# STUDENT'S SOLUTIONS MANUAL

## BEVERLY FUSFIELD

# ESSENTIALS OF COLLEGE ALGEBRA
## TENTH EDITION

## MARGARET LIAL
*American River College*

## JOHN HORNSBY
*University of New Orleans*

## DAVID SCHNEIDER
*University of Maryland*

**Addison-Wesley**
is an imprint of

Copyright © 2011, 2008 Pearson Education, Inc.
Publishing as Pearson Addison-Wesley, 75 Arlington Street, Boston, MA 02116.

ISBN-13: 978-0-321-66421-1
ISBN-10: 0-321-66421-3

1 2 3 4 5 6 BB 13 12 11 10 09

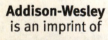

**Addison-Wesley**
is an imprint of

**PEARSON**

www.pearsonhighered.com

# CONTENTS

**R    REVIEW OF BASIC CONCEPTS**

R.1: Sets ............................................................................................1

R.2: Real Numbers and Their Properties ...........................................3

R.3: Polynomials ...............................................................................5

R.4: Factoring Polynomials ..............................................................9

R.5: Rational Expressions ...............................................................12

R.6: Rational Exponents .................................................................15

R.7: Radical Expressions ...............................................................18

Chapter R: Review Exercises .........................................................20

Chapter R: Test .............................................................................23

**1    EQUATIONS AND INEQUALITIES**

1:1 Linear Equations ......................................................................26

1.2 Applications and Modeling with Linear Equations .....................28

1.3 Complex Numbers .....................................................................32

1.4 Quadratic Equations .................................................................34

Chapter 1: Quiz (Sections 1.1−1.4) ..............................................38

1.5 Applications and Modeling with Quadratic Equations ...............38

1.6 Other Types of Equations and Applications ..............................43

Summary Exercises on Solving Equations ......................................53

1.7 Inequalities ..............................................................................55

1.8 Absolute Value Equations and Inequalities ...............................67

Chapter 1: Review Exercises .........................................................70

Chapter 1: Test ............................................................................79

**2    Graphs and Functions**

2.1 Rectangular Coordinates and Graphs .......................................84

2.2 Circles ....................................................................................88

2.3 Functions ................................................................................92

2.4 Linear Functions ......................................................................96

Chapter 2 Quiz (Sections 2.1−2.4) ..............................................100

2.5 Equations of Lines; Curve Fitting ...........................................100

Summary Exercises on Graphs, Circles, Functions, and Equations ...105

2.6 Graphs of Basic Functions ......................................................107

2.7 Graphing Techniques ..............................................................110

Chapter 2 Quiz (Sections 2.5−2.7) ..............................................116

2.8 Function Operations and Composition .....................................116

Chapter 2: Review Exercises .......................................................121

Chapter 2: Test ..........................................................................127

## 3      Graphs and Functions

3.1 Rectangular Coordinates and Graphs....................................................131
3.2 Synthetic Division..............................................................................143
3.3 Zeros of Polynomial Functions............................................................146
3.4 Polynomial Functions: Graphs, Applications, and Models .......................155
Summary Exercises on Polynomial Functions, Zeros, and Graphs...............167
3.5 Rational Functions: Graphs, Applications, and Models ...........................172
Chapter 3 Quiz (Sections 3.1–3.5)...........................................................183
3.6 Variation .........................................................................................185
Chapter 3: Review Exercises ...................................................................188
Chapter 3: Test......................................................................................197

## 4      Inverse, Exponential, and Logarithmic Functions

4.1 Inverse Functions.............................................................................204
4.2 Exponential Functions .......................................................................210
4.3 Logarithmic Functions.......................................................................217
Summary Exercises on Inverse, Exponential, and Logarithmic Functions .....222
4.4 Evaluating Logarithms and the Change-of-Base Theorem.......................224
Chapter 4 Quiz (Sections 4.1–4.4)...........................................................227
4.5 Exponential and Logarithmic Equations................................................227
4.6: Applications and Models of Exponential Growth and Decay ...................232
Summary Exercises on Functions: Domains and Defining Equations ............235
Chapter 4 Review Exercises ....................................................................237
Chapter 4 Test......................................................................................240

## 5      Systems and Matrices

5.1 Systems of Linear Equations ..............................................................243
5.2 Matrix Solution of Linear Systems ......................................................255
5.3 Determinant Solution of Linear Systems...............................................264
5.4 Partial Fractions ..............................................................................274
Chapter 5 Quiz (Sections 5.1–5.4)...........................................................280
5.5 Nonlinear Systems of Equations.........................................................282
Summary Exercises on Systems of Equations............................................290
5.6 Systems of Inequalities and Linear Programming...................................294
5.7 Properties of Matrices.......................................................................303
5.8 Matrix Inverses................................................................................309
Chapter 5 Review Exercises ...................................................................318
Chapter 5 Test......................................................................................327

# Chapter R

## REVIEW OF BASIC CONCEPTS

### Section R.1: Sets

1. The elements of the set {12, 13, 14, ..., 20} are all the natural numbers from 12 to 20 inclusive. There are 9 elements in the set, {12, 13, 14, 15, 16, 17, 18, 19, 20}.

3. Each element of the set $\left\{1, \frac{1}{2}, \frac{1}{4}, ..., \frac{1}{32}\right\}$ after the first is found by multiplying the preceding number by $\frac{1}{2}$. There are 6 elements in the set, $\left\{1, \frac{1}{2}, \frac{1}{4}, \frac{1}{8}, \frac{1}{16}, \frac{1}{32}\right\}$.

5. To find the elements of the set {17, 22, 27 ..., 47}, start with 17 and add 5 to find the next number. There are 7 elements in the set, {17, 22, 27, 32, 37, 42, 47}.

7. When you list all elements in the set {all natural numbers greater than 7 and less than 15}, you obtain {8, 9, 10, 11, 12, 13, 14}.

9. The set {4, 5, 6, ..., 15} has a limited number of elements, so it is a finite set.

11. The set $\left\{1, \frac{1}{2}, \frac{1}{4}, \frac{1}{8}, ...\right\}$ has an unlimited number of elements, so it is an infinite set.

13. The set $\{x \mid x$ is a natural number larger than 5$\}$, which can also be written as {6, 7, 8, 9, ...}, has an unlimited number of elements, so it is an infinite set.

15. There are an infinite number of fractions between 0 and 1, so $\{x \mid x$ is a fraction between 0 and 1$\}$ is an infinite set.

17. 6 is an element of the set {3, 4, 5, 6}, so we write $6 \in \{3, 4, 5, 6\}$.

19. −4 is not an element of {4, 6, 8, 10}, so we write $-4 \notin \{4, 6, 8, 10\}$.

21. 0 is an element of {2, 0, 3, 4}, so we write $0 \in \{2, 0, 3, 4\}$.

23. {3} is a subset of {2, 3, 4, 5}, not an element of {2, 3, 4, 5}, so we write $\{3\} \notin \{2, 3, 4, 5\}$.

25. {0} is a subset of {0, 1, 2, 5}, not an element of {0, 1, 2, 5}, so we write $\{0\} \notin \{0, 1, 2, 5\}$.

27. 0 is not an element of $\varnothing$, since the empty set contains no elements. Thus, $0 \notin \varnothing$.

29. $3 \in \{2, 5, 6, 8\}$
    Since 3 is not one of the elements in {2, 5, 6, 8}, the statement is false.

31. $1 \in \{3, 4, 5, 11, 1\}$
    Since 1 is one of the elements of {3, 4, 5, 11, 1}, the statement is true.

33. $9 \notin \{2, 1, 5, 8\}$
    Since 9 is not one of the elements of {2, 1, 5, 8}, the statement is true.

35. $\{2, 5, 8, 9\} = \{2, 5, 9, 8\}$
    This statement is true because both sets contain exactly the same four elements.

37. $\{5, 8, 9\} = \{5, 8, 9, 0\}$
    These two sets are not equal because {5, 8, 9, 0} contains the element 0, which is not an element of {5, 8, 9}. Therefore, the statement is false.

39. $\{x \mid x$ is a natural number less than 3$\} = \{1, 2\}$
    Since 1 and 2 are the only natural numbers less than 3, this statement is true.

41. $\{5, 7, 9, 19\} \cap \{7, 9, 11, 15\} = \{7, 9\}$
    Since 7 and 9 are the only elements belonging to both sets, the statement is true.

43. $\{2, 1, 7\} \cup \{1, 5, 9\} = \{1\}$
    $\{2, 1, 7\} \cup \{1, 5, 9\} = \{1, 2, 5, 7, 9\}$, while $\{2, 1, 7\} \cap \{1, 5, 9\} = \{1\}$. Therefore, the statement is false.

45. $\{3, 2, 5, 9\} \cap \{2, 7, 8, 10\} = \{2\}$
    Since 2 is the only element belonging to both sets, the statement is true.

47. $\{3, 5, 9, 10\} \cap \varnothing = \{3, 5, 9, 10\}$
    In order to belong to the intersection of two sets, an element must belong to both sets. Since the empty set contains no elements, $\{3, 5, 9, 10\} \cap \varnothing = \varnothing$, so the statement is false.

49. $\{1, 2, 4\} \cup \{1, 2, 4\} = \{1, 2, 4\}$
    Since the two sets are equal, their union contains the same elements, namely 1, 2, and 4. Thus, the statement is true.

1

**51.** $\varnothing \cup \varnothing = \varnothing$

Since the empty set contains no elements, the statement is true.

For Exercises 53–63, $A = \{2, 4, 6, 8, 10, 12\}$, $B = \{2, 4, 8, 10\}$, $C = \{4, 10, 12\}$, $D = \{2, 10\}$, and $U = \{2, 4, 6, 8, 10, 12, 14\}$.

**53.** $A \subseteq U$

This statement says "$A$ is a subset of $U$." Since every element of $A$ is also an element of $U$, the statement is true.

**55.** $D \subseteq B$

Since both elements of $D$, 2 and 10, are also elements of $B$, $D$ is a subset of $B$. The statement is true.

**57.** $A \subseteq B$

Set $A$ contains two elements, 6 and 12, that are not elements of $B$. Thus, $A$ is not a subset of $B$. The statement is false.

**59.** $\varnothing \subseteq A$

The empty set is a subset of every set, so the statement is true.

**61.** $\{4, 8, 10\} \subseteq B$

Since 4, 8, and 10 are all elements of $B$, $\{4, 8, 10\}$ is a subset of $B$. The statement is true.

**63.** $B \subseteq D$

Since $B$ contains two elements, 4 and 8, that are not elements of $D$, $B$ is not a subset of $D$. The statement is false.

**65.** Every element of $\{2, 4, 6\}$ is also an element of $\{3, 2, 5, 4, 6\}$, so $\{2, 4, 6\}$ is a subset of $\{3, 2, 5, 4, 6\}$.
We write $\{2, 4, 6\} \subseteq \{3, 2, 5, 4, 6\}$.

**67.** Since 0 is an element of $\{0, 1, 2\}$, but is not an element of $\{1, 2, 3, 4, 5\}$, $\{0, 1, 2\}$ is not a subset of $\{1, 2, 3, 4, 5\}$. We write $\{0, 1, 2\} \not\subseteq \{1, 2, 3, 4, 5\}$.

**69.** The empty set is a subset of every set, so $\varnothing \subseteq \{1, 4, 6, 8\}$.

For Exercises 71–93,
$U = \{0, 1, 2, 3, 4, 5, 6, 7, 8, 9, 10, 11, 12, 13\}$,
$M = \{0, 2, 4, 6, 8\}$, $N = \{1, 3, 5, 7, 9, 11, 13\}$,
$Q = \{0, 2, 4, 6, 8, 10, 12\}$, and $R = \{0, 1, 2, 3, 4\}$.

**71.** $M \cap R$

The only elements belonging to both $M$ and $R$ are 0, 2, and 4, so $M \cap R = \{0, 2, 4\}$.

**73.** $M \cup N$

The union of two sets contains all elements that belong to either set or to both sets.
$M \cup N = \{0, 1, 2, 3, 4, 5, 6, 7, 8, 9, 11, 13\}$

**75.** $M \cap N$

There are no elements which belong to both $M$ and $N$, so $M \cap N = \varnothing$. $M$ and $N$ are disjoint sets.

**77.** $N \cup R = \{0, 1, 2, 3, 4, 5, 7, 9, 11, 13\}$

**79.** $N'$

The set $N'$ is the complement of set $N$, which means the set of all elements in the universal set $U$ that do not belong to $N$.
$N' = Q$ or $\{0, 2, 4, 6, 8, 10, 12\}$

**81.** $M' \cap Q$

First form $M'$, the complement of $M$. $M'$ contains all elements of $U$ that are not elements of $M$. Thus,
$M' = \{1, 3, 5, 7, 9, 10, 11, 12, 13\}$. Now form the intersection of $M'$ and $Q$. Thus, we have $M' \cap Q = \{10, 12\}$.

**83.** $\varnothing \cap R$

Since the empty set contains no elements, there are no elements belonging to both $\varnothing$ and $R$. Thus, $\varnothing$ and $R$ are disjoint sets, and $\varnothing \cap R = \varnothing$.

**85.** $N \cup \varnothing$

Since $\varnothing$ contains no elements, the only elements belonging to $N$ or $\varnothing$ are the elements of $N$. Thus, $\varnothing$ and $N$ are disjoint sets, and $N \cup \varnothing = N$ or $\{1, 3, 5, 7, 9, 11, 13\}$.

**87.** $(M \cap N) \cup R$

First form the intersection of $M$ and $N$. Since $M$ and $N$ have no common elements (they are disjoint), $M \cap N = \varnothing$. Thus, $(M \cap N) \cup R = \varnothing \cup R$. Now, since $\varnothing$ contains no elements, the only elements belonging to $R$ or $\varnothing$ are the elements of $R$. Thus, $\varnothing$ and $R$ are disjoint sets, and $\varnothing \cup R = R$ or $\{0, 1, 2, 3, 4\}$.

**89.** $(Q \cap M) \cup R$

First form the intersection of $Q$ and $M$. We have $Q \cap M = \{0, 2, 4, 6, 8\} = M$. Now form the union of this set with $R$. We have $(Q \cap M) \cup R = M \cup R = \{0, 1, 2, 3, 4, 6, 8\}$.

**91.** $(M' \cup Q) \cap R$

First, find $M'$, the complement of $M$. We have $M' = \{1, 3, 5, 7, 9, 10, 11, 12, 13\}$. Next, form the union of $M'$ and $Q$. We have $M' \cup Q = \{0, 1, 2, 3, 4, 5, 6, 7, 8, 9, 10, 11, 12, 13\} = U$. Thus, we have $(M' \cup Q) \cap R = U \cap R = R$ or $\{0, 1, 2, 3, 4\}$.

**93.** $Q' \cap (N' \cap U)$

First, find $Q'$, the complement of $Q$. We have $Q' = \{1, 3, 5, 7, 9, 11, 13\} = N$. Now find $N'$, the complement of $N$. We have $N' = \{0, 2, 4, 6, 8, 10, 12\} = Q$. Next, form the intersection of $N'$ and $U$. We have $N' \cap U = Q \cap U = Q$ Finally, we have $Q' \cap (N' \cap U) = Q' \cap Q = \varnothing$ Since the intersection of $Q'$ and $(N' \cap U)$ is $\varnothing$, $Q'$ and $(N' \cap U)$ are disjoint sets.

**95.** $M'$ is the set of all students in this school who are not taking this course.

**97.** $N \cap P$ is the set of all students in this school who are taking both calculus and history.

**99.** $M \cup P$ is the set of all students in this school who are taking this course or history or both.

## Section R.2: Real Numbers and Their Properties

**1. (a)** 0 is a whole number. Therefore, it is also an integer, a rational number, and a real number. 0 belongs to B, C, D, F.

**(b)** 34 is a natural number. Therefore, it is also a whole number, an integer, a rational number, and a real number. 34 belongs to A, B, C, D, F.

**(c)** $-\frac{9}{4}$ is a rational number and a real number. $-\frac{9}{4}$ belongs to D, F.

**(d)** $\sqrt{36} = 6$ is a natural number. Therefore, it is also a whole number, an integer, a rational number, and a real number. $\sqrt{36}$ belongs to A, B, C, D, F.

**(e)** $\sqrt{13}$ is an irrational number and a real number. $\sqrt{13}$ belongs to E, F.

**(f)** $2.16 = \frac{216}{100} = \frac{54}{25}$ is a rational number and a real number. 2.16 belongs to D, F.

**3.** False. Positive integers are whole numbers, but negative integers are not.

**5.** False. No irrational numbers are integers.

**7.** True. Every natural number is a whole number.

**9.** True. Some rational numbers are whole numbers.

**11.** 1 and 3 are natural numbers.

**13.** $-6$, $-\frac{12}{4}$ (or $-3$), 0, 1, and 3 are integers.

**15.** $-2^4 = -(2 \cdot 2 \cdot 2 \cdot 2) = -16$

**17.** $(-2)^4 = (-2) \cdot (-2) \cdot (-2) \cdot (-2) = 16$

**19.** $(-3)^5 = (-3) \cdot (-3) \cdot (-3) \cdot (-3) \cdot (-3) = -243$

**21.** $-2 \cdot 3^4 = -2 \cdot (3 \cdot 3 \cdot 3 \cdot 3) = -2 \cdot 81 = -162$

**23.** $\begin{aligned} -2 \cdot 5 + 12 \div 3 &= -10 + 12 \div 3 \\ &= -10 + 4 = -6 \end{aligned}$

**25.** $\begin{aligned} -4(9 - 8) + (-7)(2)^3 &= -4(1) + (-7)(2)^3 \\ &= -4(1) + (-7) \cdot 8 \\ &= -4 + (-7) \cdot 8 \\ &= -4 + (-56) = -60 \end{aligned}$

**27.** $\begin{aligned} (4 - 2^3)(-2 + \sqrt{25}) &= (4 - 8)(-2 + 5) \\ &= (-4)(3) = -12 \end{aligned}$

**29.** $\begin{aligned} &\left(-\frac{2}{9} - \frac{1}{4}\right) - \left[-\frac{5}{18} - \left(-\frac{1}{2}\right)\right] \\ &= \left(-\frac{8}{36} - \frac{9}{36}\right) - \left(-\frac{5}{18} + \frac{9}{18}\right) \\ &= \left(-\frac{17}{36}\right) - \left(\frac{4}{18}\right) = -\frac{17}{36} - \frac{8}{36} = -\frac{25}{36} \end{aligned}$

**31.** $\begin{aligned} \frac{-8 + (-4)(-6) \div 12}{4 - (-3)} &= \frac{-8 + 24 \div 12}{4 + 3} \\ &= \frac{-8 + 2}{7} \\ &= \frac{-6}{7} = -\frac{6}{7} \end{aligned}$

**33.** Let $p = -4$, $q = 8$, and $r = -10$.
$\begin{aligned} 2p - 7q + r^2 &= 2(-4) - 7 \cdot 8 + (-10)^2 \\ &= 2(-4) - 7 \cdot 8 + 100 \\ &= -8 - 7 \cdot 8 + 100 \\ &= -8 - 56 + 100 \\ &= -64 + 100 = 36 \end{aligned}$

**35.** Let $p = -4$, $q = 8$, and $r = -10$.

$$\frac{q+r}{q+p} = \frac{8+(-10)}{8+(-4)} = \frac{-2}{4} = -\frac{1}{2}$$

**37.** Let $p = -4$, $q = 8$, and $r = -10$.

$$\frac{3q}{r} - \frac{5}{p} = \frac{3\cdot 8}{-10} - \frac{5}{-4} = \frac{24}{-10} - \frac{5}{-4} = -\frac{12}{5} - \frac{5}{-4}$$

$$= -\frac{12}{5} + \frac{5}{4} = -\frac{48}{20} + \frac{25}{20} = -\frac{23}{20}$$

**39.** Let $p = -4$, $q = 8$, and $r = -10$.

$$\frac{-(p+2)^2 - 3r}{2-q} = \frac{-(-4+2)^2 - 3(-10)}{2-8}$$

$$= \frac{-(-2)^2 - 3(-10)}{-6}$$

$$= \frac{-4 - 3(-10)}{-6} = \frac{-4 - (-30)}{-6}$$

$$= \frac{-4 + 30}{-6} = \frac{26}{-6} = -\frac{13}{3}$$

**41.** $A = 451$, $C = 281$, $Y = 3049$, $T = 22$, $I = 6$
Passing Rating

$$= 85.68\left(\frac{C}{A}\right) + 4.31\left(\frac{Y}{A}\right)$$

$$+ 326.42\left(\frac{T}{A}\right) - 419.07\left(\frac{I}{A}\right)$$

$$\approx 85.68\left(\frac{281}{451}\right) + 4.31\left(\frac{3049}{451}\right)$$

$$+ 326.42\left(\frac{22}{451}\right) - 419.07\left(\frac{6}{451}\right)$$

$$\approx 53.38 + 29.14 + 15.92 - 5.58 \approx 92.9$$

**43.** $A = 610$, $C = 375$, $Y = 4359$, $T = 24$, $I = 15$
Passing Rating

$$= 85.68\left(\frac{C}{A}\right) + 4.31\left(\frac{Y}{A}\right)$$

$$+ 326.42\left(\frac{T}{A}\right) - 419.07\left(\frac{I}{A}\right)$$

$$\approx 85.68\left(\frac{375}{610}\right) + 4.31\left(\frac{4359}{610}\right)$$

$$+ 326.42\left(\frac{24}{610}\right) - 419.07\left(\frac{15}{610}\right)$$

$$\approx 52.67 + 30.80 + 12.84 - 10.31 = 86.0$$

**45.** $\text{BAC} = 48 \times 3.2 \times .075 \div 190 - 2 \times .015 \approx .031$

**47.** Exercise 45:
$\text{BAC} = 48 \times 3.2 \times .075 \div 215 - 2 \times .015 \approx .024$
Exercise 46:
$\text{BAC} = 36 \times 4.0 \times .075 \div 160 - 3 \times .015 = 0.023$
The increased weight results in lower BACs.

**49.** distributive

**51.** inverse

**53.** identity

**55.** No; in general $a - b \neq b - a$. Examples will vary, i.e. if $a = 15$ and $b = 0$, then $a - b = 15 - 0 = 15$, but $b - a = 0 - 15 = -15$.

**57.** $8p - 14p = (8 - 14)p = -6p$

**59.** $-4(z - y) = -4z - (-4y) = -4z + 4y$

**61.** $\frac{10}{11}(22z) = \left(\frac{10}{11} \cdot 22\right)z = 20z$

**63.** $(m + 5) + 6 = m + (5 + 6) = m + 11$

**65.** $\frac{3}{8}\left(\frac{16}{9}y + \frac{32}{27}z - \frac{40}{9}\right)$

$$= \frac{3}{8}\left(\frac{16}{9}y\right) + \frac{3}{8}\left(\frac{32}{27}z\right) - \frac{3}{8}\left(\frac{40}{9}\right)$$

$$= \left(\frac{3}{8} \cdot \frac{16}{9}\right)y + \left(\frac{3}{8} \cdot \frac{32}{27}\right)z - \frac{5}{3}$$

$$= \frac{2}{3}y + \frac{4}{9}z - \frac{5}{3}$$

**67.** The process in your head should be the following:

$$72 \cdot 17 + 28 \cdot 17 = (72 + 28)(17)$$
$$= (100)(17) = 1700$$

**69.** The process in your head should be the following:

$$123\frac{5}{8} \cdot 1\frac{1}{2} - 23\frac{5}{8} \cdot 1\frac{1}{2} = \left(123\frac{5}{8} - 23\frac{5}{8}\right)\left(1\frac{1}{2}\right)$$

$$= (100)\left(1\frac{1}{2}\right) = (100)(1.5)$$

$$= 150$$

**71.** This statement is false since $|6 - 8| = |-2| = 2$ and $|6| - |8| = 6 - 8 = -2$. A corrected statement would be $|6 - 8| \neq |6| - |8|$ or $|6 - 8| = |8| - |6|$

**73.** This statement is true since $|-5| \cdot |6| = 5 \cdot 6 = 30$ and $|-5 \cdot 6| = |-30| = 30$.

**75.** This statement is false. For example if you let $a = 2$ and $b = 6$ then $|2 - 6| = |-4| = 4$ and $|a| - |b| = |2| - |6| = 2 - 6 = -4$. A corrected statement is $|a - b| = |b| - |a|$, if $b > a > 0$.

**77.** $|-10| = 10$

**79.** $-\left|\dfrac{4}{7}\right| = -\dfrac{4}{7}$

**81.** Let $x = -4$ and $y = 2$.
$$|x - y| = |-4 - 2| = |-6| = 6$$

**83.** Let $x = -4$ and $y = 2$.
$$|3x + 4y| = |3(-4) + 4(2)|$$
$$= |-12 + 8| = |-4| = 4$$

**85.** Let $x = -4$ and $y = 2$.
$$\frac{2|y| - 3|x|}{|xy|} = \frac{2|2| - 3|-4|}{|-4(2)|}$$
$$= \frac{2(2) - 3(4)}{|-8|} = \frac{4 - 12}{8} = \frac{-8}{8} = -1$$

**87.** Let $x = -4$ and $y = 2$.
$$\frac{|-8y + x|}{-|x|} = \frac{|-8(2) + (-4)|}{-|-4|}$$
$$= \frac{|-16 + (-4)|}{-|-4|} = \frac{|-20|}{-(4)} = \frac{20}{-4} = -5$$

**89.** Property 2

**91.** Property 3

**93.** Property 1

**95.** Since $|-3 - 5| = |-8| = 8$ and
$|5 - (-3)| = |8| = 8$, the number of strokes between their scores is 8.

**97.** $P_d = |P - 125| = |116 - 125| = |-9| = 9$

The $P_d$ value for a woman whose actual systolic pressure is 116 and whose normal value should be 125 is 9.

**99.** The absolute value of the difference in wind-chill factors for wind at 15 mph with a 30°F temperature and wind at 10 mph with a –10°F temperature is $|19° - (-28°)| = |47°| = 47°$ F.

**101.** The absolute value of the difference in wind-chill factors for wind at 30 mph with a –30°F temperature and wind at 15 mph with a –20°F temperature is
$|-67° - (-45°)| = |-22°| = 22°$ F.

**103.** $d(P, Q) = |-1 - (-4)| = |-1 + 4| = |3| = 3$ or
$d(P, Q) = |-4 - (-1)| = |-4 + 1| = |-3| = 3$

**105.** $d(Q, R) = |8 - (-1)| = |8 + 1| = |9| = 9$ or
$d(Q, R) = |-1 - 8| = |-9| = 9$

**107.** $xy > 0$ if $x$ and $y$ have the same sign.

**109.** $\dfrac{x}{y} < 0$ if $x$ and $y$ have different signs.

**111.** Since $x^3$ has the same sign as $x$, $\dfrac{x^3}{y} > 0$ if $x$ and $y$ have the same sign.

## Section R.3: Polynomials

**1.** Incorrect: $(mn)^2 = m^2 n^2$

**3.** Incorrect: $\left(\dfrac{k}{5}\right)^3 = \dfrac{k^3}{5^3} = \dfrac{k^3}{125}$

**5.** Incorrect: $4^5 \cdot 4^2 = 4^{5+2} = 4^7$

**7.** Incorrect: $cd^0 = c \cdot 1 = c$

**9.** Correct: $\left(\dfrac{1}{4}\right)^5 = \dfrac{1}{4^5}$

**11.** $9^3 \cdot 9^5 = 9^{3+5} = 9^8$

**13.** $\left(-4x^5\right)\left(4x^2\right) = (-4 \cdot 4)\left(x^5 x^2\right)$
$$= -16x^{5+2} = -16x^7$$

**15.** $n^6 \cdot n^4 \cdot n = n^{6+4+1} = n^{11}$

**17.** $\left(-3m^4\right)\left(6m^2\right)\left(-4m^5\right)$
$$= \left[(-3)(6)(-4)\right]\left(m^4 m^2 m^5\right)$$
$$= 72m^{4+2+5} = 72m^{11}$$

**19.** $(2^2)^5 = 2^{2 \cdot 5} = 2^{10}$

**21.** $\left(-6x^2\right)^3 = (-6)^3\left(x^2\right)^3 = (-6)^3 x^6 = -216x^6$

**23.** $-(4m^3n^0)^2 = -\left[4^2(m^3)^2(n^0)^2\right]$
$$= -4^2 m^{3 \cdot 2} n^{0 \cdot 2}$$
$$= -4^2 m^6 n^0 = -\left(4^2\right)m^6 \cdot 1$$
$$= -4^2 m^6 = -16m^6$$

**25.** $\left(\dfrac{r^8}{s^2}\right)^3 = \dfrac{(r^8)^3}{(s^2)^3} = \dfrac{r^{8 \cdot 3}}{s^{2 \cdot 3}} = \dfrac{r^{24}}{s^6}$

**27.** $\left(\dfrac{-4m^2}{t}\right)^4 = \dfrac{(-4)^4(m^2)^4}{t^4} = \dfrac{(-4)^4 m^8}{t^4} = \dfrac{256m^8}{t^4}$

**29.** (a) $6^0 = 1$; B      (b) $-6^0 = -1$; C

     (c) $(-6)^0 = 1$; B      (d) $-(-6)^0 = -1$; C

**31.** Answers will vary.

$x^2 + x^2 = 2x^2$

**33.** $-5x^{11}$ is a polynomial. It is a monomial since it has one term. It has degree 11 since 11 is the highest exponent.

**35.** $18p^5q + 6pq$ is a polynomial. It is a binomial since it has two terms. It has degree 6 because 6 is the sum of the exponents in the term $18p^5q$, and this term has a higher degree than the term $6pq$.

**37.** $\sqrt{2}x^2 + \sqrt{3}x^6$ is a polynomial. It is a binomial since it has two terms. It has degree 6 since 6 is the highest exponent.

**39.** $\dfrac{1}{3}r^2s^2 - \dfrac{3}{5}r^4s^2 + rs^3$ is a polynomial. It is a trinomial since it has three terms. It has degree 6 because the sum of the exponents in the term $-\dfrac{3}{5}r^4s^2$ is 6, and this term has the highest degree.

**41.** $\dfrac{5}{p} + \dfrac{2}{p^2} + \dfrac{5}{p^3}$ is not a polynomial since positive exponents in the denominator are equivalent to negative exponents in the numerator.

**43.** $\left(5x^2 - 4x + 7\right) + \left(-4x^2 + 3x - 5\right)$

$= \left[5 + (-4)\right]x^2 + (-4 + 3)x + \left[7 + (-5)\right]$

$= 1 \cdot x^2 + (-1)x + 2 = x^2 - x + 2$

**45.** $2\left(12y^2 - 8y + 6\right) - 4\left(3y^2 - 4y + 2\right)$

$= 2\left(12y^2\right) - 2(8y) + 2(6) - 4\left(3y^2\right)$

$\qquad\qquad -4(-4y) - 4 \cdot 2$

$= 24y^2 - 16y + 12 - 12y^2 + 16y - 8$

$= 12y^2 + 4$

**47.** $\left(6m^4 - 3m^2 + m\right) - \left(2m^3 + 5m^2 + 4m\right) + \left(m^2 - m\right)$

$= 6m^4 - 3m^2 + m - 2m^3 - 5m^2 - 4m + m^2 - m$

$= 6m^4 - 2m^3 + (-3 - 5 + 1)m^2 + (1 - 4 - 1)m$

$= 6m^4 - 2m^3 + (-7)m^2 + (-4)m$

$= 6m^4 - 2m^3 - 7m^2 - 4m$

**49.** $(4r - 1)(7r + 2) = 4r(7r) + 4r(2) - 1(7r) - 1(2)$

$\qquad\qquad = 28r^2 + 8r - 7r - 2$

$\qquad\qquad = 28r^2 + r - 2$

**51.** $x^2\left(3x - \dfrac{2}{3}\right)\left(5x + \dfrac{1}{3}\right)$

$= x^2\left[\left(3x - \dfrac{2}{3}\right)\left(5x + \dfrac{1}{3}\right)\right]$

$= x^2\left[(3x)(5x) + (3x)\left(\dfrac{1}{3}\right) - \dfrac{2}{3}(5x) - \dfrac{2}{3}\left(\dfrac{1}{3}\right)\right]$

$= x^2\left(15x^2 + x - \dfrac{10}{3}x - \dfrac{2}{9}\right)$

$= x^2\left(15x^2 + \dfrac{3}{3}x - \dfrac{10}{3}x - \dfrac{2}{9}\right)$

$= x^2\left(15x^2 - \dfrac{7}{3}x - \dfrac{2}{9}\right) = 15x^4 - \dfrac{7}{3}x^3 - \dfrac{2}{9}x^2$

**53.** $4x^2\left(3x^3 + 2x^2 - 5x + 1\right)$

$= 4x^2\left(3x^3\right) + 4x^2\left(2x^2\right) - 4x^2\left(5x\right) + 4x^2 \cdot 1$

$= 12x^5 + 8x^4 - 20x^3 + 4x^2$

**55.** $(2z - 1)\left(-z^2 + 3z - 4\right)$

$= (2z - 1)\left(-z^2\right) + (2z - 1)(3z) - (2z - 1)(4)$

$= 2z\left(-z^2\right) - 1\left(-z^2\right) + 2z(3z) - 1(3z)$

$\qquad\qquad - (2z)(4) - (-1)(4)$

$= -2z^3 + z^2 + 6z^2 - 3z - 8z - (-4)$

$= -2z^3 + 7z^2 - 11z + 4$

We may also multiply vertically.

$$
\begin{array}{r}
-z^2 + 3z - 4 \\
2z - 1 \\
\hline
z^2 - 3z + 4 \quad \leftarrow -1(-z^2 + 3z - 4) \\
-2z^3 + 6z^2 - 8z \qquad \leftarrow 2z(-z^2 + 3z - 4) \\
\hline
-2z^3 + 7z^2 - 11z + 4
\end{array}
$$

**57.** $(m-n+k)(m+2n-3k)$
$\quad = (m-n+k)(m) + (m-n+k)(2n)$
$\qquad\qquad\qquad\qquad - (m-n+k)(3k)$
$\quad = m^2 - mn + km + 2mn - 2n^2$
$\qquad\quad + 2kn - 3km + 3kn - 3k^2$
$\quad = m^2 + mn - 2n^2 - 2km + 5kn - 3k^2$
We may also multiply vertically.
$$\begin{array}{r} m-\ n+\ k \\ m+2n-3k \\ \hline -3km+3kn-3k^2 \\ 2mn-2n^2 \qquad +2kn \\ m^2-mn \qquad +km \\ \hline m^2+mn-2n^2-2km+5kn-3k^2 \end{array}$$

**59.** $(2m+3)(2m-3) = (2m)^2 - 3^2$
$\qquad\qquad\qquad\quad = 4m^2 - 9$

**61.** $(4x^2-5y)(4x^2+5y) = (4x^2)^2 - (5y)^2$
$\qquad\qquad\qquad\qquad\quad = 16x^4 - 25y^2$

**63.** $(4m+2n)^2 = (4m)^2 + 2(4m)(2n) + (2n)^2$
$\qquad\qquad\quad = 16m^2 + 16mn + 4n^2$

**65.** $(5r-3t^2)^2 = (5r)^2 - 2(5r)(3t^2) + (3t^2)^2$
$\qquad\qquad\quad = 25r^2 - 30rt^2 + 9t^4$

**67.** $\left[(2p-3)+q\right]^2$
$\quad = (2p-3)^2 + 2(2p-3)(q) + q^2$
$\quad = (2p)^2 - 2(2p)(3) + (3)^2 + 4pq - 6q + q^2$
$\quad = 4p^2 - 12p + 9 + 4pq - 6q + q^2$

**69.** $[(3q+5)-p][(3q+5)+p]$
$\quad = (3q+5)^2 - p^2$
$\quad = [(3q)^2 + 2(3q)(5) + 5^2] - p^2$
$\quad = 9q^2 + 30q + 25 - p^2$

**71.** $[(3a+b)-1]^2 = (3a+b)^2 - 2(3a+b)(1) + 1^2$
$\qquad\qquad = (9a^2+6ab+b^2) - 2(3a+b) + 1$
$\qquad\qquad = 9a^2 + 6ab + b^2 - 6a - 2b + 1$

**73.** $(y+2)^3 = (y+2)^2(y+2)$
$\quad = (y^2+4y+4)(y+2)$
$\quad = y^3 + 4y^2 + 4y + 2y^2 + 8y + 8$
$\quad = y^3 + 6y^2 + 12y + 8$

**75.** $(q-2)^4 = (q-2)^2(q-2)^2$
$\quad = (q^2-4q+4)(q^2-4q+4)$
$\quad = q^4 - 4q^3 + 4q^2 - 4q^3 + 16q^2$
$\qquad - 16q + 4q^2 - 16q + 16$
$\quad = q^4 - 8q^3 + 24q^2 - 32q + 16$

**77.** $(p^3-4p^2+p) - (3p^2+2p+7)$
$\quad = p^3 - 4p^2 + p - 3p^2 - 2p - 7$
$\quad = p^3 - 7p^2 - p - 7$

**79.** $(7m+2n)(7m-2n) = (7m)^2 - (2n)^2$
$\qquad\qquad\qquad\qquad = 49m^2 - 4n^2$

**81.** $-3(4q^2-3q+2) + 2(-q^2+q-4)$
$\quad = -12q^2 + 9q - 6 - 2q^2 + 2q - 8$
$\quad = -14q^2 + 11q - 14$

**83.** $p(4p-6) + 2(3p-8) = 4p^2 - 6p + 6p - 16$
$\qquad\qquad\qquad\qquad = 4p^2 - 16$

**85.** $-y(y^2-4) + 6y^2(2y-3)$
$\quad = -y^3 + 4y + 12y^3 - 18y^2$
$\quad = 11y^3 - 18y^2 + 4y$

**87.**
$$\begin{array}{r} 2x^5+\ 7x^4-\ 5x^2+7 \\ -2x^2\overline{)-4x^7-14x^6+10x^4-14x^2} \\ \underline{-4x^7}\qquad\qquad\qquad \\ -14x^6\qquad\qquad \\ \underline{-14x^6}\qquad\qquad \\ 10x^4\qquad \\ \underline{10x^4}\qquad \\ -14x^2 \\ \underline{-14x^2} \\ 0 \end{array}$$
$$\frac{-4x^7-14x^6+10x^4-14x^2}{-2x^2}$$
$$= 2x^5 + 7x^4 - 5x^2 + 7$$

**89.** 
$$-2x^6 \overline{\smash{\big)}\ 10x^8 - 16x^6 - 4x^4} \quad \frac{-5x^2 + 8}{}$$
$$\underline{10x^8}$$
$$-16x^6$$
$$\underline{-16x^6}$$
$$-4x^4$$

$$\frac{10x^8 - 16x^6 - 4x^4}{-2x^6} = -5x^2 + 8 - \frac{4x^4}{-2x^6}$$
$$= -5x^2 + 8 + \frac{2}{x^2}$$

**91.** 
$$3m + 2 \overline{\smash{\big)}\ 6m^3 + 7m^2 - 4m + 2} \quad \frac{2m^2 + m - 2}{}$$
$$\underline{6m^3 + 4m^2}$$
$$3m^2 - 4m$$
$$\underline{3m^2 + 2m}$$
$$-6m + 2$$
$$\underline{-6m - 4}$$
$$6$$

$$\frac{6m^3 + 7m^2 - 4m + 2}{3m + 2} = 2m^2 + m - 2 + \frac{6}{3m + 2}$$

**93.** 
$$3x + 3 \overline{\smash{\big)}\ 3x^4 - 0x^3 - 6x^2 + 9x - 5} \quad \frac{x^3 - x^2 - x + 4}{}$$
$$\underline{3x^4 + 3x^3}$$
$$-3x^3 - 6x^2$$
$$\underline{-3x^3 - 3x^2}$$
$$-3x^2 + 9x$$
$$\underline{-3x^2 - 3x}$$
$$12x - 5$$
$$\underline{12x + 12}$$
$$-17$$

$$\frac{3x^4 - 6x^2 + 9x - 5}{3x + 3} = x^3 - x^2 - x + 4 + \frac{-17}{3x + 3}$$
$$\text{or } x^3 - x^2 - x + 4 - \frac{17}{3x + 3}.$$

**95.** $99 \times 101 = (100 - 1)(100 + 1) = 100^2 - 1^2$
$$= 10,000 - 1 = 9999$$

**97.** $102^2 = (100 + 2)^2 = 100^2 + 2(100)(2) + 2^2$
$$= 10,000 + 400 + 4 = 10,404$$

**99. (a)** The area of the largest square is
$$s^2 = (x + y)^2.$$

**(b)** The areas of the two squares are $x^2$ and $y^2$. The area of each rectangle is $xy$. Therefore, the area of the largest square can be written as $x^2 + 2xy + y^2$.

**(c)** Answers will vary. The total area must equal the sum of the four parts.

**(d)** It reinforces the special product for squaring a binomial:
$$(x + y)^2 = x^2 + 2xy + y^2.$$

**101. (a)** The volume is
$$V = \frac{1}{3} h(a^2 + ab + b^2)$$
$$= \frac{1}{3}(200)(314^2 + 314 \times 756 + 756^2)$$
$$\approx 60,501,000 \text{ ft}^3$$

**(b)** The shape becomes a rectangular box with a square base. Its volume is given by length × width × height or $b^2 h$.

**(c)** If we let $a = b$, then
$$V = \frac{1}{3} h(a^2 + ab + b^2) \text{ becomes}$$
$$V = \frac{1}{3} h(b^2 + bb + b^2), \text{ which simplifies}$$
to $V = hb^2$. Yes, the Egyptian formula gives the same result.

**103.** $x = 1940$
$$.000020591075(1940)^3$$
$$-.1201456829(1940)^2$$
$$+233.5530856(1940)$$
$$-151,249.8184 \approx 6.2$$
The formula is .1 high.

**105.** $x = 1978$
$$.000020591075(1978)^3$$
$$-.1201456829(1978)^2$$
$$+233.5530856(1978)$$
$$-151,249.8184 \approx 2.3$$
The formula is exact.

**107.** $\left(.25^3\right)\left(400^3\right) = \left[(.25)(400)\right]^3$
$$= 100^3 = 1,000,000$$

**109.** $\dfrac{4.2^5}{2.1^5} = \left(\dfrac{4.2}{2.1}\right)^5 = 2^5 = 32$

## Section R.4: Factoring Polynomials

**1.** The greatest common factor is 12.
$$12m + 60 = 12(m) + 12(5) = 12(m+5)$$

**3.** The greatest common factor is $8k$.
$$8k^3 + 24k = 8k(k^2) + 8k(3) = 8k(k^2+3)$$

**5.** The greatest common factor is $xy$.
$$xy - 5xy^2 = xy \cdot 1 - xy(5y) = xy(1-5y)$$

**7.** The greatest common factor is $-2p^2q^4$.
$$-4p^3q^4 - 2p^2q^5$$
$$= (-2p^2q^4)(2p) + (-2p^2q^4)(q)$$
$$= -2p^2q^4(2p+q)$$

**9.** The greatest common factor is $4k^2m^3$.
$$4k^2m^3 + 8k^4m^3 - 12k^2m^4$$
$$= (4k^2m^3)(1) + (4k^2m^3)(2k^2) - (4k^2m^3)(3m)$$
$$= 4k^2m^3(1 + 2z^2 - 3m)$$

**11.** The greatest common factor is $2(a+b)$
$$2(a+b) + 4m(a+b)$$
$$= [2(a+b)](1) + [2(a+b)](2m)$$
$$= 2(a+b)(1+2m)$$

**13.** $(5r-6)(r+3) - (2r-1)(r+3)$
$$= (r+3)[(5r-6) - (2r-1)]$$
$$= (r+3)[5r-6-2r+1] = (r+3)(3r-5)$$

**15.** $2(m-1) - 3(m-1)^2 + 2(m-1)^3$
$$= (m-1)\left[2 - 3(m-1) + 2(m-1)^2\right]$$
$$= (m-1)\left[2 - 3m + 3 + 2(m^2 - 2m + 1)\right]$$
$$= (m-1)(2 - 3m + 3 + 2m^2 - 4m + 2)$$
$$= (m-1)(2m^2 - 7m + 7)$$

**17.** The completely factored form of $4x^2y^5 - 8xy^3$
is $4xy^3(xy^2 - 2)$.

**19.** $6st + 9t - 10s - 15 = (6st + 9t) - (10s + 15)$
$$= 3t(2s+3) - 5(2s+3)$$
$$= (2s+3)(3t-5)$$

**21.** $2m^4 + 6 - am^4 - 3a$
$$= (2m^4 + 6) - (am^4 + 3a)$$
$$= 2(m^4 + 3) - a(m^4 + 3) = (m^4 + 3)(2 - a)$$

**23.** $p^2q^2 - 10 - 2q^2 + 5p^2$
$$= p^2q^2 - 2q^2 + 5p^2 - 10$$
$$= q^2(p^2 - 2) + 5(p^2 - 2)$$
$$= (p^2 - 2)(q^2 + 5)$$

**25.** The positive factors of 6 could be 2 and 3, or 1 and 6. Since the middle term is negative, we know the factors of 4 must both be negative. As factors of 4, we could have $-1$ and $-4$, or $-2$ and $-2$. Try different combinations of these factors until the correct one is found.
$$6a^2 - 11a + 4 = (2a - 1)(3a - 4)$$

**27.** The positive factors of 3 are 1 and 3. Since the middle term is positive, we know the factors of 8 must both be positive. As factors of 8, we could have 1 and 8, or 2 and 4. Try different combinations of these factors until the correct one is found.
$$3m^2 + 14m + 8 = (3m + 2)(m + 4)$$

**29.** The positive factors of 15 are 1 and 15, or 3 and 5. Since the middle term is positive, we know the factors of 8 must both be positive. As factors of 8, we could have 1 and 8, or 2 and 4. Trying different combinations of these factors we find that $15p^2 + 24p + 8$ is prime.

**31.** Factor out the greatest common factor, $2a$:
$$12a^3 + 10a^2 - 42a = 2a(6a^2 + 5a - 21).$$ Now
factor the trinomial by trial and error:
$$6a^2 + 5a - 21 = (3a + 7)(2a - 3). \text{ Thus,}$$
$$12a^3 + 10a^2 - 42a = 2a(3a + 7)(2a - 3).$$

**33.** The positive factors of 6 could be 2 and 3, or 1 and 6. As factors of $-6$, we could have $-1$ and 6, $-6$ and 1, $-2$ and 3, or $-3$ and 2. Try different combinations of these factors until the correct one is found.
$$6k^2 + 5kp - 6p^2 = (2k + 3p)(3k - 2p)$$

**35.** The positive factors of 5 can only be 1 and 5. As factors of $-6$, we could have $-1$ and 6, $-6$ and 1, $-2$ and 3, or $-3$ and 2. Try different combinations of these factors until the correct one is found.
$$5a^2 - 7ab - 6b^2 = (5a + 3b)(a - 2b)$$

**37.** The positive factors of 12 could be 4 and 3, 2 and 6, or 1 and 12. The factors of $-y^2$ are $y$ and $-y$. Try different combination of these factors until the correct one is found.
$$12x^2 - xy - y^2 = (4x + y)(3x - y)$$

**39.** Factor out the greatest common factor, $2a^2$:
$$24a^4 + 10a^3b - 4a^2b^2 = 2a^2\left(12a^2 + 5ab - 2b^2\right)$$

Now factor the trinomial by trial and error:
$$12a^2 + 5ab - 2b^2 = \left(4a - b\right)\left(3a + 2b\right)$$

Thus,
$$24a^4 + 10a^3b - 4a^2b^2 = 2a^2\left(12a^2 + 5ab - 2b^2\right)$$
$$= 2a^2\left(4a - b\right)\left(3a + 2b\right)$$

**41.** $9m^2 - 12m + 4 = (3m)^2 - 12m + 2^2$
$$= (3m)^2 - 2(3m)(2) + 2^2$$
$$= (3m - 2)^2$$

**43.** $32a^2 + 48ab + 18b^2$
$$= 2(16a^2 + 24ab + 9b^2)$$
$$= 2[(4a)^2 + 24ab + (3b)^2]$$
$$= 2[(4a)^2 + 2(4a)(3b) + (3b)^2] = 2(4a + 3b)^2$$

**45.** $4x^2y^2 + 28xy + 49 = (2xy)^2 + 28xy + 7^2$
$$= (2xy)^2 + 2(2xy)(7) + 7^2$$
$$= (2xy + 7)^2$$

**47.** $(a - 3b)^2 - 6(a - 3b) + 9$
$$= (a - 3b)^2 - 6(a - 3b) + 3^2$$
$$= (a - 3b)^2 - 2(a - 3b)(3) + 3^2$$
$$= [(a - 3b) - 3]^2 = (a - 3b - 3)^2$$

**49.** **(a)** Since $(x + 5y)^2 = x^2 + 10xy + 25y^2$,
a matches B.

**(b)** Since $(x - 5y)^2 = x^2 - 10xy + 25y^2$,
b matches C.

**(c)** Since $(x + 5y)(x - 5y) = x^2 - 25y^2$,
c matches A.

**(d)** Since $(5y + x)(5y - x) = 25y^2 - x^2$,
d matches D.

**51.** $9a^2 - 16 = (3a)^2 - 4^2$
$$= (3a + 4)(3a - 4)$$

**53.** $36x^2 - \dfrac{16}{25} = \left(6x - \dfrac{4}{5}\right)\left(6x + \dfrac{4}{5}\right)$

**55.** $25s^4 - 9t^2 = (5s^2)^2 - (3t)^2$
$$= (5s^2 + 3t)(5s^2 - 3t)$$

**57.** $(a + b)^2 - 16 = (a + b)^2 - 4^2$
$$= [(a + b) + 4][(a + b) - 4]$$
$$= (a + b + 4)(a + b - 4)$$

**59.** $p^4 - 625 = (p^2)^2 - 25^2 = (p^2 + 25)(p^2 - 25)$
$$= (p^2 + 25)(p^2 - 5^2)$$
$$= (p^2 + 25)(p + 5)(p - 5)$$

Note that $p^2 + 25$ is a prime factor.

**61.** $8 - a^3 = 2^3 - a^3 = (2 - a)(2^2 + 2 \cdot a + a^2)$
$$= (2 - a)(4 + 2a + a^2)$$

**63.** $125x^3 - 27 = (5x)^3 - 3^3$
$$= (5x - 3)\left[(5x)^2 + 5x \cdot 3 + 3^2\right]$$
$$= (5x - 3)(25x^2 + 15x + 9)$$

**65.** $27y^9 + 125z^6$
$$= (3y^3)^3 + (5z^2)^3$$
$$= (3y^3 + 5z^2)\left[(3y^3)^2 - (3y^3)(5z^2) + (5z^2)^2\right]$$
$$= (3y^3 + 5z^2)(9y^6 - 15y^3z^2 + 25z^4)$$

**67.** $(r + 6)^3 - 216$
$$= (r + 6)^3 - 6^3$$
$$= \left[(r + 6) - 6\right]\left[(r + 6)^2 + (r + 6)(6) + 6^2\right]$$
$$= \left[(r + 6) - 6\right]\left[r^2 + 12r + 36 + (r + 6)(6) + 6^2\right]$$
$$= \left[r + 6 - 6\right]\left[r^2 + 12r + 36 + 6r + 36 + 36\right]$$
$$= r(r^2 + 18r + 108)$$

**69.** $27 - (m + 2n)^3$
$$= 3^3 - (m + 2n)^3$$
$$= \left[3 - (m + 2n)\right] \cdot$$
$$\left[3^2 + (3)(m + 2n) + (m + 2n)^2\right]$$
$$= \left[3 - (m + 2n)\right] \cdot$$
$$\left[3^2 + (3)(m + 2n) + m^2 + 4mn + 4n^2\right]$$
$$= (3 - m - 2n)(9 + 3m + 6n + m^2 + 4mn + 4n^2)$$

**71.** The correct complete factorization of $x^4 - 1$ is choice B: $(x^2 + 1)(x + 1)(x - 1)$. Choice A is not a complete factorization, since $x^2 - 1$ can be factored as $(x + 1)(x - 1)$. The other choices are not correct factorizations of $x^4 - 1$.

**73.** $x^6 - 1 = (x^3)^2 - 1^2$

$\qquad = (x^3 + 1)(x^3 - 1)$ or $(x^3 - 1)(x^3 + 1)$

Use the patterns for the difference of cubes and sum of cubes to factor further. Since

$x^3 - 1 = (x - 1)(x^2 + x + 1)$ and

$x^3 + 1 = (x + 1)(x^2 - x + 1)$,

we obtain the factorization

$x^6 - 1 = (x^3 - 1)(x^3 + 1)$

$\qquad = (x - 1)(x^2 + x + 1)(x + 1)(x^2 - x + 1)$

**75.** From Exercise 73, we have

$x^6 - 1 = (x - 1)(x^2 + x + 1)(x + 1)(x^2 - x + 1)$.

From Exercise 74, we have

$x^6 - 1 = (x - 1)(x + 1)(x^4 + x^2 + 1)$.

Comparing these answers, we see that

$x^4 + x^2 + 1 = (x^2 - x + 1)(x^2 + x + 1)$.

**77.** The answer in Exercise 75 and the final line in Exercise 76 are the same.

**79.** $m^4 - 3m^2 - 10$

Let $x = m^2$, so that $x^2 = (m^2)^2 = m^4$.

$x^2 - 3x - 10 = (x - 5)(x + 2)$.

Replacing $x$ with $m^2$ gives

$m^4 - 3m^2 - 10 = (m^2 - 5)(m^2 + 2)$.

**81.** Let $x = 3k - 1$. This substitution gives

$7(3k - 1)^2 + 26(3k - 1) - 8 = 7x^2 + 26x - 8$

$\qquad\qquad\qquad\qquad\qquad = (7x - 2)(x + 4)$

Replacing $x$ with $3k - 1$ gives

$7(3k - 1)^2 + 26(3k - 1) - 8$

$= [7(3k - 1) - 2][(3k - 1) + 4]$

$= (21k - 7 - 2)(3k - 1 + 4)$

$= (21k - 9)(3k + 3) = 3(7k - 3)(3)(k + 1)$

$= 9(7k - 3)(k + 1)$.

**83.** Let $x = a - 4$. This substitution gives

$9(a - 4)^2 + 30(a - 4) + 25$

$= 9x^2 + 30x + 25$

$= (3x)^2 + 2(3x)(5) + 5^2 = (3x + 5)^2$

Replacing $x$ by $a - 4$ gives

$9(a - 4)^2 + 30(a - 4) + 25$

$= [3(a - 4) + 5]^2$

$= (3a - 12 + 5)^2 = (3a - 7)^2$

**85.** $4b^2 + 4bc + c^2 - 16$

$= (4b^2 + 4bc + c^2) - 16$

$= \left[(2b)^2 + 2(2b)(c) + c^2\right] - 16$

$= (2b + c)^2 - 4^2 = [(2b + c) + 4][(2b + c) - 4]$

$= (2b + c + 4)(2b + c - 4)$

**87.** $x^2 + xy - 5x - 5y = \left(x^2 + xy\right) - (5x + 5y)$

$\qquad\qquad\qquad\qquad = x(x + y) - 5(x + y)$

$\qquad\qquad\qquad\qquad = (x + y)(x - 5)$

**89.** $p^4(m - 2n) + q(m - 2n) = (m - 2n)(p^4 + q)$

**91.** $4z^2 + 28z + 49 = (2z)^2 + 2(2z)(7) + 7^2$

$\qquad\qquad\qquad = (2z + 7)^2$

**93.** $1000x^3 + 343y^3$

$= (10x)^3 + (7y)^3$

$= (10x + 7y)\left[(10x)^2 - (10x)(7y) + (7y)^2\right]$

$= (10x + 7y)(100x^2 - 70xy + 49y^2)$

**95.** $125m^6 - 216 = (5m^2)^3 - 6^3$

$\qquad\qquad\qquad = (5m^2 - 6)\left[(5m^2)^2 + 5m^2(6) + 6^2\right]$

$\qquad\qquad\qquad = (5m^2 - 6)(25m^4 + 30m^2 + 36)$

**97.** $64 + (3x + 2)^3$

$= 4^3 + (3x + 2)^3$

$= \left[4 + (3x + 2)\right]\left[4^2 - (4)(3x + 2) + (3x + 2)^2\right]$

$= \left[4 + (3x + 2)\right]\left[\begin{array}{l} 4^2 - (4)(3x + 2) \\ \quad + 9x^2 + 12x + 4 \end{array}\right]$

$= (4 + 3x + 2)(16 - 12x - 8 + 9x^2 + 12x + 4)$

$= (3x + 6)(9x^2 + 12)$

$= 3(x + 2)(3)(3x^2 + 4) = 9(x + 2)(3x^2 + 4)$

**99.** $\dfrac{4}{25}x^2 - 49y^2 = \left(\dfrac{2}{5}x + 7y\right)\left(\dfrac{2}{5}x - 7y\right)$

**101.** $144z^2 + 121$

The sum of squares cannot be factored.

$144z^2 + 121$ is prime.

**103.** $(x + y)^2 - (x - y)^2$

$= [(x + y) + (x - y)][(x + y) - (x - y)]$

$= (x + y + x - y)(x + y - x + y)$

$= (2x)(2y) = 4xy$

**105.** Answers will vary.

**107.** $4z^2 + bz + 81 = (2z)^2 + bz + 9^2$ will be a perfect trinomial if $bz = \pm 2(2z)(9) \Rightarrow$
$bz = \pm 36z \Rightarrow b = \pm 36.$
If $b = 36$, $4z^2 + 36z + 81 = (2z + 9)^2$.
If $b = -36$, $4z^2 - 36z + 81 = (2z - 9)^2$.

**109.** $100r^2 - 60r + c = (10r)^2 - \underbrace{2(10r)(3)}_{60r} + c$

will be a perfect trinomial if $c = 3^2 = 9$.
If $c = 9$, $100r^2 - 60r + 9 = (10r - 3)^2$.

## Section R.5: Rational Expressions

**1.** In the rational expression $\dfrac{x+3}{x-6}$, the solution to the equation $x - 6 = 0$ is excluded from the domain.
$x - 6 = 0 \Rightarrow x = 6$
The domain is $\{x \mid x \neq 6\}$.

**3.** In the rational expression $\dfrac{3x+7}{(4x+2)(x-1)}$, the solution to the equation $(4x + 2)(x - 1) = 0$ is excluded from the domain.
$(4x + 2)(x - 1) = 0$
$4x + 2 = 0 \quad$ or $\quad x - 1 = 0$
$\quad 4x = -2 \qquad\qquad x = 1$
$\quad\; x = -\tfrac{1}{2}$
The domain is $\left\{ x \mid x \neq -\dfrac{1}{2}, 1 \right\}$.

**5.** In the rational expression $\dfrac{12}{x^2 + 5x + 6}$, the solution to the equation $x^2 + 5x + 6 = 0$ is excluded from the domain.
$x^2 + 5x + 6 = 0$
$(x + 3)(x + 2) = 0$
$x + 3 = 0 \quad$ or $\quad x + 2 = 0$
$\quad x = -3 \qquad\qquad x = -2$
The domain is $\{x \mid x \neq -3, -2\}$.

**7.** $x = 4, y = 2$

**(a)** $\dfrac{1}{x} + \dfrac{1}{y} = \dfrac{1}{4} + \dfrac{1}{2} = \dfrac{1}{4} + \dfrac{2}{4} = \dfrac{3}{4}$

**(b)** $\dfrac{1}{x+y} = \dfrac{1}{4+2} = \dfrac{1}{6}$.

**9.** If $x = 3, y = 5$

**(a)** $\dfrac{1}{x} - \dfrac{1}{y} = \dfrac{1}{3} - \dfrac{1}{5} = \dfrac{5}{15} - \dfrac{3}{15} = \dfrac{2}{15}$

**(b)** $\dfrac{1}{x-y} = \dfrac{1}{3-5} = \dfrac{1}{-2} = -\dfrac{1}{2}$.

**11.** $\dfrac{8k+16}{9k+18} = \dfrac{8(k+2)}{9(k+2)} = \dfrac{8}{9}$

**13.** $\dfrac{3(3-t)}{(t+5)(t-3)} = \dfrac{3(3-t)(-1)}{(t+5)(t-3)(-1)} = \dfrac{-3}{t+5}$

**15.** $\dfrac{8x^2 + 16x}{4x^2} = \dfrac{8x(x+2)}{4x^2} = \dfrac{2 \cdot 4x(x+2)}{x \cdot 4x}$
$= \dfrac{2(x+2)}{x} = \dfrac{2x+4}{x}$

**17.** $\dfrac{m^2 - 4m + 4}{m^2 + m - 6} = \dfrac{(m-2)(m-2)}{(m-2)(m+3)} = \dfrac{m-2}{m+3}$

**19.** $\dfrac{8m^2 + 6m - 9}{16m^2 - 9} = \dfrac{(2m+3)(4m-3)}{(4m+3)(4m-3)} = \dfrac{2m+3}{4m+3}$

**21.** $\dfrac{15p^3}{9p^2} \div \dfrac{6p}{10p^2} = \dfrac{15p^3}{9p^2} \cdot \dfrac{10p^2}{6p} = \dfrac{150p^5}{54p^3}$
$= \dfrac{25 \cdot 6p^5}{9 \cdot 6p^3} = \dfrac{25p^2}{9}$

**23.** $\dfrac{2k+8}{6} \div \dfrac{3k+12}{2} = \dfrac{2k+8}{6} \cdot \dfrac{2}{3k+12}$
$= \dfrac{2(k+4)(2)}{6(3)(k+4)} = \dfrac{2}{9}$

**25.** $\dfrac{x^2+x}{5} \cdot \dfrac{25}{xy+y} = \dfrac{x(x+1)}{5} \cdot \dfrac{25}{y(x+1)}$
$= \dfrac{25x(x+1)}{5y(x+1)} = \dfrac{5x}{y}$

**27.** $\dfrac{4a+12}{2a-10} \div \dfrac{a^2-9}{a^2-a-20} = \dfrac{4a+12}{2a-10} \cdot \dfrac{a^2-a-20}{a^2-9}$
$= \dfrac{4(a+3)}{2(a-5)} \cdot \dfrac{(a-5)(a+4)}{(a+3)(a-3)}$
$= \dfrac{2(a+4)}{a-3} = \dfrac{2a+8}{a-3}$

**29.** $\dfrac{p^2-p-12}{p^2-2p-15} \cdot \dfrac{p^2-9p+20}{p^2-8p+16}$
$= \dfrac{(p-4)(p+3)}{(p-5)(p+3)} \cdot \dfrac{(p-5)(p-4)}{(p-4)(p-4)} = 1$

**31.** $\dfrac{m^2+3m+2}{m^2+5m+4} \div \dfrac{m^2+5m+6}{m^2+10m+24}$

$= \dfrac{m^2+3m+2}{m^2+5m+4} \cdot \dfrac{m^2+10m+24}{m^2+5m+6}$

$= \dfrac{(m+2)(m+1)}{(m+4)(m+1)} \cdot \dfrac{(m+6)(m+4)}{(m+3)(m+2)} = \dfrac{m+6}{m+3}$

**33.** $\dfrac{xz-xw+2yz-2yw}{z^2-w^2} \cdot \dfrac{4z+4w+xz+wx}{16-x^2}$

$= \dfrac{x(z-w)+2y(z-w)}{(z+w)(z-w)} \cdot \dfrac{4(z+w)+x(z+w)}{(4+x)(4-x)}$

$= \dfrac{(z-w)(x+2y)}{(z+w)(z-w)} \cdot \dfrac{(z+w)(4+x)}{(4+x)(4-x)} = \dfrac{x+2y}{4-x}$

**35.** $\dfrac{x^3+y^3}{x^3-y^3} \cdot \dfrac{x^2-y^2}{x^2+2xy+y^2}$

$= \dfrac{(x+y)(x^2-xy+y^2)}{(x-y)(x^2+xy+y^2)} \cdot \dfrac{(x+y)(x-y)}{(x+y)(x+y)}$

$= \dfrac{x^2-xy+y^2}{x^2+xy+y^2}$

**37.** Expressions (B) and (C) are both equal to –1, since the numerator and denominator are additive inverses.

**B.** $\dfrac{-x-4}{x+4} = \dfrac{-1(x+4)}{x+4} = -1$

**C.** $\dfrac{x-4}{4-x} = \dfrac{-1(4-x)}{4-x} = -1$

**39.** $\dfrac{3}{2k}+\dfrac{5}{3k} = \dfrac{3\cdot3}{2k\cdot3}+\dfrac{5\cdot2}{3k\cdot2} = \dfrac{9}{6k}+\dfrac{10}{6k} = \dfrac{19}{6k}$

**41.** $\dfrac{1}{6m}+\dfrac{2}{5m}+\dfrac{4}{m} = \dfrac{1\cdot5}{6m\cdot5}+\dfrac{2\cdot6}{5m\cdot6}+\dfrac{4\cdot30}{m\cdot30}$

$= \dfrac{5}{30m}+\dfrac{12}{30m}+\dfrac{120}{30m} = \dfrac{137}{30m}$

**43.** $\dfrac{1}{a}-\dfrac{b}{a^2} = \dfrac{1\cdot a}{a\cdot a}-\dfrac{b}{a^2} = \dfrac{a}{a^2}-\dfrac{b}{a^2} = \dfrac{a-b}{a^2}$

**45.** $\dfrac{5}{12x^2y}-\dfrac{11}{6xy} = \dfrac{5}{12x^2y}-\dfrac{11\cdot2x}{6xy\cdot2x}$

$= \dfrac{5}{12x^2y}-\dfrac{22x}{12x^2y} = \dfrac{5-22x}{12x^2y}$

**47.** $\dfrac{17y+3}{9y+7}-\dfrac{-10y-18}{9y+7} = \dfrac{(17y+3)-(-10y-18)}{9y+7}$

$= \dfrac{17y+3+10y+18}{9y+7}$

$= \dfrac{27y+21}{9y+7}$

$= \dfrac{3(9y+7)}{9y+7} = 3$

**49.** $\dfrac{1}{x+z}+\dfrac{1}{x-z} = \dfrac{1\cdot(x-z)}{(x+z)(x-z)}+\dfrac{1\cdot(x+z)}{(x-z)(x+z)}$

$= \dfrac{(x-z)+(x+z)}{(x-z)(x+z)}$

$= \dfrac{x-z+x+z}{(x+z)(x-z)} = \dfrac{2x}{(x+z)(x-z)}$

**51.** Since $a-2=(-1)(2-a)$ we have

$\dfrac{3}{a-2}-\dfrac{1}{2-a} = \dfrac{3}{a-2}-\dfrac{1(-1)}{(2-a)(-1)}$

$= \dfrac{3}{a-2}-\dfrac{-1}{a-2} = \dfrac{3-(-1)}{a-2}$

$= \dfrac{4}{a-2}$

We may also use $2-a$ as the common denominator.

$\dfrac{3}{a-2}-\dfrac{1}{2-a} = \dfrac{3(-1)}{(a-2)(-1)}-\dfrac{1}{2-a}$

$= \dfrac{-3}{2-a}-\dfrac{1}{2-a} = \dfrac{-4}{2-a}$

The two results, $\dfrac{4}{a-2}$ and $\dfrac{-4}{2-a}$, are equivalent rational expressións.

**53.** Since $2x-y=(-1)(y-2x)$

$\dfrac{x+y}{2x-y}-\dfrac{2x}{y-2x} = \dfrac{x+y}{2x-y}-\dfrac{2x(-1)}{(y-2x)(-1)}$

$= \dfrac{x+y}{2x-y}-\dfrac{-2x}{2x-y}$

$= \dfrac{x+y-(-2x)}{2x-y}$

$= \dfrac{x+y+2x}{2x-y} = \dfrac{3x+y}{2x-y}$

We may also use $y-2x$ as the common denominator. In this case, our result will be $\dfrac{-3x-y}{y-2x}$. The two results are equivalent rational expressions.

**55.** $\dfrac{4}{x+1}+\dfrac{1}{x^2-x+1}-\dfrac{12}{x^3+1}=\dfrac{4}{x+1}+\dfrac{1}{x^2-x+1}-\dfrac{12}{(x+1)\left(x^2-x+1\right)}$

$$=\dfrac{4\left(x^2-x+1\right)}{(x+1)\left(x^2-x+1\right)}+\dfrac{1(x+1)}{(x+1)\left(x^2-x+1\right)}-\dfrac{12}{(x+1)\left(x^2-x+1\right)}$$

$$=\dfrac{4\left(x^2-x+1\right)+(x+1)-12}{(x+1)\left(x^2-x+1\right)}=\dfrac{4x^2-4x+4+x+1-12}{(x+1)\left(x^2-x+1\right)}$$

$$=\dfrac{4x^2-3x-7}{(x+1)\left(x^2-x+1\right)}=\dfrac{(4x-7)(x+1)}{(x+1)\left(x^2-x+1\right)}=\dfrac{4x-7}{x^2-x+1}$$

**57.** $\dfrac{3x}{x^2+x-12}-\dfrac{x}{x^2-16}=\dfrac{3x}{(x-3)(x+4)}-\dfrac{x}{(x-4)(x+4)}=\dfrac{3x(x-4)}{(x-3)(x+4)(x-4)}-\dfrac{x(x-3)}{(x-4)(x+4)(x-3)}$

$$=\dfrac{3x(x-4)-x(x-3)}{(x-3)(x+4)(x-4)}=\dfrac{3x^2-12x-x^2+3x}{(x-3)(x+4)(x-4)}=\dfrac{2x^2-9x}{(x-3)(x+4)(x-4)}$$

**59.** $\dfrac{1+\frac{1}{x}}{1-\frac{1}{x}}=\dfrac{x\left(1+\frac{1}{x}\right)}{x\left(1-\frac{1}{x}\right)}=\dfrac{x\cdot1+x\left(\frac{1}{x}\right)}{x\cdot1-x\left(\frac{1}{x}\right)}=\dfrac{x+1}{x-1}$

**61.** $\dfrac{\frac{1}{x+1}-\frac{1}{x}}{\frac{1}{x}}=\dfrac{x(x+1)\left(\frac{1}{x+1}-\frac{1}{x}\right)}{x(x+1)\left(\frac{1}{x}\right)}=\dfrac{x(x+1)\left(\frac{1}{x+1}\right)-x(x+1)\left(\frac{1}{x}\right)}{x(x+1)\left(\frac{1}{x}\right)}=\dfrac{x-(x+1)}{x+1}=\dfrac{x-x-1}{x+1}=\dfrac{-1}{x+1}$

**63.** $\dfrac{1+\frac{1}{1-b}}{1-\frac{1}{1+b}}=\dfrac{(1-b)(1+b)\left(1+\frac{1}{1-b}\right)}{(1-b)(1+b)\left(1-\frac{1}{1+b}\right)}=\dfrac{(1-b)(1+b)(1)+(1-b)(1+b)\left(\frac{1}{1-b}\right)}{(1-b)(1+b)(1)-(1-b)(1+b)\left(\frac{1}{1+b}\right)}=\dfrac{(1-b)(1+b)+(1+b)}{(1-b)(1+b)-(1-b)}$

$$=\dfrac{(1+b)[(1-b)+1]}{(1-b)[(1+b)-1]}=\dfrac{(1+b)(2-b)}{(1-b)b}\ \text{ or }\ \dfrac{(2-b)(1+b)}{b(1-b)}$$

**65.** $\dfrac{m-\frac{1}{m^2-4}}{\frac{1}{m+2}}=\dfrac{m-\frac{1}{(m+2)(m-2)}}{\frac{1}{m+2}}$

Multiply both numerator and denominator by the LCD of all the fractions, $(m+2)(m-2)$.

$$\dfrac{m-\frac{1}{m^2-4}}{\frac{1}{m+2}}=\dfrac{(m+2)(m-2)\left(m-\frac{1}{(m+2)(m-2)}\right)}{(m+2)(m-2)\left(\frac{1}{m+2}\right)}=\dfrac{(m+2)(m-2)(m)-(m+2)(m-2)\left(\frac{1}{(m+2)(m-2)}\right)}{(m+2)(m-2)\left(\frac{1}{m+2}\right)}$$

$$=\dfrac{m(m+2)(m-2)-1}{m-2}=\dfrac{m(m^2-4)-1}{m-2}=\dfrac{m^3-4m-1}{m-2}$$

**67.** $\dfrac{\frac{1}{x+h}-\frac{1}{x}}{h}=\dfrac{x(x+h)\left(\frac{1}{x+h}-\frac{1}{x}\right)}{x(x+h)(h)}$

$$=\dfrac{x(x+h)\left(\frac{1}{x+h}\right)-x(x+h)\left(\frac{1}{x}\right)}{x(x+h)(h)}=\dfrac{x-(x+h)}{x(x+h)(h)}=\dfrac{x-x-h}{x(x+h)(h)}=\dfrac{-h}{x(x+h)(h)}=\dfrac{-1}{x(x+h)}$$

**69.** $\dfrac{\frac{y+3}{y}-\frac{4}{y-1}}{\frac{y}{y-1}+\frac{1}{y}}=\dfrac{y(y-1)\left(\frac{y+3}{y}-\frac{4}{y-1}\right)}{y(y-1)\left(\frac{y}{y-1}+\frac{1}{y}\right)}=\dfrac{y(y-1)\left(\frac{y+3}{y}\right)-y(y-1)\left(\frac{4}{y-1}\right)}{y(y-1)\left(\frac{y}{y-1}\right)+y(y-1)\left(\frac{1}{y}\right)}=\dfrac{(y-1)(y+3)-4y}{y^2+y-1}$

$$=\dfrac{y^2+2y-3-4y}{y^2+y-1}=\dfrac{y^2-2y-3}{y^2+y-1}$$

**71.** Altitude of 1200 feet, $x = 1.2$ (thousand)
The distance from the origin is

$$\frac{7-1.2}{0.639(1.2)+1.75} \approx 2.305,$$

which represents about 2305 miles.

## Section R.6: Rational Exponents

**1.** (a)  $4^{-2} = \frac{1}{4^2} = \frac{1}{16}$; B

(b)  $-4^{-2} = -\left(4^{-2}\right) = -\frac{1}{4^2} = -\frac{1}{16}$; D

(c)  $(-4)^{-2} = \frac{1}{(-4)^2} = \frac{1}{16}$; B

(d)  $-(-4)^{-2} = -\frac{1}{(-4)^2} = -\frac{1}{16}$; D

**3.**  $(-4)^{-3} = \frac{1}{(-4)^3} = \frac{1}{-64} = -\frac{1}{64}$

**5.**  $-(-5)^{-4} = -\frac{1}{(-5)^4} = -\frac{1}{625}$

**7.**  $\left(\frac{1}{3}\right)^{-2} = \frac{1}{\left(\frac{1}{3}\right)^2} = \frac{1}{\frac{1}{9}} = 9$,  also

$\left(\frac{1}{3}\right)^{-2} = \frac{1}{\left(\frac{1}{3}\right)^2} = \left(\frac{3}{1}\right)^2 = 3^2 = 9$

**9.**  $(4x)^{-2} = \frac{1}{(4x)^2} = \frac{1}{16x^2}$

**11.**  $4x^{-2} = 4 \cdot x^{-2} = 4 \cdot \frac{1}{x^2} = \frac{4}{x^2}$

**13.**  $-a^{-3} = -\frac{1}{a^3}$

**15.**  $\frac{4^8}{4^6} = 4^{8-6} = 4^2 = 16$

**17.**  $\frac{x^{12}}{x^8} = x^{12-8} = x^4$

**19.**  $\frac{r^7}{r^{10}} = r^{7-10} = r^{-3} = \frac{1}{r^3}$

**21.**  $\frac{6^4}{6^{-2}} = 6^{4-(-2)} = 6^6$

Because $6^6$ is a relatively large number, it is generally acceptable not to simplify it to be 46,656.

**23.**  $\frac{4r^{-3}}{6r^{-6}} = \frac{4}{6} \cdot \frac{r^{-3}}{r^{-6}} = \frac{2}{3}r^{-3-(-6)} = \frac{2}{3}r^3 = \frac{2r^3}{3}$

**25.**  $\frac{16m^{-5}n^4}{12m^2n^{-3}} = \frac{16}{12} \cdot \frac{m^{-5}}{m^2} \cdot \frac{n^4}{n^{-3}} = \frac{4}{3}m^{-5-2}n^{4-(-3)}$

$= \frac{4}{3}m^{-7}n^7 = \frac{4n^7}{3m^7}$

**27.**  $-4r^{-2}\left(r^4\right)^2 = -4r^{-2}r^8 = -4r^{-2+8} = -4r^6$

**29.**  $\left(5a^{-1}\right)^4\left(a^2\right)^{-3} = 5^4a^{-4}a^{-6} = 5^4a^{-4+(-6)}$

$= 5^4a^{-10} = \frac{5^4}{a^{10}}$

Because $5^4$ is a relatively large number, it is generally acceptable not to simplify to $\frac{625}{a^{10}}$.

**31.**  $\frac{\left(p^{-2}\right)^0}{5p^{-4}} = \frac{p^0}{5p^{-4}} = \frac{1}{5} \cdot \frac{p^0}{p^{-4}}$

$= \frac{1}{5}p^{0-(-4)} = \frac{1}{5}p^4 = \frac{p^4}{5}$

**33.**  $\frac{(3pq)q^2}{6p^2q^4} = \frac{3pq^{1+2}}{6p^2q^4} = \frac{3pq^3}{6p^2q^4}$

$= \frac{3}{6} \cdot \frac{p}{p^2} \cdot \frac{q^3}{q^4} = \frac{1}{2}p^{1-2}q^{3-4}$

$= \frac{1}{2}p^{-1}q^{-1} = \frac{1}{2pq}$

**35.**  $\frac{4a^5\left(a^{-1}\right)^3}{\left(a^{-2}\right)^{-2}} = \frac{4a^5a^{-3}}{a^4} = \frac{4a^{5+(-3)}}{a^4}$

$= \frac{4a^2}{a^4} = 4a^{2-4} = 4a^{-2} = \frac{4}{a^2}$

**37.**  $169^{1/2} = 13$,  because $13^2 = 169$.

**39.**  $16^{1/4} = 2$,  because $2^4 = 16$.

**41.**  $\left(-\frac{64}{27}\right)^{1/3} = -\frac{4}{3}$,  because $\left(-\frac{4}{3}\right)^3 = -\frac{64}{27}$.

**43.** $(-4)^{1/2}$ is not a real number, because no real number, when squared, will yield a negative quantity.

**45.** **(a)** $\left(\dfrac{4}{9}\right)^{3/2} = \left[\left(\dfrac{4}{9}\right)^{1/2}\right]^3 = \left(\dfrac{2}{3}\right)^3 = \dfrac{8}{27}$ ; E

**(b)** $\left(\dfrac{4}{9}\right)^{-3/2} = \left(\dfrac{9}{4}\right)^{3/2} = \left[\left(\dfrac{9}{4}\right)^{1/2}\right]^3$

$= \left(\dfrac{3}{2}\right)^3 = \dfrac{27}{8}$ ; G

**(c)** $-\left(\dfrac{9}{4}\right)^{3/2} = -\left[\left(\dfrac{9}{4}\right)^{1/2}\right]^3$

$= -\left(\dfrac{3}{2}\right)^3 = -\dfrac{27}{8}$ ; F

**(d)** $-\left(\dfrac{4}{9}\right)^{-3/2} = -\left(\dfrac{9}{4}\right)^{3/2} = -\left[\left(\dfrac{9}{4}\right)^{1/2}\right]^3$

$= -\left(\dfrac{3}{2}\right)^3 = -\dfrac{27}{8}$ ; F

**47.** $8^{2/3} = \left(8^{1/3}\right)^2 = 2^2 = 4$

**49.** $100^{3/2} = \left(100^{1/2}\right)^3 = 10^3 = 1000$

**51.** $-81^{3/4} = -\left(81^{1/4}\right)^3 = -3^3 = -27$

**53.** $\left(\dfrac{27}{64}\right)^{-4/3} = \left(\dfrac{64}{27}\right)^{4/3} = \left[\left(\dfrac{64}{27}\right)^{1/3}\right]^4$

$= \left(\dfrac{4}{3}\right)^4 = \dfrac{256}{81}$

**55.** $3^{1/2} \cdot 3^{3/2} = 3^{1/2+3/2} = 3^{4/2} = 3^2 = 9$

**57.** $\dfrac{64^{5/3}}{64^{4/3}} = 64^{5/3-4/3} = 64^{1/3} = 4$

**59.** $y^{7/3} \cdot y^{-4/3} = y^{7/3+(-4/3)} = y^{3/3} = y^1 = y$

**61.** $\dfrac{k^{1/3}}{k^{2/3} \cdot k^{-1}} = \dfrac{k^{1/3}}{k^{2/3+(-1)}} = \dfrac{k^{1/3}}{k^{2/3+(-3/3)}}$

$= \dfrac{k^{1/3}}{k^{-1/3}} = k^{1/3-(-1/3)} = k^{2/3}$

**63.** $\dfrac{\left(x^{1/4}y^{2/5}\right)^{20}}{x^2} = \dfrac{x^5 y^8}{x^2} = x^{5-2}y^8 = x^3 y^8$

**65.** $\dfrac{\left(x^{2/3}\right)^2}{\left(x^2\right)^{7/3}} = \dfrac{x^{4/3}}{x^{14/3}} = x^{4/3-14/3} = x^{-10/3} = \dfrac{1}{x^{10/3}}$

**67.** $\left(\dfrac{16m^3}{n}\right)^{1/4}\left(\dfrac{9n^{-1}}{m^2}\right)^{1/2} = \dfrac{16^{1/4}m^{3/4}}{n^{1/4}} \cdot \dfrac{9^{1/2}n^{-1/2}}{m}$

$= \dfrac{2m^{3/4}}{n^{1/4}} \cdot \dfrac{3n^{-1/2}}{m}$

$= (2 \cdot 3)\dfrac{m^{3/4}}{m} \cdot \dfrac{n^{-1/2}}{n^{1/4}}$

$= 6m^{(3/4)-1}n^{(-1/2)-(1/4)}$

$= 6m^{(3/4)-(4/4)}n^{(-2/4)-(1/4)}$

$= 6m^{-1/4}n^{-3/4} = \dfrac{6}{m^{1/4}n^{3/4}}$

**69.** $\dfrac{p^{1/5}p^{7/10}p^{1/2}}{\left(p^3\right)^{-1/5}} = \dfrac{p^{1/5+7/10+1/2}}{p^{-3/5}}$

$= \dfrac{p^{2/10+7/10+5/10}}{p^{-6/10}} = \dfrac{p^{14/10}}{p^{-6/10}}$

$= p^{(14/10)-(-6/10)}$

$= p^{20/10} = p^2$

**71.** **(a)** Let $w = 25$

$t = \dfrac{31,293}{w^{1.5}} = \dfrac{31,293}{(25)^{3/2}} = \dfrac{31,293}{(25^{1/2})^3}$

$= \dfrac{31,293}{(5)^3} = \dfrac{31,293}{125} \approx 250$ sec

**(b)** To double the weight, replace $w$ with $2w$ to get $\dfrac{31,293}{(2w)^{1.5}} = \dfrac{31,293}{2^{3/2}w^{1.5}} = \dfrac{1}{2^{3/2}}(t)$; so the holding time changes by a factor of

$\dfrac{1}{2^{3/2}} \approx .3536$ .

**73.** $y^{5/8}(y^{3/8} - 10y^{11/8}) = y^{5/8}y^{3/8} - 10y^{5/8}y^{11/8}$

$= y^{5/8+3/8} - 10y^{5/8+11/8}$

$= y - 10y^2$

**75.** $-4k(k^{7/3} - 6k^{1/3}) = -4k^1k^{7/3} + 24k^1k^{1/3}$

$= -4k^{10/3} + 24k^{4/3}$

**77.** $(x + x^{1/2})(x - x^{1/2})$ has the form $(a-b)(a+b) = a^2 - b^2$.

$(x + x^{1/2})(x - x^{1/2}) = x^2 - (x^{1/2})^2 = x^2 - x$

**79.** $\left(r^{1/2} - r^{-1/2}\right)^2$

$= \left(r^{1/2}\right)^2 - 2\left(r^{1/2}\right)\left(r^{-1/2}\right) + \left(r^{-1/2}\right)^2$

$= r - 2r^{(1/2)+(-1/2)} + r^{-1} = r - 2r^0 + r^{-1}$

$= r - 2 \cdot 1 + r^{-1} = r - 2 + r^{-1} = r - 2 + \dfrac{1}{r}$

**81.** Factor $4k^{-1} + k^{-2}$, using the common factor
$k^{-2}$: $4k^{-1} + k^{-2} = k^{-2}(4k + 1)$

**83.** Factor $4t^{-2} + 8t^{-4}$, using the common factor
$4t^{-4}$: $4t^{-2} + 8t^{-4} = 4t^{-4}(t^2 + 2)$

**85.** Factor $9z^{-1/2} + 2z^{1/2}$, using the common
factor $z^{-1/2}$:
$9z^{-1/2} + 2z^{1/2} = z^{-1/2}(9 + 2z)$

**87.** Factor $p^{-3/4} - 2p^{-7/4}$, using the common
factor $p^{-7/4}$:
$p^{-3/4} - 2p^{-7/4} = p^{-7/4}\left(p^{4/4} - 2\right)$
$\qquad\qquad\qquad = p^{-7/4}(p - 2)$

**89.** Factor $-4a^{-2/5} + 16a^{-7/5}$, using the common
factor $4a^{-7/5}$:
$-4a^{-2/5} + 16a^{-7/5} = 4a^{-7/5}(-a + 4)$

**91.** Factor $(p+4)^{-3/2} + (p+4)^{-1/2} + (p+4)^{1/2}$,
using the common factor $(p+4)^{-3/2}$.
$(p+4)^{-3/2} + (p+4)^{-1/2} + (p+4)^{1/2}$
$= (p+4)^{-3/2} \cdot [1 + (p+4) + (p+4)^2]$
$= (p+4)^{-3/2} \cdot (1 + p + 4 + p^2 + 8p + 16)$
$= (p+4)^{-3/2}(p^2 + 9p + 21)$

**93.** Factor
$2(3x+1)^{-3/2} + 4(3x+1)^{-1/2} + 6(3x+1)^{1/2}$
using the common factor $2(3x+1)^{-3/2}$:
$2(3x+1)^{-3/2} + 4(3x+1)^{-1/2} + 6(3x+1)^{1/2}$
$= 2(3x+1)^{-3/2}\left[1 + 2(3x+1) + 3(3x+1)^2\right]$
$= 2(3x+1)^{-3/2}\left[1 + (6x+2) + 3\left(9x^2 + 6x + 1\right)\right]$
$= 2(3x+1)^{-3/2}\left[1 + 6x + 2 + 27x^2 + 18x + 3\right]$
$= 2(3x+1)^{-3/2}\left(27x^2 + 24x + 6\right)$
$= 2(3x+1)^{-3/2} \cdot 3\left(9x^2 + 8x + 2\right)$
$= 6(3x+1)^{-3/2}\left(9x^2 + 8x + 2\right)$

**95.** Factor
$4x(2x+3)^{-5/9} + 6x^2(2x+3)^{4/9} - 8x^3(2x+3)^{13/9}$
using the common factor $2x(2x+3)^{-5/9}$:
$4x(2x+3)^{-5/9} + 6x^2(2x+3)^{4/9} - 8x^3(2x+3)^{13/9}$
$= 2x(2x+3)^{-5/9}\left[2 + 3x(2x+3) - 4x^2(2x+3)^2\right]$
$= 2x(2x+3)^{-5/9}$
$\quad \cdot \left[2 + \left(6x^2 + 9x\right) - 4x^2\left(4x^2 + 12x + 9\right)\right]$
$= 2x(2x+3)^{-5/9}$
$\quad \cdot \left[2 + 6x^2 + 9x - 16x^4 - 48x^3 - 36x^2\right]$
$= 2x(2x+3)^{-5/9}\left[-16x^4 - 48x^3 - 30x^2 + 9x + 2\right]$

**97.** $\dfrac{a^{-1} + b^{-1}}{(ab)^{-1}} = \dfrac{\frac{1}{a} + \frac{1}{b}}{\frac{1}{ab}} = \dfrac{\frac{1 \cdot b}{a \cdot b} + \frac{1 \cdot a}{b \cdot a}}{\frac{1}{ab}}$

$\qquad = \dfrac{\frac{b+a}{ab}}{\frac{1}{ab}} = \dfrac{b+a}{ab} \cdot \dfrac{ab}{1} = b + a$

**99.** $\dfrac{r^{-1} + q^{-1}}{r^{-1} - q^{-1}} \cdot \dfrac{r - q}{r + q} = \dfrac{\frac{1}{r} + \frac{1}{q}}{\frac{1}{r} - \frac{1}{q}} \cdot \dfrac{r - q}{r + q}$

$\qquad = \dfrac{rq\left(\frac{1}{r} + \frac{1}{q}\right)}{rq\left(\frac{1}{r} - \frac{1}{q}\right)} \cdot \dfrac{r - q}{r + q}$

$\qquad = \dfrac{q + r}{q - r} \cdot \dfrac{r - q}{r + q} = \dfrac{r - q}{q - r}$

$\qquad = \dfrac{r - q}{-1(r - q)} = -1$

**101.** $\dfrac{x - 9y^{-1}}{\left(x - 3y^{-1}\right)\left(x + 3y^{-1}\right)} = \dfrac{x - \dfrac{9}{y}}{\left(x - \dfrac{3}{y}\right)\left(x + \dfrac{3}{y}\right)}$

$\qquad = \dfrac{x - \dfrac{9}{y}}{x^2 - \dfrac{9}{y^2}} = \dfrac{y^2\left(x - \dfrac{9}{y}\right)}{y^2\left(x^2 - \dfrac{9}{y^2}\right)}$

$\qquad = \dfrac{y^2 x - 9y}{y^2 x^2 - 9}$ or $\dfrac{y(xy - 9)}{x^2 y^2 - 9}$

**103.** $\dfrac{(x^2+1)^4(2x) - x^2(4)(x^2+1)^3(2x)}{(x^2+1)^8}$

$= \dfrac{(2x)(x^2+1)^3\left[(x^2+1) - 4x^2\right]}{(x^2+1)^8}$

$= \dfrac{(2x)\left[(x^2+1) - 4x^2\right]}{(x^2+1)^5} = \dfrac{(2x)(1 - 3x^2)}{(x^2+1)^5}$

**105.** $\dfrac{4(x^2-1)^3 + 8x(x^2-1)^4}{16(x^2-1)^3}$

$= \dfrac{4(x^2-1)^3\left[1 + 2x(x^2-1)\right]}{16(x^2-1)^3}$

$= \dfrac{1 + 2x(x^2-1)}{4} = \dfrac{1 + 2x^3 - 2x}{4}$

**107.** $\dfrac{2(2x-3)^{1/3} - (x-1)(2x-3)^{-2/3}}{(2x-3)^{2/3}}$

$= \dfrac{(2x-3)^{1/3}\left[2 - (x-1)(2x-3)^{-1}\right]}{(2x-3)^{2/3}}$

$= \dfrac{2 - (x-1)(2x-3)^{-1}}{(2x-3)^{1/3}} = \dfrac{2 - \frac{x-1}{2x-3}}{(2x-3)^{1/3}}$

$= \dfrac{2(2x-3) - (x-1)}{(2x-3)^{4/3}} = \dfrac{3x-5}{(2x-3)^{4/3}}$

**109.** $a^7 = 30 \Rightarrow (a^7)^3 = 30^3 \Rightarrow a^{21} = 27,000$

**111.** Let $x$ = length of side of cube. Then $3x$ = length of side of bigger cube (side tripled). $x^3$ is the volume of the cube, and $(3x)^3 = 3^3 x^3 = 27x^3$ is the volume of the bigger cube. The volume will change by a factor of 27.

**113.** $.2^{2/3} \cdot 40^{2/3} = (.2 \cdot 40)^{2/3} = (8^{1/3})^2 = 2^2 = 4$

**115.** $\dfrac{2^{2/3}}{2000^{2/3}} = \left(\dfrac{2}{2000}\right)^{2/3}$

$= \left(\dfrac{1}{1000}\right)^{2/3}\left[\left(\dfrac{1}{1000}\right)^{1/3}\right]^2$

$= \left(\dfrac{1}{10}\right)^2 = \dfrac{1}{100}$

## Section R.7: Radical Expressions

**1. (a)** F; $(-3x)^{1/3} = \sqrt[3]{-3x}$

**(b)** H; $(-3x)^{-1/3} = \dfrac{1}{(-3x)^{1/3}} = \dfrac{1}{\sqrt[3]{-3x}}$

**(c)** G; $(3x)^{1/3} = \sqrt[3]{3x}$

**(d)** C; $(3x)^{-1/3} = \dfrac{1}{\sqrt[3]{3x}}$

**3.** $m^{2/3} = \sqrt[3]{m^2}$ or $\left(\sqrt[3]{m}\right)^2$

**5.** $(2m+p)^{2/3} = \sqrt[3]{(2m+p)^2}$ or $\left(\sqrt[3]{2m+p}\right)^2$

**7.** $\sqrt[5]{k^2} = k^{2/5}$

**9.** $-3\sqrt{5p^3} = -3(5p^3)^{1/2} = -3 \cdot 5^{1/2} p^{3/2}$

**11.** A

**13.** It is true for all $x \geq 0$.

**15.** $\sqrt{(-5)^2} = |-5| = 5$

**17.** $\sqrt{25k^4 m^2} = \sqrt{\left(5k^2 m\right)^2} = \left|5k^2 m\right| = 5k^2 |m|$

**19.** $\sqrt{(4x-y)^2} = |4x - y|$

**21.** $\sqrt[3]{125} = 5$

**23.** This is not a real number since no real number raised to the fourth power will yield a negative quantity.

**25.** $\sqrt[3]{81} = \sqrt[3]{27 \cdot 3} = \sqrt[3]{27} \cdot \sqrt[3]{3} = 3\sqrt[3]{3}$

**27.** $-\sqrt[4]{32} = -\sqrt[4]{16 \cdot 2} = -\sqrt[4]{16} \cdot \sqrt[4]{2} = -2\sqrt[4]{2}$

**29.** $\sqrt{14} \cdot \sqrt{3pqr} = \sqrt{14 \cdot 3pqr} = \sqrt{42pqr}$

**31.** $\sqrt[3]{7x} \cdot \sqrt[3]{2y} = \sqrt[3]{7x \cdot 2y} = \sqrt[3]{14xy}$

**33.** $-\sqrt{\dfrac{9}{25}} = -\dfrac{\sqrt{9}}{\sqrt{25}} = -\dfrac{3}{5}$

**35.** $-\sqrt[3]{\dfrac{5}{8}} = -\dfrac{\sqrt[3]{5}}{\sqrt[3]{8}} = -\dfrac{\sqrt[3]{5}}{2}$

**37.** $\sqrt[4]{\dfrac{m}{n^4}} = \dfrac{\sqrt[4]{m}}{\sqrt[4]{n^4}} = \dfrac{\sqrt[4]{m}}{n}$

**39.** $3\sqrt[5]{-3125} = 3\sqrt[5]{(-5)^5} = 3(-5) = -15$

**41.** $\sqrt[3]{16(-2)^4(2)^8} = \sqrt[3]{2^4 \cdot (-2)^4 2^8} = \sqrt[3]{2^4 \cdot 2^4 \cdot 2^8}$

$= \sqrt[3]{2^{15} \cdot 2} = \sqrt[3]{2^{15}} \cdot \sqrt[3]{2}$

$= 2^5 \cdot \sqrt[3]{2} = 32\sqrt[3]{2}$

**43.** $\sqrt{8x^5 z^8} = \sqrt{2 \cdot 4 \cdot x^4 \cdot x \cdot z^8}$

$= \sqrt{4x^4 z^8} \cdot \sqrt{2x} = 2x^2 z^4 \sqrt{2x}$

**45.** $\sqrt[4]{x^4 + y^4}$ cannot be simplified further.

**47.** $\sqrt{\dfrac{2}{3x}} = \dfrac{\sqrt{2}}{\sqrt{3x}} = \dfrac{\sqrt{2}}{\sqrt{3x}} \cdot \dfrac{\sqrt{3x}}{\sqrt{3x}} = \dfrac{\sqrt{2 \cdot 3x}}{\sqrt{9x^2}} = \dfrac{\sqrt{6x}}{3x}$

**49.** $\sqrt{\dfrac{x^5 y^3}{z^2}} = \dfrac{\sqrt{x^5 y^3}}{\sqrt{z^2}} = \dfrac{\sqrt{x^4 xy^2 y}}{z}$

$= \dfrac{\sqrt{x^4 y^2} \cdot \sqrt{xy}}{z} = \dfrac{x^2 y \sqrt{xy}}{z}$

**51.** $\sqrt[3]{\dfrac{8}{x^2}} = \dfrac{\sqrt[3]{8}}{\sqrt[3]{x^2}} = \dfrac{2}{\sqrt[3]{x^2}} \cdot \dfrac{\sqrt[3]{x}}{\sqrt[3]{x}} = \dfrac{2\sqrt[3]{x}}{\sqrt[3]{x^3}} = \dfrac{2\sqrt[3]{x}}{x}$

**53.** $\sqrt[4]{\dfrac{g^3 h^5}{9r^6}} = \dfrac{\sqrt[4]{g^3 h^5}}{\sqrt[4]{9r^6}} = \dfrac{\sqrt[4]{h^4 \left(g^3 h\right)}}{\sqrt[4]{r^4 \left(9r^2\right)}} = \dfrac{h\sqrt[4]{g^3 h}}{r\sqrt[4]{9r^2}}$

$= \dfrac{h\sqrt[4]{g^3 h}}{r\sqrt[4]{3^2 r^2}} \cdot \dfrac{\sqrt[4]{3^2 r^2}}{\sqrt[4]{3^2 r^2}} = \dfrac{h\sqrt[4]{3^2 g^3 hr^2}}{r\sqrt[4]{3^4 r^4}}$

$= \dfrac{h\sqrt[4]{9g^3 hr^2}}{3r^2}$

**55.** $\sqrt[8]{3^4} = 3^{4/8} = 3^{1/2} = \sqrt{3}$

**57.** $\sqrt[3]{\sqrt{4}} = \sqrt[3]{4^{1/2}} = (4^{1/2})^{1/3} = 4^{1/6}$

$\quad = (2^2)^{1/6} = 2^{2/6} = 2^{1/3} = \sqrt[3]{2}$

**59.** $\sqrt[4]{\sqrt[3]{2}} = \sqrt[4]{2^{1/3}} = (2^{1/3})^{1/4} = 2^{1/12} = \sqrt[12]{2}$

**61.** Cannot be simplified further

**63.** $8\sqrt{2x} - \sqrt{8x} + \sqrt{72x}$

$= 8\sqrt{2x} - \sqrt{4 \cdot 2x} + \sqrt{36 \cdot 2x}$

$= 8\sqrt{2x} - 2\sqrt{2x} + 6\sqrt{2x} = 12\sqrt{2x}$

**65.** $2\sqrt[3]{3} + 4\sqrt[3]{24} - \sqrt[3]{81}$

$= 2\sqrt[3]{3} + 4\sqrt[3]{8 \cdot 3} - \sqrt[3]{27 \cdot 3}$

$= 2\sqrt[3]{3} + 4(2)\sqrt[3]{3} - 3\sqrt[3]{3}$

$= 2\sqrt[3]{3} + 8\sqrt[3]{3} - 3\sqrt[3]{3} = 7\sqrt[3]{3}$

**67.** $\sqrt[4]{81x^6 y^3} - \sqrt[4]{16x^{10} y^3}$

$= \sqrt[4]{\left(81x^4\right) x^2 y^3} - \sqrt[4]{\left(16x^8\right) x^2 y^3}$

$= 3x\sqrt[4]{x^2 y^3} - 2x^2 \left(\sqrt[4]{x^2 y^3}\right)$

**69.** This product has the pattern

$(a+b)(a-b) = a^2 - b^2$, the difference of squares.

$\left(\sqrt{2} + 3\right)\left(\sqrt{2} - 3\right) = \left(\sqrt{2}\right)^2 - 3^2 = 2 - 9 = -7$

**71.** This product has the pattern

$(a-b)(a^2 + ab + b^2) = a^3 - b^3$, the difference of cubes.

$\left(\sqrt[3]{11} - 1\right)\left(\sqrt[3]{11^2} + \sqrt[3]{11} + 1\right) = \left(\sqrt[3]{11}\right)^3 - 1^3$

$\qquad\qquad\qquad\qquad\qquad = 11 - 1 = 10.$

**73.** This product has the pattern

$(a+b)^2 = a^2 + 2ab + b^2$, the square of a binomial.

$\left(\sqrt{3} + \sqrt{8}\right)^2 = \left(\sqrt{3}\right)^2 + 2\left(\sqrt{3}\right)\left(\sqrt{8}\right) + \left(\sqrt{8}\right)^2$

$\qquad = 3 + 2\sqrt{24} + 8 = 11 + 2\sqrt{4 \cdot 6}$

$\qquad = 11 + 2(2)\sqrt{6} = 11 + 4\sqrt{6}$

**75.** This product can be found by using the FOIL method.

$\left(3\sqrt{2} + \sqrt{3}\right)\left(2\sqrt{3} - \sqrt{2}\right)$

$= 3\sqrt{2}\left(2\sqrt{3}\right) - 3\sqrt{2}\left(\sqrt{2}\right) + \sqrt{3}\left(2\sqrt{3}\right) - \sqrt{3}\sqrt{2}$

$= 6\sqrt{6} - 3 \cdot 2 + 2 \cdot 3 - \sqrt{6} = 6\sqrt{6} - 6 + 6 - \sqrt{6}$

$= 5\sqrt{6}$

**77.** $\dfrac{\sqrt[3]{mn} \cdot \sqrt[3]{m^2}}{\sqrt[3]{n^2}} = \sqrt[3]{\dfrac{mnm^2}{n^2}} = \sqrt[3]{\dfrac{m^3}{n}}$

$\qquad = \dfrac{\sqrt[3]{m^3}}{\sqrt[3]{n}} \cdot \dfrac{\sqrt[3]{n^2}}{\sqrt[3]{n^2}} = \dfrac{m\sqrt[3]{n^2}}{n}$

**79.** $\sqrt[3]{\dfrac{2}{x^6}} - \sqrt[3]{\dfrac{5}{x^9}} = \dfrac{\sqrt[3]{2}}{\sqrt[3]{x^6}} - \dfrac{\sqrt[3]{5}}{\sqrt[3]{x^9}} = \dfrac{\sqrt[3]{2}}{x^2} - \dfrac{\sqrt[3]{5}}{x^3}$

$\qquad = \dfrac{\sqrt[3]{2} \cdot x}{x^2 \cdot x} - \dfrac{\sqrt[3]{5}}{x^3} = \dfrac{x\sqrt[3]{2}}{x^3} - \dfrac{\sqrt[3]{5}}{x^3}$

$\qquad = \dfrac{x\sqrt[3]{2} - \sqrt[3]{5}}{x^3}$

**81.** $\dfrac{1}{\sqrt{2}} + \dfrac{3}{\sqrt{8}} + \dfrac{1}{\sqrt{32}} = \dfrac{1}{\sqrt{2}} + \dfrac{3}{\sqrt{4 \cdot 2}} + \dfrac{1}{\sqrt{16 \cdot 2}}$

$\qquad = \dfrac{1}{\sqrt{2}} + \dfrac{3}{2\sqrt{2}} + \dfrac{1}{4\sqrt{2}}$

$\qquad = \dfrac{4 \cdot 1}{4\sqrt{2}} + \dfrac{3 \cdot 2}{2 \cdot 2\sqrt{2}} + \dfrac{1}{4\sqrt{2}}$

$\qquad = \dfrac{4}{4\sqrt{2}} + \dfrac{6}{4\sqrt{2}} + \dfrac{1}{4\sqrt{2}}$

$\qquad = \dfrac{4 + 6 + 1}{4\sqrt{2}} = \dfrac{11}{4\sqrt{2}}$

$\qquad = \dfrac{11\sqrt{2}}{4\sqrt{2} \cdot \sqrt{2}} = \dfrac{11\sqrt{2}}{4 \cdot 2} = \dfrac{11\sqrt{2}}{8}$

**83.** $\dfrac{-4}{\sqrt[3]{3}} + \dfrac{1}{\sqrt[3]{24}} - \dfrac{2}{\sqrt[3]{81}}$

$= \dfrac{-4}{\sqrt[3]{3}} + \dfrac{1}{\sqrt[3]{8 \cdot 3}} - \dfrac{2}{\sqrt[3]{27 \cdot 3}} = \dfrac{-4}{\sqrt[3]{3}} + \dfrac{1}{2\sqrt[3]{3}} - \dfrac{2}{3\sqrt[3]{3}}$

$= \dfrac{-4 \cdot 6}{\sqrt[3]{3} \cdot 6} + \dfrac{1 \cdot 3}{2\sqrt[3]{3} \cdot 3} - \dfrac{2 \cdot 2}{3\sqrt[3]{3} \cdot 2}$

$= \dfrac{-24}{6\sqrt[3]{3}} + \dfrac{3}{6\sqrt[3]{3}} - \dfrac{4}{6\sqrt[3]{3}} = \dfrac{-24 + 3 - 4}{6\sqrt[3]{3}} = \dfrac{-25}{6\sqrt[3]{3}}$

$= \dfrac{-25}{6\sqrt[3]{3}} \cdot \dfrac{\sqrt[3]{3^2}}{\sqrt[3]{3^2}} = \dfrac{-25\sqrt[3]{9}}{6 \cdot 3} = \dfrac{-25\sqrt[3]{9}}{18}$

**85.** $\dfrac{\sqrt{3}}{\sqrt{5} + \sqrt{3}} = \dfrac{\sqrt{3}}{\sqrt{5} + \sqrt{3}} \cdot \dfrac{\sqrt{5} - \sqrt{3}}{\sqrt{5} - \sqrt{3}}$

$= \dfrac{\sqrt{3}\left(\sqrt{5} - \sqrt{3}\right)}{\left(\sqrt{5}\right)^2 - \left(\sqrt{3}\right)^2}$

$= \dfrac{\sqrt{3}\sqrt{5} - \sqrt{3}\sqrt{3}}{5 - 3} = \dfrac{\sqrt{15} - 3}{2}$

**87.** $\dfrac{\sqrt{7} - 1}{2\sqrt{7} + 4\sqrt{2}}$

$= \dfrac{\sqrt{7} - 1}{2\sqrt{7} + 4\sqrt{2}} \cdot \dfrac{2\sqrt{7} - 4\sqrt{2}}{2\sqrt{7} - 4\sqrt{2}}$

$= \dfrac{\left(\sqrt{7} - 1\right)\left(2\sqrt{7} - 4\sqrt{2}\right)}{\left(2\sqrt{7} + 4\sqrt{2}\right)\left(2\sqrt{7} - 4\sqrt{2}\right)}$

$= \dfrac{\sqrt{7} \cdot 2\sqrt{7} - \sqrt{7} \cdot 4\sqrt{2} - 1 \cdot 2\sqrt{7} + 1 \cdot 4\sqrt{2}}{\left(2\sqrt{7}\right)^2 - \left(4\sqrt{2}\right)^2}$

$= \dfrac{2 \cdot 7 - 4\sqrt{14} - 2\sqrt{7} + 4\sqrt{2}}{4 \cdot 7 - 16 \cdot 2}$

$= \dfrac{14 - 4\sqrt{14} - 2\sqrt{7} + 4\sqrt{2}}{28 - 32}$

$= \dfrac{14 - 4\sqrt{14} - 2\sqrt{7} + 4\sqrt{2}}{-4}$

$= \dfrac{-2\left(-7 + 2\sqrt{14} + \sqrt{7} - 2\sqrt{2}\right)}{-4}$

$= \dfrac{-7 + 2\sqrt{14} + \sqrt{7} - 2\sqrt{2}}{2}$

**89.** $\dfrac{p}{\sqrt{p} + 2} = \dfrac{p}{\sqrt{p} + 2} \cdot \dfrac{\sqrt{p} - 2}{\sqrt{p} - 2}$

$= \dfrac{p\left(\sqrt{p} - 2\right)}{\left(\sqrt{p}\right)^2 - 2^2} = \dfrac{p\left(\sqrt{p} - 2\right)}{p - 4}$

**91.** $\dfrac{5\sqrt{x}}{2\sqrt{x} + \sqrt{y}} = \dfrac{5\sqrt{x}\left(2\sqrt{x} - \sqrt{y}\right)}{\left(2\sqrt{x} + \sqrt{y}\right)\left(2\sqrt{x} - \sqrt{y}\right)}$

$= \dfrac{5\sqrt{x}\left(2\sqrt{x} - \sqrt{y}\right)}{4x - y}$

**93.** $\dfrac{3m}{2 + \sqrt{m + n}} = \dfrac{3m}{2 + \sqrt{m + n}} \cdot \dfrac{2 - \sqrt{m + n}}{2 - \sqrt{m + n}}$

$= \dfrac{3m\left(2 - \sqrt{m + n}\right)}{2^2 - \left(\sqrt{m + n}\right)^2}$

$= \dfrac{3m\left(2 - \sqrt{m + n}\right)}{4 - (m + n)}$

$= \dfrac{3m\left(2 - \sqrt{m + n}\right)}{4 - m - n}$

**95.** $S = 15.18\sqrt[9]{n} = 15.18\sqrt[9]{8} \approx 19.1$

The speed of the boat with an eight-man crew is approx. 19.1 ft/sec.

**97.** Windchill temperature
$= 35.74 + .6215T - 35.75V^{.16} + .4275TV^{.16}$
Windchill temperature
$= 35.74 + .6215(10) - 35.75\left(30^{.16}\right)$
$\quad + .4275(10)\left(30^{.16}\right) \approx -12.3°$

The table gives $-12°$.

**99.** $\sqrt[4]{8} \cdot \sqrt[4]{2} = \sqrt[4]{8 \cdot 2} = \sqrt[4]{16} = 2$

**101.** $\dfrac{\sqrt[5]{320}}{\sqrt[5]{10}} = \sqrt[5]{\dfrac{320}{10}} = \sqrt[5]{32} = 2$

**103.** $\dfrac{\sqrt[3]{15}}{\sqrt[3]{5}} \cdot \sqrt[3]{9} = \sqrt[3]{3} \cdot \sqrt[3]{9} = \sqrt[3]{27} = 3$

**105.** $\dfrac{355}{113} = 3.1415929\ldots$ and $\pi \approx 3.1415926\ldots$, so it gives six decimal places of accuracy.

## Chapter R: Review Exercises

**1.** The elements of the set $\{6, 8, 10, \ldots, 20\}$ are the even numbers from 6 to 20 inclusive. The elements in the set are $\{6, 8, 10, 12, 14, 16, 18, 20\}$.

**3.** True. The set of negative integers $= \{\ldots, -4, -3, -2, -1\}$, while the set of whole numbers $= \{0, 1, 2, 3, \ldots\}$. The two sets do not intersect, and so they are disjoint.

**5.** True

7. False. The two sets are not equal because they do not have the same elements.

9. True

11. True

13. $A' = \{2, 6, 9, 10\}$

15. $B \cap E = \varnothing$

17. $D \cap \varnothing = \varnothing$

19. $(C \cap D) \cup B = \{1, 3\} \cup B = \{1, 2, 3, 4, 6, 8\}$

21. $-12, -6, -\sqrt{4}$ (or $-2$), 0, and 6 are integers.

23. 0 is a whole number, an integer, a rational number, and a real number.

25. $\dfrac{4\pi}{5}$ is an irrational number and a real number.

27. Answers will vary. Sample answer: The reciprocal of a product is the product of the reciprocals.

29. Answers will vary. Sample answer: A product raised to a power is the product of each factor raised to the power.

31. commutative

33. associative

35. identity

37. In a sample of 5000 students, 38% + 21% + 16% or 75% are expected to be over 19, or 5000(0.75) = 3750 students.

39. $(-4-1)(-3-5) - 2^3 = (-5)(-8) - 8$
$$= 40 - 8 = 32$$

41. $\left(-\dfrac{5}{9} - \dfrac{2}{3}\right) - \dfrac{5}{6} = \left(-\dfrac{5}{9} - \dfrac{6}{9}\right) - \dfrac{5}{6}$
$$= \dfrac{-11}{9} - \dfrac{5}{6} = -\dfrac{22}{18} - \dfrac{15}{18} = -\dfrac{37}{18}$$

43. $\dfrac{6(-4) - 3^2(-2)^3}{-5[-2-(-6)]} = \dfrac{6(-4) - 9(-8)}{-5[-2+6]}$
$$= \dfrac{6(-4) - 9(-8)}{-5(4)}$$
$$= \dfrac{-24 - (-72)}{-20} = \dfrac{-24 + 72}{-20}$$
$$= \dfrac{48}{-20} = -\dfrac{12}{5}$$

45. Let $a = -1$, $b = -2$, $c = 4$.
$-c(2a - 5b) = -4[2(-1) - 5(-2)]$
$$= -4\left[-2 - (-10)\right] = -4(-2 + 10)$$
$$= -4(8) = -32$$

47. Let $a = -1$, $b = -2$, $c = 4$.
$\dfrac{9a + 2b}{a + b + c} = \dfrac{9(-1) + 2(-2)}{-1 + (-2) + 4}$
$$= \dfrac{-9 + (-4)}{-3 + 4} = \dfrac{-13}{1} = -13$$

49. $(3q^3 - 9q^2 + 6) + (4q^3 - 8q + 3)$
$$= 3q^3 - 9q^2 + 6 + 4q^3 - 8q + 3$$
$$= 7q^3 - 9q^2 - 8q + 9$$

51. $(8y - 7)(2y + 7) = 16y^2 + 56y - 14y - 49$
$$= 16y^2 + 42y - 49$$

53. $(3k - 5m)^2 = (3k)^2 - 2(3k)(5m) + (5m)^2$
$$= 9k^2 - 30km + 25m^2$$

55. (a) 51 million

(b) Evaluate
$.146x^4 - 2.54x^3 + 11.0x^2 + 16.6x + 51.5$
when $x = 0$.
$.146 \cdot 0^4 - 2.54 \cdot 0^3 + 11.0 \cdot 0^2 + 16.6 \cdot 0 + 51.5$
$= .146 \cdot 0 - 2.54 \cdot 0 + 11.0 \cdot 0 + 16.6 \cdot 0 + 51.5$
$= 0 - 0 + 0 + 0 + 51.5$
$= 51.5$ or approximately 52 million

(c) The approximation is 1 million high.

57. (a) 183 million

(b) Evaluate
$.146x^4 - 2.54x^3 + 11.0x^2 + 16.6x + 51.5$
when $x = 5$.
$.146 \cdot 5^4 - 2.54 \cdot 5^3 + 11.0 \cdot 5^2 + 16.6 \cdot 5 + 51.5$
$= .146 \cdot 625 - 2.54 \cdot 125 + 11.0 \cdot 25$
$\qquad\qquad + 16.6 \cdot 5 + 51.5$
$= 91.25 - 317.5 + 275 + 83 + 51.5$
$= 183.25$ or approximately 183 million

(c) They are the same.

**59.** $\dfrac{30m^3 - 9m^2 + 22m + 5}{5m + 1}$

$$
\begin{array}{r}
6m^2 - 3m + 5 \\
5m+1\overline{\smash{)}\,30m^3 - 9m^2 + 22m + 5} \\
\underline{30m^3 + 6m^2\phantom{abcdefgh}} \\
-15m^2 + 22m\phantom{abc} \\
\underline{-15m^2 - \phantom{1}3m\phantom{abc}} \\
25m + 5 \\
\underline{25m + 5} \\
0
\end{array}
$$

Thus,

$$\frac{30m^3 - 9m^2 + 22m + 5}{5m + 1} = 6m^2 - 3m + 5.$$

**61.** $\dfrac{3b^3 - 8b^2 + 12b - 30}{b^2 + 4}$

Insert the missing term in the divisor with a 0 coefficient.

$$
\begin{array}{r}
3b - 8 \\
b^2+0b+4\overline{\smash{)}\,3b^3 - 8b^2 + 12b - 30} \\
\underline{3b^3 + 0b^2 + 12b\phantom{abcd}} \\
-8b^2 + 0b - 30 \\
\underline{-8b^2 + 0b - 32} \\
2
\end{array}
$$

Thus,

$$\frac{3b^3 - 8b^2 + 12b - 30}{b^2 + 4} = 3b - 8 + \frac{2}{b^2 + 4}.$$

**63.** $3(z-4)^2 + 9(z-4)^3 = 3(z-4)^2[1 + 3(z-4)]$
$\phantom{3(z-4)^2 + 9(z-4)^3} = 3(z-4)^2(1 + 3z - 12)$
$\phantom{3(z-4)^2 + 9(z-4)^3} = 3(z-4)^2(3z - 11)$

**65.** $z^2 - 6zk - 16k^2 = (z - 8k)(z + 2k)$

**67.** $48a^8 - 12a^7b - 90a^6b^2 = 6a^6(8a^2 - 2ab - 15b^2)$
$\phantom{48a^8 - 12a^7b - 90a^6b^2} = 6a^6(4a + 5b)(2a - 3b)$

**69.** $49m^8 - 9n^2 = (7m^4)^2 - (3n)^2$
$\phantom{49m^8 - 9n^2} = (7m^4 + 3n)(7m^4 - 3n)$

**71.** $6(3r - 1)^2 + (3r - 1) - 35$

Let $x = 3r - 1$. With this substitution,

$6(3r - 1)^2 + (3r - 1) - 35$ becomes

$6x^2 + x - 35$.

Factor the trinomial by trial and error.

$6x^2 + x - 35 = (3x - 7)(2x + 5)$

Replacing $x$ with $3r - 1$ gives

$[3(3r-1) - 7][2(3r-1) + 5]$
$= (9r - 3 - 7)(6r - 2 + 5) = (9r - 10)(6r + 3)$
$= 3(9r - 10)(2r + 1).$

**73.** $(3x - 4)^2 + (x - 5)(2)(3x - 4)(3)$
$= (3x - 4)[(3x - 4) + (x - 5)(2)(3)]$
$= (3x - 4)[3x - 4 + 6x - 30] = (3x - 4)(9x - 34)$

**75.** $\dfrac{k^2 + k}{8k^3} \cdot \dfrac{4}{k^2 - 1} = \dfrac{k(k+1)(4)}{8k^3(k+1)(k-1)}$

$\phantom{\dfrac{k^2 + k}{8k^3} \cdot \dfrac{4}{k^2 - 1}} = \dfrac{4k}{8k^3(k-1)} = \dfrac{1}{2k^2(k-1)}$

**77.** $\dfrac{x^2 + x - 2}{x^2 + 5x + 6} \div \dfrac{x^2 + 3x - 4}{x^2 + 4x + 3}$

$= \dfrac{x^2 + x - 2}{x^2 + 5x + 6} \cdot \dfrac{x^2 + 4x + 3}{x^2 + 3x - 4}$

$= \dfrac{(x+2)(x-1)}{(x+3)(x+2)} \cdot \dfrac{(x+3)(x+1)}{(x+4)(x-1)} = \dfrac{x+1}{x+4}$

**79.** $\dfrac{p^2 - 36q^2}{(p-6q)^2} \cdot \dfrac{p^2 - 5pq - 6q^2}{p^2 - 6pq + 36q^2} \div \dfrac{5p}{p^3 + 216q^3}$

$= \dfrac{p^2 - 36q^2}{(p-6q)^2} \cdot \dfrac{p^2 - 5pq - 6q^2}{p^2 - 6pq + 36q^2} \cdot \dfrac{p^3 + 216q^3}{5p}$

$= \dfrac{(p+6q)(p-6q)}{(p-6q)^2} \cdot \dfrac{(p-6q)(p+q)}{p^2 - 6pq + 36q^2}$

$\phantom{=} \cdot \dfrac{(p+6q)(p^2 - 6pq + 36q^2)}{5p}$

$= \dfrac{(p+q)(p+6q)^2}{5p}$

**81.** $\dfrac{m}{4 - m} + \dfrac{3m}{m - 4} = \dfrac{m(-1)}{(4-m)(-1)} + \dfrac{3m}{m - 4}$

$= \dfrac{-m}{m - 4} + \dfrac{3m}{m - 4} = \dfrac{2m}{m - 4}$

We may also use $4 - m$ as the common denominator. In this case, the result will be

$\dfrac{-2m}{4 - m}$. The two results are equivalent rational expressions.

**83.** $\dfrac{\frac{1}{p}+\frac{1}{q}}{1-\frac{1}{pq}}=\dfrac{pq\left(\frac{1}{p}+\frac{1}{q}\right)}{pq\left(1-\frac{1}{pq}\right)}$

$\quad=\dfrac{pq\left(\frac{1}{p}\right)+pq\left(\frac{1}{q}\right)}{pq(1)-pq\left(\frac{1}{pq}\right)}=\dfrac{q+p}{pq-1}$

**85.** $2^{-6}=\dfrac{1}{2^{6}}=\dfrac{1}{64}$

**87.** $\left(-\dfrac{5}{4}\right)^{-2}=\left(-\dfrac{4}{5}\right)^{2}=\dfrac{16}{25}$

**89.** $(5z^{3})(-2z^{5})=-10z^{3+5}=-10z^{8}$

**91.** $(-6p^{5}w^{4}m^{12})^{0}=1$   Definition of $a^{0}$

**93.** $\dfrac{-8y^{7}p^{-2}}{y^{-4}p^{-3}}=-8y^{7-(-4)}p^{(-2)-(-3)}=-8y^{11}p$

**95.** $\dfrac{(p+q)^{4}(p+q)^{-3}}{(p+q)^{6}}=(p+q)^{4+(-3)-6}$

$\quad\quad\quad=(p+q)^{-5}=\dfrac{1}{(p+q)^{5}}$

**97.** $(7r^{1/2})(2r^{3/4})(-r^{1/6})=-14r^{1/2+3/4+1/6}$

$\quad\quad\quad\quad=-14r^{17/12}$

**99.** $\dfrac{y^{5/3}\cdot y^{-2}}{y^{-5/6}}=\dfrac{y^{5/3+(-2)}}{y^{-5/6}}=\dfrac{y^{5/3+(-6/3)}}{y^{-5/6}}=\dfrac{y^{-1/3}}{y^{-5/6}}$

$\quad=y^{-1/3-(-5/6)}=y^{-2/6+5/6}$

$\quad=y^{3/6}=y^{1/2}$

**101.** $2z^{1/3}(5z^{2}-2)=2z^{1/3}(5z^{2})-2z^{1/3}(2)$

$\quad\quad=10z^{1/3+2}-4z^{1/3}$

$\quad\quad=10z^{1/3+6/3}-4z^{1/3}$

$\quad\quad=10z^{7/3}-4z^{1/3}$

**103.** $\sqrt{200}=\sqrt{100\cdot2}=\sqrt{100}\cdot\sqrt{2}=10\sqrt{2}$

**105.** $\sqrt[4]{1250}=\sqrt[4]{625\cdot2}=\sqrt[4]{625}\cdot\sqrt[4]{2}=5\sqrt[4]{2}$

**107.** $-3\sqrt{\dfrac{2}{5p^{2}}}=-\dfrac{\sqrt[3]{2}}{\sqrt[3]{5p^{2}}}=-\dfrac{\sqrt[3]{2}}{\sqrt[3]{5p^{2}}}\cdot\dfrac{\sqrt[3]{25p}}{\sqrt[3]{25p}}$

$\quad=-\dfrac{\sqrt[3]{2\cdot25p}}{\sqrt[3]{125p^{3}}}=-\dfrac{\sqrt[3]{50p}}{5p}$

**109.** $\sqrt[4]{\sqrt[3]{m}}=\left(\sqrt[3]{m}\right)^{1/4}=(m^{1/3})^{1/4}=m^{1/3\cdot1/4}$

$\quad=m^{1/12}=\sqrt[12]{m}$

**111.** This product has the pattern

$\quad(a+b)(a^{2}-ab+b^{2})=a^{3}+b^{3}$, the sum of

two cubes.

$\left(\sqrt[3]{2}+4\right)\left(\sqrt[3]{2^{2}}-4\sqrt[3]{2}+16\right)$

$=\left(\sqrt[3]{2}+4\right)\left[\left(\sqrt[3]{2}\right)^{2}-\sqrt[3]{2}\cdot4+4^{2}\right]$

$=\left(\sqrt[3]{2}\right)^{3}+4^{3}=2+64=66$

**113.** $\sqrt{18m^{3}}-3m\sqrt{32m}+5\sqrt{m^{3}}$

$=\sqrt{9m^{2}\cdot2m}-3m\sqrt{16\cdot2m}+5\sqrt{m^{2}m}$

$=3m\sqrt{2m}-4\cdot3m\sqrt{2m}+5m\sqrt{m}$

$=3m\sqrt{2m}-12m\sqrt{2m}+5m\sqrt{m}$

$=-9m\sqrt{2m}+5m\sqrt{m}$ or $m\left(-9\sqrt{2m}+5\sqrt{m}\right)$

**115.** $\dfrac{6}{3-\sqrt{2}}=\dfrac{6}{3-\sqrt{2}}\cdot\dfrac{3+\sqrt{2}}{3+\sqrt{2}}$

$\quad=\dfrac{6\left(3+\sqrt{2}\right)}{9-2}=\dfrac{6\left(3+\sqrt{2}\right)}{7}$

**117.** $x(x^{2}+5)=x\cdot x^{2}+x\cdot5=x^{3}+5x$

**119.** $(m^{2})^{3}=m^{2\cdot3}=m^{6}$

**121.** $\dfrac{\left(\frac{a}{b}\right)}{2}=\left(\dfrac{a}{b}\right)\cdot\left(\dfrac{1}{2}\right)=\dfrac{a}{2b}$

**123.** One possible answer is

$\dfrac{1}{\sqrt{a}+\sqrt{b}}=\dfrac{1}{\sqrt{a}+\sqrt{b}}\cdot\dfrac{\sqrt{a}-\sqrt{b}}{\sqrt{a}-\sqrt{b}}=\dfrac{\sqrt{a}-\sqrt{b}}{a-b}.$

**125.** $4-(t+1)=4-t-1$ or $3-t$

**127.** $(-5)^{2}=(-5)(-5)=25$ or $5^{2}$

## Chapter R: Test

**1.** False. $B'=\{2,4,6,7,8\}$

**2.** True

**3.** False. $D\cap\varnothing=\varnothing$

**4.** False. $(B\cap C)\cup D=\{1\}\cup D=\{1,4\}$

**5.** True

**6.** (a) $-13,\ -\dfrac{12}{4}$ (or $-3$), $0$, and $\sqrt{49}$ (or $7$) are

integers.

**(b)** $-13, -\dfrac{12}{4}$ (or $-3$), $0, \dfrac{3}{5}, 5.9$ (or $\frac{59}{10}$), and $\sqrt{49}$ (or $7$) are rational numbers.

**(c)** All numbers in the set are real numbers.

**7.** Let $x = -2, y = -4, z = 5.$

$$\left| \frac{x^2 + 2yz}{3(x+z)} \right| = \left| \frac{(-2)^2 + 2(-4)(5)}{3(-2+5)} \right|$$

$$= \left| \frac{4 + (-40)}{3(3)} \right| = \left| \frac{-36}{9} \right| = |-4| = 4$$

**8. (a)** associative property

**(b)** commutative

**(c)** distributive

**(d)** inverse

**9.** $A = 419, C = 267, Y = 3075, T = 15, I = 10$

Rating

$$\approx 85.68 \left( \frac{C}{A} \right) + 4.31 \left( \frac{Y}{A} \right)$$

$$+ 326.42 \left( \frac{T}{A} \right) - 419.07 \left( \frac{I}{A} \right)$$

$$= 85.68 \left( \frac{267}{419} \right) + 4.31 \left( \frac{3075}{419} \right)$$

$$+ 326.42 \left( \frac{15}{419} \right) - 419.07 \left( \frac{10}{419} \right)$$

$$\approx 87.9$$

Matt Hasselbeck's rating is 87.9.

**10.** $(x^2 - 3x + 2) - (x - 4x^2) + 3x(2x+1)$

$$= x^2 - 3x + 2 - x + 4x^2 + 6x^2 + 3x$$

$$= 11x^2 - x + 2$$

**11.** $(6r - 5)^2 = (6r)^2 - 2(6r)(5) + 5^2$

$$= 36r^2 - 60r + 25$$

**12.** $(t + 2)(3t^2 - t + 4)$

$$\begin{array}{r} 3t^2 \ -t+4 \\ t+2 \\ \hline 6t^2 - 2t + 8 \\ 3t^3 - t^2 + 4t \\ \hline 3t^3 + 5t^2 + 2t + 8 \end{array}$$

**13.** $\dfrac{2x^3 - 11x^2 + 28}{x - 5}$

$$\begin{array}{r} 2x^2 \ -x \ -5 \\ x-5{\overline{\smash{\big)}\,2x^3 - 11x^2 + 0x + 28}} \\ \underline{2x^3 - 10x^2} \\ -x^2 + 0x \\ \underline{-x^2 + 5x} \\ -5x + 28 \\ \underline{-5x + 25} \\ 3 \end{array}$$

Thus,

$$\frac{2x^3 - 11x^2 + 28}{x - 5} = 2x^2 - x - 5 + \frac{3}{x - 5}.$$

**14.** $x = 3$

Adjusted poverty threshold

$$\approx 5.476x^2 + 154.3x + 7889$$

$$= 5.476 \cdot 3^2 + 154.3 \cdot 3 + 7889$$

$$= 49.284 + 462.9 + 7889 = 8401.184$$

approximately \$8401

**15.** $x = 5$

Adjusted poverty threshold

$$\approx 5.476x^2 + 154.3x + 7889$$

$$= 5.476 \cdot 5^2 + 154.3 \cdot 5 + 7889$$

$$= 136.9 + 771.5 + 7889 = 8797.4$$

approximately \$8797

**16.** $6x^2 - 17x + 7 = (3x - 7)(2x - 1)$

**17.** $x^4 - 16 = (x^2)^2 - 4^2$

$$= (x^2 + 4)(x^2 - 4)$$

$$= (x^2 + 4)(x^2 - 2^2)$$

$$= (x^2 + 4)(x + 2)(x - 2)$$

**18.** $24m^3 - 14m^2 - 24m = 2m(12m^2 - 7m - 12)$

$$= 2m(4m + 3)(3m - 4)$$

**19.** $x^3 y^2 - 9x^3 - 8y^2 + 72$

$$= (x^3 y^2 - 9x^3) - (8y^2 - 72)$$

$$= x^3(y^2 - 9) - 8(y^2 - 9) = (x^3 - 8)(y^2 - 9)$$

$$= (x^3 - 2^3)(y^2 - 3^2)$$

$$= (x - 2)(x^2 + 2x + 4)(y + 3)(y - 3)$$

**20.** $\dfrac{5x^2-9x-2}{30x^3+6x^2} \cdot \dfrac{2x^8+6x^7+4x^6}{x^4-3x^2-4}$

$= \dfrac{(5x+1)(x-2)}{6x^2(5x+1)} \cdot \dfrac{2x^6(x^2+3x+2)}{(x^2-4)(x^2+1)}$

$= \dfrac{(5x+1)(x-2)}{6x^2(5x+1)} \cdot \dfrac{2x^6(x+2)(x+1)}{(x+2)(x-2)(x^2+1)}$

$= \dfrac{2x^6(x+1)}{6x^2(x^2+1)} = \dfrac{x^4(x+1)}{3(x^2+1)}$

**21.** $\dfrac{x}{x^2+3x+2} + \dfrac{2x}{2x^2-x-3}$

$= \dfrac{x}{(x+2)(x+1)} + \dfrac{2x}{(2x-3)(x+1)}$

The least common denominator is
$(x+2)(x+1)(2x-3)$.

$= \dfrac{x(2x-3)}{(x+2)(x+1)(2x-3)} + \dfrac{2x(x+2)}{(2x-3)(x+1)(x+2)}$

$= \dfrac{2x^2-3x}{(x+2)(x+1)(2x-3)} + \dfrac{2x^2+4x}{(x+2)(x+1)(2x-3)}$

$= \dfrac{2x^2-3x+2x^2+4x}{(x+2)(x+1)(2x-3)} = \dfrac{4x^2+x}{(x+2)(x+1)(2x-3)}$

$= \dfrac{x(4x+1)}{(x+2)(x+1)(2x-3)}$

**22.** $\dfrac{a+b}{2a-3} - \dfrac{a-b}{3-2a} = \dfrac{a+b}{2a-3} - \dfrac{(a-b)(-1)}{(3-2a)(-1)}$

$= \dfrac{a+b}{2a-3} + \dfrac{a-b}{2a-3} = \dfrac{2a}{2a-3}$

If $3-2a$ is used as the common denominator,

the result will be $\dfrac{-2a}{3-2a}$. The rational

expressions $\dfrac{2a}{2a-3}$ and $\dfrac{-2a}{3-2a}$ are

equivalent.

**23.** $\dfrac{y-2}{y-\frac{4}{y}} = \dfrac{y(y-2)}{y\left(y-\frac{4}{y}\right)} = \dfrac{y^2-2y}{y^2-4}$

$= \dfrac{y(y-2)}{(y+2)(y-2)} = \dfrac{y}{y+2}$

**24.** $\left(\dfrac{x^{-2}y^{-1/3}}{x^{-5/3}y^{-2/3}}\right)^3 = \dfrac{x^{-6}y^{-1}}{x^{-5}y^{-2}}$

$= x^{-6-(-5)}y^{-1-(-2)} = x^{-1}y = \dfrac{y}{x}$

**25.** $\left(-\dfrac{64}{27}\right)^{-2/3} = \left(-\dfrac{27}{64}\right)^{2/3} = \left[\left(-\dfrac{27}{64}\right)^{1/3}\right]^2$

$= \left(-\dfrac{3}{4}\right)^2 = \dfrac{9}{16}$

**26.** $\sqrt{18x^5y^8} = \sqrt{(9x^4y^8)(2x)}$

$= \sqrt{9x^4y^8} \cdot \sqrt{2x} = 3x^2y^4\sqrt{2x}$

**27.** $\sqrt{32x} + \sqrt{2x} - \sqrt{18x}$

$= \sqrt{16 \cdot 2x} + \sqrt{2x} - \sqrt{9 \cdot 2x}$

$= 4\sqrt{2x} + \sqrt{2x} - 3\sqrt{2x} = 2\sqrt{2x}$

**28.** $\left(\sqrt{x}-\sqrt{y}\right)\left(\sqrt{x}+\sqrt{y}\right) = \left(\sqrt{x}\right)^2 - \left(\sqrt{y}\right)^2$

$= x-y$

**29.** $\dfrac{14}{\sqrt{11}-\sqrt{7}} = \dfrac{14}{\sqrt{11}-\sqrt{7}} \cdot \dfrac{\sqrt{11}+\sqrt{7}}{\sqrt{11}+\sqrt{7}}$

$= \dfrac{14\left(\sqrt{11}+\sqrt{7}\right)}{11-7} = \dfrac{14\left(\sqrt{11}+\sqrt{7}\right)}{4}$

$= \dfrac{7\left(\sqrt{11}+\sqrt{7}\right)}{2}$

**30.** Let $L = 3.5$.

$t = 2\pi\sqrt{\dfrac{L}{32}} = 2\pi\sqrt{\dfrac{3.5}{32}} \approx 2.1$

The period of a pendulum 3.5 ft long is
approximately 2.1 seconds.

# Chapter 1

## EQUATIONS AND INEQUALITIES

### Section 1.1: Linear Equations

1. Solve the equation $2x+7 = x-1$
$$2x+7 = x-1$$
$$2x+7-x = x-1-x$$
$$x+7 = -1$$
$$x+7-7 = -1-7$$
$$x = -8$$
Moreover, replacing $x$ with $-8$ in $2x+7 = x-1$ yields a true statement. Therefore, the given statement is true.

3. The equations $x^2 = 4$ and $x+2 = 4$ are not equivalent. The first has a solution set $\{-2, 2\}$ while the solution set of the second is $\{2\}$.
Since the equations do not have the same solution set, they are not equivalent. The given statement is false.

5. Answers will vary.

7. B cannot be written in the form $ax + b = 0$. A can be written as $15x - 7 = 0$ or $15x + (-7) = 0$, C can be written as $2x = 0$ or $2x + 0 = 0$, and D can be written as $-.04x - .4 = 0$ or $-.04x + (-.4) = 0$.

9. $5x+4 = 3x-4$
$$2x+4 = -4$$
$$2x = -8 \Rightarrow x = -4$$
Solution set: $\{-4\}$

11. $6(3x-1) = 8 - (10x-14)$
$$18x-6 = 8 - 10x + 14$$
$$18x-6 = 22 - 10x$$
$$28x - 6 = 22$$
$$28x = 28 \Rightarrow x = 1$$
Solution set: $\{1\}$

13. $\dfrac{5}{6}x - 2x + \dfrac{4}{3} = \dfrac{5}{3}$
$$6 \cdot \left[ \dfrac{5}{6}x - 2x + \dfrac{4}{3} \right] = 6 \cdot \dfrac{5}{3}$$
$$5x - 12x + 8 = 10$$
$$-7x + 8 = 10$$
$$-7x = 2 \Rightarrow x = -\dfrac{2}{7}$$
Solution set: $\left\{ -\dfrac{2}{7} \right\}$

15. $3x + 5 - 5(x+1) = 6x + 7$
$$3x + 5 - 5x - 5 = 6x + 7$$
$$-2x = 6x + 7$$
$$-8x = 7 \Rightarrow x = \dfrac{7}{-8} = -\dfrac{7}{8}$$
Solution set: $\left\{ -\dfrac{7}{8} \right\}$

17. $2\left[ x - (4+2x) + 3 \right] = 2x + 2$
$$2(x - 4 - 2x + 3) = 2x + 2$$
$$2(-x-1) = 2x + 2$$
$$-2x - 2 = 2x + 2$$
$$-2 = 4x + 2$$
$$-4 = 4x \Rightarrow -1 = x$$
Solution set: $\{-1\}$

19. $\dfrac{1}{14}(3x-2) = \dfrac{x+10}{10}$
$$70 \cdot \left[ \dfrac{1}{14}(3x-2) \right] = 70 \cdot \left[ \dfrac{x+10}{10} \right]$$
$$5(3x-2) = 7(x+10)$$
$$15x - 10 = 7x + 70$$
$$8x - 10 = 70$$
$$8x = 80 \Rightarrow x = 10$$
Solution set: $\{10\}$

21. $.2x - .5 = .1x + 7$
$$10(.2x - .5) = 10(.1x + 7)$$
$$2x - 5 = x + 70$$
$$x - 5 = 70 \Rightarrow x = 75$$
Solution set: $\{75\}$

**23.** $-4(2x-6)+8x=5x+24+x$

$\qquad -8x+24+8x=6x+24$

$\qquad\qquad\quad 24=6x+24$

$\qquad\qquad\quad 0=6x \Rightarrow 0=x$

Solution set: $\{0\}$

**25.** $\qquad .5x+\dfrac{4}{3}x=x+10$

$\qquad\quad \dfrac{1}{2}x+\dfrac{4}{3}x=x+10$

$\qquad 6\left(\dfrac{1}{2}x+\dfrac{4}{3}x\right)=6(x+10)$

$\qquad\qquad 3x+8x=6x+60$

$\qquad\qquad\quad 11x=6x+60$

$\qquad\qquad\quad\; 5x=60 \Rightarrow x=12$

Solution set: $\{12\}$

**27.** $\qquad\quad .08x+.06(x+12)=7.72$

$\qquad 100\big[.08x+.06(x+12)\big]=100\cdot 7.72$

$\qquad\qquad 8x+6(x+12)=772$

$\qquad\qquad\quad 8x+6x+72=772$

$\qquad\qquad\qquad 14x+72=772$

$\qquad\qquad\qquad\quad 14x=700 \Rightarrow x=50$

Solution set: $\{50\}$

**29.** $4(2x+7)=2x+22+3(2x+2)$

$\qquad 8x+28=2x+22+6x+6$

$\qquad 8x+28=8x+28$

$\qquad\quad 28=28 \Rightarrow 0=0$

identity; $\{$all real numbers$\}$

**31.** $2(x-8)=3x-16$

$\quad 2x-16=3x-16$

$\quad -16=x-16 \Rightarrow 0=x$

conditional equation; $\{0\}$

**33.** $\qquad .3(x+2)-.5(x+2)=-.2x-.4$

$\quad 10\big[.3(x+2)-.5(x+2)\big]=10[-.2x-.4]$

$\qquad\quad 3(x+2)-5(x+2)=-2x-4$

$\qquad 3x+6-5x-10=-2x-4$

$\qquad\qquad -2x-4=-2x-4$

$\qquad\qquad\qquad 0=0$

identity; $\{$all real numbers$\}$

**35.** $4(x+7)=2(x+12)+2(x+1)$

$\quad 4x+28=2x+24+2x+2$

$\quad 4x+28=4x+26$

$\qquad 28=26$

contradiction; $\varnothing$

**37.** Answers will vary. In solving an equation, you cannot multiply (or divide) both sides of an equation by zero. This is essentially what happened when the student divided both sides of the equation by $x$. To solve the equation, the student should isolate the variable term, which leads to the solution set $\{0\}$.

$\qquad\qquad 5x=4x$

$\qquad 5x-4x=4x-4x$

$\qquad\qquad\quad x=0$

**39.** $\quad V=lwh$

$\quad \dfrac{V}{wh}=\dfrac{lwh}{wh}$

$\qquad l=\dfrac{V}{wh}$

**41.** $\qquad P=a+b+c$

$\quad P-a-b=c$

$\qquad\quad c=P-a-b$

**43.** $\qquad A=\dfrac{1}{2}h(B+b)$

$\qquad 2A=2\left[\dfrac{1}{2}h(B+b)\right]$

$\qquad 2A=h(B+b)$

$\qquad 2A=Bh+bh$

$\quad 2A-bh=Bh$

$\quad \dfrac{2A-bh}{h}=\dfrac{Bh}{h}$

$\qquad B=\dfrac{2A-bh}{h}=\dfrac{2A}{h}-b$

**45.** $\qquad S=2\pi rh+2\pi r^2$

$\quad S-2\pi r^2=2\pi rh$

$\quad \dfrac{S-2\pi r^2}{2\pi r}=\dfrac{2\pi rh}{2\pi r}$

$\qquad h=\dfrac{S-2\pi r^2}{2\pi r}=\dfrac{S}{2\pi r}-r$

**47.** $\qquad S=2lw+2wh+2hl$

$\quad S-2lw=2wh+2hl$

$\quad S-2lw=(2w+2l)h$

$\quad \dfrac{S-2lw}{2w+2l}=\dfrac{(2w+2l)h}{2w+2l}$

$\qquad h=\dfrac{S-2lw}{2w+2l}$

**49.** $2(x-a)+b=3x+a$

$\quad 2x-2a+b=3x+a$

$\quad -3a+b=x$

$\qquad\quad x=-3a+b$

**51.**   $ax + b = 3(x - a)$
$ax + b = 3x - 3a$
$3a + b = 3x - ax$
$3a + b = (3 - a)x$
$\dfrac{3a + b}{3 - a} = x$
$x = \dfrac{3a + b}{3 - a}$

**53.**   $\dfrac{x}{a - 1} = ax + 3$
$(a - 1)\left[\dfrac{x}{a - 1}\right] = (a - 1)(ax + 3)$
$x = a^2 x + 3a - ax - 3$
$3 - 3a = a^2 x - ax - x$
$3 - 3a = (a^2 - a - 1)x$
$\dfrac{3 - 3a}{a^2 - a - 1} = x$
$x = \dfrac{3 - 3a}{a^2 - a - 1}$

**55.**   $a^2 x + 3x = 2a^2$
$(a^2 + 3)x = 2a^2$
$x = \dfrac{2a^2}{a^2 + 3}$

**57.**   $3x = (2x - 1)(m + 4)$
$3x = 2xm + 8x - m - 4$
$m + 4 = 2xm + 5x$
$m + 4 = (2m + 5)x$
$\dfrac{m + 4}{2m + 5} = x$
$x = \dfrac{m + 4}{2m + 5}$

**59.  (a)**   Here, $r = .08$, $P = 3150$, and
$t = \dfrac{6}{12} = \dfrac{1}{2}$ (year).
$I = Prt = 3150(.08)\left(\dfrac{1}{2}\right) = \$126$
The interest is $126.

**(b)**   The amount Miguel must pay Julio at the end of the six months is
$3150 + $126 = $3276.

**61.**   $F = \dfrac{9}{5}C + 32$
$F = \dfrac{9}{5} \cdot 40 + 32$
$F = 72 + 32$
$F = 104$
Therefore, $40°\text{C} = 104°\text{F}$.

**63.**   $C = \dfrac{5}{9}(F - 32)$
$C = \dfrac{5}{9}(59 - 32)$
$C = \dfrac{5}{9} \cdot 27$
$C = 15$
Therefore, $59°\text{F} = 15°\text{C}$.

**65.**   $C = \dfrac{5}{9}(F - 32)$
$C = \dfrac{5}{9}(100 - 32)$
$C = \dfrac{5}{9} \cdot 68$
$C \approx 37.8$
Therefore, $100°\text{F} \approx 37.8°\text{C}$.

**67.**   $C = \dfrac{5}{9}(F - 32)$
$C = \dfrac{5}{9}(867 - 32)$
$C = \dfrac{5}{9} \cdot 835$
$C \approx 463.9$
Therefore, $865°\text{F} \approx 463.9°\text{C}$.

**69.**   $C = \dfrac{5}{9}(F - 32)$
$C = \dfrac{5}{9}(7 - 32)$
$C = \dfrac{5}{9} \cdot (-25)$
$C \approx -13.9$
Therefore, $7°\text{F} \approx -14°\text{C}$.

## Section 1.2: Applications and Modeling with Linear Equations

**Connections (page 96)**

Step 1 compares to Polya's first step, Steps 2 and 3 compare to his second step, Step 4 compares to his third step, and Step 6 compares to his fourth step.

**Exercises**

1.  15 minutes is $\frac{1}{4}$ of an hour, so multiply 100 mph by $\frac{1}{4}$ to get a distance of 25 mi.

3.  4% is .04, so multiply $500 by .04 and by 2 yrs to get interest of $40

5.  Concentration A, 36%, cannot possibly be the concentration of the mixture because it exceeds both the concentrations.

7.  D

9.  In the formula $P = 2l + 2w$, let
    $P = 294$ and $w = 57$.
    $294 = 2l + 2 \cdot 57$
    $294 = 2l + 114$
    $180 = 2l \Rightarrow 90 = l$
    The length is 90 cm.

11. Let $x =$ length of shortest side.
    Then $2x =$ length of each of the longer sides.
    The perimeter of a triangle is the sum of the measures of the three sides.
    $x + 2x + 2x = 30 \Rightarrow 5x = 30 \Rightarrow x = 6$
    The length of the shortest side is 6 cm.

13. Let $x =$ length of shortest side.
    Then $2x - 200 =$ length of longest side and the length of the middle side is
    $(2x - 200) - 200 = 2x - 400$.
    The perimeter of a triangle is the sum of the measures of the three sides.
    $x + (2x - 200) + (2x - 400) = 2400$
    $\quad x + 2x - 200 + 2x - 400 = 2400$
    $\qquad\qquad\qquad 5x - 600 = 2400$
    $\qquad\qquad\qquad\quad 5x = 3000 \Rightarrow x = 600$
    The length of the shortest side is 600 ft. The middle side is $2 \cdot 600 - 400 = 1200 - 400$
    $= 800$ ft. The longest side is $2 \cdot 600 - 200$
    $= 1200 - 200 = 1000$ ft.

15. Let $l =$ the length of the book.
    Then $l - .42 =$ the width of the book
    Use the formula for the perimeter of a rectangle.
    $P = 2l + 2w$
    $5.96 = 2l + 2(l - .42)$
    $5.96 = 2l + 2l - .84$
    $5.96 = 4l - .84$
    $\;\, 6.8 = 4l \Rightarrow 1.7 = l$
    The length of the book is 1.7 cm, and the width of the book is $1.7 - .42 = 1.28$ cm.

17. Let $h =$ the height of box.
    Use the formula for the surface area of a rectangular box.
    $S = 2lw + 2wh + 2hl$
    $496 = 2 \cdot 18 \cdot 8 + 2 \cdot 8h + 2h \cdot 18$
    $496 = 288 + 16h + 36h$
    $496 = 288 + 52h$
    $208 = 52h \Rightarrow 4 = h$
    The height of the box is 4 ft.

19. Let $x =$ the time (in hours) spent on the way to the business appointment.

    |            | $r$ | $t$           | $d$                      |
    |------------|-----|---------------|--------------------------|
    | Morning    | 50  | $x$           | $50x$                    |
    | Afternoon  | 40  | $x + \frac{1}{4}$ | $40\left(x + \frac{1}{4}\right)$ |

    The distance on the way to the business appointment is the same as the return trip, so
    $50x = 40\left(x + \frac{1}{4}\right)$
    $50x = 40x + 10$
    $10x = 10 \Rightarrow x = 1$
    Since she drove 1 hr, her distance traveled would be $50 \cdot 1 = 50$ mi.

21. Let $x =$ David's speed (in mph) on bike.
    Then $x + 4.5 =$ David's speed (in mph) driving.

    |      | $r$     | $t$                          | $d$                        |
    |------|---------|------------------------------|----------------------------|
    | Car  | $x + 4.5$ | 20 min $= \frac{1}{3}$ hr | $\frac{1}{3}(x + 4.5)$     |
    | Bike | $x$     | 45 min $= \frac{3}{4}$ hr    | $\frac{3}{4}x$             |

    The distance by bike and car are the same, so
    $\frac{1}{3}(x + 4.5) = \frac{3}{4}x$
    $12\left[\frac{1}{3}(x + 4.5)\right] = 12\left[\frac{3}{4}x\right]$
    $4(x + 4.5) = 9x$
    $4x + 18 = 9x$
    $18 = 5x \Rightarrow \dfrac{18}{5} = x$

    Since his rate is $\frac{18}{5}$ (or 3.6) mph, David travels
    $\frac{3}{4}\left(\frac{18}{5}\right) = \frac{27}{10} = 2.7$ mi to work.

**23.** Let $x$ = time (in hours) it takes for Russ and Janet to be 1.5 mi apart.

|       | $r$ | $t$ | $d$ |
|-------|-----|-----|-----|
| Russ  | 7   | $x$ | $7x$ |
| Janet | 5   | $x$ | $5x$ |

Since Russ's rate is faster than Janet's, he travels farther than Janet in the same amount of time. To have the difference between Russ and Janet to be 1.5 mi, solve the following equation.

$$7x - 5x = 1.5 \Rightarrow 2x = 1.5 \Rightarrow x = .75$$

It will take .75 hr = 45 min for Russ and Janet to be 1.5 mi apart.

**25.** We need to determine how many meters are in 26 miles.

$$26 \text{ mi} \cdot \frac{5,280 \text{ ft}}{1 \text{ mi}} \cdot \frac{1 \text{ m}}{3.281 \text{ ft}} \approx 41,840.9 \text{ m}$$

Tim Montgomery's rate in the 100-m dash would be $r = \dfrac{d}{t} = \dfrac{100}{9.78}$ meters per second.

Thus, the time it would take for Tim to run the 26-mi marathon would be

$$t = \frac{d}{r} = \frac{41,840.9}{\frac{100}{9.78}} = 41,840.9 \cdot \tfrac{9.78}{100} \approx 4,092 \text{ sec.}$$

Since there is 60 seconds in one minute and $60 \cdot 60 = 3,600$ seconds in one hour,

$$4,092 \text{ sec} = 1 \cdot 3600 + 8 \cdot 60 + 12 \text{ sec}$$

or 1 hr, 8 min, 12 sec.  This is about $\frac{1}{2}$ the world record time.

**27.** Let $x$ = speed (in km/hr) of Joann's boat. When Joann is traveling upstream, the current slows her down, so we subtract the speed of the current from the speed of the boat. When she is traveling downstream, the current speeds her up, so we add the speed of the current to the speed of the boat.

|            | $r$   | $t$                          | $d$               |
|------------|-------|------------------------------|-------------------|
| Upstream   | $x-5$ | 20 min = $\frac{1}{3}$ hr    | $\frac{1}{3}(x-5)$ |
| Downstream | $x+5$ | 15 min = $\frac{1}{4}$ hr    | $\frac{1}{4}(x+5)$ |

Since the distance upstream and downstream are the same, we must solve the following equation.

$$\tfrac{1}{3}(x-5) = \tfrac{1}{4}(x+5)$$
$$12\left[\tfrac{1}{3}(x-5)\right] = 12\left[\tfrac{1}{4}(x+5)\right]$$
$$4(x-5) = 3(x+5)$$
$$4x - 20 = 3x + 15$$
$$x - 20 = 15 \Rightarrow x = 35$$

The speed of Joann's boat is 35 km per hour.

**29.** Let $x$ = the amount of 5% acid solution (in gallons).

| Strength | Gallons of Solution | Gallons of Pure Acid |
|----------|---------------------|----------------------|
| 5%       | $x$                 | $.05x$               |
| 10%      | 5                   | $.10 \cdot 5 = .5$   |
| 7%       | $x+5$               | $.07(x+5)$           |

The number of gallons of pure acid in the 5% solution plus the number of gallons of pure acid in the 10% solution must equal the number of gallons of pure acid in the 7% solution.

$$.05x + .5 = .07(x+5)$$
$$.05x + .5 = .07x + .35$$
$$.5 = .02x + .35 \Rightarrow .15 = .02x$$
$$\frac{.15}{.02} = x \Rightarrow x = 7.5 = 7\tfrac{1}{2} \text{ gal}$$

$7\frac{1}{2}$ gallons of the 5% solution should be added.

**31.** Let $x$ = the amount of 100% alcohol solution (in liters).

| Strength | Liters of Solution | Liters of Pure Alcohol |
|----------|--------------------|------------------------|
| 100%     | $x$                | $1x = x$               |
| 10%      | 7                  | $.10 \cdot 7 = .7$     |
| 30%      | $x+7$              | $.30(x+7)$             |

The number of liters of pure alcohol in the 100% solution plus the number of liters of pure alcohol in the 10% solution must equal the number of liters of pure alcohol in the 30% solution.

$$x + .7 = .30(x+7)$$
$$x + .7 = .30x + 2.1$$
$$.7x + .7 = 2.1 \Rightarrow .7x = 1.4$$
$$x = \frac{1.4}{.7} = \frac{14}{7} = 2 \text{ L}$$

2 L of the 100% solution should be added.

**33.** Let $x$ = the amount of water (in mL).

| Strength | Milliliters of Solution | Milliliters of Salt |
|----------|-------------------------|---------------------|
| 6%       | 8                       | $.06(8) = .48$      |
| 0%       | $x$                     | $0(x) = 0$          |
| 4%       | $8+x$                   | $.04(8+x)$          |

The number of milliliters of salt in the 6% solution plus the number of milliliters of salt in the water (0% solution) must equal the number of milliliters in the 4% solution.

$$.48 + 0 = .04(8 + x)$$
$$.48 = .32 + .04x$$
$$.16 = .04x$$
$$\frac{.16}{.04} = x \Rightarrow x = \frac{16}{4} = 4 \text{ mL}$$

To reduce the saline concentration to 4%, 4 mL of water should be added.

**35.** Let $x$ = amount of the short-term note. Then $240,000 - x$ = amount of the long-term note.

| Amount of Note | Interest Rate | Interest |
|---|---|---|
| $x$ | 6% | $.06x$ |
| $240,000 - x$ | 5% | $.05(240,000 - x)$ |
| $240,000$ | | $13,000$ |

The amount of interest from the 6% note plus the amount of interest from the 5% note must equal the total amount of interest.

$$.06x + .05(240,000 - x) = 13,000$$
$$.06x + 12,000 - .05x = 13,000$$
$$.01x + 12,000 = 13,000$$
$$.01x = 1,000$$
$$x = 100,000$$

The amount of the short-term note is $100,000 and the amount of the long-term note is $240,000 - $100,000 = $140,000.

**37.** Let $x$ = amount invested at 2.5%. Then $2x$ = amount invested at 3%.

| Amount in Account | Interest Rate | Interest |
|---|---|---|
| $x$ | 2.5% | $.025x$ |
| $2x$ | 3% | $.03(2x) = .06x$ |
| | | $850$ |

The amount of interest from the 2.5% account plus the amount of interest from the 3% account must equal the total amount of interest.

$$.025x + .06x = 850$$
$$.085x = 850 \Rightarrow x = \$10,000$$

Karen deposited $10,000 at 2.5% and 2($10,000) = $20,000 at 3%.

**39.** 30% of $200,000 is $60,000, so after paying her income tax, Linda had $140,000 left to invest. Let $x$ = amount invested at 1.5%. Then $140,000 - x$ = amount invested at 4%.

| Amount Invested | Interest Rate | Interest |
|---|---|---|
| $x$ | 1.5% | $.015x$ |
| $140,000 - x$ | 4% | $.04(140,000 - x)$ |
| $140,000$ | | $4350$ |

$$.015x + .04(140,000 - x) = 4350$$
$$.015x + 5600 - .04x = 4350$$
$$-.025x + 5600 = 4350$$
$$-.025x = -1250$$
$$x = \$50,000$$

Linda invested $50,000 at 1.5% and $140,000 - $50,000 = $90,000 at 4%.

**41.** **(a)** $k = \dfrac{.132B}{W} = \dfrac{.132 \cdot 20}{75} = .0352$

**(b)** $R = (.0352)(.42) \approx .015$
An individual's increased lifetime cancer risk would be 1.5%.

**(c)** Using an average life expectancy of 72 years, $\dfrac{.015}{72} \cdot 5000 \approx 1$ case of cancer each year.

**43.** **(a)** The volume would be $10 \times 10 \times 8 = 800 \text{ ft}^3$.

**(b)** Since the paneling has an area of $4 \times 8 = 32$ sq ft, it emits $32 \cdot 3365 = 107,680 \ \mu g$ of formaldehyde.

**(c)** $F = 107,680x$

**(d)** Since $33 \ \mu g / ft^3$ causes irritation, the room would need $33 \cdot 800 = 26,400 \ \mu g$ to cause irritation.
$$F = 107,680x$$
$$26,400 = 107,680x$$
$$\frac{26,400}{107,680} = x \Rightarrow x \approx .25 \text{ day}$$
or approximately 6 hours.

**45.** **(a)** In 2008, $x = 5$.
$$y = .2145x + 15.69$$
$$y = .2145 \cdot 5 + 15.69$$
$$y = 1.0725 + 15.69$$
$$y = 16.7625$$
The projected enrollment for Fall 2008 is approximately 16.8 million

**(b)**
$$y = .2145x + 15.69$$
$$17 = .2145x + 15.69$$
$$1.31 = .2145x$$
$$\frac{1.31}{.2145} = x$$
$$x \approx 6.1$$

Enrollment is projected to reach 17 million in the year 2009

**(c)** They are quite close.

**(d)**
$$y = .2145(-10) + 15.69$$
$$y = -2.145 + 15.69$$
$$y \approx 13.5$$

The enrollment would be approximately 13.5 million

**(e)** Answers will vary.

## Section 1.3: Complex Numbers

1. true

3. true

5. false (Every real number is a complex number.)

7. $-4$ is real and complex.

9. $13i$ is complex, pure imaginary and nonreal complex.

11. $5 + i$ is complex and nonreal complex.

13. $\pi$ is real and complex.

15. $\sqrt{-25} = 5i$ is complex, pure imaginary and nonreal complex.

17. $\sqrt{-25} = i\sqrt{25} = 5i$

19. $\sqrt{-10} = i\sqrt{10}$

21. $\sqrt{-288} = i\sqrt{288} = i\sqrt{144 \cdot 2} = 12i\sqrt{2}$

23. $-\sqrt{-18} = -i\sqrt{18} = -i\sqrt{9 \cdot 2} = -3i\sqrt{2}$

25. $\sqrt{-13} \cdot \sqrt{-13} = i\sqrt{13} \cdot i\sqrt{13}$
$$= i^2\left(\sqrt{13}\right)^2 = -1 \cdot 13 = -13$$

27. $\sqrt{-3} \cdot \sqrt{-8} = i\sqrt{3} \cdot i\sqrt{8} = i^2\sqrt{3 \cdot 8}$
$$= -1 \cdot \sqrt{24} = -\sqrt{4 \cdot 6} = -2\sqrt{6}$$

29. $\dfrac{\sqrt{-30}}{\sqrt{-10}} = \dfrac{i\sqrt{30}}{i\sqrt{10}} = \sqrt{\dfrac{30}{10}} = \sqrt{3}$

31. $\dfrac{\sqrt{-24}}{\sqrt{8}} = \dfrac{i\sqrt{24}}{\sqrt{8}} = i\sqrt{\dfrac{24}{8}} = i\sqrt{3}$

33. $\dfrac{\sqrt{-10}}{\sqrt{-40}} = \dfrac{i\sqrt{10}}{i\sqrt{40}} = \sqrt{\dfrac{10}{40}} = \sqrt{\dfrac{1}{4}} = \dfrac{1}{2}$

35. $\dfrac{\sqrt{-6} \cdot \sqrt{-2}}{\sqrt{3}} = \dfrac{i\sqrt{6} \cdot i\sqrt{2}}{\sqrt{3}} = i^2\sqrt{\dfrac{6 \cdot 2}{3}}$
$$= -1 \cdot \sqrt{\dfrac{12}{3}} = -\sqrt{4} = -2$$

37. $\dfrac{-6 - \sqrt{-24}}{2} = \dfrac{-6 - \sqrt{-4 \cdot 6}}{2} = \dfrac{-6 - 2i\sqrt{6}}{2}$
$$= \dfrac{2\left(-3 - i\sqrt{6}\right)}{2} = -3 - i\sqrt{6}$$

39. $\dfrac{10 + \sqrt{-200}}{5} = \dfrac{10 + \sqrt{-100 \cdot 2}}{5}$
$$= \dfrac{10 + 10i\sqrt{2}}{5} = \dfrac{5\left(2 + 2i\sqrt{2}\right)}{5}$$
$$= 2 + 2i\sqrt{2}$$

41. $\dfrac{-3 + \sqrt{-18}}{24} = \dfrac{-3 + \sqrt{-9 \cdot 2}}{24} = \dfrac{-3 + 3i\sqrt{2}}{24}$
$$= \dfrac{3\left(-1 + i\sqrt{2}\right)}{24} = \dfrac{-1 + i\sqrt{2}}{8}$$
$$= -\dfrac{1}{8} + \dfrac{\sqrt{2}}{8}i$$

43. $(3 + 2i) + (9 - 3i) = (3 + 9) + \left[2 + (-3)\right]i$
$$= 12 + (-1)i = 12 - i$$

45. $(-2 + 4i) - (-4 + 4i)$
$$= \left[-2 - (-4)\right] + (4 - 4)i$$
$$= 2 + 0i = 2$$

47. $(2 - 5i) - (3 + 4i) - (-2 + i)$
$$= \left[2 - 3 - (-2)\right] + (-5 - 4 - 1)i$$
$$= 1 + (-10)i = 1 - 10i$$

49. $-i - 2 - (6 - 4i) - (5 - 2i)$
$$= (-2 - 6 - 5) + \left[-1 - (-4) - (-2)\right]i$$
$$= -13 + 5i$$

51. $(2 + i)(3 - 2i)$
$$= 2(3) + 2(-2i) + i(3) + i(-2i)$$
$$= 6 - 4i + 3i - 2i^2 = 6 - i - 2(-1)$$
$$= 6 - i + 2 = 8 - i$$

53. $(2 + 4i)(-1 + 3i)$
$$= 2(-1) + 2(3i) + 4i(-1) + 4i(3i)$$
$$= -2 + 6i - 4i + 12i^2 = -2 + 2i + 12(-1)$$
$$= -2 + 2i - 12 = -14 + 2i$$

**55.** $(3-2i)^2 = 3^2 - 2(3)(2i) + (2i)^2$
$\qquad = 9 - 12i - 4 = 5 - 12i$

**57.** $(3+i)(3-i) = 3^2 - i^2 = 9 - (-1) = 10$

**59.** $(-2-3i)(-2+3i) = (-2)^2 - (3i)^2 = 4 - 9i^2$
$\qquad = 4 - 9(-1) = 13$

**61.** $\left(\sqrt{6}+i\right)\left(\sqrt{6}-i\right) = \left(\sqrt{6}\right)^2 - i^2$
$\qquad = 6 - (-1) = 6 + 1 = 7$

**63.** $i(3-4i)(3+4i) = i\left[(3-4i)(3+4i)\right]$
$\qquad = i\left[3^2 - (4i)^2\right]$
$\qquad = i\left[9 - 16i^2\right]$
$\qquad = i\left[9 - 16(-1)\right]$
$\qquad = i(9+16) = 25i$

**65.** $3i(2-i)^2 = 3i\left(2^2 - 2(2i) + i^2\right)$
$\qquad = 3i(4 - 4i - 1) = 3i(3 - 4i)$
$\qquad = 9i - 12i^2 = 9i - 12(-1)$
$\qquad = 12 + 9i$

**67.** $(2+i)(2-i)(4+3i) = \left[(2+i)(2-i)\right](4+3i)$
$\qquad = \left[2^2 - i^2\right](4+3i)$
$\qquad = \left[4 - (-1)\right](4+3i)$
$\qquad = 5(4+3i) = 20 + 15i$

**69.** $i^{25} = i^{24} \cdot i = \left(i^4\right)^6 \cdot i = 1^6 \cdot i = i$

**71.** $i^{22} = i^{20} \cdot i^2 = \left(i^4\right)^5 \cdot (-1) = 1^5 \cdot (-1) = -1$

**73.** $i^{23} = i^{20} \cdot i^3 = \left(i^4\right)^5 \cdot i^3 = 1^5 \cdot (-i) = -i$

**75.** $i^{32} = \left(i^4\right)^8 = 1^8 = 1$

**77.** $i^{-13} = i^{-16} \cdot i^3 = \left(i^4\right)^{-4} \cdot i^3 = 1^{-4} \cdot (-i) = -i$

**79.** $\dfrac{1}{i^{-11}} = i^{11} = i^8 \cdot i^3 = \left(i^4\right)^2 \cdot i^3 = 1^2 \cdot (-i) = -i$

**81.** Answers will vary.

**83.** $\dfrac{6+2i}{1+2i} = \dfrac{(6+2i)(1-2i)}{(1+2i)(1-2i)}$
$\qquad = \dfrac{6 - 12i + 2i - 4i^2}{1^2 - (2i)^2} = \dfrac{6 - 10i - 4(-1)}{1 - 4i^2}$
$\qquad = \dfrac{6 - 10i + 4}{1 - 4(-1)} = \dfrac{10 - 10i}{1 + 4} = \dfrac{10 - 10i}{5}$
$\qquad = \dfrac{10}{5} - \dfrac{10}{5}i = 2 - 2i$

**85.** $\dfrac{2-i}{2+i} = \dfrac{(2-i)(2-i)}{(2+i)(2-i)} = \dfrac{2^2 - 2(2i) + i^2}{2^2 - i^2}$
$\qquad = \dfrac{4 - 4i + (-1)}{4 - (-1)} = \dfrac{3 - 4i}{5} = \dfrac{3}{5} - \dfrac{4}{5}i$

**87.** $\dfrac{1-3i}{1+i} = \dfrac{(1-3i)(1-i)}{(1+i)(1-i)} = \dfrac{1 - i - 3i + 3i^2}{1^2 - i^2}$
$\qquad = \dfrac{1 - 4i + 3(-1)}{1 - (-1)} = \dfrac{1 - 4i - 3}{2}$
$\qquad = \dfrac{-2 - 4i}{2} = \dfrac{-2}{2} - \dfrac{4}{2}i = -1 - 2i$

**89.** $\dfrac{-5}{i} = \dfrac{-5(-i)}{i(-i)} = \dfrac{5i}{-i^2}$
$\qquad = \dfrac{5i}{-(-1)} = \dfrac{5i}{1} = 5i \text{ or } 0 + 5i$

**91.** $\dfrac{8}{-i} = \dfrac{8 \cdot i}{-i \cdot i} = \dfrac{8i}{-i^2}$
$\qquad = \dfrac{8i}{-(-1)} = \dfrac{8i}{1} = 8i \text{ or } 0 + 8i$

**93.** $\dfrac{2}{3i} = \dfrac{2(-3i)}{3i \cdot (-3i)} = \dfrac{-6i}{-9i^2} = \dfrac{-6i}{-9(-1)}$
$\qquad = \dfrac{-6i}{9} = -\dfrac{2}{3}i \text{ or } 0 - \dfrac{2}{3}i$

Note: In the above solution, we multiplied the numerator and denominator by the complex conjugate of $3i$, namely $-3i$. Since there is a reduction in the end, the same results can be achieved by multiplying the numerator and denominator by $-i$.

**95.** We need to show that $\left(\dfrac{\sqrt{2}}{2} + \dfrac{\sqrt{2}}{2}i\right)^2 = i$.

$$\left(\dfrac{\sqrt{2}}{2} + \dfrac{\sqrt{2}}{2}i\right)^2$$

$$= \left(\dfrac{\sqrt{2}}{2}\right)^2 + 2 \cdot \dfrac{\sqrt{2}}{2} \cdot \dfrac{\sqrt{2}}{2}i + \left(\dfrac{\sqrt{2}}{2}i\right)^2$$

$$= \dfrac{2}{4} + 2 \cdot \dfrac{2}{4}i + \dfrac{2}{4}i^2 = \dfrac{1}{2} + i + \dfrac{1}{2}i^2$$

$$= \dfrac{1}{2} + i + \dfrac{1}{2}(-1) = \dfrac{1}{2} + i - \dfrac{1}{2} = i$$

**97.** Let $z = 3 - 2i$.

$$3z - z^2 = 3(3 - 2i) - (3 - 2i)^2$$

$$= 9 - 6i - \left[3^2 - 2(6i) + (2i)^2\right]$$

$$= 9 - 6i - \left(9 - 12i + 4i^2\right)$$

$$= 9 - 6i - \left[9 - 12i + 4(-1)\right]$$

$$= 9 - 6i - \left[9 - 12i + (-4)\right]$$

$$= 9 - 6i - (5 - 12i)$$

$$= 9 - 6i - 5 + 12i = 4 + 6i$$

## Section 1.4: Quadratic Equations

**1.** $x^2 = 25$

$\quad x = \pm\sqrt{25} = \pm 5$; G

**3.** $x^2 + 5 = 0$

$\quad x^2 = -5$

$\quad\quad x = \pm\sqrt{-5} = \pm i\sqrt{5}$; C

**5.** $x^2 = -20$

$\quad x = \pm\sqrt{-20} = \pm 2i\sqrt{5}$; H

**7.** $x - 5 = 0$

$\quad x = 5$; D

**9.** D is the only one set up for direct use of the zero-factor property.

$(3x - 1)(x - 7) = 0$

$3x - 1 = 0 \quad$ or $\quad x - 7 = 0$

$\quad x = \dfrac{1}{3} \quad$ or $\quad\quad x = 7$

Solution set: $\left\{\dfrac{1}{3}, 7\right\}$

**11.** C is the only one that does not require Step 1 of the method of completing the square.

$\quad\quad x^2 + x = 12 \quad\quad$ Note:

$x^2 + x + \dfrac{1}{4} = 12 + \dfrac{1}{4} \quad \left[\dfrac{1}{2} \cdot 1\right]^2 = \left(\dfrac{1}{2}\right)^2 = \dfrac{1}{4}$

$\quad \left(x + \dfrac{1}{2}\right)^2 = \dfrac{49}{4}$

$x + \dfrac{1}{2} = \pm\sqrt{\dfrac{49}{4}}$

$x + \dfrac{1}{2} = \pm\dfrac{7}{2} \Rightarrow x = -\dfrac{1}{2} \pm \dfrac{7}{2}$

$-\dfrac{1}{2} - \dfrac{7}{2} = \dfrac{-8}{2} = -4$ and $-\dfrac{1}{2} + \dfrac{7}{2} = \dfrac{6}{2} = 3$

Solution set: $\{-4, 3\}$

**13.** $\quad x^2 - 5x + 6 = 0$

$(x - 2)(x - 3) = 0$

$x - 2 = 0 \Rightarrow x = 2 \quad$ or $\quad x - 3 = 0 \Rightarrow x = 3$

Solution set: $\{2, 3\}$

**15.** $\quad 5x^2 - 3x - 2 = 0$

$(5x + 2)(x - 1) = 0$

$5x + 2 = 0 \Rightarrow x = -\dfrac{2}{5} \quad$ or $\quad x - 1 = 0 \Rightarrow x = 1$

Solution set: $\left\{-\dfrac{2}{5}, 1\right\}$

**17.** $-4x^2 + x = -3$

$\quad 0 = 4x^2 - x - 3$

$\quad 0 = (4x + 3)(x - 1)$

$4x + 3 = 0 \Rightarrow x = -\dfrac{3}{4} \quad$ or $\quad x - 1 = 0 \Rightarrow x = 1$

Solution set: $\left\{-\dfrac{3}{4}, 1\right\}$

**19.** $x^2 = 16$

$\quad x = \pm\sqrt{16} = \pm 4$

Solution set: $\{\pm 4\}$

**21.** $27 - x^2 = 0$

$\quad\quad 27 = x^2 \Rightarrow x = \pm\sqrt{27} = \pm 3\sqrt{3}$

Solution set: $\left\{\pm 3\sqrt{3}\right\}$

**23.** $x^2 = -81$

$\quad x = \pm\sqrt{-81} = \pm 9i$

Solution set: $\{\pm 9i\}$

**25.** $(3x - 1)^2 = 12 \Rightarrow 3x - 1 = \pm\sqrt{12}$

$\quad\quad 3x = 1 \pm 2\sqrt{3} \Rightarrow x = \dfrac{1 \pm 2\sqrt{3}}{3}$

Solution set: $\left\{\dfrac{1 \pm 2\sqrt{3}}{3}\right\}$

**27.** $(x + 5)^2 = -3$

$\quad x + 5 = \pm\sqrt{-3} \Rightarrow x + 5 = \pm i\sqrt{3}$

$\quad\quad x = -5 \pm i\sqrt{3}$

Solution set: $\left\{-5 \pm i\sqrt{3}\right\}$

**29.** $(5x-3)^2 = -3 \Rightarrow 5x - 3 = \pm\sqrt{-3} \Rightarrow$

$5x - 3 = \pm i\sqrt{3} \Rightarrow 5x = 3 \pm i\sqrt{3} \Rightarrow$

$x = \dfrac{3 \pm i\sqrt{3}}{5} = \dfrac{3}{5} \pm \dfrac{\sqrt{3}}{5} i$

Solution set: $\left\{ \dfrac{3}{5} \pm \dfrac{\sqrt{3}}{5} i \right\}$

**31.** $x^2 - 4x + 3 = 0$

$x^2 - 4x + 4 = -3 + 4$ 　 Note:

$\left[ \frac{1}{2} \cdot 4 \right]^2 = 2^2 = 4$

$(x-2)^2 = 1$

$x - 2 = \pm\sqrt{1} \Rightarrow x - 2 = \pm 1$

$x = 2 \pm 1 \Rightarrow x = 3$ or $x = 1$

Solution set: $\{1, 3\}$

**33.** $2x^2 - x - 28 = 0$

$x^2 - \frac{1}{2}x - 14 = 0$ 　 Multiply by $\frac{1}{2}$.

$x^2 - \frac{1}{2}x + \frac{1}{16} = 14 + \frac{1}{16}$

Note: $\left[ \frac{1}{2} \cdot \left( -\frac{1}{2} \right) \right]^2 = \left( -\frac{1}{4} \right)^2 = \frac{1}{16}$

$\left( x - \frac{1}{4} \right)^2 = \frac{225}{16}$

$x - \frac{1}{4} = \pm\sqrt{\frac{225}{16}}$

$x - \frac{1}{4} = \pm\frac{15}{4}$

$x = \frac{1}{4} \pm \frac{15}{4}$

$\frac{1}{4} - \frac{15}{4} = \frac{-14}{4} = -\frac{7}{2}$ and $\frac{1}{4} + \frac{15}{4} = \frac{16}{4} = 4$

Solution set: $\left\{ -\frac{7}{2}, 4 \right\}$

**35.** $x^2 - 2x - 2 = 0$

$x^2 - 2x + 1 = 2 + 1$

Note: $\left[ \frac{1}{2} \cdot (-2) \right]^2 = (-1)^2 = 1$

$(x-1)^2 = 3$

$x - 1 = \pm\sqrt{3}$

$x = 1 \pm \sqrt{3}$

Solution set: $\left\{ 1 \pm \sqrt{3} \right\}$

**37.** 　 $2x^2 + x = 10$

$x^2 + \frac{1}{2}x = 5$

$x^2 + \frac{1}{2}x + \frac{1}{16} = 5 + \frac{1}{16}$ 　 Note:

$\left[ \frac{1}{2} \cdot \frac{1}{2} \right]^2 = \left( \frac{1}{4} \right)^2 = \frac{1}{16}$

$\left( x + \frac{1}{4} \right)^2 = \frac{81}{16} \Rightarrow x + \frac{1}{4} = \pm\sqrt{\frac{81}{16}}$

$x + \frac{1}{4} = \pm\frac{9}{4} \Rightarrow x = -\frac{1}{4} \pm \frac{9}{4}$

$-\frac{1}{4} - \frac{9}{4} = \frac{-10}{4} = -\frac{5}{2}$ and $-\frac{1}{4} + \frac{9}{4} = \frac{8}{4} = 2$

Solution set: $\left\{ -\frac{5}{2}, 2 \right\}$

**39.** $2x^2 - 4x - 3 = 0$

$x^2 - 2x - \frac{3}{2} = 0$

$x^2 - 2x + 1 = \frac{3}{2} + 1$ 　 Note:

$\left[ \frac{1}{2} \cdot (-2) \right]^2 = (-1)^2 = 1$

$(x-1)^2 = \frac{5}{2}$

$x - 1 = \pm\sqrt{\frac{5}{2}} = \pm\frac{\sqrt{10}}{2}$

$x = 1 \pm \frac{\sqrt{10}}{2} = \frac{2 \pm \sqrt{10}}{2}$

Solution set: $\left\{ \dfrac{2 \pm \sqrt{10}}{2} \right\}$

**41.** 　 $-4x^2 + 8x = 7$

$x^2 - 2x = -\frac{7}{4}$

$x^2 - 2x + 1 = -\frac{7}{4} + 1$ 　 Note:

$\left[ \frac{1}{2} \cdot (-2) \right]^2 = (-1)^2 = 1$

$(x-1)^2 = \frac{-3}{4}$

$x - 1 = \pm\sqrt{\frac{-3}{4}} = \pm\frac{i\sqrt{3}}{2} \Rightarrow x = 1 \pm \frac{\sqrt{3}}{2} i$

Solution set: $\left\{ 1 \pm \frac{\sqrt{3}}{2} i \right\}$

**43.** Francisco is incorrect because $c = 0$ and the

quadratic formula, $x = \dfrac{-b \pm \sqrt{b^2 - 4ac}}{2a}$, can

be evaluated with $a = 1, b = -8,$ and $c = 0$.

**45.** $x^2 - x - 1 = 0$

Let $a = 1, b = -1,$ and $c = -1$.

$x = \dfrac{-b \pm \sqrt{b^2 - 4ac}}{2a}$

$= \dfrac{-(-1) \pm \sqrt{(-1)^2 - 4(1)(-1)}}{2(1)}$

$= \dfrac{1 \pm \sqrt{1 + 4}}{2} = \dfrac{1 \pm \sqrt{5}}{2}$

Solution set: $\left\{ \dfrac{1 \pm \sqrt{5}}{2} \right\}$

**47.** $x^2 - 6x = -7 \Rightarrow x^2 - 6x + 7 = 0$

Let $a = 1, b = -6,$ and $c = 7$.

$x = \dfrac{-b \pm \sqrt{b^2 - 4ac}}{2a}$

$= \dfrac{-(-6) \pm \sqrt{(-6)^2 - 4(1)(7)}}{2(1)} = \dfrac{6 \pm \sqrt{36 - 28}}{2}$

$= \dfrac{6 \pm \sqrt{8}}{2} = \dfrac{6 \pm 2\sqrt{2}}{2} = 3 \pm \sqrt{2}$

Solution set: $\left\{ 3 \pm \sqrt{2} \right\}$

**49.** 
$$x^2 = 2x - 5$$
$$x^2 - 2x + 5 = 0$$
Let $a = 1, b = -2,$ and $c = 5.$

$$x = \frac{-b \pm \sqrt{b^2 - 4ac}}{2a}$$

$$= \frac{-(-2) \pm \sqrt{(-2)^2 - 4(1)(5)}}{2(1)}$$

$$= \frac{2 \pm \sqrt{4 - 20}}{2} = \frac{2 \pm \sqrt{-16}}{2} = \frac{2 \pm 4i}{2} = 1 \pm 2i$$

Solution set: $\{1 \pm 2i\}$

**51.** 
$$-4x^2 = -12x + 11$$
$$0 = 4x^2 - 12x + 11$$
Let $a = 4, b = -12,$ and $c = 11.$

$$x = \frac{-b \pm \sqrt{b^2 - 4ac}}{2a}$$

$$= \frac{-(-12) \pm \sqrt{(-12)^2 - 4(4)(11)}}{2(4)}$$

$$= \frac{12 \pm \sqrt{144 - 176}}{8} = \frac{12 \pm \sqrt{-32}}{8}$$

$$= \frac{12 \pm 4i\sqrt{2}}{8} = \frac{12}{8} \pm \frac{4\sqrt{2}}{8}i = \frac{3}{2} \pm \frac{\sqrt{2}}{2}i$$

Solution set: $\left\{\frac{3}{2} \pm \frac{\sqrt{2}}{2}i\right\}$

**53.** 
$$\tfrac{1}{2}x^2 + \tfrac{1}{4}x - 3 = 0$$
$$4\left(\tfrac{1}{2}x^2 + \tfrac{1}{4}x - 3\right) = 4 \cdot 0$$
$$2x^2 + x - 12 = 0$$
Let $a = 2, b = 1,$ and $c = -12.$

$$x = \frac{-b \pm \sqrt{b^2 - 4ac}}{2a} = \frac{-1 \pm \sqrt{1^2 - 4(2)(-12)}}{2(2)}$$

$$= \frac{-1 \pm \sqrt{1 + 96}}{4} = \frac{-1 \pm \sqrt{97}}{4}$$

Solution set: $\left\{\frac{-1 \pm \sqrt{97}}{4}\right\}$

**55.** 
$$.2x^2 + .4x - .3 = 0$$
$$10\left(.2x^2 + .4x - .3\right) = 10 \cdot 0$$
$$2x^2 + 4x - 3 = 0$$
Let $a = 2, b = 4,$ and $c = -3.$

$$x = \frac{-b \pm \sqrt{b^2 - 4ac}}{2a}$$

$$x = \frac{-4 \pm \sqrt{4^2 - 4(2)(-3)}}{2(2)} = \frac{-4 \pm \sqrt{16 + 24}}{4}$$

$$= \frac{-4 \pm \sqrt{40}}{4} = \frac{-4 \pm 2\sqrt{10}}{4} = \frac{-2 \pm \sqrt{10}}{2}$$

Solution set: $\left\{\frac{-2 \pm \sqrt{10}}{2}\right\}$

**57.** 
$$(4x - 1)(x + 2) = 4x$$
$$4x^2 + 7x - 2 = 4x \Rightarrow 4x^2 + 3x - 2 = 0$$
Let $a = 4, b = 3,$ and $c = -2.$

$$x = \frac{-b \pm \sqrt{b^2 - 4ac}}{2a} = \frac{-3 \pm \sqrt{3^2 - 4(4)(-2)}}{2(4)}$$

$$= \frac{-3 \pm \sqrt{9 + 32}}{8} = \frac{-3 \pm \sqrt{41}}{8}$$

Solution set: $\left\{\frac{-3 \pm \sqrt{41}}{8}\right\}$

**59.** 
$$(x - 9)(x - 1) = -16$$
$$x^2 - 10x + 9 = -16$$
$$x^2 - 10x + 25 = 0$$
Let $a = 1, b = -10,$ and $c = 25.$

$$x = \frac{-(-10) \pm \sqrt{(-10)^2 - 4(1)(25)}}{2}$$

$$= \frac{10 \pm \sqrt{100 - 100}}{2} = \frac{10 \pm 0}{2} = 5$$

Solution set: $\{5\}$

**61.** 
$$x^3 - 8 = 0 \Rightarrow x^3 - 2^3 = 0$$
$$(x - 2)(x^2 + 2x + 4) = 0$$
$$x - 2 = 0 \Rightarrow x = 2 \text{ or}$$
$$x^2 + 2x + 4 = 0$$
$$a = 1, b = 2, \text{ and } c = 4$$

$$x = \frac{-b \pm \sqrt{b^2 - 4ac}}{2a}$$

$$= \frac{-2 \pm \sqrt{2^2 - 4(1)(4)}}{2(1)}$$

$$= \frac{-2 \pm \sqrt{4 - 16}}{2} = \frac{-2 \pm \sqrt{-12}}{2}$$

$$= \frac{-2 \pm 2i\sqrt{3}}{2} = -1 \pm \sqrt{3}i$$

Solution set: $\left\{2, -1 \pm \sqrt{3}i\right\}$

**63.**
$$x^3 + 27 = 0$$
$$x^3 + 3^3 = 0$$
$$(x+3)(x^2 - 3x + 9) = 0$$
$$x + 3 = 0 \Rightarrow x = -3 \text{ or}$$
$$x^2 - 3x + 9 = 0$$
$$a = 1, b = -3, \text{ and } c = 9$$
$$x = \frac{-b \pm \sqrt{b^2 - 4ac}}{2a}$$
$$= \frac{-(-3) \pm \sqrt{(-3)^2 - 4(1)(9)}}{2(1)} = \frac{3 \pm \sqrt{9 - 36}}{2}$$
$$= \frac{3 \pm \sqrt{-27}}{2} = \frac{3 \pm 3i\sqrt{3}}{2} = \frac{3}{2} \pm \frac{3\sqrt{3}}{2}i$$
Solution set: $\left\{-3, \frac{3}{2} \pm \frac{3\sqrt{3}}{2}i\right\}$

**65.**  $s = \frac{1}{2}gt^2$
$$2s = 2\left[\frac{1}{2}gt^2\right] \Rightarrow 2s = gt^2 \Rightarrow \frac{2s}{g} = \frac{gt^2}{g} \Rightarrow$$
$$t^2 = \frac{2s}{g} \Rightarrow t = \pm\sqrt{\frac{2s}{g}} = \frac{\pm\sqrt{2s}}{\sqrt{g}} \cdot \frac{\sqrt{g}}{\sqrt{g}} = \frac{\pm\sqrt{2sg}}{g}$$

**67.**  $F = \frac{kMv^2}{r}$
$$rF = r\left[\frac{kMv^2}{r}\right] \Rightarrow Fr = kMv^2 \Rightarrow$$
$$\frac{Fr}{kM} = \frac{kMv^2}{kM} \Rightarrow v^2 = \frac{Fr}{kM} \Rightarrow v = \pm\sqrt{\frac{Fr}{kM}} \Rightarrow$$
$$v = \frac{\pm\sqrt{Fr}}{\sqrt{kM}} \cdot \frac{\sqrt{kM}}{\sqrt{kM}} = \frac{\pm\sqrt{FrkM}}{kM}$$

**69.**  $h = -16t^2 + v_0t + s_0$
$$16t^2 - v_0t + h - s_0 = 0$$
$$16t^2 - v_0t + (h - s_0) = 0. \quad a = 16, b = -v_0, \\ c = h - s_0$$
$$x = \frac{-b \pm \sqrt{b^2 - 4ac}}{2a}$$
$$= \frac{-(-v_0) \pm \sqrt{(-v_0)^2 - 4(16)(h - s_0)}}{2(16)}$$
$$= \frac{v_0 \pm \sqrt{v_0^2 - 64(h - s_0)}}{32}$$
$$= \frac{v_0 \pm \sqrt{v_0^2 - 64h + 64s_0}}{32}$$

**71.**
$$4x^2 - 2xy + 3y^2 = 2$$
$$4x^2 - 2xy + 3y^2 - 2 = 0$$
**(a)**  Solve for $x$ in terms of $y$.
$$4x^2 - (2y)x + (3y^2 - 2) = 0$$
$$a = 4, b = -2y, \text{ and } c = 3y^2 - 2$$
$$x = \frac{-b \pm \sqrt{b^2 - 4ac}}{2a}$$
$$= \frac{-(-2y) \pm \sqrt{(-2y)^2 - 4(4)(3y^2 - 2)}}{2(4)}$$
$$= \frac{2y \pm \sqrt{4y^2 - 16(3y^2 - 2)}}{8}$$
$$= \frac{2y \pm \sqrt{4y^2 - 48y^2 + 32}}{8}$$
$$= \frac{2y \pm \sqrt{32 - 44y^2}}{8} = \frac{2y \pm \sqrt{4(8 - 11y^2)}}{8}$$
$$= \frac{2y \pm 2\sqrt{8 - 11y^2}}{8} = \frac{y \pm \sqrt{8 - 11y^2}}{4}$$

**(b)**  Solve for $y$ in terms of $x$.
$$3y^2 - (2x)y + (4x^2 - 2) = 0$$
$$a = 3, b = -2x, \text{ and } c = 4x^2 - 2$$
$$y = \frac{-b \pm \sqrt{b^2 - 4ac}}{2a}$$
$$= \frac{-(-2x) \pm \sqrt{(-2x)^2 - 4(3)(4x^2 - 2)}}{2(3)}$$
$$= \frac{2x \pm \sqrt{4x^2 - 12(4x^2 - 2)}}{6}$$
$$= \frac{2x \pm \sqrt{4x^2 - 48x^2 + 24}}{6}$$
$$= \frac{2x \pm \sqrt{24 - 44x^2}}{6} = \frac{2x \pm \sqrt{4(6 - 11y^2)}}{6}$$
$$= \frac{2x \pm 2\sqrt{6 - 11y^2}}{6} = \frac{x \pm \sqrt{6 - 11y^2}}{3}$$

**73.**  $x^2 - 8x + 16 = 0$
$$a = 1, b = -8, \text{ and } c = 16$$
$$b^2 - 4ac = (-8)^2 - 4(1)(16) = 64 - 64 = 0$$
one rational solution (a double solution)

**75.** $3x^2 + 5x + 2 = 0$

$a = 3, b = 5,$ and $c = 2$

$b^2 - 4ac = 5^2 - 4(3)(2) = 25 - 24 = 1 = 1^2$

two distinct rational solutions

**77.** $4x^2 = -6x + 3$

$4x^2 + 6x - 3 = 0$

$a = 4, b = 6,$ and $c = -3$

$b^2 - 4ac = 6^2 - 4(4)(-3) = 36 + 48 = 84$

two distinct irrational solutions

**79.** $9x^2 + 11x + 4 = 0$

$a = 9, b = 11,$ and $c = 4$

$b^2 - 4ac = 11^2 - 4(9)(4) = 121 - 144 = -23$

two distinct nonreal complex solutions

**81.** $8x^2 - 72 = 0$

$a = 8, b = 0,$ and $c = -72$

$b^2 - 4ac = 0^2 - 4(8)(-72) = 2304 = 48^2$

two distinct rational solutions

**83.** It is not possible for the solution set of a quadratic equation with integer coefficients to consist of a single irrational number. Additional responses will vary.

In exercises 85–87, there are other possible answers.

**85.** $\quad x = 4 \quad$ or $\quad x = 5$

$x - 4 = 0 \quad$ or $\quad x - 5 = 0$

$(x - 4)(x - 5) = 0$

$x^2 - 5x - 4x + 20 = 0$

$x^2 - 9x + 20 = 0$

$a = 1, b = -9,$ and $c = 20$

**87.** $\quad x = 1 + \sqrt{2} \quad$ or $\quad x = 1 - \sqrt{2}$

$x - (1 + \sqrt{2}) = 0 \quad$ or $\quad x - (1 - \sqrt{2}) = 0$

$\left[ x - (1 + \sqrt{2}) \right]\left[ x - (1 - \sqrt{2}) \right] = 0$

$x^2 - x(1 - \sqrt{2}) - x(1 + \sqrt{2})$

$\qquad + (1 + \sqrt{2})(1 - \sqrt{2}) = 0$

$x^2 - x + x\sqrt{2} - x - x\sqrt{2} + \left[ 1^2 - (\sqrt{2})^2 \right] = 0$

$x^2 - 2x + (1 - 2) = 0$

$x^2 - 2x - 1 = 0$

$a = 1, b = -2,$ and $c = -1$

## Chapter 1 Quiz
### (Sections 1.1–1.4)

**1.** $3(x - 5) + 2 = 1 - (4 + 2x)$

$3x - 15 + 2 = 1 - 4 - 2x$

$3x - 13 = -3 - 2x$

$5x - 13 = -3$

$5x = 10 \Rightarrow x = 2$

Solution set $\{2\}$

**(b)** $5x - 9 = 5(-2 + x) + 1$

$5x - 9 = -10 + 5x + 1$

$5x - 9 = 5x - 9$

identity; solution set: {all real numbers} or $(-\infty, \infty)$

**(c)** $5x - 4 = 3(6 - x)$

$5x - 4 = 18 - 3x$

$8x - 4 = 18$

$8x = 22 \Rightarrow x = \frac{22}{8} = \frac{11}{4}$

conditional equation; solution set: $\left\{\frac{11}{4}\right\}$

**3.** $ay + 2x = y + 5x$

$ay - 3x = y$

$-3x = y - ay = y(1 - a)$

$3x = y(a - 1)$

$\dfrac{3x}{a - 1} = y$

**5.** Substitute 1999 for $x$ in the equation:

$y = .12(1999) - 234.42 = 5.46$

So, the model predicts that the minimum hourly wage for 1999 was $5.46. The difference between the actual minimum wage and the predicted wage is $5.46 - $5.15 = $0.31.

**7.** $\dfrac{7 - 2i}{2 + 4i} = \dfrac{7 - 2i}{2 + 4i} \cdot \dfrac{2 - 4i}{2 - 4i} = \dfrac{14 - 28i - 4i + (-8)}{4 - (-16)}$

$= \dfrac{6 - 32i}{20} = \dfrac{6}{20} - \dfrac{32}{20}i = \dfrac{3}{10} - \dfrac{8}{5}i$

**9.** $x^2 - 29 = 0 \Rightarrow x^2 = 29 \Rightarrow x = \pm\sqrt{29}$

Solution set: $\left\{\pm 29\right\}$

## Section 1.5: Applications and Modeling with Quadratic Equations

**Connections (page 125)**

**1. (a)** $d = 16t^2 = 16(5)^2 = 400$

It will fall 400 feet in 5 seconds.

**(b)** $d = 16t^2 = 16(10)^2 = 1600$

It will fall 1600 feet in 10 seconds.

No, the second answer is $2^2 = 4$ times the first because the number of seconds is squared in the formula.

**2.** Both formulas involve the number 16 times the square of time. However, in the formula for the distance an object falls, 16 is positive, while in the formula for a projected object, it is preceded by a negative sign. Also in the formula for a projected object, the initial velocity and height affect the distance.

### Exercises

**1.** The length of the parking area is $2x + 200$, while the width is $x$, so the area is $(2x + 200)x$. Set the area equal to 40,000.
$(2x + 200)x = 40{,}000$, so choice A is the correct choice.

**3.** Use the Pythagorean theorem with $a = x$, $b = 2x - 2$, and $c = x + 4$.
$$x^2 + (2x - 2)^2 = (x + 4)^2$$
The correct choice is D.

**5.** Let $x$ = the first integer. Then $x + 1$ = the next consecutive integer.
$$x(x+1) = 56 \Rightarrow x^2 + x = 56$$
$$x^2 + x - 56 = 0 \Rightarrow (x+8)(x-7) = 0$$
$$x + 8 = 0 \Rightarrow x = -8 \text{ or } x - 7 = 0 \Rightarrow x = 7$$
If $x = -8$, then $x + 1 = -7$. If $x = 7$, then $x + 1 = 8$. So the two integers are $-8$ and $-7$, or 7 and 8.

**7.** Let $x$ = the first even integer. Then $x + 2$ = the next consecutive even integer.
$$x(x+2) = 168 \Rightarrow x^2 + 2x = 168$$
$$x^2 + 2x - 168 = 0 \Rightarrow (x+14)(x-12) = 0$$
$$x + 14 = 0 \Rightarrow x = -14 \text{ or}$$
$$x - 12 = 0 \Rightarrow x = 12$$
If $x = -14$, then $x + 2 = -12$. If $x = 12$, then $x + 2 = 14$. so, the two even integers are $-14$ and $-12$, or 12 and 14.

**9.** Let $x$ = the first odd integer. Then $x + 2$ = the next consecutive odd integer.
$$x(x+2) = 63 \Rightarrow x^2 + 2x = 63 \Rightarrow$$
$$x^2 + 2x - 63 = 0 \Rightarrow (x+9)(x-7) = 0$$
$$x + 9 = 0 \Rightarrow x = -9 \text{ or}$$
$$x - 7 = 0 \Rightarrow x = 7$$
If $x = -9$, then $x + 2 = -7$. If $x = 7$, then $x + 2 = 9$. so, the two odd integers are $-9$ and $-7$, or 7 and 9.

**11.** Let $x$ = the first integer. Then $x + 1$ = the next consecutive integer.
$$x^2 + (x+1)^2 = 61$$
$$x^2 + x^2 + 2x + 1 = 61 \Rightarrow 2x^2 + 2x + 1 = 61$$
$$2x^2 + 2x - 60 = 0 \Rightarrow 2(x^2 + x - 30) = 0$$
$$x^2 + x - 30 = 0 \Rightarrow (x+6)(x-5) = 0$$
$$x + 6 = 0 \Rightarrow x = -6 \text{ or}$$
$$x - 5 = 0 \Rightarrow x = 5$$
If $x = -6$, then $x + 1 = -5$. If $x = 5$, then $x + 1 = 6$. So the two integers are $-6$ and $-5$, or 5 and 6.

**13.** Let $x$ = the first odd integer. Then $x + 2$ = the next consecutive odd integer.
$$x^2 + (x+2)^2 = 202$$
$$x^2 + x^2 + 4x + 4 = 202$$
$$2x^2 + 4x + 4 = 202 \Rightarrow 2x^2 + 4x - 198 = 0$$
$$2(x^2 + 2x - 99) = 0 \Rightarrow x^2 + 2x - 99 = 0$$
$$(x+11)(x-9) = 0$$
$$x + 11 = 0 \Rightarrow x = -11 \text{ or}$$
$$x - 9 = 0 \Rightarrow x = 9$$
If $x = -11$, then $x + 2 = -9$. If $x = 9$, then $x + 2 = 11$. So the two integers are $-11$ and $-9$, or 9 and 11.

**15.** Let $x$ = the length of one leg, $x + 2$ = the length of the other leg, and $x + 4$ = the length of the hypotenuse. (Remember that the hypotenuse is the longest side in a right triangle.) The Pythagorean theorem gives
$$x^2 + (x+2)^2 = (x+4)^2$$
$$x^2 + x^2 + 4x + 4 = x^2 + 8x + 16$$
$$x^2 - 4x - 12 = 0 \Rightarrow (x-6)(x+2) = 0$$
$$x - 6 = 0 \Rightarrow x = 6 \text{ or}$$
$$x + 2 = 0 \Rightarrow x = -2$$
Length cannot be negative, so reject that solution. If $x = 6$, then $x + 2 = 8$ and $x + 4 = 10$. The sides of the right triangle are 6, 8, and 10.

**17.** Let $x$ = the length of the side of the smaller square. Then $x + 3$ = the length of the side of the larger square.
$$(x+3)^2 + x^2 = 149$$
$$x^2 + 6x + 9 + x^2 = 149 \Rightarrow 2x^2 + 6x - 140 = 0$$
$$x^2 + 3x - 70 = 0 \Rightarrow (x-7)(x+10) = 0$$
$$x - 7 = 0 \Rightarrow x = 7 \text{ or}$$
$$x + 10 = 0 \Rightarrow x = -10$$
Length cannot be negative, so reject that solution. If $x = 7$, then $x + 3 = 10$. The length of the side of smaller square is 7 in., and the length of the side of the larger square is 10 in.

**19.** Use the figure and equation $A$ from Exercise 1.

$$x(2x + 200) = 40,000$$
$$2x^2 + 200x = 40,000$$
$$2x^2 + 200x - 40,000 = 0$$
$$x^2 + 100x - 20,000 = 0$$
$$(x - 100)(x + 200) = 0$$
$$x = 100 \text{ or } x = -200$$

The negative solution is not meaningful. If $x = 100$, then $2x + 200 = 400$. The dimensions of the lot are 100 yd by 400 yd.

**21.** Let $x =$ the width of the strip of floor around the rug.

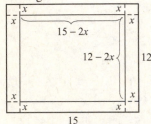

The dimensions of the carpet are $15 - 2x$ by $12 - 2x$. Since $A = lw$, the equation for the carpet area is $(15 - 2x)(12 - 2x) = 108$. Put this equation in standard form and solve by factoring.

$$(15 - 2x)(12 - 2x) = 108$$
$$180 - 30x - 24x + 4x^2 = 108$$
$$180 - 54x + 4x^2 = 108$$
$$4x^2 - 54x + 72 = 0$$
$$2x^2 - 27x + 36 = 0$$
$$(2x - 3)(x - 12) = 0$$

$$2x - 3 = 0 \Rightarrow x = \frac{3}{2}$$
$$x - 12 = 0 \Rightarrow x = 12$$

The solutions of the quadratic equation are $\frac{3}{2}$ and 12. We eliminate 12 as meaningless in this problem. If $x = \frac{3}{2}$, then $15 - 2x = 12$ and $12 - 2x = 9$. The dimensions of the carpet are 9 ft by 12 ft.

**23.** Let $x =$ the width of the metal. The dimensions of the base of the box are $x - 4$ by $x + 6$.

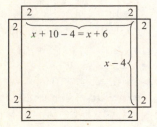

Since the formula for the volume of a box is $V = lwh$, we have

$$(x + 6)(x - 4)(2) = 832$$
$$(x + 6)(x - 4) = 416$$
$$x^2 - 4x + 6x - 24 = 416$$
$$x^2 + 2x - 24 = 416$$
$$x^2 + 2x - 440 = 0 \Rightarrow (x + 22)(x - 20) = 0$$
$$x + 22 = 0 \Rightarrow x = -22 \text{ or}$$
$$x - 20 = 0 \Rightarrow x = 20$$

The negative solution is not meaningful. If $x = 20$, then $x + 10 = 30$. The dimensions of the sheet of metal are 20 in by 30 in.

**25.** Let $h =$ height and $r =$ radius.

Area of side $= 2\pi rh$ and Area of circle $= \pi r^2$
Surface area $=$ area of side $+$ area of top $+$ area of bottom
Surface area $= 2\pi rh + \pi r^2 + \pi r^2 = 2\pi rh + 2\pi r^2$

$$371 = 2\pi r(12) + 2\pi r^2$$
$$371 = 24\pi r + 2\pi r^2$$
$$0 = 2\pi r^2 + 24\pi r - 371$$

$a = 2\pi$, $b = 24\pi$, and $c = -371$

$$r = \frac{-b \pm \sqrt{b^2 - 4ac}}{2a} r$$
$$= \frac{-24\pi \pm \sqrt{(24\pi)^2 - 4(2\pi)(-371)}}{2(2\pi)}$$
$$= \frac{-24\pi \pm \sqrt{576\pi^2 + 2968\pi}}{4\pi}$$

$r \approx -15.75$ or $r \approx 3.75$

The negative solution is not meaningful. The radius of the circular top is approximately 3.75 cm.

**27.** Let $h =$ height and $r =$ radius.

Surface area $= 2\pi rh + 2\pi r^2$

$$8\pi = 2\pi r(3) + 2\pi r^2$$
$$8\pi = 6\pi r + 2\pi r^2$$
$$0 = 2\pi r^2 + 6\pi r - 8\pi$$
$$0 = 2\pi(r^2 + 3r - 4) \Rightarrow 0 = (r + 4)(r - 1)$$

$r + 4 = 0 \Rightarrow r = -4$ or $r - 1 = 0 \Rightarrow r = 1$

The $r$ represents the radius of a cylinder, so $-4$ is not reasonable. The radius of the circular top is approximately 1 ft.

**29.** Let $x =$ length of side of square. Area $= x^2$ and perimeter $= 4x$

$$x^2 = 4x \Rightarrow x^2 - 4x = 0 \Rightarrow x(x - 4) = 0 \Rightarrow$$
$$x = 0 \text{ or } x = 4$$

We reject 0 since $x$ must be greater than 0. The side of the square measures 4 units.

**31.** Let $h$ = the height of the dock.
Then $2h + 3$ = the length of the rope from the boat to the top of the dock.
Apply the Pythagorean theorem to the triangle shown in the text.

$$h^2 + 12^2 = (2h+3)^2$$
$$h^2 + 144 = (2h)^2 + 2(6h) + 3^2$$
$$h^2 + 144 = 4h^2 + 12h + 9$$
$$0 = 3h^2 + 12h - 135$$
$$0 = h^2 + 4h - 45 \Rightarrow 0 = (h+9)(h-5)$$
$$h + 9 = 0 \Rightarrow h = -9 \text{ or } h - 5 = 0 \Rightarrow h = 5$$

The negative solution is not meaningful. The height of the dock is 5 ft.

**33.** Let $r$ = radius of circle and $x$ = length of side of square. The radius is $\frac{1}{2}$ the length of the side of the square. Area = $x^2$

$$800 = x^2 \Rightarrow x = \sqrt{800} = 20\sqrt{2} \Rightarrow$$
$$r = 10\sqrt{2}$$

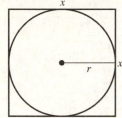

The radius is $10\sqrt{2}$ feet.

**35.** Let $x$ = length of ladder
Distance from building to ladder = $8 + 2 = 10$.
Distance from ground to window = 13
Apply the Pythagorean theorem.

$$a^2 + b^2 = c^2$$
$$10^2 + 13^2 = x^2 \Rightarrow 100 + 169 = x^2 \Rightarrow$$
$$269 = x^2 \Rightarrow \pm\sqrt{269} = x$$
$$x \approx -16.4 \text{ or } x \approx 16.4$$

The negative solution is not meaningful. The worker will need a 16.4-ft ladder.

**37.** Let $x$ = length of short leg, $x + 700$ = length of long leg, and $x + 700 + 100$ or $x + 800$ = length of hypotenuse.

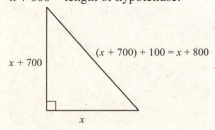

Apply the Pythagorean theorem.

$$c^2 = a^2 + b^2$$
$$(x+800)^2 = x^2 + (x+700)^2$$
$$x^2 + 1600x + 640,000$$
$$= x^2 + x^2 + 1400x + 490,000$$
$$0 = x^2 - 200x - 150,000$$
$$0 = (x+300)(x-500)$$
$$x + 300 = 0 \Rightarrow x = -300 \text{ or}$$
$$x - 500 = 0 \Rightarrow x = 500$$

The negative solution is not meaningful.
500 = length of short leg
500 + 700 = 1200 = length of long leg
1200 + 100 = 1300 = length of hypotenuse
500 + 1200 + 1300 = 3000 = length of walkway. The total length is 3000 yd.

**39. (a)** 
$$s = -16t^2 + v_0 t$$
$$s = -16t^2 + 96t$$
$$80 = -16t^2 + 96t$$
$$16t^2 - 96t + 80 = 0$$
$$a = 16, b = -96 \text{ and } c = 80$$

$$t = \frac{-b \pm \sqrt{b^2 - 4ac}}{2a}$$

$$= \frac{-(-96) \pm \sqrt{(-96)^2 - 4(16)(80)}}{2(16)}$$

$$= \frac{96 \pm \sqrt{9216 - 5120}}{32}$$

$$= \frac{96 \pm \sqrt{4096}}{32} = \frac{96 \pm 64}{32}$$

$$t = \frac{96 - 64}{32} = 1 \text{ or } t = \frac{96 + 64}{32} = 5$$

The projectile will reach 80 ft at 1 sec and 5 sec.

**(b)** 
$$s = -16t^2 + 96t$$
$$0 = -16t^2 + 96t$$
$$0 = -16t(t-6)$$
$$-16t = 0 \Rightarrow t = 0 \text{ or } t - 6 = 0 \Rightarrow t = 6$$

The projectile will return to the ground after 6 sec.

**41. (a)**

$$s = -16t^2 + v_0 t$$
$$s = -16t^2 + 32t$$
$$80 = -16t^2 + 32t$$
$$16t^2 - 32t + 80 = 0$$
$$a = 16, \, b = -32 \text{ and } c = 80$$

$$t = \frac{-b \pm \sqrt{b^2 - 4ac}}{2a}$$

$$= \frac{-(-32) \pm \sqrt{(-32)^2 - 4(16)(80)}}{2(16)}$$

$$= \frac{32 \pm \sqrt{1024 - 5120}}{32}$$

$$= \frac{32 \pm \sqrt{-4096}}{32} = \frac{32 \pm 64i}{32}$$

The projectile will not reach 80 ft.

**(b)**

$$s = -16t^2 + 32t$$
$$0 = -16t^2 + 32t \Rightarrow 0 = -16t(t - 2) \Rightarrow$$
$$-16t = 0 \Rightarrow t = 0 \text{ or } t - 2 = 0 \Rightarrow t = 2$$

The projectile will return to the ground after 2 sec.

**43.** The height of the ball is given by

$$h = -2.7t^2 + 30t + 6.5.$$

**(a)** When the ball is 12 ft above the moon's surface, $h = 12$. Set $h = 12$ and solve for $t$.

$$12 = -2.7t^2 + 30t + 6.5$$
$$2.7t^2 - 30t + 5.5 = 0$$

Use the quadratic formula with $a = 2.7$, $b = -30$, and $c = 5.5$.

$$t = \frac{30 \pm \sqrt{900 - 4(2.7)(5.5)}}{2(2.7)} = \frac{30 \pm \sqrt{840.6}}{5.4}$$

$$\frac{30 + \sqrt{840.6}}{5.4} \approx 10.92 \text{ or } \frac{30 - \sqrt{840.6}}{5.4} \approx .19$$

Therefore, the ball reaches 12 ft first after .19 sec (on the way up), then again after 10.92 sec (on the way down).

**(b)** When the ball returns to the surface, $h = 0$.

$$0 = -2.7t^2 + 30t + 6.5$$

Use the quadratic formula with $a = -2.7$, $b = 30$, and $c = 6.5$.

$$t = \frac{-30 \pm \sqrt{900 - 4(-2.7)(6.5)}}{2(-2.7)}$$

$$= \frac{-30 \pm \sqrt{970.2}}{-5.4}$$

$$\frac{-30 + \sqrt{970.2}}{-5.4} \approx -.21 \text{ or }$$

$$\frac{-30 - \sqrt{970.2}}{-5.4} \approx 11.32$$

The negative solution is not meaningful. Therefore, the ball returns to the surface after 11.32 sec.

**45. (a)** Let $x = 50$.

$$T = .00787(50)^2 - 1.528(50) + 75.89 \approx 19.2$$

The exposure time when $x = 50$ ppm is approximately 19.2 hr.

**(b)** Let $T = 3$ and solve for $x$.

$$3 = .00787x^2 - 1.528x + 75.89$$
$$.00787x^2 - 1.528x + 72.89 = 0$$

Use the quadratic formula with $a = .00787$, $b = -1.528$, and $c = 72.89$.

$$x = \frac{-(-1.528) \pm \sqrt{(-1.528)^2 - 4(.00787)(72.89)}}{2(.00787)}$$

$$= \frac{1.528 \pm \sqrt{2.334784 - 2.2945772}}{.01574}$$

$$= \frac{1.528 \pm \sqrt{.0402068}}{.01574}$$

$$\frac{1.528 + \sqrt{.0402068}}{.01574} \approx 109.8 \text{ or }$$

$$\frac{1.528 - \sqrt{.0402068}}{.01574} \approx 84.3$$

We reject the potential solution 109.8 because it is not in the interval [50, 100]. So, 84.3 ppm carbon monoxide concentration is necessary for a person to reach the 4% to 6% CoHb level in 3 hr.

**47. (a)** Let $x = 600$ and solve for $T$.

$$T = .0002x^2 - .316x + 127.9$$
$$= .0002(600)^2 - .316(600) + 127.9 = 10.3$$

The exposure time when $x = 600$ ppm is 10.3 hr.

**(b)** Let $T = 4$ and solve for $x$.

$$4 = .0002x^2 - .316x + 127.9$$
$$.0002x^2 - .316x + 123.9 = 0$$

Use the quadratic formula with $a = .0002$, $b = -.316$, and $c = 123.9$.

$$x = \frac{-(-.316) \pm \sqrt{(-.316)^2 - 4(.0002)(123.9)}}{2(.0002)}$$

$$= \frac{.316 \pm \sqrt{.099856 - .09912}}{.0004}$$

$$= \frac{.316 \pm \sqrt{.000736}}{.0004}$$

$$\frac{.316 + \sqrt{.000736}}{.0004} \approx 857.8 \text{ or}$$

$$\frac{.316 - \sqrt{.000736}}{.0004} \approx 722.2$$

857.8 is not in the interval [500, 800]. A concentration of 722.2 ppm is required.

**49.** Let $x = 4$ and solve for $y$.

$$y = .808x^2 + 2.625x + .502$$
$$= .808(4)^2 + 2.625(4) + .502$$
$$= .808(16) + 10.5 + .502$$
$$= 12.928 + 11.002 = 23.93$$

Approximately 23.93 million households are expected to pay at least one bill online each month in 2004.

**51.** For each $20 increase in rent over $300, one unit will remain vacant. Therefore, for $x$ $20 increases, $x$ units will remain vacant. Therefore, the number of rented units will be $80 - x$.

**53.** $300 + 20x$ is the rent for each apartment, and $80 - x$ is the number of apartments that will be rented at that cost. The revenue generated will then be the product of $80 - x$ and $300 + 20x$, so the correct expression is
$(80 - x)(300 + 20x)$

$$= 24,000 + 1600x - 300x - 20x^2$$
$$= 24,000 + 1300x - 20x^2.$$

**55.** $20x^2 - 1300x + 11,000 = 0$
$$x^2 - 65x + 550 = 0$$
$$(x - 10)(x - 55) = 0$$
$$x - 10 = 0 \quad \text{or} \quad x - 55 = 0$$
$$x = 10 \quad \text{or} \quad x = 55$$

If $x = 10$, $80 - x = 70$. If $x = 55$, $80 - x = 25$. Because of the restriction that at least 30 units must be rented, only $x = 10$ is valid here, and the number of units rented is 70.

**57.** Let $x$ = number of passengers in excess of 75. Then $225 - 5x$ = the cost per passenger (in dollars) and $75 + x$ = the number of passengers.

(Cost per passenger)(Number of passengers) = Revenue

$$(225 - 5x)(75 + x) = 16,000$$
$$16,875 + 225x - 375x - 5x^2 = 16,000$$
$$16,875 - 150x - 5x^2 = 16,000$$
$$0 = 5x^2 + 150x - 875$$
$$0 = x^2 + 30x - 175 \Rightarrow 0 = (x + 35)(x - 5)$$
$$x + 35 = 0 \Rightarrow x = -35 \text{ or } x - 5 = 0 \Rightarrow x = 5$$

The negative solution is not meaningful. Since there are 5 passengers in excess of 75, the total number of passengers is 80.

## Section 1.6: Other Types of Equations and Applications

**1.** $\dfrac{5}{2x+3} - \dfrac{1}{x-6} = 0$

$2x + 3 \neq 0 \Rightarrow x \neq -\frac{3}{2}$ and $x - 6 \neq 0 \Rightarrow x \neq 6$.

**3.** $\dfrac{3}{x-2} + \dfrac{1}{x+1} = \dfrac{3}{x^2 - x - 2}$

or $\dfrac{3}{x-2} + \dfrac{1}{x+1} = \dfrac{3}{(x-2)(x+1)}$

$x - 2 \neq 0 \Rightarrow x \neq 2$ and $x + 1 \neq 0 \Rightarrow x \neq -1$

**5.** $\dfrac{1}{4x} - \dfrac{2}{x} = 3$

$4x \neq 0 \Rightarrow x \neq 0$

**7.** $\dfrac{2x+5}{2} - \dfrac{3x}{x-2} = x$

The least common denominator is $2(x-2)$, which is equal to 0 if $x = 2$. Therefore, 2 cannot possibly be a solution of this equation.

$$2(x-2)\left[\frac{2x+5}{2} - \frac{3x}{x-2}\right] = 2(x-2)(x)$$
$$(x-2)(2x+5) - 2(3x) = 2x(x-2)$$
$$2x^2 + 5x - 4x - 10 - 6x = 2x^2 - 4x$$
$$-5x - 10 = -4x \Rightarrow -10 = x$$

The restriction $x \neq 2$ does not affect the result. Therefore, the solution set is $\{-10\}$.

9. $\dfrac{x}{x-3} = \dfrac{3}{x-3} + 3$

The least common denominator is $x-3$, which is equal to 0 if $x = 3$. Therefore, 3 cannot possibly be a solution of this equation.

$$(x-3)\left(\dfrac{x}{x-3}\right) = (x-3)\left[\dfrac{3}{x-3} + 3\right]$$
$$x = 3 + 3(x-3)$$
$$x = 3 + 3x - 9$$
$$x = 3x - 6 \Rightarrow -2x = -6 \Rightarrow x = 3$$

The only possible solution is 3. However, the variable is restricted to real numbers except 3. Therefore, the solution set is: $\varnothing$.

11. $\dfrac{-2}{x-3} + \dfrac{3}{x+3} = \dfrac{-12}{x^2-9}$ or
$$\dfrac{-2}{x-3} + \dfrac{3}{x+3} = \dfrac{-12}{(x+3)(x-3)}$$

The least common denominator is $(x+3)(x-3)$, which is equal to 0 if $x = -3$ or $x = 3$. Therefore, $-3$ and 3 cannot possibly be solutions of this equation.

$$(x+3)(x-3)\left[\dfrac{-2}{x-3} + \dfrac{3}{x+3}\right]$$
$$= (x+3)(x-3)\left(\dfrac{-12}{(x+3)(x-3)}\right)$$
$$-2(x+3) + 3(x-3) = -12$$
$$-2x - 6 + 3x - 9 = -12$$
$$-15 + x = -12 \Rightarrow x = 3$$

The only possible solution is 3. However, the variable is restricted to real numbers except $-3$ and 3. Therefore, the solution set is: $\varnothing$.

13. $\dfrac{4}{x^2+x-6} - \dfrac{1}{x^2-4} = \dfrac{2}{x^2+5x+6}$ or
$$\dfrac{4}{(x+3)(x-2)} - \dfrac{1}{(x+2)(x-2)} = \dfrac{2}{(x+2)(x+3)}$$

The least common denominator is $(x+3)(x-2)(x+2)$, which is equal to 0 if $x = -3$ or $x = 2$ or $x = -2$. Therefore, $-3$ and 2 and $-2$ cannot possibly be solutions of this equation.

$$(x+3)(x-2)(x+2)$$
$$\cdot\left[\dfrac{4}{(x+3)(x-2)} - \dfrac{1}{(x+2)(x-2)}\right]$$
$$= (x+3)(x-2)(x+2)\left(\dfrac{2}{(x+2)(x+3)}\right)$$

$$4(x+2) - 1(x+3) = 2(x-2)$$
$$4x + 8 - x - 3 = 2x - 4$$
$$3x + 5 = 2x - 4 \Rightarrow x + 5 = -4 \Rightarrow x = -9$$

The restrictions $x \neq -3$, $x \neq 2$, and $x \neq -2$ do not affect the result. Therefore, the solution set is $\{-9\}$.

15. $\dfrac{2x+1}{x-2} + \dfrac{3}{x} = \dfrac{-6}{x^2-2x}$ or
$$\dfrac{2x+1}{x-2} + \dfrac{3}{x} = \dfrac{-6}{x(x-2)}$$

Multiply each term in the equation by the least common denominator, $x(x-2)$, assuming $x \neq 0, 2$.

$$x(x-2)\left[\dfrac{2x+1}{x-2} + \dfrac{3}{x}\right] = x(x-2)\left(\dfrac{-6}{x(x-2)}\right)$$
$$x(2x+1) + 3(x-2) = -6$$
$$2x^2 + x + 3x - 6 = -6$$
$$2x^2 + 4x - 6 = -6$$
$$2x^2 + 4x = 0 \Rightarrow 2x(x+2) = 0$$

$$2x = 0 \Rightarrow x = 0 \quad \text{or} \quad x+2 = 0 \Rightarrow x = -2$$

Because of the restriction $x \neq 0$, the only valid solution is $-2$. The solution set is $\{-2\}$.

17. $\dfrac{x}{x-1} - \dfrac{1}{x+1} = \dfrac{2}{x^2-1}$ or
$$\dfrac{x}{x-1} - \dfrac{1}{x+1} = \dfrac{2}{(x+1)(x-1)}$$

Multiply each term in the equation by the least common denominator, $(x+1)(x-1)$, assuming $x \neq \pm 1$.

$$(x+1)(x-1)\left[\dfrac{x}{x-1} - \dfrac{1}{x+1}\right]$$
$$= (x+1)(x-1)\left(\dfrac{2}{(x+1)(x-1)}\right)$$

$$x(x+1) - (x-1) = 2 \Rightarrow x^2 + x - x + 1 = 2$$
$$x^2 + 1 = 2 \Rightarrow x^2 - 1 = 0$$
$$(x+1)(x-1) = 0 \Rightarrow x+1 = 0 \Rightarrow x = -1 \text{ or}$$
$$x - 1 = 0 \Rightarrow x = 1$$

Because of the restriction $x \neq \pm 1$, the solution set is $\varnothing$.

**19.** $\dfrac{5}{x^2} - \dfrac{43}{x} = 18$

Multiply each term in the equation by the least common denominator, $x^2$, assuming $x \neq 0$.

$$x^2\left[\dfrac{5}{x^2} - \dfrac{43}{x}\right] = x^2(18)$$

$$5 - 43x = 18x^2 \Rightarrow 0 = 18x^2 + 43x - 5$$

$$0 = (2x+5)(9x-1)$$

$$2x + 5 = 0 \Rightarrow x = -\tfrac{5}{2} \text{ or}$$

$$9x - 1 = 0 \Rightarrow x = \tfrac{1}{9}$$

The restriction $x \neq 0$ does not affect the result. Therefore, the solution set is $\left\{-\tfrac{5}{2}, \tfrac{1}{9}\right\}$.

**21.** $2 = \dfrac{3}{2x-1} + \dfrac{-1}{(2x-1)^2}$

Multiply each term in the equation by the least common denominator, $(2x-1)^2$, assuming $x \neq \tfrac{1}{2}$.

$$(2x-1)^2(2) = (2x-1)^2\left[\dfrac{3}{2x-1} + \dfrac{-1}{(2x-1)^2}\right]$$

$$2(4x^2 - 4x + 1) = 3(2x-1) - 1$$

$$8x^2 - 8x + 2 = 6x - 3 - 1$$

$$8x^2 - 8x + 2 = 6x - 4 \Rightarrow 8x^2 - 14x + 6 = 0$$

$$2(4x^2 - 7x + 3) = 0 \Rightarrow 2(4x-3)(x-1) = 0$$

$$4x - 3 = 0 \Rightarrow x = \tfrac{3}{4} \text{ or } x - 1 = 0 \Rightarrow x = 1$$

The restriction $x \neq \tfrac{1}{2}$ does not affect the result. Therefore the solution set is $\left\{\tfrac{3}{4}, 1\right\}$.

**23.** $\dfrac{2x-5}{x} = \dfrac{x-2}{3}$

Multiply each term in the equation by the least common denominator, $3x$, assuming $x \neq 0$.

$$3x\left(\dfrac{2x-5}{x}\right) = 3x\left(\dfrac{x-2}{3}\right)$$

$$3(2x-5) = x(x-2) \Rightarrow 6x - 15 = x^2 - 2x \Rightarrow$$

$$0 = x^2 - 8x + 15 = (x-3)(x-5)$$

$$x - 3 = 0 \Rightarrow x = 3 \quad \text{or} \quad x - 5 = 0 \Rightarrow x = 5$$

The restriction $x \neq 0$ does not affect the result. Therefore, the solution set is $\{3, 5\}$.

**25.** $\dfrac{2x}{x-2} = 5 + \dfrac{4x^2}{x-2}$

Multiply each term in the equation by the least common denominator, $x - 2$, assuming $x \neq 2$.

$$(x-2)\left(\dfrac{2x}{x-2}\right) = (x-2)\left[5 + \dfrac{4x^2}{x-2}\right]$$

$$2x = 5(x-2) + 4x^2$$

$$2x = 5x - 10 + 4x^2$$

$$0 = 4x^2 + 3x - 10$$

$$0 = (x+2)(4x-5)$$

$$x + 2 = 0 \Rightarrow x = -2 \text{ or } 4x - 5 = 0 \Rightarrow x = \tfrac{5}{4}$$

The restriction $x \neq 2$ does not affect the result. Therefore the solution set is $\left\{-2, \tfrac{5}{4}\right\}$.

**27.** Let $x$ = the amount of time (in hours) it takes Joe and Sam to paint the house.

|      | $r$ | $t$ | Part of the Job Accomplished |
|------|-----|-----|------------------------------|
| Joe  | $\frac{1}{3}$ | $x$ | $\frac{1}{3}x$ |
| Sam  | $\frac{1}{5}$ | $x$ | $\frac{1}{5}x$ |

Since Joe and Sam must accomplish 1 job (painting a house), we must solve the following equation.

$$\tfrac{1}{3}x + \tfrac{1}{5}x = 1 \Rightarrow 15\left[\tfrac{1}{3}x + \tfrac{1}{5}x\right] = 15 \cdot 1$$

$$5x + 3x = 15 \Rightarrow 8x = 15 \Rightarrow x = \dfrac{15}{8} = 1\tfrac{7}{8}$$

It takes Joe and Sam $1\tfrac{7}{8}$ hr working together to paint the house.

**29.** Let $x$ = the amount of time (in hours) it takes plant A to produce the pollutant. Then $2x$ = the amount of time (in hours) it takes plant B to produce the pollutant.

|                   | Rate | Time | Part of the Job Accomplished |
|-------------------|------|------|------------------------------|
| Pollution from A  | $\frac{1}{x}$  | 26 | $\frac{1}{x}(26)$  |
| Pollution from B  | $\frac{1}{2x}$ | 26 | $\frac{1}{2x}(26)$ |

Since plant A and B accomplish 1 job (producing the pollutant), we must solve the following equation.

$$\tfrac{1}{x}(26) + \tfrac{1}{2x}(26) = 1 \Rightarrow \tfrac{26}{x} + \tfrac{13}{x} = 1 \Rightarrow$$

$$x\left[\tfrac{39}{x}\right] = x \cdot 1 \Rightarrow 39 = x$$

Plant B will take $2 \cdot 39 = 78$ hr to produce the pollutant.

**31.** Let $x =$ the amount of time (in hours) to fill the pool with both pipes open.

|  | Rate | Time | Part of the Job Accomplished |
|---|---|---|---|
| Inlet pipe | $\frac{1}{5}$ | $x$ | $\frac{1}{5}x$ |
| Outlet pipe | $\frac{1}{8}$ | $x$ | $\frac{1}{8}x$ |

Filling the pool is 1 whole job, but because the outlet pipe empties the pool, its contribution should be subtracted from the contribution of the inlet pipe.

$$\frac{1}{5}x - \frac{1}{8}x = 1$$
$$40\left[\frac{1}{5}x - \frac{1}{8}x\right] = 40\cdot 1 \Rightarrow 8x - 5x = 40 \Rightarrow$$
$$3x = 40 \Rightarrow x = \frac{40}{3} = 13\frac{1}{3}\,\text{hr}$$

It took $13\frac{1}{3}$ hr to fill the pool.

**33.** Let $x =$ the amount of time (in minutes) to fill the sink with both pipes open.

|  | Rate | Time | Part of the Job Accomplished |
|---|---|---|---|
| Tap | $\frac{1}{5}$ | $x$ | $\frac{1}{5}x$ |
| Drain | $\frac{1}{10}$ | $x$ | $\frac{1}{10}x$ |

Filling the sink is 1 whole job, but because the sink is draining, its contribution should be subtracted from the contribution of the taps.

$$\frac{1}{5}x - \frac{1}{10}x = 1$$
$$10\left[\frac{1}{5}x - \frac{1}{10}x\right] = 10\cdot 1 \Rightarrow 2x - x = 10 \Rightarrow x = 10$$

It will take 10 minutes to fill the sink if Mark forgets to put in the stopper.

**35.** $x - \sqrt{2x+3} = 0$
$$x = \sqrt{2x+3} \Rightarrow x^2 = \left(\sqrt{2x+3}\right)^2$$
$$x^2 = 2x+3 \Rightarrow x^2 - 2x - 3 = 0 \Rightarrow$$
$$(x+1)(x-3) = 0 \Rightarrow x = -1 \text{ or } x = 3$$
Check $x = -1$.
$$x - \sqrt{2x+3} = 0$$
$$-1 - \sqrt{2(-1)+3} \overset{?}{=} 0$$
$$-1 - \sqrt{-2+3} = 0$$
$$-1 - \sqrt{1} = 0 \Rightarrow -1 - 1 = 0 \Rightarrow -2 = 0$$
This is a false statement. $-1$ is not a solution.

Check $x = 3$.
$$x - \sqrt{2x+3} = 0$$
$$3 - \sqrt{2(3)+3} \overset{?}{=} 0$$
$$3 - \sqrt{6+3} = 0$$
$$3 - \sqrt{9} = 0 \Rightarrow 3 - 3 = 0 \Rightarrow 0 = 0$$
This is a true statement. 3 is a solution.
Solution set: $\{3\}$

**37.** $\sqrt{3x+7} = 3x+5$
$$\left(\sqrt{3x+7}\right)^2 = (3x+5)^2$$
$$3x+7 = 9x^2 + 30x + 25$$
$$0 = 9x^2 + 27x + 18$$
$$0 = 9\left(x^2 + 3x + 2\right) = 9(x+2)(x+1)$$
$x = -2$ or $x = -1$
Check $x = -2$.
$$\sqrt{3x+7} = 3x+5$$
$$\sqrt{3(-2)+7} \overset{?}{=} 3(-2)+5$$
$$\sqrt{-6+7} = -6+5$$
$$\sqrt{1} = -1 \Rightarrow 1 = -1$$
This is a false statement. $-2$ is not a solution.
Check $x = -1$
$$\sqrt{3x+7} = 3x+5$$
$$\sqrt{3(-1)+7} \overset{?}{=} 3(-1)+5$$
$$\sqrt{-3+7} = -3+5$$
$$\sqrt{4} = 2 \Rightarrow 2 = 2$$
This is a true statement. $-1$ is a solution.
Solution set: $\{-1\}$

**39.** $\sqrt{4x+5} - 6 = 2x - 11$
$$\sqrt{4x+5} = 2x - 5$$
$$\left(\sqrt{4x+5}\right)^2 = (2x-5)^2$$
$$4x+5 = 4x^2 - 20x + 25$$
$$0 = 4x^2 - 24x + 20$$
$$0 = 4\left(x^2 - 6x + 5\right) = 4(x-1)(x-5)$$
$x = 1$ or $x = 5$
Check $x = 1$.
$$\sqrt{4x+5} - 6 = 2x - 11$$
$$\sqrt{4(1)+5} - 6 \overset{?}{=} 2(1) - 11$$
$$\sqrt{4+5} - 6 = 2 - 11$$
$$\sqrt{9} - 6 = -9$$
$$3 - 6 = -9 \Rightarrow -3 = -9$$
This is a false statement. 1 is not a solution.

Check $x = 5$.

$$\sqrt{4x+5} - 6 = 2x - 11$$

$$\sqrt{4(5)+5} - 6 \overset{?}{=} 2(5) - 11$$

$$\sqrt{20+5} - 6 = 10 - 11$$

$$\sqrt{25} - 6 = -1 \Rightarrow 5 - 6 = -1 \Rightarrow -1 = -1$$

This is a true statement. 5 is a solution.

Solution set: $\{5\}$

**41.** $\sqrt{4x} - x + 3 = 0$

$$\sqrt{4x} = x - 3$$

$$\left(\sqrt{4x}\right)^2 = (x-3)^2$$

$$4x = x^2 - 6x + 9$$

$$0 = x^2 - 10x + 9 = (x-1)(x-9)$$

$$x = 1 \text{ or } x = 9$$

Check $x = 1$.

$$\sqrt{4x} - x + 3 = 0$$

$$\sqrt{4(1)} - 1 + 3 \overset{?}{=} 0$$

$$\sqrt{4} - 1 + 3 = 0$$

$$2 - 1 + 3 = 0 \Rightarrow 4 = 0$$

This is a false statement. 1 is not a solution.

Check $x = 9$.

$$\sqrt{4x} - x + 3 = 0$$

$$\sqrt{4(9)} - 9 + 3 \overset{?}{=} 0$$

$$\sqrt{36} - 9 + 3 = 0$$

$$6 - 9 + 3 = 0 \Rightarrow 0 = 0$$

This is a true statement. 9 is a solution.

Solution set: $\{9\}$

**43.** $\sqrt{x} - \sqrt{x-5} = 1$

$$\sqrt{x} = 1 + \sqrt{x-5} \Rightarrow \left(\sqrt{x}\right)^2 = \left(1 + \sqrt{x-5}\right)^2$$

$$x = 1 + 2\sqrt{x-5} + (x-5)$$

$$x = x + 2\sqrt{x-5} - 4 \Rightarrow 4 = 2\sqrt{x-5}$$

$$2 = \sqrt{x-5} \Rightarrow 2^2 = \left(\sqrt{x-5}\right)^2$$

$$4 = x - 5 \Rightarrow 9 = x$$

Check $x = 9$.

$$\sqrt{x} - \sqrt{x-5} = 1$$

$$\sqrt{9} - \sqrt{9-5} \overset{?}{=} 1$$

$$3 - \sqrt{4} = 1$$

$$3 - 2 = 1 \Rightarrow 1 = 1$$

This is a true statement.

Solution set is: $\{9\}$

**45.** $\sqrt{x+7} + 3 = \sqrt{x-4}$

$$\left(\sqrt{x+7} + 3\right)^2 = \left(\sqrt{x-4}\right)^2$$

$$(x+7) + 6\sqrt{x+7} + 9 = x - 4$$

$$x + 6\sqrt{x+7} + 16 = x - 4 \Rightarrow 6\sqrt{x+7} = -20$$

$$3\sqrt{x+7} = -10 \Rightarrow \left(3\sqrt{x+7}\right)^2 = (-10)^2$$

$$9(x+7) = 100 \Rightarrow 9x + 63 = 100$$

$$9x = 37 \Rightarrow x = \frac{37}{9}$$

Check $x = \frac{37}{9}$.

$$\sqrt{x+7} + 3 = \sqrt{x-4}$$

$$\sqrt{\frac{37}{9}+7} + 3 \overset{?}{=} \sqrt{\frac{37}{9}-4}$$

$$\sqrt{\frac{37}{9}+\frac{63}{9}} + 3 = \sqrt{\frac{37}{9}-\frac{36}{9}}$$

$$\sqrt{\frac{100}{9}} + 3 = \sqrt{\frac{1}{9}}$$

$$\frac{10}{3} + 3 = \frac{1}{3} \Rightarrow \frac{10}{3} + \frac{9}{3} = \frac{1}{3} \Rightarrow \frac{19}{3} = \frac{1}{3}$$

This is a false statement.

Solution set: $\varnothing$

**47.** $\sqrt{x+2} - \sqrt{2x+5} = 1$

$$\sqrt{x+2} = \sqrt{2x+5} - 1$$

$$\left(\sqrt{x+2}\right)^2 = \left(\sqrt{2x+5} - 1\right)^2$$

$$x + 2 = (2x+5) - 2\sqrt{2x+5} + 1$$

$$x + 2 = 2x + 6 - 2\sqrt{2x+5}$$

$$2\sqrt{2x+5} = x + 4$$

$$\left(2\sqrt{2x+5}\right)^2 = (x+4)^2$$

$$4(2x+5) = x^2 + 8x + 16$$

$$8x + 20 = x^2 + 8x + 16$$

$$0 = x^2 - 4 \Rightarrow 0 = (x+2)(x-2)$$

$$x = \pm 2$$

Check $x = 2$.

$$\sqrt{x+2} = \sqrt{2x+5} - 1$$

$$\sqrt{2+2} \overset{?}{=} \sqrt{2(2)+5} - 1$$

$$\sqrt{4} = \sqrt{4+5} - 1 \Rightarrow 2 = \sqrt{9} - 1$$

$$2 = 3 - 1 \Rightarrow 2 = 2$$

This is a true statement. 2 is a solution.

Check $x = -2$.

$$\sqrt{x+2} = \sqrt{2x+5} - 1$$

$$\sqrt{-2+2} \overset{?}{=} \sqrt{2(-2)+5} - 1$$

$$\sqrt{0} = \sqrt{-4+5} - 1 \Rightarrow 0 = \sqrt{1} - 1$$

$$0 = 1 - 1 \Rightarrow 0 = 0$$

This is a true statement. $-2$ is a solution.

Solution set: $\{\pm 2\}$

**49.**
$$\sqrt{3x} = \sqrt{5x+1} - 1$$
$$\left(\sqrt{3x}\right)^2 = \left(\sqrt{5x+1} - 1\right)^2$$
$$3x = (5x+1) - 2\sqrt{5x+1} + 1$$
$$3x = 5x + 2 - 2\sqrt{5x+1}$$
$$2\sqrt{5x+1} = 2 + 2x \Rightarrow \sqrt{5x+1} = 1 + x$$
$$\left(\sqrt{5x+1}\right)^2 = (1+x)^2 \Rightarrow 5x+1 = 1 + 2x + x^2$$
$$0 = x^2 - 3x \Rightarrow 0 = x(x-3) \Rightarrow$$
$$x = 0 \text{ or } x = 3$$

Check $x = 0$.
$$\sqrt{3x} = \sqrt{5x+1} - 1$$
$$\sqrt{3(0)} \stackrel{?}{=} \sqrt{5(0)+1} - 1$$
$$\sqrt{0} = \sqrt{0+1} - 1 \Rightarrow 0 = \sqrt{1} - 1$$
$$0 = 1 - 1 \Rightarrow 0 = 0$$

This is a true statement. 0 is a solution.
Check $x = 3$.
$$\sqrt{3x} = \sqrt{5x+1} - 1$$
$$\sqrt{3(3)} \stackrel{?}{=} \sqrt{5(3)+1} - 1$$
$$\sqrt{9} = \sqrt{15+1} - 1 \Rightarrow 3 = \sqrt{16} - 1$$
$$3 = 4 - 1 \Rightarrow 3 = 3$$

This is a true statement. 3 is a solution.
Solution set: $\{0, 3\}$

**51.**
$$\sqrt{x+2} = 1 - \sqrt{3x+7}$$
$$\left(\sqrt{x+2}\right)^2 = (1 - \sqrt{3x+7})^2$$
$$x + 2 = 1 - 2\sqrt{3x+7} + (3x+7)$$
$$x + 2 = 3x + 8 - 2\sqrt{3x+7}$$
$$2\sqrt{3x+7} = 2x + 6$$
$$2\sqrt{3x+7} = 2(x+3)$$
$$\sqrt{3x+7} = x + 3 \Rightarrow \left(\sqrt{3x+7}\right)^2 = (x+3)^2$$
$$3x + 7 = x^2 + 6x + 9 \Rightarrow 0 = x^2 + 3x + 2$$
$$0 = (x+2)(x+1)$$
$$x = -2 \text{ or } x = -1$$

Check $x = -2$.
$$\sqrt{x+2} = 1 - \sqrt{3x+7}$$
$$\sqrt{-2+2} \stackrel{?}{=} 1 - \sqrt{3(-2)+7}$$
$$\sqrt{0} = 1 - \sqrt{-6+7}$$
$$0 = 1 - \sqrt{1}$$
$$0 = 1 - 1 \Rightarrow 0 = 0$$

This is a true statement. $-2$ is a solution.

Check $x = -1$.
$$\sqrt{x+2} = 1 - \sqrt{3x+7}$$
$$\sqrt{-1+2} \stackrel{?}{=} 1 - \sqrt{3(-1)+7}$$
$$\sqrt{1} = 1 - \sqrt{-3+7}$$
$$1 = 1 - \sqrt{4}$$
$$1 = 1 - 2 \Rightarrow 1 = -1$$

This is a false statement.
$-1$ is not a solution.
Solution set: $\{-2\}$

**53.**
$$\sqrt{2\sqrt{7x+2}} = \sqrt{3x+2}$$
$$\left(\sqrt{2\sqrt{7x+2}}\right)^2 = \left(\sqrt{3x+2}\right)^2$$
$$2\sqrt{7x+2} = 3x + 2$$
$$\left(2\sqrt{7x+2}\right)^2 = (3x+2)^2$$
$$4(7x+2) = 9x^2 + 12x + 4$$
$$28x + 8 = 9x^2 + 12x + 4$$
$$0 = 9x^2 - 16x - 4$$
$$0 = (9x+2)(x-2)$$
$$x = -\tfrac{2}{9} \text{ or } x = 2$$

Check $x = -\tfrac{2}{9}$.
$$\sqrt{2\sqrt{7x+2}} = \sqrt{3x+2}$$
$$\sqrt{2\sqrt{7\left(-\tfrac{2}{9}\right)+2}} \stackrel{?}{=} \sqrt{3\left(-\tfrac{2}{9}\right)+2}$$
$$\sqrt{2\sqrt{-\tfrac{14}{9}+2}} = \sqrt{-\tfrac{2}{3}+2}$$
$$\sqrt{2\sqrt{-\tfrac{14}{9}+\tfrac{18}{9}}} = \sqrt{-\tfrac{2}{3}+\tfrac{6}{3}}$$
$$\sqrt{2\sqrt{\tfrac{4}{9}}} = \sqrt{\tfrac{4}{3}} \Rightarrow \sqrt{2\left(\tfrac{2}{3}\right)} = \tfrac{\sqrt{4}}{\sqrt{3}}$$
$$\sqrt{\tfrac{4}{3}} = \tfrac{2}{\sqrt{3}} \cdot \tfrac{\sqrt{3}}{\sqrt{3}} \Rightarrow \tfrac{\sqrt{4}}{\sqrt{3}} = \tfrac{2\sqrt{3}}{3}$$
$$\tfrac{2}{\sqrt{3}} \cdot \tfrac{\sqrt{3}}{\sqrt{3}} = \tfrac{2\sqrt{3}}{3} \Rightarrow \tfrac{2\sqrt{3}}{3} = \tfrac{2\sqrt{3}}{3}$$

This is a true statement.
$-\tfrac{2}{9}$ is a solution.

Check $x = 2$.
$$\sqrt{2\sqrt{7x+2}} = \sqrt{3x+2}$$
$$\sqrt{2\sqrt{7(2)+2}} \stackrel{?}{=} \sqrt{3(2)+2}$$
$$\sqrt{2\sqrt{14+2}} = \sqrt{6+2}$$
$$\sqrt{2\sqrt{16}} = \sqrt{8}$$
$$\sqrt{2(4)} = 2\sqrt{2}$$
$$\sqrt{8} = 2\sqrt{2} \Rightarrow 2\sqrt{2} = 2\sqrt{2}$$

This is a true statement. 2 is a solution.
Solution set: $\left\{-\tfrac{2}{9}, 2\right\}$

**55.**
$$3 - \sqrt{x} = \sqrt{2\sqrt{x} - 3}$$
$$\left(3 - \sqrt{x}\right)^2 = \left(\sqrt{2\sqrt{x} - 3}\right)^2$$
$$9 - 6\sqrt{x} + x = 2\sqrt{x} - 3$$
$$12 + x = 8\sqrt{x}$$
$$\left(12 + x\right)^2 = \left(8\sqrt{x}\right)^2$$
$$144 + 24x + x^2 = 64x$$
$$x^2 - 40x + 144 = 0$$
$$(x - 36)(x - 4) = 0 \Rightarrow x = 36 \text{ or } x = 4$$

Check $x = 36$.
$$3 - \sqrt{x} = \sqrt{2\sqrt{x} - 3}$$
$$3 - \sqrt{36} \stackrel{?}{=} \sqrt{2\sqrt{36} - 3}$$
$$3 - 6 = \sqrt{2(6) - 3}$$
$$-3 = \sqrt{12 - 3} \Rightarrow -3 = \sqrt{9} \Rightarrow -3 = 3$$

This is a false statement. 36 is not a solution.

Check $x = 4$.
$$3 - \sqrt{x} = \sqrt{2\sqrt{x} - 3}$$
$$3 - \sqrt{4} \stackrel{?}{=} \sqrt{2\sqrt{4} - 3}$$
$$3 - 2 = \sqrt{2(2) - 3}$$
$$1 = \sqrt{4 - 3} \Rightarrow 1 = \sqrt{1} \Rightarrow 1 = 1$$

This is a true statement. 4 is a solution.
Solution set: $\{4\}$

**57.**
$$\sqrt[3]{4x + 3} = \sqrt[3]{2x - 1}$$
$$\left(\sqrt[3]{4x + 3}\right)^3 = \left(\sqrt[3]{2x - 1}\right)^3$$
$$4x + 3 = 2x - 1 \Rightarrow 2x = -4 \Rightarrow x = -2$$

Check $x = -2$.
$$\sqrt[3]{4(-2) + 3} = \sqrt[3]{2(-2) - 1}$$
$$\sqrt[3]{-5} \stackrel{?}{=} \sqrt[3]{-5} \Rightarrow -\sqrt[3]{5} = -\sqrt[3]{5}$$

This is a true statement. $-2$ is a solution.
Solution set: $\{-2\}$

**59.**
$$\sqrt[3]{5x^2 - 6x + 2} - \sqrt[3]{x} = 0$$
$$\sqrt[3]{5x^2 - 6x + 2} = \sqrt[3]{x}$$
$$\left(\sqrt[3]{5x^2 - 6x + 2}\right)^3 = \left(\sqrt[3]{x}\right)^3$$
$$5x^2 - 6x + 2 = x$$
$$5x^2 - 7x + 2 = 0$$
$$(5x - 2)(x - 1) = 0 \Rightarrow x = \tfrac{2}{5} \text{ or } x = 1$$

Check $x = \tfrac{2}{5}$.
$$\sqrt[3]{5x^2 - 6x + 2} - \sqrt[3]{x} = 0$$
$$\sqrt[3]{5\left(\tfrac{2}{5}\right)^2 - 6\left(\tfrac{2}{5}\right) + 2} - \sqrt[3]{\tfrac{2}{5}} \stackrel{?}{=} 0$$
$$\sqrt[3]{5\left(\tfrac{4}{25}\right) - \tfrac{12}{5} + 2} - \sqrt[3]{\tfrac{2}{5}} = 0$$
$$\sqrt[3]{\tfrac{4}{5} - \tfrac{12}{5} + \tfrac{10}{5}} - \sqrt[3]{\tfrac{2}{5}} = 0$$
$$\sqrt[3]{\tfrac{2}{5}} - \sqrt[3]{\tfrac{2}{5}} = 0 \Rightarrow 0 = 0$$

This is a true statement. $\tfrac{2}{5}$ is a solution.

Check $x = 1$.
$$\sqrt[3]{5x^2 - 6x + 2} - \sqrt[3]{x} = 0$$
$$\sqrt[3]{5(1)^2 - 6(1) + 2} - \sqrt[3]{1} \stackrel{?}{=} 0$$
$$\sqrt[3]{5(1) - 6 + 2} - 1 = 0$$
$$\sqrt[3]{5 - 6 + 2} - 1 = 0$$
$$\sqrt[3]{1} - 1 = 0 \Rightarrow 1 - 1 = 0 \Rightarrow 0 = 0$$

This is a true statement. 1 is a solution.
Solution set: $\left\{\tfrac{2}{5}, 1\right\}$

**61.**
$$(2x + 5)^{1/3} - (6x - 1)^{1/3} = 0$$
$$(2x + 5)^{1/3} = (6x - 1)^{1/3}$$
$$\left[(2x + 5)^{1/3}\right]^3 = \left[(6x - 1)^{1/3}\right]^3$$
$$2x + 5 = 6x - 1$$
$$5 = 4x - 1 \Rightarrow 6 = 4x$$
$$\tfrac{6}{4} = x \Rightarrow x = \tfrac{3}{2}$$

Check $x = \tfrac{3}{2}$.
$$(2x + 5)^{1/3} - (6x - 1)^{1/3} = 0$$
$$\left[2\left(\tfrac{3}{2}\right) + 5\right]^{1/3} - \left[6\left(\tfrac{3}{2}\right) - 1\right]^{1/3} \stackrel{?}{=} 0$$
$$(3 + 5)^{1/3} - (9 - 1)^{1/3} = 0$$
$$8^{1/3} - 8^{1/3} = 0$$
$$2 - 2 = 0 \Rightarrow 0 = 0$$

This is a true statement.
Solution set: $\left\{\tfrac{3}{2}\right\}$

**63.**
$$\sqrt[4]{x - 15} = 2 \Rightarrow \left(\sqrt[4]{x - 15}\right)^4 = 2^4 \Rightarrow$$
$$x - 15 = 16 \Rightarrow x = 31$$

Check $x = 31$.
$$\sqrt[4]{x - 15} = 2 \Rightarrow \sqrt[4]{31 - 15} \stackrel{?}{=} 2$$
$$\sqrt[4]{16} = 2 \Rightarrow 2 = 2$$

This is a true statement.
Solution set: $\{31\}$

**65.** $\sqrt[4]{x^2+2x}=\sqrt[4]{3}\Rightarrow\left(\sqrt[4]{x^2+2x}\right)^4=\left(\sqrt[4]{3}\right)^4$

$x^2+2x=3\Rightarrow x^2+2x-3=0$

$(x+3)(x-1)=0\Rightarrow x=-3 \text{ or } x=1$

Check $x=-3$.

$$\sqrt[4]{x^2+2x}=\sqrt[4]{3}$$

$$\sqrt[4]{(-3)^2+2(-3)}\overset{?}{=}\sqrt[4]{3}$$

$$\sqrt[4]{9-6}=\sqrt[4]{3}\Rightarrow\sqrt[4]{3}=\sqrt[4]{3}$$

This is a true statement. $-3$ is a solution.

Check $x=1$.

$$\sqrt[4]{x^2+2x}=\sqrt[4]{3}$$

$$\sqrt[4]{1^2+2(1)}\overset{?}{=}\sqrt[4]{3}\Rightarrow\sqrt[4]{1+2}=\sqrt[4]{3}\Rightarrow\sqrt[4]{3}=\sqrt[4]{3}$$

This is a true statement. 1 is a solution.

Solution set: $\{-3,1\}$

**67.** $(x^2+24x)^{1/4}=3\Rightarrow\left[\left(x^2+24x\right)^{1/4}\right]^4=3^4\Rightarrow$

$x^2+24x=81\Rightarrow x^2+24x-81=0\Rightarrow$

$(x+27)(x-3)=0\Rightarrow x+27=0\Rightarrow x=-27 \text{ or }$

$x-3=0\Rightarrow x=3$

Check $x=-27$.

$$(x^2+24x)^{1/4}=3$$

$$\left[(-27)^2+24(-27)\right]^{1/4}\overset{?}{=}3$$

$$(729-648)^{1/4}=3\Rightarrow 81^{1/4}=3\Rightarrow 3=3$$

This is a true statement. $-27$ is a solution.

Check $x=3$.

$$(x^2+24x)^{1/4}=3$$

$$\left[3^2+24(3)\right]^{1/4}\overset{?}{=}3$$

$$(9+72)^{1/4}=3\Rightarrow 81^{1/4}=3\Rightarrow 3=3$$

This is a true statement. 3 is a solution.

Solution set: $\{-27, 3\}$

**69.** $2x^4-7x^2+5=0$

Let $u=x^2$; then $u^2=x^4$. With this substitution, the equation becomes

$2u^2-7u+5=0$.

$2u^2-7u+5=0\Rightarrow(u-1)(2u-5)=0\Rightarrow$

$u=1 \text{ or } u=\frac{5}{2}$

To find $x$, replace $u$ with $x^2$.

$x^2=1\Rightarrow x=\pm1$ or

$x^2=\frac{5}{2}\Rightarrow x=\pm\sqrt{\frac{5}{2}}=\pm\frac{\sqrt{5}}{\sqrt{2}}\cdot\frac{\sqrt{2}}{\sqrt{2}}=\pm\frac{\sqrt{10}}{2}$

Solution set: $\left\{\pm1,\pm\frac{\sqrt{10}}{2}\right\}$

**71.** $x^4+2x^2-15=0$

Let $u=x^2$; then $u^2=x^4$.

$u^2+2u-15=0\Rightarrow(u-3)(u+5)=0$.

$u=3 \text{ or } u=-5$

To find $x$, replace $u$ with $x^2$.

$x^2=3\Rightarrow x=\pm\sqrt{3}$ or

$x^2=-5\Rightarrow x=\pm\sqrt{-5}=\pm i\sqrt{5}$

Solution set: $\left\{\pm\sqrt{3},\pm i\sqrt{5}\right\}$

**73.** $(2x-1)^{2/3}=x^{1/3}$

$[(2x-1)^{2/3}]^3=(x^{1/3})^3$

$(2x-1)^2=x\Rightarrow 4x^2-4x+1=x$

$4x^2-5x+1=0\Rightarrow(4x-1)(x-1)=0\Rightarrow$

$x=\frac{1}{4} \text{ or } x=1$

Check $x=\frac{1}{4}$.

$(2x-1)^{2/3}=x^{1/3}\Rightarrow\left[2\left(\frac{1}{4}\right)-1\right]^{2/3}\overset{?}{=}\left(\frac{1}{4}\right)^{1/3}$

$\left[\frac{1}{2}-1\right]^{2/3}=\frac{1}{\sqrt[3]{4}}\cdot\frac{\sqrt[3]{2}}{\sqrt[3]{2}}\Rightarrow\left[-\frac{1}{2}\right]^{2/3}=\frac{\sqrt[3]{2}}{2}\Rightarrow$

$\left[\left(-\frac{1}{2}\right)^2\right]^{1/3}=\frac{\sqrt[3]{2}}{2}\Rightarrow\left(\frac{1}{4}\right)^{1/3}=\frac{\sqrt[3]{2}}{2}$

$\frac{1}{\sqrt[3]{4}}\cdot\frac{\sqrt[3]{2}}{\sqrt[3]{2}}=\frac{\sqrt[3]{2}}{2}\Rightarrow\frac{\sqrt[3]{2}}{2}=\frac{\sqrt[3]{2}}{2}$

This is a true statement. $\frac{1}{4}$ is a solution.

Check $x=1$.

$(2x-1)^{2/3}=x^{1/3}\Rightarrow\left[2(1)-1\right]^{2/3}\overset{?}{=}(1)^{1/3}$

$[2-1]^{2/3}=1\Rightarrow 1^{2/3}=1\Rightarrow 1=1$

This is a true statement. 1 is a solution.

Solution set: $\left\{\frac{1}{4},1\right\}$

**75.** $x^{2/3}=2x^{1/3}\Rightarrow\left(x^{2/3}\right)^3=\left(2x^{1/3}\right)^3\Rightarrow$

$x^2=8x\Rightarrow x^2-8x=0\Rightarrow x(x-8)=0\Rightarrow$

$x=0 \text{ or } x=8$

Check $x=0$.

$x^{2/3}=2x^{1/3}$

$0^{2/3}\overset{?}{=}2\left(0^{1/3}\right)\Rightarrow 0=2\cdot 0\Rightarrow 0=0$

This is a true statement. 0 is a solution.

Check $x=8$.

$x^{2/3}=2x^{1/3}\Rightarrow 8^{2/3}\overset{?}{=}2\left(8^{1/3}\right)$

$\left(8^2\right)^{1/3}=2\cdot 2\Rightarrow 64^{1/3}=4\Rightarrow 4=4$

This is a true statement. 8 is a solution.

Solution set: $\{0,8\}$

**77.** $(x-1)^{2/3} + (x-1)^{1/3} - 12 = 0$

Let $u = (x-1)^{1/3}$ then,

$u^2 = \left[(x-1)^{1/3}\right]^2 = (x-1)^{2/3}$.

$u^2 + u - 12 = 0 \Rightarrow (u+4)(u-3) = 0 \Rightarrow$

$u = -4$ or $u = 3$

To find $x$, replace $u$ with $(x-1)^{1/3}$.

$(x-1)^{1/3} = -4 \Rightarrow \left[(x-1)^{1/3}\right]^3 = (-4)^3 \Rightarrow$

$\quad x-1 = -64 \Rightarrow x = -63$  or

$(x-1)^{1/3} = 3 \Rightarrow \left[(x-1)^{1/3}\right]^3 = 3^3 \Rightarrow$

$\quad x-1 = 27 \Rightarrow x = 28$

Check $x = -63$.

$(x-1)^{2/3} + (x-1)^{1/3} - 12 = 0$

$(-63-1)^{2/3} + (-63-1)^{1/3} - 12 \overset{?}{=} 0$

$(-64)^{2/3} + (-64)^{1/3} - 12 = 0$

$\left[(-64)^{1/3}\right]^2 - 4 - 12 = 0$

$(-4)^2 - 4 - 12 = 0$

$16 - 4 - 12 = 0 \Rightarrow 0 = 0$

This is a true statement. $-63$ is a solution.

Check $x = 28$.

$(x-1)^{2/3} + (x-1)^{1/3} - 12 = 0$

$(28-1)^{2/3} + (28-1)^{1/3} - 12 \overset{?}{=} 0$

$27^{2/3} + 27^{1/3} - 12 = 0$

$\left[27^{1/3}\right]^2 + 3 - 12 = 0$

$3^2 + 3 - 12 = 0$

$9 + 3 - 12 = 0 \Rightarrow 0 = 0$

This is a true statement. 28 is a solution.

Solution set: $\{-63, 28\}$

**79.** $(x+1)^{2/5} - 3(x+1)^{1/5} + 2 = 0$

Let $u = (x+1)^{1/5}$ then,

$u^2 = \left[(x+1)^{1/5}\right]^2 = (x+1)^{2/5}$.

$u^2 - 3u + 2 = 0 \Rightarrow (u-1)(u-2) = 0 \Rightarrow$

$u = 1$ or $u = 2$

To find $x$, replace $u$ with $(x+1)^{1/5}$.

$(x+1)^{1/5} = 1 \Rightarrow \left[(x+1)^{1/5}\right]^5 = 1^5 \Rightarrow$  or

$\quad x+1 = 1 \Rightarrow x = 0$

$(x+1)^{1/5} = 2 \Rightarrow \left[(x+1)^{1/5}\right]^5 = 2^5 \Rightarrow$

$\quad x+1 = 32 \Rightarrow x = 31$

Check $x = 0$.

$(x+1)^{2/5} - 3(x+1)^{1/5} + 2 = 0$

$(0+1)^{2/5} - 3(0+1)^{1/5} + 2 \overset{?}{=} 0$

$1^{2/5} - 3(1)^{1/5} + 2 = 0$

$1 - 3(1) + 2 = 0 \Rightarrow 1 - 3 + 2 = 0$

This is a true statement. 0 is a solution.

Check $x = 31$.

$(x+1)^{2/5} - 3(x+1)^{1/5} + 2 = 0$

$(31+1)^{2/5} - 3(31+1)^{1/5} + 2 \overset{?}{=} 0$

$32^{2/5} - 3(32)^{1/5} + 2 = 0$

$\left[(32)^{1/5}\right]^2 - 3(2) + 2 = 0$

$2^2 - 6 + 2 = 0$

$4 - 6 + 2 = 0 \Rightarrow 0 = 0$

This is a true statement. 31 is a solution.

Solution set: $\{0, 31\}$

**81.** $6(x+2)^4 - 11(x+2)^2 = -4$

$6(x+2)^4 - 11(x+2)^2 + 4 = 0$

Let $u = (x+2)^2$ then $u^2 = (x+2)^4$.

$6u^2 - 11u + 4 = 0 \Rightarrow (3u-4)(2u-1) = 0 \Rightarrow$

$u = \frac{4}{3}$ or $u = \frac{1}{2}$

To find $x$, replace $u$ with $(x+2)^2$.

$(x+2)^2 = \frac{4}{3} \Rightarrow x+2 = \pm\sqrt{\frac{4}{3}} = \pm\frac{2\sqrt{3}}{3}$  or

$\quad x = -2 \pm \frac{2\sqrt{3}}{3} = -\frac{6}{3} \pm \frac{2\sqrt{3}}{3} = \frac{-6 \pm 2\sqrt{3}}{3}$

$(x+2)^2 = \frac{1}{2} \Rightarrow x+2 = \pm\sqrt{\frac{1}{2}} = \pm\frac{\sqrt{2}}{2}$

$\quad x = -2 \pm \frac{\sqrt{2}}{2} = -\frac{4}{2} \pm \frac{\sqrt{2}}{2} = \frac{-4 \pm \sqrt{2}}{2}$

Solution set: $\left\{\frac{-6 \pm 2\sqrt{3}}{3}, \frac{-4 \pm \sqrt{2}}{2}\right\}$

**83.** $10x^{-2} + 33x^{-1} - 7 = 0$

Let $u = x^{-1}$; then $u^2 = x^{-2}$.

$10u^2 + 33u - 7 = 0 \Rightarrow (2u+7)(5u-1) = 0$

$u = -\frac{7}{2}$ or $u = \frac{1}{5}$

To find $x$, replace $u$ with $x^{-1}$.

$x^{-1} = -\frac{7}{2} \Rightarrow x = -\frac{2}{7}$ or $x^{-1} = \frac{1}{5} \Rightarrow x = 5$

Solution set: $\left\{-\frac{2}{7}, 5\right\}$

**85.** $x^{-2/3} + x^{-1/3} - 6 = 0$

Let $u = x^{-1/3}$; then $u^2 = \left(x^{-1/3}\right)^2 = x^{-2/3}$.

$u^2 + u - 6 = 0 \Rightarrow (u+3)(u-2) = 0$

$u = -3$ or $u = 2$

To find $x$, replace $u$ with $x^{-1/3}$.

$x^{-1/3} = -3 \Rightarrow \left(x^{-1/3}\right)^{-3} = (-3)^{-3} \Rightarrow$ or

$x = \frac{1}{(-3)^3} \Rightarrow x = -\frac{1}{27}$

$x^{-1/3} = 2 \Rightarrow \left(x^{-1/3}\right)^{-3} = 2^{-3} \Rightarrow$

$x = \frac{1}{2^3} \Rightarrow x = \frac{1}{8}$

Check $x = -\frac{1}{27}$.

$$x^{-2/3} + x^{-1/3} - 6 = 0$$

$$\left(-\frac{1}{27}\right)^{-2/3} + \left(-\frac{1}{27}\right)^{-1/3} - 6 \overset{?}{=} 0$$

$$(-27)^{2/3} + (-27)^{1/3} - 6 = 0$$

$$\left[(-27)^{1/3}\right]^2 - 3 - 6 = 0$$

$$(-3)^2 - 3 - 6 = 0$$

$$9 - 3 - 6 = 0 \Rightarrow 0 = 0$$

This is a true statement.

Check $x = \frac{1}{8}$.

$$x^{-2/3} + x^{-1/3} - 6 = 0$$

$$\left(\frac{1}{8}\right)^{-2/3} + \left(\frac{1}{8}\right)^{-1/3} - 6 \overset{?}{=} 0$$

$$8^{2/3} + 8^{1/3} - 6 = 0$$

$$\left(8^{1/3}\right)^2 + 2 - 6 = 0$$

$$2^2 + 2 - 6 = 0$$

$$4 + 2 - 6 = 0 \Rightarrow 0 = 0$$

This is a true statement.

Solution set: $\left\{-\frac{1}{27}, \frac{1}{8}\right\}$

**87.** $16x^{-4} - 65x^{-2} + 4 = 0$

Let $u = x^{-2}$; then $u^2 = x^{-4}$. Solve the resulting equation by factoring:

$16u^2 - 65u + 4 = 0 \Rightarrow (u-4)(16u-1) = 0 \Rightarrow$

$u = 4$ or $u = \frac{1}{16}$

Find $x$ by replacing $u$ with $x^{-2}$:

$x^{-2} = 4 \Rightarrow x^2 = \frac{1}{4} \Rightarrow x = \pm\frac{1}{2}$;

$x^{-2} = \frac{1}{16} \Rightarrow x^2 = 16 \Rightarrow x = \pm 4$;

Check $x = \frac{1}{2}$

$$16\left(\frac{1}{2}\right)^{-4} - 65\left(\frac{1}{2}\right)^{-2} + 4 = 0$$

$$16(2)^4 - 65(2)^2 + 4 \overset{?}{=} 0$$

$$16(16) - 65(4) + 4 = 0$$

$$256 - 260 + 4 = 0$$

$$0 = 0$$

This is a true statement, so $\frac{1}{2}$ is a solution.

Check $x = -\frac{1}{2}$

$$16\left(-\frac{1}{2}\right)^{-4} - 65\left(-\frac{1}{2}\right)^{-2} + 4 = 0$$

$$16(-2)^4 - 65(-2)^2 + 4 \overset{?}{=} 0$$

$$16(16) - 65(4) + 4 = 0$$

$$256 - 260 + 4 = 0$$

$$0 = 0$$

This is a true statement, so $-\frac{1}{2}$ is a solution.

Check $x = 4$

$$16(4)^{-4} - 65(4)^{-2} + 4 = 0$$

$$16\left(\frac{1}{4}\right)^4 - 65\left(\frac{1}{4}\right)^2 + 4 \overset{?}{=} 0$$

$$16\left(\frac{1}{256}\right) - 65\left(\frac{1}{16}\right) + 4 = 0$$

$$\frac{1}{16} - \frac{65}{16} + 4 = 0 \Rightarrow 0 = 0$$

This is a true statement, so 4 is a solution.

Check $x = -4$

$$16(-4)^{-4} - 65(-4)^{-2} + 4 = 0$$

$$16\left(-\frac{1}{4}\right)^4 - 65\left(-\frac{1}{4}\right)^2 + 4 \overset{?}{=} 0$$

$$16\left(-\frac{1}{256}\right) - 65\left(-\frac{1}{16}\right) + 4 = 0$$

$$\frac{1}{16} - \frac{65}{16} + 4 = 0 \Rightarrow 0 = 0$$

This is a true statement, so $-4$ is a solution.

Solution set: $\left\{\pm\frac{1}{2}, \pm 4\right\}$

**89.** $x - \sqrt{x} - 12 = 0$

Let $u = \sqrt{x}$; then $u^2 = x$. Solve the resulting equation by factoring.

$u^2 - u - 12 = 0 \Rightarrow (u-4)(u+3) = 0$

$u = 4$ or $u = -3$

To find $x$, replace $u$ with $\sqrt{x}$.

$\sqrt{x} = 4 \Rightarrow \left(\sqrt{x}\right)^2 = 4^2 \Rightarrow x = 16$ or

$\sqrt{x} = -3 \Rightarrow \left(\sqrt{x}\right)^2 = (-3)^2 \Rightarrow x = 9$

But $\sqrt{9} \neq -3$

So when $u = -3$, there is no solution for $x$.

Solution set: $\{16\}$

**91.** Answers will vary.

**93.** $d = k\sqrt{h}$ for $h$

$$\frac{d}{k} = \sqrt{h} \Rightarrow \frac{d^2}{k^2} = h$$

So, $h = \dfrac{d^2}{k^2}$.

**95.** $m^{3/4} + n^{3/4} = 1$ for $m$

$$m^{3/4} = 1 - n^{3/4}$$

Raise both sides to the $\frac{4}{3}$ power.

$$(m^{3/4})^{4/3} = (1 - n^{3/4})^{4/3}$$
$$m = (1 - n^{3/4})^{4/3}$$

**97.** $\dfrac{E}{e} = \dfrac{R+r}{r}$ for $e$

$$er\left(\frac{E}{e}\right) = er\left(\frac{R+r}{r}\right)$$

Multiply both sides by $er$.

$$Er = eR + er \Rightarrow Er = e(R+r)$$

$$\frac{Er}{R+r} = e$$

So, $e = \dfrac{Er}{R+r}$.

## Summary Exercises on Solving Equations

**1.** $4x - 3 = 2x + 3 \Rightarrow 2x - 3 = 3 \Rightarrow$

$\quad 2x = 6 \Rightarrow x = 3$

Solution set: $\{3\}$

**3.** $x(x+6) = 9 \Rightarrow x^2 + 6x = 9 \Rightarrow x^2 + 6x - 9 = 0$

Solve by completing the square.

$$x^2 + 6x + 9 = 9 + 9$$

$\quad$ Note: $\left[\frac{1}{2} \cdot 6\right]^2 = 3^2 = 9$

$$(x+3)^2 = 18 \Rightarrow x + 3 = \pm\sqrt{18} \Rightarrow$$
$$x + 3 = \pm 3\sqrt{2} \Rightarrow x = -3 \pm 3\sqrt{2}$$

Solve by the quadratic formula.

Let $a = 1, b = 6,$ and $c = -9$.

$$x = \frac{-b \pm \sqrt{b^2 - 4ac}}{2a}$$

$$= \frac{-6 \pm \sqrt{6^2 - 4(1)(-9)}}{2(1)}$$

$$= \frac{-6 \pm \sqrt{36 + 36}}{2} = \frac{-6 \pm \sqrt{72}}{2}$$

$$= \frac{-6 \pm 6\sqrt{2}}{2} = -3 \pm 3\sqrt{2}$$

Solution set: $\left\{-3 \pm \sqrt{2}\right\}$

**5.**

$$\sqrt{x+2} + 5 = \sqrt{x+15}$$
$$\left(\sqrt{x+2} + 5\right)^2 = \left(\sqrt{x+15}\right)^2$$
$$(x+2) + 10\sqrt{x+2} + 25 = x + 15 \Rightarrow$$
$$x + 27 + 10\sqrt{x+2} = x + 15 \Rightarrow$$
$$27 + 10\sqrt{x+2} = 15 \Rightarrow$$
$$10\sqrt{x+2} = -12 \Rightarrow 5\sqrt{x+2} = -6$$
$$\left(5\sqrt{x+2}\right)^2 = (-6)^2 \Rightarrow$$
$$25(x+2) = 36 \Rightarrow 25x + 50 = 36$$
$$25x = -14 \Rightarrow x = -\frac{14}{25}$$

Check $x = -\frac{14}{25}$.

$$\sqrt{x+2} + 5 = \sqrt{x+15}$$
$$\sqrt{-\frac{14}{25} + 2} + 5 \overset{?}{=} \sqrt{-\frac{14}{25} + 15}$$
$$\sqrt{-\frac{14}{25} + \frac{50}{25}} + 5 = \sqrt{-\frac{14}{25} + \frac{375}{25}}$$
$$\sqrt{\frac{36}{25}} + 5 = \sqrt{\frac{361}{25}} \Rightarrow \frac{6}{5} + 5 = \frac{19}{5} \Rightarrow \frac{6}{5} + \frac{25}{5} = \frac{19}{5} \Rightarrow \frac{31}{5} = \frac{19}{5}$$

This is a false statement. Solution set: $\varnothing$

**7.** $\dfrac{3x+4}{3} - \dfrac{2x}{x-3} = x$

The least common denominator is $3(x-3)$, which is equal to $0$ if $x = 3$. Therefore, $3$ cannot possibly be a solution of this equation.

$$3(x-3)\left[\frac{3x+4}{3} - \frac{2x}{x-3}\right] = 3(x-3)(x)$$
$$(x-3)(3x+4) - 3(2x) = 3x(x-3)$$
$$3x^2 + 4x - 9x - 12 - 6x = 3x^2 - 9x$$
$$3x^2 - 11x - 12 = 3x^2 - 9x$$
$$-11x - 12 = -9x \Rightarrow -12 = 2x \Rightarrow -6 = x$$

The restriction $x \neq 3$ does not affect the result. Therefore, the solution set is $\{-6\}$.

**9.** $5 - \dfrac{2}{x} + \dfrac{1}{x^2} = 0$

The least common denominator is $x^2$, which is equal to 0 if $x = 0$. Therefore, 0 cannot possibly be a solution of this equation.

$x^2 \left[ 5 - \dfrac{2}{x} + \dfrac{1}{x^2} \right] = x^2(0) \Rightarrow 5x^2 - 2x + 1 = 0$

Solve by completing the square.

$x^2 - \dfrac{2}{5}x + \dfrac{1}{5} = 0 \qquad$ Multiply by $\dfrac{1}{5}$.

$x^2 - \dfrac{2}{5}x + \dfrac{1}{25} = -\dfrac{1}{5} + \dfrac{1}{25}$

$\qquad$ Note: $\left[ \dfrac{1}{2} \cdot \left( -\dfrac{2}{5} \right) \right]^2 = \left( -\dfrac{1}{5} \right)^2 = \dfrac{1}{25}$

$\left( x - \dfrac{1}{5} \right)^2 = -\dfrac{5}{25} + \dfrac{1}{25} = \dfrac{-4}{25}$

$x - \dfrac{1}{5} = \pm\sqrt{\dfrac{-4}{25}}$

$x - \dfrac{1}{5} = \pm\dfrac{2}{5}i \Rightarrow x = \dfrac{1}{5} \pm \dfrac{2}{5}i$

Solve by the quadratic formula.
Let $a = 5$, $b = -2$, and $c = 1$.

$x = \dfrac{-b \pm \sqrt{b^2 - 4ac}}{2a}$

$= \dfrac{-(-2) \pm \sqrt{(-2)^2 - 4(5)(1)}}{2(5)}$

$= \dfrac{2 \pm \sqrt{4 - 20}}{10} = \dfrac{2 \pm \sqrt{-16}}{10}$

$= \dfrac{2 \pm 4i}{10} = \dfrac{2}{10} \pm \dfrac{4}{10}i = \dfrac{1}{5} \pm \dfrac{2}{5}i$

The restriction $x \neq 0$ does not affect the result. Therefore, the solution set is $\left\{ \dfrac{1}{5} \pm \dfrac{2}{5}i \right\}$.

**11.** $x^{-2/5} - 2x^{-1/5} - 15 = 0$

Let $u = x^{-1/5}$; then $u^2 = \left( x^{-1/5} \right)^2 = x^{-2/5}$.

$u^2 - 2u - 15 = 0 \Rightarrow (u+3)(u-5) = 0 \Rightarrow$
$u = -3$ or $u = 5$

To find $x$, replace $u$ with $x^{-1/5}$.

$x^{-1/5} = -3 \Rightarrow \left( x^{-1/5} \right)^{-5} = (-3)^{-5}$

$x = \dfrac{1}{(-3)^5} \Rightarrow x = -\dfrac{1}{243} \qquad$ or

$x^{-1/5} = 5 \Rightarrow \left( x^{-1/5} \right)^{-5} = 5^{-5} \Rightarrow x = \dfrac{1}{5^5} \Rightarrow$

$\qquad x = \dfrac{1}{3125}$

Check $x = -\dfrac{1}{243}$.

$x^{-2/5} - 2x^{-1/5} - 15 = 0$

$\left( -\dfrac{1}{243} \right)^{-2/5} - 2\left( -\dfrac{1}{243} \right)^{-1/5} - 15 = 0$ ?

$(-243)^{2/5} - 2(-243)^{1/5} - 15 = 0$

$\left[ (-243)^{1/5} \right]^2 - 2(-3) - 15 = 0$

$(-3)^2 + 6 - 15 = 0 \Rightarrow 9 + 6 - 15 = 0 \Rightarrow 0 = 0$

This is a true statement. $-\dfrac{1}{243}$ is a solution.

Check $x = \dfrac{1}{3125}$.

$x^{-2/5} - 2x^{-1/5} - 15 = 0$

$\left( \dfrac{1}{3125} \right)^{-2/5} - 2\left( \dfrac{1}{3125} \right)^{-1/5} - 15 \overset{?}{=} 0$

$(3125)^{2/5} - 2(3125)^{1/5} - 15 = 0$

$\left[ (3125)^{1/5} \right]^2 - 2(5) - 15 = 0$

$5^2 - 10 - 15 = 0$

$25 - 10 - 15 = 0 \Rightarrow 0 = 0$

This is a true statement. $\dfrac{1}{3125}$ is a solution.

Solution set: $\left\{ -\dfrac{1}{243}, \dfrac{1}{3125} \right\}$

**13.** $x^4 - 3x^2 - 4 = 0$

Let $u = x^2$; then $u^2 = x^4$.

$u^2 - 3u - 4 = 0 \Rightarrow (u+1)(u-4) = 0 \Rightarrow$
$\qquad\qquad u = -1$ or $u = 4$

To find $x$, replace $x$ with $x^2$.

$x^2 = -1 \Rightarrow x = \pm\sqrt{-1} = \pm i$ or

$x^2 = 4 \Rightarrow x = \pm\sqrt{4} = \pm 2$

Solution set: $\{ \pm i, \pm 2 \}$

**15.** $\sqrt[6]{2x+1} = \sqrt[6]{9} \Rightarrow \left( \sqrt[6]{2x+1} \right)^6 = \left( \sqrt[6]{9} \right)^6$

$\quad 2x + 1 = 9 \Rightarrow 2x = 8 \Rightarrow x = 4$

Check $x = 4$.

$\sqrt[6]{2x+1} = \sqrt[6]{9} \Rightarrow \sqrt[6]{2(4)+1} \overset{?}{=} \sqrt[6]{9}$

$\sqrt[6]{8+1} = \sqrt[6]{9} \Rightarrow \sqrt[6]{9} = \sqrt[6]{9}$

This is a true statement.
Solution set: $\{4\}$

**17.** $3[2x - (6 - 2x) + 1] = 5x$

$3(2x - 6 + 2x + 1) = 5x$

$3(4x - 5) = 5x$

$12x - 15 = 5x \Rightarrow -15 = -7x \Rightarrow$

$\dfrac{-15}{-7} = x \Rightarrow x = \dfrac{15}{7}$

Solution set: $\left\{ \dfrac{15}{7} \right\}$

**19.**
$$(14 - 2x)^{2/3} = 4$$
$$[(14 - 2x)^{2/3}]^3 = 4^3$$
$$(14 - 2x)^2 = 64$$
$$196 - 56x + 4x^2 = 64$$
$$4x^2 - 56x + 132 = 0$$
$$4(x^2 - 14x + 33) = 0$$
$$4(x - 3)(x - 11) = 0 \Rightarrow x = 3 \text{ or } x = 11$$

Check $x = 3$.
$$(14 - 2x)^{2/3} = 4$$
$$[14 - 2(3)]^{2/3} \stackrel{?}{=} 4$$
$$(14 - 6)^{2/3} = 4 \Rightarrow 8^{2/3} = 4$$
$$(8^{1/3})^2 = 4 \Rightarrow 2^2 = 4 \Rightarrow 4 = 4$$

This is a true statement.
Check $x = 11$.
$$(14 - 2x)^{2/3} = 4$$
$$[14 - 2(11)]^{2/3} \stackrel{?}{=} 4$$
$$(14 - 22)^{2/3} = 4 \Rightarrow (-8)^{2/3} = 4$$
$$[(-8)^{1/3}]^2 = 4 \Rightarrow (-2)^2 = 4 \Rightarrow 4 = 4$$

This is a true statement.
Solution set: $\{3, 11\}$

**21.** $\dfrac{3}{x - 3} = \dfrac{3}{x - 3}$

The least common denominator is $(x - 3)$ which is equal to 0 if $x = 3$. Therefore, 3 cannot possibly be a solution of this equation.
Solution set: $\{x \mid x \neq 3\}$.

## Section 1.7: Inequalities

**1.** $x < -6$
The interval includes all real numbers less than $-6$ not including $-6$. The correct interval notation is $(-\infty, -6)$, so the correct choice is F.

**3.** $-2 < x \leq 6$
The interval includes all real numbers from $-2$ to 6, not including $-2$, but including 6. The correct interval notation is $(-2, 6]$, so the correct choice is A.

**5.** $x \geq -6$
The interval includes all real numbers greater than or equal to $-6$, so it includes $-6$. The correct interval notation is $[-6, \infty)$, so the correct choice is I.

**7.** The interval shown on the number line includes all real numbers between $-2$ and 6, including $-2$, but not including 6. The correct interval notation is $[-2, 6)$, so the correct choice is B.

**9.** The interval shown on the number line includes all real numbers less than $-3$, not including $-3$, and greater than 3, not including 3. The correct interval notation is $(-\infty, -3) \cup (3, \infty)$, so the correct choice is E.

**11.** Answers will vary.

**13.** $2x + 8 \leq 16 \Rightarrow 2x + 8 - 8 \leq 16 - 8 \Rightarrow$
$$2x \leq 8 \Rightarrow \frac{2x}{2} \leq \frac{8}{2} \Rightarrow x \leq 4$$
Solution set: $(-\infty, 4]$
Graph:

**15.**
$$-2x - 2 \leq 1 + x$$
$$-2x - 2 + 2 \leq 1 + x + 2$$
$$-2x \leq x + 3 \Rightarrow -2x - x \leq 3 \Rightarrow$$
$$-3x \leq 3 \Rightarrow \frac{-3x}{-3} \geq \frac{3}{-3} \Rightarrow x \geq -1$$
Solution set: $[-1, \infty)$
Graph:

**17.** $2(x + 5) + 1 \geq 5 + 3x$
$$2x + 10 + 1 \geq 5 + 3x \Rightarrow 2x + 11 \geq 5 + 3x \Rightarrow$$
$$2x + 11 - 3x \geq 5 + 3x - 3x \Rightarrow -x + 11 \geq 5 \Rightarrow$$
$$-x + 11 - 11 \geq 5 - 11 \Rightarrow \frac{-x}{-1} \leq \frac{-6}{-1} \Rightarrow x \leq 6$$
Solution set: $(-\infty, 6]$
Graph:

**19.** $8x - 3x + 2 < 2(x + 7)$
$$5x + 2 < 2x + 14$$
$$5x + 2 - 2x < 2x + 14 - 2x$$
$$3x + 2 < 14 \Rightarrow 3x + 2 - 2 < 14 - 2 \Rightarrow$$
$$3x < 12 \Rightarrow \frac{3x}{3} < \frac{12}{3} \Rightarrow x < 4$$
Solution set: $(-\infty, 4)$
Graph:

**21.**
$$\frac{4x+7}{-3} \le 2x+5$$
$$(-3)\left(\frac{4x+7}{-3}\right) \ge (-3)(2x+5)$$
$$4x+7 \ge -6x-15$$
$$4x+7+6x \ge -6x-15+6x$$
$$10x+7 \ge -15$$
$$10x+7-7 \ge -15-7 \Rightarrow 10x \ge -22 \Rightarrow$$
$$\frac{10x}{10} \ge \frac{-22}{10} \Rightarrow x \ge -\frac{11}{5}$$

Solution set: $\left[-\frac{11}{5}, \infty\right)$

Graph:

**23.**
$$\frac{1}{3}x + \frac{2}{5}x - \frac{1}{2}(x+3) \le \frac{1}{10}$$
$$30\left[\frac{1}{3}x + \frac{2}{5}x - \frac{1}{2}(x+3)\right] \le 30\left[\frac{1}{10}\right]$$
$$10x+12x-15(x+3) \le 3$$
$$10x+12x-15x-45 \le 3$$
$$7x-45 \le 3$$
$$7x-45+45 \le 3+45$$
$$7x \le 48$$
$$\frac{7x}{7} \le \frac{48}{7} \Rightarrow x \le \frac{48}{7}$$

Solution set: $\left(-\infty, \frac{48}{7}\right]$

Graph:

**25.**
$$-5 < 5+2x < 11$$
$$-5-5 < 5+2x-5 < 11-5$$
$$-10 < 2x < 6$$
$$\frac{-10}{2} < \frac{2x}{2} < \frac{6}{2}$$
$$-5 < x < 3$$

Solution set: $(-5, 3)$

Graph:

**27.**
$$10 \le 2x+4 \le 16$$
$$10-4 \le 2x+4-4 \le 16-4$$
$$6 \le 2x \le 12$$
$$\frac{6}{2} \le \frac{2x}{2} \le \frac{12}{2}$$
$$3 \le x \le 6$$

Solution set: $[3, 6]$

Graph:

**29.**
$$-11 > -3x+1 > -17$$
$$-11-1 > -3x+1-1 > -17-1$$
$$-12 > -3x > -18$$
$$\frac{-12}{-3} < \frac{-3x}{-3} < \frac{-18}{-3}$$
$$4 < x < 6$$

Solution set: $(4, 6)$

Graph:

**31.**
$$-4 \le \frac{x+1}{2} \le 5$$
$$2(-4) \le 2\left(\frac{x+1}{2}\right) \le 2(5)$$
$$-8 \le x+1 \le 10$$
$$-8-1 \le x+1-1 \le 10-1 \Rightarrow -9 \le x \le 9$$

Solution set: $[-9, 9]$

Graph:

**33.**
$$-3 \le \frac{x-4}{-5} < 4$$
$$(-5)(-3) \ge (-5)\left(\frac{x-4}{-5}\right) > (-5)(4)$$
$$15 \ge x-4 > -20$$
$$15+4 \ge x-4+4 > -20+4$$
$$19 \ge x > -16 \Rightarrow -16 < x \le 19$$

Solution set: $(-16, 19]$

Graph:

**35.** $C = 50x + 5000; R = 60x$
The product will at least break even when $R \ge C$. Set $R \ge C$ and solve for $x$.
$$60x \ge 50x+5000 \Rightarrow 10x \ge 5000 \Rightarrow x \ge 500$$
The break-even point is at $x = 500$.
This product will at least break even if the number of units of picture frames produced is in interval $[500, \infty)$.

**37.** $C = 85x + 900; R = 105x$
The product will at least break even when $R \ge C$. Set $R \ge C$ and solve for $x$.
$$105x \ge 85x+900 \Rightarrow 20x \ge 900 \Rightarrow x \ge 45$$
The break-even point is $x = 45$.
The product will at least break even when the number of units of coffee cups produced is in the interval $[45, \infty)$.

**39.** $x^2 - x - 6 > 0$

*Step 1*: Find the values of $x$ that satisfy

$x^2 - x - 6 = 0$.

$x^2 - x - 6 = 0 \Rightarrow (x+2)(x-3) = 0$

$x + 2 = 0 \Rightarrow x = -2$ or $x - 3 = 0 \Rightarrow x = 3$

*Step 2*: The two numbers divide a number line into three regions.

| Interval A | Interval B | Interval C |
| $(-\infty, -2)$ | $(-2, 3)$ | $(3, \infty)$ |

$-2 \quad 0 \quad 3$

*Step 3*: Choose a test value to see if it satisfies the inequality, $x^2 - x - 6 > 0$.

| Interval | Test Value | Is $x^2 - x - 6 > 0$ True or False? |
|---|---|---|
| A: $(-\infty, -2)$ | $-3$ | $(-3)^2 - (-3) - 6 \overset{?}{>} 0$ <br> $6 > 0$ <br> True |
| B: $(-2, 3)$ | $0$ | $0^2 - 0 - 6 \overset{?}{>} 0$ <br> $-6 > 0$ <br> False |
| C: $(3, \infty)$ | $4$ | $4^2 - 4 - 6 \overset{?}{>} 0$ <br> $6 > 0$ <br> True |

Solution set: $(-\infty, -2) \cup (3, \infty)$

**41.** $2x^2 - 9x \le 18$

*Step 1*: Find the values of $x$ that satisfy the corresponding equation.

$2x^2 - 9x = 18$

$2x^2 - 9x - 18 = 0$

$(2x+3)(x-6) = 0$

$2x + 3 = 0 \Rightarrow x = -\frac{3}{2}$ or $x - 6 = 0 \Rightarrow x = 6$

*Step 2*: The two numbers divide a number line into three regions.

| Interval A | Interval B | Interval C |
| $(-\infty, -\frac{3}{2})$ | $(-\frac{3}{2}, 6)$ | $(6, \infty)$ |

$-\frac{3}{2} \quad 0 \quad 6$

*Step 3*: Choose a test value to see if it satisfies the inequality, $2x^2 - 9x \le 18$

| Interval | Test Value | Is $2x^2 - 9x \le 18$ True or False? |
|---|---|---|
| A: $\left(-\infty, -\frac{3}{2}\right)$ | $-2$ | $2(-2)^2 - 9(-2) \overset{?}{\le} 18$ <br> $26 \le 18$ False |

| Interval | Test Value | Is $2x^2 - 9x \le 18$ True or False? |
|---|---|---|
| B: $\left(-\frac{3}{2}, 6\right)$ | $0$ | $2(0)^2 - 9(0) \overset{?}{\le} 18$ <br> $0 \le 18$ True |
| C: $(6, \infty)$ | $7$ | $2(7)^2 - 9(7) \overset{?}{\le} 18$ <br> $35 \le 18$ False |

Solution set: $\left[-\frac{3}{2}, 6\right]$

**43.** $-x^2 - 4x - 6 \le -3$

*Step 1*: Find the values of $x$ that satisfy the corresponding equation.

$-x^2 - 4x - 6 = -3$

$x^2 + 4x + 3 = 0$

$(x+3)(x+1) = 0$

$x + 3 = 0 \Rightarrow x = -3$ or $x + 1 = 0 \Rightarrow x = -1$

*Step 2*: The two numbers divide a number line into three regions.

| Interval A | Interval B | Interval C |
| $(-\infty, -3)$ | $(-3, -1)$ | $(-1, \infty)$ |

$-3 \quad -1 \quad 0$

*Step 3*: Choose a test value to see if it satisfies the inequality, $-x^2 - 4x - 6 \le -3$

| Interval | Test Value | Is $-x^2 - 4x - 6 \le -3$ True or False? |
|---|---|---|
| A: $(-\infty, -3)$ | $-4$ | $-(-4)^2 - 4(-4) - 6 \overset{?}{\le} -3$ <br> $-6 \le -3$ <br> True |
| B: $(-3, -1)$ | $-2$ | $-(-2)^2 - 4(-2) - 6 \overset{?}{\le} -3$ <br> $-2 \le -3$ <br> False |
| C: $(-1, \infty)$ | $0$ | $-(0)^2 - 4(0) - 6 \overset{?}{\le} -3$ <br> $-6 \le -3$ <br> True |

Solution set: $(-\infty, -3] \cup [-1, \infty)$

**45.** $x(x-1) \le 6 \Rightarrow x^2 - x \le 6 \Rightarrow x^2 - x - 6 \le 0$

*Step 1*: Find the values of $x$ that satisfy

$x^2 - x - 6 = 0$.

$x^2 - x - 6 = 0 \Rightarrow (x+2)(x-3) = 0$

$x + 2 = 0 \Rightarrow x = -2$   or   $x - 3 = 0 \Rightarrow x = 3$

*Step 2*: The two numbers divide a number line into three regions.

Interval A   Interval B   Interval C
$(-\infty,-2)$   $(-2,3)$   $(3,\infty)$

$-2 \quad 0 \quad 3$

*Step 3*: Choose a test value to see if it satisfies the inequality, $x(x-1) \le 6$.

| Interval | Test Value | Is $x(x-1) \le 6$ True or False? |
|---|---|---|
| A: $(-\infty,-2)$ | $-3$ | $-3(-3-1) \overset{?}{\le} 6$ <br> $12 \le 6$ <br> False |
| B: $(-2,3)$ | $0$ | $0(0-1) \overset{?}{\le} 6$ <br> $0 \le 6$ <br> True |
| C: $(3,\infty)$ | $4$ | $4(4-1) \overset{?}{\le} 6$ <br> $12 \le 6$ <br> False |

Solution set: $[-2,3]$

**47.** $x^2 \le 9$

*Step 1*: Find the values of $x$ that satisfy $x^2 \le 9$

$x^2 = 9 \Rightarrow x^2 - 9 = 0 \Rightarrow (x+3)(x-3) = 0$

$x + 3 = 0 \Rightarrow x = -3$   or   $x - 3 = 0 \Rightarrow x = 3$

*Step 2*: The two numbers divide a number line into three regions.

Interval A   Interval B   Interval C
$(-\infty,-3)$   $(-3,3)$   $(3,\infty)$

$-3 \quad 0 \quad 3$

*Step 3*: Choose a test value to see if it satisfies the inequality, $x^2 \le 9$.

| Interval | Test Value | Is $x^2 \le 9$ True or False? |
|---|---|---|
| A: $(-\infty,-3)$ | $-4$ | $(-4)^2 \overset{?}{\le} 9$ <br> $16 \le 9$ <br> False |
| B: $(-3,3)$ | $0$ | $(0)^2 \overset{?}{\le} 9$ <br> $0 \le 9$ <br> True |

| Interval | Test Value | Is $x^2 \le 9$ True or False? |
|---|---|---|
| C: $(3,\infty)$ | $4$ | $(4)^2 \overset{?}{\le} 9$ <br> $16 \le 9$ <br> False |

Solution set: $[-3, 3]$

**49.** $x^2 + 5x - 2 < 0$

*Step 1*: Find the values of $x$ that satisfy

$x^2 + 5x - 2 = 0$.

Use the quadratic formula to solve the equation.

Let $a = 1$, $b = 5$, and $c = -2$.

$$x = \frac{-b \pm \sqrt{b^2 - 4ac}}{2a} = \frac{-5 \pm \sqrt{5^2 - 4(1)(-2)}}{2(1)}$$

$$= \frac{-5 \pm \sqrt{25 + 8}}{2} = \frac{-5 \pm \sqrt{33}}{2}$$

$$x = \frac{-5 - \sqrt{33}}{2} \approx -5.4 \text{ or}$$

$$x = \frac{-5 + \sqrt{33}}{2} \approx .4$$

*Step 2*: The two numbers divide a number line into three regions.

Interval A   Interval B   Interval C
$\left(-\infty, \frac{-5-\sqrt{33}}{2}\right)$   $\left(\frac{-5-\sqrt{33}}{2}, \frac{-5+\sqrt{33}}{2}\right)$   $\left(\frac{-5+\sqrt{33}}{2}, \infty\right)$

$-4 \quad -1 \quad 0$

*Step 3*: Choose a test value to see if it satisfies the inequality, $x^2 + 5x - 2 < 0$.

| Interval | Test Value | Is $x^2 + 5x - 2 < 0$ True or False? |
|---|---|---|
| A: $\left(-\infty, \frac{-5-\sqrt{33}}{2}\right)$ | $-6$ | $(-6)^2 + 5(-6) - 2 \overset{?}{<} 0$ <br> $4 < 0$ <br> False |
| B: $\left(\frac{-5-\sqrt{33}}{2}, \frac{-5+\sqrt{33}}{2}\right)$ | $0$ | $(0)^2 + 5(0) - 2 \overset{?}{<} 0$ <br> $-2 < 0$ <br> True |
| C: $\left(\frac{-5+\sqrt{33}}{2}, \infty\right)$ | $1$ | $(1)^2 + 5(1) - 2 \overset{?}{<} 0$ <br> $4 < 0$ <br> False |

Solution set: $\left(\frac{-5-\sqrt{33}}{2}, \frac{-5+\sqrt{33}}{2}\right)$

**51.** $x^2 - 2x \le 1 \Rightarrow x^2 - 2x - 1 \le 0$

*Step 1*: Find the values of $x$ that
satisfy $x^2 - 2x - 1 = 0$.
Use the quadratic formula to solve the
equation.
Let $a = 1$, $b = -2$, and $c = -1$.

$$x = \frac{-b \pm \sqrt{b^2 - 4ac}}{2a}$$

$$= \frac{-(-2) \pm \sqrt{(-2)^2 - 4(1)(-1)}}{2(1)}$$

$$= \frac{2 \pm \sqrt{4 + 4}}{2} = \frac{2 \pm \sqrt{8}}{2}$$

$$= \frac{2 \pm 2\sqrt{2}}{2} = 1 \pm \sqrt{2}$$

$1 - \sqrt{2} \approx -.4$ or $1 + \sqrt{2} \approx 2.4$

*Step 2*: The two numbers divide a number line
into three regions.

Interval A $(-\infty, 1-\sqrt{2})$ | Interval B $(1-\sqrt{2}, 1+\sqrt{2})$ | Interval C $(1+\sqrt{2}, \infty)$

*Step 3*: Choose a test value to see if it satisfies
the inequality, $x^2 - 2x \le 1$.

| Interval | Test Value | Is $x^2 - 2x \le 1$ True or False? |
|---|---|---|
| A: $\left(-\infty, 1-\sqrt{2}\right)$ | $-1$ | $(-1)^2 - 2(-1) \overset{?}{\le} 1$ <br> $3 \le 1$ <br> False |
| B: $\left(1-\sqrt{2}, 1+\sqrt{2}\right)$ | $0$ | $0^2 - 2(0) \overset{?}{\le} 1 \Rightarrow 0 \le 1$ <br> True |
| C: $\left(1+\sqrt{2}, \infty\right)$ | $3$ | $3^2 - 2(3) \overset{?}{\le} 1 \Rightarrow 3 \le 1$ <br> False |

Solution set: $\left[1 - \sqrt{2}, 1 + \sqrt{2}\right]$

**53.** A; $(x+3)^2$ is equal to zero when $x = -3$. For
any other real number, $(x+3)^2$ is positive.
$(x+3)^2 \ge 0$ has solution set $(-\infty, \infty)$.

**55.** $(3x - 4)(x + 2)(x + 6) = 0$
Set each factor to zero and solve.
$3x - 4 = 0 \Rightarrow x = \frac{4}{3}$ or $x + 2 = 0 \Rightarrow x = -2$ or
$x + 6 = 0 \Rightarrow x = -6$
Solution set: $\left\{\frac{4}{3}, -2, -6\right\}$

**57.**

| Interval | Test Value | Is $(3x-4)(x+2)(x+6) \le 0$ True or False? |
|---|---|---|
| A: $(-\infty, -6)$ | $-10$ | $[3(-10)-4][-10+2]$ <br> $\cdot[-10+6] \overset{?}{\le} 0$ <br> $-1088 \le 0$ <br> True |
| B: $(-6, -2)$ | $-4$ | $[3(-4)-4][-4+2]$ <br> $\cdot[-4+6] \overset{?}{\le} 0$ <br> $64 \le 0$ <br> False |
| C: $\left(-2, \frac{4}{3}\right)$ | $0$ | $[3(0)-4][0+2][0+6] \overset{?}{\le} 0$ <br> $-48 \le 0$ <br> True |
| D: $\left(\frac{4}{3}, \infty\right)$ | $4$ | $[3(4)-4][4+2][4+6] \overset{?}{\le} 0$ <br> $480 \le 0$ <br> False |

**59.** $(2x - 3)(x + 2)(x - 3) \ge 0$

*Step 1*: Solve $(2x-3)(x+2)(x-3) = 0$.
Set each factor to zero and solve.
$2x - 3 = 0 \Rightarrow x = \frac{3}{2}$ or $x + 2 = 0 \Rightarrow x = -2$ or
$x - 3 = 0 \Rightarrow x = 3$
Solution set: $\left\{-2, \frac{3}{2}, 3\right\}$

*Step 2*: Plot the solutions $-2, \frac{3}{2},$ and 3 on a
number line.

*Step 3*: Choose a test value to see if it satisfies
the inequality, $(2x+3)(x+2)(x-3) \ge 0$.

| Interval | Test Value | Is $(2x-3)(x+2)(x-3) \ge 0$ True or False? |
|---|---|---|
| A: $(-\infty, -2)$ | $-3$ | $[2(-3)-3][-3+2]$ <br> $\cdot[-3-3] \overset{?}{\ge} 0$ <br> $-54 \ge 0$ <br> False |
| B: $\left(-2, \frac{3}{2}\right)$ | $0$ | $[2(0)-3][0+2]$ <br> $\cdot[0-3] \overset{?}{\ge} 0$ <br> $18 \ge 0$ <br> True |

*(continued on next page)*

*(continued from page 59)*

| Interval | Test Value | Is $(2x-3)(x+2)(x-3) \geq 0$ True or False? |
|---|---|---|
| C: $\left(\frac{3}{2},3\right)$ | 2 | $[2(2)-3][2+2]$ $\cdot [2-3] \overset{?}{\geq} 0$ $-4 \geq 0$ False |
| D: $(3,\infty)$ | 4 | $[2(4)-3][4+2]$ $\cdot [4-3] \overset{?}{\geq} 0$ $30 \geq 0$ True |

Solution set: $\left[-2,\frac{3}{2}\right] \cup [3,\infty)$

**61.** $4x - x^3 \geq 0$

*Step 1*: Solve $4x - x^3 = 0$.
$$4x - x^3 = 0 \Rightarrow x(4-x^2) = 0 \Rightarrow$$
$$x(2+x)(2-x) = 0$$

Set each factor to zero and solve.
    $x = 0$ or $2 + x = 0 \Rightarrow x = -2$ or
$2 - x = 0 \Rightarrow x = 2$
Solution set: $\{-2, 0, 2\}$

*Step 2*: The values $-2$, $0$, and $2$ divide the number line into four intervals.

*Step 3*: Choose a test value to see if it satisfies the inequality, $4x - x^3 \geq 0$.

| Interval | Test Value | Is $4x - x^3 \geq 0$ True or False? |
|---|---|---|
| A: $(-\infty,-2)$ | $-3$ | $4(-3)-(-3)^3 \overset{?}{\geq} 0$ $15 \geq 0$ True |
| B: $(-2,0)$ | $-1$ | $4(-1)-(-1)^3 \overset{?}{\geq} 0$ $-3 \geq 0$ False |
| C: $(0,2)$ | $1$ | $4(1)-1^3 \overset{?}{\geq} 0$ $3 \geq 0$ True |
| D: $(2,\infty)$ | $3$ | $4(3)-3^3 \overset{?}{\geq} 0$ $-15 \geq 0$ False |

Solution set: $(-\infty,-2] \cup [0,2]$

**63.** $(x+1)^2 (x-3) < 0$

*Step 1*: Solve $(x+1)^2 (x-3) = 0$.
Set each distinct factor to zero and solve.
    $x+1 = 0 \Rightarrow x = -1$   or   $x-3 = 0 \Rightarrow x = 3$
Solution set: $\{-1, 3\}$

*Step 2*: The values $-1$ and $3$ divide the number line into three intervals.

*Step 3*: Choose a test value to see if it satisfies the inequality, $(x+1)^2 (x-3) < 0$.

| Interval | Test Value | Is $(x+1)^2 (x-3) < 0$ True or False? |
|---|---|---|
| A: $(-\infty,-1)$ | $-2$ | $(-2+1)^2 (-2-3) \overset{?}{<} 0$ $-5 < 0$ True |
| B: $(-1,3)$ | $0$ | $(0+1)^2 (0-3) \overset{?}{<} 0$ $-3 < 0$ True |
| C: $(3,\infty)$ | $4$ | $(4+1)^2 (4-3) \overset{?}{<} 0$ $25 < 0$ False |

Solution set: $(-\infty,-1) \cup (-1,3)$

**65.** $x^3 + 4x^2 - 9x \geq 36$

*Step 1*: Solve $x^3 + 4x^2 - 9x \geq 36$
$$x^3 + 4x^2 - 9x \geq 36$$
$$x^3 + 4x^2 - 9x - 36 = 0$$
$$x^2(x+4) - 9(x+4) = 0$$
$$(x+4)(x^2-9) = 0$$
$$(x+4)(x+3)(x-3) = 0$$

Set each factor to zero and solve.
$x+4 = 0 \Rightarrow x = -4$ or $x+3 = 0 \Rightarrow x = -3$ or
$x-3 = 0 \Rightarrow x = 3$
Solution set: $\{-4, -3, 3\}$

*Step 2*: The values $-4$, $-3$, and $3$ divide the number line into four intervals.

*Step 3*: Choose a test value to see if it satisfies the inequality, $x^3 + 4x^2 - 9x \geq 36$

| Interval | Test Value | Is $x^3 + 4x^2 - 9x \geq 36$ True or False? |
|---|---|---|
| A: $(-\infty, -4)$ | $-5$ | $(-5)^3 + 4(-5)^2 - 9(-5) \overset{?}{\geq} 36$ <br> $20 \geq 0$ <br> False |
| B: $(-4, -3)$ | $-3.5$ | $(-3.5)^3 + 4(-3.5)^2$ <br> $-9(-3.5) \overset{?}{\geq} 36$ <br> $37.625 \geq 0$ <br> True |
| C: $(-3, 3)$ | $0$ | $0^3 + 4(0)^2 - 9(0) \overset{?}{\geq} 36$ <br> $0 \geq 0$ <br> False |
| D: $(3, \infty)$ | $4$ | $4^3 + 4(4)^2 - 9(4) \overset{?}{\geq} 36$ <br> $92 \geq 36$ <br> True |

Solution set: $[-4, -3] \cup [3, \infty)$

67. $x^2(x+4)^2 \geq 0$

*Step 1*: Solve $x^2(x+4)^2 \geq 0$.

Set each distinct factor to zero and solve.

$x = 0$ or $x + 4 = 0 \Rightarrow x = -4$

Solution set: $\{-4, 0\}$

*Step 2*: The values $-4$ and $0$ divide the number line into three intervals.

*Step 3*: Choose a test value to see if it satisfies the inequality, $x^2(x+4)^2 \geq 0$.

| Interval | Test Value | Is $x^2(x+4)^2 \geq 0$ True or False? |
|---|---|---|
| A: $(-\infty, -4)$ | $-5$ | $(-5)^2(-5+4)^2 \overset{?}{\geq} 0$ <br> $25 \geq 0$  True |
| B: $(-4, 0)$ | $-1$ | $(-1)^2(-1+4)^2 \overset{?}{\geq} 0$ <br> $9 \geq 0$  True |
| C: $(0, \infty)$ | $1$ | $1^2(1+4)^2 \overset{?}{\geq} 0$ <br> $25 \geq 0$  True |

Solution set: $(-\infty, \infty)$

69. $\dfrac{x-3}{x+5} \leq 0$

Since one side of the inequality is already 0, we start with Step 2.

*Step 2*: Determine the values that will cause either the numerator or denominator to equal 0.

$x - 3 = 0 \Rightarrow x = 3$   or   $x + 5 = 0 \Rightarrow x = -5$

The values $-5$ and $3$ divide the number line into three regions. Use an open circle on $-5$ because it makes the denominator equal 0.

*Step 3*: Choose a test value to see if it satisfies the inequality, $\dfrac{x-3}{x+5} \leq 0$.

| Interval | Test Value | Is $\frac{x-3}{x+5} \leq 0$ True or False? |
|---|---|---|
| A: $(-\infty, -5)$ | $-6$ | $\frac{-6-3}{-6+5} \overset{?}{\leq} 0$ <br> $9 \leq 0$  False |
| B: $(-5, 3)$ | $0$ | $\frac{0-3}{0+5} \overset{?}{\leq} 0$ <br> $-\frac{3}{5} \leq 0$  True |
| C: $(3, \infty)$ | $4$ | $\frac{4-3}{4+5} \overset{?}{\leq} 0$ <br> $\frac{1}{9} \leq 0$  False |

Interval B satisfies the inequality. The endpoint $-5$ is not included because it makes the denominator 0.

Solution set: $(-5, 3]$

70. $\dfrac{x+1}{x-4} > 0$

Since one side of the inequality is already 0, we start with Step 2.

*Step 2*: Determine the values that will cause either the numerator or denominator to equal 0.

$x + 1 = 0 \Rightarrow x = -1$   or   $x - 4 = 0 \Rightarrow x = 4$

The values $-1$ and $4$ divide the number line into three regions.

(*continued on next page*)

*(continued from page 61)*

*Step 3:* Choose a test value to see if it satisfies

the inequality, $\dfrac{x+1}{x-4} > 0$.

| Interval | Test Value | Is $\frac{x+1}{x-4} > 0$ True or False? |
|---|---|---|
| A: $(-\infty, -1)$ | $-2$ | $\dfrac{-2+1}{-2-4} \overset{?}{>} 0$ <br> $\dfrac{1}{6} > 0$ True |
| B: $(-1, 4)$ | $0$ | $\dfrac{0+1}{0-4} \overset{?}{>} 0$ <br> $-\dfrac{1}{4} > 0$ False |
| C: $(4, \infty)$ | $5$ | $\dfrac{5+1}{5-4} \overset{?}{>} 0 \Rightarrow 6 > 0$ True |

Solution set: $(-\infty, -1) \cup (4, \infty)$

**71.** $\dfrac{1-x}{x+2} < -1$

*Step 1:* Rewrite the inequality so that 0 is on one side and there is a single fraction on the other side.

$$\dfrac{1-x}{x+2} < -1 \Rightarrow \dfrac{x-1}{x+2} > 1$$
$$\dfrac{x-1}{x+2} - 1 > 0 \Rightarrow \dfrac{x-1}{x+2} - \dfrac{x+2}{x+2} > 0$$
$$\dfrac{x-1-(x+2)}{x+2} > 0 \Rightarrow \dfrac{x-1-x-2}{x+2} > 0$$
$$\dfrac{-3}{x+2} > 0$$

*Step 2:* Since the numerator is a constant, determine the values that will cause denominator to equal 0: $x + 2 = 0 \Rightarrow x = -2$ The value $-2$ divides the number line into two regions.

| Interval A | Interval B |
|---|---|
| $(-\infty, -2)$ | $(-2, \infty)$ |

*Step 3:* Choose a test value to see if it satisfies

the inequality, $\dfrac{1-x}{x+2} < -1$

| Interval | Test Value | Is $\frac{1-x}{x+2} < -1$ True or False? |
|---|---|---|
| A: $(-\infty, -2)$ | $-3$ | $\dfrac{-1-(-3)}{-3+2} \overset{?}{<} -1$ <br> $-2 < -1$ True |
| B: $(-2, \infty)$ | $-1$ | $\dfrac{1-(-1)}{-1+2} \overset{?}{<} 1$ <br> $2 < -1$ False |

Solution set: $(-\infty, -2)$

**73.** $\dfrac{3}{x-6} \le 2$

*Step 1:* Rewrite the inequality so that 0 is on one side and there is a single fraction on the other side.

$$\dfrac{3}{x-6} - 2 \le 0 \Rightarrow \dfrac{3}{x-6} - \dfrac{2(x-6)}{x-6} \le 0$$
$$\dfrac{3-2(x-6)}{x-6} \le 0 \Rightarrow \dfrac{3-2x+12}{x-6} \le 0$$
$$\dfrac{15-2x}{x-6} \le 0$$

*Step 2:* Determine the values that will cause either the numerator or denominator to equal 0.

$$15 - 2x = 0 \Rightarrow x = \dfrac{15}{2} \quad \text{or} \quad x - 6 = 0 \Rightarrow x = 6$$

The values 6 and $\frac{15}{2}$ to divide the number line into three regions. Use an open circle on 6 because it makes the denominator equal 0.

| Interval A $(-\infty, 6)$ | Interval B $(6, \frac{15}{2})$ | Interval C $(\frac{15}{2}, \infty)$ |
|---|---|---|

*Step 3:* Choose a test value to see if it satisfies

the inequality, $\dfrac{3}{x-6} \le 2$.

| Interval | Test Value | Is $\frac{3}{x-6} \le 2$ True or False? |
|---|---|---|
| A: $(-\infty, 6)$ | $0$ | $\dfrac{3}{0-6} \overset{?}{\le} 2$ <br> $-\dfrac{1}{2} \le 2$ True |
| B: $(6, \frac{15}{2})$ | $7$ | $\dfrac{3}{7-6} \overset{?}{\le} 2$ <br> $3 \le 2$ False |
| C: $(\frac{15}{2}, \infty)$ | $8$ | $\dfrac{3}{8-6} \overset{?}{\le} 2$ <br> $\dfrac{3}{2} \le 2$ True |

Intervals A and C satisfiy the inequality. The endpoint 6 is not included because it makes the denominator 0.

Solution set: $(-\infty, 6) \cup \left[\dfrac{15}{2}, \infty\right)$

**75.** $\dfrac{-4}{1-x} < 5$

*Step 1*: Rewrite the inequality to compare a single fraction to 0:

$$\dfrac{-4}{1-x} < 5 \Rightarrow \dfrac{-4}{1-x} - 5 < 0$$

$$\dfrac{-4}{1-x} - \dfrac{5(1-x)}{1-x} < 0 \Rightarrow \dfrac{-4-5(1-x)}{1-x} < 0$$

$$\dfrac{-4-5+5x}{1-x} < 0 \Rightarrow \dfrac{-9+5x}{1-x} < 0$$

*Step 2*: Determine the values that will cause either the numerator or denominator to equal 0.

$-9 + 5x = 0 \Rightarrow x = \dfrac{9}{5}$ or $1 - x = 0 \Rightarrow x = 1$

The values 1 and $\dfrac{9}{5}$ divide the number line into three regions.

| Interval | Test Value | Is $\dfrac{-4}{1-x} < 5$ True or False? |
|---|---|---|
| A: $(-\infty, 1)$ | 0 | $\dfrac{-4}{1-0} \overset{?}{<} 5$ $-4 < 5$ True |
| B: $\left(1, \dfrac{9}{5}\right)$ | $\dfrac{6}{5}$ | $\dfrac{-4}{1-\frac{6}{5}} \overset{?}{<} 5$ $20 < 5$ False |
| C: $\left(\dfrac{9}{5}, \infty\right)$ | 2 | $\dfrac{-4}{1-2} \overset{?}{<} 5$ $4 < 5$ True |

*Step 3*: Choose a test value to see if it satisfies the inequality, $\dfrac{-4}{1-x} < 5$

Solution set: $(-\infty, 1) \cup \left(\dfrac{9}{5}, \infty\right)$

**77.** $\dfrac{10}{3+2x} \le 5$

*Step 1*: Rewrite the inequality so that 0 is on one side and there is a single fraction on the other side.

$$\dfrac{10}{3+2x} - 5 \le 0 \Rightarrow \dfrac{10}{3+2x} - \dfrac{5(3+2x)}{3+2x} \le 0$$

$$\dfrac{10-5(3+2x)}{3+2x} \le 0 \Rightarrow \dfrac{10-15-10x}{3+2x} \le 0$$

$$\dfrac{-10x-5}{3+2x} \le 0$$

*Step 2*: Determine the values that will cause either the numerator or denominator to equal 0.

$-10x - 5 = 0 \Rightarrow x = -\dfrac{1}{2}$ or $3 + 2x = 0 \Rightarrow x = -\dfrac{3}{2}$

The values $-\dfrac{3}{2}$ and $-\dfrac{1}{2}$ divide the number line into three regions. Use an open circle on $-\dfrac{3}{2}$ because it makes the denominator equal 0.

*Step 3*: Choose a test value to see if it satisfies the inequality, $\dfrac{10}{3+2x} \le 5$.

| Interval | Test Value | Is $\dfrac{10}{3+2x} \le 5$ True or False? |
|---|---|---|
| A: $\left(-\infty, -\dfrac{3}{2}\right)$ | $-2$ | $\dfrac{10}{3+2(-2)} \overset{?}{\le} 5$ $-10 \le 5$ True |
| B: $\left(-\dfrac{3}{2}, -\dfrac{1}{2}\right)$ | $-1$ | $\dfrac{10}{3+2(-1)} \overset{?}{\le} 5$ $10 \le 5$ False |
| C: $\left(-\dfrac{1}{2}, \infty\right)$ | 0 | $\dfrac{10}{3+2(0)} \overset{?}{\le} 5$ $\dfrac{10}{3} \le 5$ True |

Intervals A and C satisfy the inequality. The endpoint $-\dfrac{3}{2}$ is not included because it makes the denominator 0.

Solution set: $\left(-\infty, -\dfrac{3}{2}\right) \cup \left[-\dfrac{1}{2}, \infty\right)$

**79.** $\dfrac{7}{x+2} \ge \dfrac{1}{x+2}$

*Step 1*: Rewrite the inequality so that 0 is on one side and there is a single fraction on the other side.

$$\dfrac{7}{x+2} - \dfrac{1}{x+2} \ge 0 \Rightarrow \dfrac{6}{x+2} \ge 0$$

*Step 2*: Since the numerator is a constant, determine the value that will cause the denominator to equal 0.

$x + 2 = 0 \Rightarrow x = -2$

The value $-2$ divides the number line into two regions. Use an open circle on $-2$ because it makes the denominator equal 0.

*(continued on next page)*

(*continued from page 63*)

*Step 3*: Choose a test value to see if it satisfies the inequality, $\dfrac{7}{x+2} \geq \dfrac{1}{x+2}$.

| Interval | Test Value | Is $\dfrac{7}{x+2} \geq \dfrac{1}{x+2}$ True or False? |
|---|---|---|
| A: $(-\infty, -2)$ | $-3$ | $\dfrac{7}{-3+2} \overset{?}{\geq} \dfrac{1}{-3+2}$ <br> $-7 \geq -1$   False |
| B: $(-2, \infty)$ | $0$ | $\dfrac{7}{0+2} \overset{?}{\geq} \dfrac{1}{0+2}$ <br> $\dfrac{7}{2} \geq \dfrac{1}{2}$   True |

Interval B satisfies the inequality. The endpoint $-2$ is not included because it makes the denominator 0.

Solution set: $(-2, \infty)$

**81.** $\dfrac{3}{2x-1} > \dfrac{-4}{x}$

*Step 1*: Rewrite the inequality so that 0 is on one side and there is a single fraction on the other side.

$$\frac{3}{2x-1} + \frac{4}{x} > 0$$

$$\frac{3x}{x(2x-1)} + \frac{4(2x-1)}{x(2x-1)} > 0$$

$$\frac{3x + 4(2x-1)}{x(2x-1)} > 0$$

$$\frac{3x + 8x - 4}{x(2x-1)} > 0 \Rightarrow \frac{11x - 4}{x(2x-1)} > 0$$

*Step 2*: Determine the values that will cause either the numerator or denominator to equal 0.

$11x - 4 = 0 \Rightarrow x = \dfrac{4}{11}$  or  $x = 0$  or

$2x - 1 = 0 \Rightarrow x = \dfrac{1}{2}$

The values $0$, $\dfrac{4}{11}$, and $\dfrac{1}{2}$ divide the number line into four regions.

| Interval A $(-\infty, 0)$ | Interval B $\left(0, \frac{4}{11}\right)$ | Interval C $\left(\frac{4}{11}, \frac{1}{2}\right)$ | Interval D $\left(\frac{1}{2}, \infty\right)$ |
|---|---|---|---|

(number line marked at $0$, $\frac{4}{11}$, $\frac{1}{2}$)

*Step 3*: Choose a test value to see if it satisfies the inequality, $\dfrac{3}{2x-1} > \dfrac{-4}{x}$.

| Interval | Test Value | Is $\dfrac{3}{2x-1} > \dfrac{-4}{x}$ True or False? |
|---|---|---|
| A: $(-\infty, 0)$ | $-1$ | $\dfrac{3}{2(-1)-1} \overset{?}{>} \dfrac{-4}{-1}$ <br> $-1 > 4$   False |
| B: $\left(0, \frac{4}{11}\right)$ | $\frac{1}{11}$ | $\dfrac{3}{2\left(\frac{1}{11}\right)-1} \overset{?}{>} \dfrac{-4}{\frac{1}{11}}$  or  $-\dfrac{11}{3} \overset{?}{>} -44$ <br> $-3\frac{2}{3} > -44$   True |
| C: $\left(\frac{4}{11}, \frac{1}{2}\right)$ | $\frac{9}{22}$ | $\dfrac{3}{2\left(\frac{9}{22}\right)-1} \overset{?}{>} \dfrac{-4}{\frac{9}{22}}$  or  $-\dfrac{33}{2} \overset{?}{>} -\dfrac{88}{9}$ <br> $-16\frac{1}{2} > -9\frac{7}{9}$   False |
| D: $\left(\frac{1}{2}, \infty\right)$ | $1$ | $\dfrac{3}{2(1)-1} \overset{?}{>} \dfrac{-4}{1}$ <br> $3 > -4$   True |

Solution set: $\left(0, \frac{4}{11}\right) \cup \left(\frac{1}{2}, \infty\right)$

**83.** $\dfrac{4}{2-x} \geq \dfrac{3}{1-x}$

*Step 1*: Rewrite the inequality so that 0 is on one side and there is a single fraction on the other side.

$$\frac{4}{2-x} \geq \frac{3}{1-x}$$

$$\frac{4}{2-x} - \frac{3}{1-x} \geq 0$$

$$\frac{4(1-x)}{(2-x)(1-x)} - \frac{3(2-x)}{(1-x)(2-x)} \geq 0$$

$$\frac{4(1-x) - 3(2-x)}{(x-2)(1-x)} \geq 0$$

$$\frac{4 - 4x - 6 + 3x}{(2-x)(1-x)} \geq 0$$

$$\frac{-2 - x}{(2-x)(1-x)} \geq 0$$

*Step 2*: Determine the values that will cause either the numerator or denominator to equal 0.

$-2 - x = 0 \Rightarrow x = -2$  or  $2 - x = 0 \Rightarrow x = 2$  or

$1 - x = 0 \Rightarrow x = 1$

The values $-2$, $1$, and $2$ divide the number line into four regions. Use an open circle on 1 and 2 because they make the denominator equal 0.

| Interval A $(-\infty, -2)$ | Interval B $(-2, 1)$ | Interval C $(1, 2)$ | Interval D $(2, \infty)$ |
|---|---|---|---|

(number line marked at $-2$, $0$, $1$, $2$)

*Step 3*: Choose a test value to see if it satisfies the inequality, $\dfrac{4}{2-x} \geq \dfrac{3}{1-x}$

| Interval | Test Value | Is $\dfrac{4}{2-x} \geq \dfrac{3}{1-x}$ True or False? |
|---|---|---|
| A: $(-\infty,-2)$ | $-3$ | $\dfrac{4}{-2-(-3)} \overset{?}{\geq} \dfrac{3}{-1-(-3)}$  $4 \geq \dfrac{3}{2}$  True |
| B: $(-2,1)$ | $0$ | $\dfrac{4}{2-0} \overset{?}{\geq} \dfrac{3}{1-0}$  $2 \geq 3$  False |
| C: $(1,2)$ | $1.5$ | $\dfrac{4}{2-1.5} \overset{?}{\geq} \dfrac{3}{1-1.5}$  $8 \geq -6$  True |
| D: $(2,\infty)$ | $3$ | $\dfrac{4}{2-3} \overset{?}{\geq} \dfrac{3}{1-3}$  $-4 \geq -\dfrac{3}{2}$  False |

Intervals A and C satisfy the inequality. The endpoints 1 and 2 are not included because they make the denominator 0.

Solution set: $(-\infty,-2] \cup (1,2)$

**85.** $\dfrac{x+3}{x-5} \leq 1$

*Step 1*: Rewrite the inequality so that 0 is on one side and there is a single fraction on the other side.

$$\frac{x+3}{x-5} - 1 \leq 0 \Rightarrow \frac{x+3}{x-5} - \frac{x-5}{x-5} \leq 0$$

$$\frac{x+3-(x-5)}{x-5} \leq 0 \Rightarrow \frac{x+3-x+5}{x-5} \leq 0$$

$$\frac{8}{x-5} \leq 0$$

*Step 2*: Since the numerator is a constant, determine the value that will cause the denominator to equal 0.

$x - 5 = 0 \Rightarrow x = 5$

The value 5 divides the number line into two regions. Use an open circle on 5 because it makes the denominator equal 0.

```
Interval A          Interval B
(-∞, 5)             (5,∞)
+--+--+--+--+--+--○--+--+--+-->
   0              5
```

*Step 3*: Choose a test value to see if it satisfies the inequality, $\dfrac{x+3}{x-5} \leq 1$.

| Interval | Test Value | Is $\dfrac{x+3}{x-5} \leq 1$ True or False? |
|---|---|---|
| A: $(-\infty,5)$ | $0$ | $\dfrac{0+3}{0-5} \overset{?}{\leq} 1$  $-\dfrac{3}{5} \leq 1$  True |
| B: $(5,\infty)$ | $6$ | $\dfrac{6+3}{6-5} \overset{?}{\leq} 1$  $9 \leq 1$  False |

Interval A satisfies the inequality. The endpoint 5 is not included because it makes the denominator 0.

Solution set: $(-\infty,5)$

**87.** $\dfrac{2x-3}{x^2+1} \geq 0$

Since one side of the inequality is already 0, we start with Step 2.

*Step 2*: Determine the values that will cause either the numerator or denominator to equal 0.

$2x - 3 = 0$  or  $x^2 + 1 = 0$

$x = \dfrac{3}{2}$  has no real solutions

$\dfrac{3}{2}$ divides the number line into two intervals.

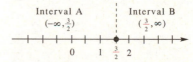

*Step 3*: Choose a test value to see if it satisfies the inequality, $\dfrac{2x-3}{x^2+1} \geq 0$.

| Interval | Test Value | Is $\dfrac{2x-3}{x^2+1} \geq 0$ True or False? |
|---|---|---|
| A: $\left(-\infty,\dfrac{3}{2}\right)$ | $0$ | $\dfrac{2(0)-3}{0^2+1} \overset{?}{\geq} 0$  $-3 \geq 0$  False |
| B: $\left(\dfrac{3}{2},\infty\right)$ | $2$ | $\dfrac{2(2)-3}{2^2+1} \overset{?}{\geq} 0$  $\dfrac{1}{5} \geq 0$  True |

Solution set: $\left[\dfrac{3}{2},\infty\right)$

**89.** $\dfrac{(5-3x)^2}{(2x-5)^3} > 0$

Since one side of the inequality is already 0, we start with Step 2.

*Step 2*: Determine the values that will cause either the numerator or denominator to equal 0.

$$5 - 3x = 0 \Rightarrow x = \tfrac{5}{3} \quad \text{or} \quad 2x - 5 = 0 \Rightarrow x = \tfrac{5}{2}$$

The values $\tfrac{5}{3}$ and $\tfrac{5}{2}$ divide the number line into three intervals.

Interval A $\;$ Interval B $\;$ Interval C
$(-\infty, \tfrac{5}{3})$ $\;$ $(\tfrac{5}{3}, \tfrac{5}{2})$ $\;$ $(\tfrac{5}{2}, \infty)$

$$\begin{array}{c|c|c|c}
 & 1 & 2 & 3
\end{array}$$

*Step 3*: Choose a test value to see if it satisfies the inequality, $\dfrac{(5-3x)^2}{(2x-5)^3} > 0$.

| Interval | Test Value | Is $\dfrac{(5-3x)^2}{(2x-5)^3} > 0$ True or False? |
|---|---|---|
| A: $\left(-\infty, \tfrac{5}{3}\right)$ | 0 | $\dfrac{(5-3\cdot 0)^2}{(2\cdot 0-5)^3} \overset{?}{>} 0$ $-\tfrac{1}{5} > 0$ False |
| B: $\left(\tfrac{5}{3}, \tfrac{5}{2}\right)$ | 2 | $\dfrac{(5-3\cdot 2)^2}{(2\cdot 2-5)^3} \overset{?}{>} 0$ $-1 > 0$ False |
| C: $\left(\tfrac{5}{2}, \infty\right)$ | 3 | $\dfrac{(5-3\cdot 3)^2}{(2\cdot 3-5)^3} \overset{?}{>} 0$ $16 > 0$ True |

Solution set: $\left(\tfrac{5}{2}, \infty\right)$

**91.** $\dfrac{(2x-3)(3x+8)}{(x-6)^3} \geq 0$

Since one side of the inequality is already 0, we start with Step 2.

*Step 2*: Determine the values that will cause either the numerator or denominator to equal 0.

$2x - 3 = 0 \quad$ or $\quad 3x + 8 = 0 \quad$ or $\quad x - 6 = 0$

$x = \tfrac{3}{2} \quad$ or $\quad x = -\tfrac{8}{3} \quad$ or $\quad x = 6$

The values $-\tfrac{8}{3}$, $\tfrac{3}{2}$, and 6 divide the number line into four intervals. Use an open circle on 6 because it makes the denominator equal 0.

Interval A $\;$ Interval B $\;$ Interval C $\;$ Interval D
$(-\infty, -\tfrac{8}{3})$ $\;$ $(-\tfrac{8}{3}, \tfrac{3}{2})$ $\;$ $(\tfrac{3}{2}, 6)$ $\;$ $(6, \infty)$

$$\begin{array}{c|c|c|c}
-\tfrac{8}{3} & 0 & \tfrac{3}{2} & 6
\end{array}$$

*Step 3*: Choose a test value to see if it satisfies the inequality, $\dfrac{(2x-3)(3x+8)}{(x-6)^3} \geq 0$.

| Interval | Test Value | Is $\dfrac{(2x-3)(3x+8)}{(x-6)^3} \geq 0$ True or False? |
|---|---|---|
| A: $\left(-\infty, -\tfrac{8}{3}\right)$ | -3 | $\dfrac{[2(-3)-3][3(-3)+8]}{(-3-6)^3} \overset{?}{\geq} 0$ $-\tfrac{1}{81} \geq 0$ False |
| B: $\left(-\tfrac{8}{3}, \tfrac{3}{2}\right)$ | 0 | $\dfrac{(2\cdot 0-3)(3\cdot 0+8)}{(0-6)^3} \overset{?}{\geq} 0$ $\tfrac{1}{9} \geq 0$ True |
| C: $\left(\tfrac{3}{2}, 6\right)$ | 2 | $\dfrac{(2\cdot 2-3)(3\cdot 2+8)}{(2-6)^3} \overset{?}{\geq} 0$ $-\tfrac{7}{32} \geq 0$ False |
| D: $(6, \infty)$ | 7 | $\dfrac{(2\cdot 7-3)(3\cdot 7+8)}{(7-6)^3} \overset{?}{\geq} 0$ $319 \geq 0$ True |

Solution set: $\left[-\tfrac{8}{3}, \tfrac{3}{2}\right] \cup (6, \infty)$

**93. (a)** Let $R = 5.3$ and then solve for $x$.
$$5.3 = 0.28944x + 3.5286$$
$$1.7714 = 0.28944x$$
$$6.1 \approx x$$
The model predicts that the receipts reach $5.3 billion about 6.1 years after 1986 which is in 1992.

**(b)** Let $R = 7$ and then solve for $x$.
$$7 = 0.28944x + 3.5286$$
$$3.4714 = 0.28944x$$
$$12.0 \approx x$$
The model predicts that the receipts reach $7 billion about 12.0 years after 1986 which is in 1998.

**95.** $-16t^2 + 220t \geq 624$

*Step 1*: Find the values of $x$ that satisfy $-16t^2 + 220t = 624$.

$$-16t^2 + 220t = 624 \Rightarrow 0 = 16t^2 - 220t + 624$$
$$0 = 4t^2 - 55t + 156$$
$$0 = (t-4)(4t-39)$$

$t - 4 = 0 \Rightarrow t = 4 \quad$ or $\quad 4t - 39 = 0 \Rightarrow t = \tfrac{39}{4} = 9.75$

*Step 2*: The two numbers divide a number line into three regions, where $t \geq 0$.

Interval A $\;$ Interval B $\;$ Interval C
$[0,4)$ $\;$ $(4, 9.75)$ $\;$ $(9.75, \infty)$

$$\begin{array}{c|c|c}
0 & 4 & 9.75
\end{array}$$

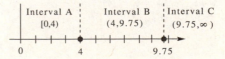

*Step 3:* Choose a test value to see if it satisfies the inequality, $-16t^2 + 220t \geq 624$.

| Interval | Test Value | Is $-16t^2 + 220t \geq 624$ True or False? |
|---|---|---|
| A: $(0,4)$ | 1 | $-16 \cdot 1^2 + 220 \cdot 1 \overset{?}{\geq} 624$<br>$204 \geq 624$ False |
| B: $(4, 9.75)$ | 5 | $-16 \cdot 5^2 + 220 \cdot 5 \overset{?}{\geq} 624$<br>$700 \geq 624$ True |
| C: $(9.75, \infty)$ | 10 | $-16 \cdot 10^2 + 220 \cdot 10 \overset{?}{\geq} 624$<br>$600 \geq 624$ False |

The projectile will be at least 624 feet above ground between 4 sec and 9.75 sec (inclusive).

**97. (a)** $1.5 \times 10^{-3} \leq R \leq 6.0 \times 10^{-3}$

$2.08 \times 10^{-5} \leq \dfrac{R}{72} \leq 8.33 \times 10^{-5}$

(Approximate values are given.)

**(b)** Let $N$ be the number of additional lung cancer deaths each year. Then $N$ would be determined by taking the annual individual risk times the total number of people. Thus, $N = (310 \times 10^6)\left(\dfrac{R}{72}\right)$. The range for $N$ would be approximately

$(310 \times 10^6)(2.08 \times 10^{-5}) \leq N \leq (310 \times 10^6)(8.33 \times 10^{-5})$
$6448 \leq N \leq 25,823$

Thus, radon gas exposure is expected by the EPA to cause approximately between 6400 and 25,800 cases of lung cancer each year in the United States.

**99.** Answers will vary. The student's answer is not correct. Taking the square root of both sides (and including a $\pm$) can be done if one is examining an equation, not an inequality. The correct solution can be found using the methods described in this section.

$x^2 \leq 144$

*Step 1:* Find the values of $x$ that satisfy $x^2 = 144$.

$x^2 = 144 \Rightarrow x^2 - 144 = 0 \Rightarrow (x+12)(x-12) = 0$
$x + 12 = 0 \Rightarrow x = -12$ or $x - 12 = 0 \Rightarrow x = 12$

*Step 2:* The two numbers divide a number line into three regions.

| Interval A<br>$(-\infty, -12)$ | Interval B<br>$(-12, 12)$ | Interval C<br>$(12, \infty)$ |

*Step 3:* Choose a test value to see if it satisfies the inequality, $x^2 \leq 144$.

| Interval | Test Value | Is $x^2 \leq 144$ True or False? |
|---|---|---|
| A: $(-\infty, -12)$ | $-20$ | $(-20)^2 \overset{?}{\leq} 144$<br>$400 \leq 144$ False |
| B: $(-12, 12)$ | 0 | $0^2 \overset{?}{\leq} 144$<br>$0 \leq 16$ True |
| C: $(12, \infty)$ | 20 | $20^2 \overset{?}{\leq} 144$<br>$400 \leq 144$ False |

Solution set: $[-12, 12]$

# Section 1.8: Absolute Value Equations and Inequalities

**1.** $|x| = 7$

The solution set includes any value of $x$ whose absolute value is 7; thus $x = 7$ or $x = -7$ are both solutions. The correct graph is F.

**3.** $|x| > -7$

The solution set is all real numbers, since the absolute value of any real number is always greater than $-7$. The correct graph is D, which shows the entire number line.

**5.** $|x| < 7$

The solution set includes any value of $x$ whose absolute value is less than 7; thus $x$ must be between $-7$ and 7, not including $-7$ or 7. The correct graph is G.

**7.** $|x| \leq 7$

The solution set includes any value of $x$ whose absolute value is less than or equal to 7; thus $x$ must be between $-7$ and 7, including $-7$ and 7. The correct graph is C.

**9.** $|3x - 1| = 2$

$3x - 1 = 2 \Rightarrow 3x = 3 \Rightarrow x = 1$ or
$3x - 1 = -2 \Rightarrow 3x = -1 \Rightarrow x = -\dfrac{1}{3}$

Solution set: $\left\{-\dfrac{1}{3}, 1\right\}$

**11.** $|5 - 3x| = 3$

$5 - 3x = 3 \Rightarrow 2 = 3x \Rightarrow \dfrac{2}{3} = x$ or
$5 - 3x = -3 \Rightarrow 8 = 3x \Rightarrow \dfrac{8}{3} = x$

Solution set: $\left\{\dfrac{2}{3}, \dfrac{8}{3}\right\}$

**13.** $\left|\dfrac{x-4}{2}\right| = 5$

$\dfrac{x-4}{2} = 5 \Rightarrow x-4 = 10 \Rightarrow x = 14$ or

$\dfrac{x-4}{2} = -5 \Rightarrow x-4 = -10 \Rightarrow x = -6$

Solution set: $\{-6, 14\}$

**15.** $\left|\dfrac{5}{x-3}\right| = 10$

$\dfrac{5}{x-3} = 10 \Rightarrow 5 = 10(x-3) \Rightarrow 5 = 10x - 30 \Rightarrow$

$\quad 35 = 10x \Rightarrow x = \dfrac{35}{10} = \dfrac{7}{2}$   or

$\dfrac{5}{x-3} = -10 \Rightarrow 5 = -10(x-3) \Rightarrow$

$\quad 5 = -10x + 30 \Rightarrow -25 = -10x \Rightarrow x = \dfrac{-25}{-10} = \dfrac{5}{2}$

Solution set: $\left\{\dfrac{5}{2}, \dfrac{7}{2}\right\}$

**17.** $\left|\dfrac{6x+1}{x-1}\right| = 3$

$\dfrac{6x+1}{x-1} = 3 \Rightarrow 6x+1 = 3(x-1) \Rightarrow$

$6x+1 = 3x-3 \Rightarrow 3x = -4 \Rightarrow x = -\dfrac{4}{3}$   or

$\dfrac{6x+1}{x-1} = -3 \Rightarrow 6x+1 = -3(x-1) \Rightarrow$

$6x+1 = -3x+3 \Rightarrow 9x = 2 \Rightarrow x = \dfrac{2}{9}$

Solution set: $\left\{-\dfrac{4}{3}, \dfrac{2}{9}\right\}$

**19.** $|2x-3| = |5x+4|$

$2x-3 = 5x+4 \Rightarrow -7 = 3x \Rightarrow -\dfrac{7}{3} = x$   or

$2x-3 = -(5x+4) \Rightarrow 2x-3 = -5x-4 \Rightarrow$

$\quad 7x = -1 \Rightarrow x = \dfrac{-1}{7} = -\dfrac{1}{7}$

Solution set: $\left\{-\dfrac{7}{3}, -\dfrac{1}{7}\right\}$

**21.** $|4-3x| = |2-3x|$

$4-3x = 2-3x \Rightarrow 4 = 2$  False  or

$4-3x = -(2-3x) \Rightarrow 4-3x = -2+3x \Rightarrow$

$\quad 6 = 6x \Rightarrow 1 = x$

Solution set: $\{1\}$

**23.** $|5x-2| = |2-5x|$

$5x-2 = 2-5x \Rightarrow 10x = 4 \Rightarrow x = \dfrac{4}{10} = \dfrac{2}{5}$   or

$5x-2 = -(2-5x) \Rightarrow 5x-2 = -2+5x \Rightarrow$

$\quad 0 = 0$  True

Solution set: $(-\infty, \infty)$

**25.** Answers will vary.

If $x$ is positive, then $-5x$ will be negative. Since the outcome of an absolute value can never be negative, a positive value of x is not possible.

**27.** $|2x+5| < 3$

$-3 < 2x+5 < 3$

$-8 < 2x < -2$

$-4 < x < -1$

Solution set: $(-4, -1)$

**29.** $|2x+5| \geq 3$

$2x+5 \leq -3 \Rightarrow 2x \leq -8 \Rightarrow x \leq -4$  or

$2x+5 \geq 3 \Rightarrow 2x \geq -2 \Rightarrow x \geq -1$

Solution set: $(-\infty, -4] \cup [-1, \infty)$

**31.** $\left|\dfrac{1}{2} - x\right| < 2$

$-2 < \dfrac{1}{2} - x < 2$

$2(-2) < 2\left(\dfrac{1}{2} - x\right) < 2(2)$

$-4 < 1 - 2x < 4$

$-5 < -2x < 3$

$\dfrac{5}{2} > x > -\dfrac{3}{2}$

Solution set: $\left(-\dfrac{3}{2}, \dfrac{5}{2}\right)$

**33.** $4|x-3| > 12 \Rightarrow |x-3| > 3$

$x-3 < -3 \Rightarrow x < 0$   or   $x-3 > 3 \Rightarrow x > 6$

Solution set: $(-\infty, 0) \cup (6, \infty)$

**35.** $|5-3x| > 7$

$5-3x < -7 \Rightarrow -3x < -12 \Rightarrow x > 4$  or

$5-3x > 7 \Rightarrow -3x > 2 \Rightarrow x < -\dfrac{2}{3}$

Solution set: $\left(-\infty, -\dfrac{2}{3}\right) \cup \left(4, \infty\right)$

**37.** $|5-3x| \leq 7$

$-7 \leq 5-3x \leq 7$

$-12 \leq -3x \leq 2$

$4 \geq x \geq -\dfrac{2}{3}$

$-\dfrac{2}{3} \leq x \leq 4$

Solution set: $\left[-\dfrac{2}{3}, 4\right]$

**39.** $\left|\frac{2}{3}x+\frac{1}{2}\right| \le \frac{1}{6}$

$-\frac{1}{6} \le \frac{2}{3}x+\frac{1}{2} \le \frac{1}{6}$

$6\left(-\frac{1}{6}\right) \le 6\left(\frac{2}{3}x+\frac{1}{2}\right) \le 6\left(\frac{1}{6}\right)$

$-1 \le 4x+3 \le 1$

$-4 \le 4x \le -2$

$-1 \le x \le -\frac{1}{2}$

Solution set: $\left[-1, -\frac{1}{2}\right]$

**41.** $|.01x+1| < .01$

$-.01 < .01x+1 < .01$

$-1 < x+100 < 1$

$-101 < x < -99$

Solution set: $(-101, -99)$

**43.** $|4x+3|-2 = -1 \Rightarrow |4x+3| = 1$

$4x+3 = 1 \Rightarrow 4x = -2 \Rightarrow x = \frac{-2}{4} = -\frac{1}{2}$ or

$4x+3 = -1 \Rightarrow 4x = -4 \Rightarrow x = -1$

Solution set: $\left\{-\frac{1}{2}, -1\right\}$

**45.** $|6-2x|+1 = 3 \Rightarrow |6-2x| = 2$

$6-2x = 2 \Rightarrow -2x = -4 \Rightarrow x = 2$ or

$6-2x = -2 \Rightarrow -2x = -8 \Rightarrow x = 4$

Solution set: $\{2, 4\}$

**47.** $|3x+1|-1 < 2 \Rightarrow |3x+1| < 3$

$-3 < 3x+1 < 3$

$-4 < 3x < 2$

$-\frac{4}{3} < x < \frac{2}{3}$

Solution set: $\left(-\frac{4}{3}, \frac{2}{3}\right)$

**49.** $\left|5x+\frac{1}{2}\right|-2 < 5 \Rightarrow \left|5x+\frac{1}{2}\right| < 7$

$-7 < 5x+\frac{1}{2} < 7$

$2(-7) < 2\left(5x+\frac{1}{2}\right) < 2(7)$

$-14 < 10x+1 < 14$

$-15 < 10x < 13$

$-\frac{15}{10} < x < \frac{13}{10} \Rightarrow -\frac{3}{2} < x < \frac{13}{10}$

Solution set: $\left(-\frac{3}{2}, \frac{13}{10}\right)$

**51.** $|10-4x|+1 \ge 5 \Rightarrow |10-4x| \ge 4$

$10-4x \le -4 \Rightarrow -4x \le -14 \Rightarrow x \ge \frac{-14}{-4} \Rightarrow x \ge \frac{7}{2}$

or

$10-4x \ge 4 \Rightarrow -4x \ge -6 \Rightarrow x \le \frac{-6}{-4} \Rightarrow x \le \frac{3}{2}$

Solution set: $\left(-\infty, \frac{3}{2}\right] \cup \left[\frac{7}{2}, \infty\right)$

**53.** $|3x-7|+1 < -2 \Rightarrow |3x-7| < -3$

An absolute value cannot be negative.

Solution set: $\varnothing$

**55.** Since the absolute value of a number is always nonnegative, the inequality $|10-4x| \ge -4$ is always true. The solution set is $(-\infty, \infty)$.

**57.** There is no number whose absolute value is less than any negative number. The solution set of $|6-3x| < -11$ is $\varnothing$.

**59.** The absolute value of a number will be 0 if that number is 0. Therefore $|8x+5| = 0$ is equivalent to $8x+5 = 0$, which has solution set $\left\{-\frac{5}{8}\right\}$.

**61.** Any number less than zero will be negative. There is no number whose absolute value is a negative number. The solution set of $|4.3x+9.8| < 0$ is $\varnothing$.

**63.** Since the absolute value of a number is always nonnegative, $|2x+1| < 0$ is never true, so $|2x+1| \le 0$ is only true when $|2x+1| = 0$.

$|2x+1| = 0 \Rightarrow 2x+1 = 0 \Rightarrow 2x = -1 \Rightarrow x = -\frac{1}{2}$

Solution set: $\left\{-\frac{1}{2}\right\}$

**65.** $|3x+2| > 0$ will be false only when $3x+2 = 0$, which occurs when $x = -\frac{2}{3}$. So the solution set for $|3x+2| > 0$ is

$\left(-\infty, -\frac{2}{3}\right) \cup \left(-\frac{2}{3}, \infty\right)$.

**67.** 6 and the opposite of 6, namely $-6$.

**69.** $x^2 - x = -6 \Rightarrow x^2 - x + 6 = 0$

The quadratic formula, $x = \dfrac{-b \pm \sqrt{b^2 - 4ac}}{2a}$, can be evaluated with $a = 1, b = -1,$ and $c = 6$.

$x = \dfrac{-b \pm \sqrt{b^2 - 4ac}}{2a}$

$= \dfrac{-(-1) \pm \sqrt{(-1)^2 - 4(1)(6)}}{2(1)} = \dfrac{1 \pm \sqrt{1-24}}{2}$

$= \dfrac{1 \pm \sqrt{-23}}{2} = \dfrac{1 \pm i\sqrt{23}}{2} = \dfrac{1}{2} \pm \dfrac{\sqrt{23}}{2}i$

Solution set: $\left\{\frac{1}{2} \pm \frac{\sqrt{23}}{2}i\right\}$

**71.** $\left|4x^2 - 23x - 6\right| = 0$

Because 0 and the opposite of 0 represent the same value, only one equation needs to be solved.

$4x^2 - 23x - 6 = 0 \Rightarrow (4x + 1)(x - 6) = 0$

$4x + 1 = 0 \quad \text{or} \quad x - 6 = 0$

$x = -\frac{1}{4} \quad \text{or} \quad x = 6$

Solution set: $\left\{-\frac{1}{4}, 6\right\}$

**73.** $\left|x^2 + 1\right| - \left|2x\right| = 0$

$\left|x^2 + 1\right| - \left|2x\right| = 0 \Rightarrow \left|x^2 + 1\right| = \left|2x\right|$

$x^2 + 1 = 2x \Rightarrow x^2 - 2x + 1 = 0$

$(x - 1)^2 = 0 \Rightarrow x - 1 = 0 \Rightarrow x = 1 \quad \text{or}$

$x^2 + 1 = -2x \Rightarrow x^2 + 2x + 1 = 0 \Rightarrow (x + 1)^2 = 0$

$x + 1 = 0 \Rightarrow x = -1$

Solution set: $\{-1, 1\}$

**75.** Any number less than zero will be negative. There is no number whose absolute value is a negative number. The solution set of $\left|x^4 + 2x^2 + 1\right| < 0$ is $\varnothing$.

**77.** $\left|\dfrac{x - 4}{3x + 1}\right| \geq 0$

This inequality will be true, except where $\frac{x-4}{3x+1}$ is undefined. This occurs when

$3x + 1 = 0$, or $x = -\frac{1}{3}$.

Solution set: $\left(-\infty, -\frac{1}{3}\right) \cup \left(-\frac{1}{3}, \infty\right)$

**79.** $\left|p - q\right| = 2$, which is equivalent to $\left|q - p\right| = 2$, indicates that the distance between $p$ and $q$ is 2 units.

**81.** "$m$ is no more than 2 units from 7" means that $m$ is 2 units or less from 7. Thus the distance between $m$ and 7 is less than or equal to 2, or $\left|m - 7\right| \leq 2$.

**83.** "$p$ is within .0001 units of 6" means that $p$ is less than .0001 units from 6. Thus the distance between $p$ and 6 is less than .0001, or $\left|p - 6\right| < .0001$.

**85.** "$r$ is no less than 1 unit from 29" means that $r$ is 1 unit or more from 29. Thus the distance between $r$ and 29 is greater than or equal to 1, or $\left|r - 29\right| \geq 1$.

**87.** Since we want $y$ to be within .002 unit of 6, we have $\left|y - 6\right| < .002$ or $\left|5x + 1 - 6\right| < .002$.

$\left|5x - 5\right| < .002$

$-.002 < 5x - 5 < .002$

$4.998 < 5x < 5.002$

$.9996 < x < 1.0004$

Values of $x$ in the interval $(.9996, 1.0004)$ would satisfy the condition.

**89.** $\left|y - 8.2\right| \leq 1.5$

$-1.5 \leq y - 8.2 \leq 1.5$

$6.7 \leq y \leq 9.7$

The range of weights, in pounds, is [6.5, 9.5].

**91.** 780 is 50 more than 730 and 680 is 50 less than 730, so all of the temperatures in the acceptable range are within 50° of 730°. That is $\left|F - 730\right| \leq 50$.

**93.** $\left|R_L - 26.75\right| \leq 1.42$

$-1.42 \leq R_L - 26.75 \leq 1.42$

$25.33 \leq R_L \leq 28.17$

$\left|R_E - 38.75\right| \leq 2.17$

$-2.17 \leq R_E - 38.75 \leq 2.17$

$36.58 \leq R_E \leq 40.92$

**95.** Answers will vary.

# Chapter 1: Review Exercises

**1.** $2x + 8 = 3x + 2 \Rightarrow 8 = x + 2 \Rightarrow 6 = x$

Solution set: $\{6\}$

**3.** $5x - 2(x + 4) = 3(2x + 1)$

$5x - 2x - 8 = 6x + 3 \Rightarrow 3x - 8 = 6x + 3 \Rightarrow$

$-8 = 3x + 3 \Rightarrow -11 = 3x \Rightarrow -\dfrac{11}{3} = x$

Solution set: $\left\{-\frac{11}{3}\right\}$

**5.** $A = \dfrac{24f}{B(p + 1)}$ for $f$ (approximate annual interest rate)

$B(p + 1)A = B(p + 1)\left(\dfrac{24f}{B(p + 1)}\right)$

$AB(p + 1) = 24f$

$\dfrac{AB(p + 1)}{24} = f$

$f = \dfrac{AB(p + 1)}{24}$

**7.** A and B cannot be equations used to find the number of pennies in a jar. The number of pennies must be a whole number.

**A.** $5x + 3 = 11 \Rightarrow 5x = 8 \Rightarrow x = \frac{8}{5}$

**B.** $12x + 6 = -4 \Rightarrow 12x = -10 \Rightarrow$
$x = -\frac{10}{12} = -\frac{5}{6}$

**C.** $100x = 50(x + 3)$
$100x = 50x + 150 \Rightarrow 50x = 150 \Rightarrow x = 3$

**D.** $6(x + 4) = x + 24$
$6x + 24 = x + 24$
$5x + 24 = 24 \Rightarrow 5x = 0 \Rightarrow x = 0$

**9.** Let $x$ = the original length of the square (in inches). Since the perimeter of a square is 4 times the length of one side, we have
$4(x - 4) = \frac{1}{2}(4x) + 10$. Solve this equation for $x$ to determine the length of each side of the original square.
$4x - 16 = 2x + 10$
$2x - 16 = 10 \Rightarrow 2x = 26 \Rightarrow x = 13$
The original square is 13 in. on each side.

**11.** Let $x$ = the amount of 100% alcohol solution (in liters).

| Strength | Liters of Solution | Liters of Pure Alcohol |
|---|---|---|
| 100% | $x$ | $1x = x$ |
| 10% | 12 | $.10 \cdot 12 = 1.2$ |
| 30% | $x + 12$ | $.30(x + 12)$ |

The number of liters of pure alcohol in the 100% solution plus the number of liters of pure alcohol in the 10% solution must equal the number of liters of pure alcohol in the 30% solution.
$x + 1.2 = .30(x + 12) \Rightarrow x + 1.2 = .30x + 3.6$
$.7x + 1.2 = 3.6 \Rightarrow .7x = 2.4$
$x = \frac{2.4}{.7} = \frac{24}{7} = 3\frac{3}{7}$ L

$3\frac{3}{7}$ L of the 100% solution should be added.

**13.** Let $x$ = average speed upriver.
Then $x + 5$ = average speed on return trip.

| | $r$ | $t$ | $d$ |
|---|---|---|---|
| Upriver | $x$ | 1.2 | $1.2x$ |
| Downriver | $x + 5$ | .9 | $.9(x + 5)$ |

Since the distance upriver and downriver are the same, we solve the following.
$1.2x = .9(x + 5)$
$1.2x = .9x + 4.5 \Rightarrow .3x = 4.5 \Rightarrow x = 15$
The average speed of the boat upriver is 15 mph.

**15. (a)** In one year, the maximum amount of lead ingested would be

$.05 \dfrac{\text{mg}}{\text{liter}} \cdot 2 \dfrac{\text{liters}}{\text{day}} \cdot 365.25 \dfrac{\text{days}}{\text{year}}$

$= 36.525 \dfrac{\text{mg}}{\text{year}}.$

The maximum amount $A$ of lead (in milligrams) ingested in $x$ years would be $A = 36.525x$.

**(b)** If $x = 72$, then
$A = 36.525(72) = 2629.8$ mg. The EPA maximum lead intake from water over a lifetime is 2629.8 mg.

**17. (a)** Using 1955 for $x = 0$, then for 1985,
$x = 30$
$y = .118x + .056$
$y = .118(30) + .056 \approx 3.60$
The minimum wage in 1985 was $3.60 according to the model. This is $0.25 more than the actual value of $3.35.

**(b)** Let $y = \$4.25$ and then solve for $x$.
$4.25 = .118x + .056$
$4.194 = .118x$
$35.5 \approx x$
The model predicts the minimum wage to be $4.25 about 3.5 years after 1955, which is mid-1990. This is consistent with the minimum wage changing to $4.25 in 1991.

**19.** $(6 - i) + (7 - 2i) = (6 + 7) + \left[-1 + (-2)\right]i$
$= 13 + (-3)i = 13 - 3i$

**21.** $15i - (3 + 2i) - 11 = (-3 - 11) + (15 - 2)i$
$= -14 + 13i$

**23.** $(5 - i)(3 + 4i) = 5(3) + 5(4i) - i(3) - i(4i)$
$= 15 + 20i - 3i - 4i^2$
$= 15 + 17i - 4(-1)$
$= 15 + 17i + 4 = 19 + 17i$

**25.** $(5 - 11i)(5 + 11i) = 5^2 - (11i)^2$  Product of the sum and difference of two terms

$= 25 - 121i^2$
$= 25 - 121(-1)$
$= 25 + 121 = 146$

**27.**
$$-5i(3-i)^2 = -5i\left[3^2 - 2(3)(i) + i^2\right]$$
$$= -5i\left[9 - 6i + (-1)\right]$$
$$= -5i(8 - 6i) = -40i + 30i^2$$
$$= -40i + 30(-1) = -40i + (-30)$$
$$= -30 - 40i$$

**29.**
$$\frac{-12-i}{-2-5i} = \frac{(-12-i)(-2+5i)}{(-2-5i)(-2+5i)}$$
$$= \frac{24 - 60i + 2i - 5i^2}{(-2)^2 - (5i)^2} = \frac{24 - 58i - 5(-1)}{4 - 25i^2}$$
$$= \frac{24 - 58i + 5}{4 - 25(-1)} = \frac{29 - 58i}{4 + 25} = \frac{29 - 58i}{29}$$
$$= \frac{29}{29} - \frac{58}{29}i = 1 - 2i$$

**31.** $i^{11} = i^8 \cdot i^3 = 1 \cdot (-i) = -i$

**33.** $i^{1001} = i^{1000} \cdot i = \left(i^4\right)^{250} \cdot i = 1^{250} \cdot i = i$

**35.** $i^{-27} = i^{-28} \cdot i = \left(i^4\right)^{-7} \cdot i = 1^{-7} \cdot i = i$

**37.** $(x+7)^2 = 5 \Rightarrow x+7 = \pm\sqrt{5} \Rightarrow x = -7 \pm \sqrt{5}$
Solution set: $\{-7 \pm \sqrt{5}\}$

**39.** $2x^2 + x - 15 = 0$
$(x+3)(2x-5) = 0$
$x+3 = 0 \Rightarrow x = -3$ or $2x - 5 = 0 \Rightarrow x = \frac{5}{2}$
Solution set: $\left\{-3, \frac{5}{2}\right\}$

**41.** $-2x^2 + 11x = -21 \Rightarrow -2x^2 + 11x + 21 = 0$
$2x^2 - 11x - 21 = 0 \Rightarrow (2x+3)(x-7) = 0$
$2x + 3 = 0 \Rightarrow x = -\frac{3}{2}$ or $x - 7 = 0 \Rightarrow x = 7$
Solution set: $\left\{-\frac{3}{2}, 7\right\}$

**43.** $(2x+1)(x-4) = x \Rightarrow 2x^2 - 8x + x - 4 = x \Rightarrow$
$2x^2 - 7x - 4 = x \Rightarrow 2x^2 - 8x - 4 = 0 \Rightarrow$
$x^2 - 4x - 2 = 0$
Solve by completing the square.
$x^2 - 4x - 2 = 0$
$x^2 - 4x + 4 = 2 + 4$
Note: $\left[\frac{1}{2} \cdot (-4)\right]^2 = (-2)^2 = 2$
$(x-2)^2 = 6 \Rightarrow x - 2 = \pm\sqrt{6} \Rightarrow x = 2 \pm \sqrt{6}$

Solve by the quadratic formula.
Let $a = 1$, $b = -4$, and $c = -2$.
$$x = \frac{-b \pm \sqrt{b^2 - 4ac}}{2a}$$
$$= \frac{-(-4) \pm \sqrt{(-4)^2 - 4(1)(-2)}}{2(1)} = \frac{4 \pm \sqrt{16 + 8}}{2}$$
$$= \frac{4 \pm \sqrt{24}}{2} = \frac{4 \pm 2\sqrt{6}}{2} = 2 \pm \sqrt{6}$$
Solution set: $\left\{2 \pm \sqrt{6}\right\}$

**45.** $x^2 - \sqrt{5}x - 1 = 0$
Using the quadratic formula would be the most direct approach.
$a = 1, b = -\sqrt{5}$, and $c = -1$.
$$x = \frac{-b \pm \sqrt{b^2 - 4ac}}{2a}$$
$$= \frac{-(-\sqrt{5}) \pm \sqrt{(-\sqrt{5})^2 - 4 \cdot 1 \cdot (-1)}}{2 \cdot 1}$$
$$= \frac{\sqrt{5} \pm \sqrt{5 + 4}}{2} = \frac{\sqrt{5} \pm \sqrt{9}}{2} = \frac{\sqrt{5} \pm 3}{2}$$
Solution set: $\left\{\frac{\sqrt{5} \pm 3}{2}\right\}$

**47.** D; $(7x+4)^2 = 11$
This equation has two real, distinct solutions since the positive number 11 has a positive square root and a negative square root.

**49.** A; $(3x-4)^2 = -9$
This equation has two imaginary solutions since the negative number −9 has two imaginary square roots.

**51.** $-6x^2 + 2x = -3 \Rightarrow -6x^2 + 2x + 3 = 0$
$a = -6, b = 2$, and $c = 3$
$b^2 - 4ac = 2^2 - 4(-6)(3) = 4 + 72 = 76$
The equation has two distinct irrational solutions since the discriminant is positive but not a perfect square.

**53.** $-8x^2 + 10x = 7 \Rightarrow 0 = 8x^2 - 10x + 7$
$a = 8, b = -10$, and $c = 7$
$b^2 - 4ac = (-10)^2 - 4(8)(7)$
$= 100 - 224 = -124$
The equation has two distinct nonreal complex solutions since the discriminant is negative.

**55.** $x(9x+6)=-1 \Rightarrow 9x^2+6x=-1 \Rightarrow$
$9x^2+6x+1=0$
$a=9,\ b=6,\ \text{and}\ c=1$
$b^2-4ac=6^2-4(9)(1)=36-36=0$
The equation has one rational solution (a double solution) since the discriminant is equal to zero.

**57.** The projectile will be 750 ft above the ground whenever $220t-16t^2=750$.
Solve this equation for $t$.
$220t-16t^2=750 \Rightarrow 0=16t^2-220t+750 \Rightarrow$
$0=8t^2-110t+375 \Rightarrow 0=(4t-25)(2t-15)$
$4t-25=0 \qquad$ or $\quad 2t-15=0$
$\quad t=\frac{25}{4}=6.25 \quad$ or $\qquad t=\frac{15}{2}=7.5$
The projectile will be 750 ft high at 6.25 sec and at 7.5 sec.

**59.** Let $x=$ width of border.
Apply the formula $A=LW$ to both the outside and inside rectangles.
Inside area= Outside area – Border area
$(12-2x)(10-2x)=12\cdot10-21$

$120-24x-20x+4x^2=120-21$
$120-44x+4x^2=99$
$4x^2-44x+120=99$
$4x^2-44x+21=0$
$(2x-21)(2x-1)=0$

$2x=21 \Rightarrow x=\frac{21}{2}=10\frac{1}{2}\quad$ or
$2x-1=0 \Rightarrow x=\frac{1}{2}$

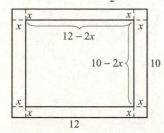

The border width cannot be $10\frac{1}{2}$ since this exceeds the width of the outside rectangle, so reject this solution. The width of the border is $\frac{1}{2}$ ft.

**61.** In 1980, $x=10$.
$y=-6.77x^2+445.34x+11,279.82$
$y=-6.77\cdot10^2+445.34\cdot10+11,279.82$
$y=-6.77\cdot100+4453.4+11,279.82$
$y=-677+4453.4+11,279.82=15,056.22$
Approximately 15,056 airports

**63.** $4x^4+3x^2-1=0$
Let $u=x^2$; then $u^2=x^4$.
With this substitution, the equation becomes
$4u^2+3u-1=0$.
Solve this equation by factoring.
$(u+1)(4u-1)=0$

$u+1 \Rightarrow u=-1$ or $4u-1=0 \Rightarrow u=\frac{1}{4}$

To find $x$, replace $u$ with $x^2$.
$x^2=-1 \Rightarrow x=\pm\sqrt{-1} \Rightarrow x=\pm i$ or
$x^2=\frac{1}{4} \Rightarrow x=\pm\sqrt{\frac{1}{4}} \Rightarrow x=\pm\frac{1}{2}$
Solution set: $\left\{\pm i,\pm\frac{1}{2}\right\}$

**65.** $\dfrac{2}{x}-\dfrac{4}{3x}=8+\dfrac{3}{x}$

$3x\left(\dfrac{2}{x}-\dfrac{4}{3x}\right)=3x\left(8+\dfrac{3}{x}\right)$
$6-4=24x+9 \Rightarrow 2=24x+9$
$-7=24x \Rightarrow -\frac{7}{24}=x$

Solution set: $\left\{-\frac{7}{24}\right\}$

**67.** $\dfrac{10}{4x-4}=\dfrac{1}{1-x} \Rightarrow \dfrac{10}{4(x-1)}=\dfrac{1}{1-x} \Rightarrow$

$\dfrac{10}{4(x-1)}=\dfrac{(-1)\cdot1}{(-1)(1-x)} \Rightarrow \dfrac{10}{4(x-1)}=\dfrac{-1}{x-1}$

Multiply each term in the equation by the least common denominator, $4(x-1)$, assuming $x\neq1$.

$4(x-1)\left[\dfrac{10}{4(x-1)}\right]=4(x-1)\left(\dfrac{-1}{x-1}\right)$
$10=-4 \Rightarrow 14=0$

This is a false statement, the solution set is $\varnothing$.
Alternate solution:
$\dfrac{10}{4x-4}=\dfrac{1}{1-x}$ or $\dfrac{10}{4(x-1)}=\dfrac{1}{1-x}$

Multiply each term in the equation by the least common denominator, $4(x-1)(1-x)$, assuming $x\neq1$.

$4(x-1)(1-x)\left[\dfrac{10}{4(x-1)}\right]$
$\qquad=4(x-1)(1-x)\left(\dfrac{1}{1-x}\right)$

$10(1-x)=4(x-1) \Rightarrow 10-10x=4x-4$
$10=14x-4 \Rightarrow 14=14x \Rightarrow 1=x$
Because of the restriction $x\neq1$, the solution set is $\varnothing$.

**69.** $\dfrac{x}{x+2} + \dfrac{1}{x} + 3 = \dfrac{2}{x^2+2x} \Rightarrow$

$\dfrac{x}{x+2} + \dfrac{1}{x} + 3 = \dfrac{2}{x(x+2)}$

Multiply each term in the equation by the least common denominator, $x(x+2)$, assuming $x \neq 0, -2$.

$x(x+2)\left[\dfrac{x}{x+2} + \dfrac{1}{x} + 3\right] = x(x+2)\left(\dfrac{2}{x(x+2)}\right)$

$x^2 + (x+2) + 3x(x+2) = 2$

$x^2 + x + 2 + 3x^2 + 6x = 2$

$4x^2 + 7x + 2 = 2$

$4x^2 + 7x = 0 \Rightarrow x(4x+7) = 0$

$x = 0$ or $4x+7 = 0 \Rightarrow x = -\dfrac{7}{4}$

Because of the restriction $x \neq 0$, the only valid solution is $-\dfrac{7}{4}$. The solution set is $\left\{-\dfrac{7}{4}\right\}$.

**71.** $(2x+3)^{2/3} + (2x+3)^{1/3} - 6 = 0$

Let $u = (2x+3)^{1/3}$. Then

$u^2 = [(2x+3)^{1/3}]^2 = (2x+3)^{2/3}$.

With this substitution, the equation becomes $u^2 + u - 6 = 0$. Solve by factoring.

$(u+3)(u-2) = 0$

$u + 3 = 0 \Rightarrow u = -3$ or $u - 2 = 0 \Rightarrow u = 2$

To find $x$, replace $u$ with $(2x+3)^{1/3}$.

$(2x+3)^{1/3} = -3 \Rightarrow \left[(2x+3)^{1/3}\right]^3 = (-3)^3 \Rightarrow$

$2x+3 = -27 \Rightarrow 2x = -30 \Rightarrow x = -15$

or

$(2x+3)^{1/3} = 2 \Rightarrow \left[(2x+3)^{1/3}\right]^3 = 2^3 \Rightarrow$

$2x+3 = 8 \Rightarrow 2x = 5 \Rightarrow x = \dfrac{5}{2}$

Check $x = -15$.

$(2x+3)^{2/3} + (2x+3)^{1/3} = 6$

$[2(-15)+3]^{2/3} + [2(-15)+3]^{1/3} = 6$

$(-30+3)^{2/3} + (-30+3)^{1/3} = 6$

$(-27)^{2/3} + (-27)^{1/3} = 6$

$\left[(-27)^{1/3}\right]^2 + (-3) = 6$

$(-3)^2 - 3 = 6$

$9 - 3 = 6 \Rightarrow 6 = 6$

This is a true statement. $-15$ is a solution.

Check $x = \dfrac{5}{2}$.

$(2x+3)^{2/3} + (2x+3)^{1/3} = 6$

$\left[2\left(\tfrac{5}{2}\right)+3\right]^{2/3} + \left[2\left(\tfrac{5}{2}\right)+3\right]^{1/3} = 6$

$(5+3)^{2/3} + (5+3)^{1/3} = 6$

$8^{2/3} + 8^{1/3} = 6$

$\left[8^{1/3}\right]^2 + 2 = 6$

$(2)^2 + 2 = 6$

$4 + 2 = 6 \Rightarrow 6 = 6$

This is a true statement. $\dfrac{5}{2}$ is a solution.

Solution set: $\left\{-15, \dfrac{5}{2}\right\}$

**73.** $\sqrt{4x-2} = \sqrt{3x+1}$

$\left(\sqrt{4x-2}\right)^2 = \left(\sqrt{3x+1}\right)^2$

$4x - 2 = 3x + 1$

$x - 2 = 1$

$x = 3$

Check $x = 3$.

$\sqrt{4x-2} = \sqrt{3x+1}$

$\sqrt{4(3)-2} = \sqrt{3(3)+1}$

$\sqrt{12-2} = \sqrt{9+1}$

$\sqrt{10} = \sqrt{10}$

This is a true statement.

Solution set: $\{3\}$

**75.** $\sqrt{x+2} - x = 2 \Rightarrow \sqrt{x+2} = 2 + x$

$\left(\sqrt{x+2}\right)^2 = (2+x)^2 \Rightarrow x + 2 = 4 + 4x + x^2$

$0 = x^2 + 3x + 2 \Rightarrow 0 = (x+2)(x+1)$

$x + 2 = 0 \Rightarrow x = -2$ or $x + 1 = 0 \Rightarrow x = -1$

Check $x = -2$.

$\sqrt{x+2} = 2 + x$

$\sqrt{-2+2} = 2 + (-2)$

$\sqrt{0} = 0 \Rightarrow 0 = 0$

This is a true statement. $-2$ is a solution.

Check $x = -1$.

$\sqrt{x+2} = 2 + x$

$\sqrt{-1+2} = 2 + (-1)$ ?

$\sqrt{1} = 1 \Rightarrow 1 = 1$

This is a true statement. $-1$ is a solution.

Solution set: $\{-2, -1\}$

**77.** $\sqrt{x+3} - \sqrt{3x+10} = 1$

$\qquad \sqrt{x+3} = 1 + \sqrt{3x+10}$

$\qquad \left(\sqrt{x+3}\right)^2 = \left(1+\sqrt{3x+10}\right)^2$

$\qquad x+3 = 1 + 2\sqrt{3x+10} + (3x+10)$

$\qquad x+3 = 3x+11+2\sqrt{3x+10}$

$\qquad -2x-8 = 2\sqrt{3x+10}$

$\qquad x+4 = -\sqrt{3x+10}$

$\qquad \left(x+4\right)^2 = \left(-\sqrt{3x+10}\right)^2$

$\qquad x^2+8x+16 = 3x+10$

$\qquad x^2+5x+6=0 \Rightarrow (x+2)(x+3)=0$

$x+3=0 \Rightarrow x=-3 \quad$ or $\quad x+2=0 \Rightarrow x=-2$

Check $x=-3$.

$\qquad \sqrt{x+3}-\sqrt{3x+10}=1$

$\qquad \sqrt{-3+3}-\sqrt{3(-3)+10}=1$

$\qquad \sqrt{0}-\sqrt{-9+10}=1$

$\qquad 0-\sqrt{1}=1$

$\qquad 0-1=1 \Rightarrow -1=1$

This is a false statement. $-3$ is not a solution.

Check $x=-2$.

$\qquad \sqrt{x+3}-\sqrt{3x+10}=1$

$\qquad \sqrt{-2+3}-\sqrt{3(-2)+10}=1$

$\qquad \sqrt{1}-\sqrt{-6+10}=1$

$\qquad 1-\sqrt{4}=1$

$\qquad 1-2=1 \Rightarrow -1=1$

This is a false statement. $-2$ is not a solution. Since neither of the proposed solutions satisfies the original equation, the equation has no solution.

Solution set: $\varnothing$

**79.** $\sqrt{x^2+3x}-2=0$

$\qquad \sqrt{x^2+3x} = 2 \Rightarrow \left(\sqrt{x^2+3x}\right)^2 = 2^2$

$\qquad x^2+3x = 4 \Rightarrow x^2+3x-4=0$

$\qquad (x-1)(x+4)=0$

$x-1=0 \Rightarrow x=1 \quad$ or $\quad x+4=0 \Rightarrow x=-4$

Check $x=-4$.

$\qquad \sqrt{x^2+3x}-2=0$

$\qquad \sqrt{(-4)^2+3(-4)}-2=0$

$\qquad \sqrt{16+(-12)}-2=0$

$\qquad \sqrt{4}-2=0 \Rightarrow 2-2=0 \Rightarrow 0=0$

This is a true statement. $-4$ is a solution.

Check $x=1$.

$\qquad \sqrt{x^2+3x}-2=0$

$\qquad \sqrt{1^2+3(1)}-2=0 \Rightarrow \sqrt{1+3}-2=0 \Rightarrow$

$\qquad \sqrt{4}-2=0 \Rightarrow 2-2=0 \Rightarrow 0=0$

This is a true statement. 1 is a solution.

Solution set: $\{-4,1\}$

**81.** $\sqrt[3]{6x+2}-\sqrt[3]{4x}=0$

$\qquad \sqrt[3]{6x+2}=\sqrt[3]{4x}$

$\qquad \left(\sqrt[3]{6x+2}\right)^3 = \left(\sqrt[3]{4x}\right)^3$

$\qquad 6x+2=4x \Rightarrow 2=-2x \Rightarrow -1=x$

Check $x=-1$.

$\qquad \sqrt[3]{6x+2}-\sqrt[3]{4x}=0$

$\qquad \sqrt[3]{6(-1)+2}-\sqrt[3]{4(-1)}=0$

$\qquad \sqrt[3]{-6+2}-\sqrt[3]{-4}=0$

$\qquad \sqrt[3]{-4}-\left(-\sqrt[3]{4}\right)=0$

$\qquad -\sqrt[3]{4}+\sqrt[3]{4}=0 \Rightarrow 0=0$

This is a true statement.

Solution set: $\{-1\}$

**83.** $-9x+3 < 4x+10 \Rightarrow -13x < 7 \Rightarrow x > -\dfrac{7}{13}$

Solution set: $\left(-\frac{7}{13}, \infty\right)$

**85.** $\quad -5x-4 \ge 3(2x-5)$

$\qquad -5x-4 \ge 6x-15$

$\qquad -11x-4 \ge -15$

$\qquad -11x \ge -11$

$\qquad x \le 1$

Solution set: $(-\infty, 1]$

**87.** $5 \le 2x-3 \le 7$

$\qquad 8 \le 2x \le 10$

$\qquad 4 \le x \le 5$

Solution set: $[4, 5]$

**89.** $x^2+3x-4 \le 0$

*Step 1*: Find the values of $x$ that satisfy $x^2+3x-4=0$.

$\qquad x^2+3x-4=0$

$\qquad (x+4)(x-1)=0$

$\qquad x+4=0 \Rightarrow x=-4 \quad$ or $\quad x-1=0 \Rightarrow x=1$

*Step 2*: The two numbers divide a number line into three regions.

| Interval A $(-\infty,-4)$ | Interval B $(-4,1)$ | Interval C $(1,\infty)$ |
|---|---|---|

$\qquad\qquad -4 \qquad\qquad 0 \quad 1$

*(continued on next page)*

(*continued from page 75*)

*Step 3*: Choose a test value to see if it satisfies the inequality, $x^2 + 3x - 4 \leq 0$.

| Interval | Test Value | Is $x^2 + 3x - 4 \leq 0$ True or False? |
|---|---|---|
| A: $(-\infty, -4)$ | $-5$ | $(-5)^2 + 3(-5) - 4 \overset{?}{\leq} 0$ $6 \leq 0$ False |
| B: $(-4, 1)$ | $0$ | $0^2 + 3(0) - 4 \overset{?}{\leq} 0$ $-4 \leq 0$ True |
| C: $(1, \infty)$ | $2$ | $2^2 + 3(2) - 4 \overset{?}{\leq} 0$ $6 \leq 0$ False |

Solution set: $[-4, 1]$

**91.** $6x^2 - 11x < 10$

*Step 1*: Find the values of $x$ that satisfy $6x^2 - 11x = 10$.

$$6x^2 - 11x = 10$$
$$6x^2 - 11x - 10 = 0$$
$$(3x + 2)(2x - 5) = 0$$

$3x + 2 = 0 \Rightarrow x = -\frac{2}{3}$ or $2x - 5 = 0 \Rightarrow x = \frac{5}{2}$

*Step 2*: The two numbers divide a number line into three regions.

Interval A $(-\infty, -\frac{2}{3})$    Interval B $(-\frac{2}{3}, \frac{5}{2})$    Interval C $(\frac{5}{2}, \infty)$

*Step 3*: Choose a test value to see if it satisfies the inequality, $6x^2 - 11x < 10$

| Interval | Test Value | Is $6x^2 - 11x < 10$ True or False? |
|---|---|---|
| A: $\left(-\infty, -\frac{2}{3}\right)$ | $-1$ | $6(-1)^2 - 11(-1) \overset{?}{<} 10$ $17 < 0$ False |
| B: $\left(-\frac{2}{3}, \frac{5}{2}\right)$ | $0$ | $6 \cdot 0^2 - 11 \cdot 0 - 10 \overset{?}{<} 10$ $-10 < 10$ True |
| C: $\left(\frac{5}{2}, \infty\right)$ | $3$ | $6 \cdot 3^2 - 11 \cdot 3 - 10 \overset{?}{<} 10$ $11 < 10$ |

Solution set: $\left(-\frac{2}{3}, \frac{5}{2}\right)$

**93.** $x^3 - 16x \leq 0$

*Step 1*: Solve $x^3 - 16x = 0$.

$$x^3 - 16x = 0 \Rightarrow x(x^2 - 16) = 0 \Rightarrow$$
$$x(x + 4)(x - 4) = 0$$

Set each factor to zero and solve.
$x = 0$ or $x + 4 = 0 \Rightarrow x = -4$ or
$x - 4 = 0 \Rightarrow x = 4$

*Step 2*: The values $-4$, $0$, and, $4$ divide the number line into four intervals.

Interval A $(-\infty, -4)$   Interval B $(-4, 0)$   Interval C $(0, 4)$   Interval D $(4, \infty)$

*Step 3*: Choose a test value to see if it satisfies the inequality, $x^3 - 16x \leq 0$.

| Interval | Test Value | Is $x^3 - 16x \leq 0$ True or False? |
|---|---|---|
| A: $(-\infty, -4)$ | $-5$ | $(-5)^3 - 16(-5) \overset{?}{\leq} 0$ $-45 \leq 0$ True |
| B: $(-4, 0)$ | $-1$ | $(-1)^3 - 16(-1) \overset{?}{\leq} 0$ $15 \leq 0$ False |
| C: $(0, 4)$ | $1$ | $1^3 - 16 \cdot 1 \overset{?}{\leq} 0$ $-15 \leq 0$ True |
| D: $(4, \infty)$ | $5$ | $5^3 - 16 \cdot 5 \overset{?}{\leq} 0$ $45 \leq 0$ False |

Solution set: $(-\infty, -4] \cup [0, 4]$

**95.** $\dfrac{3x + 6}{x - 5} > 0$

Since one side of the inequality is already 0, we start with Step 2.

*Step 2*: Determine the values that will cause either the numerator or denominator to equal 0.
$3x + 6 = 0 \Rightarrow x = -2$ or $x - 5 = 0 \Rightarrow x = 5$

The values $-2$ and 5 to divide the number line into three regions.

Interval A $(-\infty, -2)$   Interval B $(-2, 5)$   Interval C $(5, \infty)$

**Step 3:** Choose a test value to see if it satisfies the inequality, $\dfrac{3x+6}{x-5}>0$.

| Interval | Test Value | Is $\frac{3x+6}{x-5}>0$ True or False? |
|---|---|---|
| A: $(-\infty,-2)$ | $-3$ | $\frac{3(-3)+6}{-3-5}\overset{?}{>}0$ $\frac{3}{8}>0$ True |
| B: $(-2,5)$ | $0$ | $\frac{3(0)+6}{0-5}\overset{?}{>}0$ $-\frac{6}{5}>0$ False |
| C: $(5,\infty)$ | $6$ | $\frac{3(6)+6}{6-5}\overset{?}{>}0$ $24>0$ True |

Solution set: $(-\infty,-2)\cup(5,\infty)$

**97.** $\dfrac{3x-2}{x}-4>0$

**Step 1:** Rewrite the inequality so that 0 is on one side and there is a single fraction on the other side.

$$\frac{3x-2}{x}-4>0\Rightarrow\frac{3x-2}{x}-\frac{4x}{x}>0\Rightarrow$$
$$\frac{3x-2-4x}{x}>0\Rightarrow\frac{-x-2}{x}>0$$

**Step 2:** Determine the values that will cause either the numerator or denominator to equal 0.
$-x-2=0\Rightarrow x=-2$ or $x=0$

The values $-2$ and 0 divide the number line into three regions.

Interval A $(-\infty,-2)$  | Interval B $(-2,0)$  | Interval C $(0,\infty)$
$-2$ $\quad$ $0$

**Step 3:** Choose a test value to see if it satisfies the inequality, $\dfrac{3x-2}{x}-4>0$.

| Interval | Test Value | Is $\frac{3x-2}{x}-4>0$ True or False? |
|---|---|---|
| A: $(-\infty,-2)$ | $-3$ | $\frac{3(-3)-2}{-3}-4\overset{?}{>}0$ $-\frac{1}{3}>4$ False |
| B: $(-2,0)$ | $-1$ | $\frac{3(-1)-2}{-1}-4\overset{?}{>}0$ $1>4$ True |
| C: $(0,\infty)$ | $1$ | $\frac{3\cdot1-2}{1}-4\overset{?}{>}0$ $-3>0$ False |

Solution set: $(-2,0)$

**99.** $\dfrac{3}{x-1}\le\dfrac{5}{x+3}$

**Step 1:** Rewrite the inequality so that 0 is on one side and there is a single fraction on the other side.

$$\frac{3}{x-1}-\frac{5}{x+3}\le0$$
$$\frac{3(x+3)}{(x-1)(x+3)}-\frac{5(x-1)}{(x+3)(x-1)}\le0$$
$$\frac{3(x+3)-5(x-1)}{(x-1)(x+3)}\le0$$
$$\frac{3x+9-5x+5}{(x-1)(x+3)}\le0$$
$$\frac{-2x+14}{(x-1)(x+3)}\le0$$

**Step 2:** Determine the values that will cause either the numerator or denominator to equal 0.
$-2x+14=0\Rightarrow x=7$ or $x-1\Rightarrow x=1$ or $x+3=0\Rightarrow x=-3$

The values $-3$, 1 and 7 divide the number line into four regions. Use an open circle on $-3$ and 1 because they make the denominator equal 0.

Interval A $(-\infty,-3)$ | Interval B $(-3,1)$ | Interval C $(1,7)$ | Interval D $(7,\infty)$
$-3$ $\quad$ $0\ 1$ $\qquad$ $7$

**Step 3:** Choose a test value to see if it satisfies the inequality, $\dfrac{3}{x-1}\le\dfrac{5}{x+3}$.

| Interval | Test Value | Is $\frac{3}{x-1}\le\frac{5}{x+3}$ True or False? |
|---|---|---|
| A: $(-\infty,-3)$ | $-4$ | $\frac{3}{-4-1}\overset{?}{\le}\frac{5}{-4+3}$ $-\frac{3}{5}\le-5$ False |
| B: $(-3,1)$ | $0$ | $\frac{3}{0-1}\overset{?}{\le}\frac{5}{0+3}$ $-3\le\frac{5}{3}$ True |
| C: $(1,7)$ | $2$ | $\frac{3}{2-1}\overset{?}{\le}\frac{5}{2+3}$ $3\le1$ False |
| D: $(7,\infty)$ | $8$ | $\frac{3}{8-1}\overset{?}{\le}\frac{5}{8+3}$ $\frac{3}{7}\overset{?}{\le}\frac{5}{11}$ $\frac{33}{77}\le\frac{35}{77}$ True |

Intervals B and D satisfy the inequality. The endpoints $-3$ and 1 are not included because they make the denominator 0.
Solution set: $(-3,1)\cup[7,\infty)$

**101. (a)** Answers will vary.

**(b)** Let $x =$ the maximum initial concentration of ozone.
$$x - .43x \le 50 \Rightarrow .57x \le 50$$
$$x \le 87.7 \text{ (approximately)}$$
The filter will reduce ozone concentrations that don't exceed 87.7 ppb.

**103.** $s = 320 - 16t^2$

**(a)** When $s = 0$, the projectile will be at ground level.
$$0 = 320t - 16t^2 \Rightarrow 16t^2 - 320t = 0 \Rightarrow$$
$$t^2 - 20t = 0 \Rightarrow t(t - 20) = 0 \Rightarrow$$
$$t = 0 \text{ or } t = 20$$
The projectile will return to the ground after 20 sec.

**(b)** Solve $s > 576$ for $t$.
$$320t - 16t^2 > 576$$
$$0 > 16t^2 - 320t + 576$$
$$0 > t^2 - 20t + 36$$
*Step 1*: Find the values of $x$ that satisfy $t^2 - 20t + 36 = 0$.
$$t^2 - 20t + 36 = 0 \Rightarrow (t - 2)(t - 18) = 0 \Rightarrow$$
$$t - 2 = 0 \Rightarrow t = 2 \quad \text{or} \quad t - 18 = 0 \Rightarrow t = 18$$
*Step 2*: The two numbers divide a number line into three regions.

| Interval A | Interval B | Interval C |
|:---:|:---:|:---:|
| $(0,2)$ | $(2,18)$ | $(18, \infty)$ |

$$\overset{\longleftarrow \circ + + + + + + + + \circ + + \longrightarrow}{\phantom{xx}0\phantom{x}2\phantom{xxxxxxxxxxxx}18}$$

*Step 3*: Choose a test value to see if it satisfies the inequality,
$$320t - 16t^2 > 576.$$

| Interval | Test Value | Is $320t - 16t^2 > 576$ True or False? |
|:---|:---:|:---|
| A: $(0,2)$ | 1 | $320(1) - 16(1)^2 \overset{?}{>} 576$ <br> $\phantom{xxxxxxx}304 > 576$ <br> False |
| B: $(2,18)$ | 3 | $320(3) - 16(3)^2 \overset{?}{>} 576$ <br> $\phantom{xxxxxxx}816 > 576$ <br> True |
| C: $(18,\infty)$ | 20 | $320(20) - 16(20)^2 \overset{?}{>} 576$ <br> $\phantom{xxxxxxxxxx}0 > 576$ <br> False |

The projectile will be more than 576 ft above the ground between 2 and 18 sec.

**105.** Answers will vary. 3 cannot be in the solution set because when 3 is substituted into $\frac{14x+9}{x-3}$, division by zero occurs.

**107.** "at least 65" means that the number is 65 or greater; $W \ge 65$.

**109.** "as many as 100,000" means 100,000 or less; $a \le 100,000$

**111.** $|x + 4| = 7$
$$x + 4 = 7 \Rightarrow x = 3 \quad \text{or} \quad x + 4 = -7 \Rightarrow x = -11$$
Solution set: $\{-11, 3\}$

**113.** $\left|\dfrac{7}{2-3x}\right| - 9 = 0 \Rightarrow \left|\dfrac{7}{2-3x}\right| = 9$

$$\dfrac{7}{2-3x} = 9 \Rightarrow 7 = 9(2 - 3x) \Rightarrow 7 = 18 - 27x \Rightarrow$$
$$-11 = -27x \Rightarrow \dfrac{-11}{-27} = x \Rightarrow x = \dfrac{11}{27} \quad \text{or}$$

$$\dfrac{7}{2-3x} = -9 \Rightarrow 7 = -9(2 - 3x) \Rightarrow$$
$$7 = -18 + 27x \Rightarrow 25 = 27x \Rightarrow \dfrac{25}{27} = x \Rightarrow x = \dfrac{25}{27}$$

Solution set: $\left\{\dfrac{11}{27}, \dfrac{25}{27}\right\}$

**115.** $|5x - 1| = |2x + 3|$

$$5x - 1 = 2x + 3 \Rightarrow 3x - 1 = 3 \Rightarrow 3x = 4 \Rightarrow x = \dfrac{4}{3}$$
or
$$5x - 1 = -(2x + 3) \Rightarrow 5x - 1 = -2x - 3 \Rightarrow$$
$$7x - 1 = -3 \Rightarrow 7x = -2 \Rightarrow x = -\dfrac{2}{7}$$

Solution set: $\left\{-\dfrac{2}{7}, \dfrac{4}{3}\right\}$

**117.** $|2x + 9| \le 3$
$$-3 \le 2x + 9 \le 3$$
$$-12 \le 2x \le -6$$
$$-6 \le x \le -3$$
Solution set: $[-6, -3]$

**119.** $|7x - 3| > 4$
$$7x - 3 < -4 \Rightarrow 7x < -1 \Rightarrow x < -\dfrac{1}{7} \text{ or}$$
$$7x - 3 > 4 \Rightarrow 7x > 7 \Rightarrow x > 1$$
Solution set: $\left(-\infty, -\dfrac{1}{7}\right) \cup (1, \infty)$

**121.** $|3x + 7| - 5 = 0 \Rightarrow |3x + 7| = 5$
$$3x + 7 = 5 \Rightarrow 3x = -2 \Rightarrow x = -\dfrac{2}{3} \quad \text{or}$$
$$3x + 7 = -5 \Rightarrow 3x = -12 \Rightarrow x = -4$$
Solution set: $\left\{-4, -\dfrac{2}{3}\right\}$

**123.** Since the absolute value of a number is always nonnegative, the inequality $|4x-12|\geq -3$ is always true. The solution set is $(-\infty,\infty)$.

**125.** Since the absolute value of a number is always nonnegative, $|x^2+4x|<0$ is never true, so $|x^2+4x|\leq 0$ is only true when $|x^2+4x|=0$.

$|x^2+4x|=0 \Rightarrow x^2+4x=0 \Rightarrow x(x+4)=0$

$x=0 \quad$ or $\quad x+4=0$

$x-0 \quad$ or $\quad x=-4$

Solution set: $\{-4,0\}$

**127.** "$k$ is 12 units from 6 on the number line" means that the distance between $k$ and 6 is 12 units, or $|k-6|=12$ or $|6-k|=12$.

**129.** "$t$ is no less than .01 unit from 5" means that $t$ is .01 unit or more from 5. Thus, the distance between $t$ and 5 is greater than or equal to .01, or $|t-5|\geq .01$ or $|5-t|\geq .01$

## Chapter 1: Test

**1.** $3(x-4)-5(x+2)=2-(x+24)$

$3x-12-5x-10=2-x-24$

$-2x-22=-x-22$

$-22=x-22$

$0=x$

Solution set: $\{0\}$

**2.** $\frac{2}{3}x+\frac{1}{2}(x-4)=x-4$

$6\left[\frac{2}{3}x+\frac{1}{2}(x-4)\right]=6(x-4)$

$4x+3(x-4)=6x-24$

$4x+3x-12=6x-24$

$7x-12=6x-24$

$x-12=-24$

$x=-12$

Solution set: $\{-12\}$

**3.** $6x^2-11x-7=0$

$(2x+1)(3x-7)=0$

$2x+1=0 \Rightarrow x=-\frac{1}{2}$ or $3x-7=0 \Rightarrow x=\frac{7}{3}$

Solution set: $\left\{-\frac{1}{2},\frac{7}{3}\right\}$

**4.** $(3x+1)^2=8$

$3x+1=\pm\sqrt{8}=\pm2\sqrt{2}$

$3x=-1\pm2\sqrt{2} \Rightarrow x=\frac{-1\pm2\sqrt{2}}{3}$

Solution set: $\left\{\frac{-1\pm2\sqrt{2}}{3}\right\}$

**5.** $3x^2+2x=-2$

Solve by completing the square.

$3x^2+2x=-2$

$3x^2+2x+2=0$

$x^2+\frac{2}{3}x+\frac{2}{3}=0 \Rightarrow x^2+\frac{2}{3}x+\frac{1}{9}=-\frac{2}{3}+\frac{1}{9}$

Note: $\left[\frac{1}{2}\cdot\left(-\frac{2}{3}\right)\right]^2=\left(-\frac{1}{3}\right)^2=\frac{1}{9}$

$\left(x+\frac{1}{3}\right)^2=-\frac{5}{9} \Rightarrow x+\frac{1}{3}=\pm\sqrt{-\frac{5}{9}} \Rightarrow$

$x+\frac{1}{3}=\pm\frac{\sqrt{5}}{3}i \Rightarrow x=-\frac{1}{3}\pm\frac{\sqrt{5}}{3}i$

Solve by the quadratic formula.

Let $a=3$, $b=2$, and $c=2$.

$x=\frac{-b\pm\sqrt{b^2-4ac}}{2a}$

$=\frac{-2\pm\sqrt{2^2-4(3)(2)}}{2(3)}=\frac{-2\pm\sqrt{4-24}}{6}$

$=\frac{-2\pm\sqrt{-20}}{6}=\frac{-2\pm2i\sqrt{5}}{6}$

$=-\frac{2}{6}\pm\frac{2\sqrt{5}}{6}i=-\frac{1}{3}\pm\frac{\sqrt{5}}{3}i$

Solution set: $\left\{-\frac{1}{3}\pm\frac{\sqrt{5}}{3}i\right\}$

**6.** $\frac{12}{x^2-9}=\frac{2}{x-3}-\frac{3}{x+3}$

$\frac{12}{(x+3)(x-3)}+\frac{3}{x+3}=\frac{2}{x-3}$

Multiply each term in the equation by the least common denominator, $(x+3)(x-3)$ assuming $x\neq-3,3$.

$(x+3)(x-3)\left[\frac{12}{(x+3)(x-3)}+\frac{3}{x+3}\right]$

$=(x+3)(x-3)\left(\frac{2}{x-3}\right)$

$12+3(x-3)=2(x+3)$

$12+3x-9=2x+6$

$3x+3=2x+6$

$x+3=6 \Rightarrow x=3$

The only possible solution is 3. However, the variable is restricted to real numbers except $-3$ and $3$. Therefore, the solution set is $\varnothing$.

**7.** $\dfrac{4x}{x-2} + \dfrac{3}{x} = \dfrac{-6}{x^2 - 2x}$ or $\dfrac{4x}{x-2} + \dfrac{3}{x} = \dfrac{-6}{x(x-2)}$

Multiply each term in the equation by the least common denominator, $x(x-2)$, assuming $x \neq 0, 2$.

$$x(x-2)\left[\dfrac{4x}{x-2} + \dfrac{3}{x}\right] = x(x-2)\left(\dfrac{-6}{x(x-2)}\right)$$
$$4x^2 + 3(x-2) = -6 \Rightarrow 4x^2 + 3x - 6 = -6$$
$$4x^2 + 3x = 0 \Rightarrow x(4x+3) = 0$$

$x = 0$ or $4x + 3 = 0 \Rightarrow x = -\frac{3}{4}$

Because of the restriction $x \neq 0$, the only valid solution is $-\frac{3}{4}$. The solution set is $\left\{-\frac{3}{4}\right\}$.

**8.** $\sqrt{3x+4} + 5 = 2x + 1 \Rightarrow \sqrt{3x+4} = 2x - 4$

$$\left(\sqrt{3x+4}\right)^2 = (2x-4)^2$$
$$3x + 4 = 4x^2 - 16x + 16$$
$$0 = 4x^2 - 19x + 12$$
$$0 = (4x - 3)(x - 4)$$

$4x - 3 = 0 \Rightarrow x = \frac{3}{4}$ or $x - 4 = 0 \Rightarrow x = 4$

Check $x = \frac{3}{4}$.

$$\sqrt{3x+4} + 4 = 2x$$
$$\sqrt{3\left(\frac{3}{4}\right) + 4} + 4 = 2\left(\frac{3}{4}\right)$$
$$\sqrt{\tfrac{9}{4} + 4} + 4 = \tfrac{3}{2} \Rightarrow \sqrt{\tfrac{25}{4}} + 4 = \tfrac{3}{2}$$
$$\tfrac{5}{2} + 4 = \tfrac{3}{2} \Rightarrow \tfrac{13}{2} = \tfrac{3}{2}$$

This is a false statement. $\frac{3}{4}$ is a not solution.

Check $x = 4$.

$$\sqrt{3x+4} + 4 = 2x$$
$$\sqrt{3(4)+4} + 4 = 2(4)$$
$$\sqrt{12+4} + 4 = 8$$
$$\sqrt{16} + 4 = 8$$
$$4 + 4 = 8 \Rightarrow 8 = 8$$

This is a true statement. 4 is a solution.
Solution set: {4}

**9.** $\sqrt{-2x+3} + \sqrt{x+3} = 3$

$$\sqrt{-2x+3} = 3 - \sqrt{x+3}$$
$$\left(\sqrt{-2x+3}\right)^2 = \left(3 - \sqrt{x+3}\right)^2$$
$$-2x + 3 = 9 - 6\sqrt{x+3} + (x+3)$$
$$-2x + 3 = 12 + x - 6\sqrt{x+3}$$
$$-3x - 9 = -6\sqrt{x+3}$$

$$x + 3 = 2\sqrt{x+3}$$
$$(x+3)^2 = \left(2\sqrt{x+3}\right)^2$$
$$x^2 + 6x + 9 = 4(x+3)$$
$$x^2 + 6x + 9 = 4x + 12$$
$$x^2 + 2x - 3 = 0 \Rightarrow (x+3)(x-1) = 0$$

$x + 3 = 0 \Rightarrow x = -3$ or $x - 1 = 0 \Rightarrow x = 1$

Check $x = -3$.

$$\sqrt{-2x+3} + \sqrt{x+3} = 3$$
$$\sqrt{-2(-3)+3} + \sqrt{-3+3} = 3$$
$$\sqrt{6+3} + \sqrt{0} = 3$$
$$\sqrt{9} + 0 = 3$$
$$3 + 0 = 3 \Rightarrow 3 = 3$$

This is a true statement. $-3$ is a solution.

Check $x = 1$.

$$\sqrt{-2x+3} + \sqrt{x+3} = 3$$
$$\sqrt{-2(1)+3} + \sqrt{1+3} = 3$$
$$\sqrt{-2+3} + \sqrt{4} = 3$$
$$\sqrt{1} + 2 = 3$$
$$1 + 2 = 3 \Rightarrow 3 = 3$$

This is a true statement. 1 is a solution.
Solution set: $\{-3, 1\}$

**10.** $\sqrt[3]{3x-8} = \sqrt[3]{9x+4}$

$$\left(\sqrt[3]{3x-8}\right)^3 = \left(\sqrt[3]{9x+4}\right)^3$$
$$3x - 8 = 9x + 4 \Rightarrow -8 = 6x + 4 \Rightarrow$$
$$-12 = 6x \Rightarrow -2 = x$$

Check $x = -2$.

$$\sqrt[3]{3x-8} = \sqrt[3]{9x+4}$$
$$\sqrt[3]{3(-2)-8} = \sqrt[3]{9(-2)+4}$$
$$\sqrt[3]{-6-8} = \sqrt[3]{-18+4}$$
$$\sqrt[3]{-14} = \sqrt[3]{-14} \Rightarrow -\sqrt[3]{14} = -\sqrt[3]{14}$$

This is a true statement.
Solution set: $\{-2\}$

**11.** $x^4 - 17x^2 + 16 = 0$

Let $u = x^2$; then $u^2 = x^4$.

With this substitution, the equation becomes
$u^2 - 17u + 16 = 0$.

Solve this equation by factoring.

$$(u-1)(u-16) = 0$$

$u - 1 = 0 \Rightarrow u = 1$ or $u - 16 = 0 \Rightarrow u = 16$

To find $x$, replace $u$ with $x^2$.

$x^2 = 1 \Rightarrow x = \pm\sqrt{1} \Rightarrow x = \pm 1$ or

$x^2 = 16 \Rightarrow x = \pm\sqrt{16} \Rightarrow x = \pm 4$

Solution set: $\{\pm 1, \pm 4\}$

**12.** $(x+3)^{2/3} + (x+3)^{1/3} - 6 = 0$

Let $u = (x+3)^{1/3}$. Then

$u^2 = \left[(x+3)^{1/3}\right]^2 = (x+3)^{2/3}$.

$u^2 + u - 6 = 0 \Rightarrow (u+3)(u-2) = 0$

$u + 3 = 0 \Rightarrow u = -3$ or $u - 2 = 0 \Rightarrow u = 2$

To find $x$, replace $u$ with $(x+3)^{1/3}$.

$(x+3)^{1/3} = -3 \Rightarrow \left[(x+3)^{1/3}\right]^3 = (-3)^3 \Rightarrow$

$x + 3 = 27 \Rightarrow x = -30$ or

$(x+3)^{1/3} = 2 \Rightarrow \left[(x+3)^{1/3}\right]^3 = 2^3 \Rightarrow$

$x + 3 = 8 \Rightarrow x = 5$

Check $x = -30$.

$(x+3)^{2/3} + (x+3)^{1/3} - 6 = 0$

$(-30+3)^{2/3} + (-30+3)^{1/3} - 6 = 0$

$(-27)^{2/3} + (-27)^{1/3} - 6 = 0$

$\left[(-27)^{1/3}\right]^2 + (-3) - 6 = 0$

$(-3)^2 - 3 - 6 = 0$

$9 - 3 - 6 = 0 \Rightarrow 0 = 0$

This is a true statement. $-30$ is a solution.

Check $x = 5$.

$(x+3)^{2/3} + (x+3)^{1/3} - 6 = 0$

$(5+3)^{2/3} + (5+3)^{1/3} - 6 = 0$

$8^{2/3} + 8^{1/3} - 6 = 0$

$\left[8^{1/3}\right]^2 + 2 - 6 = 0$

$2^2 + 2 - 6 = 0$

$4 + 2 - 6 = 0 \Rightarrow 0 = 0$

This is a true statement. $5$ is a solution.

Solution set: $\{-30, 5\}$

**13.** $|4x+3| = 7$

$4x + 3 = 7 \Rightarrow 4x = 4 \Rightarrow x = 1$ or

$4x + 3 = -7 \Rightarrow 4x = -10 \Rightarrow x = -\frac{10}{4} = -\frac{5}{2}$

Solution set: $\left\{-\frac{5}{2}, 1\right\}$

**14.** $|2x+1| = |5-x|$

$2x + 1 = 5 - x \Rightarrow 3x + 1 = 5 \Rightarrow 3x = 4 \Rightarrow x = \frac{4}{3}$

or

$2x + 1 = -(5-x) \Rightarrow 2x + 1 = -5 + x \Rightarrow x = -6$

Solution set: $\left\{-6, \frac{4}{3}\right\}$

**15.** $\quad S = 2HW + 2LW + 2LH$

$S - 2LH = 2HW + 2LW$

$S - 2LH = W(2H + 2L)$

$\dfrac{S - 2LH}{2H + 2L} = W$

$W = \dfrac{S - 2LH}{2H + 2L}$

**16. (a)** $(9-3i) - (4+5i) = (9-4) + (-3-5)i$

$\qquad\qquad\qquad\qquad = 5 - 8i$

**(b)** $(4+3i)(-5+3i) = -20 + 12i - 15i + 9i^2$

$\qquad\qquad\qquad = -20 - 3i + 9(-1)$

$\qquad\qquad\qquad = -20 - 3i - 9 = -29 - 3i$

**(c)** $(8+3i)^2 = 8^2 + 2(8)(3i) + (3i)^2$

$\qquad\qquad = 64 + 48i + 9i^2$

$\qquad\qquad = 64 + 48i + 9(-1)$

$\qquad\qquad = 64 + 48i - 9 = 55 + 48i$

**(d)** $\dfrac{3+19i}{1+3i} = \dfrac{(3+19i)(1-3i)}{(1+3i)(1-3i)}$

$\qquad = \dfrac{3 - 9i + 19i - 57i^2}{1 - (3i)^2}$

$\qquad = \dfrac{3 + 10i - 57(-1)}{1 - 9i^2} = \dfrac{3 + 10i + 57}{1 - 9(-1)}$

$\qquad = \dfrac{60 + 10i}{1 + 9} = \dfrac{60 + 10i}{10} = 6 + i$

**17. (a)** $i^{42} = i^{40} \cdot i^2 = \left(i^4\right)^{10} \cdot (-1) = 1^{10} \cdot (-1) = -1$

**(b)** $i^{-31} = i^{-32} \cdot i = \left(i^4\right)^{-8} \cdot i = 1^{-8} \cdot i = i$

**(c)** $\frac{1}{i^{19}} = i^{-19} = i^{-20} \cdot i = \left(i^4\right)^{-5} \cdot i = 1^{-5} \cdot i = i$

**18. (a)** Minimum:

$1120 \dfrac{\text{gal}}{\text{min}} \cdot 60 \dfrac{\text{min}}{\text{hr}} \cdot 12 \dfrac{\text{hr}}{\text{day}} = 806,400 \dfrac{\text{gal}}{\text{day}}$

The equation that will calculate the minimum amount of water pumped after $x$ days would be $A = 806,400x$.

**(b)** $A = 806,400x$ when $x = 30$ would be

$A = 806,400(30) = 24,192,000$ gal.

**(c)** Since there would be $806,400\,\dfrac{\text{gal}}{\text{day}}$ minimum and each pool requires 20,000 gal, there would be a minimum of $\dfrac{806,400}{20,000} = 40.32$ pools that could be filled each day. The equation that will calculate the minimum number of pools that could be filled after $x$ days would be $P = 40.32x$. Approximately 40 pools could be filled each day.

**(d)** Solve $P = 40.32x$ where $P = 1000$.

$$1000 = 40.32x \Rightarrow x = \tfrac{1000}{40.32} \approx 24.8 \text{ days}.$$

**19.** Let $w =$ width of rectangle. Then

$2w - 20 =$ length of rectangle.

Use the formula for the perimeter of a rectangle.

$$P = 2l + 2w$$
$$620 = 2(2w - 20) + 2w$$
$$620 = 4w - 40 + 2w$$
$$620 = 6w - 40 \Rightarrow 660 = 6w \Rightarrow 110 = w$$

The width is 110 m and the length is
$$2(110) - 20 = 220 - 20 = 200 \text{ m}.$$

**20.** Let $x =$ amount of cashews (in pounds). Then $35 - x =$ amount of walnuts (in pounds).

|         | Cost per Pound | Amount of Nuts |              |
| ------- | -------------- | -------------- | ------------ |
| Cashews | 7.00           | $x$            | $7.00x$      |
| Walnuts | 5.50           | $35 - x$       | $5.50(35 - x)$ |
| Mixture | 6.50           | 35             | $35 \cdot 6.50$ |

Solve the following equation.

$$7.00x + 5.50(35 - x) = 35 \cdot 6.50$$
$$7x + 192.5 - 5.5x = 227.5$$
$$1.5x + 192.5 = 227.5$$
$$1.5x = 35$$
$$x = \tfrac{35}{1.5} = \tfrac{350}{15} = \tfrac{70}{3} = 23\tfrac{1}{3}$$

The fruit and nut stand owner should mix $23\tfrac{1}{3}$ lbs of cashews with $35 - 23\tfrac{1}{3} = 11\tfrac{2}{3}$ lbs of walnuts.

**21.** Let $x =$ time (in hours) the mother spent driving to meet plane.
Since Mary Lynn has been in the plane for 15 minutes, and 15 minutes is $\tfrac{1}{4}$ hr, she has been traveling by plane for $x + \tfrac{1}{4}$ hr.

|                     | $d$ | $r$ | $t$           |
| ------------------- | --- | --- | ------------- |
| Mary Lynn by plane  | 420 |     | $x + \tfrac{1}{4}$ |
| Mother by car       | 20  | 40  | $x$           |

The time driven by Mary Lynn's mother can be found by $20 = 40x \Rightarrow x = \tfrac{1}{2}$ hr. Mary Lynn, therefore, flew for $\tfrac{1}{2} + \tfrac{1}{4} = \tfrac{2}{4} + \tfrac{1}{4} = \tfrac{3}{4}$ hr. The rate of Mary Lynn's plane can be found by $r = \dfrac{d}{t} = \dfrac{420}{\frac{3}{4}} = 420 \cdot \tfrac{4}{3} = 560$ km per hour.

**22.** $h = -16t^2 + 96t$

**(a)** Let $h = 80$ and solve for $t$.
$$80 = -16t^2 + 96t \Rightarrow 16t^2 - 96t + 80 = 0$$
$$t^2 - 6t + 5 = 0$$
$$(t - 1)(t - 5) = 0$$
$$t - 1 = 0 \Rightarrow t = 1 \quad \text{or} \quad t - 5 = 0 \Rightarrow t = 5$$

The projectile will reach a height of 80 ft at 1 sec and 5 sec.

**(b)** Let $h = 0$ and solve for $t$.
$$0 = -16t^2 + 96t$$
$$0 = -16t(t - 6)$$
$$t = 0 \quad \text{or} \quad t - 6 = 0 \Rightarrow t = 6$$

The projectile will return to the ground at 6 sec.

**23.** The table shows each equation evaluated at the years 1975, 1994, and 2006. Equation B best models the data.

**24.** $-2(x - 1) - 12 < 2(x + 1)$
$$-2x + 2 - 12 < 2x + 2$$
$$-2x - 10 < 2x + 2$$
$$-4x - 10 < 2$$
$$-4x < 12$$
$$x > -3$$
Solution set: $(-3, \infty)$

**25.** $-3 \le \dfrac{1}{2}x + 2 \le 3$
$$2(-3) \le 2\left(\tfrac{1}{2}x + 2\right) \le 2(3)$$
$$-6 \le x + 4 \le 6$$
$$-10 \le x \le 2$$
Solution set: $[-10, 2]$

**26.** $2x^2 - x \geq 3$

*Step 1*: Find the values of $x$ that satisfy $2x^2 - x = 3$.

$$2x^2 - x = 3$$
$$2x^2 - x - 3 = 0$$
$$(x+1)(2x-3) = 0$$

$x+1 = 0 \Rightarrow x = -1$  or  $2x-3 = 0 \Rightarrow x = \frac{3}{2}$

*Step 2*:  The two numbers divide a number line into three regions.

Interval A   Interval B   Interval C
$(-\infty, -1)$   $\left(-1, \frac{3}{2}\right)$   $\left(\frac{3}{2}, \infty\right)$

$-1 \quad 0 \quad 1 \quad \frac{3}{2} \quad 2$

*Step 3*: Choose a test value to see if it satisfies the inequality, $2x^2 - x \geq 3$

| Interval | Test Value | Is $2x^2 - x \geq 3$ True or False? |
|---|---|---|
| A: $(-\infty, -1)$ | $-2$ | $2(-2)^2 - (-2) \overset{?}{\geq} 3$  $10 \geq 3$  True |
| B: $\left(-1, \frac{3}{2}\right)$ | $0$ | $2 \cdot 0^2 - 0 \overset{?}{\geq} 3$  $0 \geq 3$  False |
| C: $\left(\frac{3}{2}, \infty\right)$ | $2$ | $2 \cdot 2^2 - 2 \overset{?}{\geq} 3$  $6 \geq 3$  True |

Solution set: $(-\infty, -1] \cup \left[\frac{3}{2}, \infty\right)$

**27.** $\dfrac{x+1}{x-3} < 5$

*Step 1*: Rewrite the inequality so that 0 is on one side and there is a single fraction on the other side.

$$\frac{x+1}{x-3} < 5 \Rightarrow \frac{x+1}{x-3} - 5 < 0$$
$$\frac{x+1}{x-3} - \frac{5(x-3)}{x-3} < 0 \Rightarrow \frac{x+1-5(x-3)}{x-3} < 0$$
$$\frac{x+1-5x+15}{x-3} < 0 \Rightarrow \frac{-4x+16}{x-3} < 0$$

*Step 2*: Determine the values that will cause either the numerator or denominator to equal 0.

$-4x + 16 = 0 \Rightarrow x = 4$  or  $x - 3 = 0 \Rightarrow x = 3$

The values 3 and 4 divide the number line into three regions.

Interval A   Interval B   Interval C
$(-\infty, 3)$   $(3, 4)$   $(4, \infty)$

$2 \quad 3 \quad 4$

*Step 3*: Choose a test value to see if it satisfies the inequality, $\frac{x+1}{x-3} < 5$.

| Interval | Test Value | Is $\frac{x+1}{x-3} < 5$ True or False? |
|---|---|---|
| A: $(-\infty, 3)$ | $0$ | $\frac{0+1}{0-3} \overset{?}{<} 5$  $-\frac{1}{3} < 5$  True |
| B: $(3, 4)$ | $3.5$ | $\frac{3.5+1}{3.5-3} \overset{?}{<} 5$  $9 < 5$  False |
| C: $(4, \infty)$ | $5$ | $\frac{5+1}{5-3} \overset{?}{<} 5$  $3 < 5$  True |

Solution set: $(-\infty, 3) \cup (4, \infty)$

**28.** $|2x - 5| < 9$

$-9 < 2x - 5 < 9$
$-4 < 2x < 14$
$-2 < x < 7$

Solution set: $(-2, 7)$

**29.** $|2x + 1| - 11 \geq 0 \Rightarrow |2x + 1| \geq 11$

$2x + 1 \leq -11$  or  $2x + 1 \geq 11$
$2x \leq -12 \qquad\qquad 2x \geq 10$
$x \leq -6$  or  $x \geq 5$

Solution set: $(-\infty, -6] \cup [5, \infty)$

**30.** $|3x + 7| \leq 0 \Rightarrow 3x + 7 \leq 0 \Rightarrow x \leq -\frac{7}{3}$

However, if $x < -\frac{7}{3}$, the expression inside the absolute value bars is negative, so $x$ cannot be less than $-\frac{7}{3}$. The solution set of $|3x + 7| \leq 0$ is $\left\{-\frac{7}{3}\right\}$.

# Chapter 2

## GRAPHS AND FUNCTIONS

### Section 2.1: Rectangular Coordinates and Graphs

#### Connections (page 190)

1. Answers will vary.

2. Answers will vary.
   Latitude and longitude values pinpoint distances north or south of the equator and east or west of the prime meridian. Similarly on a Cartesian coordinate system, $x$- and $y$-coordinates give distances and directions from the $y$-axis and $x$-axis, respectively.

#### Exercises

1. False. $(-1, 3)$ lies in Quadrant II.

3. True. The origin has coordinates $(0,0)$. So, the distance from $(0,0)$ to $(a,b)$ is
$$d = \sqrt{(a-0)^2 + (b-0)^2} = \sqrt{a^2 + b^2}$$

5. True. When $x = 0$, $y = 2(0) + 4 = 4$, so the $y$-intercept is 4. When $y = 0$, $0 = 2x + 4 \Rightarrow x = -2$, so the $x$-intercept is $-2$.

7. Any three of the following:
$$(2,-5),(-1,7),(3,-9),(5,-17),(6,-21)$$

9. Any three of the following:
$$(1993,31),(1995,35),(1997,37),$$
$$(1999,35),(2001,28),(2003,25)$$

11. $P(-5, -7)$, $Q(-13, 1)$

   (a) $d(P, Q) = \sqrt{[-13-(-5)]^2 + [1-(-7)]^2}$
   $$= \sqrt{(-8)^2 + 8^2} = \sqrt{128} = 8\sqrt{2}$$

   (b) The midpoint $M$ of the segment joining points $P$ and $Q$ has coordinates
   $$\left(\frac{-5+(-13)}{2}, \frac{-7+1}{2}\right) = \left(\frac{-18}{2}, \frac{-6}{2}\right)$$
   $$= (-9, -3).$$

13. $P(8, 2)$, $Q(3, 5)$

   (a) $d(P, Q) = \sqrt{(3-8)^2 + (5-2)^2}$
   $$= \sqrt{(-5)^2 + 3^2}$$
   $$= \sqrt{25+9} = \sqrt{34}$$

   (b) The midpoint $M$ of the segment joining points $P$ and $Q$ has coordinates
   $$\left(\frac{8+3}{2}, \frac{2+5}{2}\right) = \left(\frac{11}{2}, \frac{7}{2}\right).$$

15. $P(-6, -5)$, $Q(6, 10)$

   (a) $d(P, Q) = \sqrt{[6-(-6)]^2 + [10-(-5)]^2}$
   $$= \sqrt{12^2 + 15^2} = \sqrt{144+225}$$
   $$= \sqrt{369} = 3\sqrt{41}$$

   (b) The midpoint $M$ of the segment joining points $P$ and $Q$ has coordinates
   $$\left(\frac{-6+6}{2}, \frac{-5+10}{2}\right) = \left(\frac{0}{2}, \frac{5}{2}\right) = \left(0, \frac{5}{2}\right).$$

17. $P\left(3\sqrt{2}, 4\sqrt{5}\right)$, $Q\left(\sqrt{2}, -\sqrt{5}\right)$

   (a) $d(P, Q)$
   $$= \sqrt{\left(\sqrt{2} - 3\sqrt{2}\right)^2 + \left(-\sqrt{5} - 4\sqrt{5}\right)^2}$$
   $$= \sqrt{\left(-2\sqrt{2}\right)^2 + \left(-5\sqrt{5}\right)^2}$$
   $$= \sqrt{8+125} = \sqrt{133}$$

   (b) The midpoint $M$ of the segment joining points $P$ and $Q$ has coordinates
   $$\left(\frac{3\sqrt{2}+\sqrt{2}}{2}, \frac{4\sqrt{5}+(-\sqrt{5})}{2}\right)$$
   $$= \left(\frac{4\sqrt{2}}{2}, \frac{3\sqrt{5}}{2}\right) = \left(2\sqrt{2}, \frac{3\sqrt{5}}{2}\right).$$

19. Label the points $A(-6, -4)$, $B(0, -2)$, and $C(-10, 8)$. Use the distance formula to find the length of each side of the triangle.
$$d(A, B) = \sqrt{[0-(-6)]^2 + [-2-(-4)]^2}$$
$$= \sqrt{6^2 + 2^2} = \sqrt{36+4} = \sqrt{40}$$
$$d(B, C) = \sqrt{(-10-0)^2 + [8-(-2)]^2}$$
$$= \sqrt{(-10)^2 + 10^2} = \sqrt{100+100}$$
$$= \sqrt{200}$$
$$d(A, C) = \sqrt{[-10-(-6)]^2 + [8-(-4)]^2}$$
$$= \sqrt{(-4)^2 + 12^2} = \sqrt{16+144} = \sqrt{160}$$
Since $\left(\sqrt{40}\right)^2 + \left(\sqrt{160}\right)^2 = \left(\sqrt{200}\right)^2$, triangle $ABC$ is a right triangle.

**21.** Label the points $A(-4, 1)$, $B(1, 4)$, and $C(-6, -1)$.

$$d(A, B) = \sqrt{[1-(-4)]^2 + (4-1)^2}$$
$$= \sqrt{5^2 + 3^2} = \sqrt{25+9} = \sqrt{34}$$
$$d(B, C) = \sqrt{(-6-1)^2 + (-1-4)^2}$$
$$= \sqrt{(-7)^2 + (-5)^2} = \sqrt{49+25} = \sqrt{74}$$
$$d(A, C) = \sqrt{[-6-(-4)]^2 + (-1-1)^2}$$
$$= \sqrt{(-2)^2 + (-2)^2} = \sqrt{4+4} = \sqrt{8}$$

Since $(\sqrt{8})^2 + (\sqrt{34})^2 \neq (\sqrt{74})^2$ because $8 + 34 = 42 \neq 74$, triangle $ABC$ is not a right triangle.

**23.** Label the points $A(-4, 3)$, $B(2, 5)$, and $C(-1, -6)$.

$$d(A, B) = \sqrt{[2-(-4)]^2 + (5-3)^2}$$
$$= \sqrt{6^2 + 2^2} = \sqrt{36+4} = \sqrt{40}$$
$$d(B, C) = \sqrt{(-1-2)^2 + (-6-5)^2}$$
$$= \sqrt{(-3)^2 + (-11)^2}$$
$$= \sqrt{9+121} = \sqrt{130}$$
$$d(A, C) = \sqrt{[-1-(-4)]^2 + (-6-3)^2}$$
$$= \sqrt{3^2 + (-9)^2} = \sqrt{9+81} = \sqrt{90}$$

Since $\left(\sqrt{40}\right)^2 + \left(\sqrt{90}\right)^2 = \left(\sqrt{130}\right)^2$, triangle $ABC$ is a right triangle.

**25.** Label the given points $A(0, -7)$, $B(-3, 5)$, and $C(2, -15)$. Find the distance between each pair of points.

$$d(A, B) = \sqrt{(-3-0)^2 + [5-(-7)]^2}$$
$$= \sqrt{(-3)^2 + 12^2} = \sqrt{9+144}$$
$$= \sqrt{153} = 3\sqrt{17}$$
$$d(B, C) = \sqrt{[2-(-3)]^2 + (-15-5)^2}$$
$$= \sqrt{5^2 + (-20)^2} = \sqrt{25+400}$$
$$= \sqrt{425} = 5\sqrt{17}$$
$$d(A, C) = \sqrt{(2-0)^2 + [-15-(-7)]^2}$$
$$= \sqrt{2^2 + (-8)^2} = \sqrt{68} = 2\sqrt{17}$$

Since $d(A, B) + d(A, C) = d(B, C)$ or $3\sqrt{17} + 2\sqrt{17} = 5\sqrt{17}$, the points are collinear.

**27.** Label the points $A(0, 9)$, $B(-3, -7)$, and $C(2, 19)$.

$$d(A, B) = \sqrt{(-3-0)^2 + (-7-9)^2}$$
$$= \sqrt{(-3)^2 + (-16)^2} = \sqrt{9+256}$$
$$= \sqrt{265} \approx 16.279$$
$$d(B, C) = \sqrt{[2-(-3)]^2 + [19-(-7)]^2}$$
$$= \sqrt{5^2 + 26^2} = \sqrt{25+676}$$
$$= \sqrt{701} \approx 26.476$$
$$d(A, C) = \sqrt{(2-0)^2 + (19-9)^2}$$
$$= \sqrt{2^2 + 10^2} = \sqrt{4+100}$$
$$= \sqrt{104} \approx 10.198$$

Since $d(A, B) + d(A, C) \neq d(B, C)$

or $\sqrt{265} + \sqrt{104} \neq \sqrt{701}$
$$16.279 + 10.198 \neq 26.476,$$
$$26.477 \neq 26.476,$$

the three given points are not collinear. (Note, however, that these points are very close to lying on a straight line and may appear to lie on a straight line when graphed.)

**29.** Label the points $A(-7, 4)$, $B(6, -2)$, and $C(-1, 1)$.

$$d(A, B) = \sqrt{[6-(-7)]^2 + (-2-4)^2}$$
$$= \sqrt{13^2 + (-6)^2} = \sqrt{169+36}$$
$$= \sqrt{205} \approx 14.3178$$
$$d(B, C) = \sqrt{(-1-6)^2 + [1-(-2)]^2}$$
$$= \sqrt{(-7)^2 + 3^2} = \sqrt{49+9}$$
$$= \sqrt{58} \approx 7.6158$$
$$d(A, C) = \sqrt{[-1-(-7)]^2 + (1-4)^2}$$
$$= \sqrt{6^2 + (-3)^2} = \sqrt{36+9}$$
$$= \sqrt{45} \approx 6.7082$$

Since $d(B, C) + d(A, C) \neq d(A, B)$ or
$$\sqrt{58} + \sqrt{45} \neq \sqrt{205}$$
$$7.6158 + 6.7082 \neq 14.3178$$
$$14.3240 \neq 14.3178,$$

the three given points are not collinear. (Note, however, that these points are very close to lying on a straight line and may appear to lie on a straight line when graphed.)

**31.** Midpoint (5, 8), endpoint (13, 10)

$$\frac{13+x}{2}=5 \quad \text{and} \quad \frac{10+y}{2}=8$$

$$13+x=10 \quad \text{and} \quad 10+y=16$$

$$x=-3 \quad \text{and} \quad y=6.$$

The other endpoint has coordinates $(-3, 6)$.

**33.** Midpoint (12, 6), endpoint (19, 16)

$$\frac{19+x}{2}=12 \quad \text{and} \quad \frac{16+y}{2}=6$$

$$19+x=24 \quad \text{and} \quad 16+y=12$$

$$x=5 \quad \text{and} \quad y=-4.$$

The other endpoint has coordinates $(5, -4)$.

**35.** Midpoint $(a, b)$, endpoint $(p, q)$

$$\frac{p+x}{2}=a \quad \text{and} \quad \frac{q+y}{2}=b$$

$$p+x=2a \quad \text{and} \quad q+y=2b$$

$$x=2a-p \quad \text{and} \quad y=2b-q$$

The other endpoint has coordinates $(2a-p, 2b-q)$.

**37.** The endpoints of the segment are (1990, 20.3) and (2006, 28.0).

$$M=\left(\frac{1990+2006}{2},\frac{20.3+28.0}{2}\right)$$
$$=(1998, 24.15)$$

The estimate is 24.15%. This is close to the actual figure of 24.4%.

**39.** The points to use would be (1970, 3968) and (2004, 19157). Their midpoint is

$$\left(\frac{1970+2004}{2},\frac{3968+19,157}{2}\right)$$
$$=(1987, 11562.50).$$

In 1987, the poverty level cutoff was approximately $11,563.

**41.** The midpoint M has coordinates

$$\left(\frac{x_1+x_2}{2},\frac{y_1+y_2}{2}\right).$$

$d(P,M)$

$$=\sqrt{\left(\frac{x_1+x_2}{2}-x_1\right)^2+\left(\frac{y_1+y_2}{2}-y_1\right)^2}$$

$$=\sqrt{\left(\frac{x_1+x_2}{2}-\frac{2x_1}{2}\right)^2+\left(\frac{y_1+y_2}{2}-\frac{2y_1}{2}\right)^2}$$

$$=\sqrt{\left(\frac{x_2-x_1}{2}\right)^2+\left(\frac{y_2-y_1}{2}\right)^2}$$

$$=\sqrt{\frac{\left(x_2-x_1\right)^2}{4}+\frac{\left(y_2-y_1\right)^2}{4}}$$

$$=\sqrt{\frac{\left(x_2-x_1\right)^2+\left(y_2-y_1\right)^2}{4}}$$

$$=\frac{1}{2}\sqrt{\left(x_2-x_1\right)^2+\left(y_2-y_1\right)^2}$$

$d(M,Q)$

$$=\sqrt{\left(x_2-\frac{x_1+x_2}{2}\right)^2+\left(y_2-\frac{y_1+y_2}{2}\right)^2}$$

$$=\sqrt{\left(\frac{2x_2}{2}-\frac{x_1+x_2}{2}\right)^2+\left(\frac{2y_2}{2}-\frac{y_1+y_2}{2}\right)^2}$$

$$=\sqrt{\left(\frac{x_2-x_1}{2}\right)^2+\left(\frac{y_2-y_1}{2}\right)^2}$$

$$=\sqrt{\frac{\left(x_2-x_1\right)^2}{4}+\frac{\left(y_2-y_1\right)^2}{4}}$$

$$=\sqrt{\frac{\left(x_2-x_1\right)^2+\left(y_2-y_1\right)^2}{4}}$$

$$=\frac{1}{2}\sqrt{\left(x_2-x_1\right)^2+\left(y_2-y_1\right)^2}$$

$$d(P,Q)=\sqrt{\left(x_2-x_1\right)^2+\left(y_2-y_1\right)^2}$$

Since $\frac{1}{2}\sqrt{\left(x_2-x_1\right)^2+\left(y_2-y_1\right)^2}$

$$+\frac{1}{2}\sqrt{\left(x_2-x_1\right)^2+\left(y_2-y_1\right)^2}$$

$$=\sqrt{\left(x_2-x_1\right)^2+\left(y_2-y_1\right)^2},$$

this shows $d(P,M)+d(M,Q)=d(P,Q)$ and $d(P,M)=d(M,Q)$.

In exercises 43–53, other ordered pairs are possible.

**43. (a)**

| $x$ | $y$ | |
|---|---|---|
| 0 | –2 | y-intercept: $x=0\Rightarrow$ $6y=3(0)-12\Rightarrow$ $6y=-12\Rightarrow y=-2$ |
| 4 | 0 | x-intercept: $y=0\Rightarrow$ $6(0)=3x-12\Rightarrow$ $0=3x-12\Rightarrow$ $12=3x\Rightarrow 4=x$ |
| 2 | –1 | additional point |

**(b)**

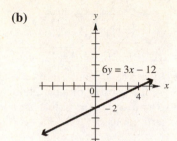

**45. (a)**

| x | y | |
|---|---|---|
| 0 | $\frac{5}{3}$ | y-intercept: $x = 0 \Rightarrow$ $2(0) + 3y = 5 \Rightarrow$ $3y = 5 \Rightarrow y = \frac{5}{3}$ |
| $\frac{5}{2}$ | 0 | x-intercept: $y = 0 \Rightarrow$ $2x + 3(0) = 5 \Rightarrow$ $2x = 5 \Rightarrow x = \frac{5}{2}$ |
| 4 | −1 | additional point |

**(b)**

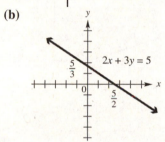

**47. (a)**

| x | y | |
|---|---|---|
| 0 | 0 | x- and y-intercept: $0 = 0^2$ |
| 1 | 1 | additional point |
| −2 | 4 | additional point |

**(b)**

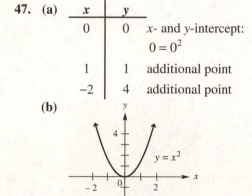

**49. (a)**

| x | y | |
|---|---|---|
| 3 | 0 | x-intercept: $y = 0 \Rightarrow$ $0 = \sqrt{x-3} \Rightarrow$ $0 = x - 3 \Rightarrow 3 = x$ |
| 4 | 1 | additional point |
| 7 | 2 | additional point |

no y-intercept:
$x = 0 \Rightarrow y = \sqrt{0-3} \Rightarrow y = \sqrt{-3}$

**(b)**

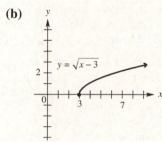

**51. (a)**

| x | y | |
|---|---|---|
| 0 | 2 | y-intercept: $x = 0 \Rightarrow$ $y = |0 - 2| \Rightarrow$ $y = |-2| \Rightarrow y = 2$ |
| 2 | 0 | x-intercept: $y = 0 \Rightarrow$ $0 = |x - 2| \Rightarrow$ $0 = x - 2 \Rightarrow 2 = x$ |
| −2 | 4 | additional point |
| 4 | 2 | additional point |

**(b)**

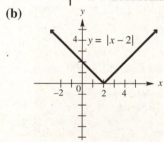

**53. (a)**

| x | y | |
|---|---|---|
| 0 | 0 | x- and y-intercept: $0 = 0^3$ |
| −1 | −1 | additional point |
| 2 | 8 | additional point |

**(b)**

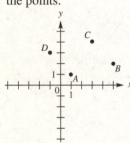

**55.** Points on the *x*-axis have *y*-coordinates equal to 0. The point on the x-axis will have the same *x*-coordinate as point (4, 3). Therefore, the line will intersect the *x*-axis at (4, 0).

**57.** Since (*a*, *b*) is in the second quadrant, *a* is negative and *b* is positive. Therefore, (*a*, – *b*) will have a negative *x* -coordinate and a negative *y*-coordinate and will lie in quadrant III. (–*a*, *b*) will have a positive *x*-coordinate and a positive *y*-coordinate and will lie in quadrant I. Also, (–*a*, – *b*) will have a positive *x* -coordinate and a negative *y*-coordinate and will lie in quadrant IV. Finally, (*b*, *a*) will have a positive *x*-coordinate and a negative *y*-coordinate and will lie in quadrant IV.

**59.** To determine which points form sides of the quadrilateral (as opposed to diagonals), plot the points.

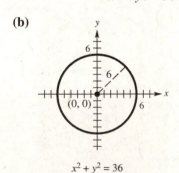

Use the distance formula to find the length of each side.

$$d(A, B) = \sqrt{(5-1)^2 + (2-1)^2}$$
$$= \sqrt{4^2 + 1^2} = \sqrt{16+1} = \sqrt{17}$$
$$d(B, C) = \sqrt{(3-5)^2 + (4-2)^2}$$
$$= \sqrt{(-2)^2 + 2^2} = \sqrt{4+4} = \sqrt{8}$$
$$d(C, D) = \sqrt{(-1-3)^2 + (3-4)^2}$$
$$= \sqrt{(-4)^2 + (-1)^2}$$
$$= \sqrt{16+1} = \sqrt{17}$$

$$d(D, A) = \sqrt{\left[1-(-1)\right]^2 + (1-3)^2}$$
$$= \sqrt{2^2 + (-2)^2} = \sqrt{4+4} = \sqrt{8}$$

Since $d(A, B) = d(C, D)$ and $d(B, C) = d(D, A)$, the points are the vertices of a parallelogram. Since $d(A, B) \neq d(B, C)$, the points are not the vertices of a rhombus.

## Section 2.2: Circles

### Connections (page 198)

Using compasses, draw circles centered at Wickenburg, Kingman, Phoenix, and Las Vegas with scaled radii of 50, 75, 105, and 180 miles respectively. The four circles should intersect at the location of Nothing.

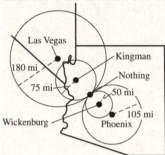

**Exercises**

**1. (a)** Center (0, 0), radius 6

$$\sqrt{(x-0)^2 + (y-0)^2} = 6$$
$$(x-0)^2 + (y-0)^2 = 6^2$$
$$x^2 + y^2 = 36$$

**(b)**

$x^2 + y^2 = 36$

**3. (a)** Center (2, 0), radius 6

$$\sqrt{(x-2)^2 + (y-0)^2} = 6$$
$$(x-2)^2 + (y-0)^2 = 6^2$$
$$(x-2)^2 + y^2 = 36$$

**(b)**

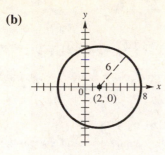

$(x-2)^2 + y^2 = 36$

**5. (a)** Center $(-2, 5)$, radius 4

$$\sqrt{\left[x-(-2)\right]^2 + (y-5)^2} = 4$$
$$[x-(-2)]^2 + (y-5)^2 = 4^2$$
$$(x+2)^2 + (y-5)^2 = 16$$

**(b)**

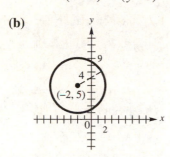

$(x+2)^2 + (y-5)^2 = 16$

**7. (a)** Center $(5, -4)$, radius 7

$$\sqrt{(x-5)^2 + \left[y-(-4)\right]^2} = 7$$
$$(x-5)^2 + [y-(-4)]^2 = 7^2$$
$$(x-5)^2 + (y+4)^2 = 49$$

**(b)**

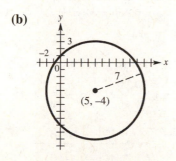

$(x-5)^2 + (y+4)^2 = 49$

**9. (a)** Center $(0, 4)$, radius 4

$$\sqrt{(x-0)^2 + (y-4)^2} = 4$$
$$x^2 + (y-4)^2 = 16$$

**(b)**

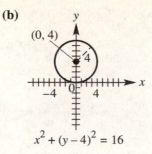

$x^2 + (y-4)^2 = 16$

**11. (a)** Center $\left(\sqrt{2}, \sqrt{2}\right)$, radius $\sqrt{2}$

$$\sqrt{\left(x-\sqrt{2}\right)^2 + \left(y-\sqrt{2}\right)^2} = \sqrt{2}$$
$$\left(x-\sqrt{2}\right)^2 + \left(y-\sqrt{2}\right)^2 = 2$$

**(b)**

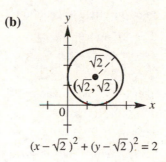

$(x-\sqrt{2})^2 + (y-\sqrt{2})^2 = 2$

**13. (a)** The center of the circle is located at the midpoint of the diameter determined by the points $(1, 1)$ and $(5, 1)$. Using the midpoint formula, we have

$$C = \left(\frac{1+5}{2}, \frac{1+1}{2}\right) = (3,1).$$ The radius is

one-half the length of the diameter:

$$r = \frac{1}{2}\sqrt{(5-1)^2 + (1-1)^2} = 2$$

The equation of the circle is

$$\left(x-3\right)^2 + \left(y-1\right)^2 = 4$$

**(b)** Expand $\left(x-3\right)^2 + \left(y-1\right)^2 = 4$ to find the equation of the circle in general form:

$$\left(x-3\right)^2 + \left(y-1\right)^2 = 4$$
$$x^2 - 6x + 9 + y^2 - 2y + 1 = 4$$
$$x^2 + y^2 - 6x - 2y + 6 = 0$$

15. **(a)** The center of the circle is located at the midpoint of the diameter determined by the points $(-2, 4)$ and $(-2, 0)$. Using the midpoint formula, we have

$$C = \left(\frac{-2+(-2)}{2}, \frac{4+0}{2}\right) = (-2, 2).$$

The radius is one-half the length of the diameter:

$$r = \frac{1}{2}\sqrt{\left[-2-(-2)\right]^2 + (4-0)^2} = 2$$

The equation of the circle is

$$(x+2)^2 + (y-2)^2 = 4$$

**(b)** Expand $(x+2)^2 + (y-2)^2 = 4$ to find the equation of the circle in general form:

$$(x+2)^2 + (y-2)^2 = 4$$
$$x^2 + 4x + 4 + y^2 - 4y + 4 = 4$$
$$x^2 + y^2 + 4x - 4y + 4 = 0$$

17. Since the center $(-3, 5)$ is in quadrant II, choice B is the correct graph.

19. $x^2 + y^2 + 6x + 8y + 9 = 0$

Complete the square on $x$ and $y$ separately.

$$\left(x^2 + 6x\right) + \left(y^2 + 8y\right) = -9$$
$$\left(x^2 + 6x + 9\right) + \left(y^2 + 8y + 16\right) = -9 + 9 + 16$$
$$(x+3)^2 + (y+4)^2 = 16$$

Yes, it is a circle. The circle has its center at $(-3, -4)$ and radius 4.

21. $x^2 + y^2 - 4x + 12y = -4$

Complete the square on $x$ and $y$ separately.

$$x^2 - 4x + y^2 + 12y = -4$$
$$\left(x^2 - 4x\right) + \left(y^2 + 12y\right) = -4$$
$$\left(x^2 - 4x + 4\right) + \left(y^2 + 12y + 36\right) = -4 + 4 + 36$$
$$(x-2)^2 + (y+6)^2 = 36$$

Yes, it is a circle. The circle has its center at $(2, -6)$ and radius 6.

23. $4x^2 + 4y^2 + 4x - 16y - 19 = 0$

Complete the square on $x$ and $y$ separately.

$$4\left(x^2 + x\right) + 4\left(y^2 - 4y\right) = 19$$
$$4\left(x^2 + x + \tfrac{1}{4}\right) + 4\left(y^2 - 4y + 4\right) = 19 + 4\left(\tfrac{1}{4}\right) + 4(4)$$

$$4\left(x+\tfrac{1}{2}\right)^2 + 4(y-2)^2 = 36$$
$$\left(x+\tfrac{1}{2}\right)^2 + (y-2)^2 = 9$$

Yes, it is a circle with center $\left(-\tfrac{1}{2}, 2\right)$ and radius 3.

25. $x^2 + y^2 + 2x - 6y + 14 = 0$

Complete the square on $x$ and $y$ separately.

$$\left(x^2 + 2x\right) + \left(y^2 - 6y\right) = -14$$
$$\left(x^2 + 2x + 1\right) + \left(y^2 - 6y + 9\right) = -14 + 1 + 9$$
$$(x+1)^2 + (y-3)^2 = -4$$

The graph is nonexistent.

27. $x^2 + y^2 - 6x - 6y + 18 = 0$

Complete the square on $x$ and $y$ separately.

$$\left(x^2 - 6x\right) + \left(y^2 - 6y\right) = -18$$
$$\left(x^2 - 6x + 9\right) + \left(y^2 - 6y + 9\right) = -18 + 9 + 9$$
$$(x-3)^2 + (y-3)^2 = 0$$

The graph is the point $(3, 3)$.

29. $9x^2 + 9y^2 + 36x = -32$

Complete the square on $x$ and $y$ separately.

$$9\left(x^2 + 4x\right) + 9y^2 = -32$$
$$9\left(x^2 + 4x + 4\right) + 9(y-0)^2 = -32 + 9(4)$$
$$9(x+2)^2 + 9(y-0)^2 = 4$$
$$(x+2)^2 + (y-0)^2 = \tfrac{4}{9} = \left(\tfrac{2}{3}\right)^2$$

Yes, it is a circle with center $(-2, 0)$ and radius $\tfrac{2}{3}$.

31. The midpoint $M$ has coordinates

$$\left(\frac{-1+5}{2}, \frac{3+(-9)}{2}\right) = \left(\frac{4}{2}, \frac{-6}{2}\right) = (2, -3).$$

33. Use points $C(2, -3)$ and $Q(5, -9)$.

$$d(C, Q) = \sqrt{(5-2)^2 + \left[-9-(-3)\right]^2}$$
$$= \sqrt{3^2 + (-6)^2} = \sqrt{9+36}$$
$$= \sqrt{45} = 3\sqrt{5}$$

The radius is $3\sqrt{5}$.

35. The center-radius form for this circle is
$$(x-2)^2 + (y+3)^2 = (3\sqrt{5})^2 \Rightarrow$$
$$(x-2)^2 + (y+3)^2 = 45.$$

**37.** The equations of the three circles are
$(x-7)^2 + (y-4)^2 = 25$,
$(x+9)^2 + (y+4)^2 = 169$, and
$(x+3)^2 + (y-9)^2 = 100$. From the graph of
the three circles, it appears that the epicenter is
located at (3, 1).

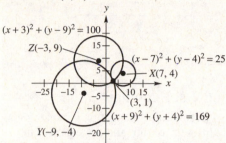

Check algebraically:
$$(x-7)^2 + (y-4)^2 = 25$$
$$(3-7)^2 + (1-4)^2 = 25$$
$$4^2 + 3^2 = 25 \Rightarrow 25 = 25$$
$$(x+9)^2 + (y+4)^2 = 169$$
$$(3+9)^2 + (1+4)^2 = 169$$
$$12^2 + 5^2 = 169 \Rightarrow 169 = 169$$
$$(x+3)^2 + (y-9)^2 = 100$$
$$(3+3)^2 + (1-9)^2 = 100$$
$$6^2 + (-8)^2 = 100 \Rightarrow 100 = 100$$

(3, 1) satisfies all three equations, so the
epicenter is at (3, 1).

**39.** From the graph of the three circles, it appears
that the epicenter is located at (−2, −2).

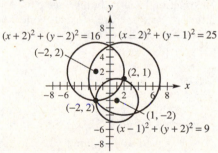

Check algebraically:
$$(x-2)^2 + (y-1)^2 = 25$$
$$(-2-2)^2 + (-2-1)^2 = 25$$
$$(-4)^2 + (-3)^2 = 25 \Rightarrow 25 = 25$$
$$(x+2)^2 + (y-2)^2 = 16$$
$$(-2+2)^2 + (-2-2)^2 = 16$$
$$0^2 + (-4)^2 = 16 \Rightarrow 16 = 16$$

$$(x-1)^2 + (y-2)^2 = 9$$
$$(-2-1)^2 + (-2-2)^2 = 9$$
$$(-3)^2 + 0^2 = 9 \Rightarrow 9 = 9$$

(−2, −2) satisfies all three equations, so the
epicenter is at (−2, −2).

**41.** The radius of this circle is the distance from
the center $C(3, 2)$ to the $x$-axis. This distance
is 2, so $r = 2$.
$$(x-3)^2 + (y-2)^2 = 2^2 \Rightarrow$$
$$(x-3)^2 + (y-2)^2 = 4$$

**43.** Label the points $P(x, y)$ and $Q(1, 3)$.
If $d(P, Q) = 4$, $\sqrt{(1-x)^2 + (3-y)^2} = 4 \Rightarrow$
$(1-x)^2 + (3-y)^2 = 16$.
If $x = y$, then we can either substitute $x$ for $y$ or
$y$ for $x$. Substituting $x$ for $y$ we solve the
following:
$$(1-x)^2 + (3-x)^2 = 16$$
$$1 - 2x + x^2 + 9 - 6x + x^2 = 16$$
$$2x^2 - 8x + 10 = 16$$
$$2x^2 - 8x - 6 = 0$$
$$x^2 - 4x - 3 = 0$$
To solve this equation, we can use the
quadratic formula with $a = 1$, $b = -4$, and
$c = -3$.
$$x = \frac{-(-4) \pm \sqrt{(-4)^2 - 4(1)(-3)}}{2(1)}$$
$$= \frac{4 \pm \sqrt{16 + 12}}{2} = \frac{4 \pm \sqrt{28}}{2}$$
$$= \frac{4 \pm 2\sqrt{7}}{2} = 2 \pm \sqrt{7}$$
Since $x = y$, the points are
$\left(2 + \sqrt{7}, 2 + \sqrt{7}\right)$ and $\left(2 - \sqrt{7}, 2 - \sqrt{7}\right)$.

**45.** Let $P(x, y)$ be a point whose distance from
$A(1, 0)$ is $\sqrt{10}$ and whose distance from
$B(5, 4)$ is $\sqrt{10}$. $d(P, A) = \sqrt{10}$, so
$$\sqrt{(1-x)^2 + (0-y)^2} = \sqrt{10} \Rightarrow$$
$(1-x)^2 + y^2 = 10$. $d(P, B) = \sqrt{10}$, so
$$\sqrt{(5-x)^2 + (4-y)^2} = \sqrt{10} \Rightarrow$$
$(5-x)^2 + (4-y)^2 = 10$.

*(continued on next page)*

*(continued from page 91)*

Thus,
$$(1-x)^2 + y^2 = (5-x)^2 + (4-y)^2$$
$$1 - 2x + x^2 + y^2 =$$
$$\qquad 25 - 10x + x^2 + 16 - 8y + y^2$$
$$1 - 2x = 41 - 10x - 8y$$
$$8y = 40 - 8x$$
$$y = 5 - x$$

Substitute $5 - x$ for $y$ in the equation $(1-x)^2 + y^2 = 10$ and solve for $x$.

$$(1-x)^2 + (5-x)^2 = 10 \Rightarrow$$
$$1 - 2x + x^2 + 25 - 10x + x^2 = 10$$
$$2x^2 - 12x + 26 = 10 \Rightarrow 2x^2 - 12x + 16 = 0$$
$$x^2 - 6x + 8 = 0 \Rightarrow (x-2)(x-4) = 0 \Rightarrow$$
$$x - 2 = 0 \quad \text{or} \quad x - 4 = 0$$
$$x = 2 \quad \text{or} \qquad x = 4$$

To find the corresponding values of $y$ use the equation $y = 5 - x$. If $x = 2$, then $y = 5 - 2 = 3$. If $x = 4$, then $y = 5 - 4 = 1$. The points satisfying the conditions are $(2, 3)$ and $(4, 1)$.

**47.** Label the points $A(3, y)$ and $B(-2, 9)$.
If $d(A, B) = 12$, then

$$\sqrt{(-2-3)^2 + (9-y)^2} = 12$$
$$\sqrt{(-5)^2 + (9-y)^2} = 12$$
$$(-5)^2 + (9-y)^2 = 12^2$$
$$25 + 81 - 18y + y^2 = 144$$
$$y^2 - 18y - 38 = 0$$

Solve this equation by using the quadratic formula with $a = 1$, $b = -18$, and $c = -38$:

$$y = \frac{-(-18) \pm \sqrt{(-18)^2 - 4(1)(-38)}}{2(1)}$$
$$= \frac{18 \pm \sqrt{324 + 152}}{2(1)} = \frac{18 \pm \sqrt{476}}{2}$$
$$= \frac{18 \pm \sqrt{4(119)}}{2} = \frac{18 \pm 2\sqrt{119}}{2} = 9 \pm \sqrt{119}$$

The values of $y$ are $9 + \sqrt{119}$ and $9 - \sqrt{119}$.

**49.** Let $P(x, y)$ be the point on the circle whose distance from the origin is the shortest. Complete the square on $x$ and $y$ separately to write the equation in center-radius form:

$$x^2 - 16x + y^2 - 14y + 88 = 0$$
$$x^2 - 16x + 64 + y^2 - 14y + 49 =$$
$$\qquad\qquad\qquad -88 + 64 + 49$$
$$(x-8)^2 + (y-7)^2 = 25$$

So, the center is $(8, 7)$ and the radius is 5.

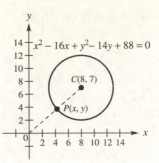

$d(P, O) = \sqrt{8^2 + 7^2} = \sqrt{113}$. Since the length

## Section 2.3: Functions

**1.** The relation is a function because for each different $x$-value there is exactly one $y$-value. This correspondence can be shown as follows.

$$\{5, 3, 4, 7\} \ x\text{-values}$$
$$\downarrow \ \downarrow \ \downarrow \ \downarrow$$
$$\{1, 2, 9, 8\} \ y\text{-values}$$

**3.** Two ordered pairs, namely $(2, 4)$ and $(2, 6)$, have the same $x$-value paired with different $y$-values, so the relation is not a function.

**5.** The relation is a function because for each different $x$-value there is exactly one $y$-value. This correspondence can be shown as follows.

$$\{-3, 4, -2\} \ x\text{-values}$$
$$\{1, 7\} \ y\text{-values}$$

**7.** The relation is a function because for each different $x$-value there is exactly one $y$-value. This correspondence can be shown as follows.

$$\{3, 7, 10\} \ x\text{-values}$$
$$\{-4\} \ y\text{-values}$$

**9.** Two sets of ordered pairs, namely $(1, 1)$ and $(1, -1)$ as well as $(2, 4)$ and $(2, -4)$, have the same $x$-value paired with different $y$-values, so the relation is not a function.
domain: $\{0, 1, 2\}$; range: $\{-4, -1, 0, 1, 4\}$

**11.** The relation is a function because for each different $x$-value there is exactly one $y$-value.
domain: $\{2, 3, 5, 11, 17\}$; range: $\{1, 7, 20\}$

**13.** The relation is a function because for each different $x$-value there is exactly one $y$-value. This correspondence can be shown as follows.

$\{0, -1, -2\}$  $x$-values

$\{0,\ \ 1,\ \ 2\}$  $y$-values

Domain: $\{0, -1, -2\}$; range: $\{0, 1, 2\}$

**15.** The relation is a function because for each different year, there is exactly one number for visitors to the Grand Canyon.
domain: $\{2001, 2002, 2003, 2004\}$
range: $\{4,400,823,\ 4,339,139,\ 4,464,400,\ 4,672,911\}$

**17.** This graph represents a function. If you pass a vertical line through the graph, one $x$-value corresponds to only one $y$-value.
domain: $(-\infty, \infty)$; range: $(-\infty, \infty)$

**19.** This graph does not represent a function. If you pass a vertical line through the graph, there are places where one value of $x$ corresponds to two values of $y$.
domain: $[3, \infty)$; range: $(-\infty, \infty)$

**21.** This graph does not represent a function. If you pass a vertical line through the graph, there are places where one value of $x$ corresponds to two values of $y$.
domain: $[-4, 4]$; range: $[-3, 3]$

**23.** $y = x^2$ represents a function since $y$ is always found by squaring $x$. Thus, each value of $x$ corresponds to just one value of $y$. $x$ can be any real number. Since the square of any real number is not negative, the range would be zero or greater.

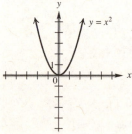

domain: $(-\infty, \infty)$; range: $[0, \infty)$

**25.** The ordered pairs $(1, 1)$ and $(1, -1)$ both satisfy $x = y^6$. This equation does not represent a function. Because $x$ is equal to the sixth power of $y$, the values of $x$ are nonnegative. Any real number can be raised to the sixth power, so the range of the relation is all real numbers.

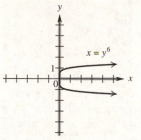

domain: $[0, \infty)$ range: $(-\infty, \infty)$

**27.** $y = 2x - 5$ represents a function since $y$ is found by multiplying $x$ by 2 and subtracting 5. Each value of $x$ corresponds to just one value of $y$. $x$ can be any real number, so the domain is all real numbers. Since $y$ is twice $x$, less 5, $y$ also may be any real number, and so the range is also all real numbers.

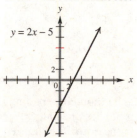

domain: $(-\infty, \infty)$; range: $(-\infty, \infty)$

**29.** By definition, $y$ is a function of $x$ if every value of $x$ leads to exactly one value of $y$. Substituting a particular value of $x$, say 1, into $x + y < 3$, corresponds to many values of $y$.
The ordered pairs $(0, 2)$ $(1, 1)$ $(1, 0)$ $(1, -1)$ and so on, all satisfy the inequality. Note that the points on the graphed line do not satisfy the inequality and only indicate the boundary of the solution set. This does not represent a function. Any number can be used for $x$ or for $y$, so the domain and range of this relation are both all real numbers.

*(continued on next page)*

(*continued from page 93*)

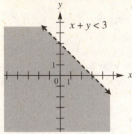

domain: $(-\infty,\infty)$; range: $(-\infty,\infty)$

31. For any choice of $x$ in the domain of $y = \sqrt{x}$, there is exactly one corresponding value of $y$, so this equation defines a function. Since the quantity under the square root cannot be negative, we have $x \geq 0$. Because the radical is nonnegative, the range is also zero or greater.

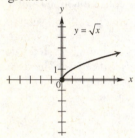

domain: $[0,\infty)$; range: $[0,\infty)$

33. Since $xy = 2$ can be rewritten as $y = \frac{2}{x}$, we can see that $y$ can be found by dividing $x$ into 2. This process produces one value of $y$ for each value of $x$ in the domain, so this equation is a function. The domain includes all real numbers except those that make the denominator equal to zero, namely $x = 0$. Values of $y$ can be negative or positive, but never zero. Therefore, the range will be all real numbers except zero.

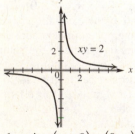

domain: $(-\infty,0) \cup (0,\infty)$;

range: $(-\infty,0) \cup (0,\infty)$

35. For any choice of $x$ in the domain of $y = \sqrt{4x+1}$ there is exactly one corresponding value of $y$, so this equation defines a function. Since the quantity under the square root cannot be negative, we have $4x+1 \geq 0 \Rightarrow 4x \geq -1 \Rightarrow x \geq -\frac{1}{4}$. Because the radical is nonnegative, the range is also zero or greater.

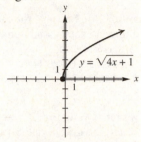

domain: $\left[-\frac{1}{4},\infty\right)$; range: $[0,\infty)$

37. Given any value in the domain of $y = \frac{2}{x-3}$, we find $y$ by subtracting 3, then dividing into 2. This process produces one value of $y$ for each value of $x$ in the domain, so this equation is a function. The domain includes all real numbers except those that make the denominator equal to zero, namely $x = 3$. Values of $y$ can be negative or positive, but never zero. Therefore, the range will be all real numbers except zero.

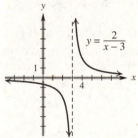

domain: $(-\infty,3) \cup (3,\infty)$;

range: $(-\infty,0) \cup (0,\infty)$

39. B

41. $f(x) = -3x + 4$
$f(0) = -3 \cdot 0 + 4 = 0 + 4 = 4$

43. $g(x) = -x^2 + 4x + 1$
$g(-2) = -(-2)^2 + 4(-2) + 1$
$= -4 + (-8) + 1 = -11$

45. $f(x) = -3x + 4$
$f\left(\frac{1}{3}\right) = -3\left(\frac{1}{3}\right) + 4 = -1 + 4 = 3$

**47.** $g(x) = -x^2 + 4x + 1$

$g\left(\frac{1}{2}\right) = -\left(\frac{1}{2}\right)^2 + 4\left(\frac{1}{2}\right) + 1$

$\qquad = -\frac{1}{4} + 2 + 1 = \frac{11}{4}$

**49.** $f(x) = -3x + 4$

$f(p) = -3p + 4$

**51.** $f(x) = -3x + 4$

$f(-x) = -3(-x) + 4 = 3x + 4$

**53.** $f(x) = -3x + 4$

$f(x+2) = -3(x+2) + 4$

$\qquad = -3x - 6 + 4 = -3x - 2$

**55.** $f(x) = -3x + 4$

$f(2m-3) = -3(2m-3) + 4$

$\qquad = -6m + 9 + 4 = -6m + 13$

**57.** (a) $f(2) = 2$ (b) $f(-1) = 3$

**59.** (a) $f(2) = 15$ (b) $f(-1) = 10$

**61.** (a) $f(2) = 3$ (b) $f(-1) = -3$

**63.** (a) $x + 3y = 12$

$\qquad 3y = -x + 12$

$\qquad y = \dfrac{-x + 12}{3}$

$\qquad y = -\frac{1}{3}x + 4 \Rightarrow f(x) = -\frac{1}{3}x + 4$

(b) $f(3) = -\frac{1}{3}(3) + 4 = -1 + 4 = 3$

**65.** (a) $y + 2x^2 = 3 - x$

$\qquad y = -2x^2 - x + 3$

$\qquad f(x) = -2x^2 - x + 3$

(b) $f(3) = -2(3)^2 - 3 + 3$

$\qquad = -2 \cdot 9 - 3 + 3 = -18$

**67.** (a) $4x - 3y = 8$

$\qquad 4x = 3y + 8$

$\qquad 4x - 8 = 3y$

$\qquad \dfrac{4x - 8}{3} = y$

$\qquad y = \frac{4}{3}x - \frac{8}{3} \Rightarrow f(x) = \frac{4}{3}x - \frac{8}{3}$

(b) $f(3) = \frac{4}{3}(3) - \frac{8}{3} = \frac{12}{3} - \frac{8}{3} = \frac{4}{3}$

**69.** $f(3) = 4$

**71.** $f(3)$ is the $y$-component of the coordinate, which is $-4$.

**73.** (a) $f(-2) = 0$ (b) $f(0) = 4$

(c) $f(1) = 2$ (d) $f(4) = 4$

**75.** (a) $f(-2) = -3$ (b) $f(0) = -2$

(c) $f(1) = 0$ (d) $f(4) = 2$

**77.** (a) $[4, \infty)$ (b) $(-\infty, -1]$

(c) $[-1, 4]$

**79.** (a) $(-\infty, 4]$ (b) $[4, \infty)$

(c) none

**81.** (a) none (b) $(-\infty, -2]; [3, \infty)$

(c) $(-2, 3)$

**83.** (a) Yes, it is the graph of a function.

(b) $[0, 24]$

(c) When $t = 8$, $y = 1200$ from the graph. At 8 A.M., approximately 1200 megawatts is being used.

(d) The most electricity was used at 17 hr or 5 P.M. The least electricity was used at 4 A.M.

(e) $f(12) = 2000$; At 12 noon, electricity use is 2000 megawatts.

(f) increasing from 4 A.M. to 5 P.M.; decreasing from midnight to 4 A.M. and from 5 P.M. to midnight

**85.** (a) At $t = 12$ and $t = 20$, $y = 55$ from the graph. Therefore, after about 12 noon until about 8 P.M. the temperature was over 55°.

(b) At $t = 5$ and $t = 22$, $y = 40$ from the graph. Therefore, until about 6 A.M. and after 10 P.M. the temperature was below 40°.

(c) The temperature at noon in Bratenahl, Ohio was 55°. Since the temperature in Greenville is 7° higher, we are looking for the time at which Bratenahl, Ohio was 55° − 7° or 48°. This occurred at approximately 10 A.M and 8:30 P.M.

## Section 2.4: Linear Functions

1. B; $f(x) = 3x + 6$ is a linear function with $y$-intercept 6.

3. C; $f(x) = -8$ is a constant function.

5. A; $f(x) = 5x$ is a linear function whose graph passes through the origin, $(0, 0)$.
   $f(0) = 2(0) = 0$.

7. $f(x) = x - 4$; Use the intercepts.
   $f(0) = 0 - 4 = -4$: $y$-intercept
   $0 = x - 4 \Rightarrow x = 4$: $x$-intercept
   Graph the line through $(0, -4)$ and $(4, 0)$.

   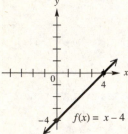

   The domain and range are both $(-\infty, \infty)$.

9. $f(x) = \frac{1}{2}x - 6$; Use the intercepts.
   $f(0) = \frac{1}{2}(0) - 6 = -6$: $y$-intercept
   $0 = \frac{1}{2}x - 6 \Rightarrow 6 = \frac{1}{2}x \Rightarrow x = 12$: $x$-intercept
   Graph the line through $(0, -6)$ and $(12, 0)$.

   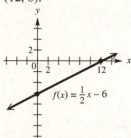

   The domain and range are both $(-\infty, \infty)$.

11. $-4x + 3y = 9$; Use the intercepts.
    $-4(0) + 3y = 9 \Rightarrow 3y = 9 \Rightarrow$
    $y = 3$: $y$-intercept
    $-4x + 3(0) = 9 \Rightarrow -4x = 9 \Rightarrow$
    $x = -\frac{9}{4}$: $x$-intercept

    Graph the line through $(0, 3)$ and $\left(-\frac{9}{4}, 0\right)$.

    The domain and range are both $(-\infty, \infty)$.

13. $3y - 4x = 0$; Use the intercepts.
    $3y - 4(0) = 0 \Rightarrow 3y = 0 \Rightarrow y = 0$: $y$-intercept
    $3(0) - 4x = 0 \Rightarrow -4x = 0 \Rightarrow x = 0$: $x$-intercept
    The graph has just one intercept. Choose an additional value, say 3, for $x$.
    $3y - 4(3) = 0 \Rightarrow 3y - 12 = 0$
    $\qquad\qquad 3y = 12 \Rightarrow y = 4$
    Graph the line through $(0, 0)$ and $(3, 4)$:

    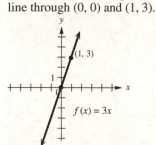

    The domain and range are both $(-\infty, \infty)$.

15. $f(x) = 3x$
    The $x$-intercept and the $y$-intercept are both zero. This gives us only one point, $(0, 0)$. If $x = 1$, $y = 3(1) = 3$. Another point is $(1, 3)$. Graph the line through $(0, 0)$ and $(1, 3)$.

    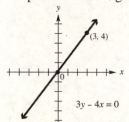

    The domain and range are both $(-\infty, \infty)$.

17. $f(x) = -4$ is a constant function.
    The graph of $f(x) = -4$ is a horizontal line with a $y$-intercept of $-4$.

    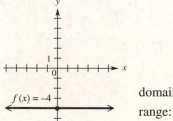

    domain: $(-\infty, \infty)$;
    range: $\{-4\}$

**19.** $x = 3$ is a vertical line, intersecting the
$x$-axis at $(3, 0)$.

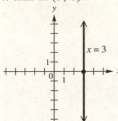

domain: $\{3\}$; range: $(-\infty, \infty)$

**21.** $2x + 4 = 0 \Rightarrow 2x = -4 \Rightarrow x = -2$ is a vertical
line intersecting the $x$-axis at $(-2, 0)$.

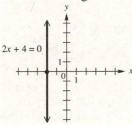

domain: $\{-2\}$; range: $(-\infty, \infty)$

**23.** $-x + 5 = 0 \Rightarrow x = 5$ is a vertical line
intersecting the $x$-axis at $(5, 0)$.

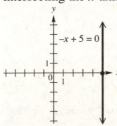

domain: $\{5\}$; range: $(-\infty, \infty)$

**25.** $y = 5$ is a horizontal line with $y$-intercept 5.
Choice A resembles this.

**27.** $x = 5$ is a vertical line with $x$-intercept 5.
Choice D resembles this.

**29.** $y = 3x + 4$; Use $Y_1 = 3X + 4$.

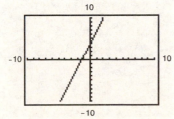

**31.** $3x + 4y = 6$; Solve for $y$.
$$3x + 4y = 6$$
$$4y = -3x + 6$$
$$y = -\tfrac{3}{4}x + \tfrac{3}{2}$$
Use $Y_1 = (-3/4)X + (3/2)$
or $Y_1 = -3/4X + 3/2$.

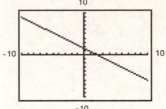

**33.** The rise is 2.5 feet while the run is 10 feet so
the slope is $\frac{2.5}{10} = .25 = 25\% = \frac{1}{4}$. So A = 0.25,
$C = \frac{2.5}{10}$, D = 25%, and $E = \frac{1}{4}$ are all
expressions of the slope.

**35.** Through $(2, -1)$ and $(-3, -3)$
Let $x_1 = 2$, $y_1 = -1$, $x_2 = -3$, and $y_2 = -3$.
Then rise $= \Delta y = -3 - (-1) = -2$ and
run $= \Delta x = -3 - 2 = -5$.
The slope is $m = \dfrac{\text{rise}}{\text{run}} = \dfrac{\Delta y}{\Delta x} = \dfrac{-2}{-5} = \dfrac{2}{5}$.

**37.** Through $(5, 9)$ and $(-2, 9)$
$$m = \frac{\Delta y}{\Delta x} = \frac{y_2 - y_1}{x_2 - x_1} = \frac{9 - 9}{-2 - 5} = \frac{0}{-7} = 0$$

**39.** Through $(5, 1)$ and $(4, 1)$
This is a horizontal line. The slope of every
horizontal line is zero, so $m = 0$.

**41.** Vertical, through $(4, -7)$
The slope of every vertical line is undefined;
$m$ is undefined.

**43.** Both B and C can be used to find the slope.

The form $m = \dfrac{y_2 - y_1}{x_2 - x_1}$ is the form that is
standardly used. If you rename points 1 and 2,
you will get the formula stated in choice B.
Choice D is incorrect because it shows a
change in $x$ to a change in $y$, which is not how
slope is defined. Choice A is incorrect because
the $y$-values are subtracted in one way, and the
$x$-values in the opposite way. This will result
in the opposite (additive inverse) of the actual
value of the slope of the line that passes
between the two points.

**45.**  $y = 3x + 5$

Find two ordered pairs that are solutions to the equation.

If $x = 0$, then $y = 3(0) + 5 \Rightarrow y = 5$.

If $x = -1$ then

$y = 3(-1) + 5 \Rightarrow y = -3 + 5 \Rightarrow y = 2$. Thus

two ordered pairs are $(0,5)$ and $(-1,2)$.

$$m = \frac{\text{rise}}{\text{run}} = \frac{y_2 - y_1}{x_2 - x_1} = \frac{2-5}{-1-0} = \frac{-3}{-1} = 3.$$

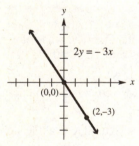

**47.**  $2y = -3x$

Find two ordered pairs that are solutions to the equation. If $x = 0$, then $2y = 0 \Rightarrow y = 0$.

If $y = -3$, then $2(-3) = -3x \Rightarrow -6 = -3x \Rightarrow$

$x = 2$. Thus two ordered pairs are $(0,0)$ and

$(2,-3)$.

$$m = \frac{\text{rise}}{\text{run}} = \frac{y_2 - y_1}{x_2 - x_1} = \frac{-3-0}{2-0} = -\frac{3}{2}.$$

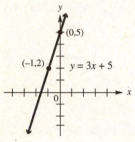

**49.**  $5x - 2y = 10$

Find two ordered pairs that are solutions to the equation. If $x = 0$, then $5(0) - 2y = 10 \Rightarrow$

$\Rightarrow y = -5$. If $y = 0$, then $5x - 2(0) = 10 \Rightarrow$

$5x = 10 \Rightarrow x = 2$.

Thus two ordered pairs are $(0,-5)$ and $(2,0)$.

$$m = \frac{\text{rise}}{\text{run}} = \frac{y_2 - y_1}{x_2 - x_1} = \frac{0-(-5)}{2-0} = \frac{5}{2}.$$

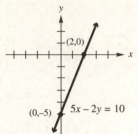

**51.**  Answers will vary.

**53.**  Through $(-1, 3)$, $m = \frac{3}{2}$

First locate the point $(-1, 3)$. Since the slope is $\frac{3}{2}$, a change of 2 units horizontally (2 units to the right) produces a change of 3 units vertically (3 units up). This gives a second point, $(1, 6)$, which can be used to complete the graph.

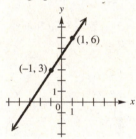

**55.**  Through $(3, -4)$, $m = -\frac{1}{3}$. First locate the point $(3, -4)$. Since the slope is $-\frac{1}{3}$, a change of 3 units horizontally (3 units to the right) produces a change of $-1$ unit vertically (1 unit down). This gives a second point, $(6, -5)$, which can be used to complete the graph.

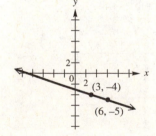

**57.** Through $\left(-\frac{1}{2}, 4\right)$,

$m = 0$.

The graph is the horizontal line through $\left(-\frac{1}{2}, 4\right)$.

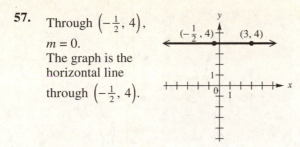

**59.** $m = \frac{1}{3}$ matches graph D because the line rises gradually as $x$ increases.

**61.** $m = 0$ matches graph A because horizontal lines have slopes of 0.

**63.** $m = 3$ matches graph E because the line rises rapidly as $x$ increases.

**65.** The average rate of change is $m = \dfrac{\Delta y}{\Delta x}$

$\dfrac{20 - 4}{0 - 4} = \dfrac{-16}{4} = -\$4$ (thousand) per year. The value of the machine is decreasing \$4000 each year during these years.

**67.** The average rate of change is $m = \dfrac{\Delta y}{\Delta x}$

$\dfrac{3 - 3}{4 - 0} = \dfrac{0}{4} = 0\%$ per year. The percent of pay raise is not changing - it is 3% each year.

**69.** For a constant function, the average rate of change is zero.

**71.** (a) Answers will vary.

(b) $m = \dfrac{12,057 - 2773}{1999 - 1950} = \dfrac{9284}{49} \approx 189.5$

This means that the average rate of change in the number of radio stations per year is an increase of about 189.5 stations.

**73.** (a) $m = \dfrac{y_2 - y_1}{x_2 - x_1} = \dfrac{21.9 - 27.6}{2004 - 1994}$

$= \dfrac{-5.7}{10} = -0.57$ million recipients per year

(b) The negative slope means the numbers of recipients *decreased* by 0.57 million each year.

**75.** $m = \dfrac{y_2 - y_1}{x_2 - x_1} = \dfrac{1.4 - 3.7}{2004 - 2000} = \dfrac{-2.3}{4}$

$= -0.575$ per year

The percent of freshman listing computer science as their probable field of study decreased an average of 0.575% per year from 2000 to 2004.

**77.** $m = \dfrac{y_2 - y_1}{x_2 - x_1} = \dfrac{19.788 - 0.315}{2006 - 1997} = \dfrac{19.473}{9}$

$\approx 2.16$ million per year. Sales of DVD players increased an average of 2.16 million each year from 1997 to 2006.

**79.** The second and third points are B(1, –3) and C(2, 0).

$m = \dfrac{0 - (-3)}{2 - 1} = \dfrac{3}{1} = 3$

**81.** The first two points are A(0, –6) and B(1, –3).

$d(A, B) = \sqrt{[-3 - (-6)]^2 + (1 - 0)^2}$

$= \sqrt{3^2 + 1^2} = \sqrt{9 + 1} = \sqrt{10}$

**83.** The first and fourth points are A(0, –6) and D(3, 3).

$d(A, D) = \sqrt{[3 - (-6)]^2 + (3 - 0)^2}$

$= \sqrt{9^2 + 3^2} = \sqrt{81 + 9}$

$= \sqrt{90} = 3\sqrt{10}$

**85.** If points A, B, and C lie on a line in that order, then the distance between A and B added to the distance between B and C is equal to the distance between A and C.

**87.** The midpoint of the segment joining E(4, 6) and F(5, 9) has coordinates $\left(\frac{4+5}{2}, \frac{6+9}{2}\right) = \left(\frac{9}{2}, \frac{15}{2}\right) = (4.5, 7.5)$. If the $x$-value 4.5 were in the table, the corresponding $y$-value would be 7.5.

**89.** (a) $C(x) = 11x + 180$

(b) $R(x) = 20x$

(c) $P(x) = R(x) - C(x)$

$= 20x - (11x + 180)$

$= 20x - 11x - 180 = 9x - 180$

(d) $C(x) = R(x)$

$11x + 180 = 20x$

$180 = 9x$

$20 = x$

20 units; produce

**91.** **(a)** $C(x) = 400x + 1650$

**(b)** $R(x) = 305x$

**(c)** $P(x) = R(x) - C(x)$
$$= 305x - (400x + 1650)$$
$$= 305x - 400x - 1650$$
$$= -95x - 1650$$

**(d)** $\quad C(x) = R(x)$
$$400x + 1650 = 305x$$
$$95x + 1650 = 0$$
$$95x = -1650$$
$$x \approx -17.37 \text{ units}$$

This result indicates a negative "break-even point," but the number of units produced must be a positive number. A calculator graph of the lines $Y_1 = 400X + 1650$ and $Y_2 = 305X$ on the same screen or solving the inequality $305x < 400x + 1650$ will show that $R(x) < C(x)$ for all positive values of $x$ (in fact whenever $x$ is greater than $-17.4$). Do not produce the product since it is impossible to make a profit.

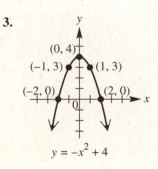

## Chapter 2 Quiz
**(Sections 2.1–2.4)**

**1.** $d(A, B) = \sqrt{(x_2 - x_1)^2 + (y_2 - y_1)^2}$
$$= \sqrt{(-8 - (-4))^2 + (-3 - 2)^2}$$
$$= \sqrt{(-4)^2 + 5^2} = \sqrt{16 + 25} = \sqrt{41}$$

**3.**

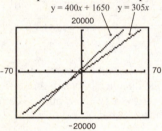

**5.** $x^2 + y^2 - 4x + 8y + 3 = 0$
Complete the square on $x$ and $y$ separately.
$$(x^2 - 4x + 4) + (y^2 + 8y + 16) = -3 + 4 + 16 \Rightarrow$$
$$(x - 2)^2 + (y + 4)^2 = 17$$
The radius is $\sqrt{17}$ and the midpoint of the circle is $(2, -4)$.

**7.** Domain: $(-\infty, \infty)$; range: $[0, \infty)$

**9.** **(a)** $m = \dfrac{11 - 5}{5 - 1} = \dfrac{6}{4} = \dfrac{3}{2}$

**(b)** $m = \dfrac{4 - 4}{-1 - (-7)} = \dfrac{0}{6} = 0$

**(c)** $m = \dfrac{-4 - 12}{6 - 6} = \dfrac{-16}{0} \Rightarrow$ the slope is undefined.

## Section 2.5: Equations of Lines; Curve Fitting

**1.** $y = \frac{1}{4}x + 2$ is graphed in D.
The slope is $\frac{1}{4}$ and the $y$-intercept is 2.

**3.** $y - (-1) = \frac{3}{2}(x - 1)$ is graphed in C. The slope is $\frac{3}{2}$ and a point on the graph is $(1, -1)$.

**5.** Through $(1, 3)$, $m = -2$.
Write the equation in point-slope form.
$$y - y_1 = m(x - x_1) \Rightarrow y - 3 = -2(x - 1)$$
Then, change to standard form.
$$y - 3 = -2x + 2 \Rightarrow 2x + y = 5$$

**7.** Through $(-5, 4)$, $m = -\frac{3}{2}$
Write the equation in point-slope form.
$$y - 4 = -\frac{3}{2}\left[x - (-5)\right]$$
Change to standard form.
$$2(y - 4) = -3(x + 5)$$
$$2y - 8 = -3x - 15$$
$$3x + 2y = -7$$

**9.** Through $(-8, 4)$, undefined slope
Since undefined slope indicates a vertical line, the equation will have the form $x = a$. The equation of the line is $x = -8$.

**11.** Through $(5, -8)$, $m = 0$
This is a horizontal line through $(5, -8)$, so the equation is $y = -8$.

**13.** Through $(-1, 3)$ and $(3, 4)$
First find $m$.
$$m = \frac{4-3}{3-(-1)} = \frac{1}{4}$$
Use either point and the point-slope form.
$$y - 4 = \tfrac{1}{4}(x - 3)$$
Change to slope-intercept form.
$$4(y - 4) = x - 3$$
$$4y - 16 = x - 3$$
$$4y = x + 13$$
$$y = \tfrac{1}{4}x + \tfrac{13}{4}$$

**15.** $x$-intercept 3, $y$-intercept $-2$
The line passes through $(3, 0)$ and $(0, -2)$. Use these points to find $m$.
$$m = \frac{-2-0}{0-3} = \frac{2}{3}$$
Using slope-intercept form we have
$$y = \tfrac{2}{3}x - 2.$$

**17.** Vertical, through $(-6, 4)$
The equation of a vertical line has an equation of the form $x = a$. Since the line passes through $(-6, 4)$, the equation is $x = -6$. (Since this slope of a vertical line is undefined, this equation cannot be written in slope-intercept form.)

**19.** Horizontal, through $(-7, 4)$
The equation of a horizontal line has an equation of the form $y = b$. Since the line passes through $(-7, 4)$, the equation is $y = 4$.

**21.** $m = 5, b = 15$
Using slope-intercept form, we have
$$y = 5x + 15.$$

**23.** $m = -\tfrac{2}{3}, b = -\tfrac{4}{5}$
Using slope-intercept form, we have
$$y = -\tfrac{2}{3}x - \tfrac{4}{5}.$$

**25.** slope 0, $y$-intercept $\tfrac{3}{2}$
These represent $m = 0$ and $b = \tfrac{3}{2}$. Using slope-intercept form we have
$$y = (0)x + \tfrac{3}{2} \Rightarrow y = \tfrac{3}{2}.$$

**27.** The line $x + 2 = 0$ has $x$-intercept $-2$. It <u>does not</u> have a $y$-intercept. The slope of this line is <u>undefined</u>.
The line $4y = 2$ has $y$-intercept $\tfrac{1}{2}$. It <u>does not</u> have an $x$-intercept. The slope of this line is <u>0</u>.

**29. (a)** The graph of $y = 2x + 3$ has a positive slope and a positive $y$-intercept. These conditions match graph B.

**(b)** The graph of $y = -2x + 3$ has a negative slope and a positive $y$-intercept. These conditions match graph D.

**(c)** The graph of $y = 2x - 3$ has a positive slope and a negative $y$-intercept. These conditions match graph A.

**(d)** The graph of $y = -2x - 3$ has a negative slope and a negative $y$-intercept. These conditions match graph C.

**31.** $y = 3x - 1$
This equation is in the slope-intercept form, $y = mx + b$.
slope: 3;
$y$-intercept: $-1$

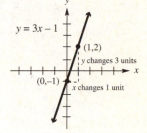

**33.** $4x - y = 7$
Solve for $y$ to write the equation in slope-intercept form.
$$-y = -4x + 7 \Rightarrow y = 4x - 7$$
slope: 4; $y$-intercept: $-7$

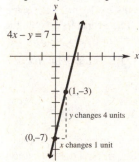

**35.** $4y = -3x$
$$y = -\tfrac{3}{4}x \text{ or}$$
$$y = -\tfrac{3}{4}x + 0$$
slope: $-\tfrac{3}{4}$;
$y$-intercept 0

**37.** $x + 2y = -4$
Solve the equation for $y$ to write the equation in slope-intercept form.
$$2y = -x - 4 \Rightarrow y = -\tfrac{1}{2}x - 2$$
slope: $-\tfrac{1}{2}$; $y$-intercept: $-2$

(*continued on next page*)

*(continued from page 101)*

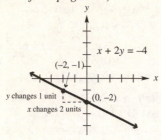

**39.** $y - \frac{3}{2}x - 1 = 0$

Solve the equation for $y$ to write the equation in slope-intercept form.

$y - \frac{3}{2}x - 1 = 0 \Rightarrow y = \frac{3}{2}x + 1$

slope: $\frac{3}{2}$; $y$-intercept: 1

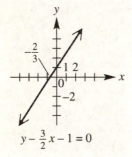

$y - \frac{3}{2}x - 1 = 0$

**41.** **(a)** The line falls 2 units each time the $x$ value increases by 1 unit. Therefore the slope is $-2$. The graph intersects the $y$-axis at the point $(0, 1)$ and intersects the $x$-axis at $\left(\frac{1}{2}, 0\right)$, so the $y$-intercept is 1 and the $x$-intercept is $\frac{1}{2}$.

**(b)** The equation defining $f$ is $y = -2x + 1$.

**43.** **(a)** The line falls 1 unit each time the $x$ value increases by 3 units. Therefore the slope is $-\frac{1}{3}$. The graph intersects the $y$-axis at the point $(0, 2)$, so the $y$-intercept is 2. The graph passes through $(3, 1)$ and will fall 1 unit when the $x$ value increases by 3, so the $x$-intercept is 6.

**(b)** The equation defining $f$ is $y = -\frac{1}{3}x + 2$.

**45.** **(a)** The line falls 200 units each time the $x$ value increases by 1 unit. Therefore the slope is $-200$. The graph intersects the $y$-axis at the point $(0, 300)$ and intersects the $x$-axis at $\left(\frac{3}{2}, 0\right)$, so the $y$-intercept is 300 and the $x$-intercept is $\frac{3}{2}$.

**(b)** The equation defining $f$ is $y = -200x + 300$.

**47.** **(a)** through $(-1, 4)$, parallel to $x + 3y = 5$
Find the slope of the line $x + 3y = 5$ by writing this equation in slope-intercept form.

$x + 3y = 5 \Rightarrow 3y = -x + 5 \Rightarrow$
$y = -\frac{1}{3}x + \frac{5}{3}$

The slope is $-\frac{1}{3}$. Since the lines are parallel, $-\frac{1}{3}$ is also the slope of the line whose equation is to be found. Substitute $m = -\frac{1}{3}$, $x_1 = -1$, and $y_1 = 4$ into the point-slope form.

$y - y_1 = m(x - x_1)$
$y - 4 = -\frac{1}{3}\left[x - (-1)\right]$
$y - 4 = -\frac{1}{3}(x + 1)$
$3y - 12 = -x - 1 \Rightarrow x + 3y = 11$

**(b)** Solve for $y$.
$3y = -x + 11 \Rightarrow y = -\frac{1}{3}x + \frac{11}{3}$

**49.** **(a)** through $(1, 6)$, perpendicular to $3x + 5y = 1$
Find the slope of the line $3x + 5y = 1$ by writing this equation in slope-intercept form.

$3x + 5y = 1 \Rightarrow 5y = -3x + 1 \Rightarrow$
$y = -\frac{3}{5}x + \frac{1}{5}$

This line has a slope of $-\frac{3}{5}$. The slope of any line perpendicular to this line is $\frac{5}{3}$, since $-\frac{3}{5}\left(\frac{5}{3}\right) = -1$. Substitute $m = \frac{5}{3}$, $x_1 = 1$, and $y_1 = 6$ into the point-slope form.

$y - 6 = \frac{5}{3}(x - 1)$
$3(y - 6) = 5(x - 1)$
$3y - 18 = 5x - 5$
$-13 = 5x - 3y$ or $5x - 3y = -13$

**(b)** Solve for $y$. $3y = 5x + 13 \Rightarrow y = \frac{5}{3}x + \frac{13}{3}$

**51.** **(a)** through $(4, 1)$, parallel to $y = -5$
Since $y = -5$ is a horizontal line, any line parallel to this line will be horizontal and have an equation of the form $y = b$. Since the line passes through $(4, 1)$, the equation is $y = 1$.

**(b)** The slope-intercept form is $y = 1$.

**53. (a)** through $(-5, 6)$, perpendicular to $x = -2$.

Since $x = -2$ is a vertical line, any line perpendicular to this line will be horizontal and have an equation of the form $y = b$. Since the line passes through $(-5, 6)$, the equation is $y = 6$.

**(b)** The slope-intercept form is $y = 6$.

**55. (a)** Find the slope of the line $3y + 2x = 6$.

$3y + 2x = 6 \Rightarrow 3y = -2x + 6 \Rightarrow$
$y = -\frac{2}{3}x + 2$

Thus, $m = -\frac{2}{3}$. A line parallel to

$3y + 2x = 6$ also has slope $-\frac{2}{3}$.

Solve for $k$ using the slope formula.
$$\frac{2 - (-1)}{k - 4} = -\frac{2}{3}$$
$$\frac{3}{k - 4} = -\frac{2}{3}$$
$$3(k - 4)\left(\frac{3}{k - 4}\right) = 3(k - 4)\left(-\frac{2}{3}\right)$$
$$9 = -2(k - 4)$$
$$9 = -2k + 8$$
$$2k = -1 \Rightarrow k = -\frac{1}{2}$$

**(b)** Find the slope of the line $2y - 5x = 1$.

$2y - 5x = 1 \Rightarrow 2y = 5x + 1 \Rightarrow$
$y = \frac{5}{2}x + \frac{1}{2}$

Thus, $m = \frac{5}{2}$. A line perpendicular to $2y$

$- 5x = 1$ will have slope $-\frac{2}{5}$, since

$\frac{5}{2}\left(-\frac{2}{5}\right) = -1$.

Solve this equation for $k$.
$$\frac{3}{k - 4} = -\frac{2}{5}$$
$$5(k - 4)\left(\frac{3}{k - 4}\right) = 5(k - 4)\left(-\frac{2}{5}\right)$$
$$15 = -2(k - 4)$$
$$15 = -2k + 8$$
$$2k = -7 \Rightarrow k = -\frac{7}{2}$$

**57.** $(1970, 43.3), (2005, 59.3)$
$$m = \frac{59.3 - 43.3}{2005 - 1970} = \frac{16}{35} \approx 0.457$$
Now use either point, say $(1970, 43.3)$, and the point-slope form to find the equation.
$y - 43.3 = 0.457(x - 1970)$
$y - 43.3 = 0.457x - 900.29$
$\quad\quad y = 0.457x - 856.99$

Let $x = 2006$
$y = 0.457(2006) - 856.99 \approx 59.8$
The percent of women in the civilian labor force is predicted to be 59.8%. This figure is very close to the actual figure.

**58.** $(1975, 46.3), (2000, 59.9)$
$$m = \frac{59.9 - 46.3}{2000 - 1975} = \frac{13.6}{25} = .544$$
Now use either point, say $(2000, 59.9)$, and the point-slope form to find the equation.
$y - 59.9 = 0.544(x - 2000)$
$y - 59.9 = 0.544x - 1088$
$\quad\quad y = 0.544x - 1028.1$

Let $x = 1996$.
$y = 0.544(1996) - 1028.1 \approx 57.7$

The percent of women in the civilian labor force is predicted to be 57.7%. This figure is reasonably close to the actual figure.

**59. (a)** $(0, 11719), (12, 22218)$
$$m = \frac{22,218 - 11,719}{12 - 0} = \frac{10,499}{12} \approx 874.9$$
From the point $(0, 11719)$, the value of $b$ is 11,719. Therefore we have
$f(x) \approx 874.9x + 11,719$

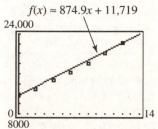

The average tuition increase is about $875 per year for the period, because this is the slope of the line.

**(b)** 2005 corresponds to $x = 11$.
$f(11) \approx 874.9(11) + 11,719 \approx \$21,343$
This is a fairly good approximation.

**(c)** From the calculator,
$f(x) \approx 877.1x + 11,322$

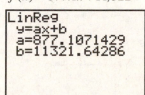

**61. (a)** The ordered pairs are $(0, 32)$ and $(100, 212)$.

The slope is $m = \dfrac{212 - 32}{100 - 0} = \dfrac{180}{100} = \dfrac{9}{5}$.

Use $(x_1, y_1) = (0, 32)$ and $m = \frac{9}{5}$ in the point-slope form.

$$y - y_1 = m(x - x_1)$$
$$y - 32 = \tfrac{9}{5}(x - 0)$$
$$y - 32 = \tfrac{9}{5}x$$
$$y = \tfrac{9}{5}x + 32 \Rightarrow F = \tfrac{9}{5}C + 32$$

**(b)** $F = \tfrac{9}{5}C + 32$
$$5F = 9(C + 32)$$
$$5F = 9C + 160 \Rightarrow 9C = 5F - 160 \Rightarrow$$
$$9C = 5(F - 32) \Rightarrow C = \tfrac{5}{9}(F - 32)$$

**(c)** $F = C \Rightarrow F = \tfrac{5}{9}(F - 32) \Rightarrow$
$$9F = 5(F - 32) \Rightarrow 9F = 5F - 160 \Rightarrow$$
$$4F = -160 \Rightarrow F = -40$$
$F = C$ when $F$ is $-40°$.

**63. (a)** Since we are wanting to find $C$ as a function of $I$, use the points $(8795, 6739)$ and $(10904, 8746)$, where the first component represents the independent variable, $I$. First find the slope of the line.

$$m = \frac{8746 - 6739}{10{,}904 - 8795} = \frac{2007}{2109} \approx 0.952$$

Now use either point, say $(8795, 6739)$, and the point-slope form to find the equation.

$$y - 6739 = 0.952(x - 8795)$$
$$y - 6739 \approx 0.952x - 8373$$
$$y \approx 0.952x - 1634$$
$$\text{or } C = 0.952I - 1634$$

**(b)** Since the slope is 0.952, the marginal propensity to consume is 0.952.

**65.** Write the equation as an equivalent equation with 0 on one side: $7x - 2x + 4 - 5 = 3x + 1 \Rightarrow$
$7x - 2x + 4 - 5 - 3x - 1 = 0$. Now graph
$Y = 7X - 2X + 4 - 5 - 3X - 1 = 0$ to find the $x$-intercept:
$Y = 7X - 2X + 4 - 5 - 3X - 1$

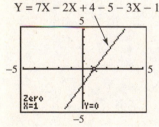

Solution set: $\{1\}$

**67.** Write the equation as an equivalent equation with 0 on one side:
$$4x - 3(4 - 2x) = 2(x - 3) + 6x + 2 \Rightarrow$$
$$4x - 3(4 - 2x) - 2(x - 3) - 6x - 2 = 0.$$
Now graph
$Y = 4X - 3(4 - 2X) - 2(X - 3) - 6X - 2$ to find the $x$-intercept:
$Y = 4X - 3(4 - 2X) - 2(X - 3) - 6X - 2$

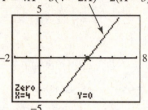

Solution set: $\{4\}$

**69. (a)** $-2(x - 5) = -x - 2$
$$-2x + 10 = -x - 2$$
$$10 = x - 2$$
$$12 = x$$
Solution set: $\{12\}$

**(b)** Answers will vary. The largest value of $x$ that is displayed in the standard viewing window is 10. As long as 12 is either a minimum or a maximum, or between the minimum and maximum, then the solution will be seen.

**71.** $d(O, P) = \sqrt{(x_1 - 0)^2 + (m_1 x_1 - 0)^2}$
$$= \sqrt{x_1^2 + m_1^2 x_1^2}$$

**73.** $d(P, Q) = \sqrt{(x_2 - x_1)^2 + (m_2 x_2 - m_1 x_1)^2}$

**75.** $-2m_1 m_2 x_1 x_2 - 2x_1 x_2 = 0$
$$-2x_1 x_2 (m_1 m_2 + 1) = 0$$

**77.** If two nonvertical lines are perpendicular, then the product of the slopes of these lines is $-1$.

**79.** Label the points as follows:
$A(-1, 5)$, $B(2, -4)$, and $C(4, -10)$.

For $A$ and $B$: $m = \dfrac{-4 - 5}{2 - (-1)} = \dfrac{-9}{3} = -3$

For $B$ and $C$, $m = \dfrac{-10 - (-4)}{4 - 2} = \dfrac{-6}{2} = -3$

For $A$ and $C$, $m = \dfrac{-10 - 5}{4 - (-1)} = \dfrac{-15}{5} = -3$

Since all three slopes are the same, the points are collinear.

**81.** $A(-1, 4)$, $B(-2, -1)$, $C(1, 14)$

For $A$ and $B$, $m = \dfrac{-1-4}{-2-(-1)} = \dfrac{-5}{-1} = 5$

For $B$ and $C$, $m = \dfrac{14-(-1)}{1-(-2)} = \dfrac{15}{3} = 5$

For $A$ and $C$, $m = \dfrac{14-4}{1-(-1)} = \dfrac{10}{2} = 5$

Since all three slopes are the same, the points are collinear.

**83.** $A(-1, -3)$, $B(-5, 12)$, $C(1, -11)$

For $A$ and $B$, $m = \dfrac{12-(-3)}{-5-(-1)} = -\dfrac{15}{4}$

For $B$ and $C$, $m = \dfrac{-11-12}{1-(-5)} = -\dfrac{23}{6}$

For $A$ and $C$, $m = \dfrac{-11-(-3)}{1-(-1)} = -\dfrac{8}{2} = -4$

Since all three slopes are not the same, the points are not collinear.

# Summary Exercises on Graphs, Functions, and Equations

**1.** $P(3, 5)$, $Q(2, -3)$

   **(a)** $d(P, Q) = \sqrt{(2-3)^2 + (-3-5)^2}$
   $= \sqrt{(-1)^2 + (-8)^2}$
   $= \sqrt{1+64} = \sqrt{65}$

   **(b)** The midpoint $M$ of the segment joining points $P$ and $Q$ has coordinates
   $\left(\dfrac{3+2}{2}, \dfrac{5+(-3)}{2}\right) = \left(\dfrac{5}{2}, \dfrac{2}{2}\right) = \left(\dfrac{5}{2}, 1\right)$.

   **(c)** First find $m$: $m = \dfrac{-3-5}{2-3} = \dfrac{-8}{-1} = 8$
   Use either point and the point-slope form.
   $y - 5 = 8(x - 3)$
   Change to slope-intercept form.
   $y - 5 = 8x - 24 \Rightarrow y = 8x - 19$

**3.** $P(-2, 2)$, $Q(3, 2)$

   **(a)** $d(P, Q) = \sqrt{[3-(-2)]^2 + (2-2)^2}$
   $= \sqrt{5^2 + 0^2} = \sqrt{25+0} = \sqrt{25} = 5$

   **(b)** The midpoint $M$ of the segment joining points $P$ and $Q$ has coordinates
   $\left(\dfrac{-2+3}{2}, \dfrac{2+2}{2}\right) = \left(\dfrac{1}{2}, \dfrac{4}{2}\right) = \left(\dfrac{1}{2}, 2\right)$.

   **(c)** First find $m$: $m = \dfrac{2-2}{3-(-2)} = \dfrac{0}{5} = 0$
   All lines that have a slope of 0 are horizontal lines. The equation of a horizontal line has an equation of the form $y = b$. Since the line passes through $(3, 2)$, the equation is $y = 2$.

**5.** $P(5, -1)$, $Q(5, 1)$

   **(a)** $d(P, Q) = \sqrt{(5-5)^2 + [1-(-1)]^2}$
   $= \sqrt{0^2 + 2^2} = \sqrt{0+4} = \sqrt{4} = 2$

   **(b)** The midpoint $M$ of the segment joining points $P$ and $Q$ has coordinates
   $\left(\dfrac{5+5}{2}, \dfrac{-1+1}{2}\right) = \left(\dfrac{10}{2}, \dfrac{0}{2}\right) = (5, 0)$.

   **(c)** First find $m$.
   $m = \dfrac{1-(-1)}{5-5} = \dfrac{2}{0} = $ undefined
   All lines that have an undefined slope are vertical lines. The equation of a vertical line has an equation of the form $x = a$. Since the line passes through $(5, 1)$, the equation is $x = 5$. (Since this slope of a vertical line is undefined, this equation cannot be written in slope-intercept form.)

**7.** $P\left(2\sqrt{3}, 3\sqrt{5}\right)$, $Q\left(6\sqrt{3}, 3\sqrt{5}\right)$

   **(a)** $d(P, Q) = \sqrt{\left(6\sqrt{3} - 2\sqrt{3}\right)^2 + \left(3\sqrt{5} - 3\sqrt{5}\right)^2}$
   $= \sqrt{\left(4\sqrt{3}\right)^2 + 0^2} = \sqrt{48} = 4\sqrt{3}$

   **(b)** The midpoint $M$ of the segment joining points $P$ and $Q$ has coordinates
   $\left(\dfrac{2\sqrt{3} + 6\sqrt{3}}{2}, \dfrac{3\sqrt{5} + 3\sqrt{5}}{2}\right)$
   $= \left(\dfrac{8\sqrt{3}}{2}, \dfrac{6\sqrt{5}}{2}\right) = \left(4\sqrt{3}, 3\sqrt{5}\right)$.

   **(c)** First find $m$: $m = \dfrac{3\sqrt{5} - 3\sqrt{5}}{6\sqrt{3} - 2\sqrt{3}} = \dfrac{0}{4\sqrt{3}} = 0$
   All lines that have a slope of 0 are horizontal lines. The equation of a horizontal line has an equation of the form $y = b$. Since the line passes through $\left(2\sqrt{3}, 3\sqrt{5}\right)$, the equation is $y = 3\sqrt{5}$.

**9.** Through $(-2, 1)$ and $(4, -1)$

First find $m$: $m = \dfrac{-1-1}{4-(-2)} = \dfrac{-2}{6} = -\dfrac{1}{3}$

Use either point and the point-slope form.

$y - (-1) = -\dfrac{1}{3}(x-4)$

Change to slope-intercept form.

$3(y+1) = -(x-4) \Rightarrow 3y+3 = -x+4 \Rightarrow$

$3y = -x+1 \Rightarrow y = -\dfrac{1}{3}x + \dfrac{1}{3}$

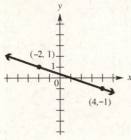

**11.** the circle with center $(2, -1)$ and radius 3

$(x-2)^2 + [y-(-1)]^2 = 3^2$

$(x-2)^2 + (y+1)^2 = 9$

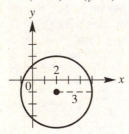

**13.** the line through $(3, -5)$ with slope $-\dfrac{5}{6}$

Write the equation in point-slope form.

$y - (-5) = -\dfrac{5}{6}(x-3)$

Change to standard form.

$6(y+5) = -5(x-3) \Rightarrow 6y+30 = -5x+15$

$6y = -5x-15 \Rightarrow y = -\dfrac{5}{6}x - \dfrac{15}{6}$

$y = -\dfrac{5}{6}x - \dfrac{5}{2}$

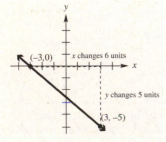

**15.** a line through $(-3, 2)$ and parallel to the line $2x+3y = 6$

First, find the slope of the line $2x+3y = 6$ by writing this equation in slope-intercept form.

$2x+3y = 6 \Rightarrow 3y = -2x+6 \Rightarrow y = -\dfrac{2}{3}x + 2$

The slope is $-\dfrac{2}{3}$. Since the lines are parallel, $-\dfrac{2}{3}$ is also the slope of the line whose equation is to be found. Substitute $m = -\dfrac{2}{3}$, $x_1 = -3$, and $y_1 = 2$ into the point-slope form.

$y - y_1 = m(x - x_1) \Rightarrow y - 2 = -\dfrac{2}{3}[x - (-3)] \Rightarrow$

$3(y-2) = -2(x+3) \Rightarrow 3y-6 = -2x-6 \Rightarrow$

$3y = -2x \Rightarrow y = -\dfrac{2}{3}x$

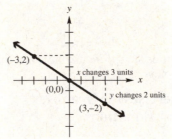

**17.** $x^2 - 4x + y^2 + 2y = 4$

Complete the square on $x$ and $y$ separately.

$\left(x^2 - 4x\right) + \left(y^2 + 2y\right) = 4$

$\left(x^2 - 4x + 4\right) + \left(y^2 + 2y + 1\right) = 4+4+1$

$(x-2)^2 + (y+1)^2 = 9$

Yes, it is a circle. The circle has its center at $(2, -1)$ and radius 3.

**19.** $x^2 - 12x + y^2 + 20 = 0$

Complete the square on $x$ and $y$ separately.

$\left(x^2 - 12x\right) + y^2 = -20$

$\left(x^2 - 12x + 36\right) + y^2 = -20 + 36$

$(x-6)^2 + y^2 = 16$

Yes, it is a circle. The circle has its center at $(6, 0)$ and radius 4.

**21.** $x^2 - 2x + y^2 + 10 = 0$

Complete the square on $x$ and $y$ separately.

$\left(x^2 - 2x\right) + y^2 = -10$

$\left(x^2 - 2x + 1\right) + y^2 = -10 + 1$

$(x-1)^2 + y^2 = -9$

No, it is not a circle.

**23.** The equation of the circle is
$(x-4)^2+(y-5)^2=4^2$. Let $y=2$ and solve
for $x$: $(x-4)^2+(2-5)^2=4^2 \Rightarrow$
$(x-4)^2+(-3)^2=4^2 \Rightarrow (x-4)^2=7 \Rightarrow$
$x-4=\pm\sqrt{7} \Rightarrow x=4\pm\sqrt{7}$
The points of intersection are $\left(4+\sqrt{7},2\right)$ and
$\left(4-\sqrt{7},2\right)$

**25.** **(a)** The equation can be rewritten as
$-4y=-x-6 \Rightarrow y=\frac{1}{4}x+\frac{6}{4} \Rightarrow y=\frac{1}{4}x+\frac{3}{2}$.
$x$ can be any real number, so the domain
is all real numbers and the range is also
all real numbers.
domain: $(-\infty,\infty)$; range: $(-\infty,\infty)$

**(b)** Each value of $x$ corresponds to just one
value of $y$. $x-4y=-6$ represents a
function.
$y=\frac{1}{4}x+\frac{3}{2} \Rightarrow f\left(x\right)=\frac{1}{4}x+\frac{3}{2}$
$f\left(-2\right)=\frac{1}{4}\left(-2\right)+\frac{3}{2}=-\frac{1}{2}+\frac{3}{2}=\frac{2}{2}=1$

**27.** **(a)** $\left(x+2\right)^2+y^2=25$ is a circle centered at
$\left(-2,0\right)$ with a radius of 5. The domain
will start 5 units to the left of –2 and end
5 units to the right of –2. The domain will
be $\left[-2-5,-2+5\right]=\left[-7,3\right]$. The range
will start 5 units below 0 and end 5 units
above 0. The range will be
$\left[0-5,0+5\right]=\left[-5,5\right]$.

**(b)** Since $\left(-2,5\right)$ and $\left(-2,-5\right)$ both satisfy
the relation, $\left(x+2\right)^2+y^2=25$ does not
represent a function.

## Section 2.6: Graphs of Basic Functions

**1.** The equation $y=x^2$ matches graph E. The
domain is $\left(-\infty,\infty\right)$.

**3.** The equation $y=x^3$ matches graph A. The
range is $\left(-\infty,\infty\right)$.

**5.** Graph F is the graph of the identity function.
Its equation is $y=x$.

**7.** The equation $y=\sqrt[3]{x}$ matches graph H. No,
there is no interval over which the function is
decreasing.

**9.** The graph in B is discontinuous at many
points. Assuming the graph continues, the
range would be $\left\{...,-3,-2,-1,0,1,2,3,...\right\}$.

**11.** The function is continuous over the entire
domain of real numbers $(-\infty,\infty)$.

**13.** The function is continuous over the interval
$\left[0,\infty\right)$.

**15.** The function has a point of discontinuity at
$x=1$. It is continuous over the interval
$(-\infty,1)$ and the interval $\left[1,\infty\right)$.

**17.** $f(x)=\begin{cases}2x & \text{if } x\le -1 \\ x-1 & \text{if } x>-1\end{cases}$

**(a)** $f(-5)=2(-5)=-10$

**(b)** $f(-1)=2(-1)=-2$

**(c)** $f(0)=0-1=-1$

**(d)** $f(3)=3-1=2$

**19.** $f(x)=\begin{cases}2+x & \text{if } x<-4 \\ -x & \text{if } -4\le x\le 2 \\ 3x & \text{if } x>2\end{cases}$

**(a)** $f(-5)=2+(-5)=-3$

**(b)** $f(-1)=-(-1)=1$

**(c)** $f(0)=-0=0$

**(d)** $f(3)=3\cdot 3=9$

**21.** $f(x)=\begin{cases}x-1 & \text{if } x\le 3 \\ 2 & \text{if } x>3\end{cases}$

Draw the graph of $y=x-1$ to the left of $x=3$,
including the endpoint at $x=3$. Draw the
graph of $y=2$ to the right of $x=3$, and note
that the endpoint at $x=3$ coincides with the
endpoint of the other ray.

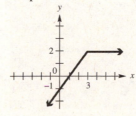

**23.** $f(x)=\begin{cases}4-x & \text{if } x<2 \\ 1+2x & \text{if } x\ge 2\end{cases}$

Draw the graph of $y=4-x$ to the left of
$x=2$, but do not include the endpoint. Draw
the graph of $y=1+2x$ to the right of $x=2$,
including the endpoint.

*(continued on next page)*

*(continued from page 107)*

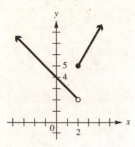

**25.** $f(x) = \begin{cases} 5x - 4 & \text{if } x \le 1 \\ x & \text{if } x > 1 \end{cases}$

Graph the line $y = 5x - 4$ to the left of $x = 1$, including the endpoint. Draw $y = x$ to the right of $x = 1$; note that the endpoint at $x = 1$ coincides with the endpoint of the other ray.

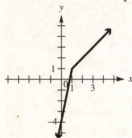

**27.** $f(x) = \begin{cases} 2 + x & \text{if } x < -4 \\ -x & \text{if } -4 \le x \le 5 \\ 3x & \text{if } x > 5 \end{cases}$

Draw the graph of $y = 2 + x$ to the left of –4, but do not include the endpoint at $x = 4$. Draw the graph of $y = -x$ between –4 and 5, including both endpoints. Draw the graph of $y = 3x$ to the right of 5, but do not include the endpoint at $x = 5$.

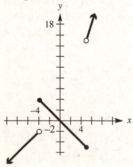

**29.** $f(x) = \begin{cases} -\frac{1}{2}x^2 + 2 & \text{if } x \le 2 \\ \frac{1}{2}x & \text{if } x > 2 \end{cases}$

Graph the curve $y = -\frac{1}{2}x^2 + 2$ to the left of $x = 2$, including the end point at $(2, 0)$. Graph the line $y = \frac{1}{2}x$ to the right of $x = 2$, but do not include the endpoint at $(2, 1)$. Notice that the endpoints of the pieces do not coincide.

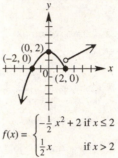

$f(x) = \begin{cases} -\frac{1}{2}x^2 + 2 & \text{if } x \le 2 \\ \frac{1}{2}x & \text{if } x > 2 \end{cases}$

**31.** $f(x) = \begin{cases} 2x & \text{if } -5 \le x < -1 \\ -2 & \text{if } -1 \le x < 0 \\ x^2 - 2 & \text{if } 0 \le x \le 2 \end{cases}$

Graph the line $y = 2x$ between $x = -5$ and $x = -1$, including the left end point at $(-5, -10)$, but not including the right endpoint at $(-1, -2)$. Graph the line $y = -2$ between $x = -1$ and $x = 0$, including the left endpoint at $(-1, -2)$ and not including the right endpoint at $(0, -2)$. Note that $(-1, -2)$ coincides with the first two sections, so it is included. Graph the curve $y = x^2 - 2$ from $x = 0$ to $x = 2$, including the endpoints at $(0, -2)$ and $(2, 2)$. Note that $(0, -2)$ coincides with the second two sections, so it is included. The graph ends at $x = -5$ and $x = 2$.

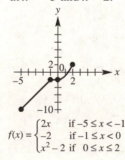

$f(x) = \begin{cases} 2x & \text{if } -5 \le x < -1 \\ -2 & \text{if } -1 \le x < 0 \\ x^2 - 2 & \text{if } 0 \le x \le 2 \end{cases}$

**33.** $f(x) = \begin{cases} x^3 + 3 & \text{if } -2 \le x \le 0 \\ x + 3 & \text{if } 0 < x < 1 \\ 4 + x - x^2 & \text{if } 1 \le x \le 3 \end{cases}$

Graph the curve $y = x^3 + 3$ between $x = -2$ and $x = 0$, including the endpoints at $(-2, -5)$ and $(0, 3)$. Graph the line $y = x + 3$ between $x = 0$ and $x = 1$, but do not include the endpoints at $(0, 3)$ and $(1, 4)$. Graph the curve $y = 4 + x - x^2$ from $x = 1$ to $x = 3$, including the endpoints at $(1, 4)$ and $(3, -2)$. The graph ends at $x = -2$ and $x = 3$.

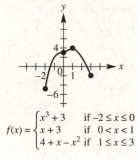

$f(x) = \begin{cases} x^3 + 3 & \text{if } -2 \le x \le 0 \\ x + 3 & \text{if } 0 < x < 1 \\ 4 + x - x^2 & \text{if } 1 \le x \le 3 \end{cases}$

**35.** The solid circle on the graph shows that the endpoint $(0, -1)$ is part of the graph, while the open circle shows that the endpoint $(0, 1)$ is not part of the graph. The graph is made up of parts of two horizontal lines. The function which fits this graph is

$f(x) = \begin{cases} -1 \text{ if } x \le 0 \\ 1 \text{ if } x > 0. \end{cases}$

domain: $(-\infty, \infty)$; range: $\{-1, 1\}$

**37.** The graph is made up of parts of two horizontal lines. The solid circle shows that the endpoint $(0, 2)$ of the one on the left belongs to the graph, while the open circle shows that the endpoint $(0, -1)$ of the one on the right does not belong to the graph. The function that fits this graph is

$f(x) = \begin{cases} 2 \text{ if } x \le 0 \\ -1 \text{ if } x > 1. \end{cases}$

domain: $(-\infty, 0] \cup (1, \infty)$; range: $\{-1, 2\}$

**39.** For $x \le 0$, that piece of the graph goes through the points $(-1, -1)$ and $(0, 0)$. The slope is 1, so the equation of this piece is $y = x$. For $x > 0$, that piece of the graph is a horizontal line passing through $(2, 2)$, so its equation is $y = 2$. We can write the function as

$f(x) = \begin{cases} x \text{ if } x \le 0 \\ 2 \text{ if } x > 0 \end{cases}$

domain: $(-\infty, \infty)$ range: $(-\infty, 0] \cup \{2\}$

**41.** For $x < 1$, that piece of the graph is a curve passes through $(-8, -2)$, $(-1, -1)$ and $(1, 1)$, so the equation of this piece is $y = \sqrt[3]{x}$. The right piece of the graph passes through $(1, 2)$ and $(2, 3)$. $m = \frac{2-3}{1-2} = 1$, and the equation of the line is $y - 2 = x - 1 \Rightarrow y = x + 1$. We can write the function as $f(x) = \begin{cases} \sqrt[3]{x} & \text{if } x < 1 \\ x + 1 & \text{if } x \ge 1 \end{cases}$

domain: $(-\infty, \infty)$ range: $(-\infty, 1) \cup [2, \infty)$

**43.** $f(x) = [\![-x]\!]$

Plot points.

| $x$ | $-x$ | $f(x) = [\![-x]\!]$ |
|---|---|---|
| $-2$ | 2 | 2 |
| $-1.5$ | 1.5 | 1 |
| $-1$ | 1 | 1 |
| $-0.5$ | 0.5 | 0 |
| 0 | 0 | 0 |
| 0.5 | $-0.5$ | $-1$ |
| 1 | $-1$ | $-1$ |
| 1.5 | $-1.5$ | $-2$ |
| 2 | $-2$ | $-2$ |

More generally, to get $y = 0$, we need $0 \le -x < 1 \Rightarrow 0 \ge x > -1 \Rightarrow -1 < x \le 0$. To get $y = 1$, we need $1 \le -x < 2 \Rightarrow -1 \ge x > -2 \Rightarrow -2 < x \le -1$. Follow this pattern to graph the step function.

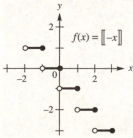

$f(x) = [\![-x]\!]$

domain: $(-\infty, \infty)$; range: $\{..., -2, -1, 0, 1, 2, ...\}$

**45.** $g(x) = \llbracket 2x - 1 \rrbracket$

To get $y = 0$, we need

$0 \le 2x - 1 < 1 \Rightarrow 1 \le 2x < 2 \Rightarrow \frac{1}{2} \le x < 1$.

To get $y = 1$, we need

$1 \le 2x - 1 < 2 \Rightarrow 2 \le 2x < 3 \Rightarrow 1 \le x < \frac{3}{2}$.

Follow this pattern to graph the step function.

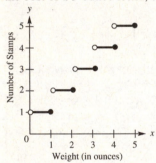

domain: $(-\infty, \infty)$; range: $\{..., 2, -1, 0, 1, 2, ...\}$

**47.** The cost of mailing a letter that weights more than 1 ounce and less than 2 ounces is the same as the cost of a 2-ounce letter, and the cost of mailing a letter that weighs more than 2 ounces and less than 3 ounces is the same as the cost of a 3-ounce letter, etc.

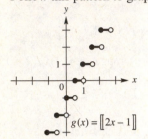

**49.**

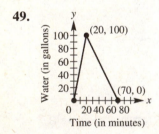

**51.** (a) For $0 \le x \le 4$, $m = \dfrac{39.2 - 42.8}{4 - 0} = -0.9$,

so $y = -0.9x + 42.8$. For $4 < x \le 8$,

$m = \dfrac{32.7 - 39.2}{8 - 4} = -1.625$, so the

equation is $y - 32.7 = -1.625(x - 8) \Rightarrow$

$y = -1.625x + 45.7$

(b) $f(x) = \begin{cases} -0.9x + 42.8 \text{ if } 0 \le x \le 4 \\ -1.625x + 45.7 \text{ if } 4 < x \le 8 \end{cases}$

**53.** (a) The initial amount is 50,000 gallons. The final amount is 30,000 gallons.

(b) The amount of water in the pool remained constant during the first and fourth days.

(c) $f(2) \approx 45,000; f(4) = 40,000$

(d) The slope of the segment between (1, 50000) and (3, 40000) is −5000, so the water was being drained at 5000 gallons per day.

**55.** (a) Since there is no charge for additional length, we use the greatest integer function. The cost is based on multiples of two feet, so $f(x) = 0.8 \llbracket \frac{x}{2} \rrbracket$ if

$6 \le x \le 18$.

(b) $f(8.5) = 0.8 \llbracket \frac{8.5}{2} \rrbracket = 0.8(4) = \$3.20$

$f(15.2) = 0.8 \llbracket \frac{15.2}{2} \rrbracket = 0.8(7) = \$5.60$

## Section 2.7: Graphing Techniques

**Connections (page 269)**

Answers will vary.

**Exercises**

**1.** (a) B; $y = (x - 7)^2$ is a shift of $y = x^2$, 7 units to the right.

(b) D; $y = x^2 - 7$ is a shift of $y = x^2$, 7 units downward.

(c) E; $y = 7x^2$ is a vertical stretch of $y = x^2$, by a factor of 7.

(d) A; $y = (x + 7)^2$ is a shift of $y = x^2$, 7 units to the left.

(e) C; $y = x^2 + 7$ is a shift of $y = x^2$, 7 units upward.

**3.** (a) B; $y = x^2 + 2$ is a shift of $y = x^2$, 2 units upward.

(b) A; $y = x^2 - 2$ is a shift of $y = x^2$, 2 units downward.

(c) G; $y = (x + 2)^2$ is a shift of $y = x^2$, 2 units to the left.

(d) C; $y = (x - 2)^2$ is a shift of $y = x^2$, 2 units to the right.

**(e)** F; $y = 2x^2$ is a vertical stretch of $y = x^2$, by a factor of 2.

**(f)** D; $y = -x^2$ is a reflection of $y = x^2$, across the $x$-axis.

**(g)** H; $y = (x-2)^2 + 1$ is a shift of $y = x^2$, 2 units to the right and 1 unit upward.

**(h)** E; $y = (x+2)^2 + 1$ is a shift of $y = x^2$, 2 units to the left and 1 unit upward.

**(i)** I; $y = (x+2)^2 - 1$ is a shift of $y = x^2$, 2 units to the left and 1 unit down.

**5.** $y = 3|x|$

| $x$ | $y = |x|$ | $y = 3|x|$ |
|---|---|---|
| −2 | 2 | 6 |
| −1 | 1 | 3 |
| 0 | 0 | 0 |
| 1 | 1 | 3 |
| 2 | 2 | 6 |

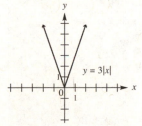

**7.** $y = \frac{2}{3}|x|$

| $x$ | $y = |x|$ | $y = \frac{2}{3}|x|$ |
|---|---|---|
| −3 | 3 | 2 |
| −2 | 2 | $\frac{4}{3}$ |
| −1 | 1 | $\frac{2}{3}$ |
| 0 | 0 | 0 |
| 1 | 1 | $\frac{2}{3}$ |
| 2 | 2 | $\frac{4}{3}$ |
| 3 | 3 | 2 |

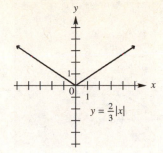

**9.** $y = -3|x|$

| $x$ | $y = |x|$ | $y = -3|x|$ |
|---|---|---|
| −2 | 2 | −6 |
| −1 | 1 | −3 |
| 0 | 0 | 0 |
| 1 | 1 | −3 |
| 2 | 2 | −6 |

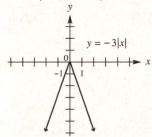

**11.** $y = \left| -\frac{1}{2}x \right|$

| $x$ | $y = |x|$ | $y = \left\| -\frac{1}{2}x \right\| = \left\| -\frac{1}{2} \right\|\|x\| = \frac{1}{2}\|x\|$ |
|---|---|---|
| −4 | 4 | 2 |
| −3 | 3 | $\frac{3}{2}$ |
| −2 | 2 | 1 |
| −1 | 1 | $\frac{1}{2}$ |
| 0 | 0 | 0 |
| 1 | 1 | $\frac{1}{2}$ |
| 2 | 2 | 1 |
| 3 | 3 | $\frac{3}{2}$ |
| 4 | 4 | 2 |

*(continued on next page)*

(*continued from page 111*)

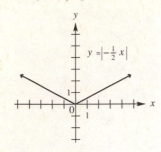

**13.** $y = \sqrt{9x}$

| $x$ | $y = \sqrt{x}$ | $y = \sqrt{4x} = 3\sqrt{x}$ |
|---|---|---|
| 0 | 0 | 0 |
| 1 | 1 | 3 |
| 2 | $\sqrt{2}$ | $3\sqrt{2}$ |
| 3 | $\sqrt{3}$ | $3\sqrt{3}$ |
| 4 | 2 | 6 |

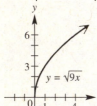

**15.** **(a)**   $y = f(x+4)$ is a horizontal translation of $f$, 4 units to the left.  The point that corresponds to $(8,12)$ on this translated function would be $(8-4,12) = (4,12)$.

   **(b)**   $y = f(x) + 4$ is a vertical translation of $f$, 4 units up. The point that corresponds to $(8,12)$ on this translated function would be $(8,12+4) = (8,16)$.

**17.** **(a)**   $y = f(4x)$ is a horizontal shrinking of $f$, by a factor of 4. The point that corresponds to $(8,12)$ on this translated function is $\left(8 \cdot \frac{1}{4},12\right) = (2,12)$.

   **(b)**   $y = f\left(\frac{1}{4}x\right)$ is a horizontal stretching of $f$, by a factor of 4. The point that corresponds to $(8,12)$ on this translated function is $(8 \cdot 4,12) = (32,12)$.

**19.** **(a)**   The point that is symmetric to $(5, -3)$ with respect to the $x$-axis is $(5, 3)$.

   **(b)**   The point that is symmetric to $(5, -3)$ with respect to the $y$-axis is $(-5, -3)$.

   **(c)**   The point that is symmetric to $(5, -3)$ with respect to the origin is $(-5, 3)$.

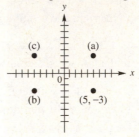

**21.** **(a)**   The point that is symmetric to $(-4, -2)$ with respect to the $x$-axis is $(-4, 2)$.

   **(b)**   The point that is symmetric to $(-4, -2)$ with respect to the $y$-axis is $(4, -2)$.

   **(c)**   The point that is symmetric to $(-4, -2)$ with respect to the origin is $(4, 2)$.

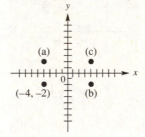

**23.**   $y = x^2 + 5$

Replace $x$ with $-x$ to obtain

$y = (-x)^2 + 5 = x^2 + 5$. The result is the same as the original equation, so the graph is symmetric with respect to the $y$-axis. Since $y$ is a function of $x$, the graph cannot be symmetric with respect to the $x$-axis. Replace $x$ with $-x$ and $y$ with $-y$ to obtain

$-y = (-x)^2 + 2 \Rightarrow -y = x^2 + 2 \Rightarrow y = -x^2 - 2$.

The result is not the same as the original equation, so the graph is not symmetric with respect to the origin. Therefore, the graph is symmetric with respect to the $y$-axis only.

**25.**   $x^2 + y^2 = 12$

Replace $x$ with $-x$ to obtain

$(-x)^2 + y^2 = 12 \Rightarrow x^2 + y^2 = 12$.

The result is the same as the original equation, so the graph is symmetric with respect to the $y$-axis. Replace $y$ with $-y$ to obtain

$x^2 + (-y)^2 = 12 \Rightarrow x^2 + y^2 = 12$

The result is the same as the original equation, so the graph is symmetric with respect to the $x$-axis. Since the graph is symmetric with respect to the $x$-axis and $y$-axis, it is also symmetric with respect to the origin.

**27.** $y = -4x^3$

Replace $x$ with $-x$ to obtain

$y = -4(-x)^3 \Rightarrow y = -4(-x^3) \Rightarrow y = 4x^3$.

The result is not the same as the original equation, so the graph is not symmetric with respect to the $y$-axis. Replace $y$ with $-y$ to obtain $-y = -4x^3 \Rightarrow y = 4x^3$.

The result is not the same as the original equation, so the graph is not symmetric with respect to the $x$-axis. Replace $x$ with $-x$ and $y$ with $-y$ to obtain

$-y = -4(-x)^3 \Rightarrow -y = -4(-x^3) \Rightarrow$
$-y = 4x^3 \Rightarrow y = -4x^3$.

The result is the same as the original equation, so the graph is symmetric with respect to the origin. Therefore, the graph is symmetric with respect to the origin only.

**29.** $y = x^2 - x + 7$

Replace $x$ with $-x$ to obtain

$y = (-x)^2 - (-x) + 7 \Rightarrow y = x^2 + x + 7$.

The result is not the same as the original equation, so the graph is not symmetric with respect to the $y$-axis. Since $y$ is a function of $x$, the graph cannot be symmetric with respect to the $x$-axis. Replace $x$ with $-x$ and $y$ with $-y$ to obtain $-y = (-x)^2 - (-x) + 7 \Rightarrow$

$-y = x^2 + x + 7 \Rightarrow y = -x^2 - x - 7$.

The result is not the same as the original equation, so the graph is not symmetric with respect to the origin. Therefore, the graph has none of the listed symmetries.

**31.** $f(x) = -x^3 + 2x$

$f(-x) = -(-x)^3 + 2(-x)$
$\qquad = x^3 - 2x = -(-x^3 + 2x) = -f(x)$

The function is odd.

**33.** $f(x) = .5x^4 - 2x^2 + 6$

$f(-x) = .5(-x)^4 - 2(-x)^2 + 6$
$\qquad = .5x^4 - 2x^2 + 6 = f(x)$

The function is even.

**35.** $f(x) = x^3 - x + 9$

$f(x) = (-x)^3 - (-x) + 9$
$\qquad = -x^3 + x + 9 = -(x^3 - x - 9) \neq -f(x)$

The function is neither.

**37.** $y = x^2 - 1$

This graph may be obtained by translating the graph of $y = x^2$, 1 unit downward.

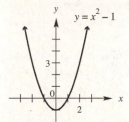

**39.** $y = x^2 + 2$

This graph may be obtained by translating the graph of $y = x^2$, 2 units upward.

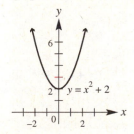

**41.** $y = (x - 4)^2$

This graph may be obtained by translating the graph of $y = x^2$, 4 units to the right.

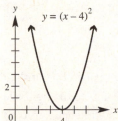

**43.** $y = (x + 2)^2$

This graph may be obtained by translating the graph of $y = x^2$, 2 units to the left.

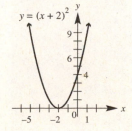

**45.** $y = |x| - 1$

The graph is obtained by translating the graph of $y = |x|$, 1 unit downward.

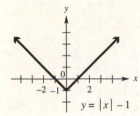

**47.** $y = -(x+1)^3$

This graph may be obtained by translating the graph of $y = x^3$, 1 unit to the left. It is then reflected across the $x$-axis.

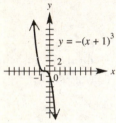

**49.** $y = 2x^2 - 1$

This graph may be obtained by translating the graph of $y = x^2$, 1 unit down. It is then stretched vertically by a factor of 2.

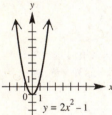

**51.** $f(x) = 2(x-2)^2 - 4$

This graph may be obtained by translating the graph of $y = x^2$, 2 units to the right and 4 units down. It is then stretched vertically by a factor of 2.

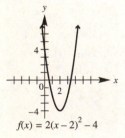

**53.** $f(x) = \sqrt{x+2}$

This graph may be obtained by translating the graph of $y = \sqrt{x}$ two units to the left.

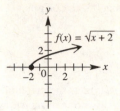

**55.** $f(x) = -\sqrt{x}$

This graph may be obtained by reflecting the graph of $y = \sqrt{x}$ across the $x$-axis.

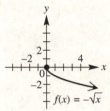

**57.** $f(x) = 2\sqrt{x} + 1$

This graph may be obtained by stretching the graph of $y = \sqrt{x}$ vertically by a factor of two and then translating the resulting graph one unit up.

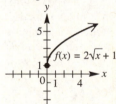

**59. (a)** $y = g(-x)$

The graph of $g(x)$ is reflected across the $y$-axis.

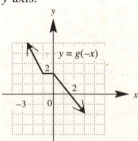

**(b)** $y = g(x - 2)$
The graph of $g(x)$ is translated to the right 2 units.

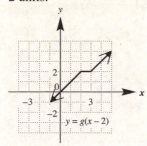

**(c)** $y = -g(x) + 2$
The graph of $g(x)$ is reflected across the $x$-axis and translated 2 units up.

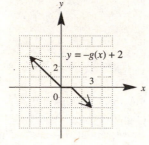

**61.** It is the graph of $f(x) = |x|$ translated 1 unit to the left, reflected across the $x$-axis, and translated 3 units up. The equation is $y = -|x + 1| + 3$.

**63.** It is the graph of $f(x) = \sqrt{x}$ translated one unit right and then three units down. The equation is $y = \sqrt{x - 1} - 3$.

**65.** It is the graph of $g(x) = \sqrt{x}$ translated 4 units to the left, stretched vertically by a factor of 2, and translated 4 units down. The equation is $y = 2\sqrt{x + 4} - 4$.

**67.** Since $f(3) = 6$, the point $(3, 6)$ is on the graph. Since the graph is symmetric with respect to the origin, the point $(-3, -6)$ is on the graph. Therefore, $f(-3) = -6$.

**69.** Since $f(3) = 6$, the point $(3, 6)$ is on the graph. Since the graph is symmetric with respect to the line $x = 6$ and since the point $(3, 6)$ is 3 units to the left of the line $x = 6$, the image point of $(3, 6)$, 3 units to the right of the line $x = 6$, is $(9, 6)$. Therefore, $f(9) = 6$.

**71.** An odd function is a function whose graph is symmetric with respect to the origin. Since $(3, 6)$ is on the graph, $(-3, -6)$ must also be on the graph. Therefore, $f(-3) = -6$.

**73.** $f(x) = 2x + 5$: Translate the graph of $f(x)$ up 2 units to obtain the graph of $t(x) = (2x + 5) + 2 = 2x + 7$.
Now translate the graph of $t(x) = 2x + 7$ left 3 units to obtain the graph of $g(x) = 2(x + 3) + 7 = 2x + 6 + 7 = 2x + 13$.
(Note that if the original graph is first translated to the left 3 units and then up 2 units, the final result will be the same.)

**75.** **(a)** Since $f(-x) = f(x)$, the graph is symmetric with respect to the $y$-axis.

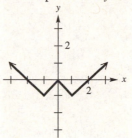

**(b)** Since $f(-x) = -f(x)$, the graph is symmetric with respect to the origin.

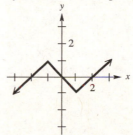

**77.** Answers will vary.
There are four possibilities for the constant, $c$.

**i)** $c > 0$  $|c| > 1$  The graph of $F(x)$ is stretched vertically by a factor of $c$.

**ii)** $c > 0$  $|c| < 1$  The graph of $F(x)$ is shrunk vertically by a factor of $c$.

**iii)** $c < 0$  $|c| > 1$  The graph of $F(x)$ is stretched vertically by a factor of $-c$ and reflected over the $x$-axis.

**iv)** $c < 0$  $|c| < 1$  The graph of $F(x)$ is shrunk vertically by a factor of $-c$ and reflected over the $x$-axis.

## Chapter 2 Quiz

**(Sections 2.5–2.7)**

1. (a) First, find the slope: $m = \dfrac{9-5}{-1-(-3)} = 2$

   Choose either point, say, $(-3, 5)$, to find the equation of the line:

   $y - 5 = 2(x - (-3)) \Rightarrow y = 2(x+3) + 5 \Rightarrow$

   $y = 2x + 11$.

   (b) To find the $x$-intercept, let $y = 0$ and solve for $x$: $0 = 2x + 11 \Rightarrow x = -\frac{11}{2}$. The

   $x$-intercept is $-\frac{11}{2}$.

3. (a) $x = -8$      (b) $y = 5$

5. (a) The highest speed limit is 55 miles per hour. The lowest speed limit is 30 miles per hour.

   (b) There are about 12 miles of highway with a speed limit of 55 miles per hour.

   (c) $f(4) = 40; f(12) = 30; f(18) = 55$

7. $f(x) = -x^3 + 1$

   Reflect the graph of $f(x) = x^3$ across the $x$-axis, and then translate the resulting graph one unit up.

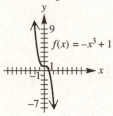

9. This is the graph of $f(x) = \sqrt{x}$, translated four units to the left, reflected across the $x$-axis, and then translated two units down. The equation is $y = -\sqrt{x+4} - 2$.

## Section 2.8: Function Operations and Composition

In Exercises 1–8, $f(x) = x^2 + 3$ and $g(x) = -2x + 6$.

1. $(f + g)(3) = f(3) + g(3)$

   $= \left[(3)^2 + 3\right] + \left[-2(3) + 6\right]$

   $= 12 + 0 = 12$

3. $(f - g)(-5) = f(-1) - g(-1)$

   $= \left[(-1)^2 + 3\right] - \left[-2(-1) + 6\right]$

   $= 4 - 8 = -4$

5. $(fg)(4) = f(4) \cdot g(4)$

   $= [4^2 + 3] \cdot [-2(4) + 6]$

   $= 19 \cdot (-2) = -38$

7. $\left(\dfrac{f}{g}\right)(-1) = \dfrac{f(-1)}{g(-1)} = \dfrac{(-1)^2 + 3}{-2(-1) + 6} = \dfrac{4}{8} = \dfrac{1}{2}$

9. $f(x) = 3x + 4, g(x) = 2x - 5$

   i) $(f + g)(x) = f(x) + g(x)$

   $= (3x + 4) + (2x - 5) = 5x - 1$

   ii) $(f - g)(x) = f(x) - g(x)$

   $= (3x + 4) - (2x - 5) = x + 9$

   iii) $(fg)(x) = f(x) \cdot g(x) = (3x + 4)(2x - 5)$

   $= 6x^2 - 15x + 8x - 20$

   $= 6x^2 - 7x - 20$

   iv) $\left(\dfrac{f}{g}\right)(x) = \dfrac{f(x)}{g(x)} = \dfrac{3x + 4}{2x - 5}$

   The domains of both $f$ and $g$ are the set of all real numbers, so the domains of $f + g$, $f - g$, and $fg$ are all $(-\infty, \infty)$. The domain of $\frac{f}{g}$ is the set of all real numbers for which $g(x) \neq 0$. This is the set of all real numbers except $\frac{5}{2}$, which is written in interval notation as $\left(-\infty, \frac{5}{2}\right) \cup \left(\frac{5}{2}, \infty\right)$.

11. $f(x) = 2x^2 - 3x, g(x) = x^2 - x + 3$

    i) $(f + g)(x) = f(x) + g(x)$

    $= (2x^2 - 3x) + (x^2 - x + 3)$

    $= 3x^2 - 4x + 3$

    ii) $(f - g)(x) = f(x) - g(x)$

    $= (2x^2 - 3x) - (x^2 - x + 3)$

    $= 2x^2 - 3x - x^2 + x - 3$

    $= x^2 - 2x - 3$

    iii) $(fg)(x) = f(x) \cdot g(x)$

    $= (2x^2 - 3x)(x^2 - x + 3)$

    $= 2x^4 - 2x^3 + 6x^2 - 3x^3 + 3x^2 - 9x$

    $= 2x^4 - 5x^3 + 9x^2 - 9x$

    iv) $\left(\dfrac{f}{g}\right)(x) = \dfrac{f(x)}{g(x)} = \dfrac{2x^2 - 3x}{x^2 - x + 3}$

    The domains of both $f$ and $g$ are the set of all real numbers, so the domains of $f + g$, $f - g$, and $fg$ are all $(-\infty, \infty)$.

The domain of $\frac{f}{g}$ is the set of all real numbers for which $g(x) \neq 0$. If $x^2 - x + 3 = 0$, then by the quadratic formula $x = \frac{1 \pm i\sqrt{11}}{2}$. The equation has no real solutions. There are no real numbers which make the denominator zero. Thus, the domain of $\frac{f}{g}$ is also $(-\infty, \infty)$.

**13.** $f(x) = \sqrt{4x-1}, \; g(x) = \dfrac{1}{x}$

**i)** $(f+g)(x) = f(x) + g(x) = \sqrt{4x-1} + \dfrac{1}{x}$

**ii)** $(f-g)(x) = f(x) - g(x) = \sqrt{4x-1} - \dfrac{1}{x}$

**iii)** $(fg)(x) = f(x) \cdot g(x)$
$$= \sqrt{4x-1}\left(\frac{1}{x}\right) = \frac{\sqrt{4x-1}}{x}$$

**iv)** $\left(\dfrac{f}{g}\right)(x) = \dfrac{f(x)}{g(x)} = \dfrac{\sqrt{4x-1}}{\frac{1}{x}} = x\sqrt{4x-1}$

Since $4x - 1 \geq 0 \Rightarrow 4x \geq 1 \Rightarrow x \geq \frac{1}{4}$, the domain of $f$ is $\left[\frac{1}{4}, \infty\right)$. The domain of $g$ is $(-\infty, 0) \cup (0, \infty)$. Considering the intersection of the domains of $f$ and $g$, the domains of $f + g$, $f - g$, and $fg$ are all $\left[\frac{1}{4}, \infty\right)$. Since $\frac{1}{x} \neq 0$ for any value of $x$, the domain of $\frac{f}{g}$ is also $\left[\frac{1}{4}, \infty\right)$.

**15.** Numerical answers may vary.
$G(1996) \approx 7.7$ and $B(1996) \approx 11.8$, thus
$T(1996) = G(1996) + B(1996)$
$\qquad = 7.7 + 11.8 = 19.5$.

**17.** Looking at the graphs of the functions, the slopes of the line segments for the period 1996–2006 are much steeper than the slopes of the corresponding line segments for the period 1991–1996. Thus, the number of sodas increased more rapidly during the period 1996–2006.

**19.** Numerical answers may vary.
$(T - S)(2000) = T(2000) - S(2000)$
$\qquad\qquad = 19 - 13 = 6$
It represents the dollars in billions spent for general science in 2000.

**21.** In space and other technologies spending was almost static in the years 1995–2000.

**23. (a)** $(f+g)(2) = f(2) + g(2)$
$\qquad\qquad = 4 + (-2) = 2$

**(b)** $(f-g)(1) = f(1) - g(1) = 1 - (-3) = 4$

**(c)** $(fg)(0) = f(0) \cdot g(0) = 0(-4) = 0$

**(d)** $\left(\dfrac{f}{g}\right)(1) = \dfrac{f(1)}{g(1)} = \dfrac{1}{-3} = -\dfrac{1}{3}$

**25. (a)** $(f+g)(-1) = f(-1) + g(-1) = 0 + 3 = 3$

**(b)** $(f-g)(-2) = f(-2) - g(-2)$
$\qquad\qquad = -1 - 4 = -5$

**(c)** $(fg)(0) = f(0) \cdot g(0) = 1 \cdot 2 = 2$

**(d)** $\left(\dfrac{f}{g}\right)(2) = \dfrac{f(2)}{g(2)} = \dfrac{3}{0} = \text{undefined}$

**27. (a)** $(f+g)(2) = f(2) + g(2) = 7 + (-2) = 5$

**(b)** $(f-g)(4) = f(4) - g(4) = 10 - 5 = 5$

**(c)** $(fg)(-2) = f(-2) \cdot g(-2) = 0 \cdot 6 = 0$

**(d)** $\left(\dfrac{f}{g}\right)(0) = \dfrac{f(0)}{g(0)} = \dfrac{5}{0} = \text{undefined}$

**29.**

| $x$ | $f(x)$ | $g(x)$ | $(f+g)(x)$ | $(f-g)(x)$ | $(fg)(x)$ | $\left(\dfrac{f}{g}\right)(x)$ |
|---|---|---|---|---|---|---|
| $-2$ | $0$ | $6$ | $0 + 6 = 6$ | $0 - 6 = -6$ | $0 \cdot 6 = 0$ | $\frac{0}{6} = 0$ |
| $0$ | $5$ | $0$ | $5 + 0 = 5$ | $5 - 0 = 5$ | $5 \cdot 0 = 0$ | $\frac{5}{0} = \text{undefined}$ |
| $2$ | $7$ | $-2$ | $7 + (-2) = 5$ | $7 - (-2) = 9$ | $7(-2) = -14$ | $\frac{7}{-2} = -3.5$ |
| $4$ | $10$ | $5$ | $10 + 5 = 15$ | $10 - 5 = 5$ | $10 \cdot 5 = 50$ | $\frac{10}{5} = 2$ |

**31.** Answers will vary.

The difference quotient, $\dfrac{f(x+h)-f(x)}{h}$,

represents the slope of the secant line which passes through points
$\bigl(x, f(x)\bigr)$ and $\bigl(x+h, f(x+h)\bigr)$. The formula is derived by applying the rule that slope represents a change in $y$ to a change in $x$.

**33.** $f(x)=2-x$

(a) $f(x+h)=2-(x+h)=2-x-h$

(b) $f(x+h)-f(x)=(2-x-h)-(2-x)$
$$=2-x-h-2+x=-h$$

(c) $\dfrac{f(x+h)-f(x)}{h}=\dfrac{-h}{h}=-1$

**35.** $f(x)=6x+2$

(a) $f(x+h)=6(x+h)+2=6x+6h+2$

(b) $f(x+h)-f(x)$
$$=(6x+6h+2)-(6x+2)$$
$$=6x+6h+2-6x-2=6h$$

(c) $\dfrac{f(x+h)-f(x)}{h}=\dfrac{6h}{h}=6$

**37.** $f(x)=-2x+5$

(a) $f(x+h)=-2(x+h)+5$
$$=-2x-2h+5$$

(b) $f(x+h)-f(x)$
$$=(-2x-2h+5)-(-2x+5)$$
$$=-2x-2h+5+2x-5$$
$$=-2h$$

(c) $\dfrac{f(x+h)-f(x)}{h}=\dfrac{-2h}{h}=-2$

**39.** $f(x)=\dfrac{1}{x}$

(a) $f(x+h)=\dfrac{1}{x+h}$

(b) $f(x+h)-f(x)$
$$=\dfrac{1}{x+h}-\dfrac{1}{x}=\dfrac{x-(x+h)}{x(x+h)}$$
$$=\dfrac{-h}{x(x+h)}$$

(c) $\dfrac{f(x+h)-f(x)}{h}=\dfrac{\frac{-h}{x(x+h)}}{h}=\dfrac{-h}{hx(x+h)}$
$$=-\dfrac{1}{x(x+h)}$$

**41.** Since $g(x)=-x+3$, $g(4)=-4+3=-1$.

Therefore, $(f\circ g)(4)=f\bigl[g(4)\bigr]=f(-1)$
$$=2(-1)-3=-2-3=-5.$$

**43.** Since $g(x)=-x+3$, $g(-2)=-(-2)+3=5$.

Therefore, $(f\circ g)(-2)=f\bigl[g(-2)\bigr]=f(5)$
$$=2(5)-3=10-3=7.$$

**45.** Since $f(x)=2x-3$,
$$f(0)=2(0)-3=0-3=-3.$$

Therefore, $(g\circ f)(0)=g\bigl[f(0)\bigr]$
$$=g(-3)=-(-3)+3=3+3=6.$$

**47.** Since $f(x)=2x-3$,
$$f(2)=2(2)-3=4-3=1.$$

Therefore, $(f\circ f)(2)=f\bigl[f(2)\bigr]$
$$=f(1)=2(1)-3=2-3=-1.$$

**49.** $(f\circ g)(2)=f[g(2)]=f(3)=1$

**51.** $(g\circ f)(3)=g[f(3)]=g(1)=9$

**53.** $(f\circ f)(4)=f[f(4)]=f(3)=1$

**55.** $(f\circ g)(1)=f[g(1)]=f(9)$

However, $f(9)$ cannot be determined from the table given.

**57.** (a) $(f\circ g)(x)=f(g(x))=f(5x+7)$
$$=-6(5x+7)+9$$
$$=-30x-42+9=-30x-33$$
The domain and range of both $f$ and $g$ are $(-\infty,\infty)$, so the domain of $f\circ g$ is $(-\infty,\infty)$.

(b) $(g\circ f)(x)=g(f(x))=g(-6x+9)$
$$=5(-6x+9)+7$$
$$=-30x+45+7=-30x+52$$
The domain of $g\circ f$ is $(-\infty,\infty)$.

**59. (a)** $(f \circ g)(x) = f(g(x)) = f(x+3) = \sqrt{x+3}$

The domain and range of $g$ are $(-\infty, \infty)$, however, the domain and range of $f$ are $[0, \infty)$. So, $x+3 \ge 0 \Rightarrow x \ge -3$.

Therefore, the domain of $f \circ g$ is $[-3, \infty)$.

**(b)** $(g \circ f)(x) = g(f(x)) = g\left(\sqrt{x}\right) = \sqrt{x} + 3$

The domain and range of $g$ are $(-\infty, \infty)$, however, the domain and range of $f$ are $[0, \infty)$. Therefore, the domain of $g \circ f$ is $[0, \infty)$.

**61. (a)** $(f \circ g)(x) = f(g(x)) = f(x^2 + 3x - 1)$
$= (x^2 + 3x - 1)^3$

The domain and range of $f$ and $g$ are $(-\infty, \infty)$, so the domain of $f \circ g$ is $(-\infty, \infty)$.

**(b)** $(g \circ f)(x) = g(f(x)) = g(x^3)$
$= \left(x^3\right)^2 + 3\left(x^3\right) - 1$
$= x^6 + 3x^3 - 1$

The domain and range of $f$ and $g$ are $(-\infty, \infty)$, so the domain of $g \circ f$ is $(-\infty, \infty)$.

**63. (a)** $(f \circ g)(x) = f(g(x)) = f(3x) = \sqrt{3x-1}$

The domain and range of $g$ are $(-\infty, \infty)$, however, the domain and range of $f$ are $[1, \infty)$. So, $3x - 1 \ge 0 \Rightarrow x \ge \frac{1}{3}$. Therefore, the domain of $f \circ g$ is $\left[\frac{1}{3}, \infty\right)$.

**(b)** $(g \circ f)(x) = g(f(x)) = g\left(\sqrt{x-1}\right)$
$= 3\sqrt{x-1}$

The domain and range of $g$ are $(-\infty, \infty)$, however, the range of $f$ is $[0, \infty)$. So $x - 1 \ge 0 \Rightarrow x \ge 1$. Therefore, the domain of $g \circ f$ is $[1, \infty)$.

**65. (a)** $(f \circ g)(x) = f(g(x)) = f(x+1) = \frac{2}{x+1}$

The domain and range of $g$ are $(-\infty, \infty)$, however, the domain of $f$ is $(-\infty, 0) \cup (0, \infty)$. So, $x+1 \ne 0 \Rightarrow x \ne -1$. Therefore, the domain of $f \circ g$ is $(-\infty, -1) \cup (-1, \infty)$.

**(b)** $(g \circ f)(x) = g(f(x)) = g\left(\frac{2}{x}\right) = \frac{2}{x} + 1$

The domain and range of $f$ is $(-\infty, 0) \cup (0, \infty)$, however, the domain and range of $g$ are $(-\infty, \infty)$. So $x \ne 0$. Therefore, the domain of $g \circ f$ is $(-\infty, 0) \cup (0, \infty)$.

**67. (a)** $(f \circ g)(x) = f(g(x)) = f\left(-\frac{1}{x}\right) = \sqrt{-\frac{1}{x} + 2}$

The domain and range of $g$ are $(-\infty, 0) \cup (0, \infty)$, however, the domain of $f$ is $[-2, \infty)$. So, $-\frac{1}{x} + 2 \ge 0 \Rightarrow$ $x < 0$ or $x \ge \frac{1}{2}$ (using test intervals). Therefore, the domain of $f \circ g$ is $(-\infty, 0) \cup \left[\frac{1}{2}, \infty\right)$.

**(b)** $(g \circ f)(x) = g(f(x)) = g\left(\sqrt{x+2}\right) = -\frac{1}{\sqrt{x+2}}$

The domain of $f$ is $[-2, \infty)$ and its range is $(-\infty, \infty)$. The domain and range of $g$ are $(-\infty, 0) \cup (0, \infty)$. So $x + 2 > 0 \Rightarrow x > -2$. Therefore, the domain of $g \circ f$ is $(-2, \infty)$.

**69. (a)** $(f \circ g)(x) = f(g(x)) = f\left(\frac{1}{x+5}\right) = \sqrt{\frac{1}{x+5}}$

The domain of $g$ is $(-\infty, -5) \cup (-5, \infty)$, and the range of $g$ is $(-\infty, 0) \cup (0, \infty)$. The domain of $f$ is $[0, \infty)$. Therefore, the domain of $f \circ g$ is $(-5, \infty)$.

**(b)** $(g \circ f)(x) = g(f(x)) = g\left(\sqrt{x}\right) = \frac{1}{\sqrt{x}+5}$

The domain and range of $f$ is $[0, \infty)$. The domain of $g$ is $(-\infty, -5) \cup (-5, \infty)$. Therefore, the domain of $g \circ f$ is $[0, \infty)$.

**71. (a)** $(f \circ g)(x) = f(g(x)) = f\left(\frac{1}{x}\right) = \frac{1}{1/x - 2} = \frac{x}{1-2x}$

The domain and range of $g$ are $(-\infty, 0) \cup (0, \infty)$. The domain of $f$ is $(-\infty, -2) \cup (-2, \infty)$, and the range of $f$ is $(-\infty, 0) \cup (0, \infty)$. So, $\frac{x}{1-2x} < 0 \Rightarrow x < 0$ or $0 < x < \frac{1}{2}$ or $x > \frac{1}{2}$ (using test intervals). Thus, $x \ne 0$ and $x \ne \frac{1}{2}$. Therefore, the domain of $f \circ g$ is $(-\infty, 0) \cup \left(0, \frac{1}{2}\right) \cup \left(\frac{1}{2}, \infty\right)$.

**(b)** $(g \circ f)(x) = g(f(x)) = g\left(\frac{1}{x-2}\right) = \frac{1}{1/(x-2)}$

$= x - 2$

The domain and range of $g$ are

$(-\infty, 0) \cup (0, \infty)$. The domain of $f$ is

$(-\infty, -2) \cup (-2, \infty)$, and the range of $f$ is

$(-\infty, 0) \cup (0, \infty)$. Therefore, the domain of

$g \circ f$ is $(-\infty, -2) \cup (-2, \infty)$.

**73.** $g\left[f(2)\right] = g(1) = 2$ and $g\left[f(3)\right] = g(2) = 5$

Since $g\left[f(1)\right] = 7$ and $f(1) = 3$, $g(3) = 7$.

| $x$ | $f(x)$ | $g(x)$ | $g\left[f(x)\right]$ |
|-----|--------|--------|----------------------|
| 1 | 3 | 2 | 7 |
| 2 | 1 | 5 | 2 |
| 3 | 2 | 7 | 5 |

**75.** Answers will vary. In general, composition of functions is not commutative.

$(f \circ g)(x) = f(2x - 3) = 3(2x - 3) - 2$

$= 6x - 9 - 2 = 6x - 11$

$(g \circ f)(x) = g(3x - 2) = 2(3x - 2) - 3$

$= 6x - 4 - 3 = 6x - 7$

Thus, $(f \circ g)(x) \neq (g \circ f)(x)$.

**77.** $(f \circ g)(x) = f\left[g(x)\right] = 4\left[\frac{1}{4}(x - 2)\right] + 2$

$= \left(4 \cdot \frac{1}{4}\right)(x - 2) + 2$

$= (x - 2) + 2 = x - 2 + 2 = x$

$(g \circ f)(x) = g\left[f(x)\right] = \frac{1}{4}\left[(4x + 2) - 2\right]$

$= \frac{1}{4}(4x + 2 - 2) = \frac{1}{4}(4x) = x$

**79.** $(f \circ g)(x) = f\left[g(x)\right] = \sqrt[3]{5\left(\frac{1}{5}x^3 - \frac{4}{5}\right) + 4}$

$= \sqrt[3]{x^3 - 4 + 4} = \sqrt[3]{x^3} = x$

$(g \circ f)(x) = g\left[f(x)\right] = \frac{1}{5}\left(\sqrt[3]{5x + 4}\right)^3 - \frac{4}{5}$

$= \frac{1}{5}(5x + 4) - \frac{4}{5} = \frac{5x}{5} + \frac{4}{5} - \frac{4}{5}$

$= \frac{5x}{5} = x$

In Exercises 81–85, we give only one of many possible ways.

**81.** $h(x) = (6x - 2)^2$

Let $g(x) = 6x - 2$ and $f(x) = x^2$.

$(f \circ g)(x) = f(6x - 2) = (6x - 2)^2 = h(x)$

**83.** $h(x) = \sqrt{x^2 - 1}$

Let $g(x) = x^2 - 1$ and $f(x) = \sqrt{x}$.

$(f \circ g)(x) = f(x^2 - 1) = \sqrt{x^2 - 1} = h(x)$.

**85.** $h(x) = \sqrt{6x} + 12$

Let $g(x) = 6x$ and $f(x) = \sqrt{x} + 12$.

$(f \circ g)(x) = f(6x) = \sqrt{6x} + 12 = h(x)$

**87.** $f(x) = 12x$, $g(x) = 5280x$

$(f \circ g)(x) = f[g(x)] = f(5280x)$

$= 12(5280x) = 63,360x$

The function $f \circ g$ computes the number of inches in $x$ miles.

**89.** $A(x) = \frac{\sqrt{3}}{4}x^2$

**(a)** $A(2x) = \frac{\sqrt{3}}{4}(2x)^2 = \frac{\sqrt{3}}{4}(4x^2) = \sqrt{3}x^2$

**(b)** $A(16) = A(2 \cdot 8) = \sqrt{3}(8)^2$

$= 64\sqrt{3}$ square units

**91. (a)** $r(t) = 4t$ and $A(r) = \pi r^2$

$(A \circ r)(t) = A[r(t)]$

$= A(4t) = \pi(4t)^2 = 16\pi t^2$

**(b)** $(A \circ r)(t)$ defines the area of the leak in terms of the time $t$, in minutes.

**(c)** $A(3) = 16\pi(3)^2 = 144\pi$ ft$^2$

**93.** Let $x$ = the number of people less than 100 people that attend.

**(a)** $x$ people fewer than 100 attend, so $100 - x$ people do attend $N(x) = 100 - x$

**(b)** The cost per person starts at $20 and increases by $5 for each of the $x$ people that do not attend. The total increase is $5x$, and the cost per person increases to $20 + $5x$. Thus, $G(x) = 20 + 5x$.

**(c)** $C(x) = N(x) \cdot G(x) = (100 - x)(20 + 5x)$

**(d)** If 80 people attend, $x = 100 - 80 = 20$.

$C(20) = (100 - 20)\left[20 + 5(20)\right]$

$= (80)(20 + 100)$

$= (80)(120) = \$9600$

## Chapter 2: Review Exercises

**1.** $P(3, -1), Q(-4, 5)$

$$d(P, Q) = \sqrt{(-4-3)^2 + [5-(-1)]^2}$$
$$= \sqrt{(-7)^2 + 6^2} = \sqrt{49+36} = \sqrt{85}$$

Midpoint:
$$\left(\frac{3+(-4)}{2}, \frac{-1+5}{2}\right) = \left(\frac{-1}{2}, \frac{4}{2}\right) = \left(-\frac{1}{2}, 2\right)$$

**3.** $A(-6, 3), B(-6, 8)$

$$d(A, B) = \sqrt{[-6-(-6)]^2 + (8-3)^2}$$
$$= \sqrt{0+5^2} = \sqrt{25} = 5$$

Midpoint:
$$\left(\frac{-6+(-6)}{2}, \frac{3+8}{2}\right) = \left(\frac{-12}{2}, \frac{11}{2}\right) = \left(-6, \frac{11}{2}\right)$$

**5.** Label the points $A(-1, 2)$, $B(-10, 5)$, and $C(-4, k)$.

$$d(A, B) = \sqrt{[-1-(-10)]^2 + (2-5)^2}$$
$$= \sqrt{9^2 + (-3)^2} = \sqrt{81+9} = \sqrt{90}$$

$$d(A, C) = \sqrt{[-4-(-1)]^2 + (k-2)^2}$$
$$= \sqrt{9 + (k-2)^2}$$

$$d(B, C) = \sqrt{[-10-(-4)]^2 + (5-k)^2}$$
$$= \sqrt{36 + (k-5)^2}$$

If segment $AB$ is the hypotenuse, then

$$\left(\sqrt{90}\right)^2 = \left[\sqrt{9+(k-2)^2}\right]^2 + \left[\sqrt{36+(k-5)^2}\right]^2.$$

$$\left(\sqrt{90}\right)^2 = \left[\sqrt{9+(k-2)^2}\right]^2 + \left[\sqrt{36+(k-5)^2}\right]^2$$
$$90 = 9 + (k-2)^2 + 36 + (k-5)^2$$
$$90 = 9 + k^2 - 4k + 4 + 36 + k^2 - 10k + 25$$
$$0 = 2k^2 - 14k - 16 \Rightarrow 0 = k^2 - 7k - 8 \Rightarrow$$
$$0 = (k-8)(k+1) \Rightarrow k = 8 \text{ or } k = -1$$

Another approach is if segment $AB$ is the hypotenuse, the product of the slopes of lines $AC$ and $BC$ is $-1$ since the product of slopes of perpendicular lines is $-1$.

$$\left(\frac{k-2}{-4-(-1)}\right) \cdot \left(\frac{k-5}{-4-(-10)}\right) = -1$$
$$\left(\frac{k-2}{-3}\right) \cdot \left(\frac{k-5}{6}\right) = -1$$
$$\frac{(k-2)(k-5)}{-18} = -1$$
$$\frac{k^2 - 5k - 2k + 10}{-18} = -1$$

$$k^2 - 7k + 10 = 18 \Rightarrow k^2 - 7k - 8 = 0 \Rightarrow$$
$$(k+8)(k-1) = 0 \Rightarrow k = 8 \text{ or } k = -1$$

We will use the second approach for investigating the other two sides of the triangle. If segment $AC$ is the hypotenuse, the product of the slopes of lines $AB$ and $BC$ is $-1$ since the product of slopes of perpendicular lines is $-1$.

$$\left(\frac{5-2}{-10-(-1)}\right) \cdot \left(\frac{k-5}{-4-(-10)}\right) = -1$$
$$\left(\frac{3}{-9}\right) \cdot \left(\frac{k-5}{6}\right) = -1$$
$$\frac{k-5}{-18} = -1$$
$$k - 5 = 18 \Rightarrow k = 23$$

If segment $BC$ is the hypotenuse, the product of the slopes of lines $AB$ and $AC$ is $-1$.

$$\left(\frac{3}{-9}\right) \cdot \left(\frac{k-2}{-4-(-1)}\right) = -1$$
$$\left(\frac{-1}{3}\right) \cdot \left(\frac{k-2}{-3}\right) = -1$$
$$\frac{k-2}{9} = -1$$
$$k - 2 = -9 \Rightarrow k = -7$$

The possible values of $k$ are $-7$, $23$, $8$, and $-1$.

**7.** Center $(-2, 3)$, radius 15

$$(x-h)^2 + (y-k)^2 = r^2$$
$$[x-(-2)]^2 + (y-3)^2 = 15^2$$
$$(x+2)^2 + (y-3)^2 = 225$$

**9.** Center $(-8, 1)$, passing through $(0, 16)$
The radius is the distance from the center to any point on the circle. The distance between $(-8, 1)$ and $(0, 16)$ is

$$r = \sqrt{(-8-0)^2 + (1-16)^2} = \sqrt{(-8)^2 + (-15)^2}$$
$$= \sqrt{64 + 225} = \sqrt{289} = 17.$$

The equation of the circle is

$$[x-(-8)]^2 + (y-1)^2 = 17^2$$
$$(x+8)^2 + (y-1)^2 = 289$$

**11.** The center of the circle is $(0, 0)$. Use the distance formula to find the radius:

$$r^2 = (3-0)^2 + (5-0)^2 = 9 + 25 = 34$$

The equation is $x^2 + y^2 = 34$.

**13.** The center of the circle is $(0, 3)$. Use the distance formula to find the radius:

$$r^2 = (-2-0)^2 + (6-3)^2 = 4 + 9 = 13$$

The equation is $x^2 + (y-3)^2 = 13$.

**15.** $x^2 - 4x + y^2 + 6y + 12 = 0$

Complete the square on $x$ and $y$ to put the equation in center-radius form.

$$\left(x^2 - 4x\right) + \left(y^2 + 6y\right) = -12$$

$$\left(x^2 - 4x + 4\right) + \left(y^2 + 6y + 9\right) = -12 + 4 + 9$$

$$\left(x - 2\right)^2 + \left(y + 3\right)^2 = 1$$

The circle has center $(2, -3)$ and radius 1.

**17.**
$$2x^2 + 14x + 2y^2 + 6y + 2 = 0$$
$$x^2 + 7x + y^2 + 3y + 1 = 0$$
$$\left(x^2 + 7x\right) + \left(y^2 + 3y\right) = -1$$
$$\left(x^2 + 7x + \tfrac{49}{4}\right) + \left(y^2 + 3y + \tfrac{9}{4}\right) = -1 + \tfrac{49}{4} + \tfrac{9}{4}$$
$$\left(x + \tfrac{7}{2}\right)^2 + \left(y + \tfrac{3}{2}\right)^2 = -\tfrac{4}{4} + \tfrac{49}{4} + \tfrac{9}{4}$$
$$\left(x + \tfrac{7}{2}\right)^2 + \left(y + \tfrac{3}{2}\right)^2 = \tfrac{54}{4}$$

The circle has center $\left(-\tfrac{7}{2}, -\tfrac{3}{2}\right)$ and radius

$$\sqrt{\tfrac{54}{4}} = \tfrac{\sqrt{54}}{\sqrt{4}} = \tfrac{\sqrt{9 \cdot 6}}{\sqrt{4}} = \tfrac{3\sqrt{6}}{2}.$$

**19.** Find all possible values of $x$ so that the distance between $(x, -9)$ and $(3, -5)$ is 6.

$$\sqrt{(3 - x)^2 + (-5 + 9)^2} = 6$$

$$\sqrt{9 - 6x + x^2 + 16} = 6$$

$$\sqrt{x^2 - 6x + 25} = 6$$

$$x^2 - 6x + 25 = 36$$

$$x^2 - 6x - 11 = 0$$

Apply the quadratic formula where $a = 1, b = -6,$ and $c = -11.$

$$x = \frac{6 \pm \sqrt{36 - 4(1)(-11)}}{2} = \frac{6 \pm \sqrt{36 + 44}}{2}$$

$$= \frac{6 \pm \sqrt{80}}{2} = \frac{6 \pm 4\sqrt{5}}{2} = \frac{2(3 \pm 2\sqrt{5})}{2}$$

$$x = 3 + 2\sqrt{5} \text{ or } x = 3 - 2\sqrt{5}$$

**21.** This is not the graph of a function because a vertical line can intersect it in two points.

domain: $[-6, 6]$;  range: $[-6, 6]$

**23.** This is not the graph of a function because a vertical line can intersect it in two points.

domain: $(-\infty, \infty)$;  range: $(-\infty, -1] \cup [1, \infty)$

**25.** This is not the graph of a function because a vertical line can intersect it in two points.

domain: $[0, \infty)$;  range: $(-\infty, \infty)$

**27.** $y = 6 - x^2$

Each value of $x$ corresponds to exactly one value of $y$, so this equation defines a function.

**29.** The equation $y = \pm\sqrt{x - 2}$ does not define $y$ as a function of $x$. For some values of $x$, there will be more than one value of $y$. For example, ordered pairs $(3, 1)$ and $(3, -1)$ satisfy the relation.

**31.** $f(x) = \dfrac{8 + x}{8 - x}$

$x$ can be any real number except 8, since this will give a denominator of zero. Thus, the domain is $(-\infty, 8) \cup (8, \infty)$.

**33.** **(a)** As $x$ is getting larger on the interval $[2, \infty)$, the value of $y$ is increasing.

**(b)** As $x$ is getting larger on the interval $(-\infty, -2]$, the value of $y$ is decreasing.

**35.** $f(x) = -2x^2 + 3x - 6$

$$f(3) = -2 \cdot 3^2 + 3 \cdot 3 - 6$$
$$= -2 \cdot 9 + 3 \cdot 3 - 6$$
$$= -18 + 9 - 6 = -15$$

**37.** $f(x) = -2x^2 + 3x - 6 \Rightarrow f(k) = -2k^2 + 3k - 6$

**39.** $2x - 5y = 5 \Rightarrow -5y = -2x + 5 \Rightarrow y = \tfrac{2}{5}x - 1$

The graph is the line with slope $\tfrac{2}{5}$ and $y$-intercept $-1$. It may also be graphed using intercepts. To do this, locate the $x$-intercept:

$x$-intercept: $y = 0$

$$2x - 5(0) = 5 \Rightarrow 2x = 5 \Rightarrow x = \tfrac{5}{2}$$

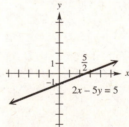

**41.** $2x + 5y = 20 \Rightarrow 5y = -2x + 20 \Rightarrow y = -\frac{2}{5}x + 4$

The graph is the line with slope of $-\frac{2}{5}$ and $y$-intercept 4. It may also be graphed using intercepts. To do this, locate the $x$-intercept: $x$-intercept: $y = 0$

$2x + 5(0) = 20 \Rightarrow 2x = 20 \Rightarrow x = 10$

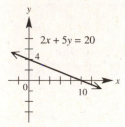

**43.** $f(x) = x$

The graph is the line with slope 1 and $y$-intercept 0, which means that it passes through the origin. Use another point such as (1, 1) to complete the graph.

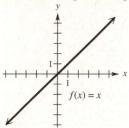

**45.** $x = -5$

The graph is the vertical line through $(-5, 0)$.

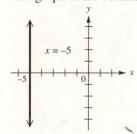

**47.** $y + 2 = 0 \Rightarrow y = -2$

The graph is the horizontal line through $(0, -2)$.

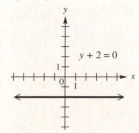

**49.** Line through (0, 5), $m = -\frac{2}{3}$

Note that $m = -\frac{2}{3} = \frac{-2}{3}$.

Begin by locating the point (0, 5).  Since the slope is $\frac{-2}{3}$, a change of 3 units horizontally (3 units to the right) produces a change of $-2$ units vertically (2 units down). This gives a second point, (3, 3), which can be used to complete the graph.

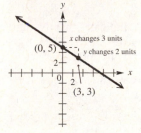

**51.** through (2, –2) and (3, –4)

$$m = \frac{y_2 - y_1}{x_2 - x_1} = \frac{-4 - (-2)}{3 - 2} = \frac{-2}{1} = -2$$

**53.** through (0, –7) and (3, –7)

$$m = \frac{-7 - (-7)}{3 - 0} = \frac{0}{3} = 0$$

**55.** $11x + 2y = 3$

Solve for $y$ to put the equation in slope-intercept form.

$$2y = -11x + 3 \Rightarrow y = -\frac{11}{2}x + \frac{3}{2}$$

Thus, the slope is $-\frac{11}{2}$.

**57.** $x - 2 = 0 \Rightarrow x = 2$

The graph is a vertical line, through (2, 0). The slope is undefined.

**59.** Initially, the car is at home. After traveling for 30 mph for 1 hr, the car is 30 mi away from home. During the second hour the car travels 20 mph until it is 50 mi away. During the third hour the car travels toward home at 30 mph until it is 20 mi away. During the fourth hour the car travels away from home at 40 mph until it is 60 mi away from home. During the last hour, the car travels 60 mi at 60 mph until it arrived home.

**61. (a)** We need to first find the slope of a line that passes between points $(0, 30.7)$ and $(5, 70.7)$

$$m = \frac{y_2 - y_1}{x_2 - x_1} = \frac{70.7 - 30.7}{5 - 0} = \frac{40}{5} = 8$$

Now use the point-slope form with $(x_1, y_1) = (0, 30.7)$ and $m = 8$. (The other point, $(5, 70.7)$, could also have been used.)

$$y - 30.7 = 8(x - 0) \Rightarrow y = 8x + 30.7$$

The slope, 8, indicates that the number of e-filing taxpayers increased by 8% each year from 2001 to 2005.

**(b)** For 2005, we evaluate the function for $x = 4$. $y = 8(4) + 30.7 = 62.7$

62.7% of the tax returns are predicted to have been filed electronically.

**63. (a)** through $(3, -5)$ with slope $-2$
Use the point-slope form.
$$y - y_1 = m(x - x_1)$$
$$y - (-5) = -2(x - 3)$$
$$y + 5 = -2(x - 3)$$
$$y + 5 = -2x + 6$$
$$y = -2x + 1$$

**(b)** Standard form: $y = -2x + 1 \Rightarrow 2x + y = 1$

**65. (a)** through $(2, -1)$ parallel to $3x - y = 1$
Find the slope of $3x - y = 1$.
$$3x - y = 1 \Rightarrow -y = -3x + 1 \Rightarrow y = 3x - 1$$
The slope of this line is 3. Since parallel lines have the same slope, 3 is also the slope of the line whose equation is to be found. Now use the point-slope form with $(x_1, y_1) = (2, -1)$ and $m = 3$.
$$y - y_1 = m(x - x_1)$$
$$y - (-1) = 3(x - 2)$$
$$y + 1 = 3x - 6 \Rightarrow y = 3x - 7$$

**(b)** Standard form:
$$y = 3x - 7 \Rightarrow -3x + y = -7 \Rightarrow 3x - y = 7$$

**67. (a)** through $(2, -10)$, perpendicular to a line with an undefined slope
A line with an undefined slope is a vertical line. Any line perpendicular to a vertical line is a horizontal line, with an equation of the form $y = b$. Since the line passes through $(2, -10)$, the equation of the line is $y = -10$.

**(b)** Standard form: $y = -10$

**69. (a)** through $(-7, 4)$, perpendicular to $y = 8$
The line $y = 8$ is a horizontal line, so any line perpendicular to it will be a vertical line. Since $x$ has the same value at all points on the line, the equation is $x = -7$. It is not possible to write this in slope-intercept form.

**(b)** Standard form: $x = -7$

**71.** $f(x) = |x| - 3$

The graph is the same as that of $y = |x|$, except that it is translated 3 units downward.

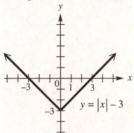

**73.** $f(x) = -|x + 1| + 3 = -|x - (-1)| + 3$

The graph of $f(x) = -|x + 1| + 3$ is a translation of the graph of $y = |x|$ to the left 1 unit, reflected over the $x$-axis and translated up 3 units.

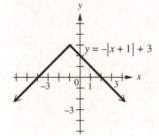

**75.** $f(x) = [\![x - 3]\!]$

To get $y = 0$, we need $0 \le x - 3 < 1 \Rightarrow 3 \le x < 4$. To get $y = 1$, we need $1 \le x - 3 < 2 \Rightarrow 4 \le x < 5$.
Follow this pattern to graph the step function.

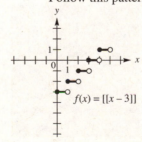

77. $f(x) = \begin{cases} -4x+2 \text{ if } x \le 1 \\ 3x-5 \quad \text{if } x > 1 \end{cases}$

Draw the graph of $y = -4x + 2$ to the left of $x = 1$, including the endpoint at $x = 1$. Draw the graph of $y = 3x - 5$ to the right of $x = 1$, but do not include the endpoint at $x = 1$. Observe that the endpoints of the two pieces coincide.

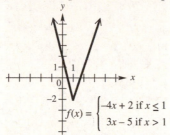

79. $f(x) = \begin{cases} |x| \quad \text{if } x < 3 \\ 6-x \text{ if } x \ge 3 \end{cases}$

Draw the graph of $y = |x|$ to the left of $x = 3$, but do not include the endpoint. Draw the graph of $y = 6 - x$ to the right of $x = 3$, including the endpoint. Observe that the endpoints of the two pieces coincide.

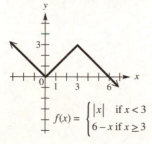

81. The graph of an even function is symmetric with respect to the $y$-axis. This statement is true.

83. If $(a, b)$ is on the graph of an even function, so is $(a, -b)$. The statement is false. For example, $f(x) = x^2$ is even, and $(2, 4)$ is on the graph but $(2, -4)$ is not.

85. The constant function $f(x) = 0$ is both even and odd. Since $f(-x) = 0 = f(x)$, the function is even. Also since $f(-x) = 0 = -0 = -f(x)$, the function is odd. This statement is true.

87. $x + y^2 = 10$

Replace $x$ with $-x$ to obtain $(-x) + y^2 = 10$. The result is not the same as the original equation, so the graph is not symmetric with respect to the $y$-axis. Replace $y$ with $-y$ to obtain $x + (-y)^2 = 10 \Rightarrow x + y^2 = 10$. The result is the same as the original equation, so the graph is symmetric with respect to the $x$-axis. Replace $x$ with $-x$ and $y$ with $-y$ to obtain $(-x) + (-y)^2 = 10 \Rightarrow (-x) + y^2 = 10$. The result is not the same as the original equation, so the graph is not symmetric with respect to the origin. The graph is symmetric with respect to the $x$-axis only.

89. $x^2 = y^3$

Replace $x$ with $-x$ to obtain $(-x)^2 = y^3 \Rightarrow x^2 = y^3$. The result is the same as the original equation, so the graph is symmetric with respect to the $y$-axis. Replace $y$ with $-y$ to obtain $x^2 = (-y)^3 \Rightarrow x^2 = -y^3$. The result is not the same as the original equation, so the graph is not symmetric with respect to the $x$-axis. Replace $x$ with $-x$ and $y$ with $-y$ to obtain $(-x)^2 = (-y)^3 \Rightarrow x^2 = -y^3$. The result is not the same as the original equation, so the graph is not symmetric with respect to the origin. Therefore, the graph is symmetric with respect to the $y$-axis only.

91. $6x + y = 4$

Replace $x$ with $-x$ to obtain $6(-x) + y = 4 \Rightarrow -6x + 7 = 4$. The result is not the same as the original equation, so the graph is not symmetric with respect to the $y$-axis. Replace $y$ with $-y$ to obtain $6x + (-y) = 4 \Rightarrow 6x - y = 4$. The result is not the same as the original equation, so the graph is not symmetric with respect to the $x$-axis.

Replace $x$ with $-x$ and $y$ with $-y$ to obtain $6(-x) + (-y) = 4 \Rightarrow -6x - y = 4$. This equation is not equivalent to the original one, so the graph is not symmetric with respect to the origin. Therefore, the graph has none of the listed symmetries.

93. To obtain the graph of $g(x) = -|x|$, reflect the graph of $f(x) = |x|$ across the $x$-axis.

**95.** To obtain the graph of $k(x) = 2|x - 4|$, translate the graph of $f(x) = |x|$ to the right 4 units and stretch vertically by a factor of 2.

**97.** If the graph of $f(x) = 3x - 4$ is reflected about the $y$-axis, we obtain a graph whose equation is $y = f(-x) = 3(-x) - 4 = -3x - 4$.

**99. (a)** To graph $y = f(x) + 3$, translate the graph of $y = f(x)$, 3 units up.

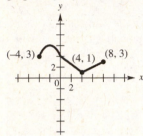

**(b)** To graph $y = f(x - 2)$, translate the graph of $y = f(x)$, 2 units to the right.

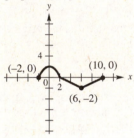

**(c)** To graph $y = f(x + 3) - 2$, translate the graph of $y = f(x)$, 3 units to the left and 2 units down.

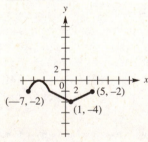

**(d)** To graph $y = |f(x)|$, keep the graph of $y = f(x)$ as it is for $y \geq 0$ and reflect the graph about the $x$-axis for $y < 0$.

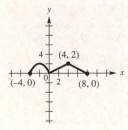

For Exercises 101–107, $f(x) = 3x^2 - 4$ and $g(x) = x^2 - 3x - 4$.

**101.** $(fg)(x) = f(x) \cdot g(x)$
$$= (3x^2 - 4)(x^2 - 3x - 4)$$
$$= 3x^4 - 9x^3 - 12x^2 - 4x^2 + 12x + 16$$
$$= 3x^4 - 9x^3 - 16x^2 + 12x + 16$$

**103.** $(f + g)(-4) = f(-4) + g(-4)$
$$= [3(-4)^2 - 4] + [(-4)^2 - 3(-4) - 4]$$
$$= [3(16) - 4] + [16 - 3(-4) - 4]$$
$$= [48 - 4] + [16 + 12 - 4]$$
$$= 44 + 24 = 68$$

**105.** $\left(\dfrac{f}{g}\right)(3) = \dfrac{f(3)}{g(3)} = \dfrac{3 \cdot 3^2 - 4}{3^2 - 3 \cdot 3 - 4} = \dfrac{3 \cdot 9 - 4}{9 - 3 \cdot 3 - 4}$
$$= \dfrac{27 - 4}{9 - 9 - 4} = \dfrac{23}{-4} = -\dfrac{23}{4}$$

**107.** The domain of $(fg)(x)$ is the intersection of the domain of $f(x)$ and the domain of $g(x)$. Both have domain $(-\infty, \infty)$, so the domain of $(fg)(x)$ is $(-\infty, \infty)$.

**109.** $f(x) = \dfrac{1}{x}, g(x) = x^2 + 1$

Since
$$(f \circ g)(x) = f[g(x)] \text{ and } (f \circ g)(x) = \dfrac{1}{x^2 + 1},$$
choices (C) and (D) are not equal to $(f \circ g)(x)$.

**111.** $f(x) = x^2 - 5x + 3$
$$f(x + h) = (x + h)^2 - 5(x + h) + 3$$
$$= x^2 + 2xh + h^2 - 5x - 5h + 3$$
$$f(x + h) - f(x)$$
$$= (x^2 + 2xh + h^2 - 5x - 5h + 3) - (x^2 - 5x + 3)$$
$$= x^2 + 2xh + h^2 - 5x - 5h + 3 - x^2 + 5x - 3$$
$$= 2xh + h^2 - 5h$$
$$\dfrac{f(x + h) - f(x)}{h} = \dfrac{2xh + h^2 - 5h}{h}$$
$$= \dfrac{h(2x + h - 5)}{h} = 2x + h - 5$$

For Exercises 113–115,
$f(x) = \sqrt{x - 2}$ and $g(x) = x^2$.

**113.** $(g \circ f)(x) = g[f(x)] = g\left(\sqrt{x - 2}\right)$
$$= \left(\sqrt{x - 2}\right)^2 = x - 2$$

**115.** Since $f(x) = \sqrt{x-2}$, $f(3) = \sqrt{3-2} = \sqrt{1} = 1$.

Therefore, $(g \circ f)(3) = g[f(3)] = g(1)$

$= 1^2 = 1$.

**117.** $(f+g)(1) = f(1) + g(1) = 7 + 1 = 8$

**119.** $(fg)(-1) = f(-1) \cdot g(-1) = 3(-2) = -6$

**121.** $(g \circ f)(-2) = g[f(-2)] = g(1) = 2$

**123.** $(f \circ g)(2) = f[g(2)] = f(2) = 1$

**125.** Let $x$ = number of yards.

$f(x) = 36x$, where $f(x)$ is the number of inches.

$g(x) = 1760x$, where $g(x)$ is the number of miles. Then

$(g \circ f)(x) = g[f(x)] = 1760(36x) = 63,360x$.

There are $63,360x$ inches in $x$ miles

**127.** If $V(r) = \frac{4}{3}\pi r^3$ and if the radius is increased by 3 inches, then the amount of volume gained is given by

$V_g(r) = V(r+3) - V(r) = \frac{4}{3}\pi(r+3)^3 - \frac{4}{3}\pi r^3$.

## Chapter 2: Test

**1. (a)** The domain of $f(x) = \sqrt{x} + 3$ occurs when $x \geq 0$. In interval notation, this correlates to the interval in D, $[0, \infty)$.

**(b)** The range of $f(x) = \sqrt{x} - 3$ is all real numbers greater than or equal to 0. In interval notation, this correlates to the interval in D, $[0, \infty)$.

**(c)** The domain of $f(x) = x^2 - 3$ is all real numbers. In interval notation, this correlates to the interval in C, $(-\infty, \infty)$.

**(d)** The range of $f(x) = x^2 + 3$ is all real numbers greater than or equal to 3. In interval notation, this correlates to the interval in B, $[3, \infty)$.

**(e)** The domain of $f(x) = \sqrt[3]{x} - 3$ is all real numbers. In interval notation, this correlates to the interval in C, $(-\infty, \infty)$.

**(f)** The range of $f(x) = \sqrt[3]{x} + 3$ is all real numbers. In interval notation, this correlates to the interval in C, $(-\infty, \infty)$.

**(g)** The domain of $f(x) = |x| - 3$ is all real numbers. In interval notation, this correlates to the interval in C, $(-\infty, \infty)$.

**(h)** The range of $f(x) = |x + 3|$ is all real numbers greater than or equal to 0. In interval notation, this correlates to the interval in D, $[0, \infty)$.

**(i)** The domain of $x = y^2$ is $x \geq 0$ since when you square any value of $y$, the outcome will be nonnegative. In interval notation, this correlates to the interval in D, $[0, \infty)$.

**(j)** The range of $x = y^2$ is all real numbers. In interval notation, this correlates to the interval in C, $(-\infty, \infty)$.

**2.** Consider the points $(-2, 1)$ and $(3, 4)$.

$$m = \frac{4-1}{3-(-2)} = \frac{3}{5}$$

**3.** We label the points $A(-2, 1)$ and $B(3, 4)$.

$$d(A, B) = \sqrt{[3-(-2)]^2 + (4-1)^2}$$
$$= \sqrt{5^2 + 3^2} = \sqrt{25+9} = \sqrt{34}$$

**4.** The midpoint has coordinates

$$\left(\frac{-2+3}{2}, \frac{1+4}{2}\right) = \left(\frac{1}{2}, \frac{5}{2}\right).$$

**5.** Use the point-slope form with $(x_1, y_1) = (-2, 1)$ and $m = \frac{3}{5}$.

$$y - y_1 = m(x - x_1)$$
$$y - 1 = \frac{3}{5}[x - (-2)]$$
$$y - 1 = \frac{3}{5}(x+2) \Rightarrow 5(y-1) = 3(x+2) \Rightarrow$$
$$5y - 5 = 3x + 6 \Rightarrow 5y = 3x + 11 \Rightarrow$$
$$-3x + 5y = 11 \Rightarrow 3x - 5y = -11$$

**6.** Solve $3x - 5y = -11$ for $y$.

$$3x - 5y = -11$$
$$-5y = -3x - 11$$
$$y = \frac{3}{5}x + \frac{11}{5}$$

Therefore, the linear function is

$$f(x) = \frac{3}{5}x + \frac{11}{5}.$$

**7. (a)** The center is at $(0, 0)$ and the radius is 2, so the equation of the circle is

$$x^2 + y^2 = 4.$$

**(b)** The center is at (1, 4) and the radius is 1, so the equation of the circle is
$$(x-1)^2+(y-4)^2=1$$

8. $x^2+y^2+4x-10y+13=0$

Complete the square on $x$ and $y$ to write the equation in standard form:
$$x^2+y^2+4x-10y+13=0$$
$$\left(x^2+4x+\quad\right)+\left(y^2-10y+\quad\right)=-13$$
$$\left(x^2+4x+4\right)+\left(y^2-10y+25\right)=-13+4+25$$
$$\left(x+2\right)^2+\left(y-5\right)^2=16$$

The circle has center $(-2, 5)$ and radius 4.

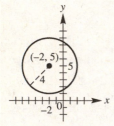

$x^2+y^2+4x-10y+13=0$

9. **(a)** This is not the graph of a function because some vertical lines intersect it in more than one point. The domain of the relation is [0, 4]. The range is [–4, 4].

   **(b)** This is the graph of a function because no vertical line intersects the graph in more than one point. The domain of the function is $(-\infty, -1)\cup(-1, \infty)$. The range is $(-\infty, 0)\cup(0, \infty)$. As $x$ is getting larger on the intervals $(-\infty, -1)$ and $(-1, \infty)$, the value of $y$ is decreasing, so the function is decreasing on these intervals. (The function is never increasing or constant.)

10. Point $A$ has coordinates $(5, -3)$.

    **(a)** The equation of a vertical line through $A$ is $x = 5$.

    **(b)** The equation of a horizontal line through $A$ is $y = -3$.

11. The slope of the graph of $y = -3x+2$ is $-3$.

    **(a)** A line parallel to the graph of $y = -3x+2$ has a slope of $-3$.
    Use the point-slope form with $(x_1, y_1) = (2, 3)$ and $m = -3$.
    $$y-y_1=m(x-x_1)$$
    $$y-3=-3(x-2)$$
    $$y-3=-3x+6\Rightarrow y=-3x+9$$

**(b)** A line perpendicular to the graph of $y=-3x+2$ has a slope of $\frac{1}{3}$ since
$$-3\left(\tfrac{1}{3}\right)=-1.$$
$$y-3=\tfrac{1}{3}(x-2)$$
$$3(y-3)=x-2\Rightarrow 3y-9=x-2\Rightarrow$$
$$3y=x+7\Rightarrow y=\tfrac{1}{3}x+\tfrac{7}{3}$$

12. **(a)** $(-\infty, -3)$

    **(b)** $(4, \infty)$

    **(c)** $[-3, 4]$

    **(d)** $(-\infty, -3);\ [-3, 4];\ (4, \infty)$

    **(e)** $(-\infty, \infty)$

    **(f)** $(-\infty, 2)$

13. To graph $y=|x-2|-1$, we translate the graph of $y=|x|$, 2 units to the right and 1 unit down.

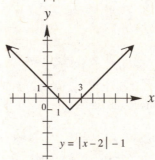

$y=|x-2|-1$

14. $f(x)=[\![x+1]\!]$

To get $y = 0$, we need $0\le x+1<1\Rightarrow -1\le x<0$. To get $y = 1$, we need $1\le x+1<2\Rightarrow 0\le x<1$. Follow this pattern to graph the step function.

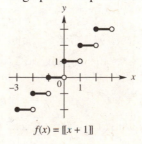

$f(x)=[\![x+1]\!]$

**15.** $f(x) = \begin{cases} 3 & \text{if } x < -2 \\ 2 - \frac{1}{2}x & \text{if } x \ge -2 \end{cases}$

For values of $x$ with $x < -2$, we graph the horizontal line $y = 3$. For values of $x$ with $x \ge -2$, we graph the line with a slope of $-\frac{1}{2}$ and a $y$-intercept of 2. Two points on this line are $(-2, 3)$ and $(0, 2)$.

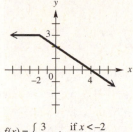

$f(x) = \begin{cases} 3 & \text{if } x < -2 \\ 2 - \frac{1}{2}x & \text{if } x \ge -2 \end{cases}$

**16. (a)** Shift $f(x)$, 2 units vertically upward.

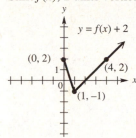

**(b)** Shift $f(x)$, 2 units horizontally to the left.

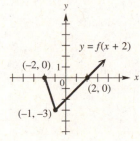

**(c)** Reflect $f(x)$, across the $x$-axis.

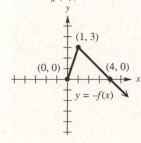

**(d)** Reflect $f(x)$, across the $y$-axis.

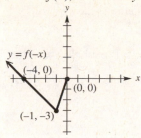

**(e)** Stretch $f(x)$, vertically by a factor of 2.

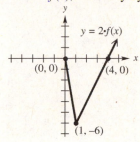

**17.** Answers will vary. Starting with $y = \sqrt{x}$, we shift it to the left 2 units and stretch it vertically by a factor of 2. The graph is then reflected over the $x$-axis and then shifted down 3 units.

**18.** $3x^2 - y^2 = 3$

**(a)** Replace $y$ with $-y$ to obtain
$3x^2 - (-y)^2 = 3 \Rightarrow 3x^2 - y^2 = 3$.
The result is the same as the original equation, so the graph is symmetric with respect to the $x$-axis.

**(b)** Replace $x$ with $-x$ to obtain
$3(-x)^2 - y^2 = 3 \Rightarrow 3x^2 - y^2 = 3$.
The result is the same as the original equation, so the graph is symmetric with respect to the $y$-axis.

**(c)** Since the graph is symmetric with respect to the $x$-axis and with respect to the $y$-axis, it must also be symmetric with respect to the origin.

**19.** $f(x) = 2x^2 - 3x + 2, \; g(x) = -2x + 1$

**(a)** $(f - g)(x) = f(x) - g(x)$
$= \left(2x^2 - 3x + 2\right) - \left(-2x + 1\right)$
$= 2x^2 - 3x + 2 + 2x - 1$
$= 2x^2 - x + 1$

**(b)** $\left(\dfrac{f}{g}\right)(x) = \dfrac{f(x)}{g(x)} = \dfrac{2x^2 - 3x + 2}{-2x + 1}$

**(c)** We must determine which values solve the equation $-2x+1=0$.

$$-2x+1=0 \Rightarrow -2x=-1 \Rightarrow x=\tfrac{1}{2}$$

Thus, $\tfrac{1}{2}$ is excluded from the domain, and the domain is $\left(-\infty,\tfrac{1}{2}\right) \cup \left(\tfrac{1}{2}, \infty\right)$..

**(d)** $f(x) = 2x^2 - 3x + 2$

$$\begin{aligned}
f(x+h) &= 2(x+h)^2 - 3(x+h) + 2 \\
&= 2\left(x^2 + 2xh + h^2\right) - 3x - 3h + 2 \\
&= 2x^2 + 4xh + 2h^2 - 3x - 3h + 2
\end{aligned}$$

$$\begin{aligned}
f(x+h) &- f(x) \\
&= \left(2x^2 + 4xh + 2h^2 - 3x - 3h + 2\right) \\
&\quad - \left(2x^2 - 3x + 2\right) \\
&= 2x^2 + 4xh + 2h^2 - 3x \\
&\quad\quad\quad -3h + 2 - 2x^2 + 3x - 2 \\
&= 4xh + 2h^2 - 3h
\end{aligned}$$

$$\begin{aligned}
\frac{f(x+h)-f(x)}{h} &= \frac{4xh + 2h^2 - 3h}{h} \\
&= \frac{h(4x + 2h - 3)}{h} \\
&= 4x + 2h - 3
\end{aligned}$$

**20. (a)**
$$\begin{aligned}
(f+g)(1) &= f(1) + g(1) \\
&= (2 \cdot 1^2 - 3 \cdot 1 + 2) + (-2 \cdot 1 + 1) \\
&= (2 \cdot 1 - 3 \cdot 1 + 2) + (-2 \cdot 1 + 1) \\
&= (2 - 3 + 2) + (-2 + 1) \\
&= 1 + (-1) = 0
\end{aligned}$$

**(b)**
$$\begin{aligned}
(fg)(2) &= f(2) \cdot g(2) \\
&= (2 \cdot 2^2 - 3 \cdot 2 + 2) \cdot (-2 \cdot 2 + 1) \\
&= (2 \cdot 4 - 3 \cdot 2 + 2) \cdot (-2 \cdot 2 + 1) \\
&= (8 - 6 + 2) \cdot (-4 + 1) \\
&= 4(-3) = -12
\end{aligned}$$

**(c)** $g(x) = -2x + 1 \Rightarrow g(0) = -2(0) + 1$
$= 0 + 1 = 1$. Therefore,
$$\begin{aligned}
(f \circ g)(0) &= f\left[g(0)\right] \\
&= f(1) = 2 \cdot 1^2 - 3 \cdot 1 + 2 \\
&= 2 \cdot 1 - 3 \cdot 1 + 2 \\
&= 2 - 3 + 2 = 1
\end{aligned}$$

**21.**
$$\begin{aligned}
(f \circ g) &= f\left(g(x)\right) = f(2x-7) \\
&= \sqrt{(2x-7)+1} = \sqrt{2x-6}
\end{aligned}$$

The domain and range of $g$ are $(-\infty, \infty)$, while the domain of $f$ is $[0, \infty)$. We need to find the values of $x$ which fit the domain of $f$: $2x - 6 \geq 0 \Rightarrow x \geq 3$. So, the domain of $f \circ g$ is $[3, \infty)$.

**22.**
$$\begin{aligned}
(g \circ f) &= g\left(f(x)\right) = g\left(\sqrt{x+1}\right) \\
&= 2\sqrt{x+1} - 7
\end{aligned}$$

The domain and range of $g$ are $(-\infty, \infty)$, while the domain of $f$ is $[0, \infty)$. We need to find the values of $x$ which fit the domain of $f$: $x + 1 \geq 0 \Rightarrow x \geq -1$. So, the domain of $g \circ f$ is $[-1, \infty)$.

**23.**
$$\begin{aligned}
f(x) &= .4[\![x]\!] + .75 \\
f(5.5) &= .4[\![5.5]\!] + .75 = .4(5) + .75 \\
&= 2 + .75 = \$2.75
\end{aligned}$$

**24. (a)** $C(x) = 3300 + 4.50x$

**(b)** $R(x) = 10.50x$

**(c)**
$$\begin{aligned}
P(x) &= R(x) - C(x) \\
&= 10.50x - (3300 + 4.50x) \\
&= 6.00x - 3300
\end{aligned}$$

**(d)**
$$\begin{aligned}
P(x) &> 0 \\
6.00x - 3300 &> 0 \\
6.00x &> 3300 \\
x &> 550
\end{aligned}$$

He must produce and sell 551 items before he earns a profit.

# Chapter 3

## POLYNOMIAL AND RATIONAL FUNCTIONS

### Section 3.1: Quadratic Functions and Models

1. $f(x) = (x+3)^2 - 4$

   (a) domain: $(-\infty, \infty)$; range: $[-4, \infty)$

   (b) vertex: $(h, k) = (-3, -4)$

   (c) axis: $x = -3$

   (d) To find the $y$-intercept, let $x = 0$.
   $$y = (0+3)^2 - 4 = 3^2 - 4 = 9 - 4 = 5$$
   $y$-intercept: 5

   (e) To find the $x$-intercepts, let $f(x) = 0$.
   $$0 = (x+3)^2 - 4$$
   $$(x+3)^2 = 4$$
   $$x+3 = \pm\sqrt{4} = \pm 2$$
   $$x = -3 \pm 2$$
   $$x = -3 - 2 = -5 \text{ or } x = -3 + 2 = -1$$
   $x$-intercepts: $-5$ and $-1$

3. $f(x) = -2(x+3)^2 + 2$

   (a) domain: $(-\infty, \infty)$; range: $(-\infty, 2]$

   (b) vertex: $(h, k) = (-3, 2)$

   (c) axis: $x = -3$

   (d) To find the $y$-intercept, let $x = 0$.
   $$y = -2(0+3)^2 + 2 = -2 \cdot 3^2 + 2$$
   $$= -2 \cdot 9 + 2 = -18 + 2 = -16$$
   $y$-intercept: $-16$

   (e) To find the $x$-intercepts, let $f(x) = 0$.
   $$0 = -2(x+3)^2 + 2$$
   $$(x+3)^2 = 1$$
   $$x+3 = \pm\sqrt{1} = \pm 1$$
   $$x = -3 \pm 1$$
   $$x = -3 - 1 = -4 \text{ or } x = -3 + 1 = -2$$
   $x$-intercepts: $-4$ and $-2$

5. $f(x) = (x-4)^2 - 3$

   Since $a > 0$, the parabola opens upward. The vertex is at $(4, -3)$. The correct graph, therefore, is B.

7. $f(x) = (x+4)^2 - 3$

   Since $a > 0$, the parabola opens upward. The vertex is at $(-4, -3)$. The correct graph, therefore, is D.

9. For parts (a), (b), (c), and (d), see the following graph.

   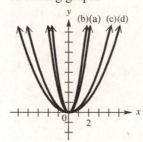

   (e) If the absolute value of the coefficient is greater than 1, it causes the graph to be stretched vertically, so it is narrower. If the absolute value of the coefficient is between 0 and 1, it causes the graph to shrink vertically, so it is broader.

11. For parts (a), (b), (c), and (d), see the following graph.

    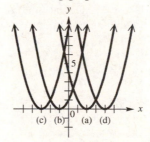

    (e) The graph of $(x-h)^2$ is translated $h$ units to the right if $h$ is positive and $|h|$ units to the left if $h$ is negative.

13. $f(x) = (x-2)^2$

    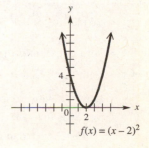

    $f(x) = (x-2)^2$

*(continued on next page)*

*(continued from page 131)*

This equation is of the form $y = (x-h)^2$, with $h = 2$. The graph opens upward and has the same shape as that of $y = x^2$. It is a horizontal translation of the graph of $y = x^2$, 2 units to the right. The vertex is (2, 0) and the axis is the vertical line $x = 2$. Additional points on the graph are $(1,1)$ and $(3,1)$. The domain is $(-\infty, \infty)$. Since the smallest value of $y$ is 0 and the graph opens upward, the range is $[0, \infty)$.

**15.** $f(x) = (x+3)^2 - 4 = \left[x - (-3)\right]^2 + (-4)$

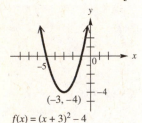

$f(x) = (x+3)^2 - 4$

This equation is of the form $y = (x-h)^2 + k$, with $h = -3$ and $k = -4$. The graph opens upward and has the same shape as $y = x^2$. The vertex is (-3, -4). It is a translation of $y = x^2$, 3 units to the left and 4 units down. The axis is the vertical line $x = -3$. Additional points on the graph are $(-4,-3)$ and $(-2,-3)$. The domain is $(-\infty, \infty)$. Since the smallest value of $y$ is -4 and the graph opens upward, the range is $[-4, \infty)$.

**17.** $f(x) = -\frac{1}{2}(x+1)^2 - 3 = -\frac{1}{2}\left[x-(-1)\right]^2 + (-3)$

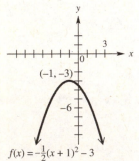

$f(x) = -\frac{1}{2}(x+1)^2 - 3$

This equation is of the form $y = a(x-h)^2 + k$, with $h = -1$, $k = -3$, and $a = -\frac{1}{2}$. The graph opens downward and is wider than $y = x^2$. The vertex is (-1, -3). It is a translation of the graph $y = -\frac{1}{2}x^2$, 1 unit to the left and 3 units down. The axis is the vertical line $x = -1$.

Additional points on the graph are $\left(-2,-3\frac{1}{2}\right)$ and $\left(0,-3\frac{1}{2}\right)$. The domain is $(-\infty, \infty)$. Since the largest value of $y$ is -3 and the graph opens downward, the range is $(-\infty, -3]$.

**19.** $f(x) = x^2 - 2x + 3$; Rewrite by completing the square on $x$.

$$f(x) = x^2 - 2x + 3 = (x^2 - 2x + 1 - 1) + 3$$
$$\text{Note: } \left[\tfrac{1}{2}(-2)\right]^2 = (-1)^2 = 1$$
$$= (x^2 - 2x + 1) - 1 + 3 = (x-1)^2 + 2$$

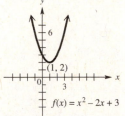

$f(x) = x^2 - 2x + 3$

This equation is of the form $y = (x-h)^2 + k$, with $h = 1$ and $k = 2$. The graph opens upward and has the same shape as $y = x^2$. The vertex is (1, 2). It is a translation of the graph $y = x^2$, 1 unit to the right and 2 units up. The axis is the vertical line $x = 1$. Additional points on the graph are $(0,3)$ and $(2,3)$. The domain is $(-\infty, \infty)$. Since the smallest value of $y$ is 2 and the graph opens upward, the range is $[2, \infty)$.

**21.** $f(x) = x^2 - 10x + 21$; Rewrite by completing the square on $x$.

$$f(x) = x^2 - 10x + 21 = (x^2 - 10x + 25 - 25) + 21$$
$$\text{Note: } \left[\tfrac{1}{2}(-10)\right]^2 = (-5)^2 = 25$$
$$= (x^2 - 10x + 25) - 25 + 21 = (x-5)^2 - 4$$
$$= (x-5)^2 + (-4)$$

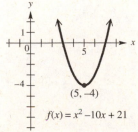

$f(x) = x^2 - 10x + 21$

This equation is of the form $y = (x-h)^2 + k$, with $h = 5$ and $k = -4$. The graph opens upward and has the same shape as $y = x^2$.

The vertex is (5, –4). It is a translation of the graph $y = x^2$, 5 units to the right and 4 units down. The axis is the vertical line $x = 5$. Additional points on the graph are $(4, -3)$ and $(6, -3)$. The domain is $(-\infty, \infty)$. Since the smallest value of $y$ is –4 and the graph opens upward, the range is $[-4, \infty)$.

**23.** $f(x) = -2x^2 - 12x - 16$; Rewrite by completing the square on $x$.

$$f(x) = -2x^2 - 12x - 16 = -2(x^2 + 6x) - 16$$
$$= -2(x^2 + 6x + 9 - 9) - 16$$

$$\text{Note: } \left[\tfrac{1}{2}(6)\right]^2 = 3^2 = 9$$

$$= -2(x^2 + 6x + 9) + 18 - 16$$
$$= -2(x + 3)^2 + 2 = -2\left[x - (-3)\right]^2 + 2$$

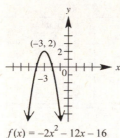

$f(x) = -2x^2 - 12x - 16$

This equation is of the form $y = a(x - h)^2 + k$, with $h = -3$, $k = 2$, and $a = -2$. The graph opens downward and is narrower than $y = x^2$. The vertex is $(-3, 2)$. It is a translation of the graph $y = -2x^2$, 3 units to the left and 2 units up. The axis is the vertical line $x = -3$. Additional points on the graph are $(-4, 0)$ and $(-2, 0)$. The domain is $(-\infty, \infty)$. Since the largest value of $y$ is 2 and the graph opens downward, the range is $(-\infty, 2]$.

**25.** $f(x) = -x^2 - 6x - 5$; Rewrite by completing the square on $x$.

$$f(x) = -x^2 - 6x - 5 = -(x^2 + 6x + 9 - 9) - 5$$

$$\text{Note: } \left[\tfrac{1}{2}(6)\right]^2 = 3^2 = 9$$

$$= -(x^2 + 6x + 9) + 9 - 5 = -(x + 3)^2 + 4$$
$$= -1 \cdot \left[x - (-3)\right]^2 + 4$$

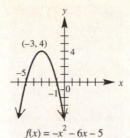

$f(x) = -x^2 - 6x - 5$

This equation is of the form $y = a(x - h)^2 + k$, with $h = -3$, $k = 4$, and $a = -1$. The graph opens downward and has the same shape as $y = x^2$. The vertex is $(-3, 4)$. It is a translation of the graph $y = -x^2$, 3 units to the left and 4 units up. The axis is the vertical line $x = -3$. Additional points on the graph are $(-2, 3)$ and $(-1, 0)$. The domain is $(-\infty, \infty)$. Since the largest value of $y$ is 4 and the graph opens downward, the range is $(-\infty, 4]$.

**27.** The minimum value of $f(x)$ is $f(-3) = 3$.

**29.** There are no real solutions to the equation $f(x) = 1$ since the value of $f(x)$ is never less than 3.

**31.** $a < 0$, $b^2 - 4ac = 0$

The correct choice is E. $a < 0$ indicates that the parabola will open downward, while $b^2 - 4ac = 0$ indicates that the graph will have exactly one $x$-intercept.

**33.** $a < 0$, $b^2 - 4ac < 0$

The correct choice is D. $a < 0$ indicates that the parabola will open downward, while $b^2 - 4ac < 0$ indicates that the graph will have no $x$-intercepts.

**35.** $a > 0$, $b^2 - 4ac > 0$

The correct choice is C. $a > 0$ indicates that the parabola will open upward, while $b^2 - 4ac > 0$ indicates that the graph will have two $x$-intercepts.

**37.** The vertex of the parabola in the figure is (2, −1) and the *y*-intercept is 0. The equation takes the form $f(x) = a(x-2)^2 - 1$. When $x = 0, f(x) = 0$, so $0 = a(0-2)^2 - 1 \Rightarrow$ $0 = 4a - 1 \Rightarrow a = \frac{1}{4}$. The equation is $f(x) = \frac{1}{4}(x-2)^2 - 1$. This function may also be written as $f(x) = \frac{1}{4}(x-2)^2 - 1$ $= \frac{1}{4}\left(x^2 - 4x + 4\right) - 1 = \frac{1}{4}x^2 - x$. Graphing this function on a graphing calculator shows that the graph matches the equation.

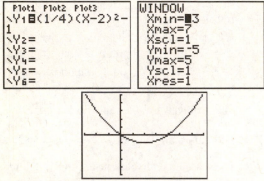

This graph can be displayed in other modes. If you press the MODE button, the following screen will appear. Toggle down to the last row and change FULL to Horiz.

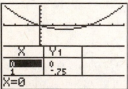

Hit 2nd then GRAPH (TABLE) to get the table to display below the graph.

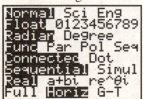

If you press the MODE button again, you can change to G-T mode (Graph-Table).

When you press the Graph button, the following screen will appear.

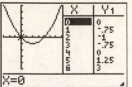

**39.** The vertex of the parabola in the figure is (1, 4) and the *y*-intercept is 2. The equation takes the form $f(x) = a(x-1)^2 + 4$. When $x = 0, f(x) = 2$, so $2 = a(0-1)^2 + 4 \Rightarrow$ $2 = a + 4 \Rightarrow a = -2$. The equation is $f(x) = -2(x-1)^2 + 4$. This function may also be written as $f(x) = -2(x^2 - 2x + 1) + 4$ $= -2x^2 + 4x - 2 + 4 = -2x^2 + 4x + 2$. Graphing this function on a graphing calculator shows that the graph matches the equation.

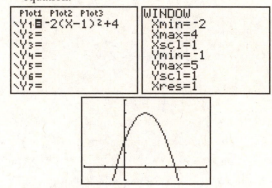

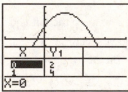

In Horizontal mode

In Graph-Table mode

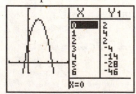

**41.** Quadratic; the points lie in a pattern that suggests a parabola opening downward, so $a < 0$.

**43.** Quadratic; the points lie in a pattern suggesting a parabola opening upward, so $a > 0$.

**45.** Linear; the points lie in pattern that suggesting a positive slope.

**47.** Let $x$ = one number. Then $12 - x$ = the other number. Now find the maximum of $x(12 - x)$ by finding the vertex of the function:

$$f(x) = x(12 - x) = 12x - x^2 = -(x^2 - 12x)$$

$$= -(x^2 - 12x + 36) + 36 \quad \text{complete the square}$$

$$= -(x - 6)^2 + 36$$

The vertex occurs at the vertex. So the two numbers are 6 and $12 - 6 = 6$. Confirm graphically:

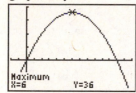

**49.** To find the number of units to be sold to minimize her costs, find the vertex of the function:

$$C(x) = x^2 - 40x + 610$$

$$= (x^2 - 40x + 400) + 610 - 400 \quad \text{complete the square}$$

$$= (x - 20)^2 + 210$$

The vertex occurs at $x = 20$, so she should sell 20 units to minimize her costs. The minimum cost is the value of the function at the vertex, $210. Confirm graphically:

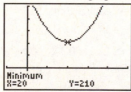

**51.** To find the amount of rainfall that will maximize the number of mosquitos, find the vertex of the function:

$$M(x) = 10x - x^2 = -(x^2 - 10x)$$

$$= -(x^2 - 10x + 25) + 25 \quad \text{complete the square}$$

$$= -(x - 5)^2 + 25$$

The vertex occurs at $x = 5$, which means that the maximum number of mosquitoes occurs when there are 5 inches of rain. The maximum number of mosquitoes is the value of the function at the vertex, 25 million. Confirm graphically:

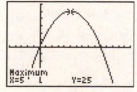

**53. (a)** Since $v_0 = 200$, and $s_0 = 50$, and

$$s(t) = -16t^2 + v_0 t + s_0 \text{ we have}$$

$$s(t) \text{ or } f(t) = -16t^2 + 200t + 50.$$

**(b)** Algebraic Solution:
Find the coordinates of the vertex of the parabola. Using the vertex formula with $a = -16$ and $b = 200$,

$$x = -\frac{b}{2a} = -\frac{200}{2(-16)} = 6.25 \text{ and}$$

$$y = -16(6.25)^2 + 200(6.25) + 50 = 675.$$

The vertex is $(6.25, 675)$. Since $a < 0$, this is the maximum point.
Graphing Calculator Solution:

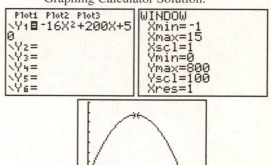

Thus, the number of seconds to reach maximum height is 6.25 seconds. The maximum height is 675 ft.

**(c)** Algebraic Solution:
To find the time interval in which the rocket will be more than 300 ft above ground level, solve the inequality
$-16t^2 + 200t + 50 > 300$ :

$$-16t^2 + 200t + 50 > 300 \Rightarrow$$

$$-16t^2 + 200t - 250 > 0 \Rightarrow$$

$$-8t^2 + 100t - 125 > 0$$

Solve the corresponding equation
$-8t^2 + 100t - 125 = 0.$
Use the quadratic formula with $a = -8$, $b = 100$, and $c = -125.$

$$t = \frac{-100 \pm \sqrt{100^2 - 4(-8)(-125)}}{2(-8)}$$

$$= \frac{-100 \pm \sqrt{10,000 - 4000}}{-16} = \frac{-100 \pm \sqrt{6000}}{-16}$$

$$t = \frac{-100 + \sqrt{6000}}{-16} \approx 1.4 \text{ or}$$

$$t = \frac{-100 - \sqrt{6000}}{-16} \approx 11.1$$

*(continued on next page)*

*(continued from page 135)*

The values 1.4 and 11.1 divide the number line into three intervals: $(-\infty, 1.4)$, $(1.4, 11.1)$, and $(11.1, \infty)$. Use a test point in each interval to determine where the inequality is satisfied.

| Interval | Test Value | Is $-16t^2 + 200t + 50 > 300$ True or False? |
|---|---|---|
| $(-\infty, 1.4)$ | 0 | $-16 \cdot 0^2 + 200 \cdot 0 + 50 \overset{?}{>} 300$<br>$50 > 300$<br>False |
| $(1.4, 11.1)$ | 2 | $-16 \cdot 2^2 + 200 \cdot 2 + 50 \overset{?}{>} 300$<br>$386 > 300$<br>True |
| $(11.1, \infty)$ | 12 | $-16 \cdot 12^2 + 200 \cdot 12 + 50 \overset{?}{>} 300$<br>$146 > 300$<br>False |

Graphing Calculator Solution:

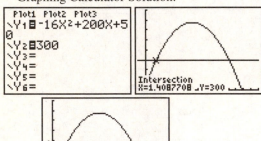

The rocket will be more than 300 ft above the ground between 1.4 sec and 11.1 sec.

**(d)** Algebraic Solution:

To find the number of seconds for the toy rocket to hit the ground, let $f(t) = 0$ and solve for $t$.

$-16t^2 + 200t + 50 = 0$

Use the quadratic formula with $a = -16$, $b = 200$, and $c = 50$.

$$t = \frac{-200 \pm \sqrt{200^2 - 4(-16)(50)}}{2(-16)}$$

$$= \frac{-200 \pm \sqrt{40,000 + 3200}}{-32}$$

$$= \frac{-200 \pm \sqrt{43,200}}{-32}$$

$$t = \frac{-200 + \sqrt{43,200}}{-32} \approx -.25 \text{ or}$$

$$t = \frac{-200 - \sqrt{43,200}}{-32} \approx 12.75$$

We reject the negative solution.

Graphing Calculator Solution:

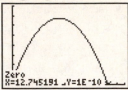

It will take approximately 12.75 seconds for the toy rocket to hit the ground.

**55. (a)** The length of the other side would be $640 - 2x$.

**(b)** In order for the two lengths to be positive, $0 < x < 320$.

**(c)** $A(x) = x(640 - 2x) = 640x - 2x^2$

$\qquad\quad = -2x^2 + 640x$

**(d)** Algebraic Solution:

Solve the inequality $30,000 < -2x^2 + 640x < 40,000$. Treat this as two inequalities,

$-2x^2 + 640x > 30,000$ and

$-2x^2 + 640x < 40,000$.

For $-2x^2 + 640x > 30,000$, we solve the corresponding equation

$-2x^2 + 640x = 30,000$.

This equation is equivalent to

$-2x^2 + 640x - 30,000 = 0$ or

$x^2 - 320x + 15,000 = 0$.

Use the quadratic formula with $a = 1$, $b = -320$, and $c = 15,000$.

$$x = \frac{-(-320) \pm \sqrt{(-320)^2 - 4(1)(15,000)}}{2(1)}$$

$$= \frac{320 \pm \sqrt{102,400 - 60,000}}{2}$$

$$= \frac{320 \pm \sqrt{42,400}}{2}$$

$$x = \frac{320 - \sqrt{42,400}}{2} \approx 57.04 \text{ or}$$

$$x = \frac{320 + \sqrt{42,400}}{2} \approx 262.96$$

The values 57.04 and 262.96 divide the number line into three intervals: $(-\infty, 57.04)$, $(57.04, 262.96)$, and $(262.96, \infty)$. Use a test point in each interval to determine where the inequality is satisfied.

| Interval | Test Value | Is $-2x^2 + 640x > 30{,}000$ True or False? |
|---|---|---|
| $(-\infty, 57.04)$ | 0 | $-2 \cdot 0^2 + 640 \cdot 0 \overset{?}{>} 30{,}000$<br>$0 > 30{,}000$<br>False |
| $(57.04, 262.96)$ | 60 | $-2 \cdot 60^2 + 640 \cdot 60 \overset{?}{>} 30{,}000$<br>$31{,}200 > 30{,}000$<br>True |
| $(262.96, \infty)$ | 300 | $-2 \cdot 300^2 + 640 \cdot 300 \overset{?}{>} 30{,}000$<br>$12{,}000 > 30{,}000$<br>False |

Thus, the first inequality is satisfied when the measure of $x$ is in the interval $(57.04, 262.96)$.

For $-2x^2 + 640x < 40{,}000$, we solve the corresponding equation
$$-2x^2 + 640x = 40{,}000.$$
This equation is equivalent to
$$-2x^2 + 640x - 40{,}000 = 0 \text{ or}$$
$$x^2 - 320x + 20{,}000 = 0$$
Use the quadratic formula with $a = 1$, $b = -320$, and $c = 20{,}000$.

$$x = \frac{-(-320) \pm \sqrt{(-320)^2 - 4(1)(20{,}000)}}{2(1)}$$
$$= \frac{320 \pm \sqrt{102{,}400 - 80{,}000}}{2}$$
$$= \frac{320 \pm \sqrt{22{,}400}}{2}$$
$$x = \frac{320 - \sqrt{22{,}400}}{2} \approx 85.17 \text{ or}$$
$$x = \frac{320 + \sqrt{22{,}400}}{2} \approx 234.83$$

The values 85.17 and 234.83 divide the number line into three intervals:
$(-\infty, 85.17)$, $(85.17, 234.83)$, and $(234.83, \infty)$. Use a test point in each interval to determine where the inequality is satisfied.

| Interval | Test Value | Is $-2x^2 + 640x < 40{,}000$ True or False? |
|---|---|---|
| $(-\infty, 85.17)$ | 0 | $-2 \cdot 0^2 + 640 \cdot 0 \overset{?}{<} 40{,}000$<br>$0 < 40{,}000$<br>True |
| $(85.17, 234.83)$ | 90 | $-2 \cdot 90^2 + 640 \cdot 90 \overset{?}{<} 40{,}000$<br>$41{,}400 < 40{,}000$<br>False |
| $(234.83, \infty)$ | 300 | $-2 \cdot 300^2 + 640 \cdot 300 \overset{?}{<} 40{,}000$<br>$12{,}000 < 40{,}000$<br>True |

Thus, the second inequality is satisfied when the measure of $x$ is in the interval $(-\infty, 85.17)$ or $(234.83, \infty)$. We must now seek the intersection of the intervals $(57.04, 262.96)$, and $(-\infty, 85.17)$ or $(234.83, \infty)$. If we use the real number line as an aid, we can see that the solution would be between 57.04 ft and 85.17 ft or 234.83 ft and 262.96 ft.

57.04  85.17                    234.83  262.96

Graphing Calculator Solution:

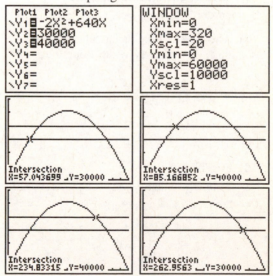

We can see from the graphs that the quadratic function lies between the lines when $x$ is between 57.04 ft and 85.17 ft or 234.83 ft and 262.96 ft.

**(e)** Algebraic Solution:

Find the coordinates of the vertex of the parabola, $A(x) = -2x^2 + 640x$. Using the vertex formula with $a = -2$ and $b = 640$. We have

$$x = -\frac{b}{2a} = -\frac{640}{2(-2)} = 160 \text{ and}$$

$$y = -2(160)^2 + 640(160) = 51,200.$$

Since $a < 0$, this is the maximum point.

Graphing Calculator Solution:

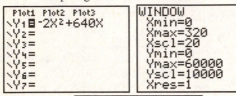

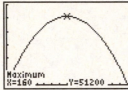

Thus, the length of the two parallel sides would be 160 ft and the third side would be $640 - 2(160) = 620 - 320 = 320$ ft.

The maximum area would be 51,200 ft².

**57. (a)** The length of the original piece of cardboard would be $2x$.

**(b)** The length of the rectangular box would be $2x - 4$ and the width would be $x - 4$, where $x > 4$.

**(c)** $V(x) = (2x - 4)(x - 4)(2)$
$= (2x^2 - 12x + 16)(2)$
$= 4x^2 - 24x + 32$

**(d)** Algebraic Solution:

Solve the equation $4x^2 - 24x + 32 = 320$.

$$4x^2 - 24x + 32 = 320$$
$$4x^2 - 24x - 288 = 0$$
$$x^2 - 6x - 72 = 0$$
$$(x + 6)(x - 12) = 0$$
$$x + 6 = 0 \Rightarrow x = -6 \text{ or}$$
$$x - 12 = 0 \Rightarrow x = 12$$

We discard the negative solution.

Graphing Calculator Solution:

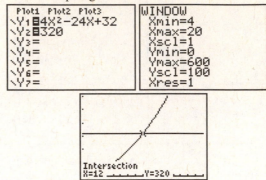

If $x = 12$, then the dimensions of the bottom of the box will be $12 - 4 = 8$ in by $2(12) - 4 = 24 - 4 = 20$ in.

**(e)** Algebraic Solution:

Solve the inequality

$400 < 4x^2 - 24x + 32 < 500$. Treat this as two inequalities, $4x^2 - 24x + 32 > 400$ and $4x^2 - 24x + 32 < 500$.

For $4x^2 - 24x + 32 > 400$, we solve the corresponding equation

$4x^2 - 24x + 32 = 400$.

This equation is equivalent to

$4x^2 - 24x - 368 = 0$ or $x^2 - 6x - 92 = 0$

Use the quadratic formula with $a = 1$, $b = -6$, and $c = -92$.

$$x = \frac{-(-6) \pm \sqrt{(-6)^2 - 4(1)(-92)}}{2(1)}$$

$$= \frac{6 \pm \sqrt{36 + 368}}{2} = \frac{6 \pm \sqrt{404}}{2}$$

$$= \frac{6 \pm 2\sqrt{101}}{2} = 3 \pm \sqrt{101}$$

$$x = 3 - \sqrt{101} \approx -7.0 \text{ or}$$
$$x = 3 + \sqrt{101} \approx 13.0$$

Since $x > 4$, we need only check the intervals: $(4, 13.0)$ and $(13.0, \infty)$. Use a test point in each interval to determine where the inequality is satisfied.

| Interval | Test Value | Is $4x^2 - 24x + 32 > 400$ True or False? |
|---|---|---|
| $(4, 13.0)$ | 5 | $4 \cdot 5^2 - 24 \cdot 5 + 32 \overset{?}{>} 400$ $12 > 400$ False |
| $(13.0, \infty)$ | 14 | $4 \cdot 14^2 - 24 \cdot 14 + 32 \overset{?}{>} 400$ $480 > 400$ True |

Thus, the first inequality is satisfied when the length, $x$, is in the interval $(13.0, \infty)$.

For $4x^2 - 24x + 32 < 500$, we solve the corresponding equation

$4x^2 - 24x + 32 = 500$. This equation is equivalent to $4x^2 - 24x - 468 = 0$ or $x^2 - 6x - 117 = 0$. Use the quadratic formula with $a = 1$, $b = -6$, and $c = -117$.

$$x = \frac{-(-6) \pm \sqrt{(-6)^2 - 4(1)(-117)}}{2(1)}$$

$$= \frac{6 \pm \sqrt{36 + 468}}{2} = \frac{6 \pm \sqrt{504}}{2}$$

$$= \frac{6 \pm 2\sqrt{126}}{2} = 3 \pm \sqrt{126}$$

$x = 3 - \sqrt{126} \approx -8.2$ or

$x = 3 + \sqrt{126} \approx 14.2$

Since $x > 4$, we need only check the intervals: $(4, 14.2)$ and $(14.2, \infty)$. Use a test point in each interval to determine where the inequality is satisfied.

| Interval | Test Value | Is $4x^2 - 24x + 32 < 500$ True or False? |
|---|---|---|
| $(4, 14.2)$ | 5 | $4 \cdot 5^2 - 24 \cdot 5 + 32 \overset{?}{<} 500$ $12 < 500$ True |
| $(14.2, \infty)$ | 15 | $4 \cdot 15^2 - 24 \cdot 15 + 32 \overset{?}{<} 500$ $572 < 500$ False |

Thus, the second inequality is satisfied when the length, $x$, is in the interval $(4, 14.2)$. We must now seek the intersection of the intervals $(13.0, \infty)$ and $(4, 14.2)$. This intersection is $(13, 14.2)$.

Graphing Calculator Solution:

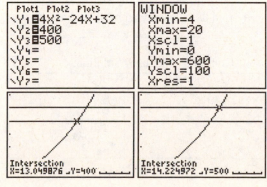

We can see from the graphs that the quadratic function lies between the lines when $x$ is between 13.0 in and 14.2 in. Thus, the volume will be between 400 in³ and 500 in³. when the original width is between 13.0 in. and 14.2 in.

**59.**  $h(x) = -.5x^2 + 1.25x + 3$

**(a)** Find $h(x)$ when $x = 2$.

$h(2) = -.5(2)^2 + 1.25(2) + 3$

$= -.5(4) + 1.25(2) + 3$

$= -2 + 2.5 + 3 = 3.5$

When the distance from the base of the stump was 2 ft, the frog was 3.5 ft high.

**(b)** Algebraic Solution:
Find $x$ when $h(x) = 3.25$.

$$3.25 = -0.5x^2 + 1.25x + 3$$

$$0.5x^2 - 1.25x + 0.25 = 0$$

$$2x^2 - 5x + 1 = 0$$

Use the quadratic formula with $a = 2$, $b = -5$, and $c = 1$.

$$x = \frac{-(-5) \pm \sqrt{(-5)^2 - 4(2)(1)}}{2(2)}$$

$$= \frac{5 \pm \sqrt{25 - 8}}{4} = \frac{5 \pm \sqrt{17}}{4}$$

$$x = \frac{5 - \sqrt{17}}{4} \approx .2 \text{ or } x = \frac{5 + \sqrt{17}}{4} \approx 2.3$$

Graphing Calculator Solution:

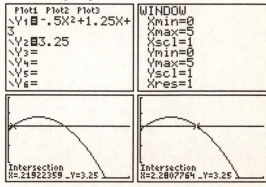

The frog was 3.25 ft above the ground when he was approximately .2 ft from the stump (on the way up) and 2.3 ft from the stump (on the way down).

**(c)** Algebraic Solution:
Since the parabola opens downward, the vertex is the maximum point. Use the vertex formula to find the $x$-coordinate of the vertex of $h(x) = -0.5x^2 + 1.25x + 3$.

$$x = -\frac{b}{2a} = -\frac{1.25}{2(-0.5)} = -\frac{1.25}{-1} = 1.25$$

Graphing Calculator Solution:

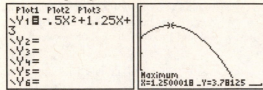

The frog reached its highest point at 1.25 ft from the stump.

**(d)** The maximum height is the $y$-coordinate of the vertex.

$$y = h(1.25)$$
$$= -.5(1.25)^2 + 1.25(1.25) + 3 = 3.78125$$

The maximum height reached by the frog was approximately 3.78 ft. This agrees with the graphing calculator solution by interpreting the $y$-coordinate of the vertex as the maximum height.

**61.** $y = \dfrac{-16x^2}{.434v^2} + 1.15x + 8$

**(a)** Let $y = 10$ and $x = 15$.

$$10 = \frac{-16(15)^2}{.434v^2} + 1.15(15) + 8$$

$$10 = \frac{-16(225)}{.434v^2} + 1.15(15) + 8$$

$$10 = \frac{-3600}{.434v^2} + 17.25 + 8$$

$$10 = \frac{-3600}{.434v^2} + 25.25$$

$$\frac{3600}{.434v^2} = 15.25$$

$$3600 = 6.6185v^2 \Rightarrow \tfrac{3600}{6.6185} = v^2$$

$$\pm\sqrt{\tfrac{3600}{6.6185}} = v \Rightarrow v \approx \pm 23.32$$

Since $v$ represents a velocity, only the positive square root is meaningful. The basketball should have an initial velocity of 23.32 ft per sec.

**(b)** $y = \dfrac{-16x^2}{.434(23.32)^2} + 1.15x + 8$

Algebraic Solution:

Since the parabola opens downward, the vertex is the maximum point. Use the vertex formula to find the $x$-coordinate of the vertex.

$$x = -\frac{b}{2a} = -\frac{1.15}{2\left(\dfrac{-16}{.434(23.32)^2}\right)} \approx 8.482$$

To find the $y$-coordinate of the vertex, evaluate the quadratic function when $x = 8.482$.

$$y = \frac{-16(8.482)^2}{.434(23.32)^2} + 1.15(8.482) + 8 \approx 12.88$$

Graphing Calculator Solution:

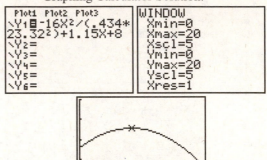

The $y$-coordinate of the vertex represents the maximum height of the basketball, which is approximately 12.88 ft.

**63.** $f(x) = .0222x^2 + .0716x + 31.8$

**(a)** The year 2009 implies $x = 16$

$$f(16) = .0222(16)^2 + .0716(16) + 31.8$$
$$\approx 38.6\%$$

Approximately 38.6% is predicted for 2005.

**(b)** No, because according to the model, the number of births to unmarried mothers should rise after 2004 since the coefficient of the $x^2$ term is positive. It is not realistic to assume they will decrease, based on the trend seen in the period 1990-2000.

**65.** $f(x) = -132.1x^2 + 1439x + 41,648$

We seek the $x$-coordinate of the vertex.

Algebraic Solution:

Since the parabola opens upward, the vertex is the minimum point. Use the vertex formula to find the $x$-coordinate of the vertex.

$$x = -\frac{b}{2a} = -\frac{1439}{2(-132.1)} = -\frac{1439}{-264.2} \approx 5.4$$

Graphing Calculator Solution:

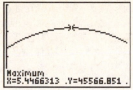

Based on the model, the median family income reach its maximum about 5.4 years after 1995, which rounds to the year 2000.

**67.**  **(a)**  Plot the 16 points given.

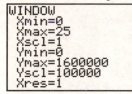 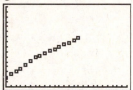

**(b)**  A quadratic function would model the data better because the data increases at a different rate each year.

**(c)**  Use the quadratic regression function on the graphing calculator to find the equation:

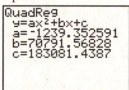

The equation is

$$f(x) = -1239x^2 + 70,792x + 183,081$$

**(d)**  Plotting the points together with $f(x)$, we see that $f$ models the data almost exactly.

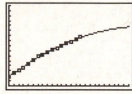

**(e)**  $x = 19$ corresponds to the year 2009, and $x = 20$ corresponds to the year 2010.

Algebraic Solution:

$$f(19) \approx -1239(19)^2 + 70,792(19) + 183,081$$
$$\approx 1,080,850$$

$$f(20) \approx -1239(20)^2 + 70,792(20) + 183,081$$
$$\approx 1,103,321$$

Graphing Calculator Solution:

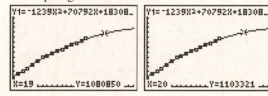

In the year 2009, approximately 1,080,850 people will have been diagnosed with AIDS since 1990. In the year 2010, approximately 1,103,321 people will have been diagnosed with AIDS since 1990.

**(f)**  The number of new cases in the year 2010 will be approximately

$$f(20) - f(19) \approx 1,103,321 - 1,080,850$$
$$\approx 22,471$$

**69.**  **(a)**  Plot the 6 points given.

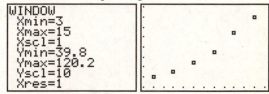

**(b)**  Let the point $(4, 50)$, be the vertex. Then,

$$f(x) = a(x - 4)^2 + 50.$$

Next let the point $(14, 110)$ lie on the graph of the function and solve for $a$.

$$f(14) = a(14 - 4)^2 + 50 = 110$$
$$a(10)^2 + 50 = 110$$
$$100a + 50 = 110$$
$$100a = 60 \Rightarrow a = \frac{60}{100} = .6$$

Thus,  $f(x) = .6(x - 4)^2 + 50.$

**(c)**  Plotting the points together with $f(x)$, we see that there is a relatively good fit.

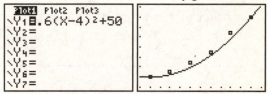

**(d)**  The quadratic regression curve is

$$g(x) = .402x^2 - 1.175x + 48.343$$

**(e)**  $x = 16$ corresponds to the year 2006.

Algebraic Solution:

$$f(16) = .6(16 - 4)^2 + 50 = 136.4$$

$$g(16) = .402(16)^2 - 1.175(16) + 48.343$$
$$\approx 132.5$$

Graphing Calculator Solution:

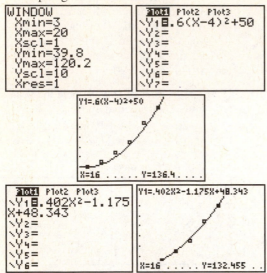

The graphing calculator agrees with the above calculations. In the year 2006, approximately 136.4 (thousand) are predicted by $f$ to be over 100 in 2006. In the year 2006, approximately 132.5 (thousand) are predicted by $g$ to be over 100 in 2006.

**71. (a)** Plot the 8 points given.

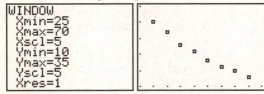

**(b)** $g(x) = .0074x^2 - 1.185x + 59.02$ models the data very well.

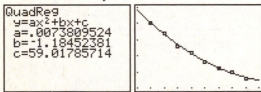

**(c)** Algebraic Solution:

$$g(70) = .0074(70)^2 - 1.185(70) + 59.02$$
$$= 12.33$$

Graphing Calculator Solution:

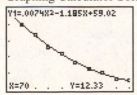

With an initial speed of 70 mph, the coast-down time would be 12.33 sec.

**(d)** A speed of approximately 39.1 mph corresponds to a coast-down time of 24 sec.

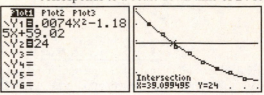

**73.** $y = x^2 - 10x + c$

An $x$-intercept occurs where $y = 0$, or

$0 = x^2 - 10x + c$. There will be exactly one $x$-intercept if this equation has exactly one solution, or if the discriminant is zero.

$$b^2 - 4ac = 0 \Rightarrow (-10)^2 - 4(1)c = 0 \Rightarrow$$
$$100 - 4c = 0 \Rightarrow 100 = 4c \Rightarrow c = 25$$

**75.** $x$-intercepts 2 and 5, and $y$-intercept 5
Since we have $x$-intercepts 2 and 5, $f$ has linear factors of $x - 2$ and $x - 5$.

$$f(x) = a(x - 2)(x - 5) = a\left(x^2 - 7x + 10\right)$$

Since the $y$-intercept is 5, $f(0) = 5$.

$$f(x) = a\left(0^2 - 7(0) + 10\right) = 5 \Rightarrow 10a = 5 \Rightarrow a = \tfrac{1}{2}$$

The required quadratic function is

$$f(x) = \tfrac{1}{2}\left(x^2 - 7x + 10\right) = \tfrac{1}{2}x^2 - \tfrac{7}{2}x + 5.$$

**77.** Use the distance formula,

$$d(P, R) = \sqrt{(x_1 - x_2)^2 + (y_1 - y_2)^2}, \text{ to find the}$$

distance between the points $P(x, 2x)$ and

$R(1, 7)$, where $P$ is any point on the line $y = 2x$.

$$d(P, R) = \sqrt{(x - 1)^2 + (2x - 7)^2}$$
$$= \sqrt{(x^2 - 2x + 1) + (4x^2 - 28x + 49)}$$
$$= \sqrt{5x^2 - 30x + 50}$$

Consider the equation $y = 5x^2 - 30x + 50$. This is the equation of a parabola opening upward, so the expression $5x^2 - 30x + 50$ has a minimum value, which is the $y$-value of the vertex. Complete the square to find the vertex.

$$y = 5x^2 - 30x + 50 = 5(x^2 - 6x) + 50$$
$$= 5(x^2 - 6x + 9 - 9) + 50$$
$$= 5(x^2 - 6x + 9) - 45 + 50 = 5(x - 3)^2 + 5$$

The vertex is $(3, 5)$, so the minimum value of $5x^2 - 30x + 50$ is 5 when $x = 3$. Thus, the minimum value of $\sqrt{5x^2 - 30x + 50}$ is $\sqrt{5}$ when $x = 3$. The point on the line $y = 2x$ for which $x = 3$ is $(3, 6)$. Thus, the closest point on the line $y = 2x$ to the point $(1, 7)$ is the point $(3, 6)$.

**79.** Graph the function $f(x) = x^2 + 2x - 8$.

Complete the square to find the vertex, $(-1, -9)$.

$$y = x^2 - 2x - 8 = (x^2 - 2x) - 8$$
$$= (x^2 - 2x + 1 - 1) - 8$$
$$= (x^2 - 2x + 1) - 1 - 8 = (x - 1)^2 - 9$$

From the graph (and table), we see that the $x$-intercepts are $-4$ and 2.

Use a table of values to find points on the graph.

| $x$ | $y$ |
|-----|-----|
| $-5$ | $7$ |
| $-4$ | $0$ |
| $-3$ | $-5$ |
| $-2$ | $-8$ |
| $-1$ | $-9$ |
| $0$ | $-8$ |
| $1$ | $-5$ |
| $2$ | $0$ |
| $3$ | $7$ |

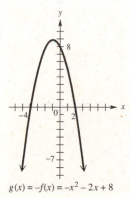

$f(x) = x^2 + 2x - 8$

**81.** Graph $g(x) = -f(x) = -x^2 - 2x + 8$. The graph of $g$ is obtained by reflecting the graph of $f$ across the $x$-axis.

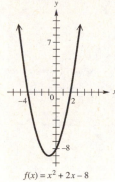

$g(x) = -f(x) = -x^2 - 2x + 8$

**83.** The two solution sets are the same, the open interval $(-4, 2)$.

## Section 3.2: Synthetic Division

### Connections (page 326)

**1.** To find $f(-2 + i)$ use synthetic division with
$$f(x) = x^3 - 4x^2 + 2x - 29i.$$

$$-2 + i \overline{)\begin{array}{cccc} 1 & -4 & 2 & -29i \end{array}}$$
$$\begin{array}{cccc} & -2 + i & 11 - 8i & -18 + 29i \end{array}$$
$$\begin{array}{cccc} \hline 1 & -6 + i & 13 - 8i & -18 \end{array}$$

The remainder is $-18$ so by the remainder theorem $f(-2 + i) = -18$.

**2.** Use synthetic division to check if $i$ is a zero of
$$f(x) = x^3 + 2ix^2 + 2x + i.$$

$$i \overline{)\begin{array}{cccc} 1 & 2i & 2 & i \end{array}}$$
$$\begin{array}{cccc} & i & -3 & -i \end{array}$$
$$\begin{array}{cccc} \hline 1 & 3i & -1 & 0 \end{array}$$

Since the remainder is 0, $f(i) = 0$ and $i$ is a zero of $f(x)$. To check $-i$, perform synthetic division.

$$-i \overline{)\begin{array}{cccc} 1 & 2i & 2 & i \end{array}}$$
$$\begin{array}{cccc} & -i & 1 & -3i \end{array}$$
$$\begin{array}{cccc} \hline 1 & i & 3 & -2i \end{array}$$

Since the remainder is $-2i$, $f(-i) = -2i$ and $-i$ is not a zero of $f(x)$.

**3.** Answers will vary.

One example is the function $f(x) = x^2 + 1$.

### Exercises

**1.** Since $x + 1$ is in the form $x + k$, $k = -1$.

$$-1 \overline{)\begin{array}{cccc} 1 & 3 & 11 & 9 \end{array}}$$
$$\begin{array}{cccc} & -1 & -2 & -9 \end{array}$$
$$\begin{array}{cccc} \hline 1 & 2 & 9 & 0 \end{array}$$

$$\frac{x^3 + 3x^2 + 11x + 9}{x + 1} = x^2 + 2x + 9$$

**3.** Express $x + 1$ in the form $x - k$ by writing it as $x - (-1)$. Thus $k = -1$.

$$-1 \overline{)\begin{array}{ccccc} 5 & 5 & 2 & -1 & -3 \end{array}}$$
$$\begin{array}{ccccc} & -5 & 0 & -2 & 3 \end{array}$$
$$\begin{array}{ccccc} \hline 5 & 0 & 2 & -3 & 0 \end{array}$$

$$\frac{5x^4 + 5x^3 + 2x^2 - x - 3}{x + 1} = 5x^3 + 2x - 3$$

**5.** Express $x + 4$ in the form $x - k$ by writing it as $x - (-4)$. Thus $k = -4$.

$$-4 \overline{)\begin{array}{ccccc} 1 & 4 & 2 & 9 & 4 \end{array}}$$
$$\begin{array}{ccccc} & -4 & 0 & -8 & -4 \end{array}$$
$$\begin{array}{ccccc} \hline 1 & 0 & 2 & 1 & 0 \end{array}$$

$$\frac{x^4 + 4x^3 + 2x^2 + 9x + 4}{x + 4} = x^3 + 2x + 1$$

**7.** Express $x + 2$ in the form $x - k$ by writing it as $x - (-2)$. Thus $k = -2$.

$$
\begin{array}{r|rrrrrr}
-2 & 1 & 3 & 2 & 2 & 3 & 1 \\
   &   & -2 & -2 & 0 & -4 & 2 \\
\hline
   & 1 & 1 & 0 & 2 & -1 & 3
\end{array}
$$

$$\frac{x^5 + 3x^4 + 2x^3 + 2x^2 + 3x + 1}{x + 2}$$
$$= x^4 + x^3 + 2x - 1 + \frac{3}{x + 2}$$

**9.** Since $x - 2$ is in the form $x - k$, $k = 2$.

$$
\begin{array}{r|rrrr}
2 & -9 & 8 & -7 & 2 \\
  &    & -18 & -20 & -54 \\
\hline
  & -9 & -10 & -27 & -52
\end{array}
$$

$$\frac{-9x^3 + 8x^2 - 7x + 2}{x - 2} = -9x^2 - 10x - 27 + \frac{-52}{x - 2}$$

**11.** Since $x - \frac{1}{3}$ is in the form $x - k$, $k = \frac{1}{3}$.

$$
\begin{array}{r|rrrr}
\frac{1}{3} & \frac{1}{3} & -\frac{2}{9} & \frac{1}{27} & 1 \\
            &   & \frac{1}{9} & -\frac{1}{27} & 0 \\
\hline
            & \frac{1}{3} & -\frac{1}{9} & 0 & 1
\end{array}
$$

$$\frac{\frac{1}{3}x^3 - \frac{2}{9}x^2 + \frac{1}{27}x + 1}{x - \frac{1}{3}} = \frac{1}{3}x^2 - \frac{1}{9}x + \frac{1}{x - \frac{1}{3}}$$

**13.** Since $x - 2$ is in the form $x - k$, $k = 2$. The constant term is missing, so include a 0.

$$
\begin{array}{r|rrrrr}
2 & 1 & -3 & -4 & 12 & 0 \\
  &   & 2 & -2 & -12 & 0 \\
\hline
  & 1 & -1 & -6 & 0 & 0
\end{array}
$$

$$\frac{x^4 - 3x^3 - 4x^2 + 12x}{x - 2} = x^3 - x^2 - 6x$$

**15.** Since $x - 1$ is in the form $x - k$, $k = 1$. The $x^2$- and $x$-terms are missing, so include 0's as their coefficients.

$$
\begin{array}{r|rrrr}
1 & 1 & 0 & 0 & -1 \\
  &   & 1 & 1 & 1 \\
\hline
  & 1 & 1 & 1 & 0
\end{array}
$$

$$\frac{x^3 - 1}{x - 1} = x^2 + x + 1$$

**17.** Express $x + 1$ in the form $x - k$ by writing it as $x - (-1)$. Thus $k = -1$. Since the $x^4$, $x^3$, $x^2$ and $x$-terms are missing, include 0's as their coefficients.

$$
\begin{array}{r|rrrrrr}
-1 & 1 & 0 & 0 & 0 & 0 & 1 \\
   &   & -1 & 1 & -1 & 1 & -1 \\
\hline
   & 1 & -1 & 1 & -1 & 1 & 0
\end{array}
$$

$$\frac{x^5 + 1}{x + 1} = x^4 - x^3 + x^2 - x + 1$$

**19.** $f(x) = 2x^3 + x^2 + x - 8$; $k = -1$
Use synthetic division to write the polynomial in the form $f(x) = (x - k)q(x) + r$.

$$
\begin{array}{r|rrrr}
-1 & 2 & 1 & 1 & -8 \\
   &   & -2 & 1 & -2 \\
\hline
   & 2 & -1 & 2 & -10
\end{array}
$$

$$f(x) = \left[x - (-1)\right](2x^2 - x + 2) - 10$$
$$= (x + 1)(2x^2 - x + 2) - 10$$

**21.** $f(x) = x^3 + 4x^2 + 5x + 2$; $k = -2$

$$
\begin{array}{r|rrrr}
-2 & 1 & 4 & 5 & 2 \\
   &   & -2 & -4 & -2 \\
\hline
   & 1 & 2 & 1 & 0
\end{array}
$$

$$f(x) = \left[x - (-2)\right](x^2 + 2x + 1) + 0$$
$$= (x + 2)(x^2 + 2x + 1) + 0$$

**23.** $f(x) = 4x^4 - 3x^3 - 20x^2 - x$; $k = 3$

$$
\begin{array}{r|rrrrr}
3 & 4 & -3 & -20 & -1 & 0 \\
  &   & 12 & 27 & 21 & 60 \\
\hline
  & 4 & 9 & 7 & 20 & 60
\end{array}
$$

$$f(x) = (x - 3)(4x^3 + 9x^2 + 7x + 20) + 60$$

**25.** $f(x) = 3x^4 + 4x^3 - 10x^2 + 15$; $k = -1$

$$
\begin{array}{r|rrrrr}
-1 & 3 & 4 & -10 & 0 & 15 \\
   &   & -3 & -1 & 11 & -11 \\
\hline
   & 3 & 1 & -11 & 11 & 4
\end{array}
$$

$$f(x) = \left[x - (-1)\right](3x^3 + x^2 - 11x + 11) + 4$$
$$= (x + 1)(3x^3 + x^2 - 11x + 11) + 4$$

**27.** $f(x) = x^2 + 5x + 6$; $k = -2$

$$
\begin{array}{r|rrr}
-2 & 1 & 5 & 6 \\
   &   & -2 & -6 \\
\hline
   & 1 & 3 & 0
\end{array}
$$

$$f(-2) = 0$$

**29.** $f(x) = 2x^2 - 3x - 3$; $k = 2$

$$2\overline{)2 \ -3 \ -3}$$
$$\underline{\phantom{2} \ \ 4 \ \ 2}$$
$$2 \ \ \ 1 \ -1$$

$f(2) = -1$

**31.** $f(x) = x^3 - 4x^2 + 2x + 1$; $k = -1$

$$-1\overline{)1 \ -4 \ \ 2 \ \ \ 1}$$
$$\underline{\phantom{-1} \ -1 \ \ 5 \ -7}$$
$$1 \ -5 \ \ 7 \ -6$$

$f(-1) = -6$

**33.** $f(x) = 2x^5 - 10x^3 - 19x^2 - 50$; $k = 3$

$$3\overline{)2 \ \ 0 \ -10 \ -19 \ \ \ 0 \ -50}$$
$$\underline{\phantom{3} \ \ \ \ 6 \ \ 18 \ \ 24 \ \ 15 \ \ 45}$$
$$2 \ \ 6 \ \ \ 8 \ \ \ \ 5 \ \ 15 \ \ -5$$

$f(3) = -5$

**35.** $f(x) = 6x^4 + x^3 - 8x^2 + 5x + 6$; $k = \frac{1}{2}$

$$\tfrac{1}{2}\overline{)6 \ \ 1 \ -8 \ \ \ 5 \ \ 6}$$
$$\underline{\phantom{\tfrac{1}{2}} \ \ \ 3 \ \ \ 2 \ -3 \ \ 1}$$
$$6 \ \ 4 \ -6 \ \ \ 2 \ \ 7$$

$f\left(\frac{1}{2}\right) = 7$

**37.** $f(x) = x^2 - 5x + 1$; $k = 2 + i$

$$2+i\overline{)1 \ -5 \ \ \ \ \ \ \ \ 1}$$
$$\underline{\phantom{2+i} \ \ 2+i \ -7-i}$$
$$1 \ -3+i \ -6-i$$

$f(2 + i) = -6 - i$

**39.** $f(x) = x^2 + 4$; $k = 2i$

$$-2i\overline{)1 \ \ \ \ 0 \ \ \ \ \ \ \ 4}$$
$$\underline{\phantom{-2i} \ \ -2i \ -4}$$
$$1 \ \ 0-2i \ \ \ 0$$

$f(2i) = 0$

**41.** To determine if $k = 2$ is a zero of
$f(x) = x^2 + 2x - 8$, divide synthetically.

$$2\overline{)1 \ \ 2 \ -8}$$
$$\underline{\phantom{2} \ \ 2 \ \ 8}$$
$$1 \ \ 4 \ \ 0$$

Yes, 2 is a zero of $f(x)$ because $f(2) = 0$.

**43.** To determine if $k = 2$ is a zero of
$f(x) = x^3 - 3x^2 + 4x - 4$, divide synthetically.

$$2\overline{)1 \ -3 \ \ \ 4 \ -4}$$
$$\underline{\phantom{2} \ \ \ 2 \ -2 \ \ \ 4}$$
$$1 \ -1 \ \ \ 2 \ \ \ 0$$

Yes, 2 is a zero of $f(x)$ because $f(2) = 0$.

**45.** To determine if $k = 1$ is a zero of
$f(x) = 2x^3 - 6x^2 - 9x + 4$, divide
synthetically.

$$1\overline{)2 \ -6 \ \ \ -9 \ \ \ \ 4}$$
$$\underline{\phantom{1} \ \ \ 2 \ \ -4 \ -13}$$
$$2 \ -4 \ -13 \ -9$$

No, 1 is not a zero of $f(x)$ because
$f(1) = -9$.

**47.** To determine if $k = 0$ is a zero of
$f(x) = x^3 + 7x^2 + 10x$, divide synthetically.

$$0\overline{)1 \ \ 7 \ \ 10 \ \ 0}$$
$$\underline{\phantom{0} \ \ 0 \ \ \ 0 \ \ 0}$$
$$1 \ \ 7 \ \ \ 2 \ \ 0$$

Yes, 0 is a zero of $f(x)$ because $f(0) = 0$.

**49.** To determine if $k = -\frac{3}{2}$ is a zero of
$f(x) = 2x^4 + 3x^3 - 8x^2 - 2x + 15$, divide
synthetically.

$$-\tfrac{3}{2}\overline{)2 \ \ 3 \ -8 \ -2 \ \ \ 15}$$
$$\underline{\phantom{-\tfrac{3}{2}} \ -3 \ \ \ 0 \ \ 12 \ -15}$$
$$2 \ \ 0 \ -8 \ \ 10 \ \ \ \ 0$$

Yes, $-\frac{3}{2}$ is a zero of $f(x)$ because
$f\left(-\frac{3}{2}\right) = 0$.

**51.** To determine if $k = \frac{2}{5}$ is a zero of
$f(x) = 5x^4 + 2x^3 - x + 3$, divide synthetically.

$$\tfrac{2}{5}\overline{)5 \ \ 2 \ \ \ 0 \ \ \ -1 \ \ \ \ \ 3}$$
$$\underline{\phantom{\tfrac{2}{5}} \ \ 2 \ \ \tfrac{8}{5} \ \ \tfrac{16}{25} \ -\tfrac{18}{125}}$$
$$5 \ \ 4 \ \ \tfrac{8}{5} \ -\tfrac{9}{25} \ \ \tfrac{357}{125}$$

No, $\frac{2}{5}$ is not a zero of $f(x)$ because
$f\left(\frac{2}{5}\right) = \frac{357}{125}$.

**53.** To determine if $k = 1 - i$ is a zero of
$f(x) = x^2 - 2x + 2$, divide synthetically.

$$1-i \overline{)1 \ -2 \qquad 2}$$
$$\phantom{1-i)1 \ } \underline{1-i \ -2}$$
$$\phantom{1-i)1 \ } 1 \ -1-i \quad 0$$

Yes, $1 - i$ is a zero of $f(x)$ because
$f(1 - i) = 0$.

**55.** To determine if $k = 2 + i$ is a zero of
$f(x) = x^2 + 3x + 4$, divide synthetically.

$$2+i \overline{)1 \ 3 \qquad 4}$$
$$\phantom{2+i)1 \ } \underline{2+i \ 9+7i}$$
$$\phantom{2+i)1 \ } 1 \ 5+i \ 13+7i$$

No, $2 + i$ is not a zero of $f(x)$ because
$f(2 + i) = 13 + 7i$.

**57.** To determine if $k = 1 + i$ is a zero of
$f(x) = x^3 + 3x^2 - x + 1$, divide synthetically.

$$1+i \overline{)1 \ 3 \quad -1 \qquad\quad 1}$$
$$\phantom{1+i)1 \ } \underline{1+i \ 3+5i \ -3+7i}$$
$$\phantom{1+i)1 \ } 1 \ 4+i \ 2+5i \ -2+7i$$

No, $1 + i$ is not a zero of $f(x)$ because
$f(1 + i) = -2 + 7i$.

**59.**
$$-2 \overline{)1 \ -2 \quad -1 \qquad 2}$$
$$\phantom{-2)1 \ } \underline{-2 \qquad 8 \quad -14}$$
$$\phantom{-2)1 \ } 1 \ -4 \qquad 7 \quad -12$$

$f(-2) = -12$

The coordinates of the corresponding point are
$(-2, -12)$.

**61.**
$$0 \overline{)1 \ -2 \quad -1 \qquad 2}$$
$$\phantom{0)1 \ } \underline{0 \qquad 0 \qquad 0}$$
$$\phantom{0)1 \ } 1 \ -2 \quad -1 \qquad 2$$

$f(0) = 2$

The coordinates of the corresponding point are
$(0, 2)$.

**63.**
$$\tfrac{3}{2} \overline{)1 \ -2 \quad -1 \qquad 2}$$
$$\phantom{\tfrac{3}{2})1 \ } \underline{\tfrac{3}{2} \quad -\tfrac{3}{4} \quad -\tfrac{21}{8}}$$
$$\phantom{\tfrac{3}{2})1 \ } 1 \ -\tfrac{1}{2} \quad -\tfrac{7}{4} \quad -\tfrac{5}{8}$$

$f\left(\tfrac{3}{2}\right) = -\tfrac{5}{8}$

The coordinates of the corresponding point are
$\left(\tfrac{3}{2}, -\tfrac{5}{8}\right)$.

**65.**
$$3 \overline{)1 \ -2 \quad -1 \qquad 2}$$
$$\phantom{3)1 \ } \underline{3 \qquad 3 \qquad 6}$$
$$\phantom{3)1 \ } 1 \ \ 1 \qquad 2 \qquad 8$$

$f(3) = 8$

The coordinates of the corresponding point are
$(3, 8)$.

## Section 3.3: Zeros of Polynomial Functions

**Connections (page 336)**

**1.** $x = \sqrt[3]{\dfrac{n}{2} + \sqrt{\left(\dfrac{n}{2}\right)^2 + \left(\dfrac{m}{3}\right)^3}} - \sqrt[3]{\dfrac{-n}{2} + \sqrt{\left(\dfrac{n}{2}\right)^2 + \left(\dfrac{m}{3}\right)^3}}$

can be used to solve a cubic equation
$x^3 + mx = n$.
For $x^3 + 9x = 26$, $m = 9$ and $n = 26$.
Substitute

$x = \sqrt[3]{\dfrac{26}{2} + \sqrt{\left(\dfrac{26}{2}\right)^2 + \left(\dfrac{9}{3}\right)^3}} - \sqrt[3]{-\dfrac{26}{2} + \sqrt{\left(\dfrac{26}{2}\right)^2 + \left(\dfrac{9}{3}\right)^3}}$

$= \sqrt[3]{13 + \sqrt{13^2 + 3^3}} - \sqrt[3]{-13 + \sqrt{13^2 + 3^3}}$

$= \sqrt[3]{13 + \sqrt{169 + 27}} - \sqrt[3]{-13 + \sqrt{169 + 27}}$

$= \sqrt[3]{13 + 14} - \sqrt[3]{-13 + 14}$

$= \sqrt[3]{27} - \sqrt[3]{1} = 3 - 1 = 2$

**Exercises**

**1.** Since $x - 1$ is a factor of
$f(x) = x^6 - x^4 + 2x^2 - 2$, we are assured
that $f(1) = 0$. This statement is justified by the
factor theorem; therefore, it is true.

**3.** For the function $f(x) = (x + 2)^4 (x - 3)$, 2 is a
zero of multiplicity 4. To find the zero, set the
factor equal to 0: $x + 2 = 0 \Rightarrow x = -2$
2 is not a zero of the function; therefore, the
statement is false. (It would be true to say that
$-2$ is a zero of multiplicity 4.)

**5.** $x^3 - 5x^2 + 3x + 1; \ x - 1$
Let $f(x) = x^3 - 5x^2 + 3x + 1$. By the factor
theorem, $x - 1$ will be a factor of $f(x)$ if and
only if $f(1) = 0$. Use synthetic division and the
remainder theorem.

$$1 \overline{)1 \ -5 \quad 3 \qquad 1}$$
$$\phantom{1)1 \ } \underline{1 \ -4 \ -1}$$
$$\phantom{1)1 \ } 1 \ -4 \ -1 \qquad 0$$

Since $f(1) = 0$, $x - 1$ is a factor of $f(x)$.

**7.** $2x^4 + 5x^3 - 8x^2 + 3x + 13;\ x + 1$

Let $f(x) = 2x^4 + 5x^3 - 8x^2 + 3x + 13$. By the factor theorem, $x + 1$ will be a factor of $f(x)$ if and only if $f(-1) = 0$. Use synthetic division and the remainder theorem.

$$-1)\overline{2\quad 5\quad -8\quad 3\quad 13}$$
$$\phantom{-1)2}\ -2\ -3\ \ 11\ -14$$
$$\overline{\phantom{-1)}2\quad 3\ -11\ \ 14\ \ -1}$$

Since the remainder is $-1$, $f(-1) = -1$, so $x + 1$ is not a factor of $f(x)$.

**9.** $-x^3 + 3x - 2;\ x + 2$

Let $f(x) = -x^3 + 3x - 2$. By the factor theorem, $x + 2$ will be a factor of $f(x)$ if and only if $f(-2) = 0$. Use synthetic division and the remainder theorem.

$$-2)\overline{-1\quad 0\quad 3\quad -2}$$
$$\phantom{-2)-1}\ \ 2\ -4\quad 2$$
$$\overline{\phantom{-2)}-1\quad 2\ -1\quad 0}$$

Since $f(-2) = 0$, $x + 2$ is a factor of $f(x)$.

**11.** $4x^2 + 2x + 54;\ x - 4$

Let $f(x) = 4x^2 + 2x + 54$. By the factor theorem, $x - 4$ will be a factor of $f(x)$ if and only if $f(4) = 0$. Use synthetic division and the remainder theorem.

$$4)\overline{4\quad 2\quad 54}$$
$$\phantom{4)4}\ 16\quad 72$$
$$\overline{\phantom{4)}4\ \ 18\ \ 126}$$

Since the remainder is 126, $f(4) = 126$, so $x - 4$ is not a factor of $f(x)$.

**13.** $x^3 + 2x^2 - 3;\ x - 1$

Let $f(x) = x^3 + 2x^2 - 1$. By the factor theorem, $x - 1$ will be a factor of $f(x)$ if and only if $f(1) = 0$. Use synthetic division and the remainder theorem.

$$1)\overline{1\quad 2\quad 0\ -3}$$
$$\phantom{1)1}\ \ 1\quad 3\quad 3$$
$$\overline{\phantom{1)}1\quad 3\quad 3\quad 0}$$

Since $f(1) = 0$, $x - 1$ is a factor of $f(x)$.

**15.** $2x^4 + 5x^3 - 2x^2 + 5x + 6;\ x + 3$

Let $f(x) = 2x^4 + 5x^3 - 2x^2 + 5x + 6$. By the factor theorem, $x + 3$ will be a factor of $f(x)$ if and only if $f(-3) = 0$. Use synthetic division and the remainder theorem.

$$-3)\overline{2\quad 5\ -2\quad 5\quad 6}$$
$$\phantom{-3)2}\ -6\quad 3\ -3\ -6$$
$$\overline{\phantom{-3)}2\ -1\quad 1\quad 2\quad 0}$$

Since $f(-3) = 0$, $x + 3$ is a factor of $f(x)$.

**17.** $f(x) = 2x^3 - 3x^2 - 17x + 30;\ k = 2$

Since 2 is a zero of $f(x)$, $x - 2$ is a factor. Divide $f(x)$ by $x - 2$.

$$2)\overline{2\ -3\ -17\quad 30}$$
$$\phantom{2)2}\ \ 4\quad 2\ -30$$
$$\overline{\phantom{2)}2\quad 1\ -15\quad 0}$$

Thus, $f(x) = (x - 2)(2x^2 + x - 15)$
$$= (x - 2)(2x - 5)(x + 3).$$

**19.** $f(x) = 6x^3 + 13x^2 - 14x + 3;\ k = -3$

Since $-3$ is a zero of $f(x)$, $x + 3$ is a factor. Divide $f(x)$ by $x + 3$.

$$-3)\overline{6\quad 13\ -14\quad 3}$$
$$\phantom{-3)6}\ -18\quad 15\ -3$$
$$\overline{\phantom{-3)}6\ -5\quad 1\quad 0}$$

Thus, $f(x) = (x + 3)(6x^2 - 5x + 1)$
$$= (x + 3)(3x - 1)(2x - 1)$$

**21.** $f(x) = 6x^3 + 25x^2 + 3x - 4;\ k = -4$

Since $-4$ is a zero of $f(x)$, $x + 4$ is a factor. Divide $f(x)$ by $x + 4$.

$$-4)\overline{6\quad 25\quad 3\ -4}$$
$$\phantom{-4)6}\ -24\ -4\quad 4$$
$$\overline{\phantom{-4)}6\quad 1\ -1\quad 0}$$

Thus, $f(x) = (x + 4)(6x^2 + x - 1)$
$$= (x + 4)(3x - 1)(2x + 1)$$

**23.** $f(x) = x^3 + (7 - 3i)x^2 + (12 - 21i)x - 36i$
$k = 3i$

Since $3i$ is a zero of $f(x)$, $x - 3i$ is a factor. Divide $f(x)$ by $x - 3i$.

$$3i)\overline{1\quad 7-3i\ \ 12-21i\ \ -36i}$$
$$\phantom{3i)1}\ \ 3i\quad 21i\quad 36i$$
$$\overline{\phantom{3i)}1\quad 7\quad 12\quad 0}$$

Thus, $f(x) = (x - 3i)(x^2 + 7x + 12)$
$$= (x - 3i)(x + 4)(x + 3)$$

**25.** $f(x) = 2x^3 + (3 - 2i)x^2 + (-8 - 5i)x + (3 + 3i)$
$k = 1 + i$

Since $1 + i$ is a zero of $f(x)$, $x - (1 + i)$ is a factor. Divide $f(x)$ by $x - (1 + i)$.

$$1 + i \overline{)2 \quad 3 - 2i \quad -8 - 5i \quad 3 + 3i}$$
$$\phantom{1 + i)2} \; 2 + 2i \quad 5 + 5i \quad -3 - 3i$$
$$\overline{\phantom{1+i)}2 \quad 5 \quad\quad -3 \quad\quad\quad 0}$$

Thus, $f(x) = \left[x - (1 + i)\right](2x^2 + 5x - 3)$
$$= \left[x - (1 + i)\right](2x - 1)(x + 3)$$

**27.** $f(x) = x^4 + 2x^3 - 7x^2 - 20x - 12; \; k = -2$
(multiplicity 2)

Since $-2$ is a zero of $f(x)$, $x + 2$ is a factor. Divide $f(x)$ by $x + 2$.

$$-2 \overline{)1 \quad 2 \quad -7 \quad -20 \quad -12}$$
$$\phantom{-2)1} \; -2 \quad 0 \quad 14 \quad 12$$
$$\overline{\phantom{-2)}1 \quad 0 \quad -7 \quad -6 \quad\quad 0}$$

Thus, $f(x) = x^4 + 2x^3 - 7x^2 - 20x - 12$
$$= (x + 2)\left(x^3 - 7x - 6\right)$$

Since $-2$ has multiplicity 2, divide the quotient polynomial by $x + 2$.

$$-2 \overline{)1 \quad 0 \quad -7 \quad -6}$$
$$\phantom{-2)1} \; -2 \quad 4 \quad 6$$
$$\overline{\phantom{-2)}1 \quad -2 \quad -3 \quad\quad 0}$$

Thus, $f(x) = (x + 2)\left(x^3 - 7x - 6\right)$
$$= (x + 2)^2 \left(x^2 - 2x - 3\right)$$
$$= (x + 2)^2 (x + 1)(x - 3)$$

**29.** $f(x) = x^3 - x^2 - 4x - 6; \; 3$

Since 3 is a zero, first divide $f(x)$ by $x - 3$.

$$3 \overline{)1 \quad -1 \quad -4 \quad -6}$$
$$\phantom{3)1} \; 3 \quad 6 \quad 6$$
$$\overline{\phantom{3)}1 \quad 2 \quad 2 \quad\quad 0}$$

This gives $f(x) = (x - 3)(x^2 + 2x + 2)$. Since $x^2 + 2x + 2$ cannot be factored, use the quadratic formula with $a = 1$, $b = 2$, and $c = 2$ to find the remaining two zeros.

$$x = \frac{-2 \pm \sqrt{4 - 4(1)(2)}}{2(1)} = \frac{-2 \pm \sqrt{4 - 8}}{2}$$
$$= \frac{-2 \pm \sqrt{-4}}{2} = \frac{-2 \pm 2i}{2} = -1 \pm i$$

The remaining zeros are $-1 \pm i$.

**31.** $f(x) = x^3 - 7x^2 + 17x - 15; \; 2 - i$

Since $2 - i$ is also a zero, first divide $f(x)$ by $x - (2 - i)$.

$$2 - i \overline{)1 \quad -7 \quad\quad 17 \quad\quad -15}$$
$$\phantom{2 - i)1} \; 2 - i \quad -11 + 3i \quad 15$$
$$\overline{\phantom{2-i)}1 \quad -5 - i \quad 6 + 3i \quad\quad 0}$$

By the conjugate zeros theorem, $2 + i$ is also a zero, so divide the quotient polynomial from the first synthetic division by $x - (2 + i)$.

$$2 + i \overline{)1 \quad -5 - i \quad 6 + 3i}$$
$$\phantom{2 + i)1} \; 2 + i \quad -6 - 3i$$
$$\overline{\phantom{2+i)}1 \quad -3 \quad\quad 0}$$

This gives
$f(x) = [x - (2 - i)][x - (2 + i)](x - 3)$. The remaining zeros are $2 + i$ and 3.

**33.** $f(x) = x^4 + 5x^2 + 4; \; -i$

Since $-i$ is a zero, first divide $f(x)$ by $x + i$.

$$-i \overline{)1 \quad 0 \quad 5 \quad 0 \quad 4}$$
$$\phantom{-i)1} \; -i \quad -1 \quad -4i \quad -4$$
$$\overline{\phantom{-i)}1 \quad -i \quad 4 \quad -4i \quad\quad 0}$$

By the conjugate zeros theorem, $i$ is also a zero, so divide the quotient polynomial from the first synthetic division by $x - i$.

$$i \overline{)1 \quad -i \quad 4 \quad -4i}$$
$$\phantom{i)1} \; i \quad 0 \quad 4i$$
$$\overline{\phantom{i)}1 \quad 0 \quad 4 \quad\quad 0}$$

The remaining zeros will be zeros of the new quotient polynomial, $x^2 + 4$. Find the remaining zeros by using the square root property.

$$x^2 + 4 = 0 \Rightarrow x^2 = -4 \Rightarrow x = \pm\sqrt{-4} \Rightarrow x = \pm 2i$$

The other zeros are $i$ and $\pm 2i$.

**35. (a)** $f(x) = x^3 - 2x^2 - 13x - 10$

$p$ must be a factor of $a_0 = -10$ and $q$ must be a factor of $a_3 = 1$. Thus, $p$ can be $\pm 1$, $\pm 2$, $\pm 5$, $\pm 10$ and $q$ can be $\pm 1$. The possible rational zeros, $\frac{p}{q}$, are $\pm 1$, $\pm 2$, $\pm 5$, $\pm 10$.

**(b)** The remainder theorem shows that $-1$ is a zero.

$$-1 \overline{)\,1\ -2\ -13\ -10\,}$$
$$\phantom{-1)\,}\ -1\ \ \ 3\ \ \ 10$$
$$\overline{\phantom{-1)}\ 1\ -3\ -10\ \ \ 0}$$

The new quotient polynomial will be
$x^2 - 3x - 10$.
$$x^2 - 3x - 10 = 0$$
$$(x+2)(x-5) = 0$$
$$x + 2 = 0 \Rightarrow x = -2 \ \text{ or } \ x - 5 = 0 \Rightarrow x = 5$$
The rational zeros are $-1$, $-2$, and $5$.

**(c)** Since the three zeros are $-1$, $-2$, and $5$, the factors are $x + 1$, $x + 2$, and $x - 5$.
$$f(x) = (x+1)(x+2)(x-5)$$

**37. (a)** $f(x) = x^3 + 6x^2 - x - 30$

$p$ must be a factor of $a_0 = -30$ and $q$ must be a factor of $a_3 = 1$. Thus, $p$ can be $\pm 1, \pm 2, \pm 3, \pm 5, \pm 6, \pm 10, \pm 15, \pm 30$ and $q$ can be $\pm 1$. The possible zeros, $\frac{p}{q}$, are $\pm 1, \pm 2, \pm 3, \pm 5, \pm 6, \pm 10, \pm 15, \pm 30$.

**(b)** The remainder theorem shows that $-5$ is a zero.

$$-5 \overline{)\,1\ \ \ 6\ -1\ -30\,}$$
$$\phantom{-5)\,}\ -5\ -5\ \ \ 30$$
$$\overline{\phantom{-5)}\ 1\ \ \ 1\ -6\ \ \ \ 0}$$

The new quotient polynomial will be
$x^2 + x - 6$.
$$x^2 + x - 6 = 0$$
$$(x+3)(x-2) = 0$$
$$x + 3 = 0 \Rightarrow x = -3 \ \text{ or } \ x - 2 = 0 \Rightarrow x = 2$$
The rational zeros are $-5$, $-3$, and $2$.

**(c)** Since the three zeros are $-5$, $-3$, and $2$, the factors are $x + 5$, $x + 3$, and $x - 2$.
$$f(x) = (x+5)(x+3)(x-2)$$

**39. (a)** $f(x) = 6x^3 + 17x^2 - 31x - 12$

$p$ must be a factor of $a_0 = -12$ and $q$ must be a factor of $a_3 = 6$. Thus, $p$ can be $\pm 1, \pm 2, \pm 3, \pm 4, \pm 6, \pm 12$ and $q$ can be $\pm 1, \pm 2, \pm 3, \pm 6$. The possible zeros, $\frac{p}{q}$, are $\pm 1, \pm 2, \pm 3, \pm 4, \pm 6, \pm 12, \pm \frac{1}{2}, \pm \frac{3}{2}, \pm \frac{1}{3}, \pm \frac{2}{3}, \pm \frac{4}{3}, \pm \frac{1}{6}$.

**(b)** The remainder theorem shows that $-4$ is a zero.

$$-4 \overline{)\,6\ \ \ \ 17\ \ \ -31\ \ \ -12\,}$$
$$\phantom{-4)\,}\ -24\ \ \ \ 28\ \ \ \ 12$$
$$\overline{\phantom{-4)}\ 6\ \ \ -7\ \ \ -3\ \ \ \ \ 0}$$

The new quotient polynomial is
$6x^2 - 7x - 3$.
$$6x^2 - 7x - 3 = 0 \Rightarrow (3x+1)(2x-3) = 0$$
$$3x + 1 = 0 \Rightarrow x = -\frac{1}{3} \ \text{ or }$$
$$2x - 3 = 0 \Rightarrow x = \frac{3}{2}$$
The rational zeros are $-4$, $-\frac{1}{3}$, and $\frac{3}{2}$.

**(c)** Since the three zeros are $-4$, $-\frac{1}{3}$, and $\frac{3}{2}$, the factors are $x + 4$, $3x + 1$, and $2x - 3$. $f(x) = (x+4)(3x+1)(2x-3)$

**41. (a)** $f(x) = 24x^3 + 40x^2 - 2x - 12$

$p$ must be a factor of $a_0 = -12$ and $q$ must be a factor of $a_3 = 24$. Thus, $p$ can be $\pm 1, \pm 2, \pm 3, \pm 4, \pm 6, \pm 12$ and $q$ can be $\pm 1, \pm 2, \pm 3, \pm 4, \pm 6, \pm 8, \pm 12, \pm 24$. The possible zeros, $\frac{p}{q}$, are $\pm 1, \pm 2, \pm 3, \pm 4, \pm 6, \pm 12, \pm \frac{1}{2}, \pm \frac{3}{2}, \pm \frac{1}{3}, \pm \frac{2}{3}, \pm \frac{4}{3}, \pm \frac{1}{4}, \pm \frac{3}{4}, \pm \frac{1}{6}, \pm \frac{1}{8}, \pm \frac{3}{8}, \pm \frac{1}{12}, \pm \frac{1}{24}$.

**(b)** The remainder theorem shows that $-\frac{3}{2}$ is a zero.

$$-\tfrac{3}{2} \overline{)\,24\ \ \ 40\ \ \ -2\ \ \ -12\,}$$
$$\phantom{-\tfrac{3}{2})\,}\ -36\ \ \ -6\ \ \ \ 12$$
$$\overline{\phantom{-\tfrac{3}{2})}\ 24\ \ \ \ 4\ \ \ -8\ \ \ \ \ 0}$$

The new quotient polynomial is
$24x^2 + 4x - 8$.
$$24x^2 + 4x - 8 = 0 \Rightarrow 4(6x^2 + x - 2) = 0$$
$$\Rightarrow 4(3x+2)(2x-1) = 0$$
$$3x + 2 = 0 \Rightarrow x = -\frac{2}{3} \ \text{ or }$$
$$2x - 1 = 0 \Rightarrow x = \frac{1}{2}$$
The rational zeros are $-\frac{3}{2}, -\frac{2}{3}$, and $\frac{1}{2}$.

**(c)** Since the three rational zeros are $-\frac{3}{2}, -\frac{2}{3},$ and $\frac{1}{2},$ the factors are $3x+2,$ $2x+3,$ and $2x-1.$

$$f(x) = 2(2x+3)(3x+2)(2x-1)$$

Note: Since $-\frac{3}{2}$ is a zero,

$$\begin{aligned} f(x) &= 24x^3 + 40x^2 - 2x - 12 \\ &= \left[x - \left(-\tfrac{3}{2}\right)\right]\left(24x^2 + 4x - 8\right) \\ &= \left(x + \tfrac{3}{2}\right)\left[4(6x^2 + x + 2)\right] \\ &= 4\left(x + \tfrac{3}{2}\right)(3x+2)(2x-1) \\ &= 2\left[2\left(x + \tfrac{3}{2}\right)\right](3x+2)(2x-1) \\ &= 2(2x+3)(3x+2)(2x-1) \end{aligned}$$

**43.** $f(x) = 7x^3 + x$

To find the zeros, let $f(x) = 0$ and factor the binomial. $7x^3 + x = 0 \Rightarrow x(7x^2 + 1) = 0$

Set each factor equal to zero and solve for $x.$

$$x = 0 \quad \text{or} \quad 7x^2 + 1 = 0$$
$$7x^2 = -1$$
$$x^2 = -\tfrac{1}{7}$$
$$x = \pm\sqrt{-\tfrac{1}{7}} = \pm\tfrac{\sqrt{7}}{7}$$

The zeros are 0 and $\pm\frac{\sqrt{7}}{7}i.$

**45.** $f(x) = 3(x-2)(x+3)(x^2 - 1)$

To find the zeros, let $f(x) = 0.$

Set each factor equal to zero and solve for $x.$

$$x - 2 = 0 \Rightarrow x = 2$$
$$x + 3 = 0 \Rightarrow x = -3$$
$$x^2 - 1 = 0 \Rightarrow x^2 = 1 \Rightarrow x = \pm 1$$

The zeros are 2, −3, 1, and −1.

**47.** $f(x) = (x^2 + x - 2)^5 (x - 1 + \sqrt{3})^2$

To find the zeros, let $f(x) = 0.$

Set each factor equal to zero and solve for $x.$

$$(x^2 + x - 2)^5 = 0 \Rightarrow x^2 + x - 2 = 0 \Rightarrow$$
$$(x+2)(x-1) = 0 \Rightarrow$$
$x = -2,$ multiplicity 5 or $x = 1,$ multiplicity 5

$$(x - 1 + \sqrt{3})^2 = 0 \Rightarrow x - 1 + \sqrt{3} = 0 \Rightarrow$$
$x = 1 - \sqrt{3},$ multiplicity 2

The zeros are −2 (multiplicity 5), 1 (multiplicity 5) and $1 - \sqrt{3}$ (multiplicity 2).

**49.** Zeros of −3, 1, and 4, $f(2) = 30$

These three zeros give

$x - (-3) = x + 3,$ $x - 1,$ and $x - 4$ as factors of $f(x).$ Since $f(x)$ is to be degree 3, these are the only possible factors by the number of zeros theorem. Therefore, $f(x)$ has the form $f(x) = a(x+3)(x-1)(x-4)$ for some real number $a.$ To find $a,$ use the fact that $f(2) = 30.$

$$f(2) = a(2+3)(2-1)(2-4) = 30 \Rightarrow$$
$$a(5)(1)(-2) = 30 \Rightarrow -10a = 30 \Rightarrow$$
$$a = -3$$

Thus, $\begin{aligned}[t] f(x) &= -3(x+3)(x-1)(x-4) \\ &= -3(x^2 + 2x - 3)(x-4) \\ &= -3(x^3 - 2x^2 - 11x + 12) \\ &= -3x^3 + 6x^2 + 33x - 36 \end{aligned}$

**51.** Zeros of −2, 1, and 0; $f(-1) = -1$

These three zeros give

$x - (-2) = x + 2,$ $x - 1,$ and $x - 0 = x$ as factors of $f(x).$ Since $f(x)$ is to be degree 3, these are the only possible factors by the number of zeros theorem. Therefore, $f(x)$ has the form $f(x) = a(x+2)(x-1)x$ for some real number $a.$ To find $a,$ use the fact that $f(-1) = -1.$

$$f(-1) = a(-1+2)(-1-1)(-1) = -1 \Rightarrow$$
$$a(1)(-2)(-1) = -1 \Rightarrow 2a = -1 \Rightarrow a = -\tfrac{1}{2}$$

Thus,

$$\begin{aligned} f(x) &= -\tfrac{1}{2}(x+2)(x-1)x \\ &= -\tfrac{1}{2}(x^2 + x - 2)x = -\tfrac{1}{2}x^3 - \tfrac{1}{2}x^2 + x \end{aligned}$$

**53.** Zero of −3 having multiplicity 3; $f(3) = 36$

These three zeros give

$x - (-3) = x + 3,$ $x - (-3) = x + 3,$ and $x - (-3) = x + 3$ as factors of $f(x).$ Since $f(x)$ is to be degree 3, these are the only possible factors by the number of zeros theorem. Therefore, $f(x)$ has the form

$f(x) = a(x+3)(x+3)(x+3) = a(x+3)^3$ for some real number $a.$ To find $a,$ use the fact that $f(3) = 36.$

$$f(3) = a(3+3)^3 = 36 \Rightarrow a(6)^3 = 36 \Rightarrow$$
$$216a = 36 \Rightarrow a = \tfrac{1}{6}$$

Thus, $f(x) = \frac{1}{6}(x+3)^3 = \frac{1}{6}(x+3)^2(x+3)$

$\qquad = \frac{1}{6}(x^2+6x+9)(x+3)$

$\qquad = \frac{1}{6}(x^3+9x^2+27x+27)$

$\qquad = \frac{1}{6}x^3 + \frac{3}{2}x^2 + \frac{9}{2}x + \frac{9}{2}$

In Exercises 55–71, we must find a polynomial of least degree with real coefficients having the given zeros. For each of these exercises, other answers are possible.

**55.** $5+i$ and $5-i$

$f(x) = \left[x-(5+i)\right]\left[x-(5-i)\right]$

$\qquad = (x-5-i)(x-5+i)$

$\qquad = \left[(x-5)-i\right]\left[(x-5)+i\right] = (x-5)^2 - i^2$

$\qquad = \left(x^2-10x+25\right)-i^2$

$\qquad = x^2-10x+25+1 = x^2-10x+26$

**57.** $2$ and $1+i$

By the conjugate zeros theorem, $1-i$ must also be a zero.

$f(x) = (x-2)\left[x-(1+i)\right]\left[x-(1-i)\right]$

$\qquad = (x-2)(x-1-i)(x-1+i)$

$\qquad = (x-2)\left[(x-1)-i\right]\left[(x-1)+i\right]$

$\qquad = (x-2)\left[(x-1)^2 - i^2\right]$

$\qquad = (x-2)\left[\left(x^2-2x+1\right)-i^2\right]$

$\qquad = (x-2)\left(x^2-2x+1+1\right)$

$\qquad = (x-2)\left(x^2-2x+2\right)$

$\qquad = x^3-4x^2+6x-4$

**59.** $1+\sqrt{2}$, $1-\sqrt{2}$, and $1$

$f(x) = \left[x-\left(1+\sqrt{2}\right)\right]\left[x-\left(1-\sqrt{2}\right)\right](x-1) = \left(x-1-\sqrt{2}\right)\left(x-1+\sqrt{2}\right)(x-1)$

$\qquad = \left[(x-1)-\sqrt{2}\right]\left[(x-1)+\sqrt{2}\right](x-1) = \left[(x-1)^2 - \left(\sqrt{2}\right)^2\right](x-1)$

$\qquad = \left(x^2-2x+1-2\right)(x-1) = \left(x^2-2x-1\right)(x-1) = x^3-3x^2+x+1$

**61.** $2+i$, $2-i$, $3$, and $-1$

$f(x) = \left[x-(2+i)\right]\left[x-(2-i)\right](x-3)(x+1) = \left[(x-2-i)(x-2+i)\right]\left[(x-3)(x+1)\right]$

$\qquad = \left[(x-2)-i\right]\left[(x-2)+i\right]\left(x^2-2x-3\right) = \left[(x-2)^2-i^2\right]\left(x^2-2x-3\right) = \left[\left(x^2-4x+4\right)-i^2\right]\left(x^2-2x-3\right)$

$\qquad = \left(x^2-4x+4+1\right)\left(x^2-2x-3\right) = \left(x^2-4x+5\right)\left(x^2-2x-3\right) = x^4-6x^3+10x^2+2x-15$

**63.** $2$ and $3+i$

By the conjugate zeros theorem, $3-i$ must also be a zero.

$f(x) = (x-2)\left[x-(3+i)\right]\left[x-(3-i)\right] = (x-2)\left[(x-3)-i\right]\left[(x-3)+i\right] = (x-2)\left[(x-3)^2-i^2\right]$

$\qquad = (x-2)\left(x^2-6x+9+1\right) = (x-2)\left(x^2-6x+10\right) = x^3-8x^2+22x-20$

**65.** $1-\sqrt{2}$, $1+\sqrt{2}$, and $1-i$

By the conjugate zeros theorem, $1+i$ must also be a zero.

$f(x) = \left[x-\left(1-\sqrt{2}\right)\right]\left[x-\left(1+\sqrt{2}\right)\right] \cdot \left[x-(1-i)\right]\left[x-(1+i)\right]$

$\qquad = \left(x-1+\sqrt{2}\right)\left(x-1-\sqrt{2}\right) \cdot (x-1+i)(x-1-i) = \left[(x-1)+\sqrt{2}\right]\left[(x-1)-\sqrt{2}\right] \cdot \left[(x-1)+i\right]\left[(x-1)-i\right]$

$\qquad = \left[(x-1)^2 - \left(\sqrt{2}\right)^2\right]\left[(x-1)^2-i^2\right] = \left(x^2-2x+1-2\right)\left[\left(x^2-2x+1\right)-i^2\right] = \left(x^2-2x-1\right)\left(x^2-2x+1+1\right)$

$\qquad = \left(x^2-2x-1\right)\left(x^2-2x+2\right) = x^4-4x^3+5x^2-2x-2$

**67.** $2-i$ and $6-3i$

By the conjugate zeros theorem, $2+i$ and $6+3i$ must also be zeros.

$$f(x)=\left[x-(2-i)\right]\left[x-(2+i)\right]\cdot\left[x-(6-3i)\right]\left[x-(6+3i)\right]=\left[(x-2+i)(x-2-i)\right]\cdot\left[(x-6+3i)(x-6-3i)\right]$$

$$=\left[(x-2)+i\right]\left[(x-2)-i\right]\cdot\left[(x-6)+3i\right]\left[(x-6)-3i\right]=\left[(x-2)^2-i^2\right]\left[(x-6)^2-(3i)^2\right]$$

$$=\left[(x^2-4x+4)-i^2\right]\cdot\left[(x^2-12x+36)-9i^2\right]=(x^2-4x+4+1)(x^2-12x+36+9)$$

$$=(x^2-4x+5)(x^2-12x+45)=x^4-16x^3+98x^2-240x+225$$

**69.** $4,\ 1-2i,$ and $3+4i$

By the conjugate zeros theorem, $1+2i$ and $3-4i$ must also be zeros.

$$f(x)=(x-4)\left[x-(1-2i)\right]\left[x-(1+2i)\right]\cdot\left[x-(3+4i)\right]\left[x-(3-4i)\right]$$

$$=(x-4)(x-1+2i)(x-1-2i)\cdot(x-3-4i)(x-3+4i)$$

$$=(x-4)\left[(x-1)+2i\right]\left[(x-1)-2i\right]\cdot\left[(x-3)-4i\right]\left[(x-3)+4i\right]$$

$$=(x-4)\left[(x-1)^2-(2i)^2\right]\left[(x-3)^2-(4i)^2\right]=(x-4)(x^2-2x+1-4i^2)\cdot(x^2-6x+9-16i^2)$$

$$=(x-4)(x^2-2x+1+4)(x^2-6x+9+16)=(x-4)(x^2-2x+5)(x^2-6x+25)$$

$$=(x-4)(x^4-8x^3+42x^2-80x+125)=x^5-12x^4+74x^3-248x^2+445x-500$$

**71.** $1+2i,\ 2$ (multiplicity 2).

By the conjugate zeros theorem, $1-2i$ must also be a zero.

$$f(x)=(x-2)^2\left[x-(1+2i)\right]\left[x-(1-2i)\right]$$

$$=(x^2-4x+4)\left[(x-1)-2i\right]\left[(x-1)+2i\right]$$

$$=(x^2-4x+4)\left[(x-1)^2-(2i)^2\right]$$

$$=(x^2-4x+4)(x^2-2x+1-4i^2)$$

$$=(x^2-4x+4)(x^2-2x+1+4)$$

$$=(x^2-4x+4)(x^2-2x+5)$$

$$=x^4-6x^3+17x^2-28x+20$$

**73.** $f(x)=2x^3-4x^2+2x+7$

$f(x)=2x^3-4x^2+2x+7$ has 2 variations in sign. $f$ has either 2 or $2-2=0$ positive real zeros.

$f(-x)=-2x^3-4x^2-2x+7$ has 1 variation in sign. $f$ has 1 negative real zero.

**75.** $f(x)=5x^4+3x^2+2x-9$

$f(x)=5x^4+3x^2+2x-9$ has 1 variation in sign. $f$ has 1 positive real zero.

$f(-x)=5x^4+3x^2-2x-9$ has 1 variation in sign. $f$ has 1 negative real zero.

**77.** $f(x)=x^5+3x^4-x^3+2x+3$

$f(x)=x^5+3x^4-x^3+2x+3$ has 2 variations in sign. $f$ has 2 or $2-2=0$ positive real zeros.

$f(-x)=-x^5+3x^4+x^3-2x+3$ has 3 variations in sign. $f$ has 3 or $3-2=1$ negative real zeros.

**79.** $f(x)=x^4+2x^3-3x^2+24x-180$

$p$ must be a factor of $a_0=-180$ and $q$ must be a factor of $a_4=1$. Thus, $p$ can be $\pm1,\ \pm2,$ $\pm3,\ \pm4,\ \pm5,\ \pm6,\ \pm9,\ \pm10,\ \pm12,\ \pm15,$ $\pm18,\ \pm20,\ \pm30,\ \pm36,\ \pm45,\ \pm60,\ \pm90,$ and $\pm180$. $q$ can be $\pm1$. The possible rational zeros, $\frac{p}{q}$, are $\pm1,\ \pm2,\ \pm3,\ \pm4,\ \pm5,\ \pm6,$ $\pm9,\ \pm10,\ \pm12,\ \pm15,\ \pm18,\ \pm20,\ \pm30,$ $\pm36,\ \pm45,\ \pm60,\ \pm90,$ and $\pm180$.

Using the remainder theorem and synthetic division, we find that one zero is $x=-5$.

$$\begin{array}{r|rrrr}
-5 & 1 & 2 & -3 & 24 & -180 \\
   &   & -5 & 15 & -60 & 180 \\
\hline
   & 1 & -3 & 12 & -36 & 0
\end{array}$$

Setting the quotient $x^3 - 3x^2 + 12x - 36$ equal to zero and factoring by grouping, we have:

$$0 = x^3 - 3x^2 + 12x - 36$$
$$= x^2(x-3) + 12(x-3) = (x^2 + 12)(x-3) \Rightarrow$$
$$x^2 + 12 = 0 \Rightarrow x^2 = -12 \Rightarrow x = \sqrt{-12} \Rightarrow$$
$$x = \pm 2i\sqrt{3} \quad \text{or} \quad x - 3 = 0 \Rightarrow x = 3$$

Thus, the zeros of $f(x)$ are $\left\{-5, 3, \pm 2i\sqrt{3}\right\}$.

**81.** $f(x) = x^4 + x^3 - 9x^2 + 11x - 4$

$p$ must be a factor of $a_0 = -4$ and $q$ must be a factor of $a_4 = 1$. Thus, $p$ can be $\pm 1$, $\pm 2$, and $\pm 4$. $q$ can be $\pm 1$. The possible rational zeros, $\frac{p}{q}$, are $\pm 1$, $\pm 2$, and $\pm 4$. Using the remainder theorem and synthetic division, we find that one zero is $x = -4$.

$$
\begin{array}{r|rrrr}
-4 & 1 & 1 & -9 & 11 & -4 \\
   &   & -4 & 12 & -12 & 4 \\
\hline
   & 1 & -3 & 3 & -1 & 0 \\
\end{array}
$$

Setting the quotient $x^3 - 3x^2 + 3x - 1$ equal to zero and factoring by grouping, we have:

$$x^3 - 3x^2 + 3x - 1 = 0$$
$$(x-1)^3 = 0 \Rightarrow x = 1 \text{ (with multiplicity 3)}$$

Thus, the zeros of $f(x)$ are $\{-4, 1, 1, 1\}$.

**83.** $f(x) = 2x^5 + 11x^4 + 16x^3 + 15x^2 + 36x$

$a_0 = 0$, so 0 is a zero of the function. Factoring, we have

$$2x^5 + 11x^4 + 16x^3 + 15x^2 + 36x$$
$$= x(2x^4 + 11x^3 + 16x^2 + 15x + 36)$$

To find a zero of the quotient, $2x^4 + 11x^3 + 16x^2 + 15x + 36$, $p$ must be a factor of $a_0 = 36$ and $q$ must be a factor of $a_4 = 2$. Thus, $p$ can be $\pm 1$, $\pm 2$, $\pm 3$, $\pm 4$, $\pm 6$, $\pm 9$, $\pm 12$, $\pm 18$, or $\pm 36$. $q$ can be $\pm 1$ or $\pm 2$. The possible rational zeros, $\frac{p}{q}$, are

$\pm \frac{1}{2}, \pm \frac{3}{2}, \pm \frac{9}{2}, \pm 1, \pm 2, \pm 3, \pm 4, \pm 6, \pm 8, \pm 9, \pm 12$
$\pm 18$, or $\pm 36$. Using the remainder theorem and synthetic division, we find that one zero is $x = -3$.

$$
\begin{array}{r|rrrrrr}
-3 & 2 & 11 & 16 & 15 & 36 & 0 \\
   &   & -6 & -15 & -3 & -36 & 0 \\
\hline
   & 2 & 5 & 1 & 12 & 0 & 0 \\
\end{array}
$$

Now find a zero of the quotient $2x^3 + 5x^2 + x + 12$. $p$ must be a factor of $a_0 = 12$ and $q$ must be a factor of $a_3 = 2$. Thus, $p$ can be $\pm 1$, $\pm 2$, $\pm 3$, $\pm 4$, $\pm 6$, or $\pm 12$. $q$ can be $\pm 1$ or $\pm 2$. The possible rational zeros, $\frac{p}{q}$, are $\pm \frac{1}{2}, \pm \frac{3}{2}, \pm 1,$

$\pm 2, \pm 3, \pm 4, \pm 6,$ or $\pm 12$. Using the remainder theorem and synthetic division, we find that one zero is $x = -3$

$$
\begin{array}{r|rrrr}
-3 & 2 & 5 & 1 & 12 \\
   &   & -6 & 3 & -12 \\
\hline
   & 2 & -1 & 4 & 0 \\
\end{array}
$$

Setting the quotient $2x^2 - x + 4$ equal to zero and using the quadratic equation with $a = 2$, $b = -1$, and $c = 4$ to solve this, we have:

$$x = \frac{-(-1) \pm \sqrt{(-1)^2 - 4(2)(4)}}{2(2)}$$
$$= \frac{1 \pm \sqrt{1 - 32}}{4} = \frac{1 \pm i\sqrt{31}}{4}$$

Thus, the zeros of $f(x)$ are

$$\left\{-3, -3, 0, \frac{1 \pm i\sqrt{31}}{4}\right\}.$$

**85.** $f(x) = x^5 - 6x^4 + 14x^3 - 20x^2 + 24x - 16$

$p$ must be a factor of $a_0 = -16$ and $q$ must be a factor of $a_3 = 2$. Thus, $p$ can be $\pm 1$, $\pm 2$, $\pm 4$, $\pm 8$, or $\pm 16$. $q$ can be $\pm 1$. The possible rational zeros, $\frac{p}{q}$, are $\pm 1$, $\pm 2$, $\pm 4$, $\pm 8$, or $\pm 16$. Using the remainder theorem and synthetic division, we find that one zero is $x = 2$

$$
\begin{array}{r|rrrrrr}
2 & 1 & -6 & 14 & -20 & 24 & -16 \\
  &   & 2 & -8 & 12 & -16 & 16 \\
\hline
  & 1 & -4 & 6 & -8 & 8 & 0 \\
\end{array}
$$

To find a zero of the quotient, $x^4 - 4x^3 + 6x^2 - 8x + 8$, $p$ must be a factor of $a_0 = 8$ and $q$ must be a factor of $a_4 = 1$. Thus, $p$ can be $\pm 1$, $\pm 2$, $\pm 4$, or $\pm 8$. $q$ can be $\pm 1$. The possible rational zeros, $\frac{p}{q}$, are $\pm 1$, $\pm 2$, $\pm 4$, or $\pm 8$. Using the remainder theorem and synthetic division, we find that one zero is $x = 2$.

(*continued on next page*)

*(continued from page 153)*

$$2\overline{)1 \quad -4 \quad 6 \quad -8 \quad 8}$$
$$\underline{\quad 2 \quad -4 \quad 4 \quad -8}$$
$$1 \quad -2 \quad 2 \quad -4 \quad 0$$

Now find a zero of the quotient
$x^3 - 2x^2 + 2x - 4$. $p$ must be a factor of
$a_0 = -4$ and $q$ must be a factor of $a_3 = 1$.
Thus, $p$ can be $\pm 1$, $\pm 2$, $\pm 3$, or $\pm 4$. $q$ can be
$\pm 1$. The possible rational zeros, $\frac{p}{q}$, are $\pm 1$,
$\pm 2$, $\pm 3$, or $\pm 4$. Using the remainder theorem
and synthetic division, we find that one zero is
$x = 2$

$$2\overline{)1 \quad -2 \quad 2 \quad -4}$$
$$\underline{\quad 2 \quad 0 \quad 4}$$
$$1 \quad 0 \quad 2 \quad 0$$

Setting the quotient $x^2 + 2$ equal to zero, we
have: $x^2 + 2 = 0 \Rightarrow x^2 = -2 \Rightarrow x = \pm i\sqrt{2}$
Thus, the zeros of $f(x)$ are $\left\{2, 2, 2, \pm i\sqrt{2}\right\}$.

**87.** $f(x) = 2x^4 - x^3 + 7x^2 - 4x - 4$

$p$ must be a factor of $a_0 = -4$ and $q$ must be a
factor of $a_4 = 2$. Thus, $p$ can be $\pm 1$, $\pm 2$,and
$\pm 4$. $q$ can be $\pm 1$ or $\pm 2$. The possible rational
zeros, $\frac{p}{q}$, are $\pm\frac{1}{2}$, $\pm 1$, $\pm 2$, and $\pm 4$. Using
the remainder theorem and synthetic division,
we find that one zero is $x = 1$.

$$1\overline{)2 \quad -1 \quad 7 \quad -4 \quad -4}$$
$$\underline{\quad 2 \quad 1 \quad 8 \quad 4}$$
$$2 \quad 1 \quad 8 \quad 4 \quad 0$$

Setting the quotient $2x^3 + x^2 + 8x + 4$ equal to
zero and factoring by grouping, we have:
$$2x^3 + x^2 + 8x + 4 = 0$$
$$x^2(2x+1) + 4(2x+1) = 0$$
$$(x^2+4)(2x+1) = 0 \Rightarrow$$
$$x^2 + 4 = 0 \Rightarrow x^2 = -4 \Rightarrow x = \pm 2i \text{ or}$$
$$2x + 1 = 0 \Rightarrow x = -\tfrac{1}{2}$$

Thus, the zeros of $f(x)$ are $\left\{-\tfrac{1}{2}, 1, \pm 2i\right\}$.

**89.** $f(x) = 5x^3 - 9x^2 + 28x + 6$

$p$ must be a factor of $a_0 = 6$ and $q$ must be a
factor of $a_3 = 5$. Thus, $p$ can be $\pm 1, \pm 2, \pm 3,$ or
$\pm 6$. $q$ can be $\pm 1$ or $\pm 5$. The possible rational
zeros, $\frac{p}{q}$, are $\pm 1, \pm 2, \pm 3, \pm 6, \pm\frac{1}{5}, \pm\frac{2}{5}, \pm\frac{3}{5},$ or
$\pm\frac{6}{5}$. Using the remainder theorem and
synthetic division, we find that one zero is
$x = -\frac{1}{5}$

$$-\tfrac{1}{5}\overline{)5 \quad -9 \quad 28 \quad 6}$$
$$\underline{\quad -1 \quad 2 \quad -6}$$
$$5 \quad -10 \quad 30 \quad 0$$

Setting the quotient
$5x^2 - 10x + 30 = 5(x^2 - 2x + 6)$ equal to zero
and using the quadratic equation with $a = 1$,
$b = -2$, and $c = 6$ to solve this, we have:

$$x = \frac{-(-2) \pm \sqrt{(-2)^2 - 4(1)(6)}}{2(1)} = \frac{2 \pm \sqrt{4 - 24}}{2}$$

$$= \frac{2 \pm \sqrt{-20}}{2} = \frac{2 \pm 2i\sqrt{5}}{2} = 1 \pm i\sqrt{5}$$

Thus, the zeros of $f(x)$ are $\left\{-\tfrac{1}{5}, 1 \pm i\sqrt{5}\right\}$.

**91.** $f(x) = x^4 + 29x^2 + 100$

Letting $u = x^2$, we have
$x^4 + 29x^2 + 100 = u^2 + 29u + 100$. Set the
function equal to zero, and solve for $u$ using
the quadratic formula with $a = 1$, $b = 29$ and
$c = 100$.

$$u = \frac{-29 \pm \sqrt{29^2 - 4(1)(100)}}{2(1)}$$

$$= \frac{-29 \pm \sqrt{441}}{2} = \frac{-29 \pm 21}{2} = -25 \text{ or } u = -4$$

$$u = -25 \Rightarrow x^2 = -25 \Rightarrow x = \pm 5i$$
$$u = -4 \Rightarrow x^2 = -4 \Rightarrow x = \pm 2i$$
Thus, the zeros of $f(x)$ are $\left\{\pm 2i, \pm 5i\right\}$.

**93.** $f(x) = x^4 + 2x^2 + 1$

Setting $f(x)$ equal to 0, then factoring to
solve for $x$, we have
$$x^4 + 2x^2 + 1 = 0$$
$$(x^2 + 1)^2 = 0$$
$$x^2 + 1 = 0 \Rightarrow x^2 = -1 \Rightarrow$$
$$x = \pm i \text{ (with multiplicity 2)}$$
Thus, the zeros of $f(x)$ are $\left\{\pm i, \pm i\right\}$.

**95.** $f(x) = x^4 - 6x^3 + 7x^2$

Setting $f(x)$ equal to zero and solving for $x$, we have $x^4 - 6x^3 + 7x^2 = 0 \Rightarrow x^2(x^2 - 6x + 7) = 0 \Rightarrow$

$$x^2 = 0 \Rightarrow$$
$$x = 0 \text{ (with multiplicity 2) or}$$
$$x^2 - 6x + 7 = 0 \Rightarrow$$

$$x = \frac{-(-6) \pm \sqrt{(-6)^2 - 4(1)(7)}}{2(1)}$$
$$= \frac{6 \pm \sqrt{36 - 28}}{2} = \frac{6 \pm \sqrt{8}}{2} = 3 \pm \sqrt{2}$$

Thus, the zeros of $f(x)$ are $\left\{0, 0, 3 \pm \sqrt{2}\right\}$.

**97.** $f(x) = x^4 - 8x^3 + 29x^2 - 66x + 72$

$p$ must be a factor of $a_0 = 72$ and $q$ must be a factor of $a_4 = 1$. Thus, $p$ can be $\pm 1, \pm 2, \pm 3,$ $\pm 4, \pm 6, \pm 8, \pm 9, \pm 12, \pm 18, \pm 24, \pm 36,$ or $\pm 72$.

$q$ can be $\pm 1$. The possible rational zeros, $\frac{p}{q}$, are $\pm 1, \pm 2, \pm 3, \pm 4, \pm 6, \pm 8, \pm 9, \pm 12, \pm 18, \pm 24,$ $\pm 36,$ or $\pm 72$.

Using the remainder theorem and synthetic division, we find that one zero is $x = 3$.

$$3)\overline{1 \quad -8 \quad 29 \quad -66 \quad 72}$$
$$\underline{\quad\quad 3 \quad -15 \quad 42 \quad -72}$$
$$1 \quad -5 \quad 14 \quad -24 \quad 0$$

Now find a zero of the quotient $x^3 - 5x^2 + 14x - 24$. $p$ must be a factor of $a_0 = -24$ and $q$ must be a factor of $a_3 = 1$. Thus, $p$ can be $\pm 1, \pm 2, \pm 3, \pm 4, \pm 6, \pm 8, \pm 12,$ or $\pm 24$, while $q$ can be $\pm 1$.

Using the remainder theorem and synthetic division, we find that one zero is $x = 3$.

$$3)\overline{1 \quad -5 \quad 14 \quad -24}$$
$$\underline{\quad\quad 3 \quad -6 \quad 24}$$
$$1 \quad -2 \quad 8 \quad 0$$

Setting the quotient $x^2 - 2x + 8$ equal to zero and solving for $x$ using the quadratic formula with $a = 1$, $b = -2$, and $c = 8$, we have

$$x = \frac{-(-2) \pm \sqrt{(-2)^2 - 4(1)(8)}}{2(1)}$$
$$= \frac{2 \pm \sqrt{4 - 32}}{2} = \frac{2 \pm \sqrt{-28}}{2} = 1 \pm i\sqrt{7}$$

Thus, the zeros of $f(x)$ are $\left\{3, 3, 1 \pm i\sqrt{7}\right\}$.

**99.** $f(x) = x^6 - 9x^4 - 16x^2 + 144$

Let $u = x^2$. Then $x^6 - 9x^4 - 16x^2 + 144 \Rightarrow$ $u^3 - 9u^2 - 16u + 144$. Set $f(u)$ equal to 0, then factor by grouping to solve for $u$:

$$u^3 - 9u^2 - 16u + 144 = 0$$
$$u^2(u - 9) - 16(u - 9) = 0$$
$$(u^2 - 16)(u - 9) = 0 \Rightarrow$$
$$u^2 - 16 = 0 \Rightarrow u^2 = 16 \Rightarrow u = \pm 4 \text{ or}$$
$$u - 9 = 0 \Rightarrow u = 9$$
$$u = 4 \Rightarrow x^2 = 4 \Rightarrow x = \pm 2$$
$$u = -4 \Rightarrow x^2 = -4 \Rightarrow x = \pm 2i$$
$$u = 9 \Rightarrow x^2 = 9 \Rightarrow x = \pm 3$$

Thus, the zeros of $f(x)$ are $\left\{\pm 2, \pm 3, \pm 2i\right\}$.

For exercises 101–103, let $c = a + bi$ and $d = m + ni$.

**101.** $\overline{c + d} = \overline{(a + bi) + (m + ni)}$
$$= \overline{(a + m) + (b + n)i} = (a + m) - (b + n)i$$
$$= a + m - bi - ni = (a - bi) + (m - ni)$$
$$= \overline{c} + \overline{d}$$

**103.** If $a$ is a real number then $a$ is of the form $a + 0 \cdot i$. $\overline{a} = \overline{a + 0 \cdot i} = a - 0i = a$. Therefore, the statement is true.

## Section 3.4: Polynomial Functions: Graphs, Applications, and Models

**1.** $y = x^3 - 3x^2 - 6x + 8$

The range of an odd-degree polynomial is $(-\infty, \infty)$. The $y$-intercept of the graph is 8. The graph fitting these criteria is A.

**3.** Since graph C crosses the $x$-axis at one point, the graph has one real zero.

**5.** A polynomial of degree 3 can have at most 2 turning points. Graphs B and D have more than 2 turning points, so they cannot be graphs of cubic polynomial functions.

**7.** Since graph B touches the $x$-axis at –5, the function has 2 real zeros of –5. The two other real zeros are where the graph crosses the $x$-axis, at 0 and 3.

$$f(x) = x^4 + 7x^3 - 5x^2 - 75x = x(x + 5)^2(x - 3)$$

**9.** $f(x) = 2x^4$ is in the form $f(x) = ax^n$.
$|a| = 2 > 1$, so the graph is narrower than
$f(x) = x^4$. It includes the points $(-2, 32)$,
$(-1, 2)$, $(0, 0)$, $(1, 2)$, and $(2, 32)$. Connect
these points with a smooth curve.

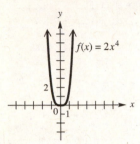

**11.** $f(x) = -\frac{2}{3}x^5$ is in the form $f(x) = ax^n$.
$|a| = \frac{2}{3} < 1$, so the graph is broader than that
of $f(x) = x^5$. Since $a = -\frac{2}{3}$ is a negative, the
graph is the reflection of $f(x) = \frac{2}{3}x^5$ about
the $x$-axis. It includes the points $\left(-2, \frac{64}{3}\right)$,
$\left(-1, \frac{2}{3}\right)$, $(0, 0)$, $\left(1, -\frac{2}{3}\right)$, and $\left(2, -\frac{64}{3}\right)$.
Connect these points with a smooth curve.

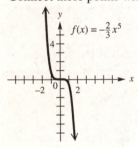

**13.** $f(x) = \frac{1}{2}x^3 + 1$ is in the form $f(x) = ax^n + k$,
with $|a| = \frac{1}{2} < 0$ and $k = 1$. The graph of
$f(x) = \frac{1}{2}x^3 + 1$ looks like $y = x^3$ but is
broader and is translated 1 unit up. The graph
includes the points $(-2, -3)$,
$\left(-1, \frac{1}{2}\right)$, $(0, 1)$, $\left(1, \frac{3}{2}\right)$, and $(2, 5)$.

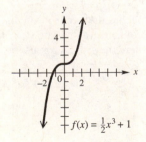

**15.** $f(x) = -(x+1)^3 = -\left[x - (-1)\right]^3$
The graph can be obtained by reflecting the
graph of $f(x) = x^3$ about the $x$-axis and then
translating it 1 unit to the left.

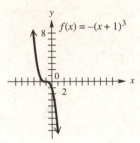

**17.** $f(x) = (x-1)^4 + 2$ This graph has the same
shape as $y = x^4$, but is translated 1 unit to the
right and 2 units up.

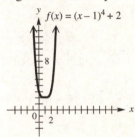

**19.** $f(x) = \frac{1}{2}(x-2)^2 + 4$ is in the form
$f(x) = a(x-h)^n + k$, with $|a| = \frac{1}{2} < 1$.
The graph is broader than that of $f(x) = x^2$.
Since $h = 2$, the graph has been translated 2
units to the right. Also, since $k = 4$, the graph
has been translated 4 units up.

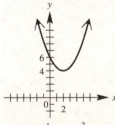

**21.**      **23.**

**25.**      **27.**

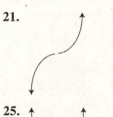

**29.** $f(x) = x^3 + 5x^2 + 2x - 8$

*Step 1:* $p$ must be a factor of $a_0 = -8$ and $q$ must be a factor of $a_3 = 1$. Thus, $p$ can be $\pm 1$, $\pm 2$, $\pm 4$, $\pm 8$ and $q$ can be $\pm 1$. The possible rational zeros, $\frac{p}{q}$, are $\pm 1$, $\pm 2$, $\pm 4$, and $\pm 8$. The remainder theorem shows that $-4$ is a zero.

$$-4 \overline{)\begin{array}{rrrr} 1 & 5 & 2 & -8 \\ & -4 & -4 & 8 \\ \hline 1 & 1 & -2 & 0 \end{array}}$$

The new quotient polynomial is $x^2 + x - 2$.
$$x^2 + x - 2 = 0$$
$$(x+2)(x-1) = 0$$
$$x + 2 = 0 \Rightarrow x = -2 \quad \text{or} \quad x - 1 = 0 \Rightarrow x = 1$$
The rational zeros are $-4$, $-2$, and $1$. Since the three zeros are $-4$, $-2$, and $1$, the factors are $x + 4$, $x + 2$, and $x - 1$ and thus
$$f(x) = (x+4)(x+2)(x-1).$$

*Step 2:* $f(0) = -8$, so plot $(0, -8)$.

*Step 3:* The $x$-intercepts divide the $x$-axis into four intervals:

| Interval | Test Point | Value of $f(x)$ | Sign of $f(x)$ | Graph Above or Below $x$-Axis |
|---|---|---|---|---|
| $(-\infty, -4)$ | $-5$ | $-18$ | Negative | Below |
| $(-4, -2)$ | $-3$ | $4$ | Positive | Above |
| $(-2, 1)$ | $0$ | $-8$ | Negative | Below |
| $(1, \infty)$ | $2$ | $24$ | Positive | Above |

Plot the $x$-intercepts, $y$-intercept and test points (the $y$-intercept is one of the test points) with a smooth curve to get the graph.

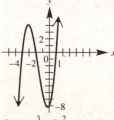

$f(x) = x^3 + 5x^2 + 2x - 8$

**31.** $f(x) = 2x(x-3)(x+2)$

*Step 1:* Set each factor equal to 0 and solve the resulting equations to find the zeros of the function.
$$2x = 0 \Rightarrow x = 0 \quad \text{or} \quad x - 3 = 0 \Rightarrow x = 3 \quad \text{or}$$
$$x + 2 = 0 \Rightarrow x = -2$$

The three zeros, $-2$, $0$, and $3$, divide the $x$-axis into four regions. Test a point in each region to find the sign of $f(x)$ in that region.

*Step 2:* $f(0) = 0$, so plot $(0, 0)$.

*Step 3:* The $x$-intercepts divide the $x$-axis into four intervals.

| Interval | Test Point | Value of $f(x)$ | Sign of $f(x)$ | Graph Above or Below $x$-Axis |
|---|---|---|---|---|
| $(-\infty, -2)$ | $-3$ | $-36$ | Negative | Below |
| $(-2, 0)$ | $-1$ | $8$ | Positive | Above |
| $(0, 3)$ | $1$ | $-12$ | Negative | Below |
| $(3, \infty)$ | $4$ | $48$ | Positive | Above |

Plot the $x$-intercepts, $y$-intercept (which is also an $x$-intercept in this exercise), and test points with a smooth curve to get the graph.

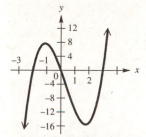

$f(x) = 2x(x-3)(x+2)$

**33.** $f(x) = x^2(x-2)(x+3)^2$

*Step 1:* Set each factor equal to 0 and solve the resulting equations to find the zeros of the function.
$$x^2 = 0 \Rightarrow x = 0 \quad \text{or} \quad x - 2 = 0 \Rightarrow x = 2 \quad \text{or}$$
$$(x+3)^2 = 0 \Rightarrow x + 3 = 0 \Rightarrow x = -3$$

The zeros are $-3$, $0$, and $2$; divide the $x$-axis into four regions. Test a point in each region to find the sign of $f(x)$ in that region.

*Step 2:* $f(0) = 0$, so plot $(0, 0)$.

*Step 3:* The $x$-intercepts divide the $x$-axis into four intervals.

| Interval | Test Point | Value of $f(x)$ | Sign of $f(x)$ | Graph Above or Below $x$-Axis |
|---|---|---|---|---|
| $(-\infty, -3)$ | $-4$ | $-96$ | Negative | Below |
| $(-3, 0)$ | $-1$ | $-12$ | Negative | Below |
| $(0, 2)$ | $1$ | $-16$ | Negative | Below |
| $(2, \infty)$ | $3$ | $324$ | Positive | Above |

*(continued on next page)*

*(continued from page 157)*

Plot the *x*-intercepts, *y*-intercept (which is also an *x*-intercept in this exercise), and test points with a smooth curve to get the graph.

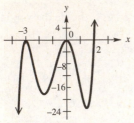

$$f(x) = x^2(x - 2)(x + 3)^2$$

**35.** $f(x) = (3x - 1)(x + 2)^2$

*Step 1:* Set each factor equal to 0 and solve the resulting equations to find the zeros of the function.

$$3x - 1 = 0 \Rightarrow x = \tfrac{1}{3} \text{ or } x + 2 = 0 \Rightarrow x = -2$$

The zeros are $-2$ and $\tfrac{1}{3}$, which divide the *x*-axis into three regions. Test a point in each region to find the sign of $f(x)$ in that region.

*Step 2:* $f(0) = -4$, so plot $(0, -4)$.

*Step 3:* The *x*-intercepts divide the *x*-axis into three intervals.

| Interval | Test Point | Value of $f(x)$ | Sign of $f(x)$ | Graph Above or Below *x*-Axis |
|---|---|---|---|---|
| $(-\infty, -2)$ | $-3$ | $-10$ | Negative | Below |
| $\left(-2, \tfrac{1}{3}\right)$ | $0$ | $-4$ | Negative | Below |
| $\left(\tfrac{1}{3}, \infty\right)$ | $1$ | $18$ | Positive | Above |

Plot the *x*-intercepts, *y*-intercept, and test points (the *y*-intercept is one of the test points) with a smooth curve to get the graph.

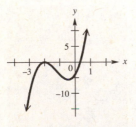

$$f(x) = (3x - 1)(x + 2)^2$$

**37.** $f(x) = x^3 + 5x^2 - x - 5$

*Step 1:* Find the zeros of the function by factoring by grouping.

$$f(x) = x^3 + 5x^2 - x - 5 = x^2(x + 5) - 1(x + 5)$$
$$= (x + 5)(x^2 - 1) = (x + 5)(x + 1)(x - 1)$$

$$x + 5 = 0 \Rightarrow x = -5 \text{ or } x + 1 = 0 \Rightarrow x = -1 \text{ or}$$
$$x - 1 = 0 \Rightarrow x = 1$$

The zeros are $-5$, $-1$, and $1$, which divide the *x*-axis into four regions. Test a point in each region to find the sign of $f(x)$ in that region.

*Step 2:* $f(0) = -5$, so plot $(0, -5)$.

*Step 3:* The *x*-intercepts divide the *x*-axis into four intervals.

| Interval | Test Point | Value of $f(x)$ | Sign of $f(x)$ | Graph Above or Below *x*-Axis |
|---|---|---|---|---|
| $(-\infty, -5)$ | $-6$ | $-35$ | Negative | Below |
| $(-5, -1)$ | $-2$ | $9$ | Positive | Above |
| $(-1, 1)$ | $0$ | $-5$ | Negative | Below |
| $(1, \infty)$ | $2$ | $21$ | Positive | Above |

Plot the *x*-intercepts, *y*-intercept, and test points (the *y*-intercept is one of the test points) with a smooth curve to get the graph.

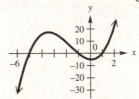

$$f(x) = x^3 + 5x^2 - x - 5$$

**39.** $f(x) = x^3 - x^2 - 2x$

*Step 1:* Find the zeros of the function by factoring out the common factor, *x*, and then factoring the resulting quadratic factor. Set each factor equal to 0 and solve the resulting equations.

$$f(x) = x^3 - x^2 - 2x$$
$$= x(x^2 - x - 2) = x(x + 1)(x - 2)$$

$$x = 0 \quad \text{or} \quad x + 1 = 0 \quad \text{or} \quad x - 2 = 0$$
$$x = -1 \qquad x = 2$$

The zeros are $-1$, $0$, and $2$, which divide the *x*-axis into four regions. Test a point in each region to find the sign of $f(x)$ in that region.

*Step 2:* $f(0) = 0$, so plot $(0, 0)$.

*Step 3:* The x-intercepts divide the x-axis into four intervals.

| Interval | Test Point | Value of $f(x)$ | Sign of $f(x)$ | Graph Above or Below x-Axis |
|---|---|---|---|---|
| $(-\infty, -1)$ | $-2$ | $-8$ | Negative | Below |
| $(-1, 0)$ | $-\frac{1}{2}$ | $\frac{5}{8}$ | Positive | Above |
| $(0, 2)$ | $1$ | $-2$ | Negative | Below |
| $(2, \infty)$ | $3$ | $12$ | Positive | Above |

Plot the x-intercepts, y-intercept (which is also an x-intercept in this exercise), and test points with a smooth curve to get the graph.

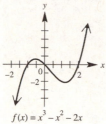

$f(x) = x^3 - x^2 - 2x$

**41.** $f(x) = 2x^3(x^2 - 4)(x - 1)$

*Step 1:* Find the zeros of the function by factoring the difference of two squares in the polynomial. Set each factor equal to 0 and solve the resulting equations.

$$f(x) = 2x^3(x^2 - 4)(x - 1)$$
$$= 2x^3(x + 2)(x - 2)(x - 1)$$

$2x^3 = 0 \Rightarrow x = 0$ or $x + 2 = 0 \Rightarrow x = -2$ or $x - 2 = 0 \Rightarrow x = 2$ or $x - 1 = 0 \Rightarrow x = 1$

The zeros are –2, 0, 1, and 2, which divide the x-axis into five regions. Test a point in each region to find the sign of $f(x)$ in that region.

*Step 2:* $f(0) = 0$, so plot $(0, 0)$.

*Step 3:* The x-intercepts divide the x-axis into four intervals.

| Interval | Test Point | Value of $f(x)$ | Sign of $f(x)$ | Graph Above or Below x-Axis |
|---|---|---|---|---|
| $(-\infty, -2)$ | $-3$ | $1080$ | Positive | Above |
| $(-2, 0)$ | $-1$ | $-12$ | Negative | Below |
| $(0, 1)$ | $\frac{1}{2}$ | $\frac{15}{32} \approx .5$ | Positive | Above |
| $(1, 2)$ | $\frac{3}{2}$ | $-\frac{189}{32} \approx -5.9$ | Negative | Below |
| $(2, \infty)$ | $3$ | $540$ | Positive | Above |

Plot the x-intercepts, y-intercept (which is also an x-intercept in this exercise), and test points with a smooth curve to get the graph.

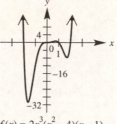

$f(x) = 2x^3(x^2 - 4)(x - 1)$

**43.** $f(x) = 2x^3 - 5x^2 - x + 6$

*Step 1:* Find the zeros of the function.
$p$ must be a factor of $a_0 = 6$ and $q$ must be a factor of $a_3 = 2$. Thus, $p$ can be ±1, ±2, ±3, or ±6 and $q$ can be ±1 or ±2. The possible rational zeros, $\frac{p}{q}$, are $\pm 1, \pm \frac{3}{2}$ ±2, ±3, or ±6.

The remainder theorem shows that 2 is a zero.

$$\begin{array}{r} 2\,\overline{)\,2 \quad -5 \quad -1 \quad 6} \\ \underline{4 \quad -2 \quad -6} \\ 2 \quad -1 \quad -3 \quad 0 \end{array}$$

The new quotient polynomial is $2x^2 - x - 3$.
$$2x^2 - x - 3 = 0$$
$$(2x - 3)(x + 1) = 0$$

$2x - 3 = 0 \Rightarrow x = \frac{3}{2}$ or $x + 1 = 0 \Rightarrow x = -1$

The rational zeros are $-1$, $\frac{3}{2}$, and 2. Since the three zeros are $-1$, $\frac{3}{2}$, and 2, the factors are

$x + 1$, $x - \frac{3}{2}$, and $x - 2$ and thus

$$f(x) = (x + 1)\left(x - \frac{3}{2}\right)(x - 2).$$

*Step 2:* $f(0) = 6$, so plot $(0, 6)$.

*Step 3:* The x-intercepts divide the x-axis into four intervals.

| Interval | Test Point | Value of $f(x)$ | Sign of $f(x)$ | Graph Above or Below x-Axis |
|---|---|---|---|---|
| $(-\infty, -1)$ | $-2$ | $-28$ | Negative | Below |
| $\left(-1, \frac{3}{2}\right)$ | $1$ | $2$ | Positive | Above |
| $\left(\frac{3}{2}, 2\right)$ | $\frac{7}{4}$ | $-\frac{11}{32}$ | Negative | Below |
| $(2, \infty)$ | $3$ | $12$ | Positive | Above |

*(continued on next page)*

*(continued from page 159)*

Plot the $x$-intercepts, $y$-intercept, and test points with a smooth curve to get the graph.

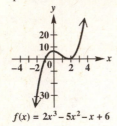

$f(x) = 2x^3 - 5x^2 - x + 6$

**45.** $f(x) = 3x^4 - 7x^3 - 6x^2 + 12x + 8$

*Step 1:* Find the zeros of the function.

$p$ must be a factor of $a_0 = 8$ and $q$ must be a factor of $a_4 = 3$. Thus, $p$ can be $\pm 1$, $\pm 2$, $\pm 4$, or $\pm 8$, and $q$ can be $\pm 1$ or $\pm 3$. The possible rational zeros, $\frac{p}{q}$, are $\pm\frac{1}{3}, \pm\frac{2}{3}, \pm 1, \pm\frac{4}{3}, \pm 2,$

$\pm\frac{8}{3}, \pm 4,$ or $\pm 8$. The remainder theorem shows that 2 is a zero.

$$
\begin{array}{r|rrrrr}
2 & 3 & -7 & -6 & 12 & 8 \\
  &   & 6 & -2 & -16 & -8 \\
\hline
  & 3 & -1 & -8 & -4 & 0 \\
\end{array}
$$

The new quotient polynomial is $3x^3 - x^2 - 8x - 4$. Find the zeros of this function. $p$ must be a factor of $a_0 = -4$ and $q$ must be a factor of $a_3 = 3$. Thus, $p$ can be $\pm 1$, $\pm 2$, or $\pm 4$, and $q$ can be $\pm 1$ or $\pm 3$.

The possible rational zeros, $\frac{p}{q}$, are $\pm\frac{1}{3}, \pm\frac{2}{3}, \pm 1, \pm\frac{4}{3}, \pm 2,$ or $\pm 4$. The remainder theorem shows that $-1$ is a zero.

$$
\begin{array}{r|rrrr}
-1 & 3 & -1 & -8 & -4 \\
   &   & -3 & 4 & 4 \\
\hline
   & 3 & -4 & -4 & 0 \\
\end{array}
$$

The new quotient polynomial is $3x^2 - 4x - 4$.

$3x^2 - 4x - 4 = 0$

$(3x + 2)(x - 2) = 0$

$3x + 2 = 0 \Rightarrow x = -\frac{2}{3}$ or $x - 2 = 0 \Rightarrow x = 2$

The rational zeros are $-1$, $-\frac{2}{3}$, 2, and 2. Since the four zeros are $-1$, $-\frac{2}{3}$, 2, and 2, the factors are $x + 1$, $x + \frac{2}{3}$, $x - 2$, and $x - 2$, and thus $f(x) = (x + 1)\left(x + \frac{2}{3}\right)(x - 2)^2$.

*Step 2:* $f(0) = 8$, so plot $(0, 8)$.

*Step 3:* The $x$-intercepts divide the $x$-axis into four intervals.

| Interval | Test Point | Value of $f(x)$ | Sign of $f(x)$ | Graph Above or Below $x$-Axis |
|---|---|---|---|---|
| $(-\infty, -1)$ | $-2$ | 64 | Positive | Above |
| $\left(-1, -\frac{2}{3}\right)$ | $-\frac{5}{6}$ | $-\frac{289}{432}$ | Negative | Below |
| $\left(-\frac{2}{3}, 2\right)$ | 1 | 10 | Positive | Above |
| $(2, \infty)$ | 3 | 44 | Positive | Above |

Plot the $x$-intercepts, $y$-intercept, and test points with a smooth curve to get the graph.

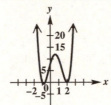

$f(x) = 3x^4 - 7x^3 - 6x^2 + 12x + 8$

**47.** $f(x) = 2x^2 - 7x + 4$; 2 and 3

Use synthetic division to find $f(2)$ and $f(3)$.

$$
\begin{array}{r|rrr} \qquad
2 & 2 & -7 & 4 \\
  &   & 4 & -6 \\
\hline
  & 2 & -3 & -2 \\
\end{array}
\qquad
\begin{array}{r|rrr}
3 & 2 & -7 & 4 \\
  &   & 6 & -3 \\
\hline
  & 2 & -1 & 1 \\
\end{array}
$$

Since $f(2) = -2$ is negative and $f(3) = 1$ is positive, there is a zero between 2 and 3.

**49.** $f(x) = 2x^3 - 5x^2 - 5x + 7$; 0 and 1

Since $f(0) = 7$ can easily be determined, use synthetic division only to find $f(1)$.

$$
\begin{array}{r|rrrr}
1 & 2 & -5 & -5 & 7 \\
  &   & 2 & -3 & -8 \\
\hline
  & 2 & -3 & -8 & -1 \\
\end{array}
$$

Since $f(0) = 7$ is positive and $f(1) = -1$ is negative, there is a zero between 0 and 1.

**51.** $f(x) = 2x^4 - 4x^2 + 4x - 8$; 1 and 2

Use synthetic division to find $f(1)$ and $f(2)$.

$$
\begin{array}{r|rrrrr} \quad
1 & 2 & 0 & -4 & 4 & -8 \\
  &   & 2 & 2 & -2 & 2 \\
\hline
  & 2 & 2 & -2 & 2 & -6 \\
\end{array}
\quad
\begin{array}{r|rrrrr}
2 & 2 & 0 & -4 & 4 & -8 \\
  &   & 4 & 8 & 8 & 24 \\
\hline
  & 2 & 4 & 4 & 12 & 16 \\
\end{array}
$$

Since $f(1) = -6$ is negative and $f(2) = 16$ is positive, there is a zero between 1 and 2.

**53.** $f(x) = x^4 + x^3 - 6x^2 - 20x - 16$; 3.2 and 3.3

Use synthetic division to find $f(3.2)$ and $f(3.3)$.

```
3.2)1    1      -6      -20      -16
         3.2    13.44   23.808   12.1856
     1   4.2    7.44    3.808    -3.8144
```

```
3.3)1    1      -6      -20      -16
         3.3    14.19   27.027   23.1891
     1   4.3    8.19    7.027    7.1891
```

Since $f(3.2) = -3.8144$ is negative and $f(3.3) = 7.1892$ is positive, there is a zero between 3.2 and 3.3.

**55.** $f(x) = x^4 - 4x^3 - 20x^2 + 32x + 12$; $-1$ and $0$

Since $f(0) = 12$ can easily be determined, use synthetic division only to find $f(-1)$.

```
-1)1    -4     -20    32     12
        -1      5     15    -47
    1   -5     -15    47    -35
```

Since $f(-1) = -35$ is negative and $f(0) = 12$ is positive, there is a zero between $-1$ and $0$.

**57.** $f(x) = x^4 - x^3 + 3x^2 - 8x + 8$; no real zero greater than 0.

Since $f(x)$ has real coefficients and the leading coefficient, 1, is positive, use the boundedness theorem. Divide $f(x)$ synthetically by $x - 2$. Since $2 > 0$ and all numbers in the last row are nonnegative, $f(x)$ has no zero greater than 2.

```
2)1   -1    3    -8    8
       2    2    10    4
  1    1    5     2   12
```

**59.** $f(x) = x^4 + x^3 - x^2 + 3$; no real zero less than $-2$.

Since $f(x)$ has real coefficients, use the boundedness theorem. Divide $f(x)$ synthetically by $x + 2 = x - (-2)$. Since $-2 < 0$ and the numbers in the last row alternate in sign, $f(x)$ has no zero less than $-2$.

```
-2)1    1    -1     0    3
       -2    2     -2    4
    1  -1    1     -2    7
```

**61.** $f(x) = 3x^4 + 2x^3 - 4x^2 + x - 1$; no real zero greater than 1.

Since $f(x)$ has real coefficients and the leading coefficient, 3, is positive, use the boundedness theorem. Divide $f(x)$ synthetically by $x - 1$. Since $1 > 0$ and all numbers in the last row are nonnegative, $f(x)$ has no zero greater than 1.

```
1)3   2   -4    1   -1
      3    5    1    2
  3   5    1    2    1
```

**63.** $f(x) = x^5 - 3x^3 + x + 2$; no real zero greater than 2.

Since $f(x)$ has real coefficients and the leading coefficient, 1, is positive, use the boundedness theorem. Divide $f(x)$ synthetically by $x - 2$. Since $2 > 0$ and all numbers in the last row are nonnegative, $f(x)$ has no zero greater than 2.

```
2)1   0   -3    0    1    2
      2   4     2    4   10
  1   2   1     2    5   12
```

**65.** The graph shows that the zeros are $-6$, 2, and 5. The polynomial function has the form $f(x) = a(x+6)(x-2)(x-5)$. Since $(0, 30)$ is on the graph, $f(0) = 30$.

$f(0) = a(0+6)(0-2)(0-5) \Rightarrow 30 = 60a \Rightarrow$
$\frac{1}{2} = a$

A cubic polynomial that has the graph shown is $f(x) = \frac{1}{2}(x+6)(x-2)(x-5)$ or

$f(x) = \frac{1}{2}x^3 - \frac{1}{2}x^2 - 16x + 30$.

**67.** The graph shows that the zeros are $-1$ and 1. There is a turning point at both points. Since the graph crosses and is tangent to the $x$-axis at both $x = -1$ and $x = 1$, these are zeros with odd multiplicity. This has to be 3 in order to find a polynomial of least degree. The polynomial function has the form $f(x) = a(x-1)^3(x+1)^3$. Since $(0, -1)$ is on the graph, $f(0) = -1$.

$f(0) = -1 = a(0-1)^3(0+1)^3 \Rightarrow a = 1$

(*continued on next page*)

*(continued from page 161)*

So the function is

$$f(x) = (x-1)^3 (x+1)^3$$
$$= (x^3 - 3x^2 + 3x - 1)(x^3 + 3x^2 + 3x + 1)$$
$$= x^6 - 3x^4 + 3x^2 - 1$$

**69.** The graph shows that the zeros are −3 and 3. The graph is tangent to the *x*-axis at both $x = -3$ and $x = 3$, so these are zeros with even multiplicity. This has to be 2 in order to find a polynomial of least degree. The polynomial function has the form

$f(x) = a(x-3)^2 (x+3)^2$. Since (0, 81) is on the graph, $f(0) = 81$.

$f(0) = 81 = a(0-3)^2 (0+3)^2 \Rightarrow 81 = 81a \Rightarrow$
$a = 1$.

So the function is

$$f(x) = (x-3)^2 (x+3)^2$$
$$= (x^2 - 6x + 9)(x^2 + 6x + 9)$$
$$= x^4 - 18x^2 + 81$$

**71.** $f(1.25) \approx -14.21875$

In order to have the calculator give the third screen, enter your function and the window, then press the GRAPH button. Press the TRACE button, then type in 1.25 then hit the ENTER key.

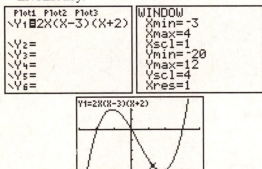

**73.** $f(1.25) \approx 29.046875$

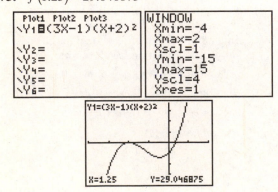

**75.** $f(x) = 2x^2 - 7x + 4$; 2 and 3

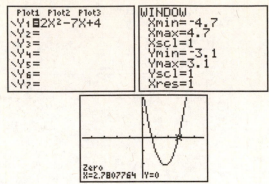

The real zero between 2 and 3 is approximately 2.7807764.

**77.** $f(x) = 2x^4 - 4x^2 + 4x - 8$; 1 and 2

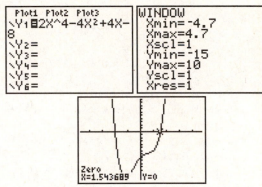

The real zero between 1 and 2 is approximately 1.543689.

**79.** $f(x) = x^3 + 3x^2 - 2x - 6$

The highest degree term is $x^3$, so the graph will have end behavior similar to the graph of $f(x) = x^3$, which is downward at the left and upward at the right. There is at least one real zero because the polynomial is of odd degree. There are at most three real zeros because the polynomial is third-degree. A graphing calculator can be used to approximate each zero.

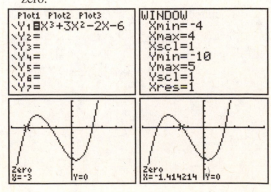

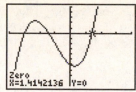

The graphs show that the zeros are approximately –3.0, –1.4 and 1.4.

**81.** $f(x) = -2x^4 - x^2 + x + 5$

The highest degree term is $-2x^4$ so the graph will have the same end behavior as the graph of $f(x) = -x^4$, which is downward at both the left and the right. Since $f(0) = 5 > 0$, the end behavior and the intermediate value theorem tell us that there must be at least one zero on each side of the $y$-axis, that is, at least one negative and one positive zero. A graphing calculator can be used to approximate each zero.

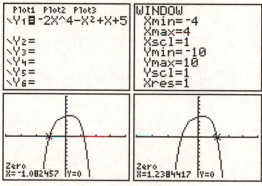

The graphs show that the only zeros are approximately –1.1 and 1.2

**83.** $f(x) = x^3 + 4x^2 - 8x - 8; [-3.8, -3]$

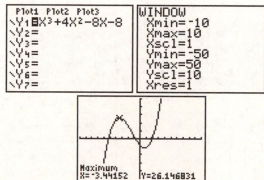

The turning point is (–3.44, 26.15).

**85.** $f(x) = 2x^3 - 5x^2 - x + 1; [-1, 0]$

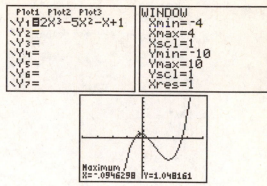

The turning point is (–.09, 1.05).

**87.** $f(x) = x^4 - 7x^3 + 13x^2 + 6x - 28; [-1, 0]$

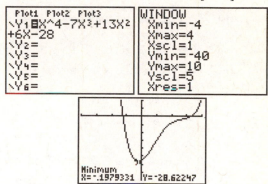

The turning point is (–.20, –28.62).

**89.** Answers will vary.

**91.** $f(x) = x^3 - 3x^2 - 6x + 8 = (x-4)(x-1)(x+2)$

*Step 1:* Set each factor equal to 0 and solve the resulting equations to find the zeros of the function.

$x - 4 = 0 \Rightarrow x = 4$ or $x - 1 = 0 \Rightarrow x = 1$ or $x + 2 = 0 \Rightarrow x = -2$

The zeros are –2, 1, and 4, which divide the $x$-axis into four regions. Test a point in each region to find the sign of $f(x)$ in that region.

*Step 2:* $f(0) = 8$, so plot $(0, 8)$.

*Step 3:* The $x$-intercepts divide the $x$-axis into four intervals.

| Interval | Test Point | Value of $f(x)$ | Sign of $f(x)$ | Graph Above or Below $x$-Axis |
|---|---|---|---|---|
| $(-\infty, -2)$ | –3 | –28 | Negative | Below |
| $(-2, 1)$ | 0 | 8 | Positive | Above |
| $(1, 4)$ | 2 | –8 | Negative | Below |
| $(4, \infty)$ | 5 | 28 | Positive | Above |

(*continued on next page*)

(*continued from page 163*)

Plot the *x*-intercepts, *y*-intercept, and test points (the *y*-intercept is one of the test points) with a smooth curve to get the graph.

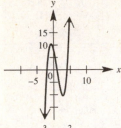

$$f(x) = x^3 - 3x^2 - 6x + 8$$
$$= (x - 4)(x - 1)(x + 2)$$

**(a)** $f(x) = 0$ has the solution set $\{-2, 1, 4\}$.

**(b)** $f(x) < 0$ has the solution set $(-\infty, -2) \cup (1, 4)$.

**(c)** $f(x) > 0$ has the solution set $(-2, 1) \cup (4, \infty)$.

**93.** $f(x) = 2x^4 - 9x^3 - 5x^2 + 57x - 45$
$$= (x - 3)^2(2x + 5)(x - 1)$$

*Step 1*: Set each factor equal to 0 and solve the resulting equations to find the zeros of the function.

$(x - 3)^2 = 0 \Rightarrow x = 3$ or $2x + 5 = 0 \Rightarrow x = -\frac{5}{2}$
or $x - 1 = 0 \Rightarrow x = 1$

The zeros are $-\frac{5}{2} = -2.5$, 1, and 3, which divide the *x*-axis into four regions. Test a point in each region to find the sign of $f(x)$ in that region.

*Step 2*: $f(0) = -45$, so plot $(0, -45)$.

*Step 3*: The *x*-intercepts divide the *x*-axis into four intervals.

| Interval | Test Point | Value of $f(x)$ | Sign of $f(x)$ | Graph Above or Below *x*-Axis |
|---|---|---|---|---|
| $(-\infty, -2.5)$ | $-3$ | 144 | Positive | Above |
| $(-2.5, 1)$ | 0 | $-45$ | Negative | Below |
| $(1, 3)$ | 2 | 9 | Positive | Above |
| $(3, \infty)$ | 4 | 39 | Positive | Above |

Plot the *x*-intercepts, *y*-intercept, and test points (the *y*-intercept is one of the test points) with a smooth curve to get the graph.

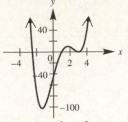

$$f(x) = 2x^4 - 9x^3 - 5x^2 + 57x - 45$$
$$= (x - 3)^2(2x + 5)(x - 1)$$

**(a)** $f(x) = 0$ has the solution set
$$\{-2.5, 1, 3 \text{ (multiplicity 2)}\}.$$

**(b)** $f(x) < 0$ has the solution set $(-2.5, 1)$.

**(c)** $f(x) > 0$ has the solution set
$$(-\infty, -2.5) \cup (1, 3) \cup (3, \infty).$$

**95.** $f(x) = -x^4 - 4x^3 + 3x^2 + 18x$
$$= x(2 - x)(x + 3)^2$$

*Step 1*: Set each factor equal to 0 and solve the resulting equations to find the zeros of the function.

$x = 0$ or $2 - x = 0 \Rightarrow 2 = x$ or
$(x + 3)^2 = 0 \Rightarrow x = -3$

The zeros are 0, 2, and $-3$, which divide the *x*-axis into four regions. Test a point in each region to find the sign of $f(x)$ in that region.

*Step 2*: $f(0) = 0$, so plot $(0, 0)$.

*Step 3*: The *x*-intercepts divide the *x*-axis into four intervals.

| Interval | Test Point | Value of $f(x)$ | Sign of $f(x)$ | Graph Above or Below *x*-Axis |
|---|---|---|---|---|
| $(-\infty, -3)$ | $-4$ | $-24$ | Negative | Below |
| $(-3, 0)$ | $-1$ | $-12$ | Negative | Below |
| $(0, 2)$ | 1 | 16 | Positive | Above |
| $(2, \infty)$ | 3 | $-108$ | Negative | Below |

Plot the *x*-intercepts, *y*-intercept (which is also an *x*-intercept in this exercise), and test points with a smooth curve to get the graph.

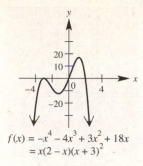

$$f(x) = -x^4 - 4x^3 + 3x^2 + 18x$$
$$= x(2-x)(x+3)^2$$

(a)  $f(x) = 0$ has the solution set
$\{-3 \text{ (multiplicity 2)}, 0, 2\}$.

(b)  $f(x) \geq 0$ has the solution set
$\{-3\} \cup [0, 2]$.

(c)  $f(x) \leq 0$ has the solution set
$(-\infty, 0] \cup [2, \infty)$.

**97.** (a)  In order for the length, width and height of the box to be positive quantities, $0 < x < 6$.

(b)  The width of the rectangular base of the box would be $12 - 2x$. The length would be $18 - 2x$ and the height would be $x$. The volume would be the product of these three measures; therefore,
$$V(x) = x(18 - 2x)(12 - 2x)$$
$$= 4x^3 - 60x^2 + 216x$$

(c)  When $x$ is approximately 2.35 in. the maximum volume will be approximately 228.16 in.$^3$.

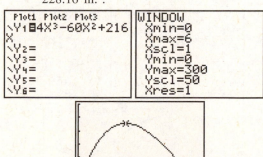

(d)  When $x$ is between .42 in. (approximately) and 5 in., the volume will be greater than 80 in.$^3$.

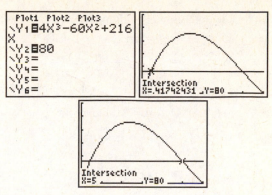

**99.** (a)  length of the leg $= x - 1$. The domain is $x > 1$ or $(1, \infty)$.

(b)  By the Pythagorean theorem,
$$a^2 + b^2 = c^2 \Rightarrow a^2 + (x-1)^2 = x^2 \Rightarrow$$
$$a^2 = x^2 - (x-1)^2 \Rightarrow a = \sqrt{x^2 - (x-1)^2}$$
Thus, the length of the other leg is
$$\sqrt{x^2 - (x-1)^2}.$$

(c)  $A = \frac{1}{2}bh \Rightarrow 84 = \frac{1}{2}(x-1)\left(\sqrt{x^2 - (x-1)^2}\right)$

Multiply by 2: $168 = (x-1)\left(\sqrt{x^2 - (x-1)^2}\right)$

Square both sides.
$$28,224 = (x-1)^2\left[x^2 - (x-1)^2\right]$$
$$28,224 = (x^2 - 2x + 1)\cdot\left[x^2 - (x^2 - 2x + 1)\right]$$
$$28,224 = (x^2 - 2x + 1)(2x - 1)$$
$$28,224 = 2x^3 - 5x^2 + 4x - 1$$
$$2x^3 - 5x^2 + 4x - 28,225 = 0$$

(d)  Solving this cubic equation graphically, we obtain $x = 25$. If $x = 25$, $x - 1 = 24$, and
$$\sqrt{x^2 - (x-1)^2} = \sqrt{625 - 576} = \sqrt{49} = 7.$$
The hypotenuse is 25 in.; the legs are 24 in. and 7 in.

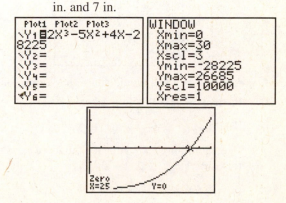

**101.** Use the following volume formulas:

$V_{cylinder} = \pi r^2 h$ and

$V_{hemisphere} = \frac{1}{2}V_{sphere} = \frac{1}{2}\left(\frac{4}{3}\pi r^3\right) = \frac{2}{3}\pi r^3$

$\pi r^2 h + 2\left(\frac{2}{3}\pi r^3\right) = $ Total volume of tank $(V)$

Let $V = 144\pi$, $h = 12$, and $r = x$.

$\pi x^2(12) + \frac{4}{3}\pi x^3 = 144\pi$

$\frac{4}{3}x^3 + 12x^2 = 144$

$\frac{4}{3}x^3 + 12x^2 - 144 = 0$

$4x^3 + 36x^2 - 432 = 0$

$x^3 + 9x^2 - 108 = 0$

The end behavior of the graph of

$f(x) = x^3 + 9x^2 - 108$, together with the negative $y$-intercept, tells us that this cubic polynomial must have one positive zero. (We are not interested in negative zeros because $x$ represents the radius.)

Algebraic Solution:

Use synthetic division or a graphing calculator to locate this zero.

```
3)1  9   0  -108
      3  36  108
   _____
   1 12  36    0
```

Note that the bottom line of the synthetic division above implies the polynomial

function $g(x) = x^2 + 12x + 36 = (x+6)^2$,

which has the zero $-6$ (multiplicity 2).

Graphing Calculator Solution:

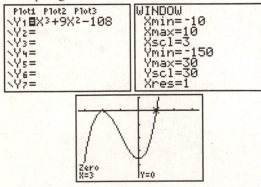

A radius of 3 feet would cause the volume of the tank to be $144\pi$ ft$^3$.

**103.** $f(x) = \frac{\pi}{3}x^3 - 5\pi x^2 + \frac{500\pi d}{3}$

Use a graphing calculator for the three parts of this exercise.

**(a)** When $d = .8$ we have

$f(x) = \frac{\pi}{3}x^3 - 5\pi x^2 + \frac{500\pi(.8)}{3}$.

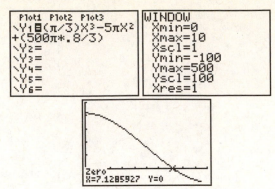

The approximate depth is 7.13 cm. The ball floats partly above the surface.

**(b)** When $d = 2.7$ we have

$f(x) = \frac{\pi}{3}x^3 - 5\pi x^2 + \frac{500\pi(2.7)}{3}$.

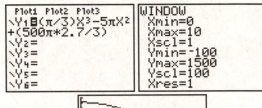

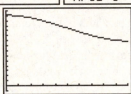

There is no $x$-intercept in this window. The sphere is more dense than water and sinks below the surface.

**(c)** When $d = 1$ we have

$f(x) = \frac{\pi}{3}x^3 - 5\pi x^2 + \frac{500\pi}{3}$.

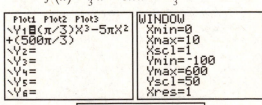

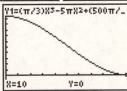

By tracing on the curve we see that the approximate depth is 10 cm. The balloon is submerged with its top even with the surface.

**105. (a)**

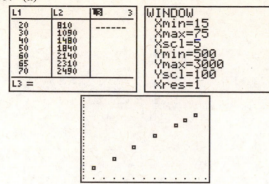

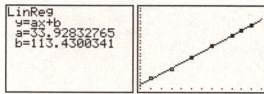

**(b)** The best-fitting linear function would be
$y = 33.93x + 113.4$.

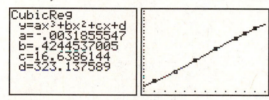

**(c)** The best-fitting cubic function would be
$y = -.0032x^3 + .4245x^2 + 16.64x + 323.1$.

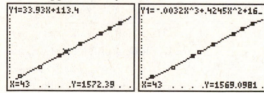

**(d)** linear: approximately 1572 ft; cubic: approximately 1569 ft

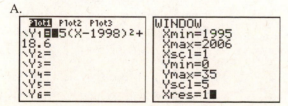

**(e)** The cubic function appears slightly better because only one data point is not on the curve.

**107.** The function in B,
$g(x) = 1.84(x - 1998) + 18.6$ provides the best model.

A.

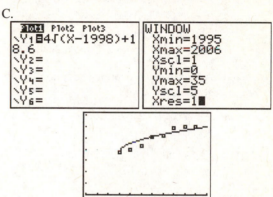

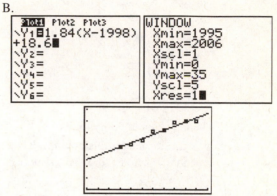

B.

C.

## Summary Exercises on Polynomial Functions, Zeros, and Graphs

**1.** $f(x) = x^4 + 3x^3 - 3x^2 - 11x - 6$

**(a)** $f(x) = x^4 \underbrace{+3x^3}_{1} - 3x^2 - 11x - 6$ has 1 variation in sign. $f$ has 1 positive real zero.

$f(-x) = \underbrace{x^4}_{} \underbrace{-3x^3}_{1} \underbrace{-3x^2}_{2} \underbrace{+11x}_{3} - 6$ has 3 variations in sign. $f$ has 3 or $3 - 2 = 1$ negative real zeros.

**(b)** $p$ must be a factor of $a_0 = -6$ and $q$ must be a factor of $a_4 = 1$. Thus, $p$ can be $\pm 1$, $\pm 2$, $\pm 3$, $\pm 6$ and $q$ can be $\pm 1$. The possible zeros, $\frac{p}{q}$, are $\pm 1$, $\pm 2$, $\pm 3$, $\pm 6$.

**(c)** The remainder theorem shows that 2 is zero.

$$2 \overline{)\ 1 \quad 3 \quad -3 \quad -11 \quad -6}$$
$$\quad \quad \underline{2 \quad 10 \quad 14 \quad 6}$$
$$\quad \ 1 \quad 5 \quad 7 \quad \quad 3 \quad \ \ 0$$

The new quotient polynomial is $x^3 + 5x^2 + 7x + 3$. Since $f$ has 1 positive real zero, try negative values. The remainder theorem shows that $-3$ is zero.

$$-3 \overline{)\ 1 \quad 5 \quad 7 \quad 3}$$
$$\quad \quad \underline{-3 \quad -6 \quad -3}$$
$$\quad \ 1 \quad 2 \quad 1 \quad \ \ 0$$

The new quotient polynomial is $x^2 + 2x + 1$. Factor this polynomial and set equal to zero to find the remaining zeros.
$$x^2 + 2x + 1 = 0 \Rightarrow (x+1)^2 = 0 \Rightarrow$$
$$x + 1 = 0 \Rightarrow x = -1$$
The rational zeros are $-3, -1$ (multiplicity 2), and 2.

**(d)** All zeros have been found, and they are all rational.

**(e)** All zeros have been found, and they are all real.

**(f)** The $x$-intercepts occur when $x = -3, -1$, and 2.

**(g)** $f(0) = -6$

**(h)**
$$4 \overline{)\ 1 \quad 3 \quad -3 \quad -11 \quad -6}$$
$$\quad \quad \underline{4 \quad 28 \quad 100 \quad 356}$$
$$\quad \ 1 \quad 7 \quad 25 \quad \ \ 89 \quad 350$$

The corresponding point on the graph is $(4, 350)$.

**(i)**

**(j)** The $x$-intercepts divide the $x$-axis into four intervals:

| Interval | Test Point | Value of $f(x)$ | Sign of $f(x)$ | Graph Above or Below $x$-Axis |
|---|---|---|---|---|
| $(-\infty, -3)$ | $-4$ | 54 | Positive | Above |
| $(-3, -1)$ | $-2$ | $-4$ | Negative | Below |
| $(-1, 2)$ | 0 | $-6$ | Negative | Below |
| $(2, \infty)$ | 3 | 96 | Positive | Above |

Plot the $x$-intercepts, $y$-intercept and test points with a smooth curve to get the graph.

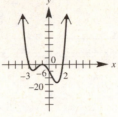

$$f(x) = x^4 + 3x^3 - 3x^2 - 11x - 6$$

**3.** $f(x) = 2x^5 - 10x^4 + x^3 - 5x^2 - x + 5$

**(a)** $f(x) = \underbrace{2x^5}\ \underbrace{-10x^4}\ +\ \underbrace{x^3}\ \underbrace{-5x^2}\ \underbrace{-x}\ +\ 5$ has
$\quad\quad\quad\ \ 1\quad\quad\ 2\quad\quad\ 3\quad\quad\ 4$
4 variations in sign. $f$ has 4 or $4 - 2 = 2$ or $2 - 2 = 0$ positive real zeros.

$$f(-x) = -2x^5 - 10x^4 - x^3 \underbrace{-5x^2}\ +\ x + 5$$
$$\quad\quad\quad\quad\quad\quad\quad\quad\quad\quad\quad\quad 1$$

has 1 variation in sign. $f$ has 1 negative real zero.

**(b)** $p$ must be a factor of $a_0 = 5$ and $q$ must be a factor of $a_5 = 2$. Thus, $p$ can be $\pm 1$, $\pm 5$ and $q$ can be $\pm 1$, $\pm 2$. The possible zeros, $\frac{p}{q}$, are $\pm 1$, $\pm 5$, $\pm \frac{1}{2}$, $\pm \frac{5}{2}$.

**(c)** The remainder theorem shows that 5 is zero.

$$5 \overline{)\ 2 \quad -10 \quad 1 \quad -5 \quad -1 \quad \ \ 5}$$
$$\quad \quad \underline{10 \quad \ \ 0 \quad 5 \quad \ \ 0 \quad -5}$$
$$\quad \ 2 \quad \ \ 0 \quad \ \ 1 \quad 0 \quad -1 \quad \ \ 0$$

The new quotient polynomial will be $2x^4 + x^2 - 1$. At this point, if we try to determine which of the possible rational zero will result in a remainder of zero, we will find that none of them work. We have found the only rational zero of our polynomial function. The rational zero is 5.

**(d)** Since the new quotient polynomial is $2x^4 + x^2 - 1$, we can factor this polynomial and set equal to zero to find the remaining zeros.
$$2x^4 + x^2 - 1 = 0 \Rightarrow \left(x^2 + 1\right)\left(2x^2 - 1\right) = 0$$
Since $x^2 + 1 = 0$ does not yield any real solutions, we will examine $2x^2 - 1 = 0$ first.

$2x^2 - 1 = 0 \Rightarrow 2x^2 = 1 \Rightarrow x^2 = \frac{1}{2} \Rightarrow$

$x = \pm\sqrt{\frac{1}{2}} = \pm\frac{\sqrt{2}}{2}$

The other real zeros are $-\frac{\sqrt{2}}{2}$ and $\frac{\sqrt{2}}{2}$.

**(e)** Examining $x^2 + 1 = 0$, we have
$x^2 = -1 \Rightarrow x = \pm\sqrt{-1} = \pm i$. Thus, the complex zeros are $-i$ and $i$.

**(f)** The $x$-intercepts occur when
$x = -\frac{\sqrt{2}}{2} \approx -.71,\ \frac{\sqrt{2}}{2} \approx .71,$ and 5.

**(g)** $f(0) = 5$

**(h)**
$$4\overline{)2 \quad -10 \quad 1 \quad -5 \quad -1 \quad 5}$$
$$\phantom{4)}\underline{8 \quad -8 \quad -28 \quad -132 \quad -532}$$
$$\phantom{4)}2 \quad -2 \quad -7 \quad -33 \quad -133 \quad -527$$

The corresponding point on the graph is $(4, -527)$.

**(i)**

**(j)** The $x$-intercepts divide the $x$-axis into four intervals:

| Interval | Test Point | Value of $f(x)$ | Sign of $f(x)$ | Graph Above or Below $x$-Axis |
|---|---|---|---|---|
| $\left(-\infty, -\frac{\sqrt{2}}{2}\right)$ | $-1$ | $-12$ | Negative | Below |
| $\left(-\frac{\sqrt{2}}{2}, \frac{\sqrt{2}}{2}\right)$ | $0$ | $5$ | Positive | Above |
| $\left(\frac{\sqrt{2}}{2}, 5\right)$ | $1$ | $-8$ | Negative | Below |
| $(5, \infty)$ | $6$ | $2627$ | Positive | Above |

Plot the $x$-intercepts, $y$-intercept and test points (the $y$-intercept is also a test point) with a smooth curve to get the graph.

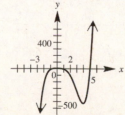

$f(x) = 2x^5 - 10x^4 + x^3 - 5x^2 - x + 5$

**5.** $f(x) = -2x^4 - x^3 + x + 2$

**(a)** $f(x) = -2x^4 \underbrace{-x^3}_{1} + x + 2$ has 1 variation in sign. $f$ has 1 positive real zero.

$f(x) = \underbrace{-2x^4}_{1} \underbrace{+x^3}_{2} \underbrace{-x}_{3} + 2$ has 3 variations in sign. $f$ has 3 or $3 - 2 = 1$ negative real zeros.

**(b)** $p$ must be a factor of $a_0 = 2$ and $q$ must be a factor of $a_4 = -2$. Thus, $p$ can be $\pm 1$, $\pm 2$ and $q$ can be $\pm 1,\ \pm 2$. The possible zeros, $\frac{p}{q}$, are $\pm 1,\ \pm 2,\ \pm\frac{1}{2}$.

**(c)** The remainder theorem shows that $-1$ is zero.

$$-1\overline{)-2 \quad -1 \quad 0 \quad 1 \quad 2}$$
$$\phantom{-1)}\underline{\phantom{-2}\ \ 2 \quad -1 \quad 1 \quad -2}$$
$$\phantom{-1)}-2 \quad 1 \quad -1 \quad 2 \quad 0$$

The new quotient polynomial will be $-2x^3 + x^2 - x + 2$. Since the signs are alternating in the bottom row of the synthetic division, we know that $-1$ is a lower bound. The remainder theorem shows that 1 is zero.

$$1\overline{)-2 \quad 1 \quad -1 \quad 2}$$
$$\phantom{1)}\underline{\phantom{-2}\ \ -2 \quad -1 \quad -2}$$
$$\phantom{1)}-2 \quad -1 \quad -2 \quad 0$$

The new quotient polynomial is $-2x^2 - x - 2 = -(2x^2 + x + 2)$. At this point, if we try to determine which of the possible rational zeros will result in a remainder of zero, we will find that none of them work. We have found the only rational zeros of our polynomial function. The rational zeros are $-1$ and 1.

**(d)** If we examine the discriminant of $2x^2 + x + 2$, namely $1^2 - 4(2)(2) = 1 - 16 = -15$, we see that it is negative, which implies there are no other real zeros.

**(e)** To find the remaining complex zeros, we must solve $2x^2 + x + 2 = 0$. To solve this equation, we use the quadratic formula with $a = 2$, $b = 1$, and $c = 2$.

$$x = \frac{-1 \pm \sqrt{1^2 - 4(2)(2)}}{2(2)} = \frac{-1 \pm \sqrt{1 - 16}}{4}$$

$$= \frac{-1 \pm \sqrt{-15}}{4} = \frac{-1 \pm i\sqrt{15}}{4}$$

Thus, the other complex zeros are

$-\frac{1}{4} + \frac{\sqrt{15}}{4}i$ and $-\frac{1}{4} - \frac{\sqrt{15}}{4}i$

**(f)** The $x$-intercepts occur when $x = -1$ and $1$.

**(g)** $f(0) = 2$

**(h)**
$$4\overline{)\,-2 \quad -1 \quad 0 \quad 1 \quad 2}$$
$$\underline{\quad\quad -8 \; -36 \; -144 \; -572}$$
$$-2 \; -9 \; -36 \; -143 \; -570$$

The corresponding point on the graph is $(4, -570)$.

**(i)**

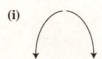

**(j)** The $x$-intercepts divide the $x$-axis into three intervals:

| Interval | Test Point | Value of $f(x)$ | Sign of $f(x)$ | Graph Above or Below $x$-Axis |
|---|---|---|---|---|
| $(-\infty, -1)$ | $-2$ | $-24$ | Negative | Below |
| $(-1, 1)$ | $0$ | $2$ | Positive | Above |
| $(1, \infty)$ | $2$ | $-36$ | Negative | Below |

Plot the $x$-intercepts, $y$-intercept and test points (the $y$-intercept is also a test point) with a smooth curve to get the graph.

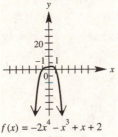

$f(x) = -2x^4 - x^3 + x + 2$

**7.** $f(x) = 3x^4 - 14x^2 - 5$

**(a)** $f(x) = \underset{1}{3x^4 - 14x^2} - 5$ has 1 variation in sign. $f$ has 1 positive real zero.

$f(-x) = \underset{1}{3x^4 - 14x^2} - 5$ has 1 variation in sign. $f$ has 1 negative real zeros.

**(b)** $p$ must be a factor of $a_0 = -5$ and $q$ must be a factor of $a_4 = 3$. Thus, $p$ can be $\pm 1$, $\pm 5$ and $q$ can be $\pm 1, \pm 3$. The possible zeros, $\frac{p}{q}$, are $\pm 1$, $\pm 5$, $\pm \frac{1}{3}$, $\pm \frac{5}{3}$.

**(c)** If we try to determine which of the possible rational zeros will result in a remainder of zero, we will find that none of them work. There are no rational zeros.

**(d)** We can factor the polynomial, $3x^4 - 14x^2 - 5$, and set equal to zero to the remaining zeros.

$$3x^4 - 14x^2 - 5 = 0$$
$$\left(3x^2 + 1\right)\left(x^2 - 5\right) = 0$$

Since $3x^2 + 1 = 0$ does not yield any real solutions, we will examine $x^2 - 5 = 0$ first.

$$x^2 - 5 = 0 \Rightarrow x^2 = 5 \Rightarrow x = \pm\sqrt{5}$$

The real zeros are $-\sqrt{5}$ and $\sqrt{5}$.

**(e)** Examining $3x^2 + 1 = 0$, we have

$$3x^2 = -1 \Rightarrow x^2 = -\frac{1}{3} \Rightarrow x = \pm\sqrt{-\frac{1}{3}} = \pm i\frac{\sqrt{3}}{3}.$$

Thus, the complex zeros are $-\frac{\sqrt{3}}{3}i$ and $\frac{\sqrt{3}}{3}i$.

**(f)** The $x$-intercepts occur when $x = -\sqrt{5} \approx -2.24$ and $\sqrt{5} \approx 2.24$.

**(g)** $f(0) = -5$

**(h)**
$$4\overline{)\,3 \quad 0 \quad -14 \quad 0 \quad -5}$$
$$\underline{\quad\quad 12 \quad 48 \; 136 \; 544}$$
$$3 \; 12 \quad 34 \; 136 \; 539$$

The corresponding point on the graph is $(4, 539)$.

**(i)**

**(j)** The $x$-intercepts divide the $x$-axis into three intervals:

| Interval | Test Point | Value of $f(x)$ | Sign of $f(x)$ | Graph Above or Below $x$-Axis |
|---|---|---|---|---|
| $\left(-\infty, -\sqrt{5}\right)$ | $-3$ | $112$ | Positive | Above |
| $\left(-\sqrt{5}, \sqrt{5}\right)$ | $0$ | $-5$ | Negative | Below |
| $\left(\sqrt{5}, \infty\right)$ | $3$ | $112$ | Positive | Above |

Plot the $x$-intercepts, $y$-intercept and test points (the $y$-intercept is also a test point) with a smooth curve to get the graph.

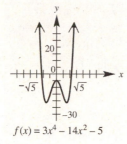

$f(x) = 3x^4 - 14x^2 - 5$

**9.** $f(x) = -3x^4 + 22x^3 - 55x^2 + 52x - 12$

**(a)** $f(x) = \underbrace{-3x^4}_{1} + \underbrace{22x^3}_{2} - \underbrace{55x^2}_{3} + \underbrace{52x}_{4} - 12$

has 4 variations in sign. $f$ has 4 or $4 - 2 = 2$ or $2 - 2 = 0$ positive real zeros.

$f(-x) = -3x^4 - 22x^3 - 55x^2 - 52x - 12$

has 0 variations in sign. $f$ has 0 negative real zeros.

**(b)** $p$ must be a factor of $a_0 = -12$ and $q$ must be a factor of $a_4 = -3$. Thus, $p$ can be $\pm 1, \pm 2, \pm 3, \pm 4, \pm 6, \pm 12$ and $q$ can be $\pm 1, \pm 3$. The possible zeros, $\frac{p}{q}$, are

$\pm 1, \ \pm 2, \ \pm 3, \ \pm 4, \ \pm 6, \ \pm 12, \ \pm \frac{1}{3},$

$\pm \frac{2}{3}, \ \pm \frac{4}{3}.$

**(c)** The remainder theorem shows that 2 is a zero.

$$2\overline{)\begin{array}{rrrrr} -3 & 22 & -55 & 52 & -12 \\ & -6 & 32 & -46 & 12 \end{array}}$$
$$\begin{array}{rrrrr} -3 & 16 & -23 & 6 & 0 \end{array}$$

The new quotient polynomial will be $-3x^3 + 16x^2 - 23x + 6$. The remainder theorem shows that 3 is zero.

$$3\overline{)\begin{array}{rrrr} -3 & 16 & -23 & 6 \\ & -9 & 21 & -6 \end{array}}$$
$$\begin{array}{rrrr} -3 & 7 & -2 & 0 \end{array}$$

The new quotient polynomial is $-3x^2 + 7x - 2 = -\left(3x^2 - 7x + 2\right)$. Factor $3x^2 - 7x + 2$ and set equal to zero to find the remaining zeros.

$3x^2 - 7x + 2 = 0 \Rightarrow (3x - 1)(x - 2) = 0$

$3x - 1 = 0 \Rightarrow x = \frac{1}{3}$ or $x - 2 = 0 \Rightarrow x = 2$

The rational zeros are $\frac{1}{3}$, 2 (multiplicity 2), and 3.

**(d)** All zeros have been found, and they are all rational.

**(e)** All zeros have been found, and they are all real.

**(f)** The $x$-intercepts occur when $x = \frac{1}{3}$, 2, and 3.

**(g)** $f(0) = -12$

**(h)** $$4\overline{)\begin{array}{rrrrr} -3 & 22 & -55 & 52 & -12 \\ & -12 & 40 & -60 & -32 \end{array}}$$
$$\begin{array}{rrrrr} -3 & 10 & -15 & -8 & -44 \end{array}$$

The corresponding point on the graph is $(4, -44)$.

**(i)**

**(j)** The $x$-intercepts divide the $x$-axis into four intervals:

| Interval | Test Point | Value of $f(x)$ | Sign of $f(x)$ | Graph Above or Below $x$-Axis |
|---|---|---|---|---|
| $\left(-\infty, \frac{1}{3}\right)$ | $0$ | $-12$ | Negative | Below |
| $\left(\frac{1}{3}, 2\right)$ | $1$ | $4$ | Positive | Above |
| $(2, 3)$ | $\frac{5}{2}$ | $\frac{13}{16}$ | Positive | Above |
| $(3, \infty)$ | $4$ | $-44$ | Negative | Below |

*(continued on next page)*

(*continued from page 171*)

Plot the $x$-intercepts, $y$-intercept and test points (the $y$-intercept is also a test point) with a smooth curve to get the graph.

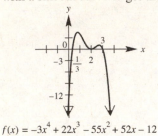

$$f(x) = -3x^4 + 22x^3 - 55x^2 + 52x - 12$$

## Section 3.5: Rational Functions: Graphs, Applications, and Models

**1.** $f(x) = \frac{1}{x}$

Domain: $(-\infty, 0) \cup (0, \infty)$

Range: $(-\infty, 0) \cup (0, \infty)$

**3.** $f(x) = \frac{1}{x}$

Increasing: nowhere

Decreasing: $(-\infty, 0) \cup (0, \infty)$

Constant: nowhere

**5.** $y = \frac{1}{x-3} + 2$

Vertical asymptote: $x = 3$

Horizontal asymptote: $y = 2$

**7.** $f(x) = \frac{1}{x^2}$ is an even function. It exhibits symmetry with respect to $y$-axis.

**9.** Graphs A, B, and C have a domain of $(-\infty, 3) \cup (3, \infty)$.

**11.** Graph A has a range of $(-\infty, 0) \cup (0, \infty)$.

**13.** Graph A has a single solution to the equation $f(x) = 3$.

**15.** Graphs A, C, and D have the $x$-axis as a horizontal asymptote.

**17.** $f(x) = \frac{2}{x}$

To obtain the graph of $f(x) = \frac{2}{x}$, stretch the graph of $f(x) = \frac{1}{x}$, vertically by a factor of 2. Just as with the graph of $f(x) = \frac{1}{x}$, $y = 0$ is the horizontal asymptote and $x = 0$ is the vertical asymptote.

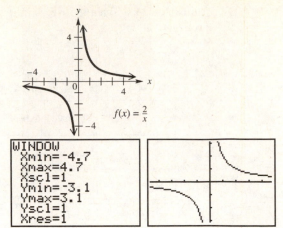

$f(x) = \frac{2}{x}$

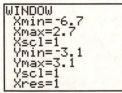

Domain: $(-\infty, 0) \cup (0, \infty)$

Range: $(-\infty, 0) \cup (0, \infty)$

**19.** $f(x) = \frac{1}{x+2}$

To obtain the graph of $f(x) = \frac{1}{x+2}$ shift the graph of $f(x) = \frac{1}{x}$, to the left 2 units. Just as with $f(x) = \frac{1}{x}$, $y = 0$ is the horizontal asymptote, but this graph has $x = -2$ as its vertical asymptote (this affects the domain).

$f(x) = \frac{1}{x+2}$

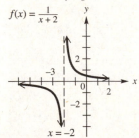

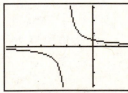

Domain: $(-\infty, -2) \cup (-2, \infty)$

Range: $(-\infty, 0) \cup (0, \infty)$

**21.** $f(x) = \frac{1}{x} + 1$

To obtain the graph of $f(x) = \frac{1}{x} + 1$, translate the graph of $f(x) = \frac{1}{x}$, 1 unit up. Just as with $f(x) = \frac{1}{x}$, $x = 0$ is the vertical asymptote, but this graph has $y = 1$ as its horizontal asymptote (this affects the range).

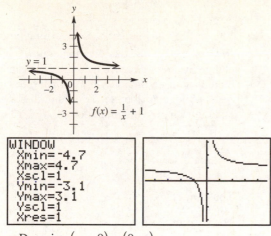

Domain: $(-\infty, 0) \cup (0, \infty)$

Range: $(-\infty, 1) \cup (1, \infty)$

**23.** $f(x) = -\frac{2}{x^2}$

To obtain the graph of $f(x) = -\frac{2}{x^2}$, stretch the graph of $f(x) = \frac{1}{x^2}$, by a factor of 2, and then reflect the graph will be reflected across the $x$-axis.

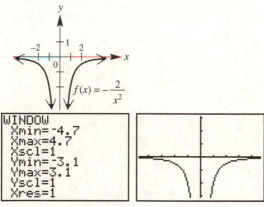

Domain: $(-\infty, 0) \cup (0, \infty)$

Range: $(-\infty, 0)$

**25.** $f(x) = \frac{1}{(x-3)^2}$

To obtain the graph of $f(x) = \frac{1}{(x-3)^2}$, shift the graph of $f(x) = \frac{1}{x^2}$, 3 units to the right. Just as with $f(x) = \frac{1}{x^2}$, $y = 0$ is the horizontal asymptote, but this graph has $x = 3$ as its vertical asymptote.

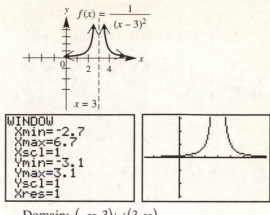

Domain: $(-\infty, 3) \cup (3, \infty)$

Range: $(0, \infty)$

**27.** $f(x) = \frac{-1}{(x+2)^2} - 3$

To obtain the graph of $f(x) = \frac{-1}{(x+2)^2} - 3$, shift the graph of $f(x) = \frac{1}{x^2}$, 2 units to the left, reflect the graph across the $x$-axis, and then shfit the graph 3 units down. Unlike the graph of $f(x) = \frac{1}{x^2}$, the vertical asymptote of $f(x) = \frac{-1}{(x+2)^2} - 3$ will be $x = -2$ and the horizontal asymptote is $y = -3$.

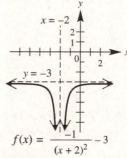

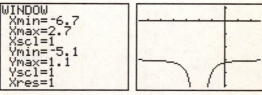

Domain: $(-\infty, -2) \cup (-2, \infty)$

Range: $(-\infty, -3)$

**29.** D. The graph of $f(x) = \frac{x+7}{x+1}$ has the vertical asymptote at $x = -1$.

**31.** G. The graph of $f(x) = \frac{1}{x+4}$ has the $x$-axis as its horizontal asymptote, and the $y$-axis is not its vertical asymptote. The line $x = -4$ is its vertical asymptote.

**33.** E. The graph of $f(x) = \frac{x^2-16}{x+4}$ has a "hole" in its graph located at $x = -4$ since
$$f(x) = \frac{x^2-16}{x+4} = \frac{(x+4)(x-4)}{x+4} = x - 4, \ x \neq -4.$$

**35.** F. The graph of $f(x) = \frac{x^2+3x+4}{x-5}$ has an oblique asymptote since $f(x) = x + 8 + \frac{44}{x-5}$.

**37.** $f(x) = \frac{3}{x-5}$. To find the vertical asymptote, set the denominator equal to zero.
$x - 5 = 0 \Rightarrow x = 5$
The equation of the vertical asymptote is $x = 5$. To find the horizontal asymptote, divide each term by the largest power of $x$ in the expression.
$$f(x) = \frac{\frac{3}{x}}{\frac{x}{x} - \frac{5}{x}} = \frac{\frac{3}{x}}{1 - \frac{5}{x}}$$
As $|x| \to \infty, \frac{1}{x}$ approaches 0, thus $f(x)$ approaches $\frac{0}{1-0} = \frac{0}{1} = 0$. The line $y = 0$ (that is, the $x$-axis) is the horizontal asymptote.

**39.** $f(x) = \frac{4-3x}{2x+1}$
To find the vertical asymptote, set the denominator equal to zero.
$2x + 1 = 0 \Rightarrow x = -\frac{1}{2}$
The equation of the vertical asymptote is $x = -\frac{1}{2}$. To find the horizontal asymptote, divide each term by the largest power of $x$ in the expression.
$$f(x) = \frac{\frac{4}{x} - \frac{3x}{x}}{\frac{2x}{x} + \frac{1}{x}} = \frac{\frac{4}{x} - 3}{2 + \frac{1}{x}}$$
As $|x| \to \infty, \frac{1}{x}$ approaches 0, so $f(x)$ approaches $\frac{0-3}{2+0} = -\frac{3}{2}$. The equation of the horizontal asymptote is $y = -\frac{3}{2}$.

**41.** $f(x) = \frac{x^2-1}{x+3}$
The vertical asymptote is $x = -3$, found by solving $x + 3 = 0$. Since the numerator is of degree exactly one more than the denominator, there is no horizontal asymptote, but there may be an oblique asymptote. To find it, divide the numerator by the denominator.

$$-3 \overline{)\begin{array}{rrr} 1 & 0 & -1 \\ & -3 & 9 \\ \hline 1 & -3 & 8 \end{array}}$$

Thus, $f(x) = \frac{x^2-1}{x+3} = x - 3 + \frac{8}{x+3}$.

For very large values of $|x|$, $\frac{8}{x+3}$ is close to 0, and the graph approaches the line $y = x - 3$.

**43.** $f(x) = \frac{(x-3)(x+1)}{(x+2)(2x-5)}$
To find the vertical asymptotes, set the denominator equal to zero and solve.
$x + 2 = 0 \Rightarrow x = -2$ and $2x - 5 = 0 \Rightarrow x = \frac{5}{2}$
Thus, the vertical asymptotes are $x = -2$ and $x = \frac{5}{2}$.
To determine the horizontal asymptote, first multiply the factors in the numerator and denominator to get $f(x) = \frac{x^2-2x-3}{2x^2-x-10}$. Divide the numerator and denominator by $x^2$.
$$f(x) = \frac{\frac{x^2}{x^2} - \frac{2x}{x^2} - \frac{3}{x^2}}{\frac{2x^2}{x^2} - \frac{x}{x^2} - \frac{10}{x^2}} = \frac{1 - \frac{2}{x} - \frac{3}{x^2}}{2 - \frac{1}{x} - \frac{10}{x^2}}$$
As $|x| \to \infty, \frac{1}{x}$ and $\frac{1}{x^2}$ approach 0, so $f(x)$ approaches $\frac{1-0-0}{2-0-0} = \frac{1}{2}$. Thus, the equation of the horizontal asymptote is $y = \frac{1}{2}$.

**45.** $f(x) = \frac{x^2+1}{x^2+9}$
To find the vertical asymptotes, set the denominator equal to zero and solve.
$x^2 + 9 = 0 \Rightarrow x^2 = -9 \Rightarrow x = \pm\sqrt{-9} = \pm 3i$
Thus, there are no vertical asymptotes.
To determine the horizontal asymptote, divide the numerator and denominator by $x^2$.
$$f(x) = \frac{\frac{x^2}{x^2} + \frac{1}{x^2}}{\frac{x^2}{x^2} + \frac{9}{x^2}} = \frac{1 + \frac{1}{x^2}}{1 + \frac{9}{x^2}}$$
As $|x| \to \infty, \frac{1}{x^2}$ approaches 0, so $f(x)$ approaches $\frac{1+0}{1+0} = \frac{1}{1} = 1$. Thus, the equation of the horizontal asymptote is $y = 1$.

**47.** **(a)** Translating $y = \frac{1}{x}$ three units to the right

gives $y = \frac{1}{x-3}$. Translating $y = \frac{1}{x-3}$ two

units up yields

$y = \frac{1}{x-3} + 2 = \frac{1}{x-3} + \frac{2x-6}{x-3} = \frac{2x-5}{x-3}$. So

$f(x) = \frac{2x-5}{x-3}$.

**(b)** $f(x) = \frac{2x-5}{x-3}$ has a zero when $2x - 5 = 0$

or $x = \frac{5}{2}$.

**(c)** $f(x) = \frac{2x-5}{x-3}$ has a horizontal asymptote

at $y = 2$ and a vertical asymptote at

$x = 3$.

**49.** **(a)** $f(x) = x+1+\frac{x^2-x}{x^4+1}$ has an oblique

asymptote of $y = x+1$.

**(b)** In order to determine when the function
intersects the oblique asymptote, we must
determine the values of that that make

$\frac{x^2-x}{x^4+1} = 0$. Since

$\frac{x^2-x}{x^4+1} = \frac{x(x-1)}{x^4+1}$, and $x(x-1) = 0$ when

$x = 0$ or $x = 1$, the function crosses its
asymptote at $x = 0$ and $x = 1$.

**(c)** For large values of $x$, $\frac{x^2-x}{x^4+1} > 0$. Thus as

$x \to \infty$, the function approaches its
asymptote from above.

**51.** Function A, because the denominator can
never be equal to 0.

**53.** From the graph, the vertical asymptote is $x = 2$,
the horizontal asymptote is $y = 4$, and there is
no oblique asymptote. The function is defined
for all real numbers $x$ such that $x \neq 2$,
therefore the domain is $(-\infty, 2) \cup (2, \infty)$.

**55.** From the graph, the vertical asymptotes are
$x = -2$ and $x = 2$, the horizontal asymptote is
$y = -4$, and there is no oblique asymptote. The
function is defined for all real numbers $x$ such
that $x \neq \pm 2$, therefore the domain is
$(-\infty, -2) \cup (-2, 2) \cup (2, \infty)$.

**57.** From the graph, there is no vertical asymptote,
the horizontal asymptote is $y = 0$, and there is
no oblique asymptote. The function is defined
for all $x$; therefore, the domain is $(-\infty, \infty)$.

**59.** From the graph, the vertical asymptote is
$x = -1$, there is no horizontal asymptote, and
the oblique asymptote passes through the
points $(0, -1)$ and $(1, 0)$. Thus, the equation of
the oblique asymptote is $y = x - 1$. The
function is defined for all $x \neq -1$, therefore
the domain is $(-\infty, -1) \cup (-1, \infty)$.

For exercises 61–99, follow the 7 steps outlined on
page 365 to sketch the graph. The solution for
exercise 61 shows all of the steps. The solutions for
exercises 63–99 are abbreviated.

**61.** *Step 1:* Find the vertical asymptote by setting
the denominator equal to 0: $x - 4 = 0 \Rightarrow x = 4$
*Step 2:* Find the horizontal asymptote. Since
the numerator and denominator have the same
degree, the horizontal asymptote has equation

$y = \frac{a_n}{b_n} = \frac{1}{1} = 1$. There is no oblique asymptote.

*Step 3:* Find the $y$-intercept: $f(0) = \frac{0+1}{0-4} = -\frac{1}{4}$

*Step 4:* Find the $x$-intercept:

$\frac{x+1}{x-4} = 0 \Rightarrow x+1 = 0 \Rightarrow x = -1$

*Step 5:* Determine whether the graph will
intersect the horizontal asymptote:

$f(x) = \frac{x+1}{x-4} = 1 \Rightarrow x+1 = x-4 \Rightarrow 0 = -5 \Rightarrow$

the graph does not intersect its horizontal
asymptote.
*Step 6:* Plot a point in each of the intervals
$(-\infty, -1), (-1, 4)$, and $(4, \infty)$.

| Interval | Test Point | Value of $f(x)$ | Sign of $f(x)$ | Graph Above or Below $x$-Axis |
|----------|-----------|-----------------|----------------|-------------------------------|
| $(-\infty, -1)$ | $-5$ | $\frac{4}{9}$ | Positive | Above |
| $(-1, 4)$ | $1$ | $-1$ | Negative | Below |
| $(4, \infty)$ | $5$ | $6$ | Positive | Above |

*Step 7:* Plot the intercepts and test points to
sketch the graph.

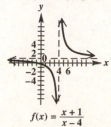

$f(x) = \frac{x+1}{x-4}$

**63.** $f(x) = \frac{x+2}{x-3}$

vertical asymptote: $x = 3$
horizontal asymptote: $y = 1$

$y$-intercept: $-\frac{2}{3}$; $x$-intercept: $-2$

$f(x)$ does not intersect the horizontal asymptote

| Interval | Test Point | Value of $f(x)$ | Sign of $f(x)$ | Graph Above or Below $x$-Axis |
|---|---|---|---|---|
| $(-\infty, -2)$ | $-5$ | $\frac{3}{8}$ | Positive | Above |
| $(-2, 3)$ | $1$ | $-\frac{3}{2}$ | Negative | Below |
| $(3, \infty)$ | $5$ | $\frac{7}{2}$ | Positive | Above |

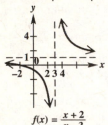

$f(x) = \frac{x+2}{x-3}$

**65.** $f(x) = \frac{4-2x}{8-x}$

vertical asymptote: $x = 8$
horizontal asymptote: $y = 2$

$y$-intercept: $\frac{1}{2}$; $x$-intercept: $2$

$f(x)$ does not intersect the horizontal asymptote

| Interval | Test Point | Value of $f(x)$ | Sign of $f(x)$ | Graph Above or Below $x$-Axis |
|---|---|---|---|---|
| $(-\infty, 2)$ | $-5$ | $\frac{14}{13}$ | Positive | Above |
| $(2, 8)$ | $4$ | $-1$ | Negative | Below |
| $(8, \infty)$ | $10$ | $8$ | Positive | Above |

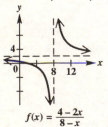

$f(x) = \frac{4-2x}{8-x}$

**67.** $f(x) = \frac{3x}{(x+1)(x-2)}$

vertical asymptotes: $x = -1$, $x = 2$
horizontal asymptote: $y = 0$

$y$-intercept: 0; $x$-intercept: 0

$f(x)$ intersects the horizontal asymptote

| Interval | Test Point | Value of $f(x)$ | Sign of $f(x)$ | Graph Above or Below $x$-Axis |
|---|---|---|---|---|
| $(-\infty, -1)$ | $-5$ | $-\frac{15}{28}$ | Negative | Below |
| $(-1, 0)$ | $-\frac{1}{2}$ | $\frac{6}{5}$ | Positive | Above |
| $(0, 2)$ | $1$ | $-\frac{3}{2}$ | Negative | Below |
| $(2, \infty)$ | $5$ | $\frac{5}{6}$ | Positive | Above |

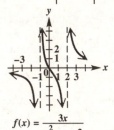

$f(x) = \frac{3x}{x^2 - x - 2}$

**69.** $f(x) = \frac{5x}{x^2-1}$

vertical asymptotes: $x = -1$, $x = 1$
horizontal asymptote: $y = 0$
$y$-intercept: 0; $x$-intercept: 0
$f(x)$ intersects the horizontal asymptote

| Interval | Test Point | Value of $f(x)$ | Sign of $f(x)$ | Graph Above or Below $x$-Axis |
|---|---|---|---|---|
| $(-\infty, -1)$ | $-2$ | $-\frac{10}{3}$ | Negative | Below |
| $(-1, 0)$ | $-\frac{1}{2}$ | $\frac{10}{3}$ | Positive | Above |
| $(0, 1)$ | $\frac{1}{2}$ | $-\frac{10}{3}$ | Negative | Below |
| $(1, \infty)$ | $2$ | $\frac{10}{3}$ | Positive | Above |

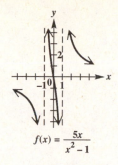

$$f(x) = \frac{5x}{x^2-1}$$

**71.** $f(x) = \frac{(x+6)(x-2)}{(x+3)(x-4)}$

vertical asymptotes: $x = -3$, $x = 4$

·horizontal asymptote: $y = 1$

$y$-intercept: 1; $x$-intercepts: $-6$, 2

$f(x)$ intersects the horizontal asymptote

| Interval | Test Point | Value of $f(x)$ | Sign of $f(x)$ | Graph Above or Below $x$-Axis |
|---|---|---|---|---|
| $(-\infty, -6)$ | $-8$ | $\frac{1}{3}$ | Positive | Above |
| $(-6, -3)$ | $-4$ | $-\frac{3}{2}$ | Negative | Below |
| $(-3, 2)$ | 1 | $\frac{7}{12}$ | Positive | Above |
| $(2, 4)$ | 3 | $-\frac{3}{2}$ | Negative | Below |
| $(4, \infty)$ | 6 | $\frac{8}{3}$ | Positive | Above |

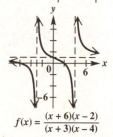

$$f(x) = \frac{(x+6)(x-2)}{(x+3)(x-4)}$$

**73.** $f(x) = \frac{3x^2+3x-6}{x^2-x-12}$

vertical asymptotes: $x = -3$, $x = 4$

horizontal asymptote: $y = 3$

$y$-intercept: $\frac{1}{2}$ ; $x$-intercepts: $-2$, 1

$f(x)$ intersects the horizontal asymptote

| Interval | Test Point | Value of $f(x)$ | Sign of $f(x)$ | Graph Above or Below $x$-Axis |
|---|---|---|---|---|
| $(-\infty, -3)$ | $-4$ | $\frac{15}{4}$ | Positive | Above |
| $(-3, -2)$ | $-\frac{5}{2}$ | $-\frac{21}{13}$ | Negative | Below |

| Interval | Test Point | Value of $f(x)$ | Sign of $f(x)$ | Graph Above or Below $x$-Axis |
|---|---|---|---|---|
| $(-2, 1)$ | $-1$ | $\frac{3}{5}$ | Positive | Above |
| $(1, 4)$ | 2 | $-\frac{6}{5}$ | Negative | Below |
| $(4, \infty)$ | 6 | $\frac{20}{3}$ | Positive | Above |

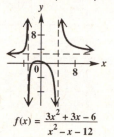

$$f(x) = \frac{3x^2 + 3x - 6}{x^2 - x - 12}$$

**75.** $f(x) = \frac{9x^2-1}{x^2-4}$

vertical asymptotes: $x = -2$, $x = 2$

horizontal asymptote: $y = 9$

$y$-intercept: $\frac{1}{4}$ ; $x$-intercepts: $-\frac{1}{3}, \frac{1}{3}$

$f(x)$ does not intersect the horizontal asymptote

| Interval | Test Point | Value of $f(x)$ | Sign of $f(x)$ | Graph Above or Below $x$-Axis |
|---|---|---|---|---|
| $(-\infty, -2)$ | $-3$ | 16 | Positive | Above |
| $\left(-2, -\frac{1}{3}\right)$ | $-1$ | $-\frac{8}{3}$ | Negative | Below |
| $\left(-\frac{1}{3}, \frac{1}{3}\right)$ | $\frac{1}{4}$ | $\frac{1}{9}$ | Positive | Above |
| $\left(\frac{1}{3}, 2\right)$ | 1 | $-\frac{8}{3}$ | Negative | Below |
| $(2, \infty)$ | 3 | 16 | Positive | Above |

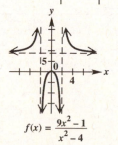

$$f(x) = \frac{9x^2 - 1}{x^2 - 4}$$

**77.** $f(x) = \frac{(x-3)(x+1)}{(x-1)^2}$

vertical asymptote: $x = 1$
horizontal asymptote: $y = 1$
$y$-intercept: $-3$; $x$-intercepts: $-1, 3$
$f(x)$ intersects the horizontal asymptote

| Interval | Test Point | Value of $f(x)$ | Sign of $f(x)$ | Graph Above or Below $x$-Axis |
|---|---|---|---|---|
| $(-\infty, -1)$ | $-2$ | $\frac{5}{9}$ | Positive | Above |
| $(-1, 1)$ | $0$ | $-3$ | Negative | Below |
| $(1, 3)$ | $2$ | $-3$ | Negative | Below |
| $(3, \infty)$ | $4$ | $\frac{5}{9}$ | Positive | Above |

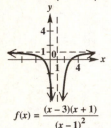

$$f(x) = \frac{(x-3)(x+1)}{(x-1)^2}$$

**79.** $f(x) = \frac{x}{x^2-9}$

vertical asymptotes: $x = -3, x = 3$
horizontal asymptote: $y = 0$
$y$-intercept: $0$; $x$-intercept: $0$
$f(x)$ intersects the horizontal asymptote

| Interval | Test Point | Value of $f(x)$ | Sign of $f(x)$ | Graph Above or Below $x$-Axis |
|---|---|---|---|---|
| $(-\infty, -3)$ | $-4$ | $-\frac{4}{7}$ | Negative | Below |
| $(-3, 0)$ | $-1$ | $\frac{1}{8}$ | Positive | Above |
| $(0, 3)$ | $1$ | $-\frac{1}{8}$ | Negative | Below |
| $(3, \infty)$ | $4$ | $\frac{4}{7}$ | Positive | Above |

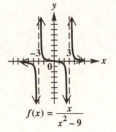

$$f(x) = \frac{x}{x^2-9}$$

**81.** $f(x) = \frac{1}{x^2+1}$

vertical asymptote: none
horizontal asymptote: $y = 0$
$y$-intercept: $1$; $x$-intercept: none
$f(x)$ does not intersect the horizontal asymptote

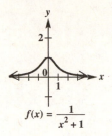

$$f(x) = \frac{1}{x^2+1}$$

**83.** $f(x) = \frac{(x+4)^2}{(x-1)(x+5)}$

vertical asymptotes: $x = -5, x = 1$
horizontal asymptote: $y = 1$
$y$-intercept: $-\frac{16}{5}$; $x$-intercept: $-4$

$f(x)$ intersects the horizontal asymptote

| Interval | Test Point | Value of $f(x)$ | Sign of $f(x)$ | Graph Above or Below $x$-Axis |
|---|---|---|---|---|
| $(-\infty, -5)$ | $-6$ | $\frac{4}{7}$ | Positive | Above |
| $(-5, -4)$ | $-\frac{9}{2}$ | $-\frac{1}{11}$ | Negative | Below |
| $(-4, 1)$ | $-2$ | $-\frac{4}{9}$ | Negative | Below |
| $(1, \infty)$ | $2$ | $\frac{36}{7}$ | Positive | Above |

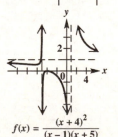

$$f(x) = \frac{(x+4)^2}{(x-1)(x+5)}$$

**85.** $f(x) = \frac{20+6x-2x^2}{8+6x-2x^2}$

vertical asymptotes: $x = -1, x = 4$
horizontal asymptote: $y = 1$
$y$-intercept: $\frac{5}{2}$; $x$-intercepts: $-2, 5$
$f(x)$ does not intersect the horizontal asymptote

| Interval | Test Point | Value of $f(x)$ | Sign of $f(x)$ | Graph Above or Below $x$-Axis |
|---|---|---|---|---|
| $(-\infty, -2)$ | $-4$ | $\frac{3}{4}$ | Positive | Above |
| $(-2, -1)$ | $-\frac{3}{2}$ | $-\frac{13}{11}$ | Negative | Below |
| $(-1, 4)$ | $2$ | $2$ | Positive | Above |
| $(4, 5)$ | $\frac{9}{2}$ | $-\frac{13}{11}$ | Negative | Below |
| $(5, \infty)$ | $6$ | $\frac{4}{7}$ | Positive | Above |

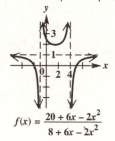

$$f(x) = \frac{20 + 6x - 2x^2}{8 + 6x - 2x^2}$$

**87.** $f(x) = \frac{x^2+1}{x+3}$

vertical asymptote: $x = -3$
oblique asymptote: $y = x - 3$

$y$-intercept: $\frac{1}{3}$; $x$-intercepts: none

$f(x)$ does not intersect the oblique asymptote

| Interval | Test Point | Value of $f(x)$ | Sign of $f(x)$ | Graph Above or Below $x$-Axis |
|---|---|---|---|---|
| $(-\infty, -3)$ | $-4$ | $-17$ | Negative | Below |
| $(-3, \infty)$ | $4$ | $\frac{17}{7}$ | Positive | Above |

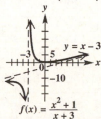

$$f(x) = \frac{x^2+1}{x+3}$$

**89.** $f(x) = \frac{x^2+2x}{2x-1}$

vertical asymptote: $x = \frac{1}{2}$

oblique asymptote: $y = \frac{x}{2} + \frac{5}{4}$

$y$-intercept: 0; $x$-intercepts: $-2$, 0
$f(x)$ does not intersect the oblique asymptote

| Interval | Test | Value | Sign of | Graph |
|---|---|---|---|---|

| Point | of $f(x)$ | $f(x)$ | Above or Below $x$-Axis |
|---|---|---|---|
| $(-\infty, -2)$ | $-3$ | $-\frac{3}{7}$ | Negative | Below |
| $(-2, 0)$ | $-1$ | $\frac{1}{3}$ | Positive | Above |
| $(0, \frac{1}{2})$ | $\frac{1}{4}$ | $-\frac{9}{8}$ | Negative | Below |
| $(\frac{1}{2}, \infty)$ | $2$ | $\frac{8}{3}$ | Positive | Above |

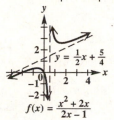

$$f(x) = \frac{x^2 + 2x}{2x - 1}$$

**91.** $f(x) = \frac{x^2-9}{x+3}$

The function degenerates into the line
$f(x) = x - 3, x \ne -3$

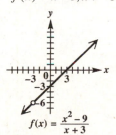

$$f(x) = \frac{x^2 - 9}{x + 3}$$

**93.** $f(x) = \frac{2x^2-5x-2}{x-2}$

vertical asymptote: $x = 2$
oblique asymptote: $y = 2x - 1$
$y$-intercept: 1; $x$-intercepts: $\approx -.4, \approx 2.9$
$f(x)$ does not intersect the oblique asymptote

| Interval | Test Point | Value of $f(x)$ | Sign of $f(x)$ | Graph Above or Below $x$-Axis |
|---|---|---|---|---|
| $(-\infty, -.4)$ | $-2$ | $-4$ | Negative | Below |
| $(-.4, 2)$ | $1$ | $5$ | Positive | Above |
| $(2, 2.9)$ | $2.5$ | $-4$ | Negative | Below |
| $(2.9, \infty)$ | $4$ | $5$ | Positive | Above |

(*continued on next page*)

*(continued from page 179)*

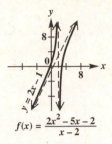

$$f(x) = \frac{2x^2 - 5x - 2}{x - 2}$$

**95.** $f(x) = \frac{x^2 - 1}{x^2 - 4x + 3}$

vertical asymptote: $x = 3$

horizontal asymptote: $y = 1$

$y$-intercept: $-\frac{1}{3}$; $x$-intercept: $-1$

$f(x)$ is not defined for $x = 1$, so there is a hole in the graph.

$f(x)$ does not intersect the horizontal asymptote

| Interval | Test Point | Value of $f(x)$ | Sign of $f(x)$ | Graph Above or Below $x$-Axis |
|---|---|---|---|---|
| $(-\infty, -1)$ | $-4$ | $\frac{3}{7}$ | Positive | Above |
| $(-1, 1)$ | $\frac{1}{2}$ | $-\frac{3}{5}$ | Negative | Below |
| $(1, 3)$ | $2$ | $-3$ | Negative | Below |
| $(3, \infty)$ | $4$ | $5$ | Positive | Above |

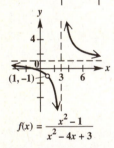

$$f(x) = \frac{x^2 - 1}{x^2 - 4x + 3}$$

**97.** $f(x) = \frac{(x^2 - 9)(2 + x)}{(x^2 - 4)(3 + x)}$

vertical asymptote: $x = 2$

horizontal asymptote: $y = 1$

$y$-intercept: $\frac{3}{2}$; $x$-intercept: $3$

$f(x)$ is not defined for $x = -3$ and $x = -2$, so there are two holes in the graph.

$f(x)$ does not intersect the horizontal asymptote

| Interval | Test Point | Value of $f(x)$ | Sign of $f(x)$ | Graph Above or Below $x$-Axis |
|---|---|---|---|---|
| $(-\infty, -3)$ | $-4$ | $\frac{7}{6}$ | Positive | Above |
| $(-3, -2)$ | $-\frac{5}{2}$ | $\frac{11}{9}$ | Positive | Above |
| $(-2, 2)$ | $1$ | $2$ | Positive | Above |
| $(2, 3)$ | $\frac{5}{2}$ | $-1$ | Negative | Below |
| $(3, \infty)$ | $4$ | $\frac{1}{2}$ | Positive | Above |

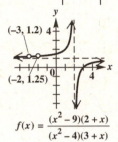

$$f(x) = \frac{(x^2 - 9)(2 + x)}{(x^2 - 4)(3 + x)}$$

**99.** $f(x) = \frac{x^4 - 20x^2 + 64}{x^4 - 10x^2 + 9}$

vertical asymptotes: $x = -3$, $x = -1$, $x = 1$, $x = 3$

horizontal asymptote: $y = 1$

$y$-intercept: $\frac{64}{9}$; $x$-intercepts: $-4, -2, 2, 4$

$f(x)$ intersects the horizontal asymptote

| Interval | Test Point | Value of $f(x)$ | Sign of $f(x)$ | Graph Above or Below $x$-Axis |
|---|---|---|---|---|
| $(-\infty, -4)$ | $-5$ | $.49$ | Positive | Above |
| $(-4, -3)$ | $-3.5$ | $-.85$ | Negative | Below |
| $(-3, -2)$ | $-2.5$ | $1.52$ | Positive | Above |
| $(-2, -1)$ | $-1.5$ | $-2.85$ | Negative | Below |
| $(-1, 1)$ | $.5$ | $9$ | Positive | Above |
| $(1, 2)$ | $1.5$ | $-2.85$ | Negative | Below |
| $(2, 3)$ | $2.5$ | $1.52$ | Positive | Above |
| $(3, 4)$ | $3.5$ | $-.85$ | Negative | Below |
| $(4, \infty)$ | $5$ | $.49$ | Positive | Above |

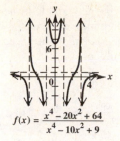

$$f(x) = \frac{x^4 - 20x^2 + 64}{x^4 - 10x^2 + 9}$$

**101.** The graph has a vertical asymptote, $x = 2$, so $x - 2$ is the denominator of the function. There is a "hole" in the graph at $x = -2$, so $x + 2$ is in the denominator and numerator also. The $x$-intercept is 3, so that when $f(x) = 0$, $x = 3$. This condition exists if $x - 3$ is a factor of the numerator. Putting these conditions together, we have a possible function $f(x) = \frac{(x-3)(x+2)}{(x-2)(x+2)}$ or $f(x) = \frac{x^2 - x - 6}{x^2 - 4}$.

**103.** The graph has vertical asymptotes at $x = 4$ and $x = 0$, so $x - 4$ and $x$ are factors in the denominator of the function. The only $x$-intercept is 2, so that when $f(x) = 0$, $x = 2$. This condition exists if $x - 2$ is a factor of the numerator. The graph has a horizontal asymptote $y = 0$, so the degree of the denominator is larger than the degree of the numerator. Putting these conditions together, we have a possible function $f(x) = \frac{x-2}{x(x-4)}$ or $f(x) = \frac{x-2}{x^2 - 4x}$.

**105.** Several answers are possible. One answer is $f(x) = \frac{(x-3)(x+1)}{(x-1)^2}$.

**107.** $f(x) = \frac{x+1}{x-4}$

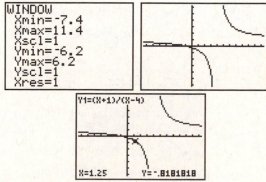

From the last screen we can see that $f(1.25) = -.\overline{81}$. If you wanted the fractional equivalent, you can go back to your homescreen and obtain $Y_1$ from the VARS menu. Thus $f(1.25) = -\frac{9}{11}$.

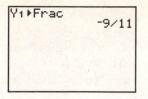

**109.** $f(x) = \frac{x^2 + 2x}{2x - 1}$

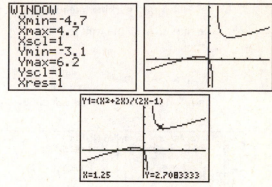

From the last screen we can see that $f(1.25) = -2.70\overline{83}$. If you wanted the fractional equivalent, you can go back to your homescreen and obtain $Y_1$ from the VARS menu. Thus $f(1.25) = \frac{65}{24}$.

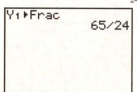

**111. (a)** $T(r) = \frac{2r - k}{2r^2 - 2kr}; k = 25 \Rightarrow$
$T(r) = \frac{2r - 25}{2r^2 - 2(25)r} = \frac{2r - 25}{2r^2 - 50r}$

Graph $Y_1 = \frac{2x - 25}{2x^2 - 50x}$ and $Y_2 = .5$ (since 30 sec = .5 min) on the same screen.

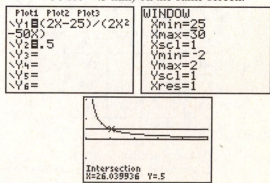

(*continued on next page*)

(*continued from page 181*)

Using the "intersect" option in the CALC menu, we find that the graphs intersect at $x \approx 26$, which represents $r \approx 26$ in the given function. Therefore, there must be an average admittance rate of 26 vehicles per minute.

**(b)** $\frac{26}{5.3} \approx 4.9$ or 5 parking attendants must be on duty to keep the wait less than 30 seconds.

**113. (a)** Graph $y = d(x) = \frac{8710x^2 - 69,400x + 470,000}{1.08x^2 - 324x + 82,200}$ and $y = 300$ on the same calculator screen.

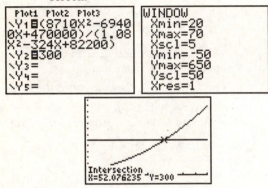

The graphs intersect when $x \approx 52.1$ miles per hour.

**(b)**

| X | Y1 | | X | Y1 |
|---|---|---|---|---|
| **20** | 33.696 | | **40** | 163.88 |
| 25 | 55.884 | | 45 | 214.66 |
| 30 | 84.776 | | 50 | 273.29 |
| 35 | 120.68 | | 55 | 340.01 |
| 40 | 163.88 | | 60 | 415.05 |
| 45 | 214.66 | | 65 | 498.59 |
| 50 | 273.29 | | 70 | 590.8 |
| X=20 | | | X=40 | |

| x | d(x) | x | d(x) |
|---|---|---|---|
| 20 | **34** | 50 | **273** |
| 25 | **56** | 55 | **340** |
| 30 | **85** | 60 | **415** |
| 35 | **121** | 65 | **499** |
| 40 | **164** | 70 | **591** |
| 45 | **215** | | |

**(c)** By comparing values in the table generated in part b, it appears that when the speed is doubled, the stopping distance is more than doubled.

**(d)** If the stopping distance doubled whenever the speed doubled, then there would be a linear relationship between the speed and the stopping distance.

**115.** $R(x) = \frac{80x - 8000}{x - 110}$

**(a)** $R(55) = \frac{80(55) - 8000}{55 - 110} \approx \$65.5$ tens of millions

**(b)** $R(60) = \frac{80(60) - 8000}{60 - 110} = \$64$ tens of millions

**(c)** $R(70) = \frac{80(70) - 8000}{70 - 110} = \$60$ tens of millions

**(d)** $R(90) = \frac{80(90) - 8000}{90 - 110} = \$40$ tens of millions

**(e)** $R(100) = \frac{80(100) - 8000}{100 - 110} = \$0$

**(f)**

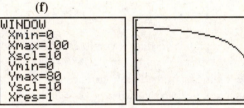

**117.** Since the degree of the numerator equals the degree of the denominator in $f(x) = \frac{x^4 - 3x^3 - 21x^2 + 43x + 60}{x^4 - 6x^3 + x^2 + 24x - 20}$, the graph has a horizontal asymptote at $y = \frac{1}{1} = 1$.

**119. (a)** Use synthetic division where $g(x) = x^4 - 6x^3 + x^2 + 24x - 20$ and $k = 1$.

$$\begin{array}{r|rrrrr} 1 & 1 & -6 & 1 & 24 & -20 \\ & & 1 & -5 & -4 & 20 \\ \hline & 1 & -5 & -4 & 20 & 0 \end{array}$$

Now use synthetic division where $h(x) = x^3 - 5x^2 - 4x + 20$ and $k = 2$.

$$\begin{array}{r|rrrr} 2 & 1 & -5 & -4 & 20 \\ & & 2 & -6 & -20 \\ \hline & 1 & -3 & -10 & 0 \end{array}$$

The resulting polynomial quotient is $x^2 - 3x - 10$, which factors to $(x + 2)(x - 5)$. Thus, the complete factorization of the denominator is $(x - 1)(x - 2)(x + 2)(x - 5)$.

**(b)** $f(x) = \frac{(x + 4)(x + 1)(x - 3)(x - 5)}{(x - 1)(x - 2)(x + 2)(x - 5)}$

**121.** Although $x - 5$ is a factor of the numerator, it will not yield an $x$-intercept because it is also a factor of the denominator. The other three factors of the numerator, namely $(x + 4)$, $(x + 1)$, and $(x - 3)$ will yield $x$-intercepts of $-4$, $-1$, and $3$ because any of these will yield $f(x) = 0$.

**123.** Although $x - 5$ is a factor of the denominator, it will not yield a vertical asymptote because it is also a factor of the numerator. The other three factors of the denominator $\left(\text{namely } (x - 1), (x - 2), \text{ and } (x + 2)\right)$ will yield vertical asymptotes of $x = 1$, $x = 2$, and $x = -2$ because any of these will yield a denominator of zero.

**125.** The vertical asymptotes are $x = 1$, $x = 2$, and $x = -2$ and the $x$-intercepts occur at $-4$, $-1$, and $3$. We should also consider the "hole" in the graph which occurs at $x = 5$. This divides the real number line into eight intervals.

| Interval | Test Point | Value of $f(x)$ | Sign of $f(x)$ | Graph Above or Below $x$-Axis |
|---|---|---|---|---|
| $(-\infty, -4)$ | $-5$ | $\frac{16}{63}$ | Positive | Above |
| $(-4, -2)$ | $-3$ | $-\frac{3}{5}$ | Negative | Below |
| $(-2, -1)$ | $-\frac{3}{2}$ | $\frac{9}{7}$ | Positive | Above |
| $(-1, 1)$ | $0$ | $-3$ | Negative | Below |
| $(1, 2)$ | $\frac{3}{2}$ | $\frac{165}{7}$ | Positive | Above |
| $(2, 3)$ | $\frac{5}{2}$ | $-\frac{91}{27}$ | Negative | Below |
| $(3, 5)$ | $4$ | $\frac{10}{9}$ | Positive | Above |
| $(5, \infty)$ | $6$ | $\frac{21}{16}$ | Positive | Above |

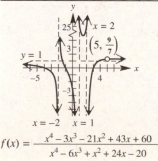

$$f(x) = \frac{x^4 - 3x^3 - 21x^2 + 43x + 60}{x^4 - 6x^3 + x^2 + 24x - 20}$$

## Chapter 3 Quiz
**(Sections 3.1–3.5)**

**1. (a)** $f(x) = -2(x + 3)^2 - 1$

This equation is of the form $y = a(x - h)^2 + k$, with $h = -3$, $k = -1$, and $a = -2$. The graph opens downward and is narrower than $y = x^2$. It is a horizontal translation of the graph of $y = -2x^2$, 3 units to the left and 1 unit down. The vertex is $(-3, -1)$. The axis is $x = -3$. The domain is $(-\infty, \infty)$. Since the largest value of $y$ is $-1$ and the graph opens downward, the range is $(-\infty, -1]$.

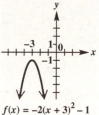

$f(x) = -2(x + 3)^2 - 1$

**(b)** $f(x) = 2x^2 - 8x + 3$; Rewrite by completing the square on $x$

$$f(x) = 2x^2 - 8x + 3$$
$$= 2(x^2 - 4x + 4) + (3 - 8)$$

$$\text{Note: } 2\left[\tfrac{1}{2}(-4)\right]^2 = 2(-2)^2 = 8$$

$$= 2(x - 2)^2 - 5$$

This equation is of the form $y = a(x - h)^2 + k$, with $h = 2$, $k = -5$, and $a = 2$. The graph opens upward and is narrower than $y = x^2$. It is a horizontal translation of the graph of $y = 2x^2$, 2 units to the right and 5 units down. The vertex is $(2, -5)$. The axis is $x = 2$. The domain is $(-\infty, \infty)$. Since the smallest value of $y$ is $-5$ and the graph opens upward, the range is $[-5, \infty)$.

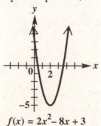

$f(x) = 2x^2 - 8x + 3$

**3.** $f(x) = 2x^4 + x^3 - 3x + 4$; $k = 2$

Use synthetic division to write the polynomial in the form $f(x) = (x - k)q(x) + r$. Since the $x^2$ is missing, include 0 as its coefficient.

```
2)2  1   0  -3   4
      4  10  20  34
   2  5  10  17  38
```

$f(x) = (x - 2)(2x^3 + 5x^2 + 10x + 17) + 38$

The remainder is 38, so $k = 2$ is not a zero of the function. $k(2) = 38$.

**5.** Since $3 - i$ is a factor, $3 + i$ is also a factor. So

$f(x)$

$= [(x - (3 - i)][(x - (3 + i)][(x - (-2)](x - 3)$

$= (x - 3 + i)(x - 3 - i)(x + 2)(x - 3)$

$= (x^2 - 6x + 1)(x^2 - x - 6)$

$= x^4 - 7x^3 + 10x^2 + 26x - 60$

**7.** $f(x) = 2x^4 - 9x^3 - 5x^2 + 57x - 45$

*Step 1*: $p$ must be a factor of $a_0 = -45$ and $q$ must be a factor of $a_4 = 2$. Thus, $p$ can be $\pm 1$, $\pm 3, \pm 5, \pm 9, \pm 15,$ or $\pm 45$, and $q$ can be $\pm 1$ or $\pm 2$. The possible rational zeros, $\frac{p}{q}$, are

$\pm \frac{1}{2}, \pm 1, \pm \frac{3}{2}, \pm \frac{5}{2}, \pm 3, \pm \frac{9}{2}, \pm 5, \pm \frac{15}{2}, \pm 15, \pm \frac{45}{2},$ or $\pm 45$. The remainder theorem shows that 1 is a zero.

```
1)2  -9  -5   57  -45
      2  -7  -12   45
   2  -7  -12  45    0
```

The new quotient polynomial is $2x^3 - 7x^2 - 12x + 45$.

$p$ must be a factor of $a_0 = -45$ and $q$ must be a factor of $a_3 = 2$. Thus, $p$ can be $\pm 1, \pm 3, \pm 5,$ $\pm 9, \pm 15,$ or $\pm 45$, and $q$ can be $\pm 1$ or $\pm 2$. The possible rational zeros, $\frac{p}{q}$, are $\pm \frac{1}{2}, \pm 1,$

$\pm \frac{3}{2}, \pm \frac{5}{2}, \pm 3, \pm \frac{9}{2}, \pm 5, \pm \frac{15}{2}, \pm 15, \pm \frac{45}{2},$ or $\pm 45$. The remainder theorem shows that 3 is a zero.

```
3)2  -7  -12   45
      6  -3  -45
   2  -1  -15    0
```

The new quotient polynomial is

$2x^2 - x - 15 = 0$

$(2x + 5)(x - 3) = 0$

$2x + 5 = 0 \Rightarrow x = -\frac{5}{2}$ or $x - 3 = 0 \Rightarrow x = 3$

The rational zeros are $-\frac{5}{2}$, 1, and 3. Note that 3 is a zero of multiplicity 2, so the graph is tangent to the x-axis at $x = 3$. The factors are $2x + 5$, $x - 1$, and $(x - 3)^2$ and thus $f(x) = (2x + 5)(x - 1)(x - 3)^2$.

*Step 2*: $f(0) = -45$, so plot $(0, -45)$.

*Step 3*: The x-intercepts divide the x-axis into four intervals:

| Interval | Test Point | Value of $f(x)$ | Sign of $f(x)$ | Graph Above or Below x-Axis |
|---|---|---|---|---|
| $\left(-\infty, -\frac{5}{2}\right)$ | $-3$ | 144 | Positive | Above |
| $\left(-\frac{5}{2}, 1\right)$ | $-1$ | $-96$ | Negative | Below |
| $(1, 3)$ | 2 | 9 | Positive | Above |
| $(3, \infty)$ | 4 | 39 | Positive | Above |

Plot the x-intercepts, y-intercept and test points (the y-intercept is one of the test points) with a smooth curve to get the graph.

$f(x) = 2x^4 - 9x^3 - 5x^2 + 57x - 45$
$= (x - 3)^2(2x + 5)(x - 1)$

**9.** $f(x) = \dfrac{3x + 1}{x^2 + 7x + 10} = \dfrac{3x + 1}{(x + 5)(x + 2)}$

*Step 1*: The graph has vertical asymptotes where the denominator equals zero, that is, at $x = -5$ and $x = -2$.

*Step 2*: Since the degree of the numerator is less than the degree of the denominator, the graph has a horizontal asymptote at $y = 0$ (the x-axis).

*Step 3*: The y-intercept is

$f(0) = \dfrac{3(0) + 1}{0^2 + 7(0) + 10} = \dfrac{1}{10}$

*Step 4*: Find any x-intercepts by solving

$f(x) = 0$: $\dfrac{3x + 1}{x^2 + 7x + 10} = \dfrac{3x + 1}{(x + 2)(x + 5)} = 0 \Rightarrow$

$3x + 1 = 0 \Rightarrow x = -\frac{1}{3}$

*Step 5*: The graph will intersect the horizontal asymptote, $y = 0$, when $f(x) = 0$. From step 4, that is at $x = -\frac{1}{3}$.

*Step 6*: Plot a point in each of the intervals determined by the x-intercept and the vertical asymptotes. There are four intervals,

$(-\infty, -5), (-5, -2), \left(-2, -\frac{1}{3}\right),$ and $\left(-\frac{1}{3}, \infty\right).$

| Interval | Test Point | Value of $f(x)$ | Sign of $f(x)$ | Graph Above or Below x-Axis |
|----------|------------|-----------------|-----------------|------------------------------|
| $(-\infty, -5)$ | –6 | –4.24 | Negative | Below |
| $(-5, -2)$ | –4 | 5.5 | Positive | Above |
| $\left(-2, -\frac{1}{3}\right)$ | –1 | –.5 | Negative | Below |
| $\left(-\frac{1}{3}, \infty\right)$ | 3 | .25 | Positive | Above |

Use the asymptotes, intercepts, and these points to sketch the graph.

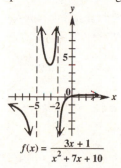

$$f(x) = \frac{3x + 1}{x^2 + 7x + 10}$$

## Section 3.6: Variation

1.  $C = 2\pi r,$ where $C$ is the circumference of a circle of radius $r$
    The circumference of a circle varies directly as (or is proportional to) its radius.

3.  $r = \frac{d}{t},$ where $r$ is the speed when traveling $d$ miles in $t$ hours
    The average speed varies directly as (or is proportional to) the distance traveled and inversely as the time.

5.  $s = kx^3,$ where $s$ is the strength of a muscle of length $x$
    The strength of a muscle varies directly as (or is proportional to) the cube of its length.

7.  $y$ varies directly as $x,$ $y = kx,$ is a straight-line model. It matches graph C.

9.  $y$ varies directly as the second power of $x,$ $y = kx^2,$ matches graph A.

11. *Step 1*: $y = kx$
    *Step 2*: Substitute $x = 4$ and $y = 20$ to find $k$.
    $20 = k(4) \Rightarrow k = 5$
    *Step 3*: $y = 5x$
    *Step 4*: Now find $y$ when $x = -6$.
    $y = 5(-6) = -30$

13. *Step 1*: $m = kxy$
    *Step 2*: Substitute $m = 10,$ $x = 2,$ and $y = 14$ to find $k$.
    $10 = k(4)(7) \Rightarrow 10 = 28k \Rightarrow k = \frac{5}{14}$
    *Step 3*: $m = \frac{5}{14} xy$
    *Step 4*: Now find $m$ when $x = 11$ and $y = 8$.
    $m = \frac{5}{14}(11)(8) = \frac{440}{14} = \frac{220}{7}$

15. *Step 1*: $y = \frac{k}{x}$
    *Step 2*: Substitute $x = 3$ and $y = 10$ to find $k$.
    $10 = \frac{k}{3} \Rightarrow k = 30$
    *Step 3*: $y = \frac{30}{x}$
    *Step 4*: Now find $y$ when $x = 20$.
    $y = \frac{30}{20} = \frac{3}{2}$

17. *Step 1*: $r = \frac{km^2}{s}$
    *Step 2*: Substitute $r = 12,$ $m = 6,$ and $s = 4$ to find $k$:
    $12 = \frac{k \cdot 6^2}{4} \Rightarrow 12 = \frac{36k}{4} \Rightarrow 12 = 9k \Rightarrow k = \frac{4}{3}$
    *Step 3*: $r = \frac{4}{3} \cdot \frac{m^2}{s} = \frac{4m^2}{3s}$
    *Step 4*: Now find $r$ when $m = 6$ and $s = 20$.
    $r = \frac{4(6^2)}{3(20)} = \frac{4(36)}{60} = \frac{144}{60} = \frac{12}{5}$

19. *Step 1*: $a = \frac{kmn^2}{y^3}$
    *Step 2*: Substitute $a = 9,$ $m = 4,$ $n = 9,$ and $y = 3$ to find $k$.
    $9 = \frac{k(4)(9^2)}{3^3} \Rightarrow 9 = \frac{k(4)(81)}{27}$
    $9 = \frac{324k}{27} \Rightarrow 9 = 12k$
    $k = \frac{3}{4}$
    *Step 3*: $a = \frac{3}{4} \cdot \frac{mn^2}{y^3} = \frac{3mn^2}{4y^3}$

(*continued on next page*)

*(continued from page 185)*

Step 4: Now find $a$ when $m = 6$, $n = 2$, and $y = 5$.

$$a = \frac{3(6)(2^2)}{4(5^3)} = \frac{3(6)(4)}{4(125)} = \frac{18}{125}$$

21. *Step 1*: Let $C$ be the circumference of the circle (in inches), $r$ is the radius of the circle (in inches). $C = kr$
    *Step 2*: Substitute $C = 43.96$ and $r = 7$ to find $k$.
    $$43.96 = k(7) \Rightarrow k = 6.28$$
    *Step 3*: $C = 6.28r$
    *Step 4*: Now find $C$ when $r = 11$.
    $$C = 6.28(11) = 69.08$$
    A radius of 11 in. yields a circumference of 69.08 in.

23. *Step 1*: Let $R$ be the resistance (in ohms); $t$ is the temperature (in degrees Kelvin, K). $R = kt$
    *Step 2*: Substitute $R = 646$ and $t = 190$ to find
    $k$: $646 = k(190) \Rightarrow k = \frac{646}{190} = \frac{17}{5}$
    *Step 3*: $R = \frac{17}{5}t$
    *Step 4*: Now find $R$ when $t = 250$.
    $$R = \frac{17}{5}(250) = 850$$
    A resistance of 850 ohms occurs at 250K.

25. *Step 1*: Let $e$ be weight on earth (in lbs). $m$ is weight on moon (in lbs). $e = km$
    *Step 2*: Substitute $e = 200$ and $m = 32$ to find
    $k$: $200 = k(32) \Rightarrow k = \frac{200}{32} = 6.25$
    *Step 3*: $e = 6.25m$
    *Step 4*: Now find $m$ when $e = 50$.
    $$60 = 6.25m \Rightarrow m = \frac{50}{6.25} = 8$$
    The dog would weigh 8 lbs on the moon.

27. *Step 1*: Let $d$ be the distance the spring stretches (in inches), $f$ is the force applied (in lbs). $d = kf$
    *Step 2*: Substitute $d = 8$ & $f = 15$ to find $k$.
    $$8 = k(15) \Rightarrow k = \frac{8}{15}$$
    *Step 3*: $d = \frac{8}{15}f$
    *Step 4*: Now find $d$ when $f = 30$.
    $$d = \frac{8}{15}(30) = 16$$
    The spring will stretch 16 in.

29. *Step 1*: Let $v$ be the speed of the pulley (in rpm). $D$ is the diameter of the pulley (in inches). $v = \frac{k}{D}$
    *Step 2*: Substitute $v = 150$ and $D = 3$ to find $k$.
    $$150 = \frac{k}{3} \Rightarrow k = 450$$
    *Step 3*: $v = \frac{450}{d}$
    *Step 4*: Now find $v$ when $D = 5$: $v = \frac{450}{5} = 90$
    The 5-in. diameter pulley will have a speed of 90 revolutions per minute.

31. *Step 1*: Let $R$ be the resistance (in ohms), $D$ is the diameter (in inches). $R = \frac{k}{D^2}$
    *Step 2*: Substitute $D = .01$ and $R = .4$ to find $k$.
    $$.4 = \frac{k}{.01^2} \Rightarrow .4 = \frac{k}{.0001} \Rightarrow k = .00004$$
    *Step 3*: $R = \frac{.00004}{D^2}$
    *Step 4*: Now find $R$ when $D = .03$.
    $$R = \frac{.00004}{.03^2} = \frac{.00004}{.0009} \approx .0444$$
    The resistance is approximately .0444 ohm.

33. *Step 1*: Let $i$ be the interest (in dollars). $p$ is the principal (in dollars). $t$ is the time (in years). $i = kpt$
    *Step 2*: Substitute $i = 110$, $p = 1000$, and $t = 2$ to find $k$.
    $$110 = k(1000)(2) \Rightarrow 110 = 2000k$$
    $$k = \frac{110}{2000} = .055 \, (5.5\%)$$
    *Step 3*: $i = .055pt$
    *Step 4*: Now find $i$ when $p = 5000$ and $t = 5$.
    $$i = .055(5000)(5) = 1375$$
    The amount of interest is \$1375.

35. *Step 1*: Let $F$ be the force of the wind (in lbs), $A$ is the area (in ft$^2$), $v$ is the velocity of the wind (in mph). $F = kAv^2$
    *Step 2*: Substitute $F = 50$, $A = \frac{1}{2}$, and $v = 40$ to find $k$.
    $$50 = k\left(\frac{1}{2}\right)\left(40^2\right) = k\left(\frac{1}{2}\right)(1600)$$
    $$50 = 800k \Rightarrow k = \frac{50}{800} = \frac{1}{16}$$
    *Step 3*: $F = \frac{1}{16}Av^2$
    *Step 4*: Now find $F$ when $v = 80$, and $A = 2$.
    $$F = \frac{1}{16}(2)\left(80^2\right) = \frac{1}{16}(2)(6400) = 800$$
    The force would be 800 pounds.

**37.** *Step 1*: Let $L$ be the load (in metric tons), $D$ is the diameter (in meters), $h$ is the height (in meters). $L = \frac{kD^4}{h^2}$

*Step 2*: Substitute $h = 9$, $D = 1$, and $L = 8$ to find $k$: $8 = \frac{k(1^4)}{9^2} \Rightarrow 8 = \frac{k}{81} \Rightarrow k = 648$

*Step 3*: $L = \frac{648D^4}{h^2}$

*Step 4*: Now find $L$ when $D = \frac{2}{3}$ and $h = 12$.

$$L = \frac{648\left(\frac{2}{3}\right)^4}{12^2} = \frac{648\left(\frac{16}{81}\right)}{144} = \frac{128}{144} = \frac{8}{9}$$

A column 12 m high and $\frac{2}{3}$ m in diameter will support $\frac{8}{9}$ metric ton.

**39.** *Step 1*: Let $p$ be the period of pendulum (in sec), $l$ is the length of pendulum (in cm), $a$ is the acceleration due to gravity $\left(\text{in } \frac{cm}{sec^2}\right)$.

$$p = \frac{k\sqrt{l}}{\sqrt{a}}$$

*Step 2*: Substitute $p = 6\pi$, $l = 289$, and $a = 980$ to find $k$.

$$6\pi = \frac{k\sqrt{289}}{\sqrt{980}} \Rightarrow 6\pi = \frac{17k}{14\sqrt{5}}$$
$$84\sqrt{5}\pi = 17k \Rightarrow k = \frac{84\sqrt{5}\pi}{17}$$

*Step 3*: $p = \frac{84\sqrt{5}\pi}{17} \cdot \frac{\sqrt{l}}{\sqrt{a}} = \frac{84\sqrt{5}\pi\sqrt{l}}{17\sqrt{a}}$

*Step 4*: Now find $p$ when $l = 121$ and $a = 980$.

$$p = \frac{84\pi\sqrt{5}\sqrt{121}}{17\sqrt{980}} = \frac{84\pi\sqrt{5}(11)}{17(14\sqrt{5})} = \frac{924\pi\sqrt{5}}{238\sqrt{5}} = \frac{66\pi}{17}$$

The period is $\frac{66\pi}{17}$ sec.

**41.** *Step 1*: Let $B$ be the BMI, $w$ is the weight (in lbs), $h$ is the height (in inches). $B = \frac{kw}{h^2}$

*Step 2*: Substitute $w = 177$, $B = 24$, and $h = 72$ (6 feet) to find $k$.

$$24 = \frac{k(177)}{72^2} \Rightarrow 24 = \frac{177k}{5184} \Rightarrow 124,416 = 177k$$
$$k = \frac{124,416}{177} = \frac{41,472}{59}$$

*Step 3*: $B = \frac{41,472}{59} \cdot \frac{w}{h^2} = \frac{41,472w}{59h^2}$

*Step 4*: Now find $B$ when $w = 130$ and $h = 66$.

$$B = \frac{41,472(130)}{59(66^2)} = \frac{5,391,360}{59(4356)} = \frac{5,391,360}{257,004} \approx 20.98$$

The BMI would be approximately 21.

**43.** *Step 1*: Let $R$ be the radiation, $t$ is the temperature (in degrees Kelvin, K). $R = kt^4$ :

*Step 2*: Substitute $R = 213.73$ and $t = 293$ to find $k$.

$$213.73 = k\left(293^4\right) \Rightarrow k = \frac{213.73}{293^4} \approx 2.9 \times 10^{-8}$$

*Step 3*: $R = (2.9 \times 10^{-8})t^4$

*Step 4*: Now find $R$ when $t = 335$.

$$R = \left(2.9 \times 10^{-8}\right) \cdot 335^4 \approx 365.24$$

The radiation of heat would be 365.24.

**45.** *Step 1*: Let $p$ be the person's pelidisi, $w$ is the person's weight (in g), $h$ is the person's sitting height (in cm). $p = \frac{k\sqrt[3]{w}}{h}$

*Step 2*: Substitute $w = 48,820$, $h = 78.7$, and $p = 100$ to find $k$.

$$100 = \frac{k\sqrt[3]{48,820}}{78.7} \Rightarrow 7870 = \sqrt[3]{48,820}k$$
$$k = \frac{7870}{\sqrt[3]{48,820}} \approx 215.33$$

*Step 3*: $p = \frac{215.33\sqrt[3]{w}}{h}$

*Step 4*: Now find $p$ when $w = 54,430$ and $h = 88.9$.

$$p = \frac{215.33\sqrt[3]{54,430}}{88.9} \approx 92$$

This person's pelidisi is 92. The individual is undernourished since his pelidisi is below 100.

**47.** For $k > 0$, if $y$ varies directly as $x$, when $x$ increases, $y$ <u>increases</u>, and when $x$ decreases, $y$ <u>decreases</u>.

**49.** $y = \frac{k}{x}$

If $x$ is doubled, then $y_1 = \frac{k}{2x} = \frac{1}{2} \cdot \frac{k}{x} = \frac{1}{2}y$. Thus, $y$ is half as large as it was before.

**51.** $y = kx$

If $x$ is replaced by $\frac{1}{3}x$, then we have $y_1 = k\left(\frac{1}{3}x\right) = \frac{1}{3} \cdot kx = \frac{1}{3}y$. Thus, $y$ is one-third as large as it was before.

**53.** $p = \frac{kr^3}{t^2}$

If $r$ is replaced by $\frac{1}{2}r$ and $t$ is replaced by $2t$, then we have the following.

$$p_1 = \frac{k\left(\frac{1}{2}r\right)^3}{(2t)^2} = \frac{k\left(\frac{1}{8}r^3\right)}{4t^2} = \frac{1}{8} \cdot \frac{kr^3}{4t^2} = \frac{kr^3}{32t^2} = \frac{1}{32} \cdot \frac{kr^3}{t^2}$$

Thus, so $p$ is $\frac{1}{32}$ as large as it was before.

## Chapter 3: Review Exercises

**1.** $f(x) = 3(x+4)^2 - 5$

Since $f(x) = 3(x+4)^2 - 5$

$= 3\left[x - (-4)\right]^2 + (-5)$, the function has the

form $f(x) = a(x-h)^2 + k$ with

$a = 3$, $h = -4$, and $k = -5$. The graph is a

parabola that opens upward with vertex

$(h, k) = (-4, -5)$. The axis is

$x = h \Rightarrow x = -4$. To find the $x$-intercepts, let

$f(x) = 0$.

$3(x+4)^2 - 5 = 0 \Rightarrow 3(x+4)^2 = 5 \Rightarrow$

$(x+4)^2 = \frac{5}{3} \Rightarrow x + 4 = \pm\sqrt{\frac{5}{3}} \Rightarrow$

$x = -4 \pm \sqrt{\frac{5}{3}} = -4 \pm \frac{\sqrt{15}}{3} = \frac{-12 \pm \sqrt{15}}{3}$

Thus, the $x$-intercepts are $\frac{-12 \pm \sqrt{15}}{3}$ or

approximately $-5.3$ and $-2.7$. To find the

$y$-intercept, let $x = 0$.

$f(0) = 3(0+4)^2 - 5 = 3(16) - 5 = 48 - 5 = 43$

The $y$-intercept is 43.

$f(x) = 3(x+4)^2 - 5$

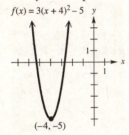

$(-4, -5)$

The domain is $(-\infty, \infty)$. Since the lowest point

on the graph is $(-4, -5)$, the range is

$[-5, \infty)$.

**3.** $f(x) = -3x^2 - 12x - 1$

Complete the square so the function has the

form $f(x) = a(x-h)^2 + k$.

$f(x) = -3x^2 - 12x - 1 = -3\left(x^2 + 4x\right) - 1$

$= -3\left(x^2 + 4x + 4 - 4\right) - 1$

$f(x) = -3(x+2)^2 + 12 - 1$

$= -3\left[x - (-2)\right]^2 + 11$

This parabola is in now in standard form with

$a = -3$, $h = -2$, and $k = 11$. It opens

downward with vertex $(h, k) = (-2, 11)$. The

axis is $x = h \Rightarrow x = -2$. To find the

$x$-intercepts, let $f(x) = 0$.

$-3(x+2)^2 + 11 = 0 \Rightarrow -3(x+2)^2 = -11$

$(x+2)^2 = \frac{11}{3} \Rightarrow x + 2 = \pm\sqrt{\frac{11}{3}}$

$x = -2 \pm \sqrt{\frac{11}{3}} = -2 \pm \frac{\sqrt{33}}{3}$

$= \frac{-6 \pm \sqrt{33}}{3}$

Thus, the $x$-intercepts are $\frac{-6 \pm \sqrt{33}}{3}$ or

approximately $-3.9$ and $-.1$.

To find the $y$-intercept, let $x = 0$.

$f(0) = -3\left(0^2\right) - 12(0) - 1 = -1$

The $y$-intercept is $-1$.

$f(x) = -3x^2 - 12x - 1$

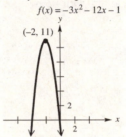

$(-2, 11)$

The domain is $(-\infty, \infty)$. Since the highest on

the graph is $(-2, 11)$, the range is $(-\infty, 11]$.

**5.** $f(x) = a(x-h)^2 + k;\ a > 0$

The graph is a parabola that opens upward.

The coordinates of the lowest point of the

graph are represented by the vertex, $(h, k)$.

**7.** For the graph to have one or more $x$-intercepts,

$f(x) = 0$ must have real number solutions.

$a(x-h)^2 + k = 0 \Rightarrow a(x-h)^2 = -k$

$(x-h)^2 = \frac{-k}{a} \Rightarrow x - h = \pm\sqrt{\frac{-k}{a}}$

$x = h \pm \sqrt{\frac{-k}{a}}$

These solutions are real only if $\frac{-k}{a} \geq 0$. Since

$a > 0$, this condition is equivalent to $-k \geq 0$ or

$k \leq 0$. The $x$-intercepts for these conditions

are given by $h \pm \sqrt{\frac{-k}{a}}$.

**9.** Let $x =$ the width of the rectangular region.

$180 - 2x =$ the length of the region.

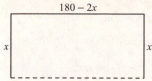

Since $A = LW$, we have

$A(x) = (180 - 2x)x = -2x^2 + 180x.$ Since the

graph of $A(x)$ is a parabola that opens

downward, the maximum area is the

$y$-coordinate of the vertex.

$$A(x) = -2x^2 + 180x = -2(x^2 - 90x)$$
$$= -2(x^2 - 90x + 2025) + 4050$$
$$= -2(x - 45)^2 + 4050$$

The $x$-coordinate of the vertex, $x = 45$, is the

width that gives the maximum area. The length

will be $L = 180 - 2(45) = 90$. Thus, the

dimensions of the region are 90 m $\times$ 45 m.

**11.** Because $V(x)$ has different equations on two

different intervals, it is a piecewise-defined

function.

$$V(x) = \begin{cases} 2x^2 - 32x + 150 & \text{if } 1 \le x < 8 \\ 31x - 226 & \text{if } 8 \le x \le 12 \end{cases}$$

Note: Since $2 \cdot 8^2 - 32 \cdot 8 + 150 = 22$ and

$31 \cdot 8 - 226 = 22$, the point (8, 22) can be

found using either rule, and the graph will

have no breaks.

**(a)** In January, $x = 1.$

$V(x) = 2x^2 - 32x + 150$

$V(1) = 2(1)^2 - 32(1) + 150 = 120$

**(b)** In May, $x = 5.$

$V(x) = 2x^2 - 32x + 150$

$V(5) = 2(5)^2 - 32(5) + 150 = 40$

**(c)** In August, $x = 8.$

$V(x) = 31x - 226$

$V(8) = 31(8) - 226 = 22$

**(d)** In October, $x = 10.$

$V(x) = 31x - 226$

$V(10) = 31(10) - 226 = 84$

**(e)** In December, $x = 12.$

$V(x) = 31x - 226$

$V(12) = 31(12) - 226 = 146$

**(f)**

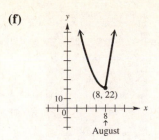

The graph shows a minimum point at

(8, 22). Thus, in August ($x = 8$) the

fewest volunteers are available.

**13.** $f(x) = -2.64x^2 + 5.47x + 3.54$

The discriminant is $b^2 - 4ac$ in the standard

quadratic equation $y = ax^2 + bx + c$. Here we

have $a = -2.64$, $b = 5.47$, and $c = 3.54$. Thus,

$b^2 - 4ac = (5.47)^2 - 4(-2.64)(3.54) = 67.3033.$

Because the discriminant is 67.3033, a positive

number, there are two $x$-intercepts.

**15. (a)** $f(x) > 0$ on the open interval

$(-.52, 2.59).$

**(b)** $f(x) < 0$ on $(-\infty, -.52) \cup (2.59, \infty).$

**17.** $\dfrac{x^3 + x^2 - 11x - 10}{x - 3}$

Since $x - 3$ is in the form $x - k$, $k = 3$.

$$\begin{array}{r} 3\overline{)1 \quad 1 \quad -11 \quad -10} \\ \underline{3 \quad 12 \quad 3} \\ 1 \quad 4 \quad 1 \quad -7 \end{array}$$

$\dfrac{x^3 + x^2 - 11x - 10}{x - 3} = x^2 + 4x + 1 + \dfrac{-7}{x - 3}$

**19.** $\dfrac{2x^3 - x + 6}{x + 4}$

Express $x + 4$ in the form $x - k$ by writing it

as $x - (-4)$. Thus we have $k = -4$. Since the

$x^2 -$ term is missing, include a 0 as its

coefficient.

$$\begin{array}{r} -4\overline{)2 \quad 0 \quad -1 \quad 6} \\ \underline{-8 \quad 32 \quad -124} \\ 2 \quad -8 \quad 31 \quad -118 \end{array}$$

$\dfrac{2x^3 - x + 6}{x + 4} = 2x^2 - 8x + 31 + \dfrac{-118}{x + 4}$

**21.** $f(x) = 5x^3 - 3x^2 + 2x - 6; \ k = 2$

Use synthetic division to write the polynomial in the form $f(x) = (x-k)q(x) + r$.

$$\begin{array}{r} 2 \overline{\smash{\big)}\ 5 -3 \ \ 2 \ -6} \\ \phantom{2)}10 \ \ 14 \ \ 32 \\ \hline \phantom{2)}5 \ \ 7 \ \ 16 \ \ 26 \end{array}$$

$f(x) = (x-2)(5x^2 + 7x + 16) + 26$

**23.** $f(x) = -x^3 + 5x^2 - 7x + 1;$ find $f(2)$.

$$\begin{array}{r} 2 \overline{\smash{\big)}\ -1 \ \ 5 \ -7 \ \ 1} \\ \phantom{2)}-2 \ \ 6 \ -2 \\ \hline \phantom{2)}-1 \ \ 3 \ -1 \ -1 \end{array}$$

The synthetic division shows that $f(2) = -1$.

**25.** $f(x) = 5x^4 - 12x^2 + 2x - 8;$ find $f(2)$.

$$\begin{array}{r} 2 \overline{\smash{\big)}\ 5 \ \ 0 \ -12 \ \ 2 \ -8} \\ \phantom{2)}10 \ \ 20 \ \ 16 \ \ 36 \\ \hline \phantom{2)}5 \ \ 10 \ \ 8 \ \ 18 \ \ 28 \end{array}$$

The synthetic division shows that $f(2) = 28$.

**27.** $f(x) = x^3 + 2x^2 + 3x + 2; \ k = -1$

$$\begin{array}{r} -1 \overline{\smash{\big)}\ 1 \ \ 2 \ \ 3 \ \ 2} \\ \phantom{-1)}-1 \ -1 \ -2 \\ \hline \phantom{-1)}1 \ \ 1 \ \ 2 \ \ 0 \end{array}$$

Since $f(-3) = 0$, we have that $-3$ is a zero of the function.

**29.** By the conjugate zeros theorem, $7 - 2i$ is also a zero.

In Exercises 31 through 33, other answers are possible.

**31.** Zeros: $-1, 4, 7$

$$\begin{aligned} f(x) &= \left[x - (-1)\right](x-4)(x-7) \\ &= (x+1)(x-4)(x-7) \\ &= (x+1)\left[(x-4)(x-7)\right] \\ &= (x+1)\left(x^2 - 11x + 28\right) \\ &= x^3 - 10x^2 + 17x + 28 \end{aligned}$$

**33.** Zeros: $\sqrt{3}, -\sqrt{3}, 2, 3$.

$$\begin{aligned} f(x) &= \left(x - \sqrt{3}\right)\left(x + \sqrt{3}\right)(x-2)(x-3) \\ &= \left[\left(x-\sqrt{3}\right)\left(x+\sqrt{3}\right)\right]\left[(x-2)(x-3)\right] \\ &= \left(x^2 - 3\right)\left(x^2 - 5x + 6\right) \\ &= x^4 - 5x^3 + 3x^2 + 15x - 18 \end{aligned}$$

**35.** $f(x) = 2x^3 - 9x^2 - 6x + 5$

p must be a factor of $a_0 = 5$ and q must be a factor of $a_3 = 2$. Thus, p can be $\pm 1, \pm 5,$ and q can be $\pm 1, \pm 2$. The possible zeros, $\frac{p}{q}$, are $\pm 1, \pm 5, \pm \frac{1}{2}, \pm \frac{5}{2}$. The remainder theorem shows that $\frac{1}{2}$ is a zero.

$$\begin{array}{r} \tfrac{1}{2} \overline{\smash{\big)}\ 2 \ -9 \ \ -6 \ \ 5} \\ \phantom{\tfrac{1}{2})}1 \ \ -4 \ -5 \\ \hline \phantom{\tfrac{1}{2})}2 \ -8 \ -10 \ \ 0 \end{array}$$

The new quotient polynomial is therefore $2x^2 - 8x - 10$.

$2x^2 - 8x - 10 = 0 \Rightarrow x^2 - 4x - 5 = 0 \Rightarrow$
$(x+1)(x-5) = 0$

$x + 1 = 0 \Rightarrow x = -1$ or $x - 5 = 0 \Rightarrow x = 5$

The rational zeros are $\frac{1}{2}, -1,$ and $5$.

**37.** $f(x) = 3x^3 - 8x^2 + x + 2$

**(a)** To show there is a zero between $-1$ and $0$, we need to find $f(-1)$ and $f(0)$.

Find $f(-1)$ by synthetic division and $f(0)$ by evaluating the function.

$$\begin{array}{r} -1 \overline{\smash{\big)}\ 3 \ -8 \ \ 1 \ \ 2} \\ \phantom{-1)}-3 \ \ 11 \ -12 \\ \hline \phantom{-1)}3 \ -11 \ \ 12 \ -10 \end{array}$$

$f(0) = 3(0^3) - 8(0^2) + (0) + 2 = 2$

Since $f(-1) = -10 < 0$ and $f(0) = 2 > 0$, there must be a zero between $-1$ and $0$.

**(b)** To show there is a zero between $2$ and $3$, we need to find $f(2)$ and $f(3)$. Find these by synthetic division.

$$\begin{array}{r} 2 \overline{\smash{\big)}\ 3 \ -8 \ \ 1 \ \ 2} \\ \phantom{2)}6 \ -4 \ -6 \\ \hline \phantom{2)}3 \ -2 \ -3 \ -4 \end{array}$$

$$\begin{array}{r} 3 \overline{\smash{\big)}\ 3 \ -8 \ \ 1 \ \ 2} \\ \phantom{3)}9 \ \ 3 \ \ 12 \\ \hline \phantom{3)}3 \ \ 1 \ \ 4 \ \ 14 \end{array}$$

Since $f(2) = -4 < 0$ and $f(3) = 14 > 0$, there must be a zero between $2$ and $3$.

**39.** $f(x) = 6x^4 + 13x^3 - 11x^2 - 3x + 5$

   **(a)** To show that there is no zero greater than 1, we must show that 1 is an upper bound.

$$\begin{array}{r|rrrrr} 1) & 6 & 13 & -11 & -3 & 5 \\ & & 6 & 19 & 8 & 5 \\ \hline & 6 & 19 & 8 & 5 & 10 \end{array}$$

   Since the 1 is positive and the bottom row has all positive numbers, there is no zero greater than 1.

   **(b)** To show that there is no zero less than $-3$, we must show that $-3$ is an lower bound.

$$\begin{array}{r|rrrrr} -3) & 6 & 13 & -11 & -3 & 5 \\ & & -18 & 15 & -12 & 45 \\ \hline & 6 & -5 & 4 & -15 & 50 \end{array}$$

   Since the $-3$ is negative and the bottom row alternates in sign, there is no zero less than $-3$.

**41.** To determine if $x+1$ is a factor of $f(x) = x^3 + 2x^2 + 3x + 2$, we can find $f(-1)$ by synthetic division

$$\begin{array}{r|rrrr} -1) & 1 & 2 & 3 & 2 \\ & & -1 & -1 & -2 \\ \hline & 1 & 1 & 2 & 0 \end{array}$$

Since $f(-1) = 0$, $x+1$ is a factor of $f(x)$.

**43.** Since $f(x) = a\left[x - (-2)\right](x-1)(x-4)$

$= a(x+2)(x-1)(x-4)$, we can solve for $a$.

$16 = a(2+2)(2-1)(2-4)$

$16 = -8a \Rightarrow a = -2$

The polynomial function is

$f(x) = -2(x+2)(x-1)(x-4)$. It can be expanded as follows.

$\begin{aligned} f(x) &= -2(x+2)(x-1)(x-4) \\ &= -2\left[(x+2)(x-1)\right](x-4) \\ &= -2\left(x^2 + x - 2\right)(x-4) \\ &= -2\left(x^3 - 3x^2 - 6x + 8\right) \\ &= -2x^3 + 6x^2 + 12x - 16 \end{aligned}$

**45.** $f(x) = 2x^4 - x^3 + 7x^2 - 4x - 4$; 1 and $-2i$ are zeros.

Use synthetic division where $k = 1$.

$$\begin{array}{r|rrrrr} 1) & 2 & -1 & 7 & -4 & -4 \\ & & 2 & 1 & 8 & 4 \\ \hline & 2 & 1 & 8 & 4 & 0 \end{array}$$

Synthetically divide $-2i$ into the quotient polynomial.

$$\begin{array}{r|rrrr} -2i) & 2 & 1 & 8 & 4 \\ & & & -4i & -8-2i & -4 \\ \hline & 2 & 1-4i & & -2i & 0 \end{array}$$

Since $-2i$ is a zero, $2i$ is also a zero. Synthetically divide $2i$ into the quotient polynomial.

$$\begin{array}{r|rrr} 2i) & 2 & 1-4i & -2i \\ & & 4i & 2i \\ \hline & 2 & 1 & 0 \end{array}$$

Since the quotient polynomial represented is $2x + 1$, we can set it equal to zero to find the remaining zero. $2x + 1 = 0 \Rightarrow x = -\frac{1}{2}$

Thus, all the zeros are 1, $-\frac{1}{2}$ and $\pm 2i$.

**47.** To determine the value of $s$ such that when the polynomial $x^3 - 3x^2 + sx - 4$ is divided by $x - 2$, the remainder is 5, we perform synthetic division.

$$\begin{array}{r|rrrr} 2) & 1 & -3 & s & -4 \\ & & 2 & -2 & 2s-4 \\ \hline & 1 & -1 & s-2 & 2s-8 \end{array}$$

To ensure that 5 is the remainder set $2s - 8$ equal to 5 and solve.

$2s - 8 = 5 \Rightarrow 2s = 13 \Rightarrow s = \frac{13}{2}$

Thus, the value of $s$ is $\frac{13}{2}$.

**49.** Any polynomial that can be factored as $a(x-b)^3$, where $a$ and $b$ are real numbers, will be a cubic polynomial function having exactly one real zero. One example is $f(x) = 2(x-1)^3$.

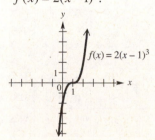

**51.** The polynomial has a leading term of $10x^7$.

   **(a)** The domain is $(-\infty, \infty)$.

   **(b)** The range is $(-\infty, \infty)$.

**(c)** As $x \to \infty, f(x) \to \infty$ and as $x \to -\infty$, $f(x) \to -\infty$.

**(d)** There are at most 7 zeros.

**(e)** There are at most 6 turning points.

**53.** $f(x) = (x-2)^2(x-5)$ is a cubic polynomial with positive $y$-values for $x > 5$, so it matches graph C.

**55.** $f(x) = (x-2)^2(x-5)^2$ is a quartic polynomial that is always greater than or equal to 0, so it matches graph E.

**57.** $f(x) = -(x-2)(x-5)$ is a quadratic polynomial that opens down, so it matches graph B.

**59.** $f(x) = (x-2)^2(x+3)$

*Step 1*: Set each factor equal to 0 and solve the resulting equations to find the zeros of the function.
$(x-2)^2 = 0 \Rightarrow x = 2$ or $x+3 = 0 \Rightarrow x = -3$
The zeros are $-3$ and 2, which divide the $x$-axis into three regions. Test a point in each region to find the sign of $f(x)$ in that region.

*Step 2*: $f(0) = (0-2)^2(0+3) = 4 \cdot 3 = 12$, so plot $(0,12)$.

*Step 3:* The $x$-intercepts divide the $x$-axis into three intervals.

| Interval | Test Point | Value of $f(x)$ | Sign of $f(x)$ | Graph Above or Below $x$-Axis |
|---|---|---|---|---|
| $(-\infty, -3)$ | $-4$ | $-36$ | Negative | Below |
| $(-3, 2)$ | 1 | 4 | Positive | Above |
| $(2, \infty)$ | 3 | 6 | Positive | Above |

Plot the $x$-intercepts, $y$-intercept, and test points with a smooth curve to get the graph.

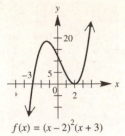

$f(x) = (x-2)^2(x+3)$

**61.** $f(x) = 2x^3 + x^2 - x$

*Step 1*: Factor the polynomial function and set each factor equal to 0 and solve the resulting equations to find the zeros of the function.
$2x^3 + x^2 - x = x(2x^2 + x - 1)$
$\qquad\qquad\qquad = x(x+1)(2x-1)$
$x = 0$ or $x+1 = 0 \Rightarrow x = -1$ or
$2x - 1 = 0 \Rightarrow x = \frac{1}{2}$

The zeros are $-1$, 0, and $\frac{1}{2}$, which divide the $x$-axis into four regions. Test a point in each region to find the sign of $f(x)$ in that region.

*Step 2*: $f(0) = 0$, so plot $(0,0)$.

*Step 3:* The $x$-intercepts divide the $x$-axis into four intervals.

| Interval | Test Point | Value of $f(x)$ | Sign of $f(x)$ | Graph Above or Below $x$-Axis |
|---|---|---|---|---|
| $(-\infty, -1)$ | $-2$ | $-10$ | Negative | Below |
| $(-1, 0)$ | $-\frac{1}{2}$ | $\frac{1}{2}$ | Positive | Above |
| $(0, \frac{1}{2})$ | $\frac{1}{4}$ | $-\frac{5}{32}$ | Negative | Below |
| $(\frac{1}{2}, \infty)$ | 1 | 2 | Positive | Above |

Plot the $x$-intercepts, $y$-intercept (the $y$-intercept is also an $x$-intercept), and test points with a smooth curve to get the graph.

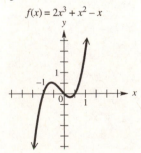

$f(x) = 2x^3 + x^2 - x$

**63.** $f(x) = x^4 + x^3 - 3x^2 - 4x - 4$

*Step 1:* The first step is to find the zeros of the polynomial function. $p$ must be a factor of $a_0 = -4$ and $q$ must be a factor of $a_4 = 1$. Thus, $p$ can be $\pm 1$, $\pm 2$, or $\pm 4$, and $q$ can be $\pm 1$. The possible zeros, $\frac{p}{q}$, are $\pm 1$, $\pm 2$, or $\pm 4$. The remainder theorem shows that 2 is a zero.

$$2\overline{)\begin{array}{ccccc} 1 & 1 & -3 & -4 & -4 \\ & 2 & 6 & 6 & 4 \\ \hline 1 & 3 & 3 & 2 & 0 \end{array}}$$

The new quotient polynomial therefore is $x^3 + 3x^2 + 3x + 2$.

$p$ must be a factor of $a_0 = 2$ and $q$ must be a factor of $a_3 = 1$. Thus, $p$ can be $\pm 1$, or $\pm 2$, and $q$ can be $\pm 1$. The possible zeros, $\frac{p}{q}$, are $\pm 1$, or $\pm 2$. The remainder theorem shows that $-2$ is a zero.

$$-2\overline{)\begin{array}{cccc} 1 & 3 & 3 & 2 \\ & -2 & -2 & -2 \\ \hline 1 & 1 & 1 & 0 \end{array}}$$

The new quotient polynomial therefore is $x^2 + x + 1$. Setting this equal to 0 and using the quadratic formula with $a = 1$, $b = 1$, and $c = 1$, we have

$$x = \frac{-1 \pm \sqrt{1^2 - 4(1)(1)}}{2(1)} = \frac{-1 \pm \sqrt{-4}}{2}$$
$$= -\tfrac{1}{2} \pm 2i$$

The rational zeros are $-2$ and $2$, which divide the $x$-axis into three regions.

*Step 2:* $f(0) = -4$, so plot $(0, -4)$.

*Step 3:* The $x$-intercepts divide the $x$-axis into three intervals.

| Interval | Test Point | Value of $f(x)$ | Sign of $f(x)$ | Graph Above or Below $x$-Axis |
|---|---|---|---|---|
| $(-\infty, -2)$ | $-3$ | 35 | Positive | Above |
| $(-2, 2)$ | $-1$ | $-3$ | Negative | Below |
| $(2, \infty)$ | 3 | 65 | Positive | Above |

Plot the $x$-intercepts, $y$-intercept, and test points with a smooth curve to get the graph.

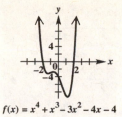

$f(x) = x^4 + x^3 - 3x^2 - 4x - 4$

**65.** Using the calculator, we find that the real zeros are 7.6533119, 1, and $-.6533119$. For the two approximations of the zeros, a higher accuracy can be found by going to the home screen and displaying the stored $x$-value.

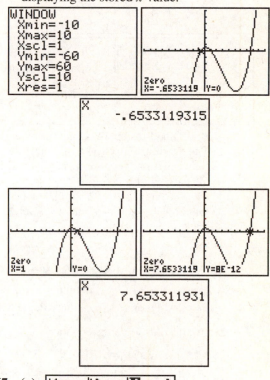

**67.** **(a)**

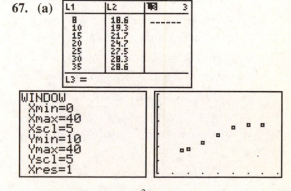

**(b)** $f(x) = -.011x^2 + .869x + 11.9$

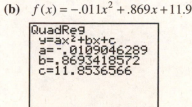

**(c)** $f(x) = -.00087x^3 + .0456x^2 - .219x + 17.8$

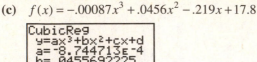

**(d)**

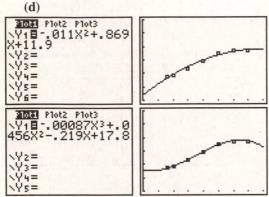

**(e)** Both functions approximate the data well. The quadratic function is probably better for prediction, because it is unlikely that the percent of out-of-pocket spending would decrease after 2025 (as the cubic function shows) unless changes were made in Medicare law.

**69.** $V_{\text{box}} = LWH$

If we let $L = x + 11$, $W = 3x$, and $H = x$, we have the volume of the rectangular box is $V(x) = (x+11)(3x)x$. Since the volume is 720 in.$^3$, we need to solve the equation

$$720 = (x+11)(3x)x \Rightarrow 720 = 3x^3 + 33x^2$$
$$0 = 3x^3 + 33x^2 - 720 \Rightarrow x^3 + 11x^2 - 240 = 0$$

Algebraic Solution:

$p$ must be a factor of $a_0 = -240$ and $q$ must be a factor of $a_3 = 1$. Including the $\pm$, there are 40 factors of $p$ and 2 factors of $q$. There are 80 possible zeros, $\frac{p}{q}$. Because this list is so long, we will simply show that 4 is a zero by the remainder theorem.

$$\begin{array}{r} 4)\overline{1 \quad 11 \quad 0 \quad -240} \\ \quad\; 4 \quad 60 \quad 240 \\ \hline 1 \quad 15 \quad 60 \quad\;\; 0 \end{array}$$

The new quotient polynomial is $x^2 + 15x + 60$. If we examine the discriminant, $b^2 - 4ac$, we will see that the quotient polynomial cannot be factored further over the reals

$$b^2 - 4ac = 15^2 - 4(1)(60)$$
$$= 225 - 240 = -15 < 0$$

Since the discriminant is negative, 4 is our only real solution to the equation $x^3 + 11x^2 - 240 = 0$.

Graphing Calculator Solution:

Graph $Y_1 = x^3 + 11x^2 - 240$ on a graphing calculator.

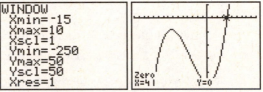

The graph will show that 4 is the only real zero. Since the only real solution is 4, the dimensions of the rectangular box are 4 in. by $3(4) = 12$ in. by $4 + 11 = 15$ in. or $12 \text{ in.} \times 4 \text{ in.} \times 15 \text{ in.}$

**71.** $f(x) = \frac{4}{x-1}$

*Step 1*: The graph has a vertical asymptote where $x - 1 = 0 \Rightarrow x = 1$

*Step 2*: Since the degree of the numerator (which is considered degree zero) is less than the degree of the denominator, the graph has a horizontal asymptote at $y = 0$ (the $x$-axis).

*Step 3*: The $y$-intercept is

$$f(0) = \frac{4}{0-1} = \frac{4}{-1} = -4.$$

*Step 4*: Any $x$-intercepts are found by solving $f(x) = 0$: $\frac{4}{x-1} = 0 \Rightarrow 4 = 0$

This is a false statement, so there is no $x$-intercept.

*Step 5*: The graph will intersect the horizontal asymptote when $\frac{4}{x-1} = 0 \Rightarrow 4 = 0$

This is a false statement, so the graph does not intersect the horizontal asymptote.

*Step 6*: Since the vertical asymptote is $x = 1$ and there is no $x$-intercept, we must determine values in two intervals.

| Interval | Test Point | Value of $f(x)$ | Sign of $f(x)$ | Graph Above or Below $x$-Axis |
|---|---|---|---|---|
| $(-\infty, 1)$ | $-2$ | $-\frac{4}{3}$ | Negative | Below |
| $(1, \infty)$ | $2$ | $4$ | Positive | Above |

*Step 7*: Use the asymptotes, intercepts, and these points to sketch the graph.

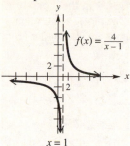

$f(x) = \frac{4}{x-1}$

**73.** $f(x) = \frac{6x}{x^2 + x - 2} = \frac{6x}{(x+2)(x-1)}$

*Step 1*: The graph has a vertical asymptote where $(x+2)(x-1) = 0$, that is, when $x = -2$ and $x = 1$.

*Step 2*: Since the degree of the numerator is less than the degree of the denominator, the graph has a horizontal asymptote at $y = 0$ (the $x$-axis).

*Step 3*: The $y$-intercept is

$f(0) = \frac{6(0)}{0^2 + 0 - 2} = \frac{0}{0+0-2} = \frac{0}{-2} = 0.$

*Step 4*: Any $x$-intercepts are found by solving $f(x) = 0.$

$\frac{6x}{x^2 + x - 2} = 0 \Rightarrow 6x = 0 \Rightarrow x = 0$

The only $x$-intercept is 0.

*Step 5*: The graph will intersect the horizontal asymptote when $\frac{6x}{x^2 + x - 2} = 0.$

$\frac{6x}{x^2 + x - 2} = 0 \Rightarrow 6x = 0 \Rightarrow x = 0$

*Step 6*: Since the vertical asymptotes are $x = -2$ and $x = 1$ and the $x$-intercept occurs at 0, we must determine values in four intervals.

| Interval | Test Point | Value of $f(x)$ | Sign of $f(x)$ | Graph Above or Below $x$-Axis |
|---|---|---|---|---|
| $(-\infty, -2)$ | $-3$ | $-\frac{9}{2}$ | Negative | Below |
| $(-2, 0)$ | $-1$ | $3$ | Positive | Above |
| $(0, 1)$ | $\frac{1}{2}$ | $-\frac{12}{5}$ | Negative | Below |
| $(1, \infty)$ | $2$ | $3$ | Positive | Above |

*Step 7*: Use the asymptotes, intercepts, and these points to sketch the graph.

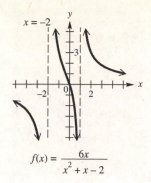

$f(x) = \frac{6x}{x^2 + x - 2}$

**75.** $f(x) = \frac{x^2 + 4}{x+2}$

*Step 1*: The graph has a vertical asymptote where $x + 2 = 0 \Rightarrow x = -2.$

*Step 2*: Since the degree of the numerator is one more than the degree of the denominator, the graph has an oblique asymptote. Divide $x^2 + 4$ by $x + 2.$

$$-2)\overline{\begin{array}{rrr} 1 & 0 & 4 \\ & -2 & 4 \\ \hline 1 & -2 & 8 \end{array}}$$

$f(x) = \frac{x^2 + 4}{x+2} = x - 2 + \frac{8}{x+3}$

The oblique asymptote is the line $y = x - 2.$

*Step 3*: The $y$-intercept is

$f(0) = \frac{0^2 + 4}{0+2} = \frac{0+4}{2} = \frac{4}{2} = 2.$

*Step 4*: Any $x$-intercepts are found by solving $f(x) = 0.$

$\frac{x^2 + 4}{x+2} = 0 \Rightarrow x^2 + 4 = 0 \Rightarrow x = \pm 2i$

There are no $x$-intercepts.

*Step 5*: The graph will intersect the oblique asymptote when $\frac{x^2 + 4}{x+2} = x - 2.$

$\frac{x^2 + 4}{x+2} = x - 2 \Rightarrow x^2 + 4 = (x+2)(x-2)$

$x^2 + 4 = x^2 - 4 \Rightarrow 4 = -4$

This is a false statement, so the graph does not intersect the oblique asymptote.

*Step 6*: Since the vertical asymptote is $x = -2$ and there are no $x$-intercepts, we must determine values in two intervals.

| Interval | Test Point | Value of $f(x)$ | Sign of $f(x)$ | Graph Above or Below $x$-Axis |
|---|---|---|---|---|
| $(-\infty, -2)$ | $-3$ | $-13$ | Negative | Below |
| $(-2, \infty)$ | $3$ | $\frac{13}{5}$ | Positive | Above |

*(continued on next page)*

(*continued from page 195*)

*Step 7*: Use the asymptotes, *y*-intercept, and these points to sketch the graph.

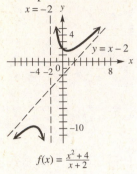

$$f(x) = \frac{x^2 + 4}{x + 2}$$

**77.** $f(x) = \frac{-2}{x^2 + 1}$

*Step 1*: The graph has a vertical asymptote where $x^2 + 1 = 0$. Since we have $x^2 + 1 = 0 \Rightarrow x = \pm i$, there are no vertical asymptotes.

*Step 2*: Since the degree of the numerator (which is considered degree zero) is less than the degree of the denominator, the graph has a horizontal asymptote at $y = 0$ (the *x*-axis).

*Step 3*: The *y*-intercept is

$$f(0) = \frac{-2}{0^2 + 1} = \frac{-2}{0 + 1} = \frac{-2}{1} = -2.$$

*Step 4*: Any *x*-intercepts are found by solving $f(x) = 0$.

$$\frac{-2}{x^2 + 1} = 0 \Rightarrow -2 = 0$$

This is a false statement, so there is no *x*-intercept.

*Step 5*: The graph will intersect the horizontal asymptote when $\frac{-2}{x^2 + 1} = 0$.

$$\frac{-2}{x^2 + 1} = 0 \Rightarrow -2 = 0$$

This is a false statement, so the graph does not intersect the horizontal asymptote.

*Step 6*: Since there are no vertical asymptotes and no *x*-intercept, we can see from the *y*-intercept that the function is always negative and, therefore, below the *x*-axis. Because of the limited information we have, we should explore symmetry and calculate a few more points on the graph.

| $x$ | $y$ |
|---|---|
| $-3$ | $-\frac{1}{5}$ |
| $\frac{1}{2}$ | $-\frac{8}{5}$ |
| $1$ | $-1$ |

Since $f(-x) = \frac{-2}{(-x)^2 + 1} = \frac{-2}{x^2 + 1} = f(x)$, the graph is symmetric about the *y*-axis.

*Step 7*: Use the asymptote, *y*-intercept, these points, and symmetry to sketch the graph.

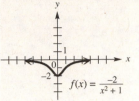

$$f(x) = \frac{-2}{x^2 + 1}$$

**79. (a)**

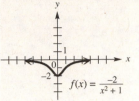

$$f(x) = \frac{(x - 2)(x - 4)}{(x - 3)^2}$$

**(b)** The graph has a vertical asymptote $x = 3$, so $x - 3$ is in the denominator of the function. The x-intercepts are 2 and 4, so that when $f(x) = 0$, $x = 2$ or $x = 4$. This would exist if $x - 2$ and $x - 4$ were factors of the numerator. The horizontal asymptote is $y = 1$, so the numerator and denominator have the same degree. Since the numerator will have degree 2, we must make the denominator also have degree 2. Putting these conditions together, we get a possible function $f(x) = \frac{(x-2)(x-4)}{(x-3)^2}$.

**81.** The graph has a vertical asymptote $x = 1$, so $x - 1$ is in the denominator of the function. The *x*-intercept is 2, so that when $f(x) = 0$, $x = 2$. This would exist if $x - 2$ was a factor of the numerator. The horizontal asymptote is $y = -3$, so the numerator and denominator have the same degree. They both have degree 1. Also, from the horizontal asymptote, we have $y = -3 = \frac{a_n}{b_n} = \frac{-3}{1}$. Putting these conditions together, we have a possible function $f(x) = \frac{-3(x-2)}{x-1}$ or $f(x) = \frac{-3x+6}{x-1}$.

**83.**  $C(x) = \frac{6.7x}{100-x}$

   **(a)**

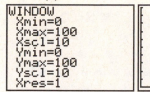 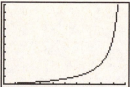

   **(b)**  To find the cost to remove 95% of the pollutant, find $C(95)$.

   $$C(95) = \frac{6.7(95)}{100-95} = 127.3$$

   This can also be found using the graphing calculator.

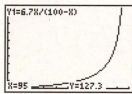

   It would cost $127.3 thousand to remove 95% of the pollutant.

**85.**  *Step 1*: $x = ky$

   *Step 2*: Substitute $x = 20$ and $y = 14$ to find

   $k$. $20 = k(14) \Rightarrow k = \frac{20}{14} = \frac{10}{7}$

   *Step 3*: $x = \frac{10}{7}y$

   *Step 4*: Now find $y$ when $x = 50$.

   $50 = \frac{10}{7}y \Rightarrow 350 = 10y \Rightarrow y = 35$

**87.**  *Step 1*: $t = \frac{k}{s}$

   *Step 2*: Substitute $t = 3$ & $s = 5$ to find $k$.

   $3 = \frac{k}{5} \Rightarrow k = 15$

   *Step 3*: $t = \frac{15}{s}$

   *Step 4*: Now find $s$ when $t = 20$.

   $20 = \frac{15}{s} \Rightarrow 20s = 15 \Rightarrow s = \frac{15}{20} = \frac{3}{4}$

**89.**  *Step 1*: $f = kg^2h$

   *Step 2*: Substitute $f = 50$, $g = 5$, and $h = 4$ to find $k$.

   $50 = k(5^2)(4) \Rightarrow 50 = 100k \Rightarrow k = \frac{50}{100} = \frac{1}{2}$

   *Step 3*: $f = \frac{1}{2}g^2h$

   *Step 4*: Now find $f$ when $g = 3$ and $h = 6$.

   $f = \frac{1}{2}(3^2)(6) = 27$

**91.**  *Step 1*: Let $f$ be the force to keep car from skidding (in lbs), $r$ is the radius of curve (in ft), $w$ is the weight of car (in lbs), $s$ is the speed of the car (in mph). $f = \frac{kws^2}{r}$

   *Step 2*: $f = 3000$, $w = 2000$, $r = 500$, and $s = 30$ to find $k$.

   $$3000 = \frac{k(2000)(30^2)}{500} = \frac{1,800,000k}{500}$$
   $$3000 = 3600k \Rightarrow k = \frac{3000}{3600} = \frac{5}{6}$$

   *Step 3*: $f = \frac{5}{6} \cdot \frac{ws^2}{r} = \frac{5ws^2}{6r}$

   *Step 4*: Now find $f$ when $w = 2000$, $r = 800$, and $s = 60$

   $$f = \frac{5(2000)(60^2)}{6(800)} = 7500$$

   A force of 7500 lb is needed.

## Chapter 3: Test

**1.**  $f(x) = -2x^2 + 6x - 3$

   Complete the square so the function has the form $f(x) = a(x-h)^2 + k$.

   $$f(x) = -2x^2 + 6x - 3 = -2\left(x^2 - 3x\right) - 3$$
   $$= -2\left(x^2 - 3x + \frac{9}{4} - \frac{9}{4}\right) - 3$$
   $$\text{Note: } \left[\frac{1}{2}(-3)\right]^2 = \frac{9}{4}$$
   $$= -2\left(x - \frac{3}{2}\right)^2 + \frac{9}{2} - 3 = -2\left(x - \frac{3}{2}\right)^2 + \frac{3}{2}$$

   This parabola is in now in standard form with $a = -2$, $h = \frac{3}{2}$, and $k = \frac{3}{2}$. It opens downward with vertex $(h, k) = \left(\frac{3}{2}, \frac{3}{2}\right)$. The axis is $x = h \Rightarrow x = \frac{3}{2}$. To find the $x$-intercepts, let $f(x) = 0$.

   $$-2\left(x - \frac{3}{2}\right)^2 + \frac{3}{2} = 0 \Rightarrow -2\left(x - \frac{3}{2}\right)^2 = -\frac{3}{2}$$
   $$\left(x - \frac{3}{2}\right)^2 = \frac{3}{4} \Rightarrow x - \frac{3}{2} = \pm\sqrt{\frac{3}{4}}$$
   $$x = \frac{3}{2} \pm \sqrt{\frac{3}{4}} = \frac{3}{2} \pm \frac{\sqrt{3}}{2} = \frac{3 \pm \sqrt{3}}{2}$$

   Thus the $x$-intercepts are $\frac{3 \pm \sqrt{3}}{2}$ or approximately .63 and 2.37.

   Note: If you use the quadratic formula to solve $-2x^2 + 6x - 3 = 0$ where $a = -2$, $b = 6$, and $c = -3$, you will arrive at the equivalent answer of $x$-intercepts: $\frac{-3 \pm \sqrt{3}}{-2}$.

   To find the $y$-intercept, let $x = 0$.

   $$f(0) = -2(0^2) + 6(0) - 3 = -3$$

   The $y$-intercept is $-3$.

(*continued on next page*)

(*continued from page 197*)

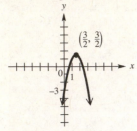

$$f(x) = -2x^2 + 6x - 3$$

The domain is $(-\infty, \infty)$. Since the highest on

the graph is $\left(\frac{3}{2}, \frac{3}{2}\right)$, the range is $\left(-\infty, \frac{3}{2}\right]$.

**2.** $f(x) = -.3857x^2 + 1.2829x + 11.329$

The year 2003 corresponds to $x = 5$ in this model.

$$f(5) = -.3857\left(5^2\right) + 1.2829(5) + 11.329$$
$$= 8.101$$

Approximately 8.1 million tickets would have been sold in 2003 according to the model.

**3.** $\dfrac{3x^3 + 4x^2 - 9x + 6}{x + 2}$

Express $x + 2$ in the form $x - k$ by writing it as $x - (-2)$. Thus $k = -2$.

$$\begin{array}{r} -2)\overline{3 \quad 4 \quad -9 \quad 6} \\ \underline{-6 \quad 4 \quad 10} \\ 3 \quad -2 \quad -5 \quad 16 \end{array}$$

Thus,

$$\frac{3x^3 + 4x^2 - 9x + 6}{x + 2} = 3x^2 - 2x - 5 + \frac{16}{x + 2}$$

**4.** $\dfrac{2x^3 - 11x^2 + 28}{x - 5}$

Since $x - 5$ is in the form $x - k$, $k = 5$.

$$\begin{array}{r} 5)\overline{2 \quad -11 \quad 0 \quad 28} \\ \underline{10 \quad -5 \quad -25} \\ 2 \quad -1 \quad -5 \quad 3 \end{array}$$

$$\frac{2x^3 - 11x^2 + 28}{x - 5} = 2x^2 - x - 5 + \frac{3}{x - 5}$$

**5.** $f(x) = 2x^3 - 9x^2 + 4x + 8; \ k = 5$

$$\begin{array}{r} 5)\overline{2 \quad -9 \quad 4 \quad 8} \\ \underline{10 \quad 5 \quad 45} \\ 2 \quad 1 \quad 9 \quad 53 \end{array}$$

Thus, $f(5) = 53$.

**6.** $6x^4 - 11x^3 - 35x^2 + 34x + 24; \ x - 3$

Let $f(x) = 6x^4 - 11x^3 - 35x^2 + 34x + 24$. By the factor theorem, $x - 3$ will be a factor of $f(x)$ only if $f(3) = 0$.

$$\begin{array}{r} 3)\overline{6 \quad -11 \quad -35 \quad 34 \quad 24} \\ \underline{18 \quad 21 \quad -42 \quad -24} \\ 6 \quad 7 \quad -14 \quad -8 \quad 0 \end{array}$$

Since $f(3) = 0$, $x - 3$ is a factor of $f(x)$. The other factor is $6x^3 + 7x^2 - 14x - 8$.

**7.** $f(x) = x^3 + 8x^2 + 25x + 26; \ -2$

Since $-2$ is a zero, first divide $f(x)$ by $x + 2$.

$$\begin{array}{r} -2)\overline{1 \quad 8 \quad 25 \quad 26} \\ \underline{-2 \quad -12 \quad -26} \\ 1 \quad 6 \quad 13 \quad 0 \end{array}$$

This gives $f(x) = (x + 2)(x^2 + 6x + 13)$. Since $x^2 + 6x + 13$ cannot be factored, use the quadratic formula with $a = 1$, $b = 6$, and $c = 13$ to find the remaining two zeros.

$$x = \frac{-6 \pm \sqrt{6^2 - 4(1)(13)}}{2(1)} = \frac{-6 \pm \sqrt{36 - 52}}{2}$$

$$= \frac{-6 \pm \sqrt{-16}}{2} = \frac{-6 \pm 4i}{2} = -3 \pm 2i$$

The zeros are $-2, -3 - 2i,$ and $-3 + 2i$.

**8.** Zeros of $-1$, 2 and $i$; $f(3) = 80$

By the conjugate zeros theorem, $-i$ is also a zero. The polynomial has the following form.

$$f(x) = a(x + 1)(x - 2)(x - i)(x + i)$$

Use the condition $f(3) = 80$ to find $a$.

$$80 = a(3 + 1)(3 - 2)(3 - i)(3 + i)$$
$$80 = a(4)(1)\left[(3 - i)(3 + i)\right]$$
$$80 = a\left[(4)(1)\right]\left[3^2 - i^2\right]$$
$$80 = a(4)\left[9 - (-1)\right]$$
$$80 = a(4)(10) \Rightarrow 80 = 40a \Rightarrow a = 2$$

Thus we have the following.

$$f(x) = 2(x + 1)(x - 2)(x - i)(x + i)$$
$$= 2\left[(x + 1)(x - 2)\right]\left[(x - i)(x + i)\right]$$
$$= 2\left(x^2 - x - 2\right)\left(x^2 + 1\right)$$
$$= 2\left(x^4 - x^3 - x^2 - x - 2\right)$$
$$= 2x^4 - 2x^3 - 2x^2 - 2x - 4$$

9. Since $f(x) = x^4 + 8x^2 + 12$

$= (x^2 + 6)(x^2 + 2)$, the zeros are $\pm i\sqrt{6}$ and

$\pm i\sqrt{2}$. Moreover, $f(x) > 0$ for all $x$; the graph never crosses or touches the $x$-axis, so $f(x)$ has no real zeros.

10. **(a)** $f(x) = x^3 - 5x^2 + 2x + 7$; find $f(1)$ and $f(2)$ synthetically.

$$1\overline{)\,1\ -5\quad 2\quad 7}\qquad 2\overline{)\,1\ -5\quad 2\quad 7}$$
$$\phantom{1)\,}\underline{\ \ \ \ 1\ -4\ -2}\qquad\phantom{2)}\underline{\ \ \ \ 2\ -6\ -8}$$
$$\phantom{1)\,}1\ -4\ -2\quad 5\qquad\phantom{2)}1\ -3\ -4\ -1$$

By the intermediate value theorem, since $f(1) = 5 > 0$ and $f(2) = -1 < 0$, there must be at least one real zero between 1 and 2.

**(b)** $f(x) = \underset{1}{x^3} - \underset{2}{5x^2} + 2x + 7$ has 2 variations in sign. $f$ has 2 or $2 - 2 = 0$ positive real zeros.

$f(-x) = -x^3 - 5x^2 \underset{1}{-2x} + 7$ has 1 variation in sign. $f$ has 1 negative real zero.

**(c)** Using the calculator, we find that the real zeros are 4.0937635, 1.8370381, and $-.9308016$. For these approximations, a higher accuracy can be found by going to the home screen and displaying the stored $x$-value.

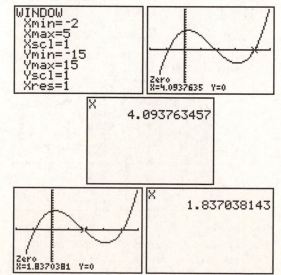

11. To obtain the graph of $f_2$, shift the graph of $f_1$ 5 units to the left, stretch by a factor of 2, reflect across the $x$-axis, and shift 3 units up.

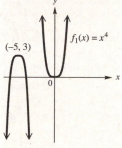

12. $f(x) = -x^7 + x - 4$

Since $f(x)$ is of odd degree and the sign of $a_n$ is negative, the end behavior is ⤵. The correct graph is C.

13. $f(x) = x^3 - 5x^2 + 3x + 9$

*Step 1*: The first step is to find the zeros of the polynomial function. $p$ must be a factor of $a_0 = 9$ and $q$ must be a factor of $a_3 = 1$. Thus, $p$ can be $\pm 1, \pm 3, \pm 9$ and $q$ can be $\pm 1$. The possible zeros, $\frac{p}{q}$, are $\pm 1, \pm 3, \pm 9$. The remainder theorem shows that 3 is a zero.

$$3\overline{)\,1\ -5\quad 3\quad 9}$$
$$\phantom{3)\,}\underline{\ \ \ \ 3\ -6\ -9}$$
$$\phantom{3)\,}1\ -2\ -3\quad 0$$

The new quotient polynomial is $x^2 - 2x - 3$.

$x^2 - 2x - 3 = 0 \Rightarrow (x+1)(x-3) = 0$

$x + 1 = 0 \Rightarrow x = -1$ or $x - 3 = 0 \Rightarrow x = 3$

The rational zeros are $-1$ and 3 (multiplicity 2) which divide the $x$-axis into three regions. Test a point in each region to find the sign of $f(x)$ in that region.

*Step 2*: $f(0) = 9$, so plot $(0, 9)$.

(*continued on next page*)

(*continued from page 199*)

*Step 3*: The *x*-intercepts divide the *x*-axis into three intervals.

| Interval | Test Point | Value of $f(x)$ | Sign of $f(x)$ | Graph Above or Below *x*-Axis |
|---|---|---|---|---|
| $(-\infty, -1)$ | $-2$ | $-25$ | Negative | Below |
| $(-1, 3)$ | $0$ | $9$ | Positive | Above |
| $(3, \infty)$ | $4$ | $5$ | Positive | Above |

Plot the *x*-intercepts, *y*-intercept, and test points (the *y*-intercept is one of the test points) with a smooth curve to get the graph.

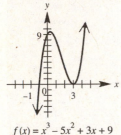

$$f(x) = x^3 - 5x^2 + 3x + 9$$

**14.** $f(x) = 2x^2(x-2)^2$

*Step 1*: Set each factor equal to 0 and solve the resulting equations to find the zeros of the function.

$$2x^2 = 0 \Rightarrow x = 0 \quad \text{or} \quad (x-2)^2 = 0 \Rightarrow x = 2$$

The zeros are 0 (multiplicity 2) and 2 (multiplicity 2), which divide the *x*-axis into three regions. Test a point in each region to find the sign of $f(x)$ in that region.

*Step 2*: $f(0) = 0$, so plot $(0, 0)$.

*Step 3*: The *x*-intercepts divide the *x*-axis into three intervals.

| Interval | Test Point | Value of $f(x)$ | Sign of $f(x)$ | Graph Above or Below *x*-Axis |
|---|---|---|---|---|
| $(-\infty, 0)$ | $-1$ | $18$ | Positive | Above |
| $(0, 2)$ | $1$ | $2$ | Positive | Above |
| $(2, \infty)$ | $3$ | $18$ | Positive | Above |

Plot the *x*-intercepts, *y*-intercept (the *y*-intercept is an *x*-intercept), and test points with a smooth curve to get the graph.

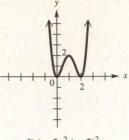

$$f(x) = 2x^2(x-2)^2$$

**15.** $f(x) = -x^3 - 4x^2 + 11x + 30$

*Step 1*: The first step is to find the zeros of the polynomial function. $p$ must be a factor of $a_0 = 30$ and $q$ must be a factor of $a_3 = -1$. Thus, $p$ can be $\pm 1$, $\pm 2$, $\pm 3$, $\pm 5$, $\pm 6$, $\pm 10$, $\pm 15$, $\pm 30$ and $q$ can be $\pm 1$. The possible zeros, $\frac{p}{q}$, are $\pm 1$, $\pm 2$, $\pm 3$, $\pm 5$, $\pm 6$, $\pm 10$, $\pm 15$, $\pm 30$. The remainder theorem shows that 3 is a zero.

$$3\overline{)\begin{array}{rrrr} -1 & -4 & 11 & 30 \\ & -3 & -21 & -30 \\ \hline -1 & -7 & -10 & 0 \end{array}}$$

The new quotient polynomial is $-x^2 - 7x - 10$.

$$-x^2 - 7x - 10 = 0 \Rightarrow x^2 + 7x + 10 = 0$$
$$(x+5)(x+2) = 0$$
$$x + 5 = 0 \Rightarrow x = -5 \quad \text{or} \quad x + 2 = 0 \Rightarrow x = -2$$

The rational zeros are $-5$, $-2$, and 3, which divide the *x*-axis into four regions. Test a point in each region to find the sign of $f(x)$ in that region.

*Step 2*: $f(0) = 30$, so plot $(0, 30)$.

*Step 3*: The *x*-intercepts divide the *x*-axis into four intervals.

| Interval | Test Point | Value of $f(x)$ | Sign of $f(x)$ | Graph Above or Below *x*-Axis |
|---|---|---|---|---|
| $(-\infty, -5)$ | $-6$ | $36$ | Positive | Above |
| $(-5, -2)$ | $-3$ | $-12$ | Negative | Below |
| $(-2, 3)$ | $0$ | $30$ | Positive | Above |
| $(3, \infty)$ | $4$ | $-54$ | Negative | Below |

Plot the *x*-intercepts, *y*-intercept, and test points (the *y*-intercept is one of the test points) with a smooth curve to get the graph.

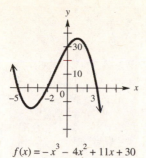

$$f(x) = -x^3 - 4x^2 + 11x + 30$$

**16.** The zeros are $-3$ and $2$. Since the graph of $f$ touches the $x$-axis at $2$, the zero $2$ has multiplicity $2$. Thus, $f(x) = a(x-2)^2(x+3)$.

Also, since the point $(0, 24)$ is on the graph, we have $f(0) = 24$.

$$24 = a(0-2)^2(0+3) \Rightarrow 24 = a(-2)^2(3)$$
$$24 = a(4)(3) \Rightarrow 24 = 12a \Rightarrow a = 2$$

The polynomial function is the following.

$$\begin{aligned} f(x) &= 2(x-2)^2(x+3) \\ &= 2(x^2 - 4x + 4)(x+3) \\ &= 2(x^3 - x^2 - 8x + 12) \\ &= 2x^3 - 2x^2 - 16x + 24 \end{aligned}$$

**17.** $f(t) = 1.06t^3 - 24.6t^2 + 180t$

**(a)** $f(2) = 1.06(2^3) - 24.6(2)^2 + 180(2)$
$$= 270.08$$

This can also be found using the graphing calculator.

 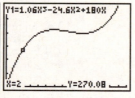

**(b)** From the graph we see that the amount of change is increasing from $t = 0$ to $t = 5.9$ and from $t = 9.5$ to $t = 15$ and decreasing from $t = 5.9$ to $t = 9.5$.

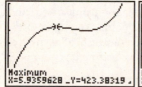 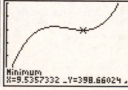

**18.** $f(x) = \frac{3x-1}{x-2}$

*Step 1*: The graph has a vertical asymptote where $x - 2 = 0$, that is, when $x = 2$.

*Step 2*: Since the degree of the numerator equals the degree of the denominator, the graph has a horizontal asymptote at $y = \frac{3}{1} = 3$.

*Step 3*: The $y$-intercept is

$$f(0) = \frac{3(0)-1}{0-2} = \frac{0-1}{-2} = \frac{-1}{-2} = \frac{1}{2}.$$

*Step 4*: Any $x$-intercepts are found by solving $f(x) = 0$.

$$\frac{3x-1}{x-2} = 0 \Rightarrow 3x - 1 = 0 \Rightarrow 3x = 1 \Rightarrow x = \frac{1}{3}$$

The only $x$-intercept is $\frac{1}{3}$.

*Step 5*: The graph will intersect the horizontal asymptote when $\frac{3x-1}{x-2} = 3$.

$$\frac{3x-1}{x-2} = 3 \Rightarrow 3x - 1 = 3(x-2)$$
$$3x - 1 = 3x - 6 \Rightarrow -1 = -6$$

This is a false statement, so the graph does not intersect the horizontal asymptote.

*Step 6*: Since the vertical asymptote is $x = 2$ and the $x$-intercept occurs at $\frac{1}{3}$, we must determine values in three intervals.

| Interval | Test Point | Value of $f(x)$ | Sign of $f(x)$ | Graph Above or Below $x$-Axis |
|---|---|---|---|---|
| $\left(-\infty, \frac{1}{3}\right)$ | $0$ | $\frac{1}{2}$ | Positive | Above |
| $\left(\frac{1}{3}, 2\right)$ | $1$ | $-2$ | Negative | Below |
| $(2, \infty)$ | $3$ | $8$ | Positive | Above |

*Step 7*: Use the asymptotes, intercepts, and test points (the $y$-intercept is one of the test points) to sketch the graph.

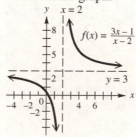

**19.** $f(x) = \frac{x^2-1}{x^2-9} = \frac{(x+1)(x-1)}{(x+3)(x-3)}$

*Step 1*: The graph has a vertical asymptote where $(x+3)(x-3) = 0$, that is, when

$x = -3$ and $x = 3$.

*Step 2*: Since the degree of the numerator equals the degree of the denominator, the graph has a horizontal asymptote at $y = \frac{1}{1} = 1$.

*Step 3*: The $y$-intercept is

$f(0) = \frac{0^2-1}{0^2-9} = \frac{-1}{-9} = \frac{1}{9}$.

*Step 4*: Any $x$-intercepts are found by solving $f(x) = 0$.

$\frac{(x+1)(x-1)}{(x+3)(x-3)} = 0 \Rightarrow (x+1)(x-1) = 0$

$x = -1$ or $x = 1$

The $x$-intercepts are $-1$ and $1$.

*Step 5*: The graph will intersect the horizontal asymptote when $\frac{x^2-1}{x^2-9} = 1$.

$\frac{x^2-1}{x^2-9} = 1 \Rightarrow x^2 - 1 = x^2 - 9 \Rightarrow -1 = -9$

This is a false statement, so the graph does not intersect the horizontal asymptote.

*Step 6*: Since the vertical asymptotes are $x = -3$ and $x = 3$, and the $x$-intercepts occur at $-1$ and $1$, we must determine values in five intervals.

| Interval | Test Point | Value of $f(x)$ | Sign of $f(x)$ | Graph Above or Below $x$-Axis |
|---|---|---|---|---|
| $(-\infty, -3)$ | $-4$ | $\frac{15}{7}$ | Positive | Above |
| $(-3, -1)$ | $-2$ | $-\frac{3}{5}$ | Negative | Below |
| $(-1, 1)$ | $0$ | $\frac{1}{9}$ | Positive | Above |
| $(1, 3)$ | $2$ | $-\frac{3}{5}$ | Negative | Below |
| $(3, \infty)$ | $4$ | $\frac{15}{7}$ | Positive | Above |

*Step 7*: Use the asymptotes, intercepts, and test points (the $y$-intercept is one of the test points) to sketch the graph. Note: This function is an even function, thus the graph is symmetric with respect to the $y$-axis.

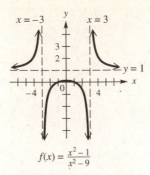

$f(x) = \frac{x^2-1}{x^2-9}$

**20.** $f(x) = \frac{2x^2 + x - 6}{x - 1}$

**(a)** Since the degree of the numerator is one more than the degree of the denominator, the graph has an oblique asymptote.

Divide $2x^2 + x - 6$ by $x - 1$.

$$\begin{array}{r} 1\overline{)2 \quad 1 \quad -6} \\ \underline{2 \quad \quad 3} \\ 2 \quad 3 \quad -3 \end{array}$$

$f(x) = \frac{2x^2+x-6}{x-1} = 2x + 3 - \frac{3}{x-1}$

The oblique asymptote is the line $y = 2x + 3$.

**(b)** To find the $x$-intercepts, let $f(x) = 0$.

$\frac{2x^2+x-6}{x-1} = 0 \Rightarrow 2x^2 + x - 6 = 0$

$(x+2)(2x-3) = 0$

$x = -2$ or $x = \frac{3}{2}$

The $x$-intercepts are $-2$ and $\frac{3}{2}$.

**(c)** The $y$-intercept is

$f(0) = \frac{2(0^2)+0-6}{0-1} = \frac{2(0)+0-6}{-1}$

$= \frac{0+0-6}{-1} = \frac{-6}{-1} = 6$.

**(d)** To find the vertical asymptote, set the denominator equal to zero and solve for $x$.

$x - 1 = 0 \Rightarrow x = 1$

The equation of the vertical asymptote is $x = 1$.

**(e)** Use the information from (a)–(d) and a few additional points to graph the function. Since the vertical asymptote is $x = 1$, and the $x$-intercepts occur at $-2$ and $\frac{3}{2}$, we must determine values in four intervals.

| Interval | Test Point | Value of $f(x)$ | Sign of $f(x)$ | Graph Above or Below $x$-Axis |
|---|---|---|---|---|
| $(-\infty, -2)$ | $-3$ | $-\frac{9}{4}$ | Negative | Below |
| $(-2, 1)$ | $0$ | $6$ | Positive | Above |
| $\left(1, \frac{3}{2}\right)$ | $\frac{5}{4}$ | $-\frac{13}{2}$ | Negative | Below |
| $\left(\frac{3}{2}, \infty\right)$ | $2$ | $4$ | Positive | Above |

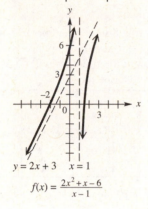

$y = 2x + 3 \quad x = 1$

$$f(x) = \frac{2x^2 + x - 6}{x - 1}$$

21. *Step 1*: $y = k\sqrt{x}$

    *Step 2*: Substitute $y = 12$ and $x = 4$ to find $k$.

    $12 = k\sqrt{4} \Rightarrow 12 = 2k \Rightarrow k = 6$

    *Step 3*: $y = 6\sqrt{x}$

    *Step 4*: Now find $y$ when $x = 100$.

    $y = 6\sqrt{100} = 6 \cdot 10 = 60$

22. *Step 1*: Let $w$ be the weight of the object (in kg), $d$ is the distance from the center of Earth (in km). $w = \frac{k}{d^2}$

    *Step 2*: Substitute $w = 90$ and $d = 6400$ to find $k$.

    $90 = \frac{k}{6400^2} \Rightarrow k = 3,686,400,000$

    *Step 3*: $w = \frac{3,686,400,000}{d^2}$

    *Step 4*: Now find $w$ when $d = 800 + 6400 = 7200$.

    $w = \frac{3,686,400,000}{7200^2} = \frac{3,686,400,000}{51,840,000} = \frac{640}{9} \approx 71.1$

    The man weighs $\frac{640}{9}$ kg or approximately 71.1 kg.

# Chapter 4

## INVERSE, EXPONENTIAL, AND LOGARITHMIC FUNCTIONS

### Section 4.1: Inverse Functions

1. Yes, it is one-to-one, because every number in the list of registered passenger cars is used only once.

3. This is a one-to-one function since every horizontal line intersects the graph in no more than one point.

5. This is a one-to-one function since every horizontal line intersects the graph in no more than one point.

7. This is not a one-to-one function since there is a horizontal line that intersects the graph in more than one point. (Here a horizontal line intersects the curve at an infinite number of points.)

9. $y = 2x - 8$

   Using the definition of a one-to-one function, we have $f(a) = f(b) \Rightarrow 2a - 8 = 2b - 8 \Rightarrow 2a = 2b \Rightarrow a = b$. So the function is one-to-one.

11. $y = \sqrt{36 - x^2}$

   If $x = 6$, $y = \sqrt{36 - 6^2} = \sqrt{36 - 36} = \sqrt{0} = 0$.
   If $x = -6$,
   $$y = \sqrt{36 - (-6)^2} = \sqrt{36 - 36} = \sqrt{0} = 0.$$
   Since two different values of $x$ lead to the same value of $y$, the function is not one-to-one.

13. $y = 2x^3 - 1$

   Looking at this function graphed on a TI-83, we can see that it appears that any horizontal line passed through the function will intersect the graph in at most one place.

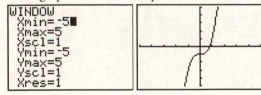

Another way of showing that a function is one-to-one is to assume that you have two equal $y$-values $\left(f(a) = f(b)\right)$ and show that they must have come from the same $x$-value $(a = b)$.

$$f(a) = f(b) \Rightarrow 2a^3 - 1 = 2b^3 - 1 \Rightarrow$$
$$2a^3 = 2b^3 \Rightarrow a^3 = b^3 \Rightarrow \sqrt[3]{a^3} = \sqrt[3]{b^3} \Rightarrow a = b$$

So, the function is one-to-one.

15. $y = -\frac{1}{x+2}$

   Looking at this function graphed on a TI-83, we can see that it appears that any horizontal line passed through the function will intersect the graph in at most one place.

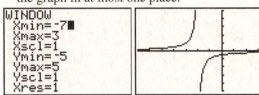

We could also show that $f(a) = f(a)$ implies $a = b$.

$$f(a) = f(b) \Rightarrow -\frac{1}{a+2} = -\frac{1}{b+2} \Rightarrow$$
$$b + 2 = a + 2 \Rightarrow b = a$$

So, the function is one-to-one.

17. $y = 2(x + 1)^2 - 6$

   Looking at this function graphed on a TI-83, we can see that it appears that any horizontal line passed through the function will intersect the graph in two places, except a horizontal line through the vertex. For example,
   $$f(0) = 2(0 + 1)^2 - 6 = -4 \text{ and}$$
   $$f(-2) = 2(-2 + 1)^2 - 6 = -4$$

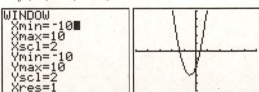

So, the function is not one-to-one.

19. In order for a function to have an inverse, it must be <u>one-to-one</u>.

21. The domain of $f$ is equal to the <u>range</u> of $f^{-1}$, and the range of $f$ is equal to the <u>domain</u> of $f^{-1}$.

**23.** It is false that if $f(x) = x^2$, then

$f^{-1}(x) = \sqrt{x}$ because $f$ is not a one-to-one function and, hence, does not have an inverse.

**25.** If a function $f$ has an inverse and $f(-3) = 6$,

then $f^{-1}(6) = \underline{-3}$.

**27.** Answers will vary. A polynomial of even degree has end behavior pointing in the same direction. This would indicate that the function would not pass the horizontal line test and cannot have an inverse.

**29.** The inverse operation of tying your shoelaces would be untying your shoelaces, since untying "undoes" tying.

**31.** The inverse operation of entering a room would be leaving a room, since leaving "undoes" entering.

**33.** The inverse operation of screwing in a light bulb would be unscrewing the light bulb.

**35.** For each point $(x, y)$ for the first function, there is a point $(y, x)$ for the second function, so $f(x)$ and $g(x)$ are inverses of each other.

**37.** The point $(3, 5)$ is on $f(x)$, but the point $(5, 3)$ is not on $g(x)$ (there is another example), so the functions are not inverses of each other.

**39.** These functions are inverses since their graphs are symmetric with respect to the line $y = x$.

**41.** $f(x) = 2x + 4, g(x) = \frac{1}{2}x - 2$

$(f \circ g)(x) = 2\left(\frac{1}{2}x - 2\right) + 4 = x - 4 + 4 = x$

$(g \circ f)(x) = \frac{1}{2}(2x + 4) - 2 = x + 2 - 2 = x$

Since $(f \circ g)(x) = x$ and $(g \circ f)(x) = x$, these functions are inverses.

**43.** $f(x) = -3x + 12, g(x) = -\frac{1}{3}x - 12$

$(f \circ g)(x) = -3\left(-\frac{1}{3}x - 12\right) + 12$

$\qquad = x + 36 + 12 = x + 48$

Since $(f \circ g)(x) \neq x$, the functions are not inverses. It is not necessary to check $(g \circ f)(x)$.

**45.** $f(x) = \frac{x+1}{x-2}, g(x) = \frac{2x+1}{x-1}$

$(f \circ g)(x) = \frac{\frac{2x+1}{x-1} + 1}{\frac{2x+1}{x-1} - 2} = \frac{\frac{2x+1+x-1}{x-1}}{\frac{2x+1-2(x-1)}{x-1}} = \frac{3x}{3} = x$

$(g \circ f)(x) = \frac{2\left(\frac{x+1}{x-2}\right) + 1}{\frac{x+1}{x-2} - 1} = \frac{\frac{2x+2}{x-2} + 1}{\frac{x+1-(x-2)}{x-2}}$

$\qquad = \frac{\frac{2x+2+x-2}{x-2}}{\frac{x+1-(x-2)}{x-2}} = \frac{3x}{3} = x$

Since $(f \circ g)(x) = x$ and $(g \circ f)(x) = x$, these functions are inverses.

**47.** $f(x) = \frac{2}{x+6}, g(x) = \frac{6x+2}{x}$

$(f \circ g)(x) = f[g(x)] = \frac{2}{\frac{6x+2}{x} + 6} = \frac{2}{\frac{6x+2+6x}{x}}$

$\qquad = \frac{2}{1} \cdot \frac{x}{12x+2} = \frac{2x}{12x+2} = \frac{x}{6x+1} \neq x$

Since $(f \circ g)(x) \neq x$, the functions are not inverses. It is not necessary to check $(g \circ f)(x)$.

**49.** $f(x) = x^2 + 3$, domain $[0, \infty)$;

$g(x) = \sqrt{x-3}$, domain $[3, \infty)$

$(f \circ g)(x) = f\left(\sqrt{x-3}\right) = \left(\sqrt{x-3}\right)^2 + 3 = x$

$(g \circ f)(x) = g\left(x^2 + 3\right) = \sqrt{x^2 + 3 - 3}$

$\qquad = \sqrt{x^2} = |x| = x$ for $[0, \infty)$

Since $(f \circ g)(x) = x$ and $(g \circ f)(x) = x$, these functions are inverses.

**51.** Since each $y$-value corresponds to only one $x$-value, this function is one-to one and has an inverse. The inverse is: $\{(6, -3), (1, 2), (8, 5)\}$.

**53.** Since the $y$-value $-3$ corresponds to two different $x$-values, this function is not one-to-one.

**55.** $y = 3x - 4$

The function, $f(x) = 3x - 4$, is one-to-one.

(a) *Step 1:* Interchange $x$ and $y$: $x = 3y - 4$

*Step 2:* Solve for $y$.

$x = 3y - 4 \Rightarrow x + 4 = 3y \Rightarrow \frac{x+4}{3} = y \Rightarrow$

$y = \frac{x+4}{3} = \frac{1}{3}x + \frac{4}{3}$

*Step 3:* Replace $y$ with $f^{-1}(x)$.

$f^{-1}(x) = \frac{1}{3}x + \frac{4}{3}$

**(b)** The graph of the original function, $f(x) = 3x - 4$, is a line with slope 3 and y-intercept –4. Since $f^{-1}(x) = \frac{1}{3}x + \frac{4}{3}$, the graph of the inverse function is a line with slope $\frac{1}{3}$ and y-intercept $\frac{4}{3}$.

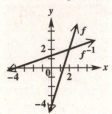

**(c)** For both $f(x)$ and $f^{-1}(x)$, the domain is $(-\infty, \infty)$ and the range is $(-\infty, \infty)$.

**57.** $f(x) = -4x + 3$

This function is one-to-one.

**(a)** *Step 1*: Replace $f(x)$ with $y$ and interchange $x$ and $y$.

$y = -4x + 3$

$x = -4y + 3$

*Step 2*: Solve for $y$.

$x = -4y + 3 \Rightarrow x - 3 = -4y \Rightarrow \frac{x-3}{-4} = y \Rightarrow$

$y = \frac{x-3}{-4} = -\frac{1}{4}x + \frac{3}{4}$

*Step 3*: Replace $y$ with $f^{-1}(x)$.

$f^{-1}(x) = -\frac{1}{4}x + \frac{3}{4}$

**(b)** The graph of the original function, $f(x) = -4x + 3$, is a line with slope –4 and y-intercept 3. Since $f^{-1}(x) = -\frac{1}{4}x + \frac{3}{4}$, the graph of the inverse function is a line with slope $-\frac{1}{4}$ and y-intercept $\frac{3}{4}$.

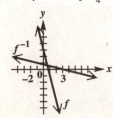

**(c)** For both $f(x)$ and $f^{-1}(x)$, the domain is $(-\infty, \infty)$ and the range is $(-\infty, \infty)$.

**59.** $f(x) = x^3 + 1$

This function is one-to-one.

**(a)** *Step 1*: Replace $f(x)$ with $y$ and interchange $x$ and $y$.

$y = x^3 + 1 \Rightarrow x = y^3 + 1$

*Step 2*: Solve for $y$.

$x = y^3 + 1 \Rightarrow x - 1 = y^3 \Rightarrow$

$\sqrt[3]{x-1} = y \Rightarrow y = \sqrt[3]{x-1}$

*Step 3*: Replace $y$ with $f^{-1}(x)$.

$f^{-1}(x) = \sqrt[3]{x-1}$

**(b)** Tables of ordered pairs will be helpful in drawing the graphs of these functions.

| $x$ | $f(x)$ | $x$ | $f^{-1}(x)$ |
|----|----|----|----|
| –2 | –7 | –7 | –2 |
| –1 | 0 | 0 | –1 |
| 0 | 1 | 1 | 0 |
| 1 | 2 | 2 | 1 |
| 2 | 9 | 9 | 2 |

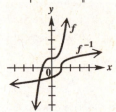

**(c)** For both $f(x)$ and $f^{-1}(x)$, the domain is $(-\infty, \infty)$ and the range is $(-\infty, \infty)$.

**61.** $y = x^2$

This is not a one-to-one function since two different x-values can correspond to the same y-value ($2^2 = 4$ and $(-2)^2 = 4$, for example), so this function is not one-to-one. Thus, the function has no inverse function.

**63.** $y = \frac{1}{x}$

The function, $f(x) = \frac{1}{x}$, is one-to-one.

**(a)** *Step 1*: Interchange $x$ and $y$.

$y = \frac{1}{x} \Rightarrow x = \frac{1}{y}$

*Step 2*: Solve for $y$.

$x = \frac{1}{y} \Rightarrow xy = 1 \Rightarrow y = \frac{1}{x}$

*Step 3*: Replace $y$ with $f^{-1}(x)$.

$f^{-1}(x) = \frac{1}{x}$

**(b)** Tables of ordered pairs will be helpful in drawing the graph of this function $\left(\text{in this case, } f(x) = f^{-1}(x)\right)$.

| $x$ | $f(x) = f^{-1}(x)$ |
|---|---|
| $-2$ | $-\frac{1}{2}$ |
| $-1$ | $-1$ |
| $-\frac{1}{2}$ | $-2$ |
| $\frac{1}{2}$ | $2$ |
| $1$ | $1$ |
| $2$ | $\frac{1}{2}$ |

**(c)** For both $f(x)$ and $f^{-1}(x)$, the domain and range are both $(-\infty, 0) \cup (0, \infty)$.

**65.** $f(x) = \frac{1}{x-3}$

This function is one-to-one.

**(a)** *Step 1*: Replace $f(x)$ with $y$ and interchange $x$ and $y$.

$y = \frac{1}{x-3} \Rightarrow x = \frac{1}{y-3}$

*Step 2*: Solve for $y$.

$x = \frac{1}{y-3} \Rightarrow x(y-3) = 1 \Rightarrow xy - 3x = 1 \Rightarrow$

$xy = 1 + 3x \Rightarrow y = \frac{1+3x}{x}$

*Step 3*: Replace $y$ with $f^{-1}(x)$.

$f^{-1}(x) = \frac{1+3x}{x}$

**(b)** To graph $f(x) = \frac{1}{x-3}$, we can determine that there are no $x$-intercepts. The $y$-intercept is $f(0) = \frac{1}{0-3} = -\frac{1}{3}$. There is a vertical aymptote when the denominator is zero, $x - 3 = 0$, which implies $x = 3$ is the vertical asymptote. Also, the horizontal asymptote is $y = 0$ since the degree of the numerator is less than the denominator. Examining the following intervals, we have test points which will be helpful in drawing the graph of $f$ as well as $f^{-1}$.

| Interval | Test Point | Value of $f(x)$ | Sign of $f(x)$ | Graph Above or Below $x$-Axis |
|---|---|---|---|---|
| $(-\infty, 3)$ | 2 | $-1$ | Negative | Below |
| $(3, \infty)$ | 4 | 1 | Positive | Above |

Plot the vertical asymptote, $y$-intercept, and test points with a smooth curve to get the graph of $f$.

To graph $f^{-1}(x) = \frac{1+3x}{x}$, we can determine that the $x$-intercept occurs when $1 + 3x = 0 \Rightarrow x = -\frac{1}{3}$. There is no $y$-intercept since 0 is not in the domain $f^{-1}$. There is a vertical aymptote when the denominator is zero, namely $x = 0$. Also, since the degree of the numerator is the same as the denominator, the horizontal asymptote is $y = \frac{3}{1} = 3$. Using this information along with the points $(-1, 2)$ and $(1, 4)$, we can sketch $f^{-1}$.

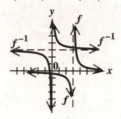

**(c)** Domain of $f$ = range of $f^{-1} = (-\infty, 3) \cup (3, \infty)$; Domain of $f^{-1}$ = range of $f = (-\infty, 0) \cup (0, \infty)$

**67.** $f(x) = \frac{x+1}{x-3}$

This function is one-to-one.

**(a)** *Step 1*: Replace $f(x)$ with $y$ and interchange $x$ and $y$.

$y = \frac{x+1}{x-3} \Rightarrow x = \frac{y+1}{y-3}$

*Step 2*: Solve for $y$.

$x = \frac{y+1}{y-3} \Rightarrow x(y-3) = y+1 \Rightarrow$

$xy - 3x = y + 1 \Rightarrow xy - y = 3x + 1 \Rightarrow$

$y(x-1) = 3x + 1 \Rightarrow y = \frac{3x+1}{x-1}$

*Step 3*: Replace $y$ with $f^{-1}(x)$.

$f^{-1}(x) = \frac{3x+1}{x-1}$

**(b)** To graph $f(x) = \frac{x+1}{x-3}$, find the

$x$-intercept:

$\frac{x+1}{x-3} = 0 \Rightarrow x + 1 = 0 \Rightarrow x = -1$. The

$y$-intercept is $f(0) = \frac{0+1}{0-3} = -\frac{1}{3}$. There is

a vertical aymptote when the denominator
is zero, $x - 3 = 0 \Rightarrow x = 3$ is the vertical
asymptote. Since the degree of the
numerator equals the degree of the
denominator, the horizontal asymptote is

$y = \frac{1}{1} = 1$. Examining the following

intervals, we have test points which will
be helpful in drawing the graph of $f$ as
well as $f^{-1}$.

| Interval | Test Point | Value of $f(x)$ | Sign of $f(x)$ | Graph Above or Below $x$-Axis |
|---|---|---|---|---|
| $\left(-\infty, -\frac{1}{3}\right)$ | $-3$ | $\frac{1}{3}$ | Positive | Above |
| $\left(-\frac{1}{3}, 1\right)$ | $\frac{1}{2}$ | $-\frac{3}{5}$ | Negative | Below |
| $(1, \infty)$ | $2$ | $-3$ | Negative | Below |

Plot the vertical asymptote, $y$-intercept,
and test points with a smooth curve to get
the graph of $f$.

To graph $f^{-1}(x) = \frac{3x+1}{x-1}$ we can

determine that the $x$-intercept occurs

when $3x + 1 = 0 \Rightarrow x = -\frac{1}{3}$. The

$y$-intercept is $f(0) = \frac{3(0)+1}{0-1} = -1$. There is

a vertical aymptote when the denominator
is zero, $x - 1 = 0 \Rightarrow x = 1$. Also, since the
degree of the numerator is the same as the
denominator, the horizontal asymptote is

$y = \frac{3}{1} = 3$. Examining the following

intervals, we have test points which will
be helpful in drawing the graph of $f$ as
well as $f^{-1}$.

| Interval | Test Point | Value of $f(x)$ | Sign of $f(x)$ | Graph Above or Below $x$-Axis |
|---|---|---|---|---|
| $\left(-\infty, -\frac{1}{3}\right)$ | $-2$ | $\frac{5}{3}$ | Positive | Above |
| $\left(-\frac{1}{3}, 1\right)$ | $\frac{1}{2}$ | $-5$ | Negative | Below |
| $(1, \infty)$ | $2$ | $7$ | Positive | Above |

Plot the vertical asymptote, $y$-intercept,
and test points with a smooth curve to get
the graph of $f^{-1}$.

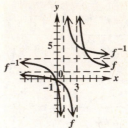

**(c)** Domain of $f$ = range of

$f^{-1} = (-\infty, 3) \cup (3, \infty)$;

Domain of $f^{-1}$ = range of

$f = (-\infty, 1) \cup (1, \infty)$

**69.** $f(x) = \sqrt{6 + x}$

This function is one-to-one.

**(a)** *Step 1*: Replace $f(x)$ with $y$ and
interchange $x$ and $y$.

$y = \sqrt{6 + x} \Rightarrow x = \sqrt{6 + y}$

*Step 2*: Solve for $y$. In this problem we must
consider that the range of $f$ will be the

domain of $f^{-1}$.

$x = \sqrt{6 + y}$

$x^2 = \left(\sqrt{6 + y}\right)^2$, for $x \geq 0$

$x^2 = 6 + y$, for $x \geq 0$

$x^2 - 6 = y$, for $x \geq 0$

*Step 3*: Replace $y$ with $f^{-1}(x)$.

$f^{-1}(x) = x^2 - 6$, for $x \geq 0$

**(b)** Tables of ordered pairs will be helpful in
drawing the graphs of these functions.

| $x$ | $f(x)$ | $x$ | $f^{-1}(x)$ |
|---|---|---|---|
| $-6$ | $0$ | $0$ | $-6$ |
| $-5$ | $1$ | $1$ | $-5$ |
| $-2$ | $2$ | $2$ | $-2$ |
| $3$ | $3$ | $3$ | $3$ |

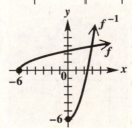

(c) Domain of $f$ = range of $f^{-1} = [-6, \infty)$;

Range of $f$ = domain of $f^{-1} = [0, \infty)$

71. Draw the mirror image of the original graph across the line $y = x$.

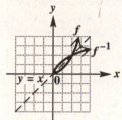

73. Carefully draw the mirror image of the original graph across the line $y = x$.

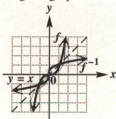

75. Draw the mirror image of the original graph across the line $y = x$.

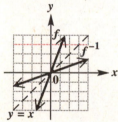

77. To find $f^{-1}(4)$, find the point with $y$-coordinate equal to 4. That point is $(4, 4)$. The graph of $f^{-1}$ contains $(4, 4)$. Hence $f^{-1}(4) = 4$.

79. To find $f^{-1}(0)$, find the point with $y$-coordinate equal to 0. That point is $(2, 0)$. The graph of $f^{-1}$ contains $(0, 2)$. Hence $f^{-1}(0) = 2$.

81. To find $f^{-1}(-3)$, find the point with $y$-coordinate equal to $-3$. That point is $(-2, -3)$. The graph of $f^{-1}$ contains $(-3, -2)$. Hence $f^{-1}(-3) = -2$.

83. $f^{-1}(1000)$ represents the number of dollars required to build 1000 cars.

85. If a line has slope $a$, the slope of its reflection in the line $y = x$ will be reciprocal of $a$, which is $\frac{1}{a}$.

87. The horizontal line test will show that this function is not one-to-one.

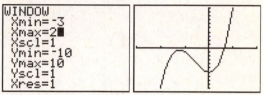

89. The horizontal line test will show that this function is one-to-one.

Find the equation of $f^{-1}$.

*Step 1*: Replace $f(x) = \frac{x-5}{x+3}$ with $y$ and interchange $x$ and $y$.

$$y = \frac{x-5}{x+3} \Rightarrow x = \frac{y-5}{y+3}$$

*Step 2*: Solve for $y$.

$$x = \frac{y-5}{y+3} \Rightarrow x(y+3) = y - 5 \Rightarrow$$
$$xy + 3x = y - 5 \Rightarrow xy - y = -5 - 3x \Rightarrow$$
$$y(x-1) = -5 - 3x \Rightarrow y = \frac{-5-3x}{x-1}$$

*Step 3*: Replace $y$ with $f^{-1}(x)$.

$$f^{-1}(x) = \frac{-5-3x}{x-1}$$

91. Given $f(x) = 3x - 2$, find $f^{-1}(x)$.

*Step 1*: Replace $f(x) = 3x - 2$ with $y$ and interchange $x$ and $y$.

$$y = 3x - 2 \Rightarrow x = 3y - 2$$

*Step 2*: Solve for $y$.

$$x = 3y - 2 \Rightarrow x + 2 = 3y \Rightarrow \frac{x+2}{3} = y$$

*Step 3*: Replace $y$ with $f^{-1}(x)$.

$$f^{-1}(x) = \frac{x+2}{3}$$

| | |
|---|---|
| 37; $f^{-1}(37) = \frac{37+2}{3}$ $= \frac{39}{3} = 13$; M | 25; $f^{-1}(25) = \frac{25+2}{3}$ $= \frac{27}{3} = 9$; I |
| 19; $f^{-1}(19) = \frac{19+2}{3}$ $= \frac{21}{3} = 7$; G | 61; $f^{-1}(61) = \frac{61+2}{3}$ $= \frac{63}{3} = 21$; U |
| 13; $f^{-1}(13) = \frac{13+2}{3}$ $= \frac{15}{3} = 5$; E | 34; $f^{-1}(34) = \frac{34+2}{3}$ $= \frac{36}{3} = 12$; L |
| 22; $f^{-1}(22) = \frac{22+2}{3}$ $= \frac{24}{3} = 8$; H | 1; $f^{-1}(1) = \frac{1+2}{3}$ $= \frac{3}{3} = 1$; A |
| 55; $f^{-1}(55) = \frac{55+2}{3}$ $= \frac{57}{3} = 19$; S | 1; $f^{-1}(1) = \frac{1+2}{3}$ $= \frac{3}{3} = 1$; A |
| 52; $f^{-1}(52) = \frac{52+2}{3}$ $= \frac{54}{3} = 18$; R | 52; $f^{-1}(52) = \frac{52+2}{3}$ $= \frac{54}{3} = 18$; R |
| 25; $f^{-1}(25) = \frac{25+2}{3}$ $= \frac{27}{3} = 9$; I | 64; $f^{-1}(64) = \frac{64+2}{3}$ $= \frac{66}{3} = 22$; V |
| 13; $f^{-1}(13) = \frac{13+2}{3}$ $= \frac{15}{3} = 5$; E | 10; $f^{-1}(10) = \frac{10+2}{3}$ $= \frac{12}{3} = 4$; D |

The message is MIGUEL HAS ARRIVED.

**93.** Given $f(x) = x^3 - 1$, we have the following.

| | |
|---|---|
| S = 19; $f(19) = 19^3 - 1$ $= 6859 - 1$ $= 6858$ | E = 5; $f(5) = 5^3$ $= 125 - 1 = 124$ |
| N = 14; $f(14) = 14^3 - 1$ $= 2744 - 1$ $= 2743$ | D = 4; $f(4) = 4^3 - 1$ $= 64 - 1 = 63$ |
| H = 8; $f(8) = 8^3 - 1$ $= 512 - 1 = 511$ | E = 5; $f(5) = 5^3 - 1$ $= 125 - 1 = 124$ |
| L = 12; $f(12) = 12^3 - 1$ $= 1728 - 1$ $= 1727$ | P = 16; $f(16) = 16^3 - 1$ $= 4096 - 1$ $= 4095$ |

Given $f(x) = x^3 - 1$, find $f^{-1}(x)$.

*Step 1:* Replace $f(x) = x^3 - 1$ with $y$ and interchange $x$ and $y$.
$$y = x^3 - 1 \Rightarrow x = y^3 - 1$$

*Step 2:* Solve for $y$.
$$x = y^3 - 1 \Rightarrow x + 1 = y^3 \Rightarrow \sqrt[3]{x+1} = y$$

*Step 3:* Replace $y$ with $f^{-1}(x)$.
$$f^{-1}(x) = \sqrt[3]{x+1}$$

**95.** Answers will vary. In the encoding/decoding process, you want to ensure that one value to be encoded will yield one encoded value (so the encoding relation itself must be a function). In order to ensure that one encoded value will yield only one decoded value, your decoding relation must also be a function. To ensure this, the original encoding function must be one-to one.

## Section 4.2: Exponential Functions
### Connections (page 427)

**1.** $e^1 \approx 1 + 1 + \frac{1^2}{2 \cdot 1} + \frac{1^3}{3 \cdot 2 \cdot 1} + \frac{1^4}{4 \cdot 3 \cdot 2 \cdot 1} + \frac{1^5}{5 \cdot 4 \cdot 3 \cdot 2 \cdot 1}$

$= 1 + 1 + \frac{1}{2} + \frac{1}{6} + \frac{1}{24} + \frac{1}{120} \approx 2.717$

```
e^(1)
        2.718281828
```

**2.** $e^{-.05} \approx 1 + (-.05) + \frac{(-.05)^2}{2 \cdot 1} + \frac{(-.05)^3}{3 \cdot 2 \cdot 1}$

$+ \frac{(-.05)^4}{4 \cdot 3 \cdot 2 \cdot 1} + \frac{(-.05)^5}{5 \cdot 4 \cdot 3 \cdot 2 \cdot 1}$

$= 1 - .05 + \frac{.0025}{2} - \frac{.000125}{6}$

$+ \frac{.00000625}{24} - \frac{.0000003125}{120}$

$\approx .9512$

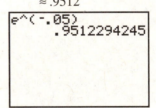

```
e^(-.05)
         .9512294245
```

**3.** $\dfrac{x^6}{6 \cdot 5 \cdot 4 \cdot 3 \cdot 2 \cdot 1}$

### Exercises

**1.** $f(x) = 3^x$
$f(2) = 3^2 = 9$

**3.** $f(x) = 3^x$
$f(-2) = 3^{-2} = \frac{1}{3^2} = \frac{1}{9}$

**5.** $g(x) = \left(\frac{1}{4}\right)^x$
$g(2) = \left(\frac{1}{4}\right)^2 = \frac{1}{16}$

**7.** $g(x) = \left(\frac{1}{4}\right)^x$
$g(-2) = \left(\frac{1}{4}\right)^{-2} = 4^2 = 16$

**9.**  $f(x) = 3^x$

$f\left(\tfrac{3}{2}\right) = 3^{3/2} \approx 5.196$

**11.**  $g(x) = \left(\tfrac{1}{4}\right)^x$

$g(2.34) = \left(\tfrac{1}{4}\right)^{2.34} \approx .039$

**13.**  The $y$-intercept of $f(x) = 3^x$ is 1, and the $x$-axis is a horizontal asymptote. Make a table of values.

| $x$ | $f(x)$ |
|---|---|
| $-2$ | $\tfrac{1}{9} \approx .1$ |
| $-1$ | $\tfrac{1}{3} \approx .3$ |
| $-\tfrac{1}{2}$ | $\approx .6$ |
| $0$ | $1$ |
| $\tfrac{1}{2}$ | $\approx 1.7$ |
| $1$ | $3$ |
| $2$ | $9$ |

Plot these points and draw a smooth curve through them. This is an increasing function. The domain is $(-\infty, \infty)$ and the range is $(0, \infty)$ and is one-to-one.

**15.**  The $y$-intercept of $f(x) = \left(\tfrac{1}{3}\right)^x$ is 1, and the $x$-axis is a horizontal asymptote. Make a table of values.

| $x$ | $f(x)$ |
|---|---|
| $-2$ | $9$ |
| $-1$ | $3$ |
| $-\tfrac{1}{2}$ | $\approx 1.7$ |
| $0$ | $1$ |
| $\tfrac{1}{2}$ | $\approx .6$ |
| $1$ | $\tfrac{1}{3} \approx .3$ |
| $2$ | $\tfrac{1}{9} \approx .1$ |

Plot these points and draw a smooth curve through them. This is a decreasing function. The domain is $(-\infty, \infty)$ and the range is $(0, \infty)$ and is one-to-one. Note: Since

$f(x) = \left(\tfrac{1}{3}\right)^x = \left(3^{-1}\right)^x = 3^{-x}$, the graph of

$f(x) = \left(\tfrac{1}{3}\right)^x$ is the reflection of the graph of

$f(x) = 3^x$ (Exercise 13) about the $y$-axis.

**17.**  The $y$-intercept of $f(x) = \left(\tfrac{3}{2}\right)^x$ is 1, and the $x$-axis is a horizontal asymptote. Make a table of values.

| $x$ | $f(x)$ |
|---|---|
| $-2$ | $\approx .4$ |
| $-1$ | $\approx .7$ |
| $-\tfrac{1}{2}$ | $\approx .8$ |
| $0$ | $1$ |
| $\tfrac{1}{2}$ | $\approx 1.2$ |
| $1$ | $1.5$ |
| $2$ | $2.25$ |

Plot these points and draw a smooth curve through them. This is an increasing function. The domain is $(-\infty, \infty)$ and the range is $(0, \infty)$ and is one-to-one.

**19.**  The $y$-intercept of $f(x) = 10^x$ is 1, and the $x$-axis is a horizontal asymptote. Make a table of values.

| $x$ | $f(x)$ |
|---|---|
| $-2$ | $.01$ |
| $-1$ | $.1$ |
| $-\tfrac{1}{2}$ | $\approx .3$ |
| $0$ | $1$ |
| $\tfrac{1}{2}$ | $\approx 3.2$ |
| $1$ | $10$ |
| $2$ | $100$ |

Plot these points and draw a smooth curve through them. This is an increasing function. The domain is $(-\infty, \infty)$ and the range is $(0, \infty)$ and is one-to-one.

**21.**  The $y$-intercept of $f(x) = 4^{-x}$ is 1, and the $x$-axis is a horizontal asymptote. Make a table of values.

| $x$ | $f(x)$ |
|---|---|
| $-2$ | $16$ |
| $-1$ | $4$ |
| $-\tfrac{1}{2}$ | $2$ |
| $0$ | $1$ |
| $\tfrac{1}{2}$ | $.5$ |
| $1$ | $.25$ |
| $2$ | $.0625$ |

*(continued on next page)*

*(continued from page 211)*

Plot these points and draw a smooth curve through them. This is a decreasing function. The domain is $(-\infty, \infty)$ and the range is $(0, \infty)$ and is one-to-one. Note: The graph of $f(x) = 4^{-x}$ is the reflection of the graph of $f(x) = 4^x$ (Exercise 14) about the $y$-axis.

**23.** The $y$-intercept of $f(x) = 2^{|x|}$ is 1, and the $x$-axis is a horizontal asymptote. Make a table of values.

| $x$ | $f(x)$ |
|---|---|
| $-2$ | 4 |
| $-1$ | 2 |
| $-\frac{1}{2}$ | $\approx 1.4$ |
| 0 | 1 |
| $\frac{1}{2}$ | $\approx 1.4$ |
| 1 | 2 |
| 2 | 4 |

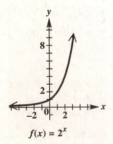

Plot these points and draw a smooth curve through them. The domain is $(-\infty, \infty)$ and the range is $[1, \infty)$ and is not one-to-one. Note: For $x < 0$, $|x| = -x$, so the graph is the same as that of $f(x) = 2^{-x}$. For $x \ge 0$, we have $|x| = x$, so the graph is the same as that of $f(x) = 2^x$. Since $|-x| = |x|$, the graph is symmetric with respect to the $y$-axis.

For Exercises 25–27, refer to the following graph of $f(x) = 2^x$.

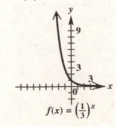

**25.** The graph of $f(x) = 2^x + 1$ is obtained by translating the graph of $f(x) = 2^x$ up 1 unit.

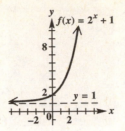

**27.** Since $f(x) = 2^{x+1} = 2^{x-(-1)}$, the graph is obtained by translating the graph of $f(x) = 2^x$ to the left 1 unit.

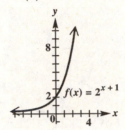

**29.** The graph of $f(x) = -2^{x+2}$ is obtained by translating the graph of $f(x) = 2^x$ to the left 2 units and then reflecting the graph across the $x$-axis.

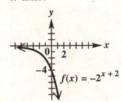

**31.** The graph of $f(x) = 2^{-x}$ is obtained by reflecting the graph across the $y$-axis.

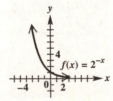

For Exercises 33–39, refer to the following graph of $f(x) = \left(\frac{1}{3}\right)^x$.

**33.** The graph of $f(x) = \left(\frac{1}{3}\right)^x - 2$ obtained by translating the graph of $f(x) = \left(\frac{1}{3}\right)^x$ down 2 units.

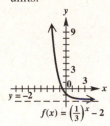

**35.** Since $f(x) = \left(\frac{1}{3}\right)^{x+2} = \left(\frac{1}{3}\right)^{x-(-2)}$, the graph is obtained by translating the graph of $f(x) = \left(\frac{1}{3}\right)^x$ 2 units to the left.

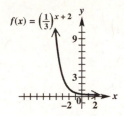

**37.** The graph of $f(x) = \left(\frac{1}{3}\right)^{-x+1}$ is obtained by translating the graph of $f(x) = \left(\frac{1}{3}\right)^x$ left one unit and then reflecting the resulting graph across the $y$-axis.

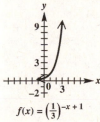

**39.** The graph of $f(x) = \left(\frac{1}{3}\right)^{-x}$ is obtained by reflecting the graph of $f(x) = \left(\frac{1}{3}\right)^x$ across the $y$-axis.

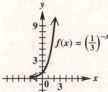

**41.** The graph of $f(x) = a^{-x}$ is the same as $g(x) = \left(\frac{1}{a}\right)^x$.

**43.** Since the horizontal asymptote is $y = -1$, the graph has been shifted down one unit. So the general form of the equation is $f(x) = a^{x+b} - 1$. The base is either 2 or 3, so try $a = 2$. Then substitute the coordinates of a point in the equation and solve for $b$:
$$1 = 2^{-2+b} - 1 \Rightarrow 2 = 2^{-2+b} \Rightarrow 2^1 = 2^{-2+b} \Rightarrow$$
$$1 = -2 + b \Rightarrow 3 = b$$
So, the equation is $f(x) = 2^{x+3} - 1$. Verify that the coordinates of other two points given satisfy the equation.

Alternate solution: Working backward and shifting the graph up one unit and right three units to transform the given graph into the graph of $y = 2^x$, it goes through the points $(3, 8)$, $(1, 2)$, and $(0, 1)$, which is the $y$-intercept. $8 = 2^3$, so $a = 2$, and the equation is $f(x) = 2^{x+3} - 1$. Verify by checking that the coordinates of the points satisfy the equation.

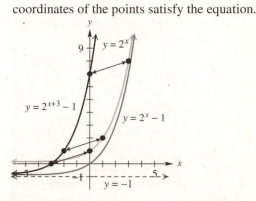

**45.** Since the horizontal asymptote is $y = 3$, the graph has been shifted up three units. The graph has also been reflected across the $x$-axis. So the general form of the equation is $f(x) = -a^{x+b} + 3$. The base is either 2 or 3, so try $a = 2$. Then substitute the coordinates of a point in the equation and solve for $b$:
$$-1 = -2^{0+b} + 3 \Rightarrow -4 = -2^b \Rightarrow 4 = 2^b \Rightarrow$$
$$2^2 = 2^b \Rightarrow 2 = b$$
So, the equation is $f(x) = -2^{x+2} + 3$. Verify that the coordinates of other two points given satisfy the equation.

Alternate solution: Working backward and shifting the graph down three units and right two units to transform the given graph into the graph of $y = -(2^x)$, it goes through the points $(0, -1)$, $(1, -2)$, and $(2, -4)$.

*(continued on next page)*

(*continued from page 213*)

The $y$-intercept is $(0, -1)$. $-2 = -\left(2^1\right)$, so $a = 2$, and the equation is $f(x) = 2^{x+1} + 3$.
Verify by checking that the coordinates of the points satisfy the equation.

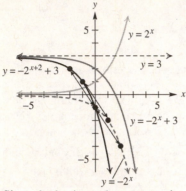

47. Since the horizontal asymptote is $y = 1$, the graph has been shifted up one unit. The graph has also been reflected across the $y$-axis. So the general form of the equation is
$f(x) = a^{-x+b} + 1$. The base is either 2 or 3, so try $a = 3$. Then substitute the coordinates of a point in the equation and solve for $b$:
$4 = 3^{-(-1)+b} + 1 \Rightarrow 3 = 3^{1+b} \Rightarrow 3^1 = 3^{1+b} \Rightarrow$
$1 = 1 + b \Rightarrow 0 = b$
So, the equation is $f(x) = 3^{-x} + 1$. Verify by checking that the coordinates of the other two points satisfy the equation.
Alternate solution: Working backward and shifting the graph down one unit to transform the given graph into the graph of $y = 3^{-x}$, it goes through the points $(-1, 3)$, $(0, 1)$, and $\left(1, \frac{1}{3}\right)$. The $y$-intercept is $(0, 1)$. $3 = 3^{-(-1)}$, so $a = 3$, and the equation is $f(x) = 3^{-x} + 1$.
Verify by checking that the coordinates of the points satisfy the equation.

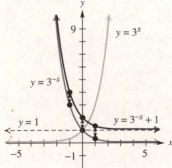

49. $4^x = 2 \Rightarrow \left(2^2\right)^x = 2^1 \Rightarrow 2^{2x} = 2^1 \Rightarrow$
$2x = 1 \Rightarrow x = \frac{1}{2}$
Solution set: $\left\{\frac{1}{2}\right\}$

51. $\left(\frac{1}{2}\right)^k = 4 \Rightarrow \left(2^{-1}\right)^k = 2^2 \Rightarrow 2^{-k} = 2^2$
$-k = 2 \Rightarrow k = -2$
Solution set: $\{-2\}$

53. $2^{3-y} = 8 \Rightarrow 2^{3-y} = 2^3 \Rightarrow 3 - y = 3 \Rightarrow$
$-y = 0 \Rightarrow y = 0$
Solution set: $\{0\}$

55. $e^{4x-1} = \left(e^2\right)^x \Rightarrow e^{4x-1} = e^{2x} \Rightarrow 4x - 1 = 2x \Rightarrow$
$-1 = -2x \Rightarrow \frac{1}{2} = x$
Solution set: $\left\{\frac{1}{2}\right\}$

57. $27^{4z} = 9^{z+1} \Rightarrow \left(3^3\right)^{4z} = \left(3^2\right)^{z+1} \Rightarrow$
$3^{3(4z)} = 3^{2(z+1)} \Rightarrow 3^{12z} = 3^{2z+2} \Rightarrow$
$12z = 2z + 2 \Rightarrow 10z = 2 \Rightarrow z = \frac{1}{5}$
Solution set: $\left\{\frac{1}{5}\right\}$

59. $4^{x-2} = 2^{3x+3} \Rightarrow \left(2^2\right)^{x-2} = 2^{3x+3} \Rightarrow$
$2^{2(x-2)} = 2^{3x+3} \Rightarrow 2^{2x-4} = 2^{3x+3} \Rightarrow$
$2x - 4 = 3x + 3 \Rightarrow -4 = x + 3 \Rightarrow -7 = x$
Solution set: $\{-7\}$

61. $\left(\frac{1}{e}\right)^{-x} = \left(\frac{1}{e^2}\right)^{x+1} \Rightarrow \left(e^{-1}\right)^{-x} = \left(e^{-2}\right)^{x+1} \Rightarrow$
$e^x = e^{-2(x+1)} \Rightarrow e^x = e^{-2x-2} \Rightarrow$
$x = -2x - 2 \Rightarrow 3x = -2 \Rightarrow x = -\frac{2}{3}$
Solution set: $\left\{-\frac{2}{3}\right\}$

63. $\left(\sqrt{2}\right)^{x+4} = 4^x \Rightarrow \left(2^{1/2}\right)^{x+4} = \left(2^2\right)^x \Rightarrow$
$2^{(1/2)(x+4)} = 2^{2x} \Rightarrow 2^{(1/2)x+2} = 2^{2x} \Rightarrow$
$\frac{1}{2}x + 2 = 2x \Rightarrow 2 = \frac{3}{2}x \Rightarrow \frac{2}{3} \cdot 2 = x \Rightarrow x = \frac{4}{3}$
Solution set: $\left\{\frac{4}{3}\right\}$

65. $\frac{1}{27} = b^{-3} \Rightarrow 3^{-3} = b^{-3} \Rightarrow b = 3$
Alternate solution:
$\frac{1}{27} = b^{-3} \Rightarrow \frac{1}{27} = \frac{1}{b^3} \Rightarrow 27 = b^3 \Rightarrow b = \sqrt[3]{27} = 3$
Solution set: $\{3\}$

**67.** $r^{2/3} = 4 \Rightarrow \left(r^{2/3}\right)^{3/2} = 4^{3/2} \Rightarrow r = \left(\pm\sqrt{4}\right)^3 \Rightarrow$

$r = \left(\pm 2\right)^3 \Rightarrow r = \pm 8$

Recall from Chapter 1 that it is necessary to check all proposed solutions in the original equation when you raise both sides to a power.

Check $r = -8$.

$\overline{4 = r^{2/3}}$

$4 \overset{?}{=} \left(-8\right)^{2/3}$

$4 = \left(\sqrt[3]{-8}\right)^2 \Rightarrow 4 = \left(-2\right)^2 \Rightarrow 4 = 4$

This is a true statement. $-8$ is a solution.

Check $r = 8$.

$\overline{4 = r^{2/3}}$

$4 \overset{?}{=} 8^{2/3}$

$4 = \left(\sqrt[3]{8}\right)^2 \Rightarrow 4 = 2^2 \Rightarrow 4 = 4$

This is a true statement. 8 is a solution.

Solution set: $\{-8, 8\}$

**69.** $x^{5/3} = -243 \Rightarrow \left(x^{5/3}\right)^{3/5} = \left(-243\right)^{3/5} \Rightarrow$

$x = \left(-3\right)^3 = -27$

Recall from Chapter 1 that it is necessary to check all proposed solutions in the original equation when you raise both sides to a power.

Check $z = -27$.

$\overline{x^{5/3} = -243}$

$\left(-27\right)^{5/3} \overset{?}{=} -243$

$\left(-3\right)^5 = -243$

This is a true statement. $-27$ is a solution.

Solution set: $\{-27\}$

**71. (a)** Use the compound interest formula to find the future value, $A = P\left(1 + \frac{r}{n}\right)^{tn}$, given $n = 2$, $P = 8906.54$, $r = .05$, and $t = 9$.

$A = P\left(1 + \frac{r}{n}\right)^{tn} = \left(8906.54\right)\left(1 + \frac{.05}{2}\right)^{9(2)}$

$= \left(8906.54\right)\left(1 + .025\right)^{18} \approx 13{,}891.16276$

Rounding to the nearest cent, the future value is \$13,891.16. The amount of interest would be

$\$13{,}891.16 - \$8906.54 = \$4984.62$.

**(b)** Use the continuous compounding interest formula to find the future value, $A = Pe^{rt}$, given $P = 8906.54$, $r = .05$, and $t = 9$.

$A = Pe^{rt} = 8906.54e^{.05(9)} = 8906.54e^{.45}$

$\approx 8906.54\left(1.568312\right) \approx 13{,}968.23521$

Rounding to the nearest cent, the future value is \$13,968.24. The amount of interest would be

$\$13{,}968.24 - \$8906.54 = \$5061.70$.

**73.** Use the compound interest formula to find the present amount, $A = P\left(1 + \frac{r}{n}\right)^{tn}$, given $n = 4$, $A = 25{,}000$, $r = .06$, and $t = \frac{11}{4}$.

$A = P\left(1 + \frac{r}{n}\right)^{tn}$

$25{,}000 = P\left(1 + \frac{.06}{4}\right)^{(11/4)(4)}$

$25{,}000 = P\left(1.015\right)^{11}$

$P = \dfrac{25{,}000}{\left(1.015\right)^{11}} \approx \$21{,}223.33083$

Rounding to the nearest cent, the present value is \$21,223.33.

**75.** Use the compound interest formula to find the present value, $A = P\left(1 + \frac{r}{n}\right)^{tn}$, given $n = 4$, $A = 5{,}000$, $r = .035$, and $t = 10$.

$A = P\left(1 + \frac{r}{n}\right)^{tn}$

$5{,}000 = P\left(1 + \frac{.035}{4}\right)^{10(4)}$

$5{,}000 = P\left(1.00875\right)^{40}$

$P = \dfrac{5{,}000}{\left(1.00875\right)^{40}} \approx \$3528.808535$

Rounding to the nearest cent, the present value is \$3528.81.

**77.** Use the compound interest formula to find the interest rate, $A = P\left(1 + \frac{r}{n}\right)^{tn}$, given $n = 4$, $A = 1500$, $P = 1200$, and $t = 5$.

$A = P\left(1 + \frac{r}{n}\right)^{tn}$

$1500 = 1200\left(1 + \frac{r}{4}\right)^{5(4)}$

$1500 = 1200\left(1 + \frac{r}{4}\right)^{20}$

$1.25 = \left(1 + \frac{r}{4}\right)^{20}$

$\left(1.25\right)^{\frac{1}{20}} = 1 + \frac{r}{4} \Rightarrow \left(1.25\right)^{\frac{1}{20}} - 1 = \frac{r}{4} \Rightarrow$

$4\left[\left(1.25\right)^{\frac{1}{20}} - 1\right] = r \Rightarrow r \approx .044878604$

The interest rate, to the nearest tenth, is 4.5%.

**79.** For each bank we need to calculate $\left(1+\frac{r}{n}\right)^{n}$.

Since the base, $1+\frac{r}{n}$, is greater than 1, we need only compare the three values calculated to determine which bank will yield the least amount of interest. It is understood that the amount of time, $t$, and the principal, $P$, are the same for all three banks.

Bank A: Calculate $\left(1+\frac{r}{n}\right)^{n}$ where $n=1$ and $r=.064$.

$\left(1+\frac{.064}{1}\right)^{1}=\left(1+.064\right)^{1}=\left(1.064\right)^{1}=1.064$

Bank B: Calculate $\left(1+\frac{r}{n}\right)^{n}$ where $n=12$ and $r=.063$.

$\left(1+\frac{.063}{12}\right)^{12}=\left(1+.00525\right)^{12}=\left(1.00525\right)^{12}$
$\approx 1.064851339$

Bank C: Calculate $\left(1+\frac{r}{n}\right)^{n}$ where $n=4$ and $r=.0635$.

$\left(1+\frac{.0635}{4}\right)^{4}=\left(1+.015875\right)^{4}=\left(1.015875\right)^{4}$
$\approx 1.06502816$

Bank A will charge you the least amount of interest, even though it has the highest stated rate.

**81. (a)**

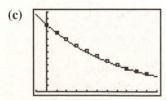

**(b)** From the graph above, we can see that the data are not linear but exponentially decreasing.

**(c)**

**(d)**    $P(x)=1013e^{-.0001341x}$
$P(1500)=1013e^{-.0001341(1500)}$
$\approx 1013\left(.817790\right)\approx 828$
$P(11,000)=1013e^{-.0001341(11,000)}$
$\approx 1013\left(.228756\right)\approx 232$

When the altitude is 1500 m, the function $P$ gives a pressure of 828 mb, which is less than the actual value of 846 mb. When the altitude is 11,000 m, the function $P$ gives a pressure of 232 mb, which is more than the actual value of 227 mb.

**83. (a)** Evaluate $T=50,000\left(1+.06\right)^{n}$ where $n=4$.

$T=50,000\left(1+.06\right)^{4}$
$=50,000\left(1.06\right)^{4}\approx 63,123.848$

Total population after 4 years is about 63,000.

**(b)** Evaluate $T=30,000\left(1+.12\right)^{n}$ where $n=3$.

$T=30,000\left(1+.12\right)^{3}$
$=30,000\left(1.12\right)^{3}=42,147.84$

There would be about 42,000 deer after 3 years.

**(c)** Evaluate $T=45,000(1+.08)^{n}$ where $n=5$.

$T=45,000\left(1+.08\right)^{5}$
$=45,000\left(1.08\right)^{5}\approx 66,119.76346$

There would be about 66,000 deer after 5 years. Thus, we can expect about $66,000-45,000=21,000$ additional deer after 5 years.

**85.** $5e^{3x}=75$

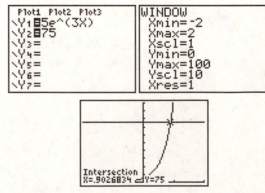

Since $5e^{3x}=75\Rightarrow e^{3x}=15$ we could also do the following.

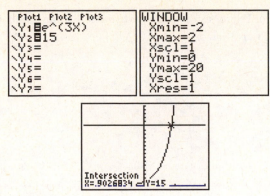

Solution set: $\{.9\}$

**87.** $3x + 2 = 4^x$

We can see from the following screens that there are two solutions.

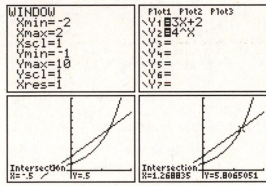

Solution set: $\{-.5, 1.3\}$

**89.** Answers will vary.

**91.** If the graph of the exponential function $f(x) = a^x$ contains the point $(3, 8)$, we have $a^3 = 8$. This implies $a = \sqrt[3]{8} = 2$. Thus, the equation which satisfies the given condition is $f(x) = 2^x$.

**93.** If the graph of the exponential function $f(x) = a^x$ contains the point $(-3, 64)$, we have $a^{-3} = 64$. This implies $a^3 = \frac{1}{64} \Rightarrow a = \sqrt[3]{\frac{1}{64}} = \frac{1}{4}$. Thus, the equation which satisfies the given condition is $f(x) = \left(\frac{1}{4}\right)^x$.

**95.** $f(t) = 3^{2t+3} = 3^{2t} \cdot 3^3 = 27 \cdot \left(3^2\right)^t = 27 \cdot 9^t$

**97.** $f(t) = \left(\frac{1}{3}\right)^{1-2t} = \left(\frac{1}{3}\right)^1 \left(\frac{1}{3}\right)^{-2t}$
$= \left(\frac{1}{3}\right)(3)^{2t} = \left(\frac{1}{3}\right)\left(3^2\right)^t = \left(\frac{1}{3}\right)9^t$

**99.** Answers will vary.

**101.** Yes; $f(x) = a^x$ is a one-to-one function. Therefore, an inverse function exists for $f$.

**103.** Since $f(x) = a^x$ has an inverse, we find it as follows: $y = a^x \Rightarrow x = a^y$

**105.** If $a = e$, the equation for $f^{-1}(x)$ will be given by $x = e^y$.

## Section 4.3: Logarithmic Functions

### Connections (page 441)

**1.**  $\log_{10} 458.3 \approx 2.661149857$
$+ \log_{10} 294.6 \approx 2.469232743$
$\qquad\qquad\qquad \approx 5.130382600$
$10^{5.130382600} \approx 135,015.18$
A calculator gives
$(458.3)(294.6) = 135,015.18$.

**2.** Answers will vary.

### Exercises

**1.** **(a)**  C; $\log_2 16 = 4$ because $2^4 = 16$.

 **(b)**  A; $\log_3 1 = 0$ because $3^0 = 1$.

 **(c)**  E; $\log_{10} .1 = -1$ because $10^{-1} = .1$.

 **(d)**  B; $\log_2 \sqrt{2} = \frac{1}{2}$ because $2^{1/2} = \sqrt{2}$.

 **(e)**  F; $\log_e\left(\frac{1}{e^2}\right) = -2$ because $e^{-2} = \frac{1}{e^2}$.

 **(f)**  D; $\log_{1/2} 8 = -3$ because $\left(\frac{1}{2}\right)^{-3} = 8$.

**3.** $3^4 = 81$ is equivalent to $\log_3 81 = 4$.

**5.** $\left(\frac{2}{3}\right)^{-3} = \frac{27}{8}$ is equivalent to $\log_{2/3} \frac{27}{8} = -3$.

**7.** $\log_6 36 = 2$ is equivalent to $6^2 = 36$.

**9.** $\log_{\sqrt{3}} 81 = 8$ is equivalent to $\left(\sqrt{3}\right)^8 = 81$.

**11.** Answers will vary.

**13.** $x = \log_5 \frac{1}{625} \Rightarrow 5^x = \frac{1}{625} \Rightarrow 5^x = \frac{1}{5^4} \Rightarrow$
$5^x = 5^{-4} \Rightarrow x = -4$
Solution set: $\{-4\}$

**15.** $\log_x \frac{1}{32} = 5 \Rightarrow x^5 = \frac{1}{32} \Rightarrow x^5 = \frac{1}{2^5} = \left(\frac{1}{2}\right)^5 \Rightarrow$

$\quad x = \frac{1}{2}$

Solution set: $\left\{\frac{1}{2}\right\}$

**17.** $x = \log_8 \sqrt[4]{8} \Rightarrow 8^x = \sqrt[4]{8} \Rightarrow 8^x = 8^{1/4} \Rightarrow x = \frac{1}{4}$

Solution set: $\left\{\frac{1}{4}\right\}$

**19.** $x = 3^{\log_3 8}$

Writing as a logarithmic equation, we have
$\log_3 8 = \log_3 x \Rightarrow x = 8$
Using the Theorem of Inverses on page 440,
we can directly state that $x = 8$.
Solution set: $\{8\}$

**21.** $x = 2^{\log_2 9}$

Writing as a logarithmic equation, we have
$\log_2 9 = \log_2 x \Rightarrow x = 9$
Using the Theorem of Inverses on page 440,
we can directly state that $x = 9$.
Solution set: $\{9\}$

**23.** $\quad \log_x 25 = -2 \Rightarrow x^{-2} = 25 \Rightarrow x^{-2} = 5^2 \Rightarrow$

$\left(x^{-2}\right)^{-1/2} = \left(5^2\right)^{-1/2} \Rightarrow x = 5^{-1}$

Do not include a $\pm$ since the base,
$x$, cannot be negative.

$x = \frac{1}{5}$

Solution set: $\left\{\frac{1}{5}\right\}$

**25.** $\log_4 x = 3 \Rightarrow 4^3 = x \Rightarrow 64 = x$

Solution set: $\{64\}$

**27.** $\quad x = \log_4 \sqrt[3]{16} \Rightarrow 4^x = \sqrt[3]{16} \Rightarrow 4^x = (16)^{1/3} \Rightarrow$

$4^x = \left(4^2\right)^{1/3} \Rightarrow 4^x = 4^{2/3} \Rightarrow x = \frac{2}{3}$

Solution set: $\left\{\frac{2}{3}\right\}$

**29.** $\log_9 x = \frac{5}{2} \Rightarrow 9^{5/2} = x \Rightarrow (3)^5 = 243 = x$

Note that we do not include $\sqrt{9} = -3$ because
logarithms are not defined for negative
numbers.
Solution set: $\{243\}$

**31.** Answers will vary.

For Exercises 33–35, refer to the following graph of
$f(x) = \log_2 x$.

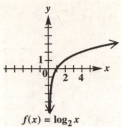

$f(x) = \log_2 x$

**33.** The graph of $f(x) = \left(\log_2 x\right) + 3$ is obtained
by translating the graph of $f(x) = \log_2 x$ up 3
units.

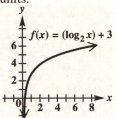

Domain: $(0, \infty)$; range: $(-\infty, \infty)$

**35.** To find the graph of $f(x) = \left|\log_2(x+3)\right|$,
translate the graph of $f(x) = \log_2 x$ to the left
3 units to obtain the graph of $\log_2(x+3)$. (See
Exercise 34.) For the portion of the graph
where $f(x) \geq 0$, that is, where $x \geq -2$, use the
same graph as in 34. For the portion of the
graph in 34 where $f(x) < 0$, $-3 < x < -2$,
reflect the graph about the $x$-axis. In this way,
each negative value of $f(x)$ on the graph in
34 is replaced by its opposite, which is
positive. The graph has a vertical asymptote at
$x = -3$.

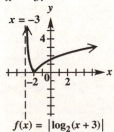

$f(x) = \left|\log_2(x+3)\right|$

Domain: $(-3, \infty)$; range: $[0, \infty)$

For Exercise 37, refer to the following graph of
$f(x) = \log_{1/2} x$.

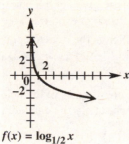

$f(x) = \log_{1/2} x$

**37.** The graph of $f(x) = \log_{1/2}(x - 2)$ is obtained by translating the graph of $f(x) = \log_{1/2} x$ to the right 2 units. The graph has a vertical asymptote at $x = 2$.

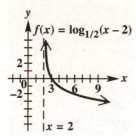

Domain: $(2, \infty)$; range: $(-\infty, \infty)$

**39.** Because $f(x) = \log_2 x$ has a vertical asymptote, which is the $y$-axis (the line $x = 0$), $x$-intercept of 1, and is increasing, the correct choice is the graph in E.

**41.** Because $f(x) = \log_2 \frac{1}{x} = \log_2 x^{-1} = -\log_2 x$, it has a vertical asymptote, which is the $y$-axis (the line $x = 0$), has an $x$-intercept 1, and is the reflection of $f(x) = \log_2 x$ across the $x$-axis, it is decreasing and the correct choice is the graph in B

**43.** Because $f(x) = \log_2(x - 1)$ represents the horizontal shift of $f(x) = \log_2 x$ to the right 1 unit, the function has a vertical asymptote which is the line $x = 1$, has an $x$-intercept when $x - 1 = 1 \Rightarrow x = 2$, and is increasing, the correct choice is the graph in F.

**45.** $f(x) = \log_5 x$

Since $f(x) = y = \log_5 x$, we can write the exponential form as $x = 5^y$ to find ordered pairs that satisfy the equation. It is easier to choose values for $y$ and find the corresponding values of $x$. Make a table of values.

| $x$ | $y = \log_5 x$ | |
|---|---|---|
| $\frac{1}{25} = .04$ | $-2$ | |
| $\frac{1}{5} = .2$ | $-1$ | |
| $1$ | $0$ | |
| $5$ | $1$ | |
| $25$ | $2$ | |

The graph can also be found by reflecting the graph of $f(x) = 5^x$ about the line $y = x$. The graph has the $y$-axis as a vertical asymptote.

**47.** $f(x) = \log_{1/2}(1 - x)$

Since $f(x) = y = \log_{1/2}(1 - x)$, we can write the exponential form as
$1 - x = \left(\frac{1}{2}\right)^y \Rightarrow x = 1 - \left(\frac{1}{2}\right)^y$ to find ordered pairs that satisfy the equation. It is easier to choose values for $y$ and find the corresponding values of $x$. Make a table of values.

| $x$ | $y = \log_{1/2}(1 - x)$ | |
|---|---|---|
| $-3$ | $-2$ | |
| $-1$ | $-1$ | |
| $0$ | $0$ | |
| $\frac{1}{2} = .5$ | $1$ | |
| $\frac{3}{4} = .75$ | $2$ | |

The graph has the line $x = 1$ as a vertical asymptote.

**49.** $f(x) = \log_3(x - 1)$

Since $f(x) = y = \log_3(x - 1)$, we can write the exponential form as
$x - 1 = 3^y \Rightarrow x = 3^y + 1$ to find ordered pairs that satisfy the equation. It is easier to choose values for $y$ and find the corresponding values of $x$. Make a table of values.

| $x$ | $y = \log_3(x - 1)$ | |
|---|---|---|
| $\frac{10}{9} \approx 1.1$ | $-2$ | |
| $\frac{4}{3} \approx 1.3$ | $-1$ | |
| $2$ | $0$ | |
| $4$ | $1$ | |
| $10$ | $2$ | |

The vertical asymptote will be $x = 1$.

**51.** Since the vertical asymptote is $x = -1$, the graph has been shifted left one unit. So the general form of the equation is
$f(x) = \log_a(x+1) + k$. The base is either 2 or 3, so try $a = 2$. Then substitute the coordinates of a point in the equation and solve for $k$:
$-2 = \log_2(1+1) + k \Rightarrow -2 - k = \log_2 2 \Rightarrow$
$2^{-2-k} = 2 \Rightarrow 2^{-2-k} = 2^1 \Rightarrow$
$-2 - k = 1 \Rightarrow -3 = k$
So, the equation is $f(x) = \log_2(x+1) - 3$.
Verify that the coordinates of other two points given satisfy the equation.
Alternate solution: Working backward and shifting the graph up three units and right one unit to transform the given graph into the graph of $y = \log_2 x$, it goes through the points $(1, 0)$, which is the $x$-intercept, $(2, 1)$, and $(8, 3)$. $3 = \log_2 8$, so $a = 2$, and the equation is $f(x) = \log_2(x+1) - 3$. Verify by checking that the coordinates of the points shown on the graph satisfy the equation.

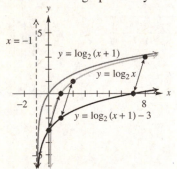

**53.** The graph has been reflected across the $y$-axis, so the general form of the equation is
$f(x) = \log_a(-x-b) + k$. Since the vertical asymptote of the graph is $x = 3$, the graph has been shifted right three units, and $b = -3$. So the general form of the equation is
$f(x) = \log_a(-x+3) + k$. The base is either 2 or 3, so try $a = 2$. Then substitute the coordinates of a point in the equation and solve for $k$:
$-1 = \log_2(-1+3) + k \Rightarrow -1 - k = \log_2 2 \Rightarrow$
$2^{-1-k} = 2 \Rightarrow 2^{-1-k} = 2^1 \Rightarrow$
$-1 - k = 1 \Rightarrow -2 = k$
So, the equation is $f(x) = \log_2(-x+3) - 2$.
Verify that the coordinates of other two points given satisfy the equation.

Alternate solution: Working backward and shifting the graph up two units and left three units to transform the given graph into the graph of $y = \log_2 x$, it goes through the points $(-1, 0)$, which is the $x$-intercept, $(-2, 1)$, and $(-4, 2)$. $2 = \log_2[-(-4)]$, so $a = 2$, and the equation is $f(x) = \log_2(-x+3) - 2$. Verify by checking that the coordinates of the points shown on the graph satisfy the equation.

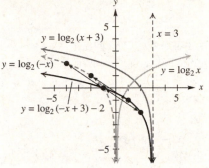

**55.** The graph has been reflected across the $x$-axis, so the general form of the equation is
$f(x) = -\log_a(x-b) + k$. Since the vertical asymptote of the original graph is $x = 1$, the graph has been shifted right one unit and $b = 1$. So the general form of the equation is
$f(x) = -\log_a(x-1) + k$. The base is either 2 or 3, so try $a = 3$. Then substitute the coordinates of a point in the equation and solve for $k$:
$-1 = -\log_3(4-1) + k \Rightarrow$
$-(-1-k) = \log_3 3 \Rightarrow 3^{1+k} = 3 \Rightarrow$
$3^{1+k} = 3^1 \Rightarrow 1 + k = 1 \Rightarrow k = 0$
So, the equation is $f(x) = -\log_3(x-1)$.
Verify that the coordinates of other two points given satisfy the equation.
Alternate solution: Working backward and shifting the graph left one unit, it goes through the points $(1, 0)$, which is the $x$-intercept, $\left(\frac{1}{3}, 1\right)$, and $(3, -1)$. $-1 = -\log_3 3$, so $a = 3$, and the equation is $f(x) = -\log_3(x-1)$. Verify by checking that the coordinates of the points shown on the graph satisfy the equation.

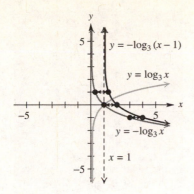

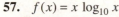

$y = -\log_3(x-1)$

$y = \log_3 x$

$y = -\log_3 x$

$x = 1$

**57.**  $f(x) = x \log_{10} x$

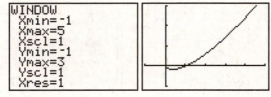

**59.**  $\log_2 \frac{6x}{y} = \log_2 6x - \log_2 y$
$$= \log_2 6 + \log_2 x - \log_2 y$$

**61.**  $\log_5 \frac{5\sqrt{7}}{3} = \log_5 5\sqrt{7} - \log_5 3$
$$= \log_5 5 + \log_5 \sqrt{7} - \log_5 3$$
$$= 1 + \log_5 7^{1/2} - \log_5 3$$
$$= 1 + \tfrac{1}{2}\log_5 7 - \log_5 3$$

**63.**  $\log_4(2x + 5y)$

Since this is a sum, none of the logarithm properties apply, so this expression cannot be simplified.

**65.**  $\log_m \sqrt{\frac{5r^3}{z^5}} = \log_m \left(\frac{5r^3}{z^5}\right)^{1/2} = \tfrac{1}{2}\log_m \frac{5r^3}{z^5}$
$$= \tfrac{1}{2}\left(\log_m 5r^3 - \log_m z^5\right)$$
$$= \tfrac{1}{2}\left(\log_m 5 + \log_m r^3 - \log_m z^5\right)$$
$$= \tfrac{1}{2}\left(\log_m 5 + 3\log_m r - 5\log_m z\right)$$

**67.**  $\log_2 \frac{ab}{cd} = \log_2(ab) - \log_2(cd)$
$$= \log_2 a + \log_2 b - \left(\log_2 c + \log_2 d\right)$$
$$= \log_2 a + \log_2 b - \log_2 c - \log_2 d$$

**69.**  $\log_3 \frac{\sqrt{x}\sqrt[3]{y}}{w^2\sqrt{z}}$
$$= \log_3\left(x^{1/2}y^{1/3}\right) - \log_3\left(w^2 z^{1/2}\right)$$
$$= \log_3 x^{1/2} + \log_3 y^{1/3}$$
$$\qquad - \left(\log_3 w^2 + \log_3 z^{1/2}\right)$$
$$= \tfrac{1}{2}\log_3 x + \tfrac{1}{3}\log_3 y - \left(2\log_3 w + \tfrac{1}{2}\log_3 z\right)$$
$$= \tfrac{1}{2}\log_3 x + \tfrac{1}{3}\log_3 y - 2\log_3 w - \tfrac{1}{2}\log_3 z$$

**71.**  $\log_a x + \log_a y - \log_a m$
$$= \log_a xy - \log_a m = \log_a \tfrac{xy}{m}$$

**73.**  $\log_a m - \log_a n - \log_a t$
$$= \log_a m - \left(\log_a n + \log_a t\right)$$
$$= \log_a m - \log_a(nt) = \log_a \tfrac{m}{nt}$$

**75.**  $2\log_m a - 3\log_m b^2 = \log_m a^2 - \log_m \left(b^2\right)^3$
$$= \log_m a^2 - \log_m b^6$$
$$= \log_m \tfrac{a^2}{b^6}$$

**77.**  $2\log_a(z+1) + \log_a(3z+2)$
$$= \log_a(z+1)^2 + \log_a(3z+2)$$
$$= \log_a\left[(z+1)^2(3z+2)\right]$$

**79.**  $-\tfrac{2}{3}\log_5 5m^2 + \tfrac{1}{2}\log_5 25m^2$
$$= \log_5\left(5m^2\right)^{-2/3} + \log_5\left(25m^2\right)^{1/2}$$
$$= \log_5\left[\left(5m^2\right)^{-2/3}\cdot\left(25m^2\right)^{1/2}\right]$$
$$= \log_5\left(5^{-2/3}m^{-4/3}\cdot 5m\right)$$
$$= \log_5\left(5^{-2/3}\cdot 5^1\cdot m^{-4/3}\cdot m^1\right)$$
$$= \log_5\left(5^{1/3}\cdot m^{-1/3}\right) = \log_5 \tfrac{5^{1/3}}{m^{1/3}} = \log_5 \sqrt[3]{\tfrac{5}{m}}$$

**81.**  $\log_{10} 6 = \log_{10}(2\cdot 3) = \log_{10} 2 + \log_{10} 3$
$$= .3010 + .4771 = .7781$$

**83.**  $\log_{10} \tfrac{3}{2} = \log_{10} 3 - \log_{10} 2$
$$= .4771 - .3010 = .1761$$

**85.**  $\log_{10} \tfrac{9}{4} = \log_{10} 9 - \log_{10} 4 = \log_{10} 3^2 - \log_{10} 2^2$
$$= 2\log_{10} 3 - 2\log_{10} 2$$
$$= 2(.4771) - 2(.3010) = .9542 - .6020$$
$$= .3522$$

**87.**  $\log_{10} \sqrt{30} = \log_{10} 30^{1/2} = \tfrac{1}{2}\log_{10} 30$
$$= \tfrac{1}{2}\log_{10}(10\cdot 3)$$
$$= \tfrac{1}{2}\left(\log_{10} 10 + \log_{10} 3\right) = \tfrac{1}{2}\left(1 + .4771\right)$$
$$= \tfrac{1}{2}(1.4771) \approx .7386$$

**89. (a)** If the $x$-values are representing years, 3 months is $\frac{3}{12} = \frac{1}{4} = .25$ yr and 6 months is $\frac{6}{12} = \frac{1}{2} = .5$ yr.

| L1 | L2 | L3 | 3 |
|----|----|----|---|
| .25 | .83 | ------ | |
| .5 | .91 | | |
| 2 | 1.35 | | |
| 5 | 2.46 | | |
| 10 | 3.54 | | |
| 30 | 4.58 | | |
| ------ | ------ | | |

L3 =

```
WINDOW
Xmin=0
Xmax=35
Xscl=5
Ymin=0
Ymax=5
Yscl=1
Xres=1
```

**(b)** A logarithmic function will model the data best.

**91.** $f(x) = \log_a x$ and $f(3) = 2$

$2 = \log_a 3 \Rightarrow a^2 = 3 \Rightarrow \left(a^2\right)^{1/2} = 3^{1/2} \Rightarrow a = \sqrt{3}$

(There is no $\pm$ because $a$ must be positive and not equal to 1.) We now have $f(x) = \log_{\sqrt{3}} x$.

**(a)** $f\left(\frac{1}{9}\right) = \log_{\sqrt{3}} \frac{1}{9} \Rightarrow y = \log_{\sqrt{3}} \frac{1}{9} \Rightarrow$

$\left(\sqrt{3}\right)^y = \frac{1}{9} \Rightarrow \left(3^{1/2}\right)^y = \frac{1}{3^2} \Rightarrow$

$3^{y/2} = 3^{-2} \Rightarrow \frac{y}{2} = -2 \Rightarrow y = -4$

**(b)** $f(27) = \log_{\sqrt{3}} 27 \Rightarrow y = \log_{\sqrt{3}} 27 \Rightarrow$

$\left(\sqrt{3}\right)^y = 27 \Rightarrow \left(3^{1/2}\right)^y = 3^3 \Rightarrow$

$3^{y/2} = 3^3 \Rightarrow \frac{y}{2} = 3 \Rightarrow y = 6$

**(c)** $f(9) = \log_{\sqrt{3}} 9 \Rightarrow y = \log_{\sqrt{3}} 9 \Rightarrow$

$\left(\sqrt{3}\right)^y = 9 \Rightarrow \left(3^{1/2}\right)^y = 3^2 \Rightarrow$

$3^{y/2} = 3^2 \Rightarrow \frac{y}{2} = 2 \Rightarrow y = 4$

**(d)** $f\left(\frac{\sqrt{3}}{3}\right) = \log_{\sqrt{3}}\left(\frac{\sqrt{3}}{3}\right) \Rightarrow y = \log_{\sqrt{3}}\left(\frac{\sqrt{3}}{3}\right) \Rightarrow$

$\left(\sqrt{3}\right)^y = \frac{\sqrt{3}}{3} \Rightarrow \left(3^{1/2}\right)^y = \frac{3^{1/2}}{3^1} \Rightarrow$

$\left(3^{1/2}\right)^y = 3^{1/2-1} = 3^{-1/2} \Rightarrow$

$3^{y/2} = 3^{-1/2} \Rightarrow \frac{y}{2} = -\frac{1}{2} \Rightarrow y = -1$

**93.** In the formula $A = P\left(1+\frac{r}{n}\right)^{tn}$ we substitute $A = 2P$ since we want the present value to be doubled in the future. Thus, we need to solve for $t$ in the equation $2P = P\left(1+\frac{r}{n}\right)^{tn}$.

$2P = P\left(1+\frac{r}{n}\right)^{tn} \Rightarrow 2 = \left(1+\frac{r}{n}\right)^{tn}$

$2^{\frac{1}{tn}} = \left[\left(1+\frac{r}{n}\right)^{tn}\right]^{\frac{1}{tn}} \Rightarrow 2^{\frac{1}{tn}} = \left(1+\frac{r}{n}\right)$

$\log_2\left(1+\frac{r}{n}\right) = \frac{1}{tn} \Rightarrow tn = \frac{1}{\log_2\left(1+\frac{r}{n}\right)}$

$t = \frac{1}{n\log_2\left(1+\frac{r}{n}\right)} \Rightarrow t = \frac{1}{\log_2\left(1+\frac{r}{n}\right)^n}$

**95.** $\log_{10} x = x - 2$

The $x$-coordinates of the intersection points will be the solutions to the given equation. There are two points of intersection, hence there are two possible solutions.

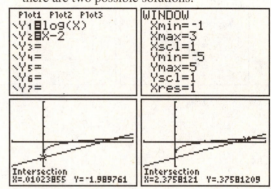

Solution set: $\{.01, 2.38\}$

**97.** Prove that $\log_a \frac{x}{y} = \log_a x - \log_a y$.

Let $m = \log_a x$ and $n = \log_a y$. Changing to exponential form we have $a^m = x$ and $a^n = y$. Since $\frac{x}{y} = \frac{a^m}{a^n}$ we have $\frac{x}{y} = a^{m-n}$. Changing to logarithmic form, we have $\log_a \frac{x}{y} = m - n$. Substituting for $m$ and $n$ we have $\log_a \frac{x}{y} = \log_a x - \log_a y$.

## Summary Exercises on Inverse, Exponential, and Logarithmic Functions

**1.** $f(x) = 3x - 4$, $g(x) = \frac{1}{3}x + \frac{4}{3}$

$(f \circ g)(x) = f[g(x)] = 3\left(\frac{1}{3}x + \frac{4}{3}\right) - 4$ Since
$\qquad = x + 4 - 4 = x$

$(g \circ f)(x) = g[f(x)] = \frac{1}{3}(3x - 4) + \frac{4}{3}$
$\qquad = \frac{3x}{3} = x$

$(f \circ g)(x) = x$ and $(g \circ f)(x) = x$, these functions are inverses.

**3.** $f(x) = 1 + \log_2 x, \ g(x) = 2^{x-1}$

$$(f \circ g)(x) = f[g(x)] = f(2^{x-1}) = 1 + \log_2 2^{x-1}$$
$$= 1 + (x-1)\log_2 2 = 1 + x - 1 = x$$
$$(g \circ f)(x) = g[f(x)] = g(1 + \log_2 x)$$
$$= 2^{(1+\log_2 x - 1)} = 2^{\log_2 x} = x$$

Since $(f \circ g)(x) = x$ and $(g \circ f)(x) = x$, these functions are inverses.

**5.** Since any horiztontal line when passed through the graph of this function will touch the graph in at most one place, the function is one-to-one. A sketch of the graph of the inverse function is as follows.

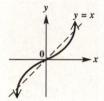

**7.** Since a horizontal line when passed through the graph of this function will touch the graph in more than one place, the function is not one-to-one.

**9.** Because $f(x) = \log_3(x+2)$ has a vertical asymptote when $x + 2 = 0 \Rightarrow x = -2$, $x$-intercept when $x + 2 = 1 \Rightarrow x = -1$, and is increasing, the correct choice is the graph in B.

**11.** Because $f(x) = \log_2(5-x)$ has a vertical asymptote when $5 - x = 0 \Rightarrow x = 5$, $x$-intercept when $5 - x = 1 \Rightarrow x = 4$. The correct choice is the graph in C.

**13.** The functions in Exercises 9 and 12 are inverses. The functions in Exercises 10 and 11 are inverses.

**15.** $f(x) = 3x - 6$

This function is one-to-one.

*Step 1*: Replace $f(x)$ with $y$ and interchange $x$ and $y$. $y = 3x - 6 \Rightarrow x = 3y - 6$

*Step 2*: Solve for $y$.
$$x = 3y - 6 \Rightarrow x + 6 = 3y \Rightarrow \tfrac{x+6}{3} = y$$

*Step 3*: Replace $y$ with $f^{-1}(x)$:
$$f^{-1}(x) = \tfrac{x+6}{3} = \tfrac{1}{3}x + 2$$

For both $f(x)$ and $f^{-1}(x)$, the domain is $(-\infty, \infty)$ and the range is $(-\infty, \infty)$.

**17.** $f(x) = 3x^2$

If $x = 1$, $f(1) = 3(1)^2 = 3(1) = 3$. Also if $x = -1$, $f(-1) = 3(-1)^2 = 3(1) = 3$.

Since two different values of $x$ lead to the same value of $y$, the function is not one-to-one.

**19.** $f(x) = \sqrt[3]{5 - x^4}$

If $x = 1$, $f(1) = \sqrt[3]{5 - 1^4} = \sqrt[3]{5 - 1} = \sqrt[3]{4}$. Also if $x = -1$, $f(-1) = \sqrt[3]{5 - (-1)^4} = \sqrt[3]{5 - 1} = \sqrt[3]{4}$.

Since two different values of $x$ lead to the same value of $y$, the function is not one-to-one.

**21.** $\left(\tfrac{1}{10}\right)^{-3} = 1000$ is equivalent to $\log_{1/10} 1000 = -3$.

**23.** $\left(\sqrt{3}\right)^4 = 9$ is equivalent to $\log_{\sqrt{3}} 9 = 4$.

**25.** $2^x = 32$ is equivalent to $\log_2 32 = x$.

**27.** $3x = 7^{\log_7 6} \Rightarrow 3x = 6 \Rightarrow x = 2$

Solution set: $\{2\}$

**29.** $x = \log_6 \tfrac{1}{216} \Rightarrow 6^x = \tfrac{1}{216} \Rightarrow 6^x = 6^{-3} \Rightarrow$ $x = -3$

Solution set: $\{-3\}$

**31.** $\log_{10} .01 = x$
$$10^x = .01 \Rightarrow 10^x = 10^{-2} \Rightarrow x = -2$$

Solution set: $\{-2\}$

**33.** $\log_x 1 = 0 \Rightarrow x^0 = 1$

This is a true statement for all real numbers greater than 0, excluding 1.

Solution set: $(0,1) \cup (1, \infty)$

**35.** $\log_x \sqrt[3]{5} = \tfrac{1}{3} \Rightarrow x^{1/3} = \sqrt[3]{5} \Rightarrow x^{1/3} = 5^{1/3} \Rightarrow$
$$\left(x^{1/3}\right)^3 = \left(5^{1/3}\right)^3 \Rightarrow x = 5$$

Recall from Chapter 1 that it is necessary to check all proposed solutions in the original equation when you raise both sides to a power.

Check $x = 5$.

$\log_x \sqrt[3]{5} = \tfrac{1}{3}$

$\log_5 \sqrt[3]{5} \overset{?}{=} \tfrac{1}{3}$

$\log_5 5^{1/3} = \tfrac{1}{3} \Rightarrow \tfrac{1}{3}\log_5 5 = \tfrac{1}{3} \Rightarrow \tfrac{1}{3} \cdot 1 = \tfrac{1}{3} \Rightarrow \tfrac{1}{3} = \tfrac{1}{3}$

This is a true statement.

Solution set: $\{5\}$

**37.** $\log_{10}\left(\log_2 2^{10}\right) = x \Rightarrow$
$\log_{10}\left(10\log_2 2\right) = x \Rightarrow \log_{10}\left(10 \cdot 1\right) = x \Rightarrow$
$\log_{10} 10 = x \Rightarrow x = 1$
Solution set: $\{1\}$

**39.** $2x - 1 = \log_6 6^x \Rightarrow 2x - 1 = x \Rightarrow$
$-1 = -x \Rightarrow 1 = x$
Solution set: $\{1\}$

**41.** $2^x = \log_2 16 \Rightarrow 2^{\left(2^x\right)} = 16 \Rightarrow 2^{\left(2^x\right)} = 2^4 \Rightarrow$
$2^x = 4 \Rightarrow 2^x = 2^2 \Rightarrow x = 2$
Solution set: $\{2\}$

**43.** $\left(\frac{1}{3}\right)^{x+1} = 9^x \Rightarrow \left(3^{-1}\right)^{x+1} = \left(3^2\right)^x \Rightarrow$
$3^{(-1)(x+1)} = 3^{2x} \Rightarrow 3^{-x-1} = 3^{2x} \Rightarrow$
$-x - 1 = 2x \Rightarrow x = -\frac{1}{3}$
Solution set: $\left\{-\frac{1}{3}\right\}$

## Section 4.4: Evaluating Logarithms and the Change-of-Base Theorem

**1.** For $f(x) = a^x$, where $a > 0$, the function is _increasing_ over its entire domain.

**3.** $f(x) = 5^x$
This function is one-to-one.
*Step 1*: Replace $f(x)$ with $y$ and interchange $x$ and $y$. $y = 5^x \Rightarrow x = 5^y$
*Step 2*: Solve for $y$.
$x = 5^y \Rightarrow y = \log_5 x$
*Step 3*: Replace $y$ with $f^{-1}(x)$.
$f^{-1}(x) = \log_5 x$

**5.** A base $e$ logarithm is called a _natural_ logarithm, while a base 10 logarithm is called a _common_ logarithm.

**7.** $\log_2 0$ is undefined because there is no power of 2 that yields a result of 0. In other words, the equation $2^x = 0$ has no solution.

**9.** $\log 8 = .90308999$

**11.** $\log 53 \approx 1.7243$

**13.** $\log .0013 \approx -2.8861$

**15.** $\ln 53 \approx 3.9703$

**17.** $\ln .0013 \approx -6.6454$

**19.** $\log\left(3.1 \times 10^4\right) \approx 4.4914$

**21.** $\log\left(5.0 \times 10^{-6}\right) \approx -5.3010$

**23.** $\ln\left(6 \times e^4\right) \approx 5.7918$

**25.** $\log\left(387 \times 23\right) \approx 3.9494$

**27.** $\log\left(387 \times 23\right) = \log 387 + \log 23$ by the product property of logarithms.

**29.** Grapefruit, $f(x) = \sqrt[3]{2x - 7}$
$pH = -\log\left[H_3O^+\right] = -\log\left(6.3 \times 10^{-4}\right)$
$\qquad = -\left(\log 6.3 + \log 10^{-4}\right) = -(.7793 - 4)$
$\qquad = -.7993 + 4 \approx 3.2$

```
-log(6.3*10^(-4)
)
            3.200659451
```

The answer is rounded to the nearest tenth because it is customary to round pH values to the nearest tenth. The pH of grapefruit is 3.2.

**31.** Crackers, $f(x) = \sqrt[3]{2x - 7}$
$pH = -\log\left[H_3O^+\right] = -\log\left(3.9 \times 10^{-9}\right)$
$\qquad = -\left(\log 3.9 + \log 10^{-9}\right) = -(.59106 - 9)$
$\qquad = -(-8.409) \approx 8.4$

```
-log(3.9*10^(-9)
)
            8.408935393
```

The answer is rounded to the nearest tenth because it is customary to round pH values to the nearest tenth. The pH of crackers is 8.4.

**33.** Soda pop, 2.7
$pH = -\log\left[H_3O^+\right]$
$2.7 = -\log\left[H_3O^+\right]$
$-2.7 = \log\left[H_3O^+\right] \Rightarrow \left[H_3O^+\right] = 10^{-2.7}$

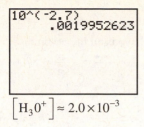

$$\left[H_3O^+\right] \approx 2.0 \times 10^{-3}$$

**35.** Beer, 4.8

$$pH = -\log\left[H_3O^+\right]$$
$$4.8 = -\log\left[H_3O^+\right]$$
$$-4.8 = \log\left[H_3O^+\right] \Rightarrow \left[H_3O^+\right] = 10^{-4.8}$$

```
10^(-4.8)
   1.584893192E-5
```

$$\left[H_3O^+\right] \approx 1.6 \times 10^{-5}$$

**37.** Wetland, $2.49 \times 10^{-5}$

$$pH = -\log\left[H_3O^+\right] = -\log\left(2.49 \times 10^{-5}\right)$$
$$= -\left(\log 2.49 + \log 10^{-5}\right)$$
$$= -\log 2.49 - (-5) = -\log 2.49 + 5$$
$$pH \approx 4.6$$

Since the pH is between 4.0 and 6.0, it is a poor fen.

**39.** Wetland, $2.49 \times 10^{-2}$

$$pH = -\log\left[H_3O^+\right] = -\log\left(2.49 \times 10^{-2}\right)$$
$$= -\left(\log 2.49 + \log 10^{-2}\right)$$
$$= -\log 2.49 - (-2) = -\log 2.49 + 2$$
$$pH \approx 1.6$$

Since the pH is 3.0 or less, it is a bog.

**41.** Wetland, $2.49 \times 10^{-7}$

$$pH = -\log\left[H_3O^+\right] = -\log\left(2.49 \times 10^{-7}\right)$$
$$= -\left(\log 2.49 + \log 10^{-7}\right)$$
$$= -\log 2.49 - (-7) = -\log 2.49 + 7$$
$$pH \approx 6.6$$

Since the pH is greater than 6.0, it is a rich fen.

**43.** **(a)** $\log 398.4 \approx 2.60031933$

**(b)** $\log 39.84 \approx 1.60031933$

**(c)** $\log 3.984 \approx .6003193298$

**(d)** The whole number parts will vary, but the decimal parts will be the same.

**45.** $d = 10\log\frac{I}{I_0}$, where $d$ is the decibel rating.

**(a)** $d = 10\log\frac{100 I_0}{I_0}$
$$= 10\log_{10} 100 = 10(2) = 20$$

**(b)** $d = 10\log\frac{1000 I_0}{I_0}$
$$= 10\log_{10} 1000 = 10(3) = 30$$

**(c)** $d = 10\log\frac{100,000 I_0}{I_0}$
$$= 10\log_{10} 100,000 = 10(5) = 50$$

**(d)** $d = 10\log\frac{1,000,000 I_0}{I_0}$
$$= 10\log_{10} 1,000,000 = 10(6) = 60$$

**(e)** $I = 2I_0$
$$d = 10\log\frac{2I_0}{I_0} = 10\log 2 \approx 3.0103$$

The described rating is increased by about 3 decimals.

**47.** $r = \log_{10}\frac{I}{I_0}$, where $r$ is the Richter scale rating of an earthquake.

**(a)** $r = \log_{10}\frac{1000 I_0}{I_0} = \log_{10} 1000 = 3$

**(b)** $r = \log_{10}\frac{1,000,000 I_0}{I_0} = \log_{10} 1,000,000 = 6$

**(c)** $r = \log_{10}\frac{100,000,000 I_0}{I_0}$
$$= \log_{10} 100,000,000 = 8$$

**49.** From exercise 47, the magnitude of an earthquake, measured on the Richter scale, is $r = \log_{10}\frac{I}{I_0}$, where $I$ is the amplitude registered on a seismograph 100 km fro the epicenter of the earthquake, and $I_0$ is the amplitude of an earthquake of a certain small size. So, $8.6 = \log_{10}\frac{I}{I_0} \Rightarrow \frac{I}{I_0} = 10^{8.6} \Rightarrow$

$$I = 10^{8.6} I_0 \approx 398,107,171 I_0$$

**51.** The year 2009 is represented by 109.
$$f(109) = -85.4 + 32.4\ln 109 \approx 66.6 \text{ million}$$
We must assume that the model continues to be logarithmic.

**53.** If $a = .36$, then

$$S(n) = a\ln\left(1 + \frac{n}{a}\right) = .36\ln\left(1 + \frac{n}{.36}\right).$$

**(a)** $S(100) = .36\ln\left(1 + \frac{100}{.36}\right) \approx 2.0269 \approx 2$

**(b)** $S(200) = .36 \ln\left(1 + \frac{200}{.36}\right) \approx 2.2758 \approx 2$

**(c)** $S(150) = .36 \ln\left(1 + \frac{150}{.36}\right) \approx 2.1725 \approx 2$

**(d)** $S(10) = .36 \ln\left(1 + \frac{10}{.36}\right) \approx 1.2095 \approx 1$

**55.** The index of diversity $H$ for 2 species is given by $H = -\left[P_1 \log_2 P_1 + P_2 \log_2 P_2\right]$. When $P_1 = \frac{50}{100} = .5$ and $P_2 = \frac{50}{100} = .5$ we have

$H = -\left[.5 \log_2 .5 + .5 \log_2 .5\right].$

Since $\log_2 .5 = \log_2 \frac{1}{2} = \log_2 2^{-1} = -1$, we have

$H = -\left[.5(-1) + .5(-1)\right] = -(-1) = 1$. Thus, the index of diversity is 1.

**57.** $T(k) = 1.03k \ln \frac{C}{C_0}$

Since $10 \le k \le 16$ and $\frac{C}{C_0} = 2$, the range for $T = 1.03k \ln \frac{C}{C_0}$ will be between $T(10)$ and $T(16)$. Since $T(10) = 1.03(10)\ln 2 \approx 7.1$ and $T(16) = 1.03(16)\ln 2 \approx 11.4$, the predicted increased global temperature due to the greenhouse effect from a doubling of the carbon dioxide in the atmosphere is between 7°F and 11°F.

**59.** $t = \left(1.26 \times 10^9\right)\frac{\ln\left[1 + 8.33(.103)\right]}{\ln 2} \approx 1.126 \times 10^9$

The rock sample is approximately 1.126 billion years old.

For Exercises 61–71, as noted on page 452 of the text, the solutions will be evaluated at the intermediate steps to four decimal places. However, the final answers are obtained without rounding the intermediate steps.

**61.** $\log_2 5 = \frac{\ln 5}{\ln 2} \approx \frac{1.6094}{.6931} \approx 2.3219$

We could also have used the common logarithm. $\log_2 5 = \frac{\log 5}{\log 2} \approx \frac{.6990}{.3010} \approx 2.3219$

**63.** $\log_8 .59 = \frac{\log .59}{\log 8} \approx \frac{-.2291}{.9031} \approx -.2537$

We could also have used the natural logarithm. $\log_8 .59 = \frac{\ln .59}{\ln 8} \approx \frac{-.5276}{2.0794} \approx -.2537$

**65.** $\log_{1/2} 3 = \frac{\log 3}{\log[1/2]} \approx \frac{.4771}{-.3010} \approx -1.5850$

We could also have used the natural logarithm. $\log_{1/2} 3 = \frac{\ln 3}{\ln[1/2]} \approx \frac{1.0986}{-.6931} \approx -1.5850$

**67.** $\log_\pi e = \frac{\ln e}{\ln \pi} \approx \frac{1}{1.1447} \approx .8736$

We could also have used the common logarithm. $\log_\pi e = \frac{\log e}{\log \pi} \approx \frac{.4343}{.4971} \approx .8736$

**69.** Since $\sqrt{13} = 13^{1/2}$, we have

$\log_{\sqrt{13}} 12 = \frac{\ln 12}{\ln \sqrt{13}} = \frac{\ln 12}{\frac{1}{2}\ln 13} \approx \frac{2.4849}{1.2825} \approx 1.9376.$

The required logarithm can also be found by entering $\ln \sqrt{13}$ directly into the calculator. We could also have used the common logarithm.

$\log_{\sqrt{13}} 12 = \frac{\log 12}{\log \sqrt{13}} = \frac{\log 12}{\frac{1}{2}\log 13} \approx \frac{1.0792}{.5570} \approx 1.9376.$

**71.** $\log_{.32} 5 = \frac{\log 5}{\log .32} \approx \frac{.6990}{-.4949} \approx -1.4125$

We could also have used the natural logarithm. $\log_{.32} 5 = \frac{\ln 5}{\ln .32} \approx \frac{1.6094}{-1.1394} \approx -1.4125$

**73.** $\ln\left(b^4 \sqrt{a}\right) = \ln\left(b^4 a^{1/2}\right) = \ln b^4 + \ln a^{1/2}$
$= 4 \ln b + \frac{1}{2}\ln a = 4v + \frac{1}{2}u$

**75.** $\ln \sqrt{\frac{a^3}{b^5}} = \ln\left(\frac{a^3}{b^5}\right)^{1/2} = \ln\left(\frac{a^{3/2}}{b^{5/2}}\right) = \ln a^{3/2} - \ln b^{5/2}$
$= \frac{3}{2}\ln a - \frac{5}{2}\ln b = \frac{3}{2}u - \frac{5}{2}v$

**77.** $g(x) = e^x$

**(a)** $g(\ln 4) = e^{\ln 4} = 4$

**(b)** $g\left[\ln\left(5^2\right)\right] = e^{\ln 5^2} = 5^2$ or 25

**(c)** $g\left[\ln\left(\frac{1}{e}\right)\right] = e^{\ln(1/e)} = \frac{1}{e}$

**79.** $f(x) = \ln x$

**(a)** $f\left(e^6\right) = \ln e^6 = 6$

**(b)** $f\left(e^{\ln 3}\right) = \ln e^{\ln 3} = \ln 3$

**(c)** $f\left(e^{2\ln 3}\right) = \ln e^{2\ln 3} = 2\ln 3$ or $\ln 9$.

**81.** $2\ln 3x = \ln(3x)^2 = \ln\left(3^2 x^2\right) = \ln 9x^2$

It is equivalent to D.

**83.** $f(x) = \ln|x|$

The domain of $f$ is all real numbers except 0: $(-\infty, 0) \cup (0, \infty)$ and the range is $(-\infty, \infty)$.

Since $f(-x) = \ln|-x| = \ln|x| = f(x)$, this is an even function and symmetric with respect to the $y$-axis.

**85.** $f(x) = \ln e^2 x = \ln e^2 + \ln x = 2 + \ln x$

$f(x) = \ln e^2 x$ is a vertical shift of the graph of $g(x) = \ln x$, 2 units up.

**87.** $f(x) = \ln \frac{x}{e^2} = \ln x - 2\ln e = \ln x - 2$

$f(x) = \ln \frac{x}{e^2}$ is a vertical shift of the graph of $g(x) = \ln x$, 2 units down.

## Chapter 4 Quiz
### (Sections 4.1–4.4)

**1.** *Step 1*: Replace $f(x)$ with $y$ and interchange $x$ and $y$: $y = \sqrt[3]{3x-6} \Rightarrow x = \sqrt[3]{3y-6}$

*Step 2*: Solve for $y$.

$x = \sqrt[3]{3y-6} \Rightarrow x^3 = 3y-6 \Rightarrow \frac{x^3+6}{3} = y$

*Step 3*: Replace $y$ with $f^{-1}(x)$.

$f^{-1}(x) = \frac{x^3+6}{3}$

**3.**

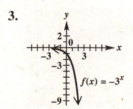

Domain: $(-\infty, \infty)$; range: $(-\infty, 0)$

**5.** Use the compound interest formula to find the amount in the account, $A = P\left(1 + \frac{r}{n}\right)^{tn}$, given $n = 4$, $P = 2000$, $r = .036$, and $t = 2$.

$A = P\left(1 + \frac{r}{n}\right)^{tn} = (2000)\left(1 + \frac{.036}{4}\right)^{2(4)}$

$= (2000)(1 + .009)^8 \approx 2148.62$

Rounding to the nearest cent, there will be $2148.62 after 2 years.

**7.** The expression $\log_6 25$ represents the exponent to which 6 must be raised in order to obtain 25.

**9.** $\log_3 \frac{\sqrt{x} \cdot y}{pq^4} = \log_3\left(\sqrt{x} \cdot y\right) - \log_3\left(pq^4\right)$

$= \log_3\left(x^{1/2}\right) + \log_3 y$

$\qquad - \left(\log_3 p + \log_3 q^4\right)$

$= \frac{1}{2}\log_3 x + \log_3 y$

$\qquad - \log_3 p - 4\log_3 q$

**11.** $\log_3 40 = \frac{\log 40}{\log 3} \approx \frac{1.6021}{.4771} \approx 3.3578$

We could also have used the natural logarithm. $\log_3 40 = \frac{\ln 40}{\ln 3} \approx \frac{3.689}{1.0986} \approx 3.3578$

## Section 4.5: Exponential and Logarithmic Equations

**1.** Since $x$ is the exponent to which 7 must be raised in order to obtain 19, the solution is $\log_7 19$ or $\frac{\log 19}{\log 7}$ or $\frac{\ln 19}{\ln 7}$.

**3.** Since $x$ is the exponent to which $\frac{1}{2}$ must be raised in order to obtain 12, the solution is $\log_{1/2} 12$ or $\frac{\log 12}{\log\left(\frac{1}{2}\right)}$ or $\frac{\ln 12}{\ln\left(\frac{1}{2}\right)}$.

**5.** $3^x = 7$

$\ln 3^x = \ln 7$

$x \ln 3 = \ln 7$

$x = \frac{\ln 7}{\ln 3} \approx 1.771$

Solution set: $\{1.771\}$

**7.** $\left(\frac{1}{2}\right)^x = 5 \Rightarrow \ln\left(\frac{1}{2}\right)^x = \ln 5 \Rightarrow x \ln \frac{1}{2} = \ln 5$

$x = \frac{\ln 5}{\ln \frac{1}{2}} \approx -2.322$

Solution set: $\{-2.322\}$

**9.** $.8^x = 4 \Rightarrow \ln\left(.8^x\right) = \ln 4 \Rightarrow x \ln .8 = \ln 4 \Rightarrow$

$x = \frac{\ln 4}{\ln .8} \approx -6.213$

Solution set: $\{-6.213\}$

**11.** $\qquad 4^{x-1} = 3^{2x} \Rightarrow \ln\left(4^{x-1}\right) = \ln\left(3^{2x}\right) \Rightarrow$

$(x-1)\ln 4 = 2x \ln 3$

$x \ln 4 - \ln 4 = 2x \ln 3$

$x \ln 4 - 2x \ln 3 = \ln 4 \Rightarrow x(\ln 4 - 2\ln 3) = \ln 4 \Rightarrow$

$x = \frac{\ln 4}{\ln 4 - 2\ln 3} \approx -1.710$

Solution set: $\{-1.710\}$

**13.** $6^{x+1} = 4^{2x-1} \Rightarrow \ln\left(6^{x+1}\right) = \ln\left(4^{2x-1}\right)$

$(x+1)\ln 6 = (2x-1)\ln 4$

$x\ln 6 + \ln 6 = 2x\ln 4 - \ln 4$

$\ln 6 + \ln 4 = 2x\ln 4 - x\ln 6$

$\ln 6 + \ln 4 = x(2\ln 4 - \ln 6)$

$x = \frac{\ln 6 + \ln 4}{2\ln 4 - \ln 6} \approx 3.240$

Solution set: $\{3.240\}$

**15.** $e^{x^2} = 100 \Rightarrow \ln\left(e^{x^2}\right) = \ln 100$

$x^2 = \ln 100 \Rightarrow x = \pm\sqrt{\ln 100} = \pm 2.146$

Solution set: $\{\pm 2.146\}$

**17.** $e^{3x-7} \cdot e^{-2x} = 4e \Rightarrow e^{x-7} = 4e$

$\ln\left(e^{x-7}\right) = \ln(4e) \Rightarrow (x-7)\ln e = \ln 4 + \ln e$

$x - 7 = \ln 4 + 1 \Rightarrow x = \ln 4 + 8 \approx 9.386$

Solution set: $\{9.386\}$

**19.** $\left(\frac{1}{3}\right)^x = -3$ has no solution since $\frac{1}{3}$ raised to any power is positive.
Solution set: $\varnothing$

**21.** $.05(1.15)^x = 5 \Rightarrow 1.15^x = \frac{5}{.05} = 100 \Rightarrow$

$\log\left(1.15^x\right) = \log 100 \Rightarrow x\log 1.15 = \log 100 \Rightarrow$

$x = \frac{\log 100}{\log 1.15} = \frac{2}{\log 1.15} \approx 32.950$

Solution set: $\{32.950\}$

**23.** $3(2)^{x-2} + 1 = 100 \Rightarrow 3(2)^{x-2} = 99 \Rightarrow$

$2^{x-2} = 33 \Rightarrow \ln\left(2^{x-2}\right) = \ln 33 \Rightarrow$

$(x-2)\ln 2 = \ln 33 \Rightarrow x = \frac{\ln 33}{\ln 2} + 2 \approx 7.044$

Solution set: $\{7.044\}$

**25.** $2(1.05)^x + 3 = 10 \Rightarrow 2(1.05)^x = 7 \Rightarrow$

$1.05^x = 3.5 \Rightarrow \ln\left(1.05^x\right) = \ln 3.5 \Rightarrow$

$x\ln 1.05 = \ln 3.5 \Rightarrow x = \frac{\ln 3.5}{\ln 1.05} \approx 25.677$

Solution set: $\{25.677\}$

**27.** $5(1.015)^{x-1980} = 8 \Rightarrow 1.015^{(x-1980)} = 1.6$

$\ln\left(1.015^{(x-1980)}\right) = \ln 1.6$

$(x-1980)\ln 1.015 = \ln 1.6$

$x = \frac{\ln 1.6}{\ln 1.015} + 1980 \approx 2011.568$

Solution set: $\{2011.568\}$

**29.** $5\ln x = 10 \Rightarrow \ln x = 2 \Rightarrow e^2 = x$
Solution set: $\left\{e^2\right\}$

**31.** $\ln(4x) = 1.5 \Rightarrow 4x = e^{1.5} \Rightarrow x = \frac{e^{1.5}}{4}$

Solution set: $\left\{\frac{e^{1.5}}{4}\right\}$

**33.** $\log(2-x) = .5 \Rightarrow 2 - x = 10^{.5} \Rightarrow 2 - x = \sqrt{10}$

$x = 2 - \sqrt{10}$

Solution set: $\left\{2 - \sqrt{10}\right\}$

**35.** $\log_6(2x+4) = 2 \Rightarrow 2x + 4 = 6^2$

$2x + 4 = 36 \Rightarrow 2x = 32 \Rightarrow x = 16$

Solution set: $\{16\}$

**37.** $\log_4\left(x^3 + 37\right) = 3 \Rightarrow x^3 + 37 = 4^3$

$x^3 = 27 \Rightarrow x = 3$

Solution set: $\{3\}$

**39.** $\ln x + \ln x^2 = 3 \Rightarrow \ln\left(x \cdot x^2\right) = 3 \Rightarrow \ln x^3 = 3$

$x^3 = e^3 \Rightarrow x = e$

Solution set: $\{e\}$

**41.** $\log x + \log(x - 21) = 2 \Rightarrow \log\left[x(x-21)\right] = 2$

$\log\left(x^2 - 21x\right) = 2 \Rightarrow x^2 - 21x = 10^2$

$x^2 - 21x - 100 = 0 \Rightarrow (x - 25)(x + 4) = 0 \Rightarrow$

$x = 25$ or $x = -4$

Since the negative solution $(x = -4)$ is not in the domain of $\log x$, it must be discarded.
Solution set: $\{25\}$

**43.** $\log(x + 25) = 1 + \log(2x - 7)$

$\log(x + 25) - \log(2x - 7) = 1$

$\log_{10}\frac{x+25}{2x-7} = 1 \Rightarrow \frac{x+25}{2x-7} = 10^1$

$x + 25 = 10(2x - 7)$

$x + 25 = 20x - 70$

$25 = 19x - 70$

$95 = 19x \Rightarrow 5 = x$

Solution set: $\{5\}$

**45.** $\ln(4x - 2) - \ln 4 = -\ln(x - 2)$

$\ln\frac{4x-2}{4} = -\ln(x-2) \Rightarrow \frac{4x-2}{4} = \frac{1}{x-2}$

$(4x - 2)(x - 2) = 4 \Rightarrow 4x^2 - 10x + 4 = 4$

$4x^2 - 10x = 0 \Rightarrow 2x(2x - 5) = 0 \Rightarrow$

$2x = 0 \Rightarrow x = 0$ or $2x - 5 = 0 \Rightarrow x = \frac{5}{2} = 2.5$

Since $x = 0$ is not in the domain of $\ln(x - 2)$, it must be discarded.
Solution set: $\{2.5\}$

**47.** $\log_5(x+2) + \log_5(x-2) = 1$

$\qquad \log_5[(x+2)(x-2)] = 1 \Rightarrow x^2 - 4 = 5^1$

$\qquad\qquad x^2 - 9 = 0$

$\qquad\qquad (x-3)(x+3) = 0 \Rightarrow x = \pm 3$

$-3$ is not in the domain, so reject it.

Solution set: $\{3\}$

**49.** $\log_2(2x-3) + \log_2(x+1) = 1$

$\qquad \log_2[(2x-3)(x+1)] = 1$

$\qquad\qquad 2x^2 - x - 3 = 2^1$

$\qquad\qquad 2x^2 - x - 5 = 0$

$$x = \frac{-(-1) \pm \sqrt{(-1)^2 - 4(2)(-5)}}{2(2)} = \frac{1 \pm \sqrt{41}}{4}$$

Since the negative solution $\left(x = \frac{1-\sqrt{41}}{4}\right)$ is not

in the domain of $\log(x+1)$, it must be discarded.

Solution set: $\left\{\frac{1+\sqrt{41}}{4}\right\}$

**51.** $\ln e^x - 2\ln e = \ln e^4 \Rightarrow x - 2 = 4 \Rightarrow x = 6$

Solution set: $\{6\}$

**53.** $\log_2(\log_2 x) = 1 \Rightarrow \log_2 x = 2^1 \Rightarrow$

$\log_2 x = 2 \Rightarrow x = 2^2 \Rightarrow x = 4$

Solution set: $\{4\}$

**55.** $\log x^2 = (\log x)^2 \Rightarrow 2\log x = (\log x)^2 \Rightarrow$

$(\log x)^2 - 2\log x = 0 \Rightarrow \log x(\log x - 2) = 0$

$\quad \log_{10} x = 0 \qquad$ or $\log_{10} x - 2 = 0$

$\qquad x = 10^0 \qquad\qquad \log_{10} x = 2$

$\qquad x = 1 \qquad\qquad\quad x = 10^2 = 100$

Solution set: $\{1, 100\}$

**57.** Answers will vary. One should not immediately reject a negative answer when solving equations involving logarithms. One must examine what happens to an answer when substituted back into the original equation. An answer (whether negative, positive, or zero) must not allow a nonpositive value in the argument of the logarithm. If it does, regardless of its sign, it must be rejected.

**59.** $p = a + \dfrac{k}{\ln x}$, for $x$

$\qquad p - a = \dfrac{k}{\ln x}$

$\quad (\ln x)(p-a) = k$

$\qquad\quad \ln x = \dfrac{k}{p-a}$

$\qquad\qquad x = e^{k/(p-a)}$

**61.** $T = T_0 + (T_1 - T_0)10^{-kt}$, for $t$

$\qquad T - T_0 = (T_1 - T_0)10^{-kt}$

$\qquad \dfrac{T - T_0}{T_1 - T_0} = 10^{-kt}$

$\qquad \log_{10}\left(\dfrac{T-T_0}{T_1-T_0}\right) = -kt$

$\qquad \dfrac{\log\left(\frac{T-T_0}{T_1-T_0}\right)}{-k} = \dfrac{-kt}{-k}$

$\qquad\qquad t = -\dfrac{1}{k}\log\left(\dfrac{T-T_0}{T_1-T_0}\right)$

**63.** $I = \dfrac{E}{R}\left(1 - e^{-Rt/2}\right)$, for $t$

$\qquad RI = R\left[\dfrac{E}{R}\left(1 - e^{-Rt/2}\right)\right]$

$\qquad RI = E\left(1 - e^{-Rt/2}\right)$

$\qquad RI = E - Ee^{-Rt/2}$

$\qquad RI + Ee^{-Rt/2} = E$

$\qquad Ee^{-Rt/2} = E - RI$

$\qquad \dfrac{Ee^{-Rt/2}}{E} = \dfrac{E-RI}{E}$

$\qquad e^{-Rt/2} = 1 - \dfrac{RI}{E}$

$\qquad \ln e^{-Rt/2} = \ln\left(1 - \dfrac{RI}{E}\right)$

$\qquad -\dfrac{Rt}{2} = \ln\left(1 - \dfrac{RI}{E}\right)$

$\qquad -\dfrac{2}{R}\left(-\dfrac{Rt}{2}\right) = -\dfrac{2}{R}\ln\left(1 - \dfrac{RI}{E}\right)$

$\qquad t = -\dfrac{2}{R}\ln\left(1 - \dfrac{RI}{E}\right)$

**65.** $y = A + B\left(1 - e^{-Cx}\right)$, for $x$

$\qquad y - A = B\left(1 - e^{-Cx}\right)$

$\qquad \dfrac{y-A}{B} = 1 - e^{-Cx}$

$\qquad \dfrac{y-A}{B} - 1 = \dfrac{y-A-B}{B} = -e^{-Cx}$

$\qquad \dfrac{A+B-y}{B} = e^{-Cx}$

$\qquad \ln\left(\dfrac{A+B-y}{B}\right) = -Cx$

$\qquad \dfrac{\ln\left(\frac{A+B-y}{B}\right)}{-C} = x$

**67.** $\log A = \log B - C\log x$, for $A$

$\qquad \log A = \log B - C\log x$

$\qquad \log A = \log\dfrac{B}{x^C}$

$\qquad A = \dfrac{B}{x^C}$

**69.** $A = P\left(1 + \dfrac{r}{n}\right)^{nt}$, for $t$

$\qquad \dfrac{A}{P} = \left(1 + \dfrac{r}{n}\right)^{nt}$

$\qquad \log\left(\dfrac{A}{P}\right) = nt\log\left(1 + \dfrac{r}{n}\right)$

$\qquad \dfrac{\log\left(\frac{A}{P}\right)}{n\log\left(1 + \frac{r}{n}\right)} = t$

**71.** $A = P\left(1 + \frac{r}{n}\right)^{tn}$

To solve for $A$, substitute $P = 10,000$, $r = .03$, $n = 4$, and $t = 5$.

$A = 10,000\left(1 + \frac{.03}{4}\right)^{(5)(4)}$

$A = 10,000(1.0075)^{20} \approx 11,611.84$

There will be $11,611.84 in the account.

**73.** $A = P\left(1 + \frac{r}{n}\right)^{tn}$

To solve for $t$, substitute $A = 30,000$, $P = 27,000$, $r = .04$, and $n = 4$.

$30,000 = 27,000\left(1 + \frac{.04}{4}\right)^{t(4)}$

$\frac{30,000}{27,000} = (1 + .01)^{4t} \Rightarrow \frac{10}{9} = 1.01^{4t}$

$\ln\frac{10}{9} = \ln\left(1.01^{4t}\right) \Rightarrow \ln\frac{10}{9} = 4t\ln 1.01$

$t = \frac{\ln\frac{10}{9}}{4\ln 1.01} \approx 2.6$

To the nearest tenth of a year, Tom will be ready to buy a car in 2.6 yr.

**75.** $A = P\left(1 + \frac{r}{n}\right)^{tn}$

To solve for $r$, substitute $A = 2500$, $P = 2000$, $t = 3.5$, and $n = 2$.

$2500 = 2000\left(1 + \frac{r}{2}\right)^{(3.5)(2)}$

$1.25 = \left(1 + \frac{r}{2}\right)^{7} \Rightarrow \sqrt[7]{1.25} = 1 + \frac{r}{2}$

$\sqrt[7]{1.25} - 1 = \frac{r}{2} \Rightarrow r = 2\left(\sqrt[7]{1.25} - 1\right)$

$r \approx .0648$

The interest rate is about 6.48%.

**77. (a)** $f(3000) = 86.3\ln 3000 - 680 \approx 10.9$

At 3000 ft, about 10.9% of the moisture falls as snow.

**(b)** $f(4000) = 86.3\ln 4000 - 680 \approx 35.8$

At 4000 ft, about 35.8% of the moisture falls as snow.

**(c)** $f(7000) = 86.3\ln 7000 - 680 \approx 84.1$

At 7000 ft, about 84.1% of the moisture falls as snow.

**79.** Double the 2000 value is $2(7,990) = 15,980$.

$f(x) = 8160(1.06)^x$

$15,980 = 8160(1.06)^x$

$\frac{15,980}{8160} = 1.06^x \Rightarrow \frac{47}{24} = 1.06^x$

$\ln\frac{47}{24} = \ln 1.06^x \Rightarrow \ln\frac{47}{24} = x\ln 1.06$

$\frac{\ln\frac{47}{24}}{\ln 1.06} = x$

$x \approx 11.53$

During 2011, the cost of a year's tuition, room and board, and fees at a public university will be double the cost in 2000.

**81.** $f(x) = \dfrac{25}{1 + 1364.3e^{-x/9.316}}$

**(a)** In 1997, $x = 97$.

$f(97) = \dfrac{25}{1 + 1364.3e^{-97/9.316}} \approx 24$

In 1997, about 24% of U.S. children lived in a home without a father.

**(b)** $10 = \dfrac{25}{1 + 1364.3e^{-x/9.316}}$

$10\left(1 + 1364.3e^{-x/9.316}\right) = 25$

$10 + 13,643e^{-x/9.316} = 25$

$13,643e^{-x/9.316} = 15$

$e^{-x/9.316} = \frac{15}{13,643}$

$-\frac{x}{9.316} = \ln\frac{15}{13,643}$

$x = -9.316\ln\frac{15}{13,643}$

$\approx 63.47$

During 1963, 10% of U.S. children lived in a home without a father.

**83.** $\ln(1 - P) = -.0034 - .0053T$

**(a)** Change this equation to exponential form, then isolate $P$.

$1 - P = e^{-.0034 - .0053T}$

$P(T) = 1 - e^{-.0034 - .0053T}$

**(b)**

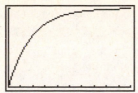

From the graph one can see that initially there is a rapid reduction of carbon dioxide emissions. However, after a while there is little benefit in raising taxes further.

**(c)**  $P(60) = 1 - e^{-.00340 - .0053(60)}$

$\approx .275 \text{ or } 27.5\%$

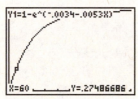

The reduction in carbon emissions from a tax of \$60 per ton of carbon is 27.5%.

**(d)**  We must determine $T$ when $P = .5$.

$$.5 = 1 - e^{-.0034 - .0053T}$$
$$.5 - 1 = -e^{-.0034 - .0053T}$$
$$-.5 = -e^{-.0034 - .0053T}$$
$$.5 = e^{-.0034 - .0053T}$$
$$\ln .5 = -.0034 - .0053T$$
$$\ln .5 + .0034 = -.0053T$$
$$T = \frac{\ln .5 + .0034}{-.0053} \approx 130.14$$

The value $T = \$130.14$ will give a 50% reduction in carbon emissions.

**85.**  *Step 1*: Replace $f(x) = e^{x+1} - 4$ with $y$ and interchange $x$ and $y$.

$$y = e^{x+1} - 4 \Rightarrow x = e^{y+1} - 4$$

*Step 2*: Solve for $y$.

$$x = e^{y+1} - 4 \Rightarrow x + 4 = e^{y+1}$$
$$\ln(x+4) = y + 1 \Rightarrow \ln(x+4) - 1 = y$$

*Step 3*: Replace $y$ with $f^{-1}(x)$.

$$f^{-1}(x) = \ln(x+4) - 1$$

Domain: $(-4, \infty)$; range: $(-\infty, \infty)$

**87.**  $e^x + \ln x = 5$

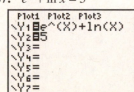

The two graphs intersect at approximately $(1.52, 5)$. The $x$-coordinate of this point is the solution of the equation.
Solution set: $\{1.52\}$

**89.**  $2e^x + 1 = 3e^{-x}$

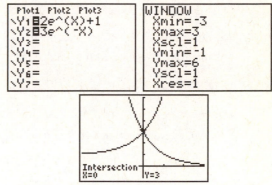

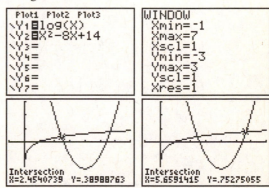

The two curves intersect at the point $(0, 3)$. The $x$-coordinate of this point is the solution of the equation.
Solution set: $\{0\}$

**91.**  $\log x = x^2 - 8x + 14$

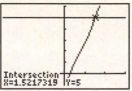

The two graphs intersect at approximately $(2.45, .39)$ and $(5.66, .75)$. The $x$-coordinate of these points represent the solution of the equation.
Solution set: $\{2.45, 5.66\}$

**93.**  The expression $e^{2x}$ is equivalent to $\left(e^x\right)^2$ by the power rule for exponents, $\left(a^m\right)^n = a^{mn}$.

**95.** $(e^x - 1)(e^x - 3) = 0$

$e^x - 1 = 0 \Rightarrow e^x = 1 \Rightarrow \ln e^x = \ln 1 \Rightarrow x = 0$ or
$e^x - 3 = 0 \Rightarrow e^x = 3 \Rightarrow \ln e^x = \ln 3 \Rightarrow x = \ln 3$
Solution set: $\{0, \ln 3\}$

## Section 4.6: Applications and Models of Exponential Growth and Decay

**1.** The equation $2e^{.02x} = 6$ represents an initial amount of 2 and a final amount 6 (triple that of 2). The correct choice is B.

**3.** The equation $y = 2e^{.02(3)}$ represents the amount after 3 yr. The correct choice is C.

**5.** $A(t) = 500e^{-.032t}$

(a) $t = 4 \Rightarrow A(4) = 500e^{-.032(4)} \approx 440$
After 4 years, about 440 g remain.

(b) $t = 8 \Rightarrow A(8) = 500e^{-.032(8)} \approx 387$
After 8 years, about 387 g remain.

(c) $t = 20 \Rightarrow A(20) = 500e^{-.032(20)} \approx 264$
After 20 years, about 264 g remain.

(d) Find $t$ when $A(t) = 250$.

$250 = 500e^{-.032t}$
$.5 = e^{-.032t}$
$\ln .5 = \ln e^{-.032t} \Rightarrow \ln .5 = -.032t$
$t = \frac{\ln .5}{-.032} \approx 21.66$
The half-life is about 21.66 yr.

**7.** $A(t) = A_0 e^{-.00043t}$
Find $t$ when $A(t) = \frac{1}{2} A_0$.

$\frac{1}{2} A_0 = A_0 e^{-.00043t}$
$\frac{1}{2} = e^{-.00043t} \Rightarrow \ln \frac{1}{2} = \ln e^{-.00043t}$
$\ln \frac{1}{2} = -.00043t \Rightarrow t = \frac{\ln \frac{1}{2}}{-.00043} \approx 1611.97$
The half-life is about 1611.97 yr.

**9.** First find the given values to find $y_0$ and then
$k$: $12 = y_0 e^{k(0)} \Rightarrow 12 = y_0$

$y = 12e^{kt} \Rightarrow 6 = 12e^{4k} \Rightarrow .5 = e^{4k} \Rightarrow$
$\ln .5 = 4k \Rightarrow \frac{\ln .5}{4} = k \Rightarrow k \approx -.173$
The exponential decay equation is
$y = 12e^{-.173t}$

To find the amount present after 7 years, let
$t = 7$. $y = 12e^{-.173(7)} \approx 3.57$
After 7 years, about 3.57 g of the substance will be present.

**11.** From Example 5, we have the amount of radiocarbon present after $t$ years is given by
$y = y_0 e^{-(\ln 2)(1/5700)t}$, where $y_0$ is the initial amount present. Letting $y = \frac{1}{3} y_0$, solve for $t$.

$\frac{1}{3} y_0 = y_0 e^{-(\ln 2)(1/5700)t}$
$\frac{1}{3} = e^{-(\ln 2)(1/5700)t}$
$\ln \frac{1}{3} = \ln e^{-(\ln 2)(1/5700)t}$
$\ln \frac{1}{3} = -\frac{\ln 2}{5700} t \Rightarrow -\frac{5700}{\ln 2} \ln \frac{1}{3} = t$
$-\frac{5700}{\ln 2}(\ln 1 - \ln 3) = t \Rightarrow t = \frac{5700 \ln 3}{\ln 2} \approx 9034.29$
The Egyptian died about 9000 yr ago.

**13.** Since $y = y_0 e^{-(\ln 2)(1/5700)t}$, where $y_0$ is the initial amount present. If we let $y = .15 y_0$, we can solve for $t$.

$.15 y_0 = y_0 e^{-(\ln 2)(1/5700)t}$
$.15 = e^{-(\ln 2)(1/5700)t}$
$\ln .15 = \ln e^{-(\ln 2)(1/5700)t}$
$\ln .15 = -\frac{\ln 2}{5700} t \Rightarrow -\frac{5700}{\ln 2} \ln .15 = t$
$t = -\frac{5700 \ln .15}{\ln 2} \approx 15,600.70$
The paintings are about 15,600 yr old.

**15.** (a) A point on the graph of $f(x) = A_0 a^{x-1950}$ is $(1950, .05)$.
Since $f(1950) = A_0 a^{1950-1950} = .05$, we have $A_0 a^0 = .05 \Rightarrow A_0 = .05$. Thus, we have the function $f(x) = .05 a^{x-1950}$.
Since the point $(2000, .35)$ is also on the graph of the function, we have
$f(2000) = .05 a^{2000-1950} = .35 \Rightarrow$
$a^{50} = 7 \Rightarrow a = \sqrt[50]{7} \approx 1.04$
Finally, we have the function
$f(x) = .05(1.04)^{x-1950}$.

**(b)** If you think of $f(x) = .05(1.04)^{x-1950}$ as

a function in the form of $A = P\left(1 + \frac{r}{n}\right)^{tn}$

where $n = 1$ (compounded annually), we have

$$f(x) = .05(1+.04)^{x-1950} = P(1+r)^t.$$

Thus, the average annual percent increase would be about 4%.

17. From Example 6, we have the temperature, $f(t)$, of a body at time t after being introduced into an environment having constant temperature $T_0$ is $f(t) = T_0 + Ce^{-kt}$, where $C$ and $k$ are constants. From the given information, when $t = 0$, $T_0 = 0$, and the temperature of the water is $f(0) = 100$.

$$100 = 0 + Ce^{-0k} \Rightarrow 100 = C$$

Thus, we have $f(t) = 0 + 100e^{-kt} = 100e^{-kt}$.

Also, when $t = \frac{24}{60} = \frac{2}{5}$ hr, $f\left(\frac{2}{5}\right) = 50$. Using this information, we have

$$50 = 100e^{-(2/5)k} \Rightarrow \tfrac{1}{2} = e^{-(2/5)k}$$

$$\ln\tfrac{1}{2} = \ln e^{-(2/5)k} \Rightarrow \ln\tfrac{1}{2} = -\tfrac{2}{5}k$$

$$\ln 1 - \ln 2 = -\tfrac{2}{5}k \Rightarrow \ln 2 = \tfrac{2}{5}k$$

$$k = \tfrac{5}{2}\ln 2 \approx 1.733$$

Thus, the model is $f(t) = 100e^{-1.733t}$.

To find the temperature after $\frac{96}{60} = \frac{8}{5}$ hrs, we find $f\left(\frac{8}{5}\right)$. Since

$f\left(\frac{8}{5}\right) = 100e^{-1.733(8/5)} \approx 6.25$, the temperature after 96 minutes is about $6.25°\,C$. Note: We could have used the exact value of $k$ to perform the calculations.

$$f\left(\tfrac{8}{5}\right) = 100e^{-\left(\frac{5}{2}\ln 2\right)\left(\frac{8}{5}\right)} = 100e^{-4\ln 2}$$

$$= 100e^{\ln\frac{1}{16}} = 100 \cdot \tfrac{1}{16} = 6.25$$

19. Given $P = 60,000$ and $t = 5$, substitute $r = .07$ and $n = 4$ into the compound interest formula, $A = P\left(1 + \frac{r}{n}\right)^{tn}$. We have

$$A = 60,000\left(1 + \tfrac{.07}{4}\right)^{5(4)}$$

$$= 60,000(1.0175)^{20} \approx 84,886.692.$$

The interest from this investment would be $84,886.69 – $60,000 = $24,886.69. Given $P = 60,000$ and $t = 5$, substitute $r = .0675$ into the continuous compounding formula, $A = Pe^{rt}$.

$$A = 60,000e^{.0675(5)} = 60,000e^{.3375} \approx 84,086.377$$

The interest from this investment would be $84,086.38 – $60,000 = $24,086.38.

**(a)** The investment that offers 7% compounded quarterly will earn more interest than the investment that offers 6.75% compounded continuously.

**(b)** Note that $24,886.69 – $24,086.38 = $800.31. The investment that offers 7% compounded quarterly will earn $800.32 more in interest. Note: Keeping all digits in the calculator rather than rounding the two amounts of interest to the nearest cent before subtracting them will yield a final answer of 800.315242 ≈ $800.32. This discrepancy is insignificant.

21. $$A = Pe^{rt}$$
$$2P = Pe^{.025t}$$
$$2 = e^{.025t}$$
$$\ln 2 = .025t$$
$$27.73 \approx t$$

The doubling time is about 27.73 yr if interest is compounded continuously.

23. $$A = Pe^{rt}$$
$$3P = Pe^{.05t} \Rightarrow 3 = e^{.05t} \Rightarrow \ln 3 = \ln e^{.05t}$$
$$\ln 3 = .05t \Rightarrow t = \tfrac{\ln 3}{.05} \approx 21.97$$

It will take about 21.97 years for the investment to triple.

25. **(a)** 1969 is represented by $t = 4$
$$M(4) = 434e^{-.08(4)} \approx 315$$
There were 315 continuously serving members in 1969.

**(b)** 1973 is represented by $t = 8$
$$M(8) = 434e^{-.08(8)} \approx 229$$
There were 229 continuously serving members in 1973.

**(c)** 1979 is represented by $t = 14$
$$M(14) = 434e^{-.08(14)} \approx 142$$
There were 142 continuously serving members in 1973.

**27.** **(a)** A point associated with the graph of
$f(x) = P_0 a^{x-2000}$ is $(2000, 1)$. Since
$f(2000) = P_0 a^{2000-2000} = 1$, we have
$P_0 a^0 = 1 \Rightarrow P_0 = 1$. Thus, we have
$f(x) = a^{x-2000}$. Since the point
$(2025, 1.4)$ is projected to be on the
graph, we have the following.
$f(2025) = a^{2025-2000} = 1.4$
$a^{25} = 1.4 \Rightarrow a = \sqrt[25]{1.4} \approx 1.01355$

**(b)** From part (a) we have
$f(x) = (1.01355)^{x-2000}$. To find the
population projected for 2010, we must
find $f(2010)$. Thus
$f(2010) = (1.01355)^{2010-2000}$
$= (1.01355)^{10} \approx 1.14$ billion is
the estimated population in 2010.

**(c)** We must solve $(1.01355)^{x-2000} = 1.5$ for $x$.
$(1.01355)^{x-2000} = 1.5$
$\ln(1.01355)^{x-2000} = \ln 1.5$
$(x-2000)\ln(1.01355) = \ln 1.5$
$x - 2000 = \frac{\ln 1.5}{\ln(1.01355)}$
$x = 2000 + \frac{\ln 1.5}{\ln(1.01355)}$
$\approx 2030.13$
In 2030, it is projected that the population
will reach 1.5 billion.

**29.** $f(x) = 14.621e^{.141x}$

**(a)** $t = 4$ represents 2004.
$f(4) = 14.621e^{.141(4)} \approx 25.7$
In 2004, gaming revenues were about
$25.7 billion.

**(b)** Find $t$ when $f(t) = 22.3$.
$22.3 = 14.621e^{.141x} \Rightarrow \frac{22.3}{14.621} = e^{.141x}$
$\ln\left(\frac{22.3}{14.621}\right) = .141x \Rightarrow \frac{\ln\left(\frac{22.3}{14.621}\right)}{.141} = x$
$x \approx 3$
Revenues reached $22.3 billion in 2003.

**31.** $L = 9 + 2e^{.15t}$

**(a)** In 1988, $t = 6$, so $L = 9 + 2e^{.15(6)} \approx 13.92$

**(b)** In 1998, $t = 16$, so
$L = 9 + 2e^{.15(16)} \approx 31.05$

**(c)** In 2008, $t = 26$, so
$L = 9 + 2e^{.15(26)} \approx 107.8$

**33.** $f(t) = 15,000e^{-.05t}$

**(a)** At the beginning of the epidemic, $t = 0$.
$f(0) = 15,000e^{-.05(0)} = 15,000$
At the beginning of the epidemic, 15,000
people were susceptible.

**(b)** After 10 days, $t = 10$.
$f(0) = 15,000e^{-.05(10)} \approx 9098$
After 10 days, approximately 9098 people
were susceptible.

**(c)** After 3 weeks, $t = 21$.
$f(21) = 15,000e^{-.05(21)} \approx 5249$
After three weeks, approximately 5249
people were susceptible.

**35.** $f(t) = 500e^{.1t}$

**(a)** $f(2) = 500e^{.1(2)} \approx 611$
At two days, the bacteria count is
approximately 611 million.

**(b)** $f(4) = 500e^{.1(4)} \approx 746$
At four days, the bacteria count is
approximately 746 million.

**(c)** $f(7) = 500e^{.1(7)} \approx 1007$
At one week (seven days), the bacteria
count is approximately 1007 million.

**37.** $f(t) = 200(.90)^{t-1}$
Find $t$ when $f(t) = 50$.
$50 = 200(.90)^{t-1}$
$.25 = (.90)^{t-1}$
$\ln .25 = \ln\left[(.90)^{t-1}\right]$
$\ln .25 = (t-1)\ln .90$
$t - 1 = \frac{\ln .25}{\ln .90} \Rightarrow t = 1 + \frac{\ln .25}{\ln .90} \approx 14.2$
The dose will reach a level of 50 mg in about
14.2 hr.

**39.** $A(t) = 100e^{.024t}$
We want to find the year in which the CPI will
be 175.
$175 = 100e^{.024t} \Rightarrow 1.75 = e^{.024t}$
$\ln 1.75 = \ln e^{.024t} \Rightarrow \ln 1.75 = .024t$
$t = \frac{\ln 1.75}{.024} \approx 23.3$
Twenty-three years after 1990, or in 2013,
costs were 75% than in 1990.

**41.** $S(t) = 50,000e^{-.1t}$

Find $t$ when $S(t) = 25,000$.

$25,000 = 50,000e^{-.1t} \Rightarrow .5 = e^{-.1t}$

$\ln .5 = \ln e^{-.1t} \Rightarrow \ln .5 = -.1t \Rightarrow t = \frac{\ln .5}{-.1} \approx 6.9$

It will take about 6.9 yr for sales to fall to half the initial sales.

**43.** Use the formula for continuous compounding with $r = .06$.

$A = Pe^{rt} \Rightarrow 2P = Pe^{.06t} \Rightarrow 2 = e^{.06t}$

$\ln 2 = \ln e^{.06t} \Rightarrow \ln 2 = .06t \Rightarrow t = \frac{\ln 2}{.06} \approx 11.6$

It will take about 11.6 yr before twice as much electricity is needed.

**45.** $f(x) = \frac{.9}{1 + 271e^{-.122x}}$

**(a)** $f(25) = \frac{.9}{1 + 271e^{-.122(25)}} = \frac{.9}{1 + 271e^{-3.05}} \approx .065$

$f(65) = \frac{.9}{1 + 271e^{-.122(65)}} = \frac{.9}{1 + 271e^{-7.93}} \approx .820$

Among people age 25, 6.5% have some CHD, while among people age 65, 82% have some CHD.

**(b)** $.50 = \frac{.9}{1 + 271e^{-.122x}}$

$.50(1 + 271e^{-.122x}) = .9$

$.5 + 135.5e^{-.122x} = .9$

$135.5e^{-.122x} = .4$

$e^{-.122x} = \frac{.4}{135.5}$

$\ln e^{-.122x} = \ln \frac{.4}{135.5}$

$-.122x = \ln \frac{.4}{135.5}$

$x = \frac{\ln \frac{.4}{135.5}}{-.122} \approx 47.75$

At about 48, the likelihood of coronary heart disease is 50%.

## Summary Exercises on Functions: Domains and Defining Equations

**1.** $f(x) = 3x - 6$

Domain: $(-\infty, \infty)$

**3.** $f(x) = |x + 4|$

Domain: $(-\infty, \infty)$

**5.** $f(x) = \frac{-2}{x^2 + 7}$

The domain is the set of all real numbers such that $x^2 + 7 \neq 0 \Rightarrow$ there is no real solution.

Domain: $(-\infty, \infty)$

**7.** $f(x) = \frac{x^2 + 7}{x^2 - 9}$

The domain is the set of all real numbers such that $x^2 - 9 \neq 0 \Rightarrow x \neq -3$ or $x \neq 3$

Domain: $(-\infty, -3) \cup (-3, 3) \cup (3, \infty)$

**9.** $f(x) = \log_5(16 - x^2)$

The domain is the set of all real numbers such that $16 - x^2 > 0 \Rightarrow 16 > x^2 \Rightarrow 4 > x$ and $-4 < x$.

Domain: $(-4, 4)$

**11.** $f(x) = \sqrt{x^2 - 7x - 8}$

The domain is the set of all real numbers such that $x^2 - 7x - 8 \geq 0$. Solve the equation to find the test intervals: $x^2 - 7x - 8 = 0 \Rightarrow (x - 8)(x + 1) = 0 \Rightarrow x = 8$ or $x = -1$

| Interval | Test Point | Value of $x^2 - 7x - 8$ | Sign of $x^2 - 7x - 8$ |
|---|---|---|---|
| $(-\infty, -1)$ | $-2$ | 10 | Positive |
| $(-1, 8)$ | 0 | $-8$ | Negative |
| $(8, \infty)$ | 10 | 22 | Positive |

Domain: $(-\infty, -1] \cup [8, \infty)$

**13.** $f(x) = \frac{1}{2x^2 - x + 7}$

The domain is the set of all real numbers such that $2x^2 - x + 7 \neq 0$

$x = \frac{-(-1) \pm \sqrt{(-1)^2 - 4(2)(7)}}{2(2)} = \frac{1 \pm \sqrt{-55}}{4} \Rightarrow$ there are no real solutions. Domain: $(-\infty, \infty)$

**15.** $f(x) = \sqrt{x^3 - 1}$

The domain is the set of all real numbers such that $x^3 - 1 \geq 0 \Rightarrow x^3 \geq 1 \Rightarrow x \geq 1$

Domain: $[1, \infty)$

**17.** $f(x) = e^{x^2 + x + 4}$

The domain is the set of values such that $x^2 + x + 4$ is real. Domain: $(-\infty, \infty)$

**19.** $f(x) = \sqrt{\frac{-1}{x^3 - 1}}$

The domain is the set of all real numbers such that $\sqrt{\frac{-1}{x^3 - 1}}$ is defined or $\frac{-1}{x^3 - 1} \geq 0$. $\frac{-1}{x^3 - 1}$ is defined for $x^3 - 1 \neq 0 \Rightarrow x^3 \neq 1 \Rightarrow x \neq 1$

*(continued on next page)*

*(continued from page 235)*

| Interval | Test Point | Value of $\frac{-1}{x^3-1}$ | Sign of $\frac{-1}{x^3-1}$ |
|---|---|---|---|
| $(-\infty, 1)$ | 0 | 1 | Positive |
| $(1, \infty)$ | 2 | $-\frac{1}{7}$ | Negative |

Domain: $(-\infty, 1)$

**21.** $f(x) = \ln\left(x^2 + 1\right)$

The domain is the set of all real numbers such that $x^2 + 1 > 0$. Domain: $(-\infty, \infty)$

**23.** $f(x) = \log\left(\frac{x+2}{x-3}\right)^2$

Since $\left(\frac{x+2}{x-3}\right)^2 \geq 0$ for all real numbers, the domain of $f(x)$ is the set of all real numbers such that $\frac{x+2}{x-3} \neq 0$. $\frac{x+2}{x-3} = 0$ when $x = -2$, and $\frac{x+2}{x-3}$ is undefined when $x = 3$.
Domain: $(-\infty, -2) \cup (-2, 3) \cup (3, \infty)$

**25.** $f(x) = e^{|1/x|}$

The domain is the set of all real numbers such that $\frac{1}{x}$ is defined, or $x \neq 0$
Domain: $(-\infty, 0) \cup (0, \infty)$

**27.** $f(x) = x^{100} - x^{50} + x^2 + 5$; Domain: $(-\infty, \infty)$

**29.** $f(x) = \sqrt[4]{16 - x^4}$

The domain is the set of all real numbers such that $16 - x^4 \geq 0$. Solve the equation to find the test intervals: $16 - x^4 = 0 \Rightarrow$
$(2 - x)(2 + x)(4 + x^2) = 0 \Rightarrow x = 2$ or $x = -2$

| Interval | Test Point | Value of $16 - x^4$ | Sign of $16 - x^4$ |
|---|---|---|---|
| $(-\infty, -2)$ | $-3$ | $-65$ | Negative |
| $(-2, 2)$ | 0 | 16 | Positive |
| $(2, \infty)$ | 3 | $-65$ | Negative |

Domain: $[-2, 2]$

**31.** $f(x) = \sqrt{\frac{x^2 - 2x - 63}{x^2 + x - 12}}$

The domain is the set of real numbers such that $\frac{x^2 - 2x - 63}{x^2 + x - 12} \geq 0$. $\frac{x^2 - 2x - 63}{x^2 + x - 12}$ is not defined for $x^2 + x - 12 = 0 \Rightarrow (x + 4)(x - 3) = 0 \Rightarrow x \neq -4$ or $x \neq 3$. Solve $\frac{x^2 - 2x - 63}{x^2 + x - 12} = 0$ to find the test intervals: $\frac{x^2 - 2x - 63}{x^2 + x - 12} = 0 \Rightarrow$
$x^2 - 2x - 63 = 0 \Rightarrow (x - 9)(x + 7) = 0 \Rightarrow x = 9$ or $x = -7$.

| Interval | Test Point | Value of $\frac{x^2 - 2x - 63}{x^2 + x - 12}$ | Sign |
|---|---|---|---|
| $(-\infty, -7)$ | $-10$ | $\frac{19}{26}$ | Positive |
| $(-7, -4)$ | $-5$ | $-\frac{7}{2}$ | Negative |
| $(-4, 3)$ | 0 | $\frac{21}{4}$ | Positive |
| $(3, 9)$ | 5 | $-\frac{8}{3}$ | Negative |
| $(9, \infty)$ | 10 | $\frac{17}{98}$ | Positive |

Domain: $(-\infty, -7] \cup (-4, 3) \cup [9, \infty)$

**33.** $f(x) = \left|\sqrt{5 - x}\right|$

The domain is the set of real numbers such that $5 - x \geq 0 \Rightarrow 5 \geq x$
Domain: $(-\infty, 5]$

**35.** $f(x) = \log\left|\frac{1}{4 - x}\right|$

The domain is the set of real numbers such that $\left|\frac{1}{4-x}\right| > 0 \Rightarrow \frac{1}{4-x} > 0 \Rightarrow 4 - x > 0 \Rightarrow 4 > x$ or $-\frac{1}{4-x} < 0 \Rightarrow -4 + x < 0 \Rightarrow x < 4$
Domain: $(-\infty, 4) \cup (4, \infty)$

**37.** $f(x) = 6^{\sqrt{x^2 - 25}}$

The domain is the set of real numbers such that $\sqrt{x^2 - 25}$ is a real number.
$x^2 - 25 \geq 0 \Rightarrow x^2 \geq 25 \Rightarrow x \geq 5$ or $x \leq -5$
Domain: $(-\infty, -5] \cup [5, \infty)$

**39.** $f(x) = \ln\left(\frac{-3}{(x+2)(x-6)}\right)$

The domain is the set of real numbers such that $\frac{-3}{(x+2)(x-6)} > 0$ and $(x + 2)(x - 6) \neq 0 \Rightarrow x \neq -2$ or $x \neq 6$.

| Interval | Test Point | Value of $\frac{-3}{(x+2)(x-6)}$ | Sign of $\frac{-3}{(x+2)(x-6)}$ |
|---|---|---|---|
| $(-\infty, -2)$ | $-3$ | $-\frac{1}{3}$ | Negative |
| $(-2, 6)$ | $0$ | $\frac{1}{4}$ | Positive |
| $(6, \infty)$ | $7$ | $-\frac{1}{3}$ | Negative |

Domain: $(-2, 6)$

**41.** Choice A can be written as a function of $x$.

$3x + 2y = 6 \Rightarrow y = f(x) = -\frac{3}{2}x + 3$

**43.** Choice C can be written as a function of $x$.

$x^3 + y^3 = 5 \Rightarrow y = f(x) = \sqrt[3]{5 - x^3}$

**45.** Choice A can be written as a function of $x$.

$x = \frac{2-y}{y+3} \Rightarrow y = f(x) = \frac{2-3x}{x+1}$

**47.** Choice D can be written as a function of $x$.

$2x = \frac{1}{y^3} \Rightarrow y = f(x) = \sqrt[3]{\frac{1}{2x}}$

**49.** Choice C can be written as a function of $x$.

$\frac{x}{4} - \frac{y}{9} = 0 \Rightarrow y = f(x) = \frac{9x}{4}$

## Chapter 4: Review Exercises

**1.** This is not a one-to-one function since a horizontal line can intersect the graph in more than one point.

**3.** $y = 5x - 4$

Looking at this function graphed on a TI-83, we can see that it appears that any horizontal line passed through the function will intersect the graph in at most one place.

 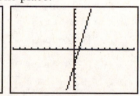

If we attempt to find the inverse function, we see that this function can be found.

*Step 1*: Interchange $x$ and $y$. $x = 5y - 4$

*Step 2*: Solve for $y$.

$x = 5y - 4 \Rightarrow x + 4 = 5y \Rightarrow y = \frac{x+4}{5}$

*Step 3*: Replace $y$ with $f^{-1}(x)$. $f^{-1}(x) = \frac{x+4}{5}$

Also, an acceptable way of showing that a function is one-to-one is to assume that you have two equal $y$-values $\left(f(x_1) = f(x_2)\right)$ and show that they must have come from the same $x$-value $(x_1 = x_2)$.

$f(x_1) = f(x_2) \Rightarrow 5x_1 - 4 = 5x_2 - 4 \Rightarrow$
$\qquad 5x_1 = 5x_2 \Rightarrow x_1 = x_2$

So, the function is one-to-one.

**5.** $y = (x + 3)^2$

If $x = -2$, $y = (-2 + 3)^2 = 1^2 = 1$.

If $x = -4$, $y = (-4 + 3)^2 = (-1)^2 = 1$.

Since two different values of $x$ lead to the same value of $y$, the function is not one-to-one.

**7.** $f(x) = x^3 - 3$

This function is one-to-one.

*Step 1*: Replace $f(x)$ with $y$ and interchange $x$ and $y$. $y = x^3 - 3 \Rightarrow x = y^3 - 3$

*Step 2*: Solve for $y$.

$x = y^3 - 3 \Rightarrow x + 3 = y^3 \Rightarrow y = \sqrt[3]{x+3}$

*Step 3*: Replace $y$ with $f^{-1}(x)$.

$f^{-1}(x) = \sqrt[3]{x+3}$

**9.** $f^{-1}(\$50,000)$ represents the number of years after 2004 required for the investment to reach $50,000.

**11.** To have an inverse, a function must be a <u>one-to-one</u> function.

**13.** $y = \log_{.3} x$

The point $(1, 0)$ is on the graph of every function of the form $y = \log_a x$, so the correct choice must be either B or C. Since the base is $a = .3$ and $0 < .3 < 1$, $y = \log_{.3} x$ is a decreasing function, and so the correct choice must be B.

**15.** $y = \ln x = \log_e x$

The point $(1, 0)$ is on the graph of every function of the form $y = \log_a x$, so the correct choice must be either B or C. Since the base is $a = e$ and $e > 1$, $y = \ln x$ is an increasing function, and so the correct choice must be C.

**17.** $2^5 = 32$ is written in logarithmic form as $\log_2 32 = 5$.

**19.** $\left(\frac{3}{4}\right)^{-1} = \frac{4}{3}$ is written in logarithmic form as $\log_{3/4} \frac{4}{3} = -1$.

**21.** $\log_3 4$ is the logarithm with the base 3 of 4. ($\log_4 3$ would be the logarithm with the base 4 of 3.)

**23.** $\log 1000 = 3$ is written in exponential form as $10^3 = 1000$.

**25.** Let $f(x) = \log_a x$ be the required function. Then $f(81) = 4 \Rightarrow \log_a 81 = 4 \Rightarrow a^4 = 81 \Rightarrow a^4 = 3^4 \Rightarrow a = 3$. The base is 3.

**27.** $\log_3 \frac{mn}{5r} = \log_3 mn - \log_3 5r$
$= \log_3 m + \log_3 n - (\log_3 5 + \log_3 r)$
$= \log_3 m + \log_3 n - \log_3 5 - \log_3 r$

**29.** $\log_7(7k + 5r^2)$
Since this is the logarithm of a sum, this expression cannot be simplified.

**31.** $\log .0411 \approx -1.3862$

**33.** $\ln 144{,}000 \approx 11.8776$

**35.** To find $\log_{2/3} \frac{5}{8}$, use the change-of-base theorem. We have
$\log_{2/3} \frac{5}{8} = \frac{\log \frac{5}{8}}{\log \frac{2}{3}} = \frac{\ln \frac{5}{8}}{\ln \frac{2}{3}} \approx 1.1592$.

**37.** $16^{x+4} = 8^{3x-2} \Rightarrow \left(2^4\right)^{x+4} = \left(2^3\right)^{3x-2} \Rightarrow$
$2^{4x+16} = 2^{9x-6} \Rightarrow 4x + 16 = 9x - 6 \Rightarrow$
$22 = 5x \Rightarrow \frac{22}{5} = x$
Solution set: $\left\{\frac{22}{5}\right\}$

**39.** $3^{2x-5} = 13 \Rightarrow \ln 3^{2x-5} = \ln 13 \Rightarrow$
$(2x-5)\ln 3 = \ln 13 \Rightarrow 2x - 5 = \frac{\ln 13}{\ln 3} \Rightarrow$
$2x = 5 + \frac{\ln 13}{\ln 3}$
$x = \frac{1}{2}\left(5 + \frac{\ln 13}{\ln 3}\right) \approx 3.667$ or

$3^{2x-5} = 13 \Rightarrow \ln 3^{2x-5} = \ln 13$
$(2x - 5)\ln 3 = \ln 13$
$2x \ln 3 - 5 \ln 3 = \ln 13$
$x \ln 3^2 - \ln 3^5 = \ln 13$
$x \ln 9 - \ln 243 = \ln 13$
$x \ln 9 = \ln 13 + \ln 243$
$x \ln 9 = \ln 3159 \Rightarrow x = \frac{\ln 3159}{\ln 9} \approx 3.667$
Solution set: $\{3.667\}$

**41.** $6^{x+3} = 4^x \Rightarrow \ln 6^{x+3} = \ln 4^x \Rightarrow$
$(x+3)\ln 6 = x \ln 4$
$x \ln 6 + 3 \ln 6 = x \ln 4$
$x \ln 6 - x \ln 4 = -3 \ln 6$
$x(\ln 6 - \ln 4) = -3 \ln 6 \Rightarrow x\left(\ln \frac{6}{4}\right) = -\ln 6^3 \Rightarrow$
$x\left(\ln \frac{3}{2}\right) = -\ln 216$
$x = \frac{-\ln 216}{\ln \frac{3}{2}} \approx -13.257$
Solution set: $\{-13.257\}$

**43.** $e^{2-x} = 12 \Rightarrow \ln e^{2-x} = \ln 12 \Rightarrow$
$2 - x = \ln 12 \Rightarrow -x = -2 + \ln 12 \Rightarrow$
$x = 2 - \ln 12 \approx -.485$
Solution set: $\{-.485\}$

**45.** $10e^{3x-7} = 5 \Rightarrow e^{3x-7} = \frac{1}{2} \Rightarrow$
$\ln e^{3x-7} = \ln \frac{1}{2} \Rightarrow 3x - 7 = \ln \frac{1}{2} \Rightarrow$
$3x = \ln \frac{1}{2} + 7 \Rightarrow x = \frac{1}{3}\left(\ln \frac{1}{2} + 7\right) \approx 2.102$
Solution set: $\{2.102\}$

**47.** $6^{x-3} = 3^{4x+1}$
$\ln 6^{x-3} = \ln 3^{4x+1}$
$(x-3)\ln 6 = (4x+1)\ln 3$
$x \ln 6 - 3 \ln 6 = 4x \ln 3 + \ln 3$
$x \ln 6 - \ln 6^3 = x \ln 3^4 + \ln 3$
$x \ln 6 - \ln 216 = x \ln 81 + \ln 3$
$x \ln 6 - x \ln 81 = \ln 3 + \ln 216$
$x(\ln 6 - \ln 81) = \ln 3 + \ln 216$
$x = \frac{\ln 3 + \ln 216}{\ln 6 - \ln 81} = \frac{\ln(3 \cdot 216)}{\ln \frac{6}{81}} = \frac{\ln 648}{\ln \frac{2}{27}}$
$x \approx -2.487$
Solution set: $\{-2.487\}$

**49.** $e^{6x} \cdot e^x = e^{21} \Rightarrow e^{7x} = e^{21} \Rightarrow 7x = 21 \Rightarrow x = 3$
Solution set: $\{3\}$

**51.** $3 \ln x = 13 \Rightarrow \ln x = \frac{13}{3} \Rightarrow x = e^{13/3}$
Solution set: $\left\{e^{13/3}\right\}$

**53.** $\log(2x+7) = .25 \Rightarrow 2x+7 = 10^{.25} = \sqrt[4]{10} \Rightarrow$

$x = \frac{\sqrt[4]{10}-7}{2}$

Solution set: $\left\{\frac{\sqrt[4]{10}-7}{2}\right\}$

**55.** $\log x + \log(13-3x) = 1 \Rightarrow$

$\log[x(13-3x)] = 1 \Rightarrow 13x-3x^2 = 10^1 \Rightarrow$

$13x-3x^2 = 10 \Rightarrow 3x^2 -13x+10 = 0 \Rightarrow$

$(3x-10)(x-1) = 0 \Rightarrow x = \frac{10}{3}$ or $x=1$

Solution set: $\left\{1, \frac{10}{3}\right\}$

**57.** $\ln(6x) - \ln(x+1) = \ln 4$

$\ln \frac{6x}{x+1} = \ln 4 \Rightarrow \frac{6x}{x+1} = 4$

$6x = 4(x+1) \Rightarrow$

$6x = 4x+4 \Rightarrow 2x = 4 \Rightarrow x = 2$

Solution set: $\{2\}$

**59.** $\ln\left[\ln\left(e^{-x}\right)\right] = \ln 3$

$\ln(-x) = \ln 3 \Rightarrow -x = 3 \Rightarrow x = -3$

Solution set: $\{-3\}$

**61.** $\frac{d}{10} = \log\left(\frac{I}{I_0}\right) \Rightarrow 10^{d/10} = \frac{I}{I_0} \Rightarrow$

$I_0\left(10^{d/10}\right) = I \Rightarrow I_0 = \frac{I}{10^{d/10}}$

**63.**

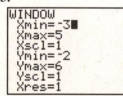 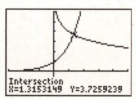

Solution set: $\{1.315\}$

**65. (a)** $8.3 = \log_{10}\frac{I}{I_0} \Rightarrow \frac{I}{I_0} = 10^{8.3} \Rightarrow$

$I = 10^{8.3}I_0 \approx 199,526,231.5I_0$

The magnitude was about $200,000,000I_0$.

**(b)** $7.1 = \log_{10}\frac{I}{I_0} \Rightarrow \frac{I}{I_0} = 10^{7.1} \Rightarrow$

$I = 10^{7.1}I_0 \approx 12,589,254.12I_0$

The magnitude was about $13,000,000I_0$.

**(c)** $\frac{200,000,000I_0}{13,000,000I_0} = \frac{200}{13} \approx 15.38$

The 1906 earthquake had a magnitude more than 15 times greater than the 1989 earthquake. Note: If the more precise values found in parts (a) and (b) were used, the 1906 earthquake had a magnitude of almost 16 times greater than the 1989 earthquake.

**67.** Substitute $A = 5760$, $P = 3500$, $t = 10$, $n = 1$ into the formula $A = P\left(1+\frac{r}{n}\right)^{tn}$.

$5760 = 3500\left(1+\frac{r}{1}\right)^{10(1)}$

$\frac{288}{175} = (1+r)^{10}$

$\left(\frac{288}{175}\right)^{1/10} = 1+r$

$\left(\frac{288}{175}\right)^{1/10} - 1 = r$

$r \approx .051$

The annual interest rate, to the nearest tenth, is 5.1%.

**69.** First, substitute $P = 10,000$, $r = .08$, $t = 12$, and $n = 1$ into the formula $A = P\left(1+\frac{r}{n}\right)^{tn}$.

$A = 10,000\left(1+\frac{.08}{1}\right)^{12(1)}$

$= 10,000(1.08)^{12} \approx 25,181.70$

After the first 12 yr, there would be about $25,181.70 in the account. To finish off the 21-year period, substitute $P = 25,181.70$, $r = .10$, $t = 9$, and $n = 2$ into the formula $A = P\left(1+\frac{r}{n}\right)^{tn}$.

$A = 25,181.70\left(1+\frac{10}{2}\right)^{9(2)} = 25,181.70(1+.05)^{18}$

$= 25,181.70(1.05)^{18} \approx 60,602.76$

At the end of the 21-year period, about $60,606.76 would be in the account. Note: If it was possible to transfer the money accrued after the 12 years to the new account without rounding, the amount after the 21-year period would be $60,606.77. The difference is not significant.

**71.** To find $t$, substitute $a = 2$, $P = 1$, and $r = .04$ into $A = Pe^{rt}$ and solve.

$2 = 1 \cdot e^{.04t} \Rightarrow 2 = e^{.04t} \Rightarrow \ln 2 = \ln e^{.04t} \Rightarrow$

$\ln 2 = .04t \Rightarrow t = \frac{\ln 2}{.04} \approx 17.3$

It would take about 17.3 yr.

**73.** Double the 2003 total payoff value is $2(152.7) = 305.4$. Using the function $f(x) = 93.54e^{.16x}$, we solve for $x$ when $f(x) = 305.4$.

$93.54e^{.16x} = 305.4 \Rightarrow e^{.16x} = \frac{305.4}{93.54} \Rightarrow$

$\ln e^{.16x} = \ln\frac{305.4}{93.54} \Rightarrow .16x = \ln\frac{305.4}{93.54} \Rightarrow$

$x = \frac{\ln\frac{305.4}{93.54}}{.16} \approx 7.40$

Since $x$ represents the number of years since 2000, in 2007 the total payoff value will be double of 2003.

**75.** $f(x) = \log_4\left(2x^2 - x\right)$

(a) Use the change-of-base theorem with base $e$ to write the function as

$$f(x) = \frac{\ln\left(2x^2 - x\right)}{\ln 4}.$$

(b)

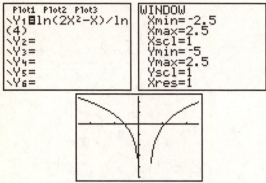

(c) From the graph, the $x$-intercepts are $-\frac{1}{2}$ and 1.

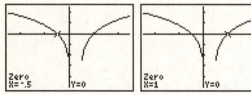

(d) From the graph, the vertical asymptotes are $x = 0$ and $x = \frac{1}{2}$. This be can verified algebraically. The vertical asymptotes will occur when $2x^2 - x = 0$.

$$2x^2 - x = 0 \Rightarrow x(2x - 1) = 0 \Rightarrow$$
$$x = 0 \text{ or } x = \frac{1}{2}$$

(e) To make a $y$-intercept, $x = 0$ must be in the domain, which is not the case here.

## Chapter 4: Test

**1. (a)** $f(x) = \sqrt[3]{2x - 7}$

Since it is a cube root, $2x - 7$ may be any real number. Domain: $(-\infty, \infty)$

Since the cube root of any real number is also any real number. Range: $(-\infty, \infty)$

**(b)** $f(x) = \sqrt[3]{2x - 7}$

The graph of $f$ passes the horizontal line test, and thus is a one-to-one function.

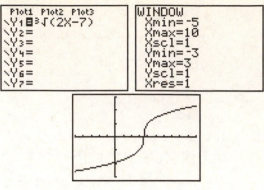

(c) *Step 1*: Replace $f(x)$ with $y$ and interchange $x$ and $y$.

$$y = \sqrt[3]{2x - 7} \Rightarrow x = \sqrt[3]{2y - 7}$$

*Step 2*: Solve for $y$.

$$x = \sqrt[3]{2y - 7} \Rightarrow x^3 = \left(\sqrt[3]{2y - 7}\right)^3 \Rightarrow$$
$$x^3 = 2y - 7 \Rightarrow x^3 + 7 = 2y \Rightarrow \frac{x^3 + 7}{2} = y$$

*Step 3*: Replace $y$ with $f^{-1}(x)$.

$$f^{-1}(x) = \frac{x^3 + 7}{2}$$

(d) Since the domain and range of $f$ are $(-\infty, \infty)$, the domain and range of $f^{-1}$ are also $(-\infty, \infty)$.

(e)

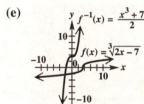

These graphs are reflections of each other across the line $y = x$.

**2. (a)** $y = \log_{1/3} x$

The point $(1, 0)$ is on the graph of every function of the form $y = \log_a x$, so the correct choice must be either B or C. Since the base is $a = \frac{1}{3}$ and $0 < \frac{1}{3} < 1$, $y = \log_{1/3} x$ is a decreasing function, and so the correct choice must be B.

**(b)** $y = e^x$

The point $(0, 1)$ is on the graph since $e^0 = 1$, so the correct choice must be either A or D. Since the base is $e$ and $e > 1$, $y = e^x$ is an increasing function, and so the correct choice must be A.

**(c)** $y = \ln x$ or $y = \log_e x$

The point $(1, 0)$ is on the graph of every function of the form $y = \log_a x$, so the correct choice must be B or C. Since the base is $a = e$ and $e > 1$, $y = \ln x$ is an increasing function, and the correct choice must be C.

**(d)** $y = \left(\frac{1}{3}\right)^x$

The point $(0, 1)$ is on the graph since $\left(\frac{1}{3}\right)^0 = 1$, so the correct choice must be either A or D. Since the base is $\frac{1}{3}$ and $0 < \frac{1}{3} < 1$, $y = \left(\frac{1}{3}\right)^x$ is a decreasing function, and so the correct choice must be D.

**3.** $\left(\frac{1}{8}\right)^{2x-3} = 16^{x+1} \Rightarrow \left(2^{-3}\right)^{2x-3} = \left(2^4\right)^{x+1} \Rightarrow$
$2^{-3(2x-3)} = 2^{4(x+1)} \Rightarrow 2^{-6x+9} = 2^{4x+4} \Rightarrow$
$-6x+9 = 4x+4 \Rightarrow -10x+9 = 4 \Rightarrow$
$-10x = -5 \Rightarrow x = \frac{1}{2}$

Solution set: $\left\{\frac{1}{2}\right\}$

**4. (a)** $4^{3/2} = 8$ is written in logarithmic form as $\log_4 8 = \frac{3}{2}$.

**(b)** $\log_8 4 = \frac{2}{3}$ is written in exponential form as $8^{2/3} = 4$.

**5.** They are inverses of each other.

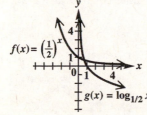

**6.** $\log_7 \frac{x^2 \sqrt[4]{y}}{z^3} = \log_7 x^2 + \log_7 \sqrt[4]{y} - \log_7 z^3$
$= \log_7 x^2 + \log_7 y^{1/4} - \log_7 z^3$
$= 2\log_7 x + \frac{1}{4}\log_7 y - 3\log_7 z$

**7.** $\log 2388 \approx 3.3780$

**8.** $\ln 2388 \approx 7.7782$

**9.** $\log_9 13 = \frac{\ln 13}{\ln 9} = \frac{\log 13}{\log 9} \approx 1.1674$

**10.** $\log_x \frac{9}{16} = 2 \Rightarrow x^2 = \frac{9}{16} \Rightarrow x = \pm\frac{3}{4}$

Since the negative solution is not in the domain it must be discarded.

Solution set: $\left\{\frac{3}{4}\right\}$

**11.** $9^x = 4 \Rightarrow x\log 9 = \log 4 \Rightarrow x = \frac{\log 4}{\log 9} \approx .631$

Solution set: $\{.631\}$

**12.**
$$2^{x+1} = 3^{x-4}$$
$$(x+1)\log 2 = (x-4)\log 3$$
$$x\log 2 + \log 2 = x\log 3 - 4\log 3$$
$$x\log 2 - x\log 3 = -\log 2 - 4\log 3$$
$$x(\log 2 - \log 3) = -\log 2 - 4\log 3$$
$$x = \frac{-\log 2 - 4\log 3}{\log 2 - \log 3} \approx 12.548$$

Solution set: $\{12.548\}$

**13.**
$$e^{.4x} = 4^{x-2} \Rightarrow .4x = (x-2)\ln 4$$
$$.4x = x\ln 4 - 2\ln 4$$
$$.4x - x\ln 4 = -2\ln 4 \Rightarrow x(.4 - \ln 4) = -2\ln 4$$
$$x = \frac{-2\ln 4}{.4 - \ln 4} \approx 2.811$$

Solution set: $\{2.811\}$

**14.** $\log_2 x + \log_2(x+2) = 3$
$$\log_2[x(x+2)] = 3$$
$$x^2 + 2x = 2^3 \Rightarrow x^2 + 2x - 8 = 0$$
$$(x+4)(x-2) = 0 \Rightarrow x = -4 \text{ or } x = 2$$

Since the negative solution is not in the domain it must be discarded.
Solution set: $\{2\}$

**15.** $\ln x - 4\ln 3 = \ln\left(\frac{1}{5}x\right)$
$$\ln x - \ln 3^4 = \ln\frac{x}{5}$$
$$\ln x - \ln 81 = \ln\frac{x}{5} \Rightarrow \ln\frac{x}{81} = \ln\frac{x}{5}$$
$$\frac{x}{81} = \frac{x}{5} \Rightarrow \frac{5}{81} = 1 \Rightarrow \text{ there is no}$$

solution. Solution set: $\varnothing$

**16.** $\log_3(x+1) - \log_3(x-3) = 2 \Rightarrow \log_3\frac{x+1}{x-3} = 2$
$$\frac{x+1}{x-3} = 3^2 \Rightarrow \frac{x+1}{x-3} = 9$$
$$x+1 = 9(x-3)$$
$$x+1 = 9x - 27$$
$$-8x = -28 \Rightarrow x = \frac{28}{8} = \frac{7}{2}$$

Solution set: $\left\{\frac{7}{2}\right\}$

**17.** Answers will vary. $\log_5 27$ is the exponent to which 5 must be raised in order to obtain 27. To approximate $\log_5 27$ on your calculator, use the change-of-base formula;
$\log_5 27 = \frac{\log 27}{\log 5} = \frac{\ln 27}{\ln 5} \approx 2.048$.

**18.** $v(t) = 176(1 - e^{-.18t})$

Find the time $t$ at which $v(t) = 147$.

$147 = 176(1 - e^{-.18t}) \Rightarrow \frac{147}{176} = 1 - e^{-.18t} \Rightarrow$

$-e^{-.18t} = \frac{147}{176} - 1 \Rightarrow -e^{-.18t} = -\frac{29}{176}$

$e^{-.18t} = \frac{29}{176} \Rightarrow \ln e^{-.18t} = \ln \frac{29}{176} \Rightarrow$

$-.18t = \ln \frac{29}{176} \Rightarrow t = \frac{\ln \frac{29}{176}}{-.18} \approx 10.02$

It will take the skydiver about 10 sec to attain the speed of 147 ft per sec (100 mph).

**19.  (a)**  Substitute $P = 5000$, $A = 18,000$, $r = .068$, and $n = 12$ into the formula

$A = P\left(1 + \frac{r}{n}\right)^{tn}$.

$18,000 = 5000\left(1 + \frac{.068}{12}\right)^{t(12)}$

$3.6 = \left(1 + \frac{.068}{12}\right)^{12t}$

$\ln 3.6 = \ln\left(1 + \frac{.068}{12}\right)^{12t}$

$\ln 3.6 = 12t \ln\left(1 + \frac{.068}{12}\right)$

$t = \frac{\ln 3.6}{12 \ln\left(1 + \frac{.068}{12}\right)} \approx 18.9$

It will take about 18.9 years.

**(b)**  Substitute $P = 5000$, $A = 18,000$, and $r = .068$, and into the formula $A = Pe^{rt}$.

$18,000 = 5000e^{.068t} \Rightarrow 3.6 = e^{.068t} \Rightarrow$

$\ln 3.6 = \ln e^{.068t} \Rightarrow \ln 3.6 = .068t \Rightarrow$

$t = \frac{\ln 3.6}{.068} \approx 18.8$

It will take about 18.8 years.

**20.**  Substitute $A = 3P$ and $r = .068$ into the continuous compounding formula $A = Pe^{rt}$, then solve for $t$:

$A = Pe^{rt}$

$3P = Pe^{.068t} \Rightarrow 3 = e^{.068t} \Rightarrow \ln 3 = \ln e^{.068t} \Rightarrow$

$\ln 3 = .068t \Rightarrow \frac{\ln 3}{.068} = t \Rightarrow t \approx 16.16$

It will take about 16.2 years for any amount of money to triple at 6.8% annual interest.

**21.**  $A(t) = 600e^{-.05t}$

**(a)**  $A(12) = 600e^{-.05(12)} = 600e^{-.6} \approx 329.3$

The amount of radioactive material present after 12 days is about 329.3 g.

**(b)**  Since $A(0) = 600e^{-.05(0)} = 600e^0 = 600$ g is the amount initially present, we seek to find $t$ when $A(t) = \frac{1}{2}(600) = 300$ g is present.

$300 = 600e^{-.05t} \Rightarrow .5 = e^{-.05t}$

$\ln .5 = \ln e^{-.05t} \Rightarrow \ln .5 = -.05t$

$t = \frac{\ln .5}{-.05} \approx 13.9$

The half-life of the material is about 13.9 days.

**22.**  Let $x = 0$ correspond to the year 2005. The population of New York (in millions) can be approximated by $y = 19.26e^{.0021x}$. The population of Florida (in millions) can be approximated by $y = 17.79e^{.0131x}$.

Graphing Calculator Solution:
Graph the two functions on the same screen and use the "intersect" option to find the $x$-coordinate of the intersection.

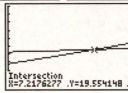

Algebraic Solution:
Set the two $y$-values equal to each other and solve for $x$.

$19.26e^{.0021x} = 17.79e^{.0131x} \Rightarrow \frac{19.26}{17.79} = \frac{e^{.0131x}}{e^{.0021x}}$

$\frac{19.26}{17.79} = e^{.0131x - .0021x} \Rightarrow \frac{19.26}{17.79} = e^{.011x}$

$\ln \frac{19.26}{17.79} = \ln e^{.011x} \Rightarrow \ln \frac{19.26}{17.79} = .011x$

$x = \frac{\ln \frac{19.26}{17.79}}{.011} \approx 7.2$

$x \approx 7.2$ corresponds to the middle of the year 2012. During 2012 the populations will be equal.

# Chapter 5

## SYSTEMS AND MATRICES

### Section 5.1: Systems of Linear Equations

1. In 2005, the trend "Unadjusted for Weather" shows a higher ozone level than the trend "Adjusted for Weather."

3. {(1998, .060), (2000, .059), (2001, .058), (2003, .059), (2005, .056)}

5. $t$ would represent time in years and $y$ would represent the ozone level in ppm.

7. 
$$4x + 3y = -13 \quad (1)$$
$$-x + y = 5 \quad (2)$$
Solve equation (2) for $y$.
$$-x + y = 5 \quad (2)$$
$$y = x + 5 \quad (3)$$
Replace $y$ with $x + 5$ in equation (1), and solve for $x$.
$$4x + 3y = -13 \Rightarrow 4x + 3(x + 5) = -13 \Rightarrow$$
$$4x + 3x + 15 = -13 \Rightarrow 7x + 15 = -13 \Rightarrow$$
$$7x = -28 \Rightarrow x = -4$$
Replace $x$ with $-4$ in equation (3) to obtain $y = -4 + 5 = 1$.
*Check:*

| $4x + 3y = -13$ (1) | $-x + y = 5$ (2) |
|---|---|
| $4(-4) + 3(1) = -13$ ? | $-(-4) + 1 = 5$ ? |
| $-16 + 3 = -13$ | $4 + 1 = 5$ |
| $-13 = -13$ True | $5 = 5$ True |

Solution set: $\{(-4, 1)\}$

9. 
$$x - 5y = 8 \quad (1)$$
$$x = 6y \quad (2)$$
Replace $x$ with $6y$ in equation (1), and solve for $y$: $x - 5y = 8 \Rightarrow 6y - 5y = 8 \Rightarrow y = 8$
Replace $y$ with 8 in equation (2) to obtain $x = 6(8) = 48$.

*Check:*

| $x - 5y = 8$ (1) | $x = 6y$ (2) |
|---|---|
| $48 - 5(8) = 8$ ? | $48 = 6(8)$ ? |
| $48 - 40 = 8$ | $48 = 48$ True |
| $8 = 8$ True | |

Solution set: $\{(48, 8)\}$

11. 
$$8x - 10y = -22 \quad (1)$$
$$3x + y = 6 \quad (2)$$
Solve equation (2) for $y$.
$$3x + y = 6 \quad (2)$$
$$y = -3x + 6 \quad (3)$$
Replace $y$ with $-3x + 6$ in equation (1), and solve for $x$.
$$8x - 10y = -22$$
$$8x - 10(-3x + 6) = -22$$
$$8x + 30x - 60 = -22$$
$$38x - 60 = -22$$
$$38x = 38$$
$$x = 1$$
Replace $x$ with 1 in equation (3) to obtain $y = -3(1) + 6 = 3$.
*Check:*

| $8x - 10y = -22$ (1) | $3x + y = 6$ (2) |
|---|---|
| $8(1) - 10(3) = -22$ ? | $3(1) + 3 = 6$ ? |
| $8 - 30 = -22$ | $3 + 3 = 6$ |
| $-22 = -22$ True | $6 = 6$ True |

Solution set: $\{(1, 3)\}$

13. 
$$7x - y = -10 \quad (1)$$
$$3y - x = 10 \quad (2)$$
Solve equation (1) for $y$.
$$7x - y = -10 \quad (1)$$
$$y = 7x + 10 \quad (3)$$
Replace $y$ with $7x + 10$ in equation (2), and solve for $x$.
$$3y - x = 10 \Rightarrow 3(7x + 10) - x = 10 \Rightarrow$$
$$21x + 30 - x = 10 \Rightarrow 20x + 30 = 10 \Rightarrow$$
$$20x = -20 \Rightarrow x = -1$$
Replace $x$ with $-1$ in equation (3) to obtain $y = 7(-1) + 10 = 3$.
*Check:*

| $7x - y = -10$ (1) | $3y - x = 10$ (2) |
|---|---|
| $7(-1) - 3 = -10$ ? | $3(3) - (-1) = 10$ ? |
| $-7 - 3 = -10$ | $9 + 1 = 10$ |
| $-10 = -10$ True | $10 = 10$ True |

Solution set: $\{(-1, 3)\}$

**15.** $-2x = 6y + 18$ (1)
$-29 = 5y - 3x$ (2)

Solve equation (1) for $x$.
$-2x = 6y + 18$ (1)
$x = -3y - 9$ (3)

Replace $x$ with $-3y - 9$ in equation (2), and solve for $y$.
$-29 = 5y - 3(-3y - 9)$
$-29 = 5y + 9y + 27 \Rightarrow -29 = 14y + 27 \Rightarrow$
$-56 = 14y \Rightarrow -4 = y$

Replace $y$ with $-4$ in equation (3) to obtain
$x = -3(-4) - 9 = 12 - 9 = 3$.

*Check*: $-2x = 6y + 18$ (1)
$-2(3) = 6(-4) + 18$ ?
$-6 = -24 + 18 \Rightarrow -6 = -6$ True
$-29 = 5y - 3x$ (2)
$-29 = 5(-4) - 3(3)$ ?
$-29 = -20 - 9 \Rightarrow -29 = -29$ True

Solution set: $\{(3, -4)\}$

**17.** $3y = 5x + 6$ (1)
$x + y = 2$ (2)

Solve equation (2) for $y$. (You could solve equation (2) for $x$ just as easily.)
$x + y = 2$ (2)
$y = 2 - x$ (3)

Replace $y$ with $2 - x$ in equation (1), and solve for $x$.
$3y = 5x + 6 \Rightarrow 3(2 - x) = 5x + 6 \Rightarrow$
$6 - 3x = 5x + 6 \Rightarrow 6 = 8x + 6 \Rightarrow$
$0 = 8x \Rightarrow 0 = x$

Replace $x$ with 0 in equation (3) to obtain $y = 2 - 0 = 2$.

*Check*: $3y = 5x + 6$ (1) | $x + y = 2$ (2)
$3(2) = 5(0) + 6$ ? | $0 + 2 = 2$ ?
$6 = 0 + 6$ | $2 = 2$ True
$6 = 6$ True

Solution set: $\{(0, 2)\}$

**19.** $3x - y = -4$ (1)
$x + 3y = 12$ (2)

Multiply equation (2) by $-3$ and add the result to equation (1).
$3x - y = -4$
$-3x - 9y = -36$
_____
$-10y = -40 \Rightarrow y = 4$

Substitute 4 for $y$ in equation (2) and solve for $x$: $x + 3(4) = 12 \Rightarrow x + 12 = 12 \Rightarrow x = 0$

*Check*:
$3x - y = -4$ (1) | $x + 3y = 12$ (2)
$3(0) - 4 = -4$ ? | $0 + 3(4) = 12$ ?
$0 - 4 = -4$ | $0 + 12 = 12$
$-4 = -4$ True | $12 = 12$ True

Solution set: $\{(0, 4)\}$

**21.** $2x - 3y = -7$ (1)
$5x + 4y = 17$ (2)

Multiply equation (1) by 4 and equation (2) by 3 and then add the resulting equations.
$8x - 12y = -28$
$15x + 12y = 51$
_____
$23x = 23 \Rightarrow x = 1$

Substitute 1 for $x$ in equation (2) and solve for $y$.
$5(1) + 4y = 17 \Rightarrow 5 + 4y = 17 \Rightarrow$
$4y = 12 \Rightarrow y = 3$

*Check*:
$2x - 3y = -7$ (1) | $5x + 4y = 17$ (2)
$2(1) - 3(3) = -7$ ? | $5(1) + 4(3) = 17$ ?
$2 - 9 = -7$ | $5 + 12 = 17$
$-7 = -7$ True | $17 = 17$ True

Solution set: $\{(1, 3)\}$

**23.** $5x + 7y = 6$ (1)
$10x - 3y = 46$ (2)

Multiply equation (1) by $-2$ and add to equation (2).

$-10x - 14y = -12$
$10x - 3y = 46$
_____
$-17y = 34 \Rightarrow y = -2$

Substitute $-2$ for $y$ in equation (2) and solve for $x$.
$10x - 3(-2) = 46 \Rightarrow 10x + 6 = 46$
$10x = 40 \Rightarrow x = 4$

*Check*: $5x + 7y = 6$ (1)
$5(4) + 7(-2) = 6$ ?
$20 - 14 = 6$
$6 = 6$ True

$10x - 3y = 46$ (2)
$10(4) - 3(-2) = 46$ ?
$40 + 6 = 46$
$46 = 46$ True

Solution set: $\{(4, -2)\}$

**25.**  $6x + 7y + 2 = 0$   (1)
$7x - 6y - 26 = 0$   (2)

Multiply equation (1) by 6 and equation (2) by 7 and then add the resulting equations.

$36x + 42y + 12 = 0$
$\underline{49x - 42y - 182 = 0}$
$85x \qquad - 170 = 0 \Rightarrow 85x = 170 \Rightarrow x = 2$

Substitute 2 for $x$ in equation (1).

$6(2) + 7y + 2 = 0$
$12 + 7y + 2 = 0$
$7y + 14 = 0$
$\qquad 7y = -14 \Rightarrow y = -2$

*Check*:   $6x + 7y + 2 = 0$   (1)
$6(2) + 7(-2) + 2 = 0$   ?
$12 - 14 + 2 = 0 \Rightarrow 0 = 0$ True
$7x - 6y - 26 = 0$   (2)
$7(2) - 6(-2) - 26 = 0$   ?
$14 + 12 - 26 = 0 \Rightarrow 0 = 0$ True

Solution set: $\{(2, -2)\}$

**27.**  $\frac{x}{2} + \frac{y}{3} = 4$   (1)
$\frac{3x}{2} + \frac{3y}{2} = 15$   (2)

To clear denominators, multiply equation (1) by 6 and equation (2) by 2.

$6\left(\frac{x}{2} + \frac{y}{3}\right) = 6(4)$
$2\left(\frac{3x}{2} + \frac{3y}{2}\right) = 2(15)$

This gives the system
$3x + 2y = 24$   (3)
$3x + 3y = 30$   (4).

Multiply equation (3) by $-1$ and add to equation (4).

$-3x - 2y = -24$
$\underline{3x + 3y = 30}$
$\qquad y = 6$

Substitute 6 for $y$ in equation (1).

$\frac{x}{2} + \frac{6}{3} = 4 \Rightarrow \frac{x}{2} + 2 = 4 \Rightarrow \frac{x}{2} = 2 \Rightarrow x = 4$

*Check*: $\frac{x}{2} + \frac{y}{3} = 4$   (1)   $\bigg|$   $\frac{3x}{2} + \frac{3y}{2} = 15$   (2)
$\frac{4}{2} + \frac{6}{3} = 4$   ?   $\bigg|$   $\frac{3(4)}{2} + \frac{3(6)}{2} = 15$   ?
$2 + 2 = 4$   $\bigg|$   $6 + 9 = 15$
$4 = 4$ True   $\bigg|$   $15 = 15$ True

Solution set: $\{(4, 6)\}$

**29.**  $\frac{2x-1}{3} + \frac{y+2}{4} = 4$ (1)
$\frac{x+3}{2} - \frac{x-y}{3} = 3$ (2)

Multiply equation (1) by 12 and equation (2) by 6 to clear denominators. Also, remove parentheses and combine like terms.

$12\left(\frac{2x-1}{3} + \frac{y+2}{4}\right) = 12(4)$
$4(2x - 1) + 3(y + 2) = 48$
$8x - 4 + 3y + 6 = 48 \Rightarrow 8x + 3y = 46$  (3)
$6\left(\frac{x+3}{2} - \frac{x-y}{3}\right) = 6(3)$
$3(x + 3) - 2(x - y) = 18$
$3x + 9 - 2x + 2y = 18 \Rightarrow x + 2y = 9$  (4)

Multiply equation (4) by $-8$ and then add the result to equation (3).

$8x + 3y = 46$
$\underline{-8x - 16y = -72}$
$\qquad -13y = -26 \Rightarrow y = 2$

Substitute 2 for $y$ into equation (4) and solve for $x$: $x + 2(2) = 9 \Rightarrow x + 4 = 9 \Rightarrow x = 5$

*Check*:

$\frac{2x-1}{3} + \frac{y+2}{4} = 4$   (1)   $\bigg|$   $\frac{x+3}{2} - \frac{x-y}{3} = 3$ (2)
$\frac{2(5)-1}{3} + \frac{2+2}{4} = 4$   ?   $\bigg|$   $\frac{5+3}{2} - \frac{5-2}{3} = 3$   ?
$\frac{10-1}{3} + \frac{4}{4} = 4$   $\bigg|$   $\frac{8}{2} - \frac{3}{3} = 3$
$\frac{9}{3} + 1 = 4$   $\bigg|$   $4 - 1 = 3$
$3 + 1 = 4$   $\bigg|$   $3 = 3$ True
$4 = 4$ True

Solution set: $\{(5, 2)\}$

**31.**  $9x - 5y = 1$ (1)
$-18x + 10y = 1$ (2)

Multiply equation (1) by 2 and add the result to equation (2).

$18x - 10y = 2$
$\underline{-18x + 10y = 1}$
$\qquad 0 = 3$

This is a false statement. The solution set is $\varnothing$, and the system is inconsistent.

**33.**  $4x - y = 9$ (1)
$-8x + 2y = -18$ (2)

Multiply equation (1) by 2 and add the result to equation (2).

$8x - 2y = 18$
$\underline{-8x + 2y = -18}$
$\qquad 0 = 0$

(*continued on next page*)

*(continued from page 245)*

This is a true statement. There are infinitely many solutions. We will now express the solution set with $y$ as the arbitrary variable. Solve equation (1) for $x$.

$$4x - y = 9 \Rightarrow 4x = y + 9 \Rightarrow x = \tfrac{y+9}{4}$$

Solution set: $\left\{ \left( \tfrac{y+9}{4}, y \right) \right\}$

**35.** $5x - 5y - 3 = 0$ (1)
    $x - y - 12 = 0$ (2)

Multiply equation (2) by $-5$ and then add the resulting equations.

$$\begin{aligned} 5x - 5y - \phantom{0}3 &= 0 \\ -5x + 5y + 60 &= 0 \\ \hline 57 &= 0 \end{aligned}$$

This is a false statement. The solution set is $\varnothing$, and the system is inconsistent.

**37.** $7x + 2y = \phantom{0}6$ (1)
    $14x + 4y = 12$ (2)

Multiply equation (1) by $-2$ and add the result to equation (2).

$$\begin{aligned} -14x - 4y &= -12 \\ 14x + 4y &= \phantom{-}12 \\ \hline 0 &= \phantom{-}0 \end{aligned}$$

This is a true statement. There are infinitely many solutions. We will express the solution set with $y$ as the arbitrary variable. Solve equation (1) for $x$.

$$7x + 2y = 6 \Rightarrow 7x = 6 - 2y \Rightarrow x = \tfrac{6-2y}{7}$$

Solution set: $\left\{ \left( \tfrac{6-2y}{7}, y \right) \right\}$

**39.** $4x - 5y = -11$
    $\phantom{4x} -5y = -4x - 11$
    $\phantom{4x} y = \tfrac{4}{5}x + \tfrac{11}{5}$
    $2x + y = 5$
    $\phantom{2x} y = -2x + 5$

Screen A is the correct choice.

**41.** Since $y = ax + b$ and the line passes through $(2, 0)$ and $(0, 3)$, we have the equations
$0 = a(2) + b$ and $3 = a(0) + b$
These becomes the following system.
    $2a + b = 0$ (1)
    $\phantom{2a} b = 3$ (2)

Substitute $b = 3$ into equation (1) to solve for $a$: $2a + 3 = 0 \Rightarrow 2a = -3 \Rightarrow a = -\tfrac{3}{2}$. Thus, the equation is $y = -\tfrac{3}{2}x + 3 \Rightarrow 3x + 2y = 6$.

The other line goes through the points $(0, 1)$ and $(-3, 0)$, so we have the equations
$1 = a(0) + b$ and $0 = a(-3) + b$. These become the following system:
    $b = 1$ (3)
    $-3a + b = 0$ (4)

Substitute $b = 1$ into equation (4), then solve for $a$:
$$-3a + 1 = 0 \Rightarrow a = \tfrac{1}{3}$$
Thus, the equation of this line is
$$y = \tfrac{1}{3}x + 1 \Rightarrow 3y = x + 3 \Rightarrow x - 3y = -3$$

**43.** Solve each equation for $y$.
$\tfrac{11}{3}x + y = .5 \Rightarrow y = .5 - \tfrac{11}{3}x$ and
$.6x - y = 3 \Rightarrow -y = 3 - .6x \Rightarrow y = .6x - 3$

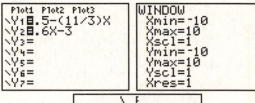

Solution set: $\{(.820, -2.508)\}$

**45.** Solve each equation for $y$.
$$\sqrt{7}x + \sqrt{2}y - 3 = 0 \Rightarrow \sqrt{2}y = 3 - \sqrt{7}x \Rightarrow y = \tfrac{3-\sqrt{7}x}{\sqrt{2}}$$
and $\sqrt{6}x - y - \sqrt{3} = 0 \Rightarrow y = \sqrt{6}x - \sqrt{3}$

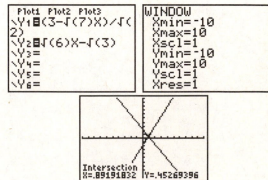

Solution set: $\{(.892, .453)\}$

**47.**  $x + y + z = 2$  (1)
$2x + y - z = 5$  (2)
$x - y + z = -2$ (3)

Eliminate $z$ by adding equations (1) and (2) to get $3x + 2y = 7$ (4).

Eliminate $z$ by adding equations (2) and (3) to get $3x = 3 \Rightarrow x = 1$ (5).

Using $x = 1$, find $y$ from equation (4) by substitution.

$3(1) + 2y = 7 \Rightarrow 3 + 2y = 7 \Rightarrow 2y = 4 \Rightarrow y = 2$

Substitute 1 for $x$ and 2 for $y$ in equation (1) to find $z$: $1 + 2 + z = 2 \Rightarrow 3 + z = 2 \Rightarrow z = -1$

Verify that the ordered triple $(1, 2, -1)$ satisfies all three equations.

*Check*:

$\begin{array}{l|l} x + y + z = 2 \;\; (1) & 2x + y - z = 5 \;\; (2) \\ 1 + 2 + (-1) = 2 \;\; ? & 2(1) + 2 - (-1) = 5 \;\; ? \\ \qquad 2 = 2 \text{ True} & \quad 2 + 2 + 1 = 5 \\ & \qquad 5 = 5 \text{ True} \end{array}$

$x - y + z = -2$ (3)
$1 - 2 + (-1) = -2$  ?
$-1 - 1 = -2$
$-2 = -2$ True

Solution set: $\{(1, 2, -1)\}$

**49.**  $x + 3y + 4z = 14$ (1)
$2x - 3y + 2z = 10$ (2)
$3x - y + z = 9$ (3)

Eliminate $y$ by adding equations (1) and (2) to get $3x + 6z = 24$ (4).

Multiply equation (3) by 3 and add the result to equation (1).

$\begin{array}{r} x + 3y + 4z = 14 \\ 9x - 3y + 3z = 27 \\ \hline 10x + \quad 7z = 41 \;\; (5) \end{array}$

Multiply equation (4) by 10 and equation (5) by $-3$ and add in order to eliminate $y$.

$\begin{array}{r} 30x + 60z = \quad 240 \\ -30x - 21z = -123 \\ \hline 39z = \quad 117 \Rightarrow z = 3 \end{array}$

Using $z = 3$, find $x$ from equation (5) by substitution.

$10x + 7(3) = 41 \Rightarrow 10x + 21 = 41 \Rightarrow$
$10x = 20 \Rightarrow x = 2$

Substitute 2 for $x$ and 3 for $z$ in equation (1) to find $y$.

$2 + 3y + 4(3) = 14 \Rightarrow 3y = 0 \Rightarrow y = 0$

Verify that the ordered triple $(2, 0, 3)$ satisfies all three equations.

*Check*:

$x + 3y + 4z = 14$  (1)
$2 + 3(0) + 4(3) = 14$  ?
$2 + 0 + 12 = 14$
$14 = 14$ True

$2x - 3y + 2z = 10$  (2)
$2(2) - 3(0) + 2(3) = 10$  ?
$4 - 0 + 6 = 10$
$10 = 10$ True

$3x - y + z = 9$  (3)
$3(2) - 0 + 3 = 9$  ?
$6 - 0 + 3 = 9$
$9 = 9$ True

Solution set: $\{(2, 0, 3)\}$

**51.**  $x + 4y - z = 6$  (1)
$2x - y + z = 3$  (2)
$3x + 2y + 3z = 16$ (3)

Eliminate $z$ by adding equations (1) and (2) to get $3x + 3y = 9$ or $x + y = 3$ (4).

Multiply equation (1) by 3 and add the result to equation (3).

$\begin{array}{r} 3x + 12y - 3z = 18 \\ 3x + \;\; 2y + 3z = 16 \\ \hline 6x + 14y \quad\;\; = 34 \;\; (5) \end{array}$

Multiply equation (4) by $-6$ and then add the result to equation (5) in order to eliminate $x$.

$\begin{array}{r} -6x - \;\; 6y = -18 \\ 6x + 14y = \;\; 34 \\ \hline 8y = \;\; 16 \Rightarrow y = 2 \end{array}$

Using $y = 2$, find $x$ from equation (4) by substitution: $x + 2 = 3 \Rightarrow x = 1$

Substitute 1 for $x$ and 2 for $y$ in equation (1) to find $z$.

$1 + 4(2) - z = 6 \Rightarrow 1 + 8 - z = 6 \Rightarrow$
$9 - z = 6 \Rightarrow -z = -3 \Rightarrow z = 3$

Verify that the ordered triple $(1, 2, 3)$ satisfies all three equations.

*Check*:

$\begin{array}{l|l} x + 4y - z = 6 \;\; (1) & 2x - y + z = 3 \;\; (2) \\ 1 + 4(2) - 3 = 6 \;\; ? & 2(1) - 2 + 3 = 3 \;\; ? \\ 1 + 8 - 3 = 6 & 2 - 2 + 3 = 3 \\ \quad 6 = 6 \text{ True} & \quad 3 = 3 \text{ True} \end{array}$

$3x + 2y + 3z = 16$  (3)
$3(1) + 2(2) + 3(3) = 16$  ?
$3 + 4 + 9 = 16$
$16 = 16$ True

Solution set: $\{(1, 2, 3)\}$

**53.**  $x - 3y - 2z = -3$  (1)
$3x + 2y - z = 12$  (2)
$-x - y + 4z = 3$  (3)

Eliminate $x$ by adding equations (1) and (3) to get $-4y + 2z = 0$ (4).

Eliminate $x$ by multiplying equation (3) by 3 and add to equation (2).

$-3x - 3y + 12z = 9$
$\underline{3x + 2y - z = 12}$
$-y + 11z = 21$  (5)

Multiply equation (5) by $-4$ and add to equation (4) in order to eliminate $y$.

$-4y + 2z = 0$
$\underline{4y - 44z = -84}$
$-42z = -84 \Rightarrow z = 2$

Using $z = 2$, find $y$ from equation (4) by substitution.

$-4y + 2(2) = 0 \Rightarrow -4y + 4 = 0 \Rightarrow$
$-4y = -4 \Rightarrow y = 1$

Substitute 1 for $y$ and 2 for $z$ in equation (1) to find $x$.

$x - 3(1) - 2(2) = -3 \Rightarrow x - 3 - 4 = -3 \Rightarrow$
$x - 7 = -3 \Rightarrow x = 4$

Verify that the ordered triple $(4, 1, 2)$ satisfies all three equations.

*Check*:   $x - 3y - 2z = -3$  (1)
$4 - 3(1) - 2(2) = -3$  ?
$4 - 3 - 4 = -3$
$-3 = -3$ True

$3x + 2y - z = 12$  (2)
$3(4) + 2(1) - 2 = 12$  ?
$12 + 2 - 2 = 12$
$12 = 12$ True

$-x - y + 4z = 3$  (3)
$-4 - 1 + 4(2) = 3$  ?
$-4 - 1 + 8 = 3$
$3 = 3$ True

Solution set: $\{(4, 1, 2)\}$

**55.**  $2x + 6y - z = 6$  (1)
$4x - 3y + 5z = -5$ (2)
$6x + 9y - 2z = 11$ (3)

Eliminate $y$ by multiplying equation (2) by 2 and add to equation (1).

$2x + 6y - z = 6$
$\underline{8x - 6y + 10z = -10}$
$10x + 9z = -4$  (4)

Eliminate $y$ by multiplying equation (2) by 3 and add to equation (3).

$6x + 9y - 2z = 11$
$\underline{12x - 9y + 15z = -15}$
$18x + 13z = -4$  (5)

Multiply equation (4) by 9 and equation (5) by $-5$ and add in order to eliminate $x$.

$90x + 81z = -36$
$\underline{-90x - 65z = 20}$
$16z = -16 \Rightarrow z = -1$

Using $z = -1$, find $x$ from equation (4) by substitution.

$10x + 9(-1) = -4 \Rightarrow 10x - 9 = -4$
$10x = 5 \Rightarrow x = \frac{1}{2}$

Substitute $\frac{1}{2}$ for $x$ and $-1$ for $z$ in equation (1) to find $y$.

$2\left(\frac{1}{2}\right) + 6y - (-1) = 6 \Rightarrow 1 + 6y + 1 = 6 \Rightarrow$
$6y + 2 = 6 \Rightarrow 6y = 4 \Rightarrow y = \frac{2}{3}$

Verify that the ordered triple $\left(\frac{1}{2}, \frac{2}{3}, -1\right)$ satisfies all three equations.

*Check*:
$2x + 6y - z = 6$  (1)
$2\left(\frac{1}{2}\right) + 6\left(\frac{2}{3}\right) - (-1) = 6$   ?
$1 + 4 + 1 = 6$
$6 = 6$ True

$4x - 3y + 5z = -5$  (2)
$4\left(\frac{1}{2}\right) - 3\left(\frac{2}{3}\right) + 5(-1) = -5$   ?
$2 - 2 - 5 = -5$
$-5 = -5$ True

$6x + 9y - 2z = 11$ (3)
$6\left(\frac{1}{2}\right) + 9\left(\frac{2}{3}\right) - 2(-1) = 11$   ?
$3 + 6 + 2 = 11$
$11 = 11$ True

Solution set: $\left\{\left(\frac{1}{2}, \frac{2}{3}, -1\right)\right\}$

**57.**  $2x - 3y + 2z - 3 = 0$ (1)
$4x + 8y + z - 2 = 0$ (2)
$-x - 7y + 3z - 14 = 0$ (3)

Eliminate $x$ by multiplying equation (3) by 2 and add to equation (1).

$2x - 3y + 2z - 3 = 0$
$\underline{-2x - 14y + 6z - 28 = 0}$
$-17y + 8z - 31 = 0$ (4)

Eliminate $x$ by multiplying equation (3) by 4 and add to equation (2).

$$\begin{array}{r} 4x+ 8y+ z- 2=0 \\ -4x-28y+12z-56=0 \\ \hline -20y+13z-58=0 \ (5) \end{array}$$

Multiply equation (5) by 17, equation (4) by $-20$, and add in order to eliminate $y$.

$$\begin{array}{r} 340y-160z+620=0 \\ -340y+221z-986=0 \\ \hline 61z-366=0 \Rightarrow z=6 \end{array}$$

Using $z=6$, find $y$ from equation (5) by substitution.

$$-20y+13(6)-58=0 \Rightarrow -20y+78-58=0$$
$$-20y+20=0 \Rightarrow -20y=-20 \Rightarrow y=1$$

Substitute 1 for $y$ and 6 for $z$ in equation (1) to find $x$.

$$2x-3(1)+2(6)-3=0 \Rightarrow 2x-3+12-3=0 \Rightarrow$$
$$2x+6=0 \Rightarrow 2x=-6 \Rightarrow x=-3$$

Verify that the ordered triple $(-3,1,6)$ satisfies all three equations.

*Check:*
$$2x-3y+2z-3=0 \ \ (1)$$
$$2(-3)-3(1)+2(6)-3=0 \ \ ?$$
$$-6-3+12-3=0$$
$$0=0 \ \text{True}$$

$$4x+8y+z-2=0 \ \ (2)$$
$$4(-3)+8(1)+6-2=0$$
$$-12+8+6-2=0 \ \ ?$$
$$0=0 \ \text{True}$$

$$-x-7y+3z-14=0 \ \ (3)$$
$$-(-3)-7(1)+3(6)-14=0 \ \ ?$$
$$3-7+18-14=0$$
$$0=0 \ \text{True}$$

Solution set: $\{(-3,1,6)\}$

**59.**  $x-2y+3z=6$ (1)
$2x-y+2z=5$ (2)

Multiply equation (2) by $-2$ and add to equation (1) in order to eliminate $y$.

$$\begin{array}{r} x-2y+3z= 6 \\ -4x+2y-4z=-10 \\ \hline -3x \quad -z= -4 \ (3) \end{array}$$

Solve equation (3) for $z$.
$$-3x-z=-4 \Rightarrow -z=3x-4 \Rightarrow z=-3x+4$$

Express $y$ in terms of $x$ by solving equation (1) for $y$ and substituting $-3x+4$ for $z$.

$$x-2y+3z=6 \Rightarrow -2y=-x-3z+6 \Rightarrow$$
$$y=\frac{-x-3z+6}{-2}=\frac{x+3z-6}{2} \Rightarrow$$
$$y=\frac{x+3(-3x+4)-6}{2}$$
$$=\frac{x-9x+12-6}{2}=\frac{-8x+6}{2}=-4x+3$$

With $x$ arbitrary, the solution set is of the form $\{(x,-4x+3,-3x+4)\}$.

**61.**  $5x-4y+z= 9$ (1)
$x+ y \quad =15$ (2)

Multiply equation (2) by 4 and add to equation (1) in order to eliminate $y$.

$$\begin{array}{r} 5x-4y+z= 9 \\ 4x+4y \quad = 60 \\ \hline 9x \quad +z=69 \ (3) \end{array}$$

Solve equation (3) for $z$.
$$9x+z=69 \Rightarrow z=-9x+69$$

Express $y$ in terms of $x$ by solving equation (1) for $y$ and substituting $-9x+69$ for $z$.

$$5x-4y+z=9 \Rightarrow -4y=-5x-z+9 \Rightarrow$$
$$y=\frac{-5x-z+9}{-4}=\frac{5x+z-9}{4}$$
$$=\frac{5x+(-9x+69)-9}{4}=\frac{-4x+60}{4}=-x+15$$

With $x$ arbitrary, the solution set is of the form $\{(x,-x+15,-9x+69)\}$.

**63.**  $3x+4y-z=13$ (1)
$x+ y+2z=15$ (2)

Multiply equation (2) by $-4$ and add to equation (1) in order to eliminate $y$.

$$\begin{array}{r} 3x+4y- z= 13 \\ -4x-4y-8z=-60 \\ \hline -x \quad -9z=-47 \ (3) \end{array}$$

Solve equation (3) for $z$.
$$-x-9z=-47 \Rightarrow -9z=-47+x \Rightarrow$$
$$z=\frac{-47+x}{-9}=\frac{47-x}{9}$$

Express $y$ in terms of $x$ by solving equation (2) for $y$ and substituting $\frac{47-x}{9}$ for $z$.

$$x+y+2z=15 \Rightarrow y=-x-2z+15 \Rightarrow$$
$$y=-x-2\left(\frac{47-x}{9}\right)+15$$
$$y=\frac{-9x-2(47-x)+135}{9}$$
$$=\frac{-9x-94+2x+135}{9}=\frac{41-7x}{9}$$

With $x$ arbitrary, the solution set is of the form $\left\{\left(x,\frac{41-7x}{9},\frac{47-x}{9}\right)\right\}$.

**65.** $\quad 3x + \;\;5y - \;z = -2 \;(1)$
$\quad\;\; 4x - \;\;\;y + 2z = \;\;1 \;(2)$
$\;\; -6x - 10y + 2z = \;\;0 \;(3)$

Multiply equation (1) by 2 and add the result to equation (3).

$\quad\;\; 6x + 10y - 2z = -4$
$\;\; \underline{-6x - 10y + 2z = \;\;0}$
$\quad\qquad\qquad\qquad 0 = -4$

We obtain a false statement. The solution set is $\varnothing,$ and the system is inconsistent.

**67.** $\quad\;\; 5x - 4y + \;z = 0 \;(1)$
$\quad\;\;\;\; x + \;\;y \qquad = 0 \;(2)$
$\;\; -10x + 8y - 2z = 0 \;(3)$

Multiply equation (2) by 4 and add the result to equation (1) in order to eliminate $y$.

$\quad 5x - 4y + z = 0$
$\quad \underline{4x + 4y \qquad = 0}$
$\quad 9x \qquad\;\; + z = 0 \quad (4)$

Multiply equation (2) by $-8$ and add the result to equation (3) in order to eliminate $y$.

$\quad\;\; -8x - 8y \qquad = 0$
$\quad \underline{-10x + 8y - 2z = 0}$
$\quad -18x \qquad\; - 2z = 0 \;\; (5)$

Multiply equation (4) by 2 and add the result to equation (5).

$\quad\; 18x + 2z = 0$
$\quad \underline{-18x - 2z = 0}$
$\qquad\qquad\;\; 0 = 0$

This is a true statement and the system has infinitely many solutions.
Solve equation (4) for $x$.

$9x + z = 0 \Rightarrow 9x = -z \Rightarrow x = -\frac{z}{9}$

Express $y$ in terms of $z$ by solving equation (1) for $y$ and substituting $-\frac{z}{9}$ for $x$.

$5x - 4y + z = 0 \Rightarrow -4y = -5x - z \Rightarrow$

$\quad y = \frac{-5x - z}{-4} = \frac{5x + z}{4} \Rightarrow$

$\quad y = \frac{5\left(-\frac{z}{9}\right) + z}{4} = \frac{-5z + 9z}{36} = \frac{4z}{36} = \frac{z}{9}$

With $z$ arbitrary, the solution set is of the form $\left\{\left(-\frac{z}{9}, \frac{z}{9}, z\right)\right\}.$

**69.** $\quad \frac{2}{x} + \frac{1}{y} = \frac{3}{2} \;\;(1)$
$\quad \frac{3}{x} - \frac{1}{y} = 1 \;\;(2)$

Let $\frac{1}{x} = t$ and $\frac{1}{y} = u.$ With these substitutions, the system becomes

$2t + u = \frac{3}{2} \quad (3)$
$3t - u = 1 \quad (4)$

Add these equations, eliminate $u,$ and solve for $t.$

$5t = \frac{5}{2} \Rightarrow t = \frac{1}{2}$

Substitute $\frac{1}{2}$ for $t$ in equation (3) and solve for $u.$

$2\left(\frac{1}{2}\right) + u = \frac{3}{2} \Rightarrow 1 + u = \frac{3}{2} \Rightarrow u = \frac{1}{2}$

Now find the values of $x$ and $y,$ the variables in the original system. So, $\frac{1}{x} = t,\, tx = 1,$ and $x = \frac{1}{t}.$ Likewise $y = \frac{1}{u}.$

$x = \frac{1}{t} = \frac{1}{\frac{1}{2}} = 2$ and $y = \frac{1}{u} = \frac{1}{\frac{1}{2}} = 2$

Solution set: $\left\{(2, 2)\right\}$

**71.** $\quad \frac{2}{x} + \frac{1}{y} = 11 \;(1)$
$\quad \frac{3}{x} - \frac{5}{y} = 10 \;(2)$

Let $\frac{1}{x} = t$ and $\frac{1}{y} = u.$ With these substitutions, the system becomes

$2t + u = 11 \quad (3)$
$3t - 5u = 10 \quad (4)$

Multiply equation (3) by 5 and add to equation (4), eliminating $u,$ and solve for $t.$

$\quad 10t + 5u = 55$
$\quad \underline{3t - 5u = 10}$
$\quad 13t \qquad = 65 \Rightarrow t = 5$

Substitute 5 for $t$ in equation (3) and solve for $u: 2(5) + u = 11 \Rightarrow 10 + u = 11 \Rightarrow u = 1$

Now find the values of $x$ and $y,$ the variables in the original system. So, $x = \frac{1}{t} = \frac{1}{5}$ and $y = \frac{1}{u} = \frac{1}{1} = 1.$

Solution set: $\left\{\left(\frac{1}{5}, 1\right)\right\}$

**73.** $\quad \frac{2}{x} + \frac{3}{y} - \frac{2}{z} = -1 \;(1)$
$\quad \frac{8}{x} - \frac{12}{y} + \frac{5}{z} = \;\;5 \;(2)$
$\quad \frac{6}{x} + \frac{3}{y} - \frac{1}{z} = \;\;1 \;(3)$

Let $\frac{1}{x} = t,\; \frac{1}{y} = u,$ and $\frac{1}{z} = v.$ With these substitutions, the system becomes

$\quad 2t + \;\;3u - 2v = -1 \;(4)$
$\quad 8t - 12u + 5v = \;\;5 \;(5)$
$\quad 6t + \;\;3u - \;\;v = \;\;1 \;(6)$

Eliminate $v$ by multiplying equation (6) by 5 and add to equation (5).

$\quad\;\; 8t - 12u + 5v = 5$
$\quad \underline{30t + 15u - 5v = 5}$
$\quad 38t + \;\;3u \qquad = 10 \;(7)$

Eliminate $v$ by multiplying equation (6) by $-2$ and add to equation (4).

$$2t + 3u - 2v = -1$$
$$\underline{-12t - 6u + 2v = -2}$$
$$-10t - 3u \qquad = -3 \ (8)$$

Add equations (7) and (8) in order to eliminate $u$.

$$38t + 3u = 10$$
$$\underline{-10t - 3u = -3}$$
$$28t \qquad = 7 \Rightarrow t = \frac{1}{4}$$

Using $t = \frac{1}{4}$, find $u$ from equation (7) by substitution.

$$38\left(\tfrac{1}{4}\right) + 3u = 10 \Rightarrow \tfrac{19}{2} + 3u = 10 \Rightarrow$$
$$3u = \tfrac{20}{2} - \tfrac{19}{2} = \tfrac{1}{2} \Rightarrow u = \tfrac{1}{6}$$

Substitute $\frac{1}{4}$ for $t$ and $\frac{1}{6}$ for $u$ in equation (4) to find $v$.

$$2\left(\tfrac{1}{4}\right) + 3\left(\tfrac{1}{6}\right) - 2v = -1 \Rightarrow \tfrac{1}{2} + \tfrac{1}{2} - 2v = -1 \Rightarrow$$
$$1 - 2v = -1 \Rightarrow -2v = -2 \Rightarrow v = 1$$

Now find the values of $x$, $y$, and $z$, the variables in the original system. So,

$$x = \tfrac{1}{t} = \tfrac{1}{\frac{1}{4}} = 4, \quad y = \tfrac{1}{u} = \tfrac{1}{\frac{1}{6}} = 6, \text{ and}$$

$$z = \tfrac{1}{v} = \tfrac{1}{1} = 1.$$

Solution set: $\left\{(4, 6, 1)\right\}$

**75.** Answers will vary.

   **(a)**   $x + y + z = 4$
             $x + 2y + z = 5$
             $2x - y + 3z = 4$

      This system has exactly one solution, namely $(4, 1, -1)$. (There are other equations that would do the same.)

   **(b)**   $x + y + z = 4$
             $x + y + z = 5$
             $2x - y + 3z = 4$

      This system has no solution, since no ordered triple can satisfy the first two equations simultaneously. (There are other equations that would do the same.)

   **(c)**   $x + y + z = 4$
             $2x + 2y + 2z = 8$
             $2x - y + 3z = 4$

      This system has infinitely many solutions, since all the ordered triples that satisfy the first equation will also satisfy the second equation. (There are other equations that would do the same.)

**77.** Since $y = ax^2 + bx + c$ and the parabola passes through the points $(2, 3)$, $(-1, 0)$, and $(-2, 2)$, we have the following equations.

$$3 = a(2)^2 + b(2) + c \quad \Rightarrow 4a + 2b + c = 3 \ (1)$$
$$0 = a(-1)^2 + b(-1) + c \Rightarrow a - b + c = 0 \ (2)$$
$$2 = a(-2)^2 + b(-2) + c \Rightarrow 4a - 2b + c = 2 \ (3)$$

Multiply equation (2) by $-1$ and add the result to equation (1) in order to eliminate $c$.

$$4a + 2b + c = 3$$
$$\underline{-a + b - c = 0}$$
$$3a + 3b \qquad = 3 \ (4)$$

Multiplying equation (4) by $-1$ and then adding the result to equation (5).

$$-3a - 3b = -3$$
$$\underline{3a - b = 2}$$
$$-4b = -1 \Rightarrow b = \tfrac{1}{4}$$

Multiply equation (2) by $-1$ and add the result to equation (3) in order to eliminate $c$.

$$-a + b - c = 0$$
$$\underline{4a - 2b + c = 2}$$
$$3a - b \qquad = 2 \ (5)$$

Substitute this value into equation (5).

$$3a - \tfrac{1}{4} = 2 \Rightarrow 3a = \tfrac{8}{4} + \tfrac{1}{4} = \tfrac{9}{4} \Rightarrow a = \tfrac{3}{4}$$

Substitute $a = \frac{3}{4}$ and $b = \frac{1}{4}$ into equation (1) in order to solve for $c$.

$$4\left(\tfrac{3}{4}\right) + 2\left(\tfrac{1}{4}\right) + c = 3$$
$$3 + \tfrac{1}{2} + c = 3 \Rightarrow c = -\tfrac{1}{2}$$

The equation of the parabola is

$$y = \tfrac{3}{4}x^2 + \tfrac{1}{4}x - \tfrac{1}{2}.$$

**79.** Since $y = ax + b$ and the line passes through the points $(2, 5)$ and $(-1, -4)$, we have the equations $5 = a(2) + b$ and $-4 = a(-1) + b$. This becomes the following system.

$$2a + b = 5 \ (1)$$
$$-a + b = -4. \ (2)$$

Multiply equation (2) by $-1$ and add the result to equation (1) in order to eliminate $b$.

$$2a + b = 5$$
$$\underline{a - b = 4}$$
$$3a \qquad = 9 \Rightarrow a = 3$$

Substitute this value into equation (1).

$$2(3) + b = 5 \Rightarrow 6 + b = 5 \Rightarrow b = -1$$

The equation is $y = 3x - 1$.

**81.** Since $y = ax^2 + bx + c$ and the parabola passes through the points $(-2, -3.75)$, $(4, -3.75)$, and $(-1, -1.25)$, we have the following equations.

$$-3.75 = a(-2)^2 + b(-2) + c$$
$$4a - 2b + c = -3.75 \qquad (1)$$
$$-3.75 = a(4)^2 + b(4) + c$$
$$16a + 4b + c = -3.75 \qquad (2)$$
$$-1.25 = a(-1)^2 + b(-1) + c$$
$$a - b + c = -1.25 \qquad (3)$$

Multiply equation (2) by $-1$ and add the result to equation (1) in order to eliminate $c$.

$$4a - 2b + c = -3.75$$
$$\underline{-16a - 4b - c = 3.75}$$
$$-12a - 6b \quad = \quad 0 \text{ or } 2a + b = 0 \quad (4)$$

Multiply equation (2) by $-1$ and add the result to equation (3) in order to eliminate $c$.

$$-16a - 4b - c = 3.75$$
$$\underline{a - b + c = -1.25}$$
$$-15a - 5b \quad = \quad 2.50 \text{ or } -3a - b = .5 \quad (5)$$

Add equations (4) and (5) in order to solve for $b$.

$$2a + b = 0$$
$$\underline{-3a - b = .5}$$
$$-a \quad = .5 \Rightarrow a = -.5$$

Substitute $a = -.5$ into equation (4) in order to solve for $b$.

$$2(-.5) + b = 0 \Rightarrow -1 + b = 0 \Rightarrow b = 1$$

Substitute $a = -.5$ and $b = 1$ into equation (3) in order to solve for $c$.

$$-.5 - 1 + c = -1.25$$
$$-1.5 + c = -1.25 \Rightarrow c = .25$$

The equation of the parabola is

$$y = -.5x^2 + x + .25 \text{ or } y = -\tfrac{1}{2}x^2 + x + \tfrac{1}{4}.$$

**83.** Since $x^2 + y^2 + ax + by + c = 0$ and the circle passes through the points $(-1, 3)$, $(6, 2)$, and $(-2, -4)$, we have the following equations.

$$(-1)^2 + 3^2 + a(-1) + b(3) + c = 0$$
$$-a + 3b + c = -10 \quad (1)$$
$$6^2 + 2^2 + a(6) + b(2) + c = 0 \Rightarrow$$
$$6a + 2b + c = -40 \quad (2)$$
$$(-2)^2 + (-4)^2 + a(-2) + b(-4) + c = 0$$
$$-2a - 4b + c = -20 \quad (3)$$

Multiply equation (1) by $-2$ and adding the result to equation (3) in order to eliminate $a$.

$$2a - 6b - 2c = 20$$
$$\underline{-2a - 4b + c = -20}$$
$$-10b - c = 0 \,(4)$$

Multiply equation (1) by 6 and adding the result to equation (2) in order to eliminate $a$.

$$-6a + 18b + 6c = -60$$
$$\underline{6a + 2b + c = -40}$$
$$20b + 7c = -100 \,(5)$$

Multiply equation (4) by 7 and adding the result to equation (5) in order to eliminate $c$.

$$-70b - 7c = 0$$
$$\underline{20b + 7c = -100}$$
$$-50b \quad = -100 \Rightarrow b = 2$$

We substitute this value into equation (4) in order to solve for $c$.

$$-10(2) - c = 0 \Rightarrow -20 - c = 0 \Rightarrow -20 = c$$

Substitute $b = 2$ and $c = -20$ into equation (1).

$$-a + 3(2) - 20 = -10 \Rightarrow -a + 6 - 20 = -10 \Rightarrow$$
$$-a - 14 = -10 \Rightarrow -a = 4 \Rightarrow a = -4$$

The equation of the circle is

$$x^2 + y^2 - 4x + 2y - 20 = 0.$$

**85.** Since $x^2 + y^2 + ax + by + c = 0$ and the circle passes through the points $(2, 1)$, $(-1, 0)$, and $(3, 3)$, we have the following equations.

$$2^2 + 1^2 + a(2) + b(1) + c = 0$$
$$2a + b + c = -5 \quad (1)$$
$$(-1)^2 + 0^2 + a(-1) + b(0) + c = 0$$
$$-a + c = -1 \quad (2)$$
$$3^2 + 3^2 + a(3) + b(3) + c = 0$$
$$3a + 3b + c = -18 \quad (3)$$

Multiply equation (1) by $-3$ and add the result to equation (3) in order to eliminate $b$.

$$-6a - 3b - 3c = 15$$
$$\underline{3a + 3b + c = -18}$$
$$-3a - 2c = -3 \quad (4)$$

Multiply equation (2) by 2 and add the result to equation (4) in order to eliminate $c$.

$$-2a + 2c = -2$$
$$\underline{-3a - 2c = -3}$$
$$-5a \quad = -5 \Rightarrow a = 1$$

We substitute this value into equation (2) in order to solve for $c$.

$$-1 + c = -1 \Rightarrow c = 0$$

Substitute $a = 1$ and $c = 0$ into equation (1).

$$2(1) + b + 0 = -5 \Rightarrow 2 + b = -5 \Rightarrow b = -7$$

The equation of the circle is

$$x^2 + y^2 + x - 7y = 0.$$

**87. (a)** Since $C = at^2 + bt + c$ and we have the ordered pairs $(0, 318)$, $(20, 341)$, and $(40, 371)$, we have the following equations.

$$318 = a(0)^2 + b(0) + c$$
$$c = 318 \qquad (1)$$
$$341 = a(20)^2 + b(20) + c$$
$$400a + 20b + c = 341 \qquad (2)$$
$$371 = a(40)^2 + b(40) + c$$
$$1600a + 40b + c = 371 \qquad (3)$$

Since $c = 318$, we substitute this value into equations (2) and (3) to obtain the following system.

$$400a + 20b + 318 = 341$$
$$400a + 20b = 23 \qquad (4)$$
$$1600a + 40b + 318 = 371$$
$$1600a + 40b = 53 \qquad (5)$$

Multiply equation (4) by $-2$ and add the result to equation (5) in order to eliminate $b$.

$$-800a - 40b = -46$$
$$\underline{1600a + 40b = \phantom{0}53}$$
$$800a \phantom{+ 40b} = \phantom{00}7 \Rightarrow a = \tfrac{7}{800}$$

Substitute $a = \tfrac{7}{800}$ into equation (5) in order to solve for $b$.

$$1600\left(\tfrac{7}{800}\right) + 40b = 53$$
$$14 + 40b = 53$$
$$40b = 39 \Rightarrow b = \tfrac{39}{40}$$

The constants are $a = \tfrac{7}{800}$, $b = \tfrac{39}{40}$, and $c = 318$ and the relationship is $C = \tfrac{7}{800}t^2 + \tfrac{39}{40}t + 318$.

**(b)** Since $t = 0$ corresponds to 1962, the amount of carbon dioxide will be double its 1962 level when $\tfrac{7}{800}t^2 + \tfrac{39}{40}t + 318 = 2(318)$.

Solve this equation.

$$\tfrac{7}{800}t^2 + \tfrac{39}{40}t + 318 = 2(318)$$
$$\tfrac{7}{800}t^2 + \tfrac{39}{40}t - 318 = 0$$
$$7t^2 + 780t - 254,400 = 0$$

Use the quadratic formula, where $a = 7, b = 780,$ and $c = -254,400$.

$$t = \frac{-780 \pm \sqrt{780^2 - 4(7)(-254,400)}}{2(7)}$$
$$= \frac{-780 \pm \sqrt{608,400 + 7,123,200}}{14}$$
$$= \frac{-780 \pm \sqrt{7,731,600}}{14}$$
$$= \frac{-780 - \sqrt{7,731,600}}{14} \approx -254.3 \text{ or}$$
$$t = \frac{-780 + \sqrt{7,731,600}}{14} \approx 142.9$$

We reject the first proposed solution because time cannot be negative. If $t$ is approximately 142.9 then the year is $1962 + 142.9 = 2104.9$ (near the end of 2104).

**89.** Let $x =$ one number; let $y =$ the other number We have the following system of equations.

$$x + y = 47 \quad (1)$$
$$x - y = 1 \quad (2)$$

Add the two equations to eliminate $y$. $2x = 48 \Rightarrow x = 24$. Substitute this value into equation (2) and solve for $y$: $24 + y = 47 \Rightarrow y = 23$.

The two numbers are 23 and 24.

**91.** Let $x =$ the FCI for football; let $y =$ the FCI for baseball. We have the following system of equations.

$$\tfrac{x+y}{2} = 250.51 \Rightarrow x + y = 501.02 \quad (1)$$
$$x - y = 158.62 \quad (2)$$

Add the two equations to eliminate $y$. $2x = 659.64 \Rightarrow x = 329.82$ Substitute this value into equation (2) and solve for $y$:

$$329.82 - y = 158.62 \Rightarrow y = 171.20$$

Verify that $(329.82, 171.20)$ satisfies both equations in the original system. The FCI for football was \$329.82, and the FCI for baseball was \$171.20.

**93.** Let $x =$ the number of \$3.00 gallons; $y =$ the number of \$4.50 gallons; $z =$ the number of \$9.00 gallons. We have the following equations.

$$x + \phantom{0}y + \phantom{0}z = 300 \qquad (1)$$
$$2x = y \Rightarrow$$
$$2x - \phantom{0}y \phantom{+ 00z} = \phantom{00}0 \qquad (2)$$
$$3.00x + 4.50y + 9.00z = 6.00(300) \Rightarrow$$
$$3x + 4.5y + 9z = 1800 \qquad (3)$$

Multiply equation (1) by $-9$ and add to equation (3) in order to eliminate $z$.

$$-9x - \phantom{0}9y - 9z = -2700$$
$$\underline{\phantom{-}3x + 4.5y + 9z = \phantom{-}1800}$$
$$-6x - 4.5y \phantom{+ 9z} = -\phantom{0}900 \quad (4)$$

*(continued on next page)*

(*continued from page 253*)

Multiply equation (2) by 3 and add to equation (4) in order to eliminate $x$.

$$
\begin{aligned}
6x - 3y &= 0 \\
-6x - 4.5y &= -900 \\
\hline
-7.5y &= -900 \Rightarrow y = 120
\end{aligned}
$$

Substitute this value into equation (2) to solve for $x$.

$2x - 120 = 0 \Rightarrow 2x = 120 \Rightarrow x = 60$

Substitute $x = 60$ and $y = 120$ into equation (1) and solve for $y$: $60 + 120 + z = 300 \Rightarrow 180 + z = 300 \Rightarrow z = 120$. She should use 60 gal of the $3.00 water, 120 gal of the $4.50 water, and 120 gal of the $9.00 water.

**95.** Let $x$ = the length of the shortest side;
$y$ = the length of the medium side;
$z$ = the length of the longest side.
We have the following equations.

$$
\begin{aligned}
x + y + z &= 59 \quad (1) \\
z = y + 11 \Rightarrow \quad -y + z &= 11 \quad (2) \\
y = x + 3 \Rightarrow -x + y \quad &= 3 \quad (3)
\end{aligned}
$$

Add equations (1) and (3) together in order to eliminate $x$.

$$
\begin{aligned}
x + y + z &= 59 \\
-x + y \quad &= 3 \\
\hline
2y + z &= 62 \quad (4)
\end{aligned}
$$

Multiply equation (2) by 2 and then add the result to equation (4) to solve for $z$.

$$
\begin{aligned}
-2y + 2z &= 22 \\
2y + z &= 62 \\
\hline
3z &= 84 \Rightarrow z = 28
\end{aligned}
$$

Substitute this value into equation (2) in order to solve for $y$.

$-y + 28 = 11 \Rightarrow -y = -17 \Rightarrow y = 17$

Substitute $y = 17$ into equation (3) in order to solve for $x$: $-x + 17 = 3 \Rightarrow -x = -14 \Rightarrow x = 14$

The lengths of the sides of the triangle are 14 inches, 17 inches, and 28 inches.
Note: $14 + 17 + 28 = 59$

**97.** Let $x$ = the amount invested in real estate (at 3%); $y$ = the amount invested in a money market account (at 2.5%); $z$ = the amount invested in CDs (at 1.5%).
Completing the table we have the following.

| | Amount Invested | Rate (as a decimal) | Annual Interest |
|---|---|---|---|
| Real Estate | $x$ | .03 | $.03x$ |
| Money Market | $y$ | .025 | $.025y$ |
| CDs | $z$ | .015 | $.015z$ |

We have the following equations.

$$
\begin{aligned}
x + y + z &= 200,000 \quad (1) \\
z = x + y - 80,000 &\Rightarrow \\
x + y - z &= 80,000 \quad (2) \\
.03x + .025y + .015z = 4900 &\Rightarrow \\
30x + 25y + 15z &= 4,900,000 \quad (3)
\end{aligned}
$$

Add equations (1) and (2) in order to eliminate $z$.

$$
\begin{aligned}
x + y + z &= 200,000 \\
x + y - z &= 80,000 \\
\hline
2x + 2y &= 280,000 \text{ or } x + y = 140,000 \quad (4)
\end{aligned}
$$

Multiply equation (2) by 15 and add the result to equation (3) in order to eliminate $z$.

$$
\begin{aligned}
15x + 15y - 15z &= 1,200,000 \\
30x + 25y + 15z &= 4,900,000 \\
\hline
45x + 40y &= 6,100,000 \text{ or} \\
9x + 8y &= 1,220,000 \quad (5)
\end{aligned}
$$

Multiply equation (4) by $-8$ and add the result to equation (5).

$$
\begin{aligned}
-8x - 8y &= -1,120,000 \\
9x + 8y &= 1,220,000 \\
\hline
x &= 100,000
\end{aligned}
$$

Substitute this value into equation (4) in order to solve for $y$.

$100,000 + y = 140,000 \Rightarrow y = 40,000$

Substitute $x = 100,000$ and $y = 40,000$ into equation (1) in order to solve for $z$.

$100,000 + 40,000 + z = 200,000$
$140,000 + z = 200,000 \Rightarrow z = 60,000$

The amounts invested were $100,000 at 3% (real estate), $40,000 at 2.5% (money market), and $60,000 at 1.5% (CDs).

**99.** $p = 16 - \frac{5}{4}q$

(a) $p = 16 - \frac{5}{4}(0) = 16 - 0 = 16$
The price is $16.

(b) $p = 16 - \frac{5}{4}(4) = 16 - 5 = 16 - 5 = 11$
The price is $11.

(c) $p = 16 - \frac{5}{4}(8) = 16 - 10 = 6$
The price is $6.

**101.** See answer to Exercise 103.

**103.**

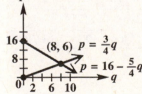

**105.**  $25x + 40y + 20z = 2200$ (4)
$\quad\;\; 4x + \;\;2y + \;\;3z = 280$  (5)
$\quad\;\; 3x + \;\;2y + \quad z = 180$  (6)

Eliminate $z$ by multiplying equation (6) by $-3$ and add to equation (5).
$$\begin{array}{r} -9x - 6y - 3z = -540 \\ 4x + 2y + 3z = \;\;280 \\ \hline -5x - 4y \qquad\; = -260 \;(7) \end{array}$$

Eliminate $z$ by multiplying equation (6) by $-20$ and add to equation (4).
$$\begin{array}{r} -60x - 40y - 20z = -3600 \\ 25x + 40y + 20z = \;\;2200 \\ \hline -35x \qquad\qquad\quad = -1400 \Rightarrow x = 40 \end{array}$$

Substitute this value in equation (7) in order to solve for $y$.
$$-5(40) - 4y = -260 \Rightarrow -200 - 4y = -260 \Rightarrow$$
$$-4y = -60 \Rightarrow y = 15$$

Substitute $x = 40$ and $y = 15$ into equation (6) to solve for $z$.
$$3(40) + 2(15) + z = 180 \Rightarrow$$
$$120 + 30 + z = 180 \Rightarrow$$
$$150 + z = 180 \Rightarrow z = 30$$
Solution set: $\{(40, 15, 30)\}$

**107.**  Let $x =$ the number of pounds of Arabian Mocha Sanai; $y =$ the number of pounds of Organic Shade-Grown Mexico; $z =$ the number of pounds of Guatemala Antigua. Completing the table we have the following.

| | Number of Pounds | Cost per Pound | Total Cost |
|---|---|---|---|
| Arabian Mocha | $x$ | 15.99 | $15.99x$ |
| Organic Mexico | $y$ | 12.99 | $12.99y$ |
| Guatemala Antigua | $z$ | 10.19 | $10.19z$ |
| Total | 50 | 12.37 | $50(12.37) =$ 618.50 |

We have the following equations.
$$x + y + z = 50 \qquad\qquad (1)$$
$$z = 2x \qquad\qquad\qquad (2)$$
$$15.99x + 12.99y + 10.19z = 618.50 \qquad (3)$$

Substitute $z = 2x$ into equations (1) and (3):
$$x + y + 2x = 50 \Rightarrow$$
$$3x + y = 50 \qquad\qquad (4)$$
$$15.99x + 12.99y + 10.19(2x) = 618.50 \Rightarrow$$
$$36.37x + 12.99y = 618.50 \qquad (5)$$

Solve equation (4) for $y$, then substitute that value into equation (5) and solve for $x$:

$$3x + y = 50 \Rightarrow y = -3x + 50$$
$$36.37x + 12.99(-3x + 50) = 618.50 \Rightarrow$$
$$36.37x - 38.97x + 649.50 = 618.50 \Rightarrow$$
$$-2.6x = -31 \Rightarrow$$
$$x \approx 11.9231 \approx 11.92$$

Substitute $x = 11.9231$ into equation (2) to solve for $z$:  $z \approx 2(11.9231) \approx 23.85$

Substitute the values for $x$ and $z$ into equation (1) to solve for $y$:
$$11.92 + y + 23.85 = 50 \Rightarrow y \approx 14.23$$

11.92 pounds of Arabian Mocha Sanani, 14.23 pounds of Organic Shade-Grown Mexico, and 23.85 pounds of Guatemala Antigua are needed. (Answers are approximations.)

## Section 5.2: Matrix Solution of Linear Systems

### Connections (page 517)

**1.**  $T(n) = \frac{2}{3}n^3 + \frac{3}{2}n^2 - \frac{7}{6}n$

Continuing in this manner we have the following.

| $n$ | $T$ |
|---|---|
| 3 | 28 |
| 6 | 191 |
| 10 | 805 |
| 29 | 17,487 |
| 100 | 681,550 |
| 200 | 5,393,100 |
| 400 | 42,906,200 |
| 1000 | 668,165,500 |
| 5000 | $8.3 \times 10^{10}$ |
| 10,000 | $6.7 \times 10^{11}$ |
| 100,000 | $6.7 \times 10^{14}$ |

**2.**  Using the tables he would have to do 17,487 operations. Yes, this is too many to do by hand.

**3.**  If the number of variables doubles, the number of operations increases by a factor of 8. 100 variables require 681,550 operations and 200 variables requires 5,393,100 operations.

The ratio $\frac{5,393,100}{681,550} \approx 7.91 \approx 8$.

**4.**  A system of 100,000 variables has $6.7 \times 10^{14}$ operations. If a Cray-T90 does 60 billion $= 6 \times 10^{10}$ operations per second, then the system would take $\frac{6.7 \times 10^{14}}{6 \times 10^{10}} \approx 11,166.67$ sec.

Since there are 3600 seconds in 1 hr, the system would take $\frac{11,166.67}{3600} \approx 3.1$ hr.

**Exercises**

**1.** $\begin{bmatrix} 2 & 4 \\ 4 & 7 \end{bmatrix}$; $-2$ times row 1 added to row 2

$$\begin{bmatrix} 2 & 4 \\ 4+(-2)(2) & 7+(-2)(4) \end{bmatrix} = \begin{bmatrix} 2 & 4 \\ 0 & -1 \end{bmatrix}$$

**3.** $\begin{bmatrix} 1 & 5 & 6 \\ -2 & 3 & -1 \\ 4 & 7 & 0 \end{bmatrix}$; 2 times row 1 added to row 2

$$\begin{bmatrix} 1 & 5 & 6 \\ -2+2(1) & 3+2(5) & -1+2(6) \\ 4 & 7 & 0 \end{bmatrix} = \begin{bmatrix} 1 & 5 & 6 \\ 0 & 13 & 11 \\ 4 & 7 & 0 \end{bmatrix}$$

**5.** $2x+3y=11$
$\quad x+2y=8$

The augmented matrix is $\begin{bmatrix} 2 & 3 & | & 11 \\ 1 & 2 & | & 8 \end{bmatrix}$. The

size is $2 \times 3$.

**7.** $2x+\ y+\ z-3=0$
$\quad 3x-4y+2z+7=0$
$\quad x+\ y+\ z-2=0$

Each equation of a linear system must have the constant term isolated on one side of the equal sign. Rewriting this system, we have
$2x+\ y+\ z=\ 3$
$\quad 3x-4y+2z=-7$
$\quad x+\ y+\ z=\ 2$

The augmented matrix is $\begin{bmatrix} 2 & 1 & 1 & | & 3 \\ 3 & -4 & 2 & | & -7 \\ 1 & 1 & 1 & | & 2 \end{bmatrix}$.

The size is $3 \times 4$.

**9.** $\begin{bmatrix} 3 & 2 & 1 & | & 1 \\ 0 & 2 & 4 & | & 22 \\ -1 & -2 & 3 & | & 15 \end{bmatrix}$ is associated with the

system $\begin{array}{r} 3x+2y+z=1 \\ 2y+4z=22 \\ -x-2y+3z=15. \end{array}$

**11.** $\begin{bmatrix} 1 & 0 & 0 & | & 2 \\ 0 & 1 & 0 & | & 3 \\ 0 & 0 & 1 & | & -2 \end{bmatrix}$ is associated with the system

$x=\ 2$
$y=\ 3$
$z=-2.$

**13.** $\begin{bmatrix} 1 & 1 & 0 & | & 3 \\ 0 & 2 & 1 & | & -4 \\ 1 & 0 & -1 & | & 5 \end{bmatrix}$ is associated with the

system $\begin{array}{r} x+y=\ 3 \\ 2y+z=-4 \\ x-z=\ 5. \end{array}$

**15.** $x+y=5$
$\quad x-y=-1$

This system has the augmented matrix

$\begin{bmatrix} 1 & 1 & | & 5 \\ 1 & -1 & | & -1 \end{bmatrix}$.

$\begin{bmatrix} 1 & 1 & | & 5 \\ 0 & -2 & | & -6 \end{bmatrix} \begin{array}{l} \\ -1R1+R2 \end{array} \Rightarrow$

$\begin{bmatrix} 1 & 1 & | & 5 \\ 0 & 1 & | & 3 \end{bmatrix} \begin{array}{l} \\ -\frac{1}{2}R2 \end{array} \Rightarrow \begin{bmatrix} 1 & 0 & | & 2 \\ 0 & 1 & | & 3 \end{bmatrix} \begin{array}{l} -1R2+R1 \\ \\ \end{array}$

Solution set: $\{(2,3)\}$

**17.** $3x+2y=-9$
$\quad 2x-5y=-6$

This system has the augmented matrix

$\begin{bmatrix} 3 & 2 & | & -9 \\ 2 & -5 & | & -6 \end{bmatrix}$.

$\begin{bmatrix} 1 & 7 & | & -3 \\ 2 & -5 & | & -6 \end{bmatrix} \begin{array}{l} -1R2+R1 \\ \\ \end{array} \Rightarrow$

$\begin{bmatrix} 1 & 7 & | & -3 \\ 0 & -19 & | & 0 \end{bmatrix} \begin{array}{l} \\ -2R1+R2 \end{array} \Rightarrow$

$\begin{bmatrix} 1 & 7 & | & -3 \\ 0 & 1 & | & 0 \end{bmatrix} \begin{array}{l} \\ -\frac{1}{19}R2 \end{array} \Rightarrow$

$\begin{bmatrix} 1 & 0 & | & -3 \\ 0 & 1 & | & 0 \end{bmatrix} \begin{array}{l} -7R2+R1 \\ \\ \end{array}$

Solution set: $\{(-3,0)\}$

**19.** $6x+y-5=0$
$\quad 5x+y-3=0$

Rewrite the system as $\begin{array}{l} 6x+y=5 \\ 5x+y=3 \end{array}$

The augmented matrix is $\begin{bmatrix} 6 & 1 & | & 5 \\ 5 & 1 & | & 3 \end{bmatrix}$.

$\begin{bmatrix} 1 & \frac{1}{6} & | & \frac{5}{6} \\ 5 & 1 & | & 3 \end{bmatrix} \begin{array}{l} \frac{1}{6}R1 \\ \\ \end{array} \Rightarrow \begin{bmatrix} 1 & \frac{1}{6} & | & \frac{5}{6} \\ 0 & \frac{1}{6} & | & -\frac{7}{6} \end{bmatrix} \begin{array}{l} \\ -5R1+R2 \end{array} \Rightarrow$

$\begin{bmatrix} 1 & \frac{1}{6} & | & \frac{5}{6} \\ 0 & 1 & | & -7 \end{bmatrix} \begin{array}{l} \\ 6R2 \end{array} \Rightarrow \begin{bmatrix} 1 & 0 & | & 2 \\ 0 & 1 & | & -7 \end{bmatrix} \begin{array}{l} -\frac{1}{6}R2+R1 \\ \\ \end{array}$

or

$$\begin{bmatrix} 1 & 0 & | & 2 \\ 5 & 1 & | & 3 \end{bmatrix} \begin{matrix} -1R2+R1 \\ \\ \end{matrix} \Rightarrow \begin{bmatrix} 1 & 0 & | & 2 \\ 0 & 1 & | & -7 \end{bmatrix} \begin{matrix} \\ -5R1+R2 \end{matrix}$$

Solution set: $\{(2,-7)\}$

**21.** $2x - y = 6$
$\quad\; 4x - 2y = 0$

This system has the augmented matrix

$$\begin{bmatrix} 2 & -1 & | & 6 \\ 4 & -2 & | & 0 \end{bmatrix}.$$

$$\begin{bmatrix} 1 & -\frac{1}{2} & | & 3 \\ 4 & -2 & | & 0 \end{bmatrix} \begin{matrix} \frac{1}{2}R1 \\ \\ \end{matrix} \Rightarrow$$

$$\begin{bmatrix} 1 & -\frac{1}{2} & | & 3 \\ 0 & 0 & | & -12 \end{bmatrix} \begin{matrix} \\ -4R1+R2 \end{matrix}$$

The second row of the augmented matrix corresponds to the equation $0x + 0y = -12$, which has no solution. Thus, the solution set is $\varnothing$.

**23.** $\quad 3x - 4y = 7$
$\quad\; -6x + 8y = -14$

This system has the augmented matrix

$$\begin{bmatrix} 3 & -4 & | & 7 \\ -6 & 8 & | & -14 \end{bmatrix}$$

$$\begin{bmatrix} 1 & -\frac{4}{3} & | & \frac{7}{3} \\ -6 & 8 & | & -14 \end{bmatrix} \begin{matrix} \frac{1}{3}R1 \\ \\ \end{matrix} \Rightarrow$$

$$\begin{bmatrix} 1 & -\frac{4}{3} & | & \frac{7}{3} \\ 0 & 0 & | & 0 \end{bmatrix} \begin{matrix} \\ 6R1+R2 \end{matrix}$$

It is impossible to go further. The equation that corresponds to the first row in the final matrix is $x - \frac{4}{3}y = \frac{7}{3} \Rightarrow x = \frac{4}{3}y + \frac{7}{3}$

Solution set: $\left\{ \frac{4}{3}y + \frac{7}{3}, y \right\}$

**25.** $\quad x + y - 5z = -18$
$\quad\; 3x - 3y + z = 6$ ; This system has the augmented matrix $\begin{bmatrix} 1 & 1 & -5 & | & -18 \\ 3 & -3 & 1 & | & 6 \\ 1 & 3 & -2 & | & -13 \end{bmatrix}$.
$\quad\; x + 3y - 2z = -13$

$$\begin{bmatrix} 1 & 1 & -5 & | & -18 \\ 0 & -6 & 16 & | & 60 \\ 1 & 3 & -2 & | & -13 \end{bmatrix} \begin{matrix} \\ -3R1+R2 \\ \\ \end{matrix} \Rightarrow \begin{bmatrix} 1 & 1 & -5 & | & -18 \\ 0 & -6 & 16 & | & 60 \\ 0 & 2 & 3 & | & 5 \end{bmatrix} \begin{matrix} \\ \\ -1R1+R3 \end{matrix} \Rightarrow \begin{bmatrix} 1 & 1 & -5 & | & -18 \\ 0 & 1 & -\frac{8}{3} & | & -10 \\ 0 & 2 & 3 & | & 5 \end{bmatrix} \begin{matrix} \\ -\frac{1}{6}R2 \\ \\ \end{matrix}$$

$$\begin{bmatrix} 1 & 0 & -\frac{7}{3} & | & -8 \\ 0 & 1 & -\frac{8}{3} & | & -10 \\ 0 & 2 & 3 & | & 5 \end{bmatrix} \begin{matrix} -1R2+R1 \\ \\ \\ \end{matrix} \Rightarrow \begin{bmatrix} 1 & 0 & -\frac{7}{3} & | & -8 \\ 0 & 1 & -\frac{8}{3} & | & -10 \\ 0 & 0 & \frac{25}{3} & | & 25 \end{bmatrix} \begin{matrix} \\ \\ -2R2+R3 \end{matrix} \Rightarrow \begin{bmatrix} 1 & 0 & -\frac{7}{3} & | & -8 \\ 0 & 1 & -\frac{8}{3} & | & -10 \\ 0 & 0 & 1 & | & 3 \end{bmatrix} \begin{matrix} \\ \\ \frac{3}{25}R3 \end{matrix}$$

$$\begin{bmatrix} 1 & 0 & 0 & | & -1 \\ 0 & 1 & -\frac{8}{3} & | & -10 \\ 0 & 0 & 1 & | & 3 \end{bmatrix} \begin{matrix} \frac{7}{3}R3+R1 \\ \\ \\ \end{matrix} \Rightarrow \begin{bmatrix} 1 & 0 & 0 & | & -1 \\ 0 & 1 & 0 & | & -2 \\ 0 & 0 & 1 & | & 3 \end{bmatrix} \begin{matrix} \\ \frac{8}{3}R3+R2 \\ \\ \end{matrix}$$

Solution set: $\{(-1,-2,3)\}$

**27.** $\quad x + y - z = 6$
$\quad\; 2x - y + z = -9$ ; This system has the augmented matrix $\begin{bmatrix} 1 & 1 & -1 & | & 6 \\ 2 & -1 & 1 & | & -9 \\ 1 & -2 & 3 & | & 1 \end{bmatrix}$.
$\quad\; x - 2y + 3z = 1$

$$\begin{bmatrix} 1 & 1 & -1 & | & 6 \\ 0 & -3 & 3 & | & -21 \\ 1 & -2 & 3 & | & 1 \end{bmatrix} \begin{matrix} \\ -2R1+R2 \\ \\ \end{matrix} \Rightarrow \begin{bmatrix} 1 & 1 & -1 & | & 6 \\ 0 & -3 & 3 & | & -21 \\ 0 & -3 & 4 & | & -5 \end{bmatrix} \begin{matrix} \\ \\ -1R1+R3 \end{matrix} \Rightarrow \begin{bmatrix} 1 & 1 & -1 & | & 6 \\ 0 & 1 & -1 & | & 7 \\ 0 & -3 & 4 & | & -5 \end{bmatrix} \begin{matrix} \\ -\frac{1}{3}R2 \\ \\ \end{matrix}$$

$$\begin{bmatrix} 1 & 0 & 0 & | & -1 \\ 0 & 1 & -1 & | & 7 \\ 0 & -3 & 4 & | & -5 \end{bmatrix} \begin{matrix} -1R2+R1 \\ \\ \\ \end{matrix} \Rightarrow \begin{bmatrix} 1 & 0 & 0 & | & -1 \\ 0 & 1 & -1 & | & 7 \\ 0 & 0 & 1 & | & 16 \end{bmatrix} \begin{matrix} \\ \\ 3R2+R3 \end{matrix} \Rightarrow \begin{bmatrix} 1 & 0 & 0 & | & -1 \\ 0 & 1 & 0 & | & 23 \\ 0 & 0 & 1 & | & 16 \end{bmatrix} \begin{matrix} \\ R3+R2 \\ \\ \end{matrix}$$

Solution set: $\{(-1,23,16)\}$

**29.**
$$\begin{array}{l} x - z = -3 \\ y + z = 9 \\ x + z = 7 \end{array}$$ ; This system has the augmented matrix $\begin{bmatrix} 1 & 0 & -1 & | & -3 \\ 0 & 1 & 1 & | & 9 \\ 1 & 0 & 1 & | & 7 \end{bmatrix}$.

$\begin{bmatrix} 1 & 0 & -1 & | & -3 \\ 0 & 1 & 1 & | & 9 \\ 0 & 0 & 2 & | & 10 \end{bmatrix}\begin{matrix} \\ \\ -1\text{R}1+\text{R}3 \end{matrix} \Rightarrow \begin{bmatrix} 1 & 0 & -1 & | & -3 \\ 0 & 1 & 1 & | & 9 \\ 0 & 0 & 1 & | & 5 \end{bmatrix}\begin{matrix} \\ \\ \frac{1}{2}\text{R}3 \end{matrix} \Rightarrow \begin{bmatrix} 1 & 0 & 0 & | & 2 \\ 0 & 1 & 1 & | & 9 \\ 0 & 0 & 1 & | & 5 \end{bmatrix}\begin{matrix} \text{R}3+\text{R}1 \\ \\ \end{matrix} \Rightarrow \begin{bmatrix} 1 & 0 & 0 & | & 2 \\ 0 & 1 & 0 & | & 4 \\ 0 & 0 & 1 & | & 5 \end{bmatrix}\begin{matrix} \\ -1\text{R}3+\text{R}2 \\ \end{matrix}$

Solution set: $\{(2,4,5)\}$

**31.**
$$\begin{array}{l} y = -2x - 2z + 1 \\ x = -2y - z + 2 \\ z = x - y \end{array}$$ ; Rewrite the system as $\begin{array}{l} 2x + y + 2z = 1 \\ x + 2y + z = 2 \\ x - y - z = 0. \end{array}$ The augmented matrix is $\begin{bmatrix} 2 & 1 & 2 & | & 1 \\ 1 & 2 & 1 & | & 2 \\ 1 & -1 & -1 & | & 0 \end{bmatrix}$.

$\begin{bmatrix} 2 & 1 & 2 & | & 1 \\ 1 & 2 & 1 & | & 2 \\ 1 & -1 & -1 & | & 0 \end{bmatrix}\begin{matrix} \\ \text{R}1\leftrightarrow\text{R}2 \\ \end{matrix} \Rightarrow \begin{bmatrix} 1 & 2 & 1 & | & 2 \\ 0 & -3 & 0 & | & -3 \\ 1 & -1 & -1 & | & 0 \end{bmatrix}\begin{matrix} \\ -2\text{R}1+\text{R}2 \\ \end{matrix} \Rightarrow \begin{bmatrix} 1 & 2 & 1 & | & 2 \\ 0 & -3 & 0 & | & -3 \\ 0 & -3 & -2 & | & -2 \end{bmatrix}\begin{matrix} \\ \\ -1\text{R}1+\text{R}3 \end{matrix}$

$\begin{bmatrix} 1 & 2 & 1 & | & 2 \\ 0 & 1 & 0 & | & 1 \\ 0 & -3 & -2 & | & -2 \end{bmatrix}\begin{matrix} \\ -\frac{1}{3}\text{R}2 \\ \end{matrix} \Rightarrow \begin{bmatrix} 1 & 0 & 1 & | & 0 \\ 0 & 1 & 0 & | & 1 \\ 0 & -3 & -2 & | & -2 \end{bmatrix}\begin{matrix} -2\text{R}2+\text{R}1 \\ \\ \end{matrix} \Rightarrow \begin{bmatrix} 1 & 0 & 1 & | & 0 \\ 0 & 1 & 0 & | & 1 \\ 0 & 0 & -2 & | & 1 \end{bmatrix}\begin{matrix} \\ \\ 3\text{R}2+\text{R}3 \end{matrix}$

$\begin{bmatrix} 1 & 0 & 1 & | & 0 \\ 0 & 1 & 0 & | & 1 \\ 0 & 0 & 1 & | & -\frac{1}{2} \end{bmatrix}\begin{matrix} \\ \\ -\frac{1}{2}\text{R}3 \end{matrix} \Rightarrow \begin{bmatrix} 1 & 0 & 0 & | & \frac{1}{2} \\ 0 & 1 & 0 & | & 1 \\ 0 & 0 & 1 & | & -\frac{1}{2} \end{bmatrix}\begin{matrix} -1\text{R}3+\text{R}1 \\ \\ \end{matrix}$

Solution set: $\left\{\left(\frac{1}{2}, 1, -\frac{1}{2}\right)\right\}$

**33.**
$$\begin{array}{l} 2x - y + 3z = 0 \\ x + 2y - z = 5 \\ 2y + z = 1 \end{array}$$ ; This system has the augmented matrix $\begin{bmatrix} 2 & -1 & 3 & | & 0 \\ 1 & 2 & -1 & | & 5 \\ 0 & 2 & 1 & | & 1 \end{bmatrix}$.

$\begin{bmatrix} 1 & 2 & -1 & | & 5 \\ 2 & -1 & 3 & | & 0 \\ 0 & 2 & 1 & | & 1 \end{bmatrix}\begin{matrix} \text{R}1\leftrightarrow\text{R}2 \\ \\ \end{matrix} \Rightarrow \begin{bmatrix} 1 & 2 & -1 & | & 5 \\ 0 & -5 & 5 & | & -10 \\ 0 & 2 & 1 & | & 1 \end{bmatrix}\begin{matrix} \\ -2\text{R}1+\text{R}2 \\ \end{matrix} \Rightarrow \begin{bmatrix} 1 & 2 & -1 & | & 5 \\ 0 & 1 & -1 & | & 2 \\ 0 & 2 & 1 & | & 1 \end{bmatrix}\begin{matrix} \\ -\frac{1}{5}\text{R}2 \\ \end{matrix}$

$\begin{bmatrix} 1 & 0 & 1 & | & 1 \\ 0 & 1 & -1 & | & 2 \\ 0 & 2 & 1 & | & 1 \end{bmatrix}\begin{matrix} -2\text{R}2+\text{R}1 \\ \\ \end{matrix} \Rightarrow \begin{bmatrix} 1 & 0 & 1 & | & 1 \\ 0 & 1 & -1 & | & 2 \\ 0 & 0 & 3 & | & -3 \end{bmatrix}\begin{matrix} \\ \\ -2\text{R}2+\text{R}3 \end{matrix} \Rightarrow \begin{bmatrix} 1 & 0 & 1 & | & 1 \\ 0 & 1 & -1 & | & 2 \\ 0 & 0 & 1 & | & -1 \end{bmatrix}\begin{matrix} \\ \\ \frac{1}{3}\text{R}3 \end{matrix}$

$\begin{bmatrix} 1 & 0 & 0 & | & 2 \\ 0 & 1 & -1 & | & 2 \\ 0 & 0 & 1 & | & -1 \end{bmatrix}\begin{matrix} -1\text{R}3+\text{R}1 \\ \\ \end{matrix} \Rightarrow \begin{bmatrix} 1 & 0 & 0 & | & 2 \\ 0 & 1 & 0 & | & 1 \\ 0 & 0 & 1 & | & -1 \end{bmatrix}\begin{matrix} \\ \text{R}3+\text{R}2 \\ \end{matrix}$

Solution set: $\{(2,1,-1)\}$

**35.**
$$\begin{array}{l} 3x + 5y - z + 2 = 0 \\ 4x - y + 2z - 1 = 0 \\ -6x - 10y + 2z = 0 \end{array}$$ ; Rewrite the system as $\begin{array}{l} 3x + 5y - z = -2 \\ 4x - y + 2z = 1 \\ -6x - 10y + 2z = 0. \end{array}$

The augmented matrix is $\begin{bmatrix} 3 & 5 & -1 & | & -2 \\ 4 & -1 & 2 & | & 1 \\ -6 & -10 & 2 & | & 0 \end{bmatrix}$.

$$\begin{bmatrix} 1 & \frac{5}{3} & -\frac{1}{3} & -\frac{2}{3} \\ 4 & -1 & 2 & 1 \\ -6 & -10 & 2 & 0 \end{bmatrix} \frac{1}{3}R1 \Rightarrow \begin{bmatrix} 1 & \frac{5}{3} & -\frac{1}{3} & -\frac{2}{3} \\ 0 & -\frac{23}{3} & \frac{10}{3} & \frac{11}{3} \\ -6 & -10 & 2 & 0 \end{bmatrix} \begin{matrix} \\ -4R1+R2 \\ \end{matrix} \Rightarrow \begin{bmatrix} 1 & \frac{5}{3} & -\frac{1}{3} & -\frac{2}{3} \\ 0 & -\frac{23}{3} & \frac{10}{3} & \frac{11}{3} \\ 0 & 0 & 0 & -4 \end{bmatrix} \begin{matrix} \\ \\ 6R1+R3 \end{matrix}$$

The last row indicates that there is no solution. The solution set is $\varnothing$.

**37.** $x - 8y + z = 4$
$3x - y + 2z = -1$

This system has the augmented matrix $\begin{bmatrix} 1 & -8 & 1 & 4 \\ 3 & -1 & 2 & -1 \end{bmatrix}$.

$$\begin{bmatrix} 1 & -8 & 1 & 4 \\ 0 & 23 & -1 & -13 \end{bmatrix} \begin{matrix} \\ -3R1+R2 \end{matrix} \Rightarrow \begin{bmatrix} 1 & -8 & 1 & 4 \\ 0 & 1 & -\frac{1}{23} & -\frac{13}{23} \end{bmatrix} \frac{1}{23}R2 \Rightarrow \begin{bmatrix} 1 & 0 & \frac{15}{23} & -\frac{12}{23} \\ 0 & 1 & -\frac{1}{23} & -\frac{13}{23} \end{bmatrix} 8R2+R1$$

The equations that correspond to the final matrix are $x + \frac{15}{23}z = -\frac{12}{23}$ and $y - \frac{1}{23}z = -\frac{13}{23}$.

This system has infinitely many solutions. We will express the solution set with $z$ as the arbitrary variable.

Therefore, $x = -\frac{15}{23}z - \frac{12}{23}$ and $y = \frac{1}{23}z - \frac{13}{23}$.

Solution set: $\left\{ \left( -\frac{15}{23}z - \frac{12}{23}, \frac{1}{23}z - \frac{13}{23}, z \right) \right\}$

**39.** $x - y + 2z + w = 4$
$y + z = 3$
$z - w = 2$
$x - y = 0$

This system has the augmented matrix $\begin{bmatrix} 1 & -1 & 2 & 1 & 4 \\ 0 & 1 & 1 & 0 & 3 \\ 0 & 0 & 1 & -1 & 2 \\ 1 & -1 & 0 & 0 & 0 \end{bmatrix}$

$$\begin{bmatrix} 1 & -1 & 2 & 1 & 4 \\ 0 & 1 & 1 & 0 & 3 \\ 0 & 0 & 1 & -1 & 2 \\ 0 & 0 & -2 & -1 & -4 \end{bmatrix} \begin{matrix} \\ \\ \\ -1R1+R4 \end{matrix} \Rightarrow \begin{bmatrix} 1 & 0 & 3 & 1 & 7 \\ 0 & 1 & 1 & 0 & 3 \\ 0 & 0 & 1 & -1 & 2 \\ 0 & 0 & -2 & -1 & -4 \end{bmatrix} \begin{matrix} R2+R1 \\ \\ \\ \end{matrix} \Rightarrow$$

$$\begin{bmatrix} 1 & 0 & 0 & 4 & 1 \\ 0 & 1 & 1 & 0 & 3 \\ 0 & 0 & 1 & -1 & 2 \\ 0 & 0 & -2 & -1 & -4 \end{bmatrix} \begin{matrix} -3R3+R1 \\ \\ \\ \end{matrix} \Rightarrow \begin{bmatrix} 1 & 0 & 0 & 4 & 1 \\ 0 & 1 & 0 & 1 & 1 \\ 0 & 0 & 1 & -1 & 2 \\ 0 & 0 & -2 & -1 & -4 \end{bmatrix} \begin{matrix} \\ -1R3+R2 \\ \\ \end{matrix} \Rightarrow$$

$$\begin{bmatrix} 1 & 0 & 0 & 4 & 1 \\ 0 & 1 & 0 & 1 & 1 \\ 0 & 0 & 1 & -1 & 2 \\ 0 & 0 & 0 & -3 & 0 \end{bmatrix} \begin{matrix} \\ \\ \\ 2R3+R4 \end{matrix} \Rightarrow \begin{bmatrix} 1 & 0 & 0 & 4 & 1 \\ 0 & 1 & 0 & 1 & 1 \\ 0 & 0 & 1 & -1 & 2 \\ 0 & 0 & 0 & 1 & 0 \end{bmatrix} \begin{matrix} \\ \\ \\ -\frac{1}{3}R4 \end{matrix} \Rightarrow$$

$$\begin{bmatrix} 1 & 0 & 0 & 0 & 1 \\ 0 & 1 & 0 & 1 & 1 \\ 0 & 0 & 1 & -1 & 2 \\ 0 & 0 & 0 & 1 & 0 \end{bmatrix} \begin{matrix} -4R4+R1 \\ \\ \\ \end{matrix} \Rightarrow \begin{bmatrix} 1 & 0 & 0 & 0 & 1 \\ 0 & 1 & 0 & 0 & 1 \\ 0 & 0 & 1 & -1 & 2 \\ 0 & 0 & 0 & 1 & 0 \end{bmatrix} \begin{matrix} \\ -1R4+R2 \\ \\ \end{matrix} \Rightarrow \begin{bmatrix} 1 & 0 & 0 & 0 & 1 \\ 0 & 1 & 0 & 0 & 1 \\ 0 & 0 & 1 & 0 & 2 \\ 0 & 0 & 0 & 1 & 0 \end{bmatrix} \begin{matrix} \\ \\ R4+R3 \\ \end{matrix}$$

Solution set: $\left\{ (1, 1, 2, 0) \right\}$

**41.**
$$\begin{aligned} x + 3y - 2z - w &= 9 \\ 4x + y + z + 2w &= 2 \\ -3x - y + z - w &= -5 \\ x - y - 3z - 2w &= 2 \end{aligned}$$ ; This system has the augmented matrix $\begin{bmatrix} 1 & 3 & -2 & -1 & | & 9 \\ 4 & 1 & 1 & 2 & | & 2 \\ -3 & -1 & 1 & -1 & | & -5 \\ 1 & -1 & -3 & -2 & | & 2 \end{bmatrix}$.

$$\begin{bmatrix} 1 & 3 & -2 & -1 & | & 9 \\ 0 & -11 & 9 & 6 & | & -34 \\ -3 & -1 & 1 & -1 & | & -5 \\ 1 & -1 & -3 & -2 & | & 2 \end{bmatrix} \begin{matrix} \\ -4R1+R2 \\ \\ \end{matrix} \Rightarrow \begin{bmatrix} 1 & 3 & -2 & -1 & | & 9 \\ 0 & -11 & 9 & 6 & | & -34 \\ 0 & 8 & -5 & -4 & | & 22 \\ 1 & -1 & -3 & -2 & | & 2 \end{bmatrix} \begin{matrix} \\ \\ 3R1+R3 \\ \end{matrix} \Rightarrow$$

$$\begin{bmatrix} 1 & 3 & -2 & -1 & | & 9 \\ 0 & -11 & 9 & 6 & | & -34 \\ 0 & 8 & -5 & -4 & | & 22 \\ 0 & -4 & -1 & -1 & | & -7 \end{bmatrix} \begin{matrix} \\ \\ \\ -1R1+R4 \end{matrix} \Rightarrow \begin{bmatrix} 1 & 3 & -2 & -1 & | & 9 \\ 0 & 1 & -\frac{9}{11} & -\frac{6}{11} & | & \frac{34}{11} \\ 0 & 8 & -5 & -4 & | & 22 \\ 0 & -4 & -1 & -1 & | & -7 \end{bmatrix} \begin{matrix} \\ -\frac{1}{11}R2 \\ \\ \end{matrix} \Rightarrow$$

$$\begin{bmatrix} 1 & 0 & \frac{5}{11} & \frac{7}{11} & | & -\frac{3}{11} \\ 0 & 1 & -\frac{9}{11} & -\frac{6}{11} & | & \frac{34}{11} \\ 0 & 8 & -5 & -4 & | & 22 \\ 0 & -4 & -1 & -1 & | & -7 \end{bmatrix} \begin{matrix} -3R2+R1 \\ \\ \\ \end{matrix} \Rightarrow \begin{bmatrix} 1 & 0 & \frac{5}{11} & \frac{7}{11} & | & -\frac{3}{11} \\ 0 & 1 & -\frac{9}{11} & -\frac{6}{11} & | & \frac{34}{11} \\ 0 & 0 & \frac{17}{11} & \frac{4}{11} & | & -\frac{30}{11} \\ 0 & -4 & -1 & -1 & | & -7 \end{bmatrix} \begin{matrix} \\ \\ -8R2+R3 \\ \end{matrix} \Rightarrow$$

$$\begin{bmatrix} 1 & 0 & \frac{5}{11} & \frac{7}{11} & | & -\frac{3}{11} \\ 0 & 1 & -\frac{9}{11} & -\frac{6}{11} & | & \frac{34}{11} \\ 0 & 0 & \frac{17}{11} & \frac{4}{11} & | & -\frac{30}{11} \\ 0 & 0 & -\frac{47}{11} & -\frac{35}{11} & | & \frac{59}{11} \end{bmatrix} \begin{matrix} \\ \\ \\ 4R2+R4 \end{matrix} \Rightarrow \begin{bmatrix} 1 & 0 & \frac{5}{11} & \frac{7}{11} & | & -\frac{3}{11} \\ 0 & 1 & -\frac{9}{11} & -\frac{6}{11} & | & \frac{34}{11} \\ 0 & 0 & 1 & \frac{4}{17} & | & -\frac{30}{17} \\ 0 & 0 & -\frac{47}{11} & -\frac{35}{11} & | & \frac{59}{11} \end{bmatrix} \begin{matrix} \\ \\ \frac{11}{17}R3 \\ \end{matrix} \Rightarrow$$

$$\begin{bmatrix} 1 & 0 & 0 & \frac{9}{17} & | & \frac{9}{17} \\ 0 & 1 & -\frac{9}{11} & -\frac{6}{11} & | & \frac{34}{11} \\ 0 & 0 & 1 & \frac{4}{17} & | & -\frac{30}{17} \\ 0 & 0 & -\frac{47}{11} & -\frac{35}{11} & | & \frac{59}{11} \end{bmatrix} \begin{matrix} -\frac{5}{11}R3+R1 \\ \\ \\ \end{matrix} \Rightarrow \begin{bmatrix} 1 & 0 & 0 & \frac{9}{17} & | & \frac{9}{17} \\ 0 & 1 & 0 & -\frac{6}{17} & | & \frac{28}{71} \\ 0 & 0 & 1 & \frac{4}{17} & | & -\frac{30}{17} \\ 0 & 0 & -\frac{47}{11} & -\frac{35}{11} & | & \frac{59}{11} \end{bmatrix} \begin{matrix} \\ \frac{9}{11}R3+R2 \\ \\ \end{matrix} \Rightarrow$$

$$\begin{bmatrix} 1 & 0 & 0 & \frac{9}{17} & | & \frac{9}{17} \\ 0 & 1 & 0 & -\frac{6}{17} & | & \frac{28}{17} \\ 0 & 0 & 1 & \frac{4}{17} & | & -\frac{30}{17} \\ 0 & 0 & 0 & -\frac{37}{17} & | & -\frac{37}{17} \end{bmatrix} \begin{matrix} \\ \\ \\ \frac{47}{11}R3+R4 \end{matrix} \Rightarrow \begin{bmatrix} 1 & 0 & 0 & \frac{9}{17} & | & \frac{9}{17} \\ 0 & 1 & 0 & -\frac{6}{17} & | & \frac{28}{17} \\ 0 & 0 & 1 & \frac{4}{17} & | & -\frac{30}{17} \\ 0 & 0 & 0 & 1 & | & 1 \end{bmatrix} \begin{matrix} \\ \\ \\ -\frac{17}{37}R4 \end{matrix} \Rightarrow$$

$$\begin{bmatrix} 1 & 0 & 0 & 0 & | & 0 \\ 0 & 1 & 0 & -\frac{6}{17} & | & \frac{28}{17} \\ 0 & 0 & 1 & \frac{4}{17} & | & -\frac{30}{17} \\ 0 & 0 & 0 & 1 & | & 1 \end{bmatrix} \begin{matrix} -\frac{9}{17}R4+R1 \\ \\ \\ \end{matrix} \Rightarrow \begin{bmatrix} 1 & 0 & 0 & 0 & | & 0 \\ 0 & 1 & 0 & 0 & | & 2 \\ 0 & 0 & 1 & \frac{4}{17} & | & -\frac{30}{17} \\ 0 & 0 & 0 & 1 & | & 1 \end{bmatrix} \begin{matrix} \\ \frac{6}{17}R4+R2 \\ \\ \end{matrix} \Rightarrow$$

$$\begin{bmatrix} 1 & 0 & 0 & 0 & | & 0 \\ 0 & 1 & 0 & 0 & | & 2 \\ 0 & 0 & 1 & 0 & | & -2 \\ 0 & 0 & 0 & 1 & | & 1 \end{bmatrix} \begin{matrix} \\ \\ -\frac{4}{17}R4+R3 \\ \end{matrix}$$

Solution set: $\{(0, 2, -2, 1)\}$

**43.** $\quad .3x + 2.7y - \sqrt{2}z = 3$
$\quad\quad \sqrt{7}x - 20y + 12z = -2$
$\quad\quad 4x + \sqrt{3}y - 1.2z = \frac{3}{4}$

This system has the augmented matrix

$$\begin{bmatrix} .3 & 2.7 & -\sqrt{2} & 3 \\ \sqrt{7} & -20 & 12 & -2 \\ 4 & \sqrt{3} & -1.2 & \frac{3}{4} \end{bmatrix}.$$

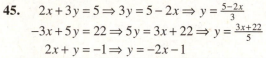

Using a graphing calculator. we obtain the approximate solution set.
Solution set: {(.571, 7.041, 11.442)}.

**45.** $\quad 2x + 3y = 5 \Rightarrow 3y = 5 - 2x \Rightarrow y = \frac{5-2x}{3}$
$\quad -3x + 5y = 22 \Rightarrow 5y = 3x + 22 \Rightarrow y = \frac{3x+22}{5}$
$\quad\quad 2x + y = -1 \Rightarrow y = -2x - 1$

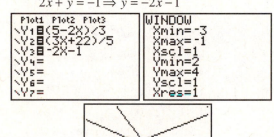

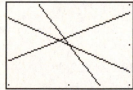

There are no solutions since the three lines do not intersect at one point.

**47.** $\quad \dfrac{1}{(x-1)(x+1)} = \dfrac{A}{x-1} + \dfrac{B}{x+1}$

If we add the rational expression on the right, we get the following.

$$\frac{1}{(x-1)(x+1)} = \frac{A(x+1)}{(x-1)(x+1)} + \frac{B(x-1)}{(x-1)(x+1)}$$

$$\frac{1}{(x-1)(x+1)} = \frac{A(x+1) + B(x-1)}{(x-1)(x+1)}$$

Since the denominators are equal, the numerators must be equal. Thus we have the equation $1 = A(x+1) + B(x-1)$.

$1 = A(x+1) + B(x-1) \Rightarrow$
$1 = Ax + A + Bx - B$
$1 = (A-B) + (Ax + Bx) \Rightarrow$
$1 = (A-B) + (A+B)x$

Equating the coefficients of like powers of $x$ gives the following system of equations.
$A + B = 0$
$A - B = 1$

Solve this system by the Gauss-Jordan method.

$$\begin{bmatrix} 1 & 1 & 0 \\ 1 & -1 & 1 \end{bmatrix} \Rightarrow \begin{bmatrix} 1 & 1 & 0 \\ 0 & -2 & 1 \end{bmatrix} \begin{array}{c} \\ -R1 + R2 \end{array} \Rightarrow$$

$$\begin{bmatrix} 1 & 1 & 0 \\ 0 & 1 & -\frac{1}{2} \end{bmatrix} -\frac{1}{2}R2 \Rightarrow$$

$$\begin{bmatrix} 1 & 0 & \frac{1}{2} \\ 0 & 1 & -\frac{1}{2} \end{bmatrix} -1R2 + R1$$

Thus, $A = \frac{1}{2}$ and $B = -\frac{1}{2}$.

**49.** $\quad \dfrac{x}{(x-a)(x+a)} = \dfrac{A}{x-a} + \dfrac{B}{x+a}$

If we add the rational expression on the right, we get the following.

$$\frac{x}{(x-a)(x+a)} = \frac{A(x+a)}{(x-a)(x+a)} + \frac{B(x-a)}{(x-a)(x+a)}$$

$$\frac{x}{(x-a)(x+a)} = \frac{A(x+a) + B(x-a)}{(x-a)(x+a)}$$

Since the denominators are equal, the numerators must be equal. Thus we have the equation $x = A(x+a) + B(x-a)$.

$x = A(x+a) + B(x-a) \Rightarrow$
$x = Ax + Aa + Bx - Ba \Rightarrow$
$x = (A+B)x + (A-B)a$

Equating the coefficients of like powers of $x$ gives the following system of equations.
$A + B = 1$
$A - B = 0$

Solve this system by the Gauss-Jordan method.

$$\begin{bmatrix} 1 & 1 & 1 \\ 1 & -1 & 0 \end{bmatrix} \Rightarrow \begin{bmatrix} 1 & 1 & 1 \\ 0 & -2 & -1 \end{bmatrix} \begin{array}{c} \\ -R1 + R2 \end{array} \Rightarrow$$

$$\begin{bmatrix} 1 & 1 & 1 \\ 0 & 1 & \frac{1}{2} \end{bmatrix} -\frac{1}{2}R2 \Rightarrow \begin{bmatrix} 1 & 0 & \frac{1}{2} \\ 0 & 1 & \frac{1}{2} \end{bmatrix} -1R2 + R1$$

Thus, $A = \frac{1}{2}$ and $B = \frac{1}{2}$.

**51.** Let $x$ = the daily wage for a day laborer
$y$ = the daily wage for a concrete finisher
From the information, we can write the system
$$7x + 2y = 1384$$
$$x + 5y = 952$$
Solve this system by the Gauss-Jordan method:

$$\begin{bmatrix} 7 & 2 & | & 1384 \\ 1 & 5 & | & 952 \end{bmatrix} \Rightarrow$$

$$\begin{bmatrix} 1 & 5 & | & 952 \\ 7 & 2 & | & 1384 \end{bmatrix} \begin{matrix} R1 \leftrightarrow R2 \end{matrix} \Rightarrow$$

$$\begin{bmatrix} 1 & 5 & | & 952 \\ 0 & -33 & | & -5280 \end{bmatrix} \begin{matrix} \\ -7R1 + R2 \end{matrix} \Rightarrow$$

$$\begin{bmatrix} 1 & 5 & | & 952 \\ 0 & 1 & | & 160 \end{bmatrix} -\frac{1}{33}R2 \Rightarrow$$

$$\begin{bmatrix} 1 & 0 & | & 152 \\ 0 & 1 & | & 160 \end{bmatrix} -5R2 + R1$$

From the final matrix, $x = 152$ and $y = 160$, so the day laborers earn \$152 per day and the concreter finishers earn \$160 per day.

**53.** Let $x$ = the first number; $y$ = the second number; $z$ = the third number.
From the information, we can write the system
$$x + y + z = 20$$
$$x = 3(y - z) \Rightarrow x = 3y - 3z \Rightarrow x - 3y + 3z = 0$$
$$y = 2 + 2z \Rightarrow \qquad\qquad y - 2z = 2$$
Solve this system by the Gauss-Jordan method:

$$\begin{bmatrix} 1 & 1 & 1 & | & 20 \\ 1 & -3 & 3 & | & 0 \\ 0 & 1 & -2 & | & 2 \end{bmatrix} \Rightarrow$$

$$\begin{bmatrix} 1 & 1 & 1 & | & 20 \\ 0 & 1 & -\frac{1}{2} & | & 5 \\ 0 & 1 & -2 & | & 2 \end{bmatrix} \begin{matrix} \\ \frac{1}{4}(R1 - R2) \\ \\ \end{matrix} \Rightarrow$$

$$\begin{bmatrix} 1 & 1 & 1 & | & 20 \\ 0 & 1 & -\frac{1}{2} & | & 5 \\ 0 & 0 & 1 & | & 2 \end{bmatrix} \begin{matrix} \\ \\ \frac{2}{3}(R2 - R3) \end{matrix} \Rightarrow$$

$$\begin{bmatrix} 1 & 1 & 1 & | & 20 \\ 0 & 1 & 0 & | & 6 \\ 0 & 0 & 1 & | & 2 \end{bmatrix} \begin{matrix} \\ R2 + \frac{1}{2}R3 \\ \\ \end{matrix} \Rightarrow$$

$$\begin{bmatrix} 1 & 0 & 1 & | & 14 \\ 0 & 1 & 0 & | & 6 \\ 0 & 0 & 1 & | & 2 \end{bmatrix} \begin{matrix} R1 - R2 \\ \\ \\ \end{matrix} \Rightarrow$$

$$\begin{bmatrix} 1 & 0 & 0 & | & 12 \\ 0 & 1 & 0 & | & 6 \\ 0 & 0 & 1 & | & 2 \end{bmatrix} \begin{matrix} R1 - R3 \\ \\ \\ \end{matrix}$$

From the final matrix, $x = 12$, $y = 6$, and $z = 2$. Thus, the numbers are 12, 6, and 2.

**55.** Let $x$ = number of cubic centimeters of the 2% solution; $y$ = number of cubic centimeters of the 7% solution.
From the information we can write the system
$$x + y = 40 \qquad \Rightarrow \qquad x + y = 40$$
$$.02x + .07y = .032(40) \Rightarrow .02x + .07y = 1.28$$
Solve this system by the Gauss-Jordan method.

$$\begin{bmatrix} 1 & 1 & | & 40 \\ .02 & .07 & | & 1.28 \end{bmatrix} \Rightarrow \begin{bmatrix} 1 & 1 & | & 40 \\ 0 & .05 & | & .48 \end{bmatrix} -.02R1 + R2$$

$$\begin{bmatrix} 1 & 1 & | & 40 \\ 0 & 1 & | & 9.6 \end{bmatrix} \frac{1}{.05}R2 \Rightarrow \begin{bmatrix} 1 & 0 & | & 30.4 \\ 0 & 1 & | & 9.6 \end{bmatrix} -1R2 + R1$$

From the final matrix, $x = 30.4$ and $y = 9.6$, so the chemist should mix 30.4 $cm^3$ of the 2% solution with 9.6 $cm^3$ of the 7% solution.

**57.** Answers will vary.
If the condition that correlates to the third row in the augmented matrix,

$$\begin{bmatrix} 1 & 1 & 1 & | & 25,000 \\ 8 & 10 & 9 & | & 222,000 \\ -1 & 2 & 0 & | & 4000 \end{bmatrix},$$ was dropped, then

that row would be dropped. We would have

$$\begin{bmatrix} 1 & 1 & 1 & | & 25,000 \\ 8 & 10 & 9 & | & 222,000 \end{bmatrix}.$$

Solve by the Gauss-Jordan method.

$$\begin{bmatrix} 1 & 1 & 1 & | & 25,000 \\ 8 & 10 & 9 & | & 222,000 \end{bmatrix} \Rightarrow$$

$$\begin{bmatrix} 1 & 1 & 1 & | & 25,000 \\ 0 & 2 & 1 & | & 22,000 \end{bmatrix} -8R1 + R2 \Rightarrow$$

$$\begin{bmatrix} 1 & 1 & 1 & | & 25,000 \\ 0 & 1 & \frac{1}{2} & | & 11,000 \end{bmatrix} \frac{1}{2}R2 \Rightarrow$$

$$\begin{bmatrix} 1 & 0 & \frac{1}{2} & | & 14,000 \\ 0 & 1 & \frac{1}{2} & | & 11,000 \end{bmatrix} -1R2 + R1$$

We can see it is not possible to find a unique solution.

**59.** Let $x$ = number of grams of food A;
$y$ = number of grams of food B;
$z$ = number of grams of food C.
Completing the table we have the following.

| Food Group | A | B | C | Total |
|---|---|---|---|---|
| Grams/Meal | $x$ | $y$ | $z$ | 400 |

From the information given in the exercise, we can write the system.

$$x + y + z = 400$$
$$x = \tfrac{1}{3}y \Rightarrow 3x - y = 0$$
$$x + z = 2y \Rightarrow x - 2y + z = 0$$

This system has the augmented matrix

$$\begin{bmatrix} 1 & 1 & 1 & 400 \\ 3 & -1 & 0 & 0 \\ 1 & -2 & 1 & 0 \end{bmatrix}$$

Solve by the Gauss-Jordan method.

$$\begin{bmatrix} 1 & 1 & 1 & 400 \\ 0 & -4 & -3 & -1200 \\ 1 & -2 & 1 & 0 \end{bmatrix} \begin{matrix} \\ -3R1 + R2 \\ \end{matrix} \Rightarrow$$

$$\begin{bmatrix} 1 & 1 & 1 & 400 \\ 0 & -4 & -3 & -1200 \\ 0 & -3 & 0 & -400 \end{bmatrix} \begin{matrix} \\ \\ -1R1 + R3 \end{matrix} \Rightarrow$$

$$\begin{bmatrix} 1 & 1 & 1 & 400 \\ 0 & 1 & \tfrac{3}{4} & 300 \\ 0 & -3 & 0 & -400 \end{bmatrix} \begin{matrix} \\ -\tfrac{1}{4}R2 \\ \end{matrix} \Rightarrow$$

$$\begin{bmatrix} 1 & 0 & \tfrac{1}{4} & 100 \\ 0 & 1 & \tfrac{3}{4} & 300 \\ 0 & -3 & 0 & -400 \end{bmatrix} \begin{matrix} -1R2 + R1 \\ \\ \end{matrix} \Rightarrow$$

$$\begin{bmatrix} 1 & 0 & \tfrac{1}{4} & 100 \\ 0 & 1 & \tfrac{3}{4} & 300 \\ 0 & 0 & \tfrac{9}{4} & 500 \end{bmatrix} \begin{matrix} \\ \\ 3R2 + R3 \end{matrix} \Rightarrow$$

$$\begin{bmatrix} 1 & 0 & \tfrac{1}{4} & 100 \\ 0 & 1 & \tfrac{3}{4} & 300 \\ 0 & 0 & 1 & \tfrac{2000}{9} \end{bmatrix} \begin{matrix} \\ \\ \tfrac{4}{9}R3 \end{matrix} \Rightarrow$$

$$\begin{bmatrix} 1 & 0 & 0 & \tfrac{400}{9} \\ 0 & 1 & \tfrac{3}{4} & 300 \\ 0 & 0 & 1 & \tfrac{2000}{9} \end{bmatrix} \begin{matrix} -\tfrac{1}{4}R3 + R1 \\ \\ \end{matrix} \Rightarrow$$

$$\begin{bmatrix} 1 & 0 & 0 & \tfrac{400}{9} \\ 0 & 1 & 0 & \tfrac{400}{3} \\ 0 & 0 & 1 & \tfrac{2000}{9} \end{bmatrix} \begin{matrix} \\ -\tfrac{3}{4}R3 + R2 \\ \end{matrix}$$

From the final matrix, we

have $x = \tfrac{400}{9} \approx 44.4$, $y = \tfrac{400}{3} \approx 133.3$, and

$z = \tfrac{2000}{9} \approx 222.2$. The diet should include 44.4 g of food A, 133.3 g of food B, and 222.2 g of food C.

**61. (a)** For the 65 or older group:
With $x = 0$ representing 2005 and $x = 45$ representing 2050, we have information that correlates to the points $(0, .124)$ and $(45, .207)$.

$$m = \frac{.207 - .124}{45 - 0} = \frac{.083}{45} = .00184$$

Since we have the point $(0, .124)$, the equation in slope-intercept form is
$y = .00184x + .124$

For the 25–34 group:
With $x = 0$ representing 2005 and $x = 45$ representing 2050, we have information that correlates to the points $(0, .134)$ and $(50, .126)$.

$$m = \frac{.126 - .134}{45 - 0} = \frac{-.008}{45} = -.000178$$

(three significant figures)
Since we have the point $(0, .134)$, the equation in slope-intercept form is
$y = -.000178x + .134$

**(b)** We need to solve the following system.

$$\begin{cases} y = .00184x + .124 \\ y = -.000178x + .134 \end{cases} \Rightarrow$$

$$\begin{cases} -.00184x + y = .124 \\ .000178x + y = .134 \end{cases} \Rightarrow$$

$$-1840x + 1,000,000y = 124,000$$
$$178x + 1,000,000y = 134,000$$

This system has the augmented matrix

$$\begin{bmatrix} -1840 & 1,000,000 & 124,000 \\ 178 & 1,000,000 & 134,000 \end{bmatrix}$$

Solve by the Gauss-Jordan method.

$$\begin{bmatrix} 1 & -\tfrac{1,000,000}{1840} & -\tfrac{124,000}{1840} \\ 1 & \tfrac{1,000,000}{178} & \tfrac{134,000}{178} \end{bmatrix} \begin{matrix} -\tfrac{1}{1840}R1 \\ \tfrac{1}{178}R2 \end{matrix}$$

Reducing the fractions, we have

$$\begin{bmatrix} 1 & -\tfrac{12,500}{23} & -\tfrac{1550}{23} \\ 1 & \tfrac{500000}{89} & \tfrac{67,000}{89} \end{bmatrix} \Rightarrow$$

$$\begin{bmatrix} 1 & -\tfrac{12,500}{23} & -\tfrac{1550}{23} \\ 0 & -\tfrac{12,612,500}{2047} & -\tfrac{1,678,950}{2047} \end{bmatrix} \begin{matrix} \\ R1 - R2 \end{matrix} \Rightarrow$$

$$\begin{bmatrix} 1 & -\tfrac{12,500}{23} & -\tfrac{1550}{23} \\ 0 & 1 & \tfrac{1,678,950}{12,612,500} \end{bmatrix} \begin{matrix} \\ -\tfrac{2047}{12,612,500}R2 \end{matrix} \Rightarrow$$

$$\begin{bmatrix} 1 & 0 & \tfrac{5000}{1009} \\ 0 & 1 & \tfrac{33,579}{252,250} \end{bmatrix} \begin{matrix} \tfrac{12,500}{23}R2 + R1 \\ \end{matrix}$$

*(continued on next page)*

(*continued from page 263*)

The solution set is

$$\left\{\left(\frac{5000}{1009}, \frac{33,579}{252,250}\right)\right\} \approx \left\{(4.9554, .1331)\right\}$$

$x = \frac{5000}{1009} \approx 4.9554$ represents the year

2009 and $y = \frac{33,579}{252,250} \approx .1331 \approx 13.31\%.$

The two groups will have the same percent of the population in 2009, about 13.3%

**63. (a)** A height of $6'11''$ is $83''$.
If $W = 7.46H - 374$, then
$W = 7.46(83) - 374 = 245.18$.
Using the first equation, the predicted weight is approximately 245 pounds.
If $W = 7.93H - 405$, then
$W = 7.93(83) - 405 = 253.19$
Using the second equation, the predicted weight is approximately 253 pounds.

**(b)** For the first model $W = 7.46H - 374$, a 1-inch increase in height results in a 7.46-pound increase in weight.
For the second model $W = 7.93H - 405$, a 1-inch increase in height results in a 7.93-pound increase in weight. In each case, the change is given by the slope of the line that is the graph of the given equation.

**(c)** $W - 7.46H = -374$
$W - 7.93H = -405$
This system has the augmented matrix

$$\begin{bmatrix} 1 & -7.46 & | & -374 \\ 1 & -7.93 & | & -405 \end{bmatrix}.$$

Solve this system by the Gauss-Jordan method.

$$\begin{bmatrix} 1 & -7.46 & | & -374 \\ 0 & -.47 & | & -31 \end{bmatrix} \begin{matrix} \\ -1R1 + R2 \end{matrix} \Rightarrow$$

$$\begin{bmatrix} 1 & -7.46 & | & -374 \\ 0 & 1 & | & 65.957 \end{bmatrix} \begin{matrix} \\ -\frac{1}{.47}R2 \end{matrix} \Rightarrow$$

$$\begin{bmatrix} 1 & 0 & | & 118.043 \\ 0 & 1 & | & 65.957 \end{bmatrix} 7.46R2 + R1$$

From the last matrix, we have $W \approx 118$ and $H \approx 66$. The two models agree at a height of 66 inches and a weight of 118 pounds.

**65.** $F = a + bA + cP + dW$
Substituting the values, we have the following system of equations.
$a + 871b + 11.5c + 3d = 239$
$a + 847b + 12.2c + 2d = 234$
$a + 685b + 10.6c + 5d = 192$
$a + 969b + 14.2c + 1d = 343.$

**67.** Using these values,
$F = -714.457 + .34756A + 48.6585P$
$+ 30.71951W$

## Section 5.3: Determinant Solution of Linear Systems

**1.** $\begin{vmatrix} -5 & 9 \\ 4 & -1 \end{vmatrix} = -5(-1) - 4 \cdot 9 = 5 - 36 = -31$

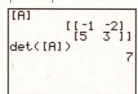

**3.** $\begin{vmatrix} -1 & -2 \\ 5 & 3 \end{vmatrix} = -1 \cdot 3 - 5(-2) = -3 - (-10) = 7$

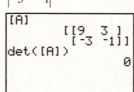

**5.** $\begin{vmatrix} 9 & 3 \\ -3 & -1 \end{vmatrix} = 9(-1) - (-3) \cdot 3 = -9 - (-9) = 0$

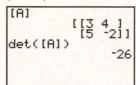

**7.** $\begin{vmatrix} 3 & 4 \\ 5 & -2 \end{vmatrix} = 3(-2) - 5 \cdot 4 = -6 - 20 = -26$

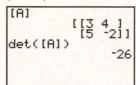

**9.** $\begin{vmatrix} -7 & 0 \\ 3 & 0 \end{vmatrix} = -7 \cdot 0 - 3 \cdot 0 = 0 - 0 = 0$

```
[A]
            [[-7 0]
             [3  0]]
det([A])
                    0
```

**11.** $\begin{vmatrix} -2 & 0 & 1 \\ 1 & 2 & 0 \\ 4 & 2 & 1 \end{vmatrix}$

To find the cofactor of 1, we have
$i = 2, j = 1,$

$M_{21} = \begin{vmatrix} 0 & 1 \\ 2 & 1 \end{vmatrix} = 0 \cdot 1 - 2 \cdot 1 = 0 - 2 = -2$

Thus, the cofactor is

$(-1)^{2+1} (-2) = (-1)^3 (-2) = (-1)(-2) = 2.$

To find the cofactor of 2, we have
$i = 2, j = 2,$

$M_{22} = \begin{vmatrix} -2 & 1 \\ 4 & 1 \end{vmatrix} = -2 \cdot 1 - 4 \cdot 1 = -2 - 4 = -6$

Thus, the cofactor is

$(-1)^{2+2} (-6) = (-1)^4 (-6) = 1(-6) = -6.$

To find the cofactor of 0, we have
$i = 2, j = 3,$

$M_{23} = \begin{vmatrix} -2 & 0 \\ 4 & 2 \end{vmatrix} = -2 \cdot 2 - 4 \cdot 0 = -4 - 0 = -4$

Thus, the cofactor is

$(-1)^{2+3} (-4) = (-1)^5 (-4) = (-1)(-4) = 4.$

**13.** $\begin{vmatrix} 1 & 2 & -1 \\ 2 & 3 & -2 \\ -1 & 4 & 1 \end{vmatrix}$

To find the cofactor of 2, we have $i = 2, j = 1,$

$M_{21} = \begin{vmatrix} 2 & -1 \\ 4 & 1 \end{vmatrix} = 2 \cdot 1 - 4(-1) = 2 - (-4) = 6$

Thus, the cofactor is

$(-1)^{2+1} (6) = (-1)^3 (6) = (-1)(6) = -6.$

To find the cofactor of 3, we have $i = 2, j = 2,$

$M_{22} = \begin{vmatrix} 1 & -1 \\ -1 & 1 \end{vmatrix} = 1 \cdot 1 - (-1)(-1) = 1 - 1 = 0$

Thus, the cofactor is

$(-1)^{2+2} (0) = (-1)^4 (0) = 1(0) = 0.$

To find the cofactor of $-2,$ we have
$i = 2, j = 3,$

$M_{23} = \begin{vmatrix} 1 & 2 \\ -1 & 4 \end{vmatrix} = 1 \cdot 4 - (-1) \cdot 2 = 4 - (-2) = 6$

Thus, the cofactor is

$(-1)^{2+3} (6) = (-1)^5 (6) = (-1)(6) = -6.$

For Exercises 15–27, an answer can be arrived at by expanding on any row or column as noted on page 526 of your text. In the solutions, we will expand on a row or column that allows a minimum number of calculations. Any row or column containing zero, will reduce the number of calculations as noted in the same paragraph.

**15.** $\begin{vmatrix} 4 & -7 & 8 \\ 2 & 1 & 3 \\ -6 & 3 & 0 \end{vmatrix}$

If we expand by the third row, we will need to find $a_{31} \cdot A_{31} + a_{32} \cdot A_{32} + a_{33} \cdot A_{33}.$ However, we do not need to calculate $A_{33},$ since $a_{33} = 0.$

$A_{31} = (-1)^{3+1} \begin{vmatrix} -7 & 8 \\ 1 & 3 \end{vmatrix} = (-1)^4 (-7 \cdot 3 - 1 \cdot 8)$

$= 1(-21 - 8) = 1(-29) = -29$

$A_{32} = (-1)^{3+2} \begin{vmatrix} 4 & 8 \\ 2 & 3 \end{vmatrix} = (-1)^5 (4 \cdot 3 - 2 \cdot 8)$

$= -1(12 - 16) = -1(-4) = 4$

$a_{31} \cdot A_{31} + a_{32} \cdot A_{32} + a_{33} \cdot A_{33}$
$= a_{31} \cdot A_{31} + a_{32} \cdot A_{32} + 0 \cdot A_{33}$
$= -6(-29) + 3(4) + 0 = 174 + 12 = 186$

**17.** $\begin{vmatrix} 1 & 2 & 0 \\ -1 & 2 & -1 \\ 0 & 1 & 4 \end{vmatrix}$

If we expand by the third row, we will need to find $a_{31} \cdot A_{31} + a_{32} \cdot A_{32} + a_{33} \cdot A_{33}.$ However, we do not need to calculate $A_{31},$ since $a_{31} = 0.$

$A_{32} = (-1)^{3+2} \begin{vmatrix} 1 & 0 \\ -1 & -1 \end{vmatrix}$

$= (-1)^5 \left[ 1(-1) - (-1) \cdot 0 \right]$

$= -1(-1 - 0) = -1(-1) = 1$

*(continued on next page)*

(*continued from page 265*)

$$A_{33} = (-1)^{3+3} \begin{vmatrix} 1 & 2 \\ -1 & 2 \end{vmatrix}$$
$$= (-1)^6 \left[ 1 \cdot 2 - (-1) \cdot 2 \right]$$
$$= 1\left[ 2 - (-2) \right] = 1(4) = 4$$

$$a_{31} \cdot A_{31} + a_{32} \cdot A_{32} + a_{33} \cdot A_{33}$$
$$= 0 \cdot A_{31} + a_{32} \cdot A_{32} + a_{33} \cdot A_{33}$$
$$= 0 + 1(1) + 4(4) = 1 + 16 = 17$$

For Exercises 19–27, we will use the sign checkerboard as described in the margin of page 526 in calculations.

**19.** $$\begin{vmatrix} 10 & 2 & 1 \\ -1 & 4 & 3 \\ -3 & 8 & 10 \end{vmatrix}$$

If we expand by the first row, we will need to find $a_{11} \cdot M_{11} - a_{12} \cdot M_{12} + a_{13} \cdot M_{13}$.

$$M_{11} = \begin{vmatrix} 4 & 3 \\ 8 & 10 \end{vmatrix} = 4 \cdot 10 - 8 \cdot 3 = 40 - 24 = 16,$$

$$M_{12} = \begin{vmatrix} -1 & 3 \\ -3 & 10 \end{vmatrix} = -1 \cdot 10 - (-3) \cdot 3$$
$$= -10 - (-9) = -1, \text{ and}$$

$$M_{13} = \begin{vmatrix} -1 & 4 \\ -3 & 8 \end{vmatrix} = -1 \cdot 8 - (-3) \cdot 4 = -8 - (-12) = 4$$

$$a_{11} \cdot M_{11} - a_{12} \cdot M_{12} + a_{13} \cdot M_{13}$$
$$= 10 \cdot 16 - 2(-1) + 1 \cdot 4$$
$$= 160 - (-2) + 4 = 166$$

**21.** $$\begin{vmatrix} 1 & -2 & 3 \\ 0 & 0 & 0 \\ 1 & 10 & -12 \end{vmatrix}$$

If we expand by the second row, we will need to find $-a_{21} \cdot M_{21} + a_{22} \cdot M_{22} - a_{23} \cdot M_{23}$. Since $a_{21} = a_{22} = a_{23} = 0$, the result will be zero.

**23.** $$\begin{vmatrix} 3 & 3 & -1 \\ 2 & 6 & 0 \\ -6 & -6 & 2 \end{vmatrix}$$

If we expand by the second row, we will need to find $-a_{21} \cdot M_{21} + a_{22} \cdot M_{22} - a_{23} \cdot M_{23}$. However, we do not need to calculate $M_{23}$, since $a_{23} = 0$.

$$M_{21} = \begin{vmatrix} 3 & -1 \\ -6 & 2 \end{vmatrix} = 3 \cdot 2 - (-6)(-1) = 6 - 6 = 0$$

and

$$M_{22} = \begin{vmatrix} 3 & -1 \\ -6 & 2 \end{vmatrix} = 3 \cdot 2 - (-6)(-1) = 6 - 6 = 0$$

$$a_{21} \cdot M_{21} - a_{22} \cdot M_{22} + a_{23} \cdot M_{23}$$
$$= a_{21} \cdot M_{21} - a_{22} \cdot M_{22} + 0 \cdot M_{23}$$
$$= 2 \cdot 0 - 6 \cdot 0 + 0 = 0 - 0 = 0$$

**25.** $$\begin{vmatrix} 1 & 0 & 0 \\ 0 & 1 & 0 \\ 0 & 0 & 1 \end{vmatrix}$$

If we expand by the first row, we will need to find $a_{11} \cdot M_{11} - a_{12} \cdot M_{12} + a_{13} \cdot M_{13}$. However, we do not need to calculate $M_{12}$ or $M_{13}$, since $a_{12} = a_{13} = 0$.

$$M_{11} = \begin{vmatrix} 1 & 0 \\ 0 & 1 \end{vmatrix} = 1 \cdot 1 - 0 \cdot 0 = 1 - 0 = 1$$

$$a_{11} \cdot M_{11} - a_{12} \cdot M_{12} + a_{13} \cdot M_{13}$$
$$= a_{11} \cdot M_{11} - 0 \cdot M_{12} + 0 \cdot M_{13}$$
$$= 1 \cdot 1 - 0 + 0 = 1$$

**27.** $$\begin{vmatrix} -2 & 0 & 1 \\ 0 & 1 & 0 \\ 0 & 0 & -1 \end{vmatrix}$$

If we expand by the first column, we will need to find $a_{11} \cdot M_{11} - a_{21} \cdot M_{21} + a_{31} \cdot M_{31}$. However, we do not need to calculate $M_{21}$ or $M_{31}$, since $a_{21} = a_{31} = 0$.

$$M_{11} = \begin{vmatrix} 1 & 0 \\ 0 & -1 \end{vmatrix} = 1(-1) - 0 \cdot 0 = -1 - 0 = -1$$

$$a_{11} \cdot M_{11} - a_{21} \cdot M_{21} + a_{31} \cdot M_{31}$$
$$= a_{11} \cdot M_{11} - 0 \cdot M_{21} + 0 \cdot M_{31}$$
$$= -2(-1) - 0 + 0 = 2$$

**29.** $$\begin{vmatrix} .4 & -.8 & .6 \\ .3 & .9 & .7 \\ 3.1 & 4.1 & -2.8 \end{vmatrix}$$

The determinant is −5.5.

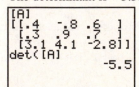

**31.**
$$\begin{vmatrix} a_{11} & a_{12} & a_{13} \\ a_{21} & a_{22} & a_{23} \\ a_{31} & a_{32} & a_{33} \end{vmatrix}\begin{matrix} a_{11} & a_{12} \\ a_{21} & a_{22} \\ a_{31} & a_{32} \end{matrix}$$

$d_1 = a_{11}a_{22}a_{33}; d_2 = a_{12}a_{23}a_{31}; d_3 = a_{13}a_{21}a_{32}$
$d_4 = a_{13}a_{22}a_{31}; d_5 = a_{11}a_{23}a_{32}; d_6 = a_{12}a_{21}a_{33}$

$\left(d_1 + d_2 + d_3\right) - \left(d_4 + d_5 + d_6\right)$
$\quad = \left(a_{11}a_{22}a_{33} + a_{12}a_{23}a_{31} + a_{13}a_{21}a_{32}\right)$
$\qquad - \left(a_{13}a_{22}a_{31} + a_{11}a_{23}a_{32} + a_{12}a_{21}a_{33}\right)$
$\quad = \left(a_{11}a_{22}a_{33} + a_{12}a_{23}a_{31} + a_{13}a_{21}a_{32}\right)$
$\qquad - \left(a_{31}a_{22}a_{13} + a_{32}a_{23}a_{11} + a_{33}a_{21}a_{12}\right)$

**33.**
$$\begin{vmatrix} 1 & 3 & 2 \\ 0 & 2 & 6 \\ 7 & 1 & 5 \end{vmatrix}$$

If we expand by the first column, we will need to find $a_{11} \cdot M_{11} - a_{21} \cdot M_{21} + a_{31} \cdot M_{31}$.

$(-1)^{1+1} \cdot 1 \cdot \begin{vmatrix} 2 & 6 \\ 1 & 5 \end{vmatrix} - (-1)^{1+2} \cdot 0 \cdot \begin{vmatrix} 3 & 2 \\ 1 & 5 \end{vmatrix}$

$\qquad + (-1)^{1+3} \cdot 7 \cdot \begin{vmatrix} 3 & 2 \\ 2 & 6 \end{vmatrix}$

$= 1(10-6) + 0 + 7(18-4) = 4 + 98 = 102$

Both methods give the same determinant.

**35.** To solve the equation $\begin{vmatrix} 5 & x \\ -3 & 2 \end{vmatrix} = 6$, we need to

solve $5 \cdot 2 - (-3) \cdot x = 6$.

$5 \cdot 2 - (-3) \cdot x = 6 \Rightarrow 10 + 3x = 6 \Rightarrow$
$\qquad 3x = -4 \Rightarrow x = -\frac{4}{3}$

Verifying $\begin{vmatrix} 5 & -\frac{4}{3} \\ -3 & 2 \end{vmatrix} = 6$, we have

$5 \cdot 2 - (-3)\left(-\frac{4}{3}\right) = 10 - 4 = 6.$

Solution set: $\left\{-\frac{4}{3}\right\}$

**37.** To solve the equation $\begin{vmatrix} x & 3 \\ x & x \end{vmatrix} = 4$, we need to

solve
$\quad x \cdot x - 3x = 4 \Rightarrow x^2 - 3x = 4 \Rightarrow$
$x^2 - 3x - 4 = 0 \Rightarrow (x+1)(x-4) = 0$
$x + 1 = 0 \Rightarrow x = -1 \ \text{ or } \ x - 4 = 0 \Rightarrow x = 4$

Verify $x = -1$.
$\begin{vmatrix} -1 & 3 \\ -1 & -1 \end{vmatrix} = -1(-1) - (-1) \cdot 3 = 1 - (-3) = 4$

Verify $x = 4$.
$\begin{vmatrix} 4 & 3 \\ 4 & 4 \end{vmatrix} = 4 \cdot 4 - 4 \cdot 3 = 16 - 12 = 4$

Solution set: $\{-1, 4\}$

**39.** To solve the equation $\begin{vmatrix} -2 & 0 & 1 \\ -1 & 3 & x \\ 5 & -2 & 0 \end{vmatrix} = 3$, expand

by the first row. In order to do this, we will need to find $a_{11} \cdot M_{11} - a_{12} \cdot M_{12} + a_{13} \cdot M_{13}$. However, we do not need to calculate $M_{12}$, since $a_{12} = 0$.

$M_{11} = \begin{vmatrix} 3 & x \\ -2 & 0 \end{vmatrix} = 3 \cdot 0 - (-2) \cdot x = 0 - (-2x) = 2x$

$M_{13} = \begin{vmatrix} -1 & 3 \\ 5 & -2 \end{vmatrix} = (-1)(-2) - 5 \cdot 3 = 2 - 15 = -13$

$a_{11} \cdot M_{11} - a_{12} \cdot M_{12} + a_{13} \cdot M_{13}$
$\quad = a_{11} \cdot M_{11} - 0 \cdot M_{12} + a_{13} \cdot M_{13}$
$\quad = -2(2x) - 0 + 1(-13)$
$\quad = -4x + (-13) = -4x - 13$

Set this equal to 3 and solve to get
$-4x - 13 = 3 \Rightarrow -4x = 16 \Rightarrow x = -4.$

Verify $\begin{vmatrix} -2 & 0 & 1 \\ -1 & 3 & -4 \\ 5 & -2 & 0 \end{vmatrix} = 3.$

Since
$M_{11} = \begin{vmatrix} 3 & -4 \\ -2 & 0 \end{vmatrix} = 3 \cdot 0 - (-2)(-4) = 0 - 8 = -8$

and $M_{13} = \begin{vmatrix} -1 & 3 \\ 5 & -2 \end{vmatrix} = -13$, we have

$a_{11} \cdot M_{11} - a_{12} \cdot M_{12} + a_{13} \cdot M_{13}$
$\quad = a_{11} \cdot M_{11} - 0 \cdot M_{12} + a_{13} \cdot M_{13}$
$\quad = -2(-8) - 0 + 1(-13) = 16 + (-13) = 3$

Solution set: $\{-4\}$

**41.** To solve the equation $\begin{vmatrix} 5 & 3x & -3 \\ 0 & 2 & -1 \\ 4 & -1 & x \end{vmatrix} = -7$,

expand by the second row. We will need to find $-a_{21} \cdot M_{21} + a_{22} \cdot M_{22} - a_{23} \cdot M_{23}$. However, we do not need to calculate $M_{21}$, since $a_{21} = 0$.

*(continued on next page)*

(*continued from page 267*)

$$M_{22} = \begin{vmatrix} 5 & -3 \\ 4 & x \end{vmatrix} = 5x - 4(-3)$$
$$= 5x - (-12) = 5x + 12$$

and $M_{23} = \begin{vmatrix} 5 & 3x \\ 4 & -1 \end{vmatrix} = 5(-1) - 4 \cdot 3x = -5 - 12x$

$$-a_{21} \cdot M_{21} + a_{22} \cdot M_{22} - a_{23} \cdot M_{23}$$
$$= -0 \cdot M_{21} + a_{22} \cdot M_{22} - a_{23} \cdot M_{23}$$
$$= 0 + 2(5x + 12) - (-1)(-5 - 12x)$$
$$= 2(5x + 12) + (-5 - 12x)$$
$$= 10x + 24 - 5 - 12x = 19 - 2x$$

Set this equal to $-7$ and solve to get
$$19 - 2x = -7 \Rightarrow -2x = -26 \Rightarrow x = 13.$$

Verify $\begin{vmatrix} 5 & 3(13) & -3 \\ 0 & 2 & -1 \\ 4 & -1 & 13 \end{vmatrix} = \begin{vmatrix} 5 & 39 & -3 \\ 0 & 2 & -1 \\ 4 & -1 & 13 \end{vmatrix} = -7.$

Since
$$M_{22} = \begin{vmatrix} 5 & -3 \\ 4 & 13 \end{vmatrix} = 5 \cdot 13 - 4(-3) = 65 - (-12) = 77$$

and
$$M_{23} = \begin{vmatrix} 5 & 39 \\ 4 & -1 \end{vmatrix} = 5(-1) - 4 \cdot 39 = -5 - 156 = -161,$$

we have the following.
$$-a_{21} \cdot M_{21} + a_{22} \cdot M_{22} - a_{23} \cdot M_{23}$$
$$= -0 \cdot M_{21} + a_{22} \cdot M_{22} - a_{23} \cdot M_{23}$$
$$= 0 + 2 \cdot 77 - (-1)(-161) = 154 - 161 = -7$$

Solution set: $\{13\}$

**43.** $P(0,0), Q(0,2), R(1,4)$

Find $D = \dfrac{1}{2} \begin{vmatrix} x_1 & y_1 & 1 \\ x_2 & y_2 & 1 \\ x_3 & y_3 & 1 \end{vmatrix}$, where

$P = (x_1, y_1) = (0,0), Q(x_2, y_2) = (0,2),$ and
$R = (x_3, y_3) = (1, 4)$

Expanding by the first row, we have the following

$$D = \frac{1}{2} \begin{vmatrix} 0 & 0 & 1 \\ 0 & 2 & 1 \\ 1 & 4 & 1 \end{vmatrix} = \frac{1}{2} \left[ 0 \begin{vmatrix} 2 & 1 \\ 4 & 1 \end{vmatrix} - 0 \begin{vmatrix} 0 & 1 \\ 1 & 1 \end{vmatrix} + 1 \begin{vmatrix} 0 & 2 \\ 1 & 4 \end{vmatrix} \right]$$
$$= \frac{1}{2} \left[ 0(2 - 4) - 0(0 - 1) + 1(0 - 2) \right]$$
$$= \frac{1}{2} \left[ 0(-2) - 0(-1) + 1(-2) \right]$$
$$= \frac{1}{2} \left[ 0 - 0 + (-2) \right] = \frac{1}{2}(-2) = -1$$

Area of triangle $= |D| = |-1| = 1.$

**45.** $P(2,5), Q(-1,3), R(4,0)$

Find $D = \dfrac{1}{2} \begin{vmatrix} x_1 & y_1 & 1 \\ x_2 & y_2 & 1 \\ x_3 & y_3 & 1 \end{vmatrix}$, where

$P = (x_1, y_1) = (2,5), Q(x_2, y_2) = (-1,3),$ and
$R = (x_3, y_3) = (4,0).$

Expanding by the third row, we have the following.

$$D = \frac{1}{2} \begin{vmatrix} 2 & 5 & 1 \\ -1 & 3 & 1 \\ 4 & 0 & 1 \end{vmatrix}$$
$$= \frac{1}{2} \left[ 4 \begin{vmatrix} 5 & 1 \\ 3 & 1 \end{vmatrix} - 0 \begin{vmatrix} 2 & 1 \\ -1 & 1 \end{vmatrix} + 1 \begin{vmatrix} 2 & 5 \\ -1 & 3 \end{vmatrix} \right]$$
$$= \frac{1}{2} \left[ 4(5 - 3) - 0(2 + 1) + 1(6 + 5) \right]$$
$$= \frac{1}{2} \left[ 4(2) - 0(3) + 1(11) \right] = \frac{1}{2}(8 - 0 + 11)$$
$$= \frac{1}{2}(19) = \frac{19}{2} = 9.5$$

Area of triangle $= |D| = |9.5| = 9.5.$

**47.** $(101.3, 52.7), (117.2, 253.9), (313.1, 301.6)$

Label the points as follows.
$P = (x_1, y_1) = (101.3, 52.7),$
$Q(x_2, y_2) = (117.2, 253.9),$ and
$R = (x_3, y_3) = (313.1, 301.6)$

Since

$$D = \frac{1}{2} \begin{vmatrix} x_1 & y_1 & 1 \\ x_2 & y_2 & 1 \\ x_3 & y_3 & 1 \end{vmatrix} = \frac{1}{2} \begin{vmatrix} 101.3 & 52.7 & 1 \\ 117.2 & 253.9 & 1 \\ 313.1 & 301.6 & 1 \end{vmatrix},$$

we will enter the $3 \times 3$ as $\begin{bmatrix} 101.3 & 52.7 & 1 \\ 117.2 & 253.9 & 1 \\ 313.1 & 301.6 & 1 \end{bmatrix}$

and perform the calculations as shown below.

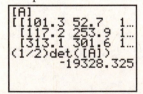

Area of triangular lot is $|-19,328.325|$ ft$^2$ or

approximately $19,328.3$ ft$^2$.

**49.** $\begin{vmatrix} 1 & 0 & 0 \\ 1 & 0 & 1 \\ 3 & 0 & 0 \end{vmatrix}$

By Theorem 1, the determinant is 0 since every entry in column 2 is 0.

**51.** $\begin{vmatrix} 6 & 8 & -12 \\ -1 & 0 & 2 \\ 4 & 0 & -8 \end{vmatrix}$

Given $\begin{bmatrix} 6 & 8 & -12 \\ -1 & 0 & 2 \\ 4 & 0 & -8 \end{bmatrix}$, add 2 times column 1

to column 3 to obtain $\begin{bmatrix} 6 & 8 & 0 \\ -1 & 0 & 0 \\ 4 & 0 & 0 \end{bmatrix}$. By

Theorem 6 the following statement is true.

$\begin{vmatrix} 6 & 8 & -12 \\ -1 & 0 & 2 \\ 4 & 0 & -8 \end{vmatrix} = \begin{vmatrix} 6 & 8 & 0 \\ -1 & 0 & 0 \\ 4 & 0 & 0 \end{vmatrix}$

By Theorem 1, the determinant is 0 since every entry in column 3 is 0.

**53.** $\begin{vmatrix} -4 & 1 & 4 \\ 2 & 0 & 1 \\ 0 & 2 & 4 \end{vmatrix}$

Since $\begin{bmatrix} -4 & 1 & 4 \\ 2 & 0 & 1 \\ 0 & 2 & 4 \end{bmatrix} \Rightarrow \begin{bmatrix} 0 & 1 & 6 \\ 2 & 0 & 1 \\ 0 & 2 & 4 \end{bmatrix} 2R2 + R1$,

by Theorem 6 we have the following.

$\begin{vmatrix} -4 & 1 & 4 \\ 2 & 0 & 1 \\ 0 & 2 & 4 \end{vmatrix} = \begin{vmatrix} 0 & 1 & 6 \\ 2 & 0 & 1 \\ 0 & 2 & 4 \end{vmatrix}$

Expanding by the first column, we have the following.

$\begin{vmatrix} 0 & 1 & 6 \\ 2 & 0 & 1 \\ 0 & 2 & 4 \end{vmatrix} = 0 \begin{vmatrix} 0 & 1 \\ 2 & 4 \end{vmatrix} - 2 \begin{vmatrix} 1 & 6 \\ 2 & 4 \end{vmatrix} + 0 \begin{vmatrix} 1 & 6 \\ 0 & 1 \end{vmatrix}$

$= 0 - 2(4 - 12) + 0 = -2(-8) = 16$

As noted on page 526 of the text, the array of signs can be extended for determinants of $4 \times 4$ matrices. For Exercises 55–57, use the following array of signs.

*For $4 \times 4$ matrices*

$\begin{matrix} + & - & + & - \\ - & + & - & + \\ + & - & + & - \\ - & + & - & + \end{matrix}$

Like previous exercises, you can arrive at a solution by expanding on any row or column. We will also use determinant theorems to reduce the number of calculations.

**55.** $\begin{vmatrix} 3 & -6 & 5 & -1 \\ 0 & 2 & -1 & 3 \\ -6 & 4 & 2 & 0 \\ -7 & 3 & 1 & 1 \end{vmatrix}$

Add 3 times row 1 to row 2 and add row 1 to row 4 to obtain

$\begin{vmatrix} 3 & -6 & 5 & -1 \\ 0 & 2 & -1 & 3 \\ -6 & 4 & 2 & 0 \\ -7 & 3 & 1 & 1 \end{vmatrix} = \begin{vmatrix} 3 & -6 & 5 & -1 \\ 9 & -16 & 14 & 0 \\ -6 & 4 & 2 & 0 \\ -4 & -3 & 6 & 0 \end{vmatrix}.$

Expanding by the fourth column, we have the following.

$\begin{vmatrix} 3 & -6 & 5 & -1 \\ 9 & -16 & 14 & 0 \\ -6 & 4 & 2 & 0 \\ -4 & -3 & 6 & 0 \end{vmatrix} = -(-1) \begin{vmatrix} 9 & -16 & 14 \\ -6 & 4 & 2 \\ -4 & -3 & 6 \end{vmatrix}$

$= \begin{vmatrix} 9 & -16 & 14 \\ -6 & 4 & 2 \\ -4 & -3 & 6 \end{vmatrix}$

Adding $-7$ times row 2 to row 1 and $-3$ times row 2 added to row 3 we have the following.

$\begin{vmatrix} 9 & -16 & 14 \\ -6 & 4 & 2 \\ -4 & -3 & 6 \end{vmatrix} = \begin{vmatrix} 51 & -44 & 0 \\ -6 & 4 & 2 \\ 14 & -15 & 0 \end{vmatrix}$

Expanding by the third column, we have

$-2 \begin{vmatrix} 51 & -44 \\ 14 & -15 \end{vmatrix} = -2(-765 + 616)$

$= -2(-149) = 298$

**57.** $\begin{vmatrix} 4 & 0 & 0 & 2 \\ -1 & 0 & 3 & 0 \\ 2 & 4 & 0 & 1 \\ 0 & 0 & 1 & 2 \end{vmatrix}$

Expanding by the second column, we have

$\begin{vmatrix} 4 & 0 & 0 & 2 \\ -1 & 0 & 3 & 0 \\ 2 & 4 & 0 & 1 \\ 0 & 0 & 1 & 2 \end{vmatrix} = -4 \begin{vmatrix} 4 & 0 & 2 \\ -1 & 3 & 0 \\ 0 & 1 & 2 \end{vmatrix}.$

*(continued on next page)*

(*continued from page 269*)

Using the definition of the determinant in the text, we have the following.

$$-4\begin{vmatrix} 4 & 0 & 2 \\ -1 & 3 & 0 \\ 0 & 1 & 2 \end{vmatrix}$$

$$= -4\left(\begin{bmatrix} 4(3)(2) + 0(0)(0) + 2(-1)(1) \end{bmatrix} \\ -\begin{bmatrix} 0(3)(2) + 1(0)(4) + 2(-1)(0) \end{bmatrix}\right)$$

$$= -4\begin{bmatrix} (24 + 0 - 2) - (0 + 0 + 0) \end{bmatrix}$$

$$= -4(22 - 0) = -4(22) = -88$$

**59.** For $\begin{aligned} 4x + 3y - 2z &= 1 \\ 7x - 4y + 3z &= 2 \\ -2x + y - 8z &= 0 \end{aligned}$, we have the following.

$$D = \begin{vmatrix} 4 & 3 & -2 \\ 7 & -4 & 3 \\ -2 & 1 & -8 \end{vmatrix}, D_x = \begin{vmatrix} 1 & 3 & -2 \\ 2 & -4 & 3 \\ 0 & 1 & -8 \end{vmatrix},$$

$$D_y = \begin{vmatrix} 4 & 1 & -2 \\ 7 & 2 & 3 \\ -2 & 0 & -8 \end{vmatrix}, \text{ and } D_z = \begin{vmatrix} 4 & 3 & 1 \\ 7 & -4 & 2 \\ -2 & 1 & 0 \end{vmatrix}$$

(a)  D    (b)  A

(c)  C    (d)  B

**61.** $\begin{aligned} x + y &= 4 \\ 2x - y &= 2 \end{aligned}$

$$D = \begin{vmatrix} 1 & 1 \\ 2 & -1 \end{vmatrix} = 1(-1) - 2(1) = -1 - 2 = -3,$$

$$D_x = \begin{vmatrix} 4 & 1 \\ 2 & -1 \end{vmatrix} = 4(-1) - 2(1) = -4 - 2 = -6,$$

$$D_y = \begin{vmatrix} 1 & 4 \\ 2 & 2 \end{vmatrix} = 1(2) - 2(4) = 2 - 8 = -6 \Rightarrow$$

$$x = \frac{D_x}{D} = \frac{-6}{-3} = 2 \text{ and } y = \frac{D_y}{D} = \frac{-6}{-3} = 2.$$

Solution set: $\{(2, 2)\}$

**63.** $\begin{aligned} 4x + 3y &= -7 \\ 2x + 3y &= -11 \end{aligned}$

$$D = \begin{vmatrix} 4 & 3 \\ 2 & 3 \end{vmatrix} = 4(3) - 2(3) = 12 - 6 = 6,$$

$$D_x = \begin{vmatrix} -7 & 3 \\ -11 & 3 \end{vmatrix} = -7(3) - (-11)(3)$$

$$= -21 + 33 = 12,$$

$$D_y = \begin{vmatrix} 4 & -7 \\ 2 & -11 \end{vmatrix} = 4(-11) - 2(-7)$$

$$= -44 + 14 = -30 \Rightarrow$$

$$x = \frac{D_x}{D} = \frac{12}{6} = 2 \text{ and } y = \frac{D_y}{D} = \frac{-30}{6} = -5.$$

Solution set: $\{(2, -5)\}$

**65.** $\begin{aligned} 5x + 4y &= 10 \\ 3x - 7y &= 6 \end{aligned}$

$$D = \begin{vmatrix} 5 & 4 \\ 3 & -7 \end{vmatrix} = 5(-7) - 3(4) = -35 - 12 = -47,$$

$$D_x = \begin{vmatrix} 10 & 4 \\ 6 & -7 \end{vmatrix} = 10(-7) - 6(4) = -70 - 24 = -94,$$

$$D_y = \begin{vmatrix} 5 & 10 \\ 3 & 6 \end{vmatrix} = 5(6) - 3(10) = 30 - 30 = 0 \Rightarrow$$

$$x = \frac{D_x}{D} = \frac{-94}{-47} = 2 \text{ and } y = \frac{D_y}{D} = \frac{0}{-47} = 0.$$

Solution set: $\{(2, 0)\}$

**67.** $\begin{aligned} 1.5x + 3y &= 5 \quad (1) \\ 2x + 4y &= 3 \quad (2) \end{aligned}$

$$D = \begin{vmatrix} 1.5 & 3 \\ 2 & 4 \end{vmatrix} = (1.5)(4) - 2(3) = 6 - 6 = 0$$

Because $D = 0$, Cramer's rule does not apply. To determine whether the system is inconsistent or has infinitely many solutions, use the elimination method.

$\quad 6x + 12y = 20$   Multiply equation (1) by 4.

$\underline{-6x - 12y = -9}$   Multiply equation (2) by $-3$.

$\qquad\qquad 0 = 11$   False

The system is inconsistent.
Solution set: $\varnothing$

**69.** $\begin{aligned} 3x + 2y &= 4 \ (1) \\ 6x + 4y &= 8 \ (2) \end{aligned}$

$$D = \begin{vmatrix} 3 & 2 \\ 6 & 4 \end{vmatrix} = 3(4) - 6(2) = 12 - 12 = 0$$

Because $D = 0$, Cramer's rule does not apply. To determine whether the system is inconsistent or has infinitely many solutions, use the elimination method.

$\quad -6x - 4y = -8$   Multiply equation (1) by $-2$.

$\underline{\quad 6x + 4y = \quad 8}$

$\qquad\qquad 0 = \quad 0$   True

This shows that equations (1) and (2) are dependent. To write the solution set with $y$ as the arbitrary variable, solve equation (1) for $x$ in terms of $y$.

$$3x + 2y = 4 \Rightarrow 3x = 4 - 2y \Rightarrow x = \frac{4 - 2y}{3}$$

Solution set: $\left\{\left(\frac{4 - 2y}{3}, y\right)\right\}$

**71.** $\frac{1}{2}x + \frac{1}{3}y = 2$

$\frac{3}{2}x - \frac{1}{2}y = -12$

$D = \begin{vmatrix} \frac{1}{2} & \frac{1}{3} \\ \frac{3}{2} & -\frac{1}{2} \end{vmatrix} = \frac{1}{2}\left(-\frac{1}{2}\right) - \frac{3}{2}\left(\frac{1}{3}\right)$

$= -\frac{1}{4} - \frac{1}{2} = -\frac{1}{4} - \frac{2}{4} = -\frac{3}{4}$

$D_x = \begin{vmatrix} 2 & \frac{1}{3} \\ -12 & -\frac{1}{2} \end{vmatrix} = 2\left(-\frac{1}{2}\right) - (-12)\left(\frac{1}{3}\right)$ and

$= -1 + 4 = 3$

$D_y = \begin{vmatrix} \frac{1}{2} & 2 \\ \frac{3}{2} & -12 \end{vmatrix} = \frac{1}{2}(-12) - \frac{3}{2}(2) = -6 - 3 = -9$

Thus, we have $x = \frac{D_x}{D} = \frac{3}{-\frac{3}{4}} = 3\left(-\frac{4}{3}\right) = -4$ and

$y = \frac{D_y}{D} = \frac{-9}{-\frac{3}{4}} = -9\left(-\frac{4}{3}\right) = 12$

Solution set: $\{(-4, 12)\}$

In Exercises 73–83, we will be using the Determinant Theorems on page 532 of the text to reduce the number of calculations necessary in simplifying determinants.

**73.** $2x - y + 4z + 2 = 0 \qquad 2x - y + 4z = -2$

$3x + 2y - z + 3 = 0 \Rightarrow 3x + 2y - z = -3$

$x + 4y + 2z - 17 = 0 \qquad x + 4y + 2z = 17$

Adding 2 times row 1 to row 2 and 4 times row 1 to row 3 we have

$D = \begin{vmatrix} 2 & -1 & 4 \\ 3 & 2 & -1 \\ 1 & 4 & 2 \end{vmatrix} = \begin{vmatrix} 2 & -1 & 4 \\ 7 & 0 & 7 \\ 9 & 0 & 18 \end{vmatrix}.$

Expanding by column two, we have

$D = -(-1)\begin{vmatrix} 7 & 7 \\ 9 & 18 \end{vmatrix} = 126 - 63 = 63.$

Adding 2 times row 1 to row 2 and 4 times row 1 to row 3, we have

$D_x = \begin{vmatrix} -2 & -1 & 4 \\ -3 & 2 & -1 \\ 17 & 4 & 2 \end{vmatrix} = \begin{vmatrix} -2 & -1 & 4 \\ -7 & 0 & 7 \\ 9 & 0 & 18 \end{vmatrix}.$

Expanding about column two, we have

$D_x = -(-1)\begin{vmatrix} -7 & 7 \\ 9 & 18 \end{vmatrix} = -126 - 63 = -189.$

Adding column 2 to column 1, we have

$D_y = \begin{vmatrix} 2 & -2 & 4 \\ 3 & -3 & -1 \\ 1 & 17 & 2 \end{vmatrix} = \begin{vmatrix} 0 & -2 & 4 \\ 0 & -3 & -1 \\ 18 & 17 & 2 \end{vmatrix}.$

Expanding by column one, we have

$D_y = 18\begin{vmatrix} -2 & 4 \\ -3 & -1 \end{vmatrix} = 18(2 + 12) = 18(14) = 252.$

Adding column 3 to column 1, we have

$D_z = \begin{vmatrix} 2 & -1 & -2 \\ 3 & 2 & -3 \\ 1 & 4 & 17 \end{vmatrix} = \begin{vmatrix} 0 & -1 & -2 \\ 0 & 2 & -3 \\ 18 & 4 & 17 \end{vmatrix}.$

Expanding by column one we have

$D_z = 18\begin{vmatrix} -1 & -2 \\ 2 & -3 \end{vmatrix} = 18(3 + 4) = 18(7) = 126.$

Thus, we have $x = \frac{D_x}{D} = \frac{-189}{63} = -3,$

$y = \frac{D_y}{D} = \frac{252}{63} = 4,$ and $z = \frac{D_z}{D} = \frac{126}{63} = 2.$

Solution set: $\{(-3, 4, 2)\}$

**75.** $4x - 3y + z = -1$

$5x + 7y + 2z = -2$

$3x - 5y - z = 1$

Adding row 3 to row 1 and 2 times row 3 to row 2, we have

$D = \begin{vmatrix} 4 & -3 & 1 \\ 5 & 7 & 2 \\ 3 & -5 & -1 \end{vmatrix} = \begin{vmatrix} 7 & -8 & 0 \\ 11 & -3 & 0 \\ 3 & -5 & -1 \end{vmatrix}.$

Expanding by column three, we have

$D = -1\begin{vmatrix} 7 & -8 \\ 11 & -3 \end{vmatrix} = -1(-21 + 88) = -1(67) = -67$

Adding column 1 to column 3 we have

$D_x = \begin{vmatrix} -1 & -3 & 1 \\ -2 & 7 & 2 \\ 1 & -5 & -1 \end{vmatrix} = \begin{vmatrix} -1 & -3 & 0 \\ -2 & 7 & 0 \\ 1 & -5 & 0 \end{vmatrix}.$

Since we have a column of zeros, $D_x = 0.$

Adding column 2 to column 3, we have

$D_y = \begin{vmatrix} 4 & -1 & 1 \\ 5 & -2 & 2 \\ 3 & 1 & -1 \end{vmatrix} = \begin{vmatrix} 4 & -1 & 0 \\ 5 & -2 & 0 \\ 3 & 1 & 0 \end{vmatrix}.$

Since we have a column of zeros, $D_y = 0.$

Adding row 3 to row 1 and 2 times row 3 to row 2, we have

$D_z = \begin{vmatrix} 4 & -3 & -1 \\ 5 & 7 & -2 \\ 3 & -5 & 1 \end{vmatrix} = \begin{vmatrix} 7 & -8 & 0 \\ 11 & -3 & 0 \\ 3 & -5 & 1 \end{vmatrix}.$

*(continued on next page)*

*(continued from page 271)*

Expanding by the third column, we have

$$D_z = 1 \begin{vmatrix} 7 & -8 \\ 11 & -3 \end{vmatrix} = -21 + 88 = 67.$$

Thus, we have $x = \frac{D_x}{D} = \frac{0}{-67} = 0,$

$y = \frac{D_y}{D} = \frac{0}{-67} = 0,$ and $z = \frac{D_z}{D} = \frac{67}{-67} = -1.$

Solution set: $\{(0, 0, -1)\}$

**77.** $x + 2y + 3z = 4$ (1)
$4x + 3y + 2z = 1$ (2)
$-x - 2y - 3z = 0$ (3)

Adding row 1 to row 3, we have

$$D = \begin{vmatrix} 1 & 2 & 3 \\ 4 & 3 & 2 \\ -1 & -2 & -3 \end{vmatrix} = \begin{vmatrix} 1 & 2 & 3 \\ 4 & 3 & 2 \\ 0 & 0 & 0 \end{vmatrix}.$$

Since we have a row of zeros, $D = 0$ and we cannot use Cramer's rule.
Using the elimination method, we can add equations (1) and (3).

$$\begin{array}{r} x + 2y + 3z = 4 \\ -x - 2y - 3z = 0 \\ \hline 0 = 4 \text{ False} \end{array}$$

The system is inconsistent.
Solution set: $\varnothing$

**79.** $-2x - 2y + 3z = 4$ (1)
$5x + 7y - z = 2$ (2)
$2x + 2y - 3z = -4$ (3)

Adding row 1 to row 3, we have

$$D = \begin{vmatrix} -2 & -2 & 3 \\ 5 & 7 & -1 \\ 2 & 2 & -3 \end{vmatrix} = \begin{vmatrix} -2 & -2 & 3 \\ 5 & 7 & -1 \\ 0 & 0 & 0 \end{vmatrix}.$$

Since we have a row of zeros, $D = 0$ and we cannot use Cramer's rule. Using the elimination method, we can add equations (1) and (3).

$$\begin{array}{r} -2x - 2y + 3z = 4 \\ 2x + 2y - 3z = -4 \\ \hline 0 = 0 \text{ True} \end{array}$$

This system will have infinitely many solutions.
Solve the system made up of equations (2) and (3) in terms of the arbitrary variable $z$. To eliminate $x$, multiply equation (2) by $-2$ and equation (3) by 5 and add the results

$$\begin{array}{r} -10x - 14y + 2z = -4 \\ 10x + 10y - 15z = -20 \\ \hline -4y - 13z = -24 \end{array}$$

Solve for $y$ in terms of $z$.
$-4y - 13z = -24 \Rightarrow -4y = -24 + 13z \Rightarrow$
$y = \frac{-24 + 13z}{-4} \Rightarrow y = \frac{24 - 13z}{4}$

Express $x$ also in terms of $z$ by solving equation (3) for $x$ and substituting $\frac{24 - 13z}{4}$ for $y$.

$2x + 2y - 3z = -4 \Rightarrow 2x = -2y + 3z - 4 \Rightarrow$
$$x = \frac{-2y + 3z - 4}{2}$$

$$x = \frac{-2\left(\frac{24-13z}{4}\right) + 3z - 4}{2} = \frac{\frac{-24+13z}{2} + 3z - 4}{2}$$

$$= \frac{-24 + 13z + 6z - 8}{4} = \frac{-32 + 19z}{4}$$

Solution set (with $z$ arbitrary):
$$\left\{ \left( \frac{-32+19z}{4}, \frac{24-13z}{4}, z \right) \right\}$$

**81.** $5x - y = -4$
$3x + 2z = 4$
$4y + 3z = 22$

Adding 4 times row 1 to row 3, we have

$$D = \begin{vmatrix} 5 & -1 & 0 \\ 3 & 0 & 2 \\ 0 & 4 & 3 \end{vmatrix} = \begin{vmatrix} 5 & -1 & 0 \\ 3 & 0 & 2 \\ 20 & 0 & 3 \end{vmatrix}.$$

Expanding by column two, we have

$$D = -(-1) \begin{vmatrix} 3 & 2 \\ 20 & 3 \end{vmatrix} = 9 - 40 = -31.$$

Adding 4 times row 1 to row 3, we have

$$D_x = \begin{vmatrix} -4 & -1 & 0 \\ 4 & 0 & 2 \\ 22 & 4 & 3 \end{vmatrix} = \begin{vmatrix} -4 & -1 & 0 \\ 4 & 0 & 2 \\ 6 & 0 & 3 \end{vmatrix}.$$

Expand by column two, we have

$$D_x = -(-1) \begin{vmatrix} 4 & 2 \\ 6 & 3 \end{vmatrix} = 12 - 12 = 0.$$

Adding column 2 to column 1, we have

$$D_y = \begin{vmatrix} 5 & -4 & 0 \\ 3 & 4 & 2 \\ 0 & 22 & 3 \end{vmatrix} = \begin{vmatrix} 1 & -4 & 0 \\ 7 & 4 & 2 \\ 22 & 22 & 3 \end{vmatrix}.$$

Adding 4 times column 1 to column 2, we

now have $D_y = \begin{vmatrix} 1 & 0 & 0 \\ 7 & 32 & 2 \\ 22 & 110 & 3 \end{vmatrix}.$

Expanding by row one, we have

$$D_y = 1 \begin{vmatrix} 32 & 2 \\ 110 & 3 \end{vmatrix} = 96 - 220 = -124.$$

Adding 4 times row 1 to row 3, we have

$$D_z = \begin{vmatrix} 5 & -1 & -4 \\ 3 & 0 & 4 \\ 0 & 4 & 22 \end{vmatrix} = \begin{vmatrix} 5 & -1 & -4 \\ 3 & 0 & 4 \\ 20 & 0 & 6 \end{vmatrix}.$$

Expanding by column two, we have

$$D_z = -(-1)\begin{vmatrix} 3 & 4 \\ 20 & 6 \end{vmatrix} = 18 - 80 = -62.$$

Thus, we have

$$x = \frac{D_x}{D} = \frac{0}{-31} = 0, \ y = \frac{D_y}{D} = \frac{-124}{-31} = 4, \text{ and}$$

$$z = \frac{D_z}{D} = \frac{-62}{-31} = 2.$$

Solution set: $\{(0, 4, 2)\}$

**83.** $x + 2y = 10$
$3x + 4z = 7$
$-y - z = 1$

Adding $-3$ times row 1 to row 2, we have

$$D = \begin{vmatrix} 1 & 2 & 0 \\ 3 & 0 & 4 \\ 0 & -1 & -1 \end{vmatrix} = \begin{vmatrix} 1 & 2 & 0 \\ 0 & -6 & 4 \\ 0 & -1 & -1 \end{vmatrix}.$$

Expanding by column one, we have

$$D = 1\begin{vmatrix} -6 & 4 \\ -1 & -1 \end{vmatrix} = 6 + 4 = 10.$$

Adding column 1 to column 2 and column 1 to column 3, we have

$$D_x = \begin{vmatrix} 10 & 2 & 0 \\ 7 & 0 & 4 \\ 1 & -1 & -1 \end{vmatrix} = \begin{vmatrix} 10 & 12 & 10 \\ 7 & 7 & 11 \\ 1 & 0 & 0 \end{vmatrix}.$$

Expanding by row three, we have

$$D_x = 1\begin{vmatrix} 12 & 10 \\ 7 & 11 \end{vmatrix} = 132 - 70 = 62.$$

Adding column 2 to column 3, we have

$$D_y = \begin{vmatrix} 1 & 10 & 0 \\ 3 & 7 & 4 \\ 0 & 1 & -1 \end{vmatrix} = \begin{vmatrix} 1 & 10 & 10 \\ 3 & 7 & 11 \\ 0 & 1 & 0 \end{vmatrix}.$$

Expanding by row three, we have

$$D_y = -1\begin{vmatrix} 1 & 10 \\ 3 & 11 \end{vmatrix} = -1(11 - 30) = -1(-19) = 19.$$

Adding column 3 to column 2, we have

$$D_z = \begin{vmatrix} 1 & 2 & 10 \\ 3 & 0 & 7 \\ 0 & -1 & 1 \end{vmatrix} = \begin{vmatrix} 1 & 12 & 10 \\ 3 & 7 & 7 \\ 0 & 0 & 1 \end{vmatrix}.$$

Expanding by row three, we have

$$D_z = 1\begin{vmatrix} 1 & 12 \\ 3 & 7 \end{vmatrix} = 7 - 36 = -29.$$

Thus, we have $x = \frac{D_x}{D} = \frac{62}{10} = \frac{31}{5}$, $y = \frac{D_y}{D} = \frac{19}{10}$,

and $z = \frac{D_z}{D} = \frac{-29}{10} = -\frac{29}{10}$.

Solution set: $\left\{\left(\frac{31}{5}, \frac{19}{10}, -\frac{29}{10}\right)\right\}$

**85.** $\frac{\sqrt{3}}{2}(W_1 + W_2) = 100$
$W_1 - W_2 = 0$

Using the distributive property, we have the following system.

$$\frac{\sqrt{3}}{2}W_1 + \frac{\sqrt{3}}{2}W_2 = 100$$
$$W_1 - W_2 = 0$$

Using Cramer's rule, we have

$$D = \begin{vmatrix} \frac{\sqrt{3}}{2} & \frac{\sqrt{3}}{2} \\ 1 & -1 \end{vmatrix} = -\frac{\sqrt{3}}{2} - \frac{\sqrt{3}}{2} = -\sqrt{3},$$

$$D_{W_1} = \begin{vmatrix} 100 & \frac{\sqrt{3}}{2} \\ 0 & -1 \end{vmatrix} = -100 - 0 = -100, \text{ and}$$

$$D_{W_2} = \begin{vmatrix} \frac{\sqrt{3}}{2} & 100 \\ 1 & 0 \end{vmatrix} = 0 - 100 = -100. \text{ This yields}$$

the following solution.

$$W_1 = \frac{D_{W_1}}{D} = \frac{-100}{-\sqrt{3}} = \frac{100}{\sqrt{3}} = \frac{100\sqrt{3}}{3} \approx 58 \text{ and}$$

$$W_2 = \frac{D_{W_2}}{D} = \frac{-100}{-\sqrt{3}} \approx 58$$

Both $W_1$ and $W_2$ are approximately 58 lb.

**87.** $bx + y = a^2$
$ax + y = b^2$

Using Cramer's rule, we have

$$D = \begin{vmatrix} b & 1 \\ a & 1 \end{vmatrix} = b - a, \ D_x = \begin{vmatrix} a^2 & 1 \\ b^2 & 1 \end{vmatrix} = a^2 - b^2,$$

and $D_y = \begin{vmatrix} b & a^2 \\ a & b^2 \end{vmatrix} = b^3 - a^3.$

$$x = \frac{D_x}{D} = \frac{a^2 - b^2}{b - a} = \frac{(a+b)(a-b)}{b-a}$$
$$= \frac{(a+b)(a-b)}{-(a-b)} = -(a+b) = -a - b$$

$$y = \frac{D_y}{D} = \frac{b^3 - a^3}{b - a} = \frac{(b-a)(b^2 + ab + a^2)}{b-a}$$
$$= b^2 + ab + a^2$$

Solution set: $\left\{\left(-a - b, a^2 + ab + b^2\right)\right\}$

**89.** $b^2x + a^2y = b^2$
$ax + by = a$

Using Cramer's rule, we have

$$D = \begin{vmatrix} b^2 & a^2 \\ a & b \end{vmatrix} = b^3 - a^3,$$

$$D_x = \begin{vmatrix} b^2 & a^2 \\ a & b \end{vmatrix} = b^3 - a^3, \text{ and}$$

$$D_y = \begin{vmatrix} b^2 & b^2 \\ a & a \end{vmatrix} = ab^2 - ab^2 = 0.$$

$$x = \frac{D_x}{D} = \frac{b^3 - a^3}{b^3 - a^3} = 1$$

$$y = \frac{D_y}{D} = \frac{0}{b^3 - a^3} = 0$$

Note: In order for $D \neq 0$, we also assumed $a \neq b$.

Solution set: $\{(1, 0)\}$

## Section 5.4: Partial Fractions

**1.** $\dfrac{5}{3x(2x+1)} = \dfrac{A}{3x} + \dfrac{B}{2x+1}$

Multiply both sides by $3x(2x+1)$ to get

$5 = A(2x+1) + B(3x)$. (1)

First substitute 0 for $x$ to get
$5 = A(2 \cdot 0 + 1) + B(3 \cdot 0) \Rightarrow A = 5$.

Replace $A$ with 5 in equation (1) and substitute $-\frac{1}{2}$ for $x$ to get the following.

$5 = 5\left[2\left(-\frac{1}{2}\right) + 1\right] + B\left[3\left(-\frac{1}{2}\right)\right]$
$= 5(-1 + 1) - \frac{3}{2}B = 5(0) - \frac{3}{2}B \Rightarrow$
$5 = -\frac{3}{2}B \Rightarrow -\frac{10}{3} = B$

Thus, we have $\dfrac{5}{3x(2x+1)} = \dfrac{5}{3x} + \dfrac{-\frac{10}{3}}{2x+1} \Rightarrow$

$\dfrac{5}{3x(2x+1)} = \dfrac{5}{3x} + \dfrac{-10}{3(2x+1)}$

**3.** $\dfrac{4x+2}{(x+2)(2x-1)} = \dfrac{A}{x+2} + \dfrac{B}{2x-1}$

Multiply both sides by $(x+2)(2x-1)$ to get

$4x + 2 = A(2x-1) + B(x+2)$. (1)

First substitute $-2$ for $x$ to get the following.

$4(-2) + 2 = A[2(-2) - 1] + B(-2+2)$
$-8 + 2 = A(-4-1) + B(0) = -6 = -5A$
$A = \frac{6}{5}$

Replace $A$ with $\frac{6}{5}$ in equation (1) and substitute $\frac{1}{2}$ for $x$ to get the following.

$4 \cdot \frac{1}{2} + 2 = \frac{6}{5}\left(2 \cdot \frac{1}{2} - 1\right) + B\left(\frac{1}{2} + 2\right)$
$2 + 2 = \frac{6}{5}(1-1) + B \cdot \frac{5}{2} \Rightarrow 4 = \frac{6}{5}(0) + \frac{5}{2}B \Rightarrow$
$4 = \frac{5}{2}B \Rightarrow B = \frac{8}{5}$

Thus, we have

$\dfrac{4x+2}{(x+2)(2x-1)} = \dfrac{\frac{6}{5}}{(x+2)} + \dfrac{\frac{8}{5}}{(2x-1)} \Rightarrow$

$\dfrac{4x+2}{(x+2)(2x-1)} = \dfrac{6}{5(x+2)} + \dfrac{8}{5(2x-1)}$

**5.** $\dfrac{x}{x^2+4x-5} = \dfrac{x}{(x+5)(x-1)} = \dfrac{A}{x+5} + \dfrac{B}{x-1}$

Multiply both sides by $(x+5)(x-1)$ to get

$x = A(x-1) + B(x+5)$. (1)

First substitute $-5$ for $x$ to get

$-5 = A(-5-1) + B(-5+5) \Rightarrow$
$-5 = A(-6) + B(0) \Rightarrow -5 = -6A \Rightarrow A = \frac{5}{6}$

Replace $A$ with $\frac{5}{6}$ in equation (1) and substitute 1 for $x$ to get

$1 = \frac{5}{6}(1-1) + B(1+5) \Rightarrow 1 = \frac{5}{6}(0) + B(6) \Rightarrow$
$1 = 6B \Rightarrow B = \frac{1}{6}$

Thus, we have

$\dfrac{x}{(x+5)(x-1)} = \dfrac{\frac{5}{6}}{x+5} + \dfrac{\frac{1}{6}}{x-1}$

$= \dfrac{5}{6(x+5)} + \dfrac{1}{6(x-1)}$

**7.** $\dfrac{2x}{(x+1)(x+2)^2} = \dfrac{A}{x+1} + \dfrac{B}{x+2} + \dfrac{C}{(x+2)^2}$

Multiply both sides by $(x+1)(x+2)^2$ to get

$2x = A(x+2)^2 + B(x+1)(x+2) + C(x+1)$ (1)

First substitute $-1$ for $x$ to get the following.

$2(-1) = A(-1+2)^2 + B(-1+1)(-1+2)$
$\qquad\qquad + C(-1+1)$
$-2 = A(1)^2 + B(0)(1) + C(0) \Rightarrow -2 = A$

Replace $A$ with $-2$ in equation (1) and substitute $-2$ for $x$ to get the following.

$2(-2) = -2(-2+2)^2 + B(-2+1)(-2+2)$
$\qquad\qquad + C(-2+1)$
$-4 = 0 + 0 - C \Rightarrow C = 4$

$2x = -2(x+2)^2 + B(x+1)(x+2) + 4(x+1)$ (2)

Now substitute 0 (arbitrary choice) for $x$ in equation (2) to get the following.

$$2(0) = -2(0+2)^2 + B(0+1)(0+2) + 4(0+1)$$
$$0 = -2(2)^2 + B(1)(2) + 4(1)$$
$$0 = -2(4) + 2B + 4 \Rightarrow 0 = -8 + 2B + 4$$
$$0 = -4 + 2B \Rightarrow 4 = 2B \Rightarrow B = 2$$

Thus we have

$$\frac{2x}{(x+1)(x+2)^2} = \frac{-2}{x+1} + \frac{2}{x+2} + \frac{4}{(x+2)^2}.$$

**9.** $\dfrac{4}{x(1-x)} = \dfrac{A}{x} + \dfrac{B}{1-x}$

Multiply both sides by $x(1-x)$ to get

$$4 = A(1-x) + Bx. \ (1)$$

First substitute 0 for $x$ to get the following.

$$4 = A(1-0) + B(0) \Rightarrow 4 = A + 0 \Rightarrow A = 4$$

Replace $A$ with 4 in equation (1) and substitute 1 for $x$ to get the following.

$$4 = 4(1-1) + B(1) \Rightarrow 4 = 0 + B \Rightarrow B = 4$$

Thus, we have $\dfrac{4}{x(1-x)} = \dfrac{4}{x} + \dfrac{4}{1-x}.$

**11.** $\dfrac{2x+1}{(x+2)^3} = \dfrac{A}{x+2} + \dfrac{B}{(x+2)^2} + \dfrac{C}{(x+2)^3}$

Multiply both sides by $(x+2)^3$ to get

$$2x+1 = A(x+2)^2 + B(x+2) + C. \ (1)$$

First substitute $-2$ for $x$ to get the following.

$$2(-2) + 1 = A(-2+2)^2 + B(-2+2) + C$$
$$-4 + 1 = 0 + 0 + C \Rightarrow C = -3$$

Replace $C$ with $-3$ in equation (1) and substitute 0 (arbitrary choice) for $x$ to get the following.

$$2(0) + 1 = A(0+2)^2 + B(0+2) - 3$$
$$1 = 4A + 2B - 3 \Rightarrow 4A + 2B = 4$$
$$2A + B = 2 \quad (3)$$

Replace $C$ with $-3$ in equation (1) and substitute $-1$ (arbitrary choice) for $x$ to get the following.

$$2(-1) + 1 = A(-1+2)^2 + B(-1+2) - 3$$
$$-1 = A + B - 3 \Rightarrow A + B = 2 \quad (4)$$

Solve the system of equations by multiplying equation (4) by $-1$ and adding to equation 3.

$$\begin{array}{r} 2A + B = \ \ 2 \\ -A - B = -2 \\ \hline A \qquad = \ \ 0 \end{array}$$

Substituting 0 for $A$ in equation (4) we obtain

$$0 + B = 2 \Rightarrow B = 2.$$

Thus, we have

$$\frac{2x+1}{(x+2)^3} = \frac{0}{x+2} + \frac{2}{(x+2)^2} + \frac{-3}{(x+2)^3}$$
$$= \frac{2}{(x+2)^2} + \frac{-3}{(x+2)^3}$$

**13.** $\dfrac{x^2}{x^2 + 2x + 1}$

This is not a proper fraction; the numerator has degree greater than or equal to that of the denominator. Divide the numerator by the denominator.

$$\begin{array}{r} 1 \phantom{00000} \\ x^2 + 2x + 1 \overline{\smash{\big)}\ x^2 \phantom{00000}} \\ \underline{x^2 + 2x + 1} \\ -2x - 1 \end{array}$$

Find the partial fraction decomposition for

$$\frac{-2x-1}{x^2 + 2x + 1} = \frac{-2x-1}{(x+1)^2}.$$

$$\frac{-2x-1}{(x+1)^2} = \frac{A}{x+1} + \frac{B}{(x+1)^2}$$

Multiply both sides by $(x+1)^2$ to get

$$-2x - 1 = A(x+1) + B. \ (1)$$

First substitute $-1$ for $x$ to get the following.

$$-2(-1) - 1 = A(-1+1) + B$$
$$2 - 1 = 0 + B \Rightarrow 1 = B$$

Replace $B$ with 1 in equation (1) and substitute 2 (arbitrary choice) for $x$ to get the following.

$$-2(2) - 1 = A(2+1) + 1 \Rightarrow -5 = 3A + 1$$
$$-6 = 3A \Rightarrow A = -2$$

Thus, we have

$$\frac{x^2}{x^2 + 2x + 1} = 1 + \frac{-2}{x+1} + \frac{1}{(x+1)^2}.$$

**15.** $\dfrac{2x^5 + 3x^4 - 3x^3 - 2x^2 + x}{2x^2 + 5x + 2}$

The degree of the numerator is greater than the degree of the denominator, so first find the quotient.

$$\begin{array}{r} x^3 - x^2 \phantom{00000000} \\ 2x^2 + 5x + 2 \overline{\smash{\big)}\ 2x^5 + 3x^4 - 3x^3 - 2x^2 + x} \\ \underline{2x^5 + 5x^4 + 2x^3} \phantom{0000000} \\ -2x^4 - 5x^3 - 2x^2 + x \\ \underline{-2x^4 - 5x^3 - 2x^2} \phantom{00} \\ x \end{array}$$

*(continued on next page)*

*(continued from page 275)*

Find the partial fraction decomposition for

$$\frac{x}{2x^2+5x+2}=\frac{x}{(2x+1)(x+2)}.$$

$$\frac{x}{(2x+1)(x+2)}=\frac{A}{2x+1}+\frac{B}{x+2}$$

Multiply both sides by $(2x+1)(x+2)$ to get

$$x=A(x+2)+B(2x+1). \quad (1)$$

First substitute $-\frac{1}{2}$ for $x$ to get the following.

$$-\frac{1}{2}=A\left(-\frac{1}{2}+2\right)+B\left[2\left(-\frac{1}{2}\right)+1\right]$$

$$-\frac{1}{2}=A\left(\frac{3}{2}\right)+B(0)\Rightarrow-\frac{1}{2}=\frac{3}{2}A\Rightarrow A=-\frac{1}{3}$$

Replace $A$ with $-\frac{1}{3}$ in equation (1) and substitute $-2$ for $x$ to get the following.

$$-2=-\frac{1}{3}(-2+2)+B\left[2(-2)+1\right]$$

$$-2=-\frac{1}{3}(0)+B(-3)\Rightarrow-2=-3B\Rightarrow B=\frac{2}{3}$$

Thus, we have

$$\frac{2x^5+3x^4-3x^3-2x^2+x}{2x^2+5x+2}$$

$$=x^3-x^2+\frac{-1}{3(2x+1)}+\frac{2}{3(x+2)}$$

**17.** $\dfrac{x^3+4}{9x^3-4x}$

Find the quotient since the degrees of the numerator and denominator are the same.

$$\begin{array}{r} \frac{1}{9} \\ 9x^3-4x\overline{\smash{\big)}\,x^3+0x^2+0x+4} \\ \underline{x^3\phantom{+0x^2}-\frac{4}{9}x\phantom{+4}} \\ \frac{4}{9}x+4 \end{array}$$

Find the partial fraction decomposition for

$$\frac{\frac{4}{9}x+4}{9x^3-4x}.$$

$$\frac{\frac{4}{9}x+4}{9x^3-4x}=\frac{\frac{4}{9}x+4}{x(9x^2-4)}=\frac{\frac{4}{9}x+4}{x(3x+2)(3x-2)}$$

$$=\frac{A}{x}+\frac{B}{3x+2}+\frac{C}{3x-2}$$

Multiply both sides of

$$\frac{\frac{4}{9}x+4}{9x^3-4x}=\frac{A}{x}+\frac{B}{3x+2}+\frac{C}{3x-2} \text{ by}$$

$x(3x+2)(3x-2)$ to get the following.

$$\frac{4}{9}x+4=A(3x+2)(3x-2)+Bx(3x-2)$$
$$+Cx(3x+2) \quad (1)$$

First substitute 0 for $x$ to get the following.

$$\frac{4}{9}(0)+4$$

$$=A\left[3(0)+2\right]\left[3(0)-2\right]+B(0)\left[3(0)-2\right]$$
$$+C(0)\left[3(0)+2\right]$$

$$0+4=A(2)(-2)+0+0\Rightarrow4=-4A\Rightarrow A=-1$$

Replace $A$ with $-1$ in equation (1) and substitute $-\frac{2}{3}$ for $x$ to get the following.

$$\frac{4}{9}\left(-\frac{2}{3}\right)+4=-\left[3\left(-\frac{2}{3}\right)+2\right]\left[3\left(-\frac{2}{3}\right)-2\right]$$
$$+B\left(-\frac{2}{3}\right)\left[3\left(-\frac{2}{3}\right)-2\right]$$
$$+C\left(-\frac{2}{3}\right)\left[3\left(-\frac{2}{3}\right)+2\right]$$

$$-\frac{8}{27}+4=0+B\left(-\frac{2}{3}\right)(-4)+0$$

$$\frac{100}{27}=\frac{8}{3}B\Rightarrow B=\frac{25}{18}$$

$$\frac{4}{9}x+4=-(3x+2)(3x-2)$$
$$+\frac{25}{18}x(3x-2)+Cx(3x+2) \quad (2)$$

Substitute $\frac{2}{3}$ in equation (2) for $x$ to get the following.

$$\frac{4}{9}\left(\frac{2}{3}\right)+4=-\left[3\left(\frac{2}{3}\right)+2\right]\left[3\left(\frac{2}{3}\right)-2\right]$$
$$+\frac{25}{18}\left(\frac{2}{3}\right)\left[3\left(\frac{2}{3}\right)-2\right]+C\left(\frac{2}{3}\right)\left[3\left(\frac{2}{3}\right)+2\right]$$

$$\frac{8}{27}+4=0+0+C\left(\frac{2}{3}\right)(4)$$

$$\frac{116}{27}=\frac{8}{3}C\Rightarrow C=\frac{29}{18}$$

Thus, we have

$$\frac{x^3+4}{9x^3-4x}=\frac{1}{9}+\frac{-1}{x}+\frac{25}{18(3x+2)}+\frac{29}{18(3x-2)}.$$

**19.** $\dfrac{-3}{x^2\left(x^2+5\right)}=\dfrac{A}{x}+\dfrac{B}{x^2}+\dfrac{Cx+D}{x^2+5}$ Multiply

both sides by $x^2\left(x^2+5\right)$ to get the following.

$$-3=Ax\left(x^2+5\right)+B\left(x^2+5\right)+(Cx+D)x^2$$

Distributing and combining like terms on the right side of the equation, we have the following.

$$-3=Ax\left(x^2+5\right)+B\left(x^2+5\right)+(Cx+D)x^2$$

$$-3=Ax^3+5Ax+Bx^2+5B+Cx^3+Dx^2$$

$$-3=(A+C)x^3+(B+D)x^2+(5A)x+5B$$

Equate the coefficients of like powers of $x$ on the two sides of the equation.

For the $x^3$- term, we have $0=A+C$.

For the $x^2$- term, we have $0=B+D$.

For the $x$-term, we have $0=5A\Rightarrow A=0$.

For the constant term, we have

$$-3=5B\Rightarrow B=-\frac{3}{5}.$$

Since $A = 0$ and $0 = A + C$, we have $C = 0$.

Since $B = -\frac{3}{5}$ and $0 = B + D$, we have

$D = \frac{3}{5}$.

Thus, we have $\dfrac{-3}{x^2\left(x^2+5\right)} = \dfrac{-3}{5x^2} + \dfrac{3}{5\left(x^2+5\right)}$.

**21.** $\dfrac{3x-2}{(x+4)\left(3x^2+1\right)} = \dfrac{A}{x+4} + \dfrac{Bx+C}{3x^2+1}$

Multiply both sides by $(x+4)\left(3x^2+1\right)$ to get the following.

$3x - 2 = A\left(3x^2+1\right) + (Bx+C)(x+4)$ (1)

First substitute $-4$ for $x$ to get the following.

$3(-4) - 2 = A\left[3(-4)^2 + 1\right]$

$\qquad\qquad + \left[B(-4) + C\right](-4+4)$

$-12 - 2 = A(48+1) + 0$

$-14 = 49A \Rightarrow -\frac{14}{49} = A \Rightarrow A = -\frac{2}{7}$

Replace $A$ with $-\frac{2}{7}$ in equation (1) and substitute 0 for $x$ to get the following.

$3(0) - 2 = -\frac{2}{7}\left[3(0)^2 + 1\right] + \left[B(0) + C\right](0+4)$

$-2 = -\frac{2}{7} + 4C \Rightarrow -\frac{12}{7} = 4C \Rightarrow C = -\frac{3}{7}$

$3x - 2 = -\frac{2}{7}\left(3x^2+1\right) + \left(Bx - \frac{3}{7}\right)(x+4)$ (2)

Substitute 1 (arbitrary value) in equation (2) for $x$ to get the following.

$3(1) - 2 = -\frac{2}{7}\left[3(1)^2 + 1\right] + \left[B(1) - \frac{3}{7}\right](1+4)$

$3 - 2 = -\frac{2}{7}(4) + \left[B - \frac{3}{7}\right](5)$

$1 = -\frac{8}{7} + 5B - \frac{15}{7}$

$1 = 5B - \frac{23}{7} \Rightarrow \frac{30}{7} = 5B \Rightarrow B = \frac{6}{7}$

Thus, we have

$\dfrac{3x-2}{(x+4)\left(3x^2+1\right)} = \dfrac{-\frac{2}{7}}{x+4} + \dfrac{\frac{6}{7}x + \left(-\frac{3}{7}\right)}{3x^2+1}$

$\qquad\qquad = \dfrac{-2}{7(x+4)} + \dfrac{6x-3}{7\left(3x^2+1\right)}$

**23.** $\dfrac{1}{x(2x+1)\left(3x^2+4\right)} = \dfrac{A}{x} + \dfrac{B}{2x+1} + \dfrac{Cx+D}{3x^2+4}$

Multiply both sides by $x(2x+1)\left(3x^2+4\right)$ to get the following.

$1 = A(2x+1)\left(3x^2+4\right) + Bx\left(3x^2+4\right)$
$\qquad\qquad + (Cx+D)(x)(2x+1)$

Expanding and combining like terms on the right side of the equation, we have the following.

$1 = A(2x+1)\left(3x^2+4\right) + Bx\left(3x^2+4\right)$
$\qquad\qquad + (Cx+D)(x)(2x+1)$

$1 = A\left(6x^3 + 3x^2 + 8x + 4\right) + B\left(3x^3 + 4x\right)$
$\qquad\qquad + C\left(2x^3 + x^2\right)$
$\qquad\qquad + D\left(2x^2 + x\right)$

$1 = 6Ax^3 + 3Ax^2 + 8Ax + 4A + 3Bx^3$
$\qquad\qquad + 4Bx + 2Cx^3 + Cx^2 + 2Dx^2 + Dx$

$1 = (6A + 3B + 2C)x^3 + (3A + C + 2D)x^2$
$\qquad\qquad + (8A + 4B + D)x + 4A$

Equate the coefficients of like powers of $x$ on the two sides of the equation.

For the $x^3$- term, we have $0 = 6A + 3B + 2C$.

For the $x^2$- term, we have $0 = 3A + C + 2D$.

For the $x$-term, we have $0 = 8A + 4B + D$.

For the constant term, we have $1 = 4A$.

If we use the Gauss-Jordan method, we begin with the following augmented matrix.

$\begin{bmatrix} 6 & 3 & 2 & 0 & | & 0 \\ 3 & 0 & 1 & 2 & | & 0 \\ 8 & 4 & 0 & 1 & | & 0 \\ 4 & 0 & 0 & 0 & | & 1 \end{bmatrix}$

$\begin{bmatrix} 6 & 3 & 2 & 0 & | & 0 \\ 3 & 0 & 1 & 2 & | & 0 \\ 8 & 4 & 0 & 1 & | & 0 \\ 1 & 0 & 0 & 0 & | & \frac{1}{4} \end{bmatrix} \frac{1}{4}\text{R4} \qquad \Rightarrow$

$\begin{bmatrix} 1 & 0 & 0 & 0 & | & \frac{1}{4} \\ 3 & 0 & 1 & 2 & | & 0 \\ 8 & 4 & 0 & 1 & | & 0 \\ 6 & 3 & 2 & 0 & | & 0 \end{bmatrix} \text{R1} \leftrightarrow \text{R4} \qquad \Rightarrow$

*(continued on next page)*

*(continued from page 277)*

$$\begin{bmatrix} 1 & 0 & 0 & 0 & \frac{1}{4} \\ 0 & 0 & 1 & 2 & -\frac{3}{4} \\ 0 & 4 & 0 & 1 & -2 \\ 0 & 3 & 2 & 0 & -\frac{3}{2} \end{bmatrix} \begin{matrix} \\ -3R1+R2 \\ -8R1+R3 \\ -6R1+R4 \end{matrix} \Rightarrow$$

$$\begin{bmatrix} 1 & 0 & 0 & 0 & \frac{1}{4} \\ 0 & 0 & 1 & 2 & -\frac{3}{4} \\ 0 & 1 & 0 & \frac{1}{4} & -\frac{1}{2} \\ 0 & 3 & 2 & 0 & -\frac{3}{2} \end{bmatrix} \begin{matrix} \\ \\ \frac{1}{4}R3 \\ \\ \end{matrix} \Rightarrow$$

$$\begin{bmatrix} 1 & 0 & 0 & 0 & \frac{1}{4} \\ 0 & 1 & 0 & \frac{1}{4} & -\frac{1}{2} \\ 0 & 0 & 1 & 2 & -\frac{3}{4} \\ 0 & 3 & 2 & 0 & -\frac{3}{2} \end{bmatrix} \begin{matrix} \\ R2 \leftrightarrow R3 \\ \\ \end{matrix} \Rightarrow$$

$$\begin{bmatrix} 1 & 0 & 0 & 0 & \frac{1}{4} \\ 0 & 1 & 0 & \frac{1}{4} & -\frac{1}{2} \\ 0 & 0 & 1 & 2 & -\frac{3}{4} \\ 0 & 0 & 2 & -\frac{3}{4} & 0 \end{bmatrix} \begin{matrix} \\ \\ \\ -3R2+R4 \end{matrix} \Rightarrow$$

$$\begin{bmatrix} 1 & 0 & 0 & 0 & \frac{1}{4} \\ 0 & 1 & 0 & \frac{1}{4} & -\frac{1}{2} \\ 0 & 0 & 1 & 2 & -\frac{3}{4} \\ 0 & 0 & 0 & -\frac{19}{4} & \frac{3}{2} \end{bmatrix} \begin{matrix} \\ \\ \\ -2R3+R4 \end{matrix} \Rightarrow$$

$$\begin{bmatrix} 1 & 0 & 0 & 0 & \frac{1}{4} \\ 0 & 1 & 0 & \frac{1}{4} & -\frac{1}{2} \\ 0 & 0 & 1 & 2 & -\frac{3}{4} \\ 0 & 0 & 0 & 1 & -\frac{6}{19} \end{bmatrix} \begin{matrix} \\ \\ \\ -\frac{4}{19}R4 \end{matrix} \Rightarrow$$

$$\begin{bmatrix} 1 & 0 & 0 & 0 & \frac{1}{4} \\ 0 & 1 & 0 & 0 & -\frac{8}{19} \\ 0 & 0 & 1 & 0 & -\frac{9}{76} \\ 0 & 0 & 0 & 1 & -\frac{6}{19} \end{bmatrix} \begin{matrix} \\ -\frac{1}{4}R4+R2 \\ -2R4+R3 \\ \\ \end{matrix}$$

Thus, we have

$$\frac{1}{x(2x+1)(3x^2+4)}$$

$$= \frac{\frac{1}{4}}{x} + \frac{-\frac{8}{19}}{2x+1} + \frac{-\frac{9}{76}x+\left(-\frac{6}{19}\right)}{3x^2+4}$$

$$= \frac{\frac{1}{4}}{x} + \frac{-\frac{8}{19}}{2x+1} + \frac{-\frac{9}{76}x-\frac{24}{76}}{3x^2+4}$$

$$= \frac{1}{4x} + \frac{-8}{19(2x+1)} + \frac{-9x-24}{76(3x^2+4)}$$

**25.** $\dfrac{3x-1}{x\left(2x^2+1\right)^2} = \dfrac{A}{x} + \dfrac{Bx+C}{2x^2+1} + \dfrac{Dx+E}{\left(2x^2+1\right)^2}$

Multiply both sides by $x\left(2x^2+1\right)^2$ to get the following.

$$3x-1 = A\left(2x^2+1\right)^2 + (Bx+C)(x)\left(2x^2+1\right) + (Dx+E)x$$

Expanding and combining like terms on the right side of the equation, we have the following.

$$3x-1 = A\left(2x^2+1\right)^2 + (Bx+C)(x)\left(2x^2+1\right) + (Dx+E)x$$

$$= A\left(4x^4+4x^2+1\right) + B\left(2x^4+x^2\right) + C\left(2x^3+x\right) + Dx^2 + Ex$$

$$= 4Ax^4 + 4Ax^2 + A + 2Bx^4 + Bx^2 + 2Cx^3 + Cx + Dx^2 + Ex$$

$$3x-1 = (4A+2B)x^4 + 2Cx^3 + (4A+B+D)x^2 + (C+E)x + A$$

Equate the coefficients of like powers of $x$ on the two sides of the equation.

For the $x^4$- term, we have $0 = 4A+2B$.
For the $x^3$- term, we have $0 = 2C \Rightarrow C = 0$.
For the $x^2$- term, we have $0 = 4A+B+D$.
For the $x$-term, we have $3 = C+E$.
For the constant term, we have $-1 = A$.
Since $A=-1$ and $0=4A+2B$, we have $B=2$. Since $A=-1$, $B=2$ and $0=4A+B+D$, we have $D=2$. Since $C=0$ and $3=C+E$, we have $E=3$.
Thus, we have

$$\frac{3x-1}{x\left(2x^2+1\right)^2} = \frac{-1}{x} + \frac{2x+0}{2x^2+1} + \frac{2x+3}{\left(2x^2+1\right)^2}$$

$$= \frac{-1}{x} + \frac{2x}{2x^2+1} + \frac{2x+3}{\left(2x^2+1\right)^2}$$

**27.** $\dfrac{-x^4-8x^2+3x-10}{(x+2)\left(x^2+4\right)^2}$

$$= \frac{A}{x+2} + \frac{Bx+C}{x^2+4} + \frac{Dx+E}{\left(x^2+4\right)^2}$$

Multiply both sides by $(x+2)\left(x^2+4\right)^2$ to get the following.

$$-x^4 - 8x^2 + 3x - 10$$
$$= A\left(x^2 + 4\right)^2 + (Bx + C)(x + 2)\left(x^2 + 4\right)$$
$$+ (Dx + E)(x + 2) \quad (1)$$

Substitute then expand and combine like terms on the right side of the equation.
Substitute $-2$ in equation (1) for $x$ to get the following.

$$-(-2)^4 - 8(-2)^2 + 3(-2) - 10$$
$$= A\left[(-2)^2 + 4\right]^2$$
$$+ \left[B(-2) + C\right](-2 + 2)\left[(-2)^2 + 4\right]$$
$$+ \left[D(-2) + E\right](-2 + 2)$$
$$-16 - 32 + (-6) - 10 = A(4 + 4)^2 + 0 + 0$$
$$-64 = 64A \Rightarrow A = -1$$

Substituting $-1$ for $A$ in equation (1) and expanding and combining like terms, we have the following.

$$-x^4 - 8x^2 + 3x - 10$$
$$= -1\left(x^2 + 4\right)^2 + (Bx + C)(x + 2)\left(x^2 + 4\right)$$
$$+ (Dx + E)(x + 2)$$
$$= -1\left(x^4 + 8x^2 + 16\right)$$
$$+ (Bx + C)\left(x^3 + 2x^2 + 4x + 8\right)$$
$$+ D\left(x^2 + 2x\right) + E(x + 2)$$
$$= -x^4 - 8x^2 - 16 + Bx^4 + 2Bx^3 + 4Bx^2 + 8Bx$$
$$+ Cx^3 + 2Cx^2 + 4Cx + 8C + Dx^2 + 2Dx$$
$$+ Ex + 2E$$
$$= (-1 + B)x^4 + (2B + C)x^3$$
$$+ (-8 + 4B + 2C + D)x^2$$
$$+ (8B + 4C + 2D + E)x$$
$$+ (-16 + 8C + 2E)$$

Equate the coefficients of like powers of $x$ on the two sides of the equation.

For the $x^4$- term, we have
$-1 = -1 + B \Rightarrow B = 0.$

For the $x^3$- term, we have $0 = 2B + C.$

For the $x^2$- term, we have
$-8 = -8 + 4B + 2C + D \Rightarrow 0 = 4B + 2C + D.$

For the $x$-term, we have
$3 = 8B + 4C + 2D + E.$

For the constant term, we have
$-10 = -16 + 8C + 2E \Rightarrow 6 = 8C + 2E.$

Since $B = 0$ and $0 = 2B + C$, we have $C = 0.$
Since $B = 0$, $C = 0$, and $0 = 4B + 2C + D$, we have $D = 0.$ Since $C = 0$ and $6 = 8C + 2E$, we have $E = 3.$

Thus we have,

$$\frac{-x^4 - 8x^2 + 3x - 10}{(x + 2)\left(x^2 + 4\right)^2}$$
$$= \frac{-1}{x + 2} + \frac{0 \cdot x + 0}{x^2 + 4} + \frac{0 \cdot x + 3}{\left(x^2 + 4\right)^2}$$
$$= \frac{-1}{x + 2} + \frac{3}{\left(x^2 + 4\right)^2}$$

**29.** $\dfrac{5x^5 + 10x^4 - 15x^3 + 4x^2 + 13x - 9}{x^3 + 2x^2 - 3x}$

Since the degree of the numerator is higher than the degree of the denominator, first find the quotient.

$$
\begin{array}{r}
5x^2 \phantom{aaaaaaaaaaaaaaaaaa} \\
x^3 + 2x^2 - 3x \overline{\smash{)}\,5x^5 + 10x^4 - 15x^3 + 4x^2 + 13x - 9} \\
\underline{5x^5 + 10x^4 - 15x^3} \phantom{aaaaaaaaaa} \\
4x^2 + 13x - 9
\end{array}
$$

Find the partial fraction decomposition of $\dfrac{4x^2 + 13x - 9}{x^3 + 2x^2 - 3x}$.

$$\frac{4x^2 + 13x - 9}{x^3 + 2x^2 - 3x} = \frac{4x^2 + 13x - 9}{x\left(x^2 + 2x - 3\right)}$$
$$= \frac{4x^2 + 13x - 9}{x(x + 3)(x - 1)}$$
$$= \frac{A}{x} + \frac{B}{x + 3} + \frac{C}{x - 1}$$

Multiply by $x(x + 3)(x - 1)$ to obtain

$$4x^2 + 13x - 9$$
$$= A(x + 3)(x - 1) + Bx(x - 1) + Cx(x + 3) \quad (1)$$

First substitute 0 for $x$ in equation (1) to get the following.

$$4(0)^2 + 13(0) - 9$$
$$= A(0 + 3)(0 - 1) + B(0)(0 - 1)$$
$$+ C(0)(0 + 3)$$
$$-9 = -3A \Rightarrow A = 3$$

Replace $A$ with 3 in equation (1) and substitute $-3$ for $x$ to get the following.

$$4(-3)^2 + 13(-3) - 9$$
$$= 3(-3 + 3)(-3 - 1) + B(-3)(-3 - 1)$$
$$+ C(-3)(-3 + 3)$$
$$-12 = 12B \Rightarrow B = -1$$

$$4x^2 + 13x - 9$$
$$= 3(x + 3)(x - 1) - x(x - 1) + Cx(x + 3) \quad (2)$$

*(continued on next page)*

(*continued from page 279*)

Now substitute 1 (arbitrary choice) for $x$ in equation (2) to get the following.

$$4(1)^2 + 13(1) - 9$$
$$= 3(1+3)(1-1) - 1(1-1) + C(1)(1+3)$$
$$8 = 4C \Rightarrow C = 2$$

Thus, we have

$$\frac{5x^5 + 10x^4 - 15x^3 + 4x^2 + 13x - 9}{x^3 + 2x^2 - 3x}$$
$$= 5x^2 + \frac{3}{x} + \frac{-1}{x+3} + \frac{2}{x-1}$$

**31.** $\dfrac{4x^2 - 3x - 4}{x^3 + x^2 - 2x} = \dfrac{2}{x} + \dfrac{-1}{x-1} + \dfrac{3}{x+2}$

The graphs coincide. The partial fraction decomposition is correct.

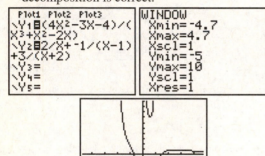

**33.** $\dfrac{x^3 - 2x}{\left(x^2 + 2x + 2\right)^2} = \dfrac{x-2}{x^2 + 2x + 2} + \dfrac{2}{\left(x^2 + 2x + 2\right)^2}$

The graphs do not coincide. The partial fraction decomposition is not correct.

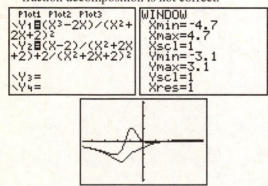

## Chapter 5 Quiz
**(Sections 5.1–5.4)**

**1.** $\quad 2x + y = -4 \quad$ (1)
$\quad -x + 2y = 2 \quad$ (2)

Solve equation (1) for $y$:
$\quad 2x + y = -4 \Rightarrow y = -2x - 4 \quad$ (3)

Replace $y$ in equation (2) with $-2x - 4$ and solve for $x$:
$$-x + 2y = 2 \Rightarrow -x + 2(-2x - 4) = 2$$
$$-5x - 8 = 2 \Rightarrow x = -2$$

Substitute $-2$ for $x$ in equation (3) and solve for $y$: $\; y = -2(-2) - 4 = 0$

Verify that the ordered pair $(-2, 0)$ satisfies both equations.
*Check:*

$$\begin{array}{c|c}
2x + y = -4 \; (1) & -x + 2y = 2 \quad (2) \\
2(-2) + 0 = -4 \;\; ? & -(-2) + 2(0) = 2 \;\; ? \\
-4 = -4 \;\; \text{True} & 2 = 2 \quad \text{True}
\end{array}$$

Solution set: $\{(-2, 0)\}$

**3.** $\quad x - y = 6 \quad$ (1)
$\quad x - y = 4 \quad$ (2)

Multiply equation (2) by $-1$, then add the result to equation (1):

$$\begin{array}{r}
x - y = 6 \\
-x + y = -4 \\
\hline
0 = 2
\end{array}$$

This is a false statement. The system is inconsistent and the solution set is $\varnothing$.

**5.** $\quad 3x + 5y = -5$
$\quad -2x + 3y = 16$

This system has the augmented matrix

$$\begin{bmatrix} 3 & 5 & | & -5 \\ -2 & 3 & | & 16 \end{bmatrix}.$$

$$\begin{bmatrix} 1 & 8 & | & 11 \\ -2 & 3 & | & 16 \end{bmatrix} \begin{array}{l} R1 + R2 \end{array} \Rightarrow$$

$$\begin{bmatrix} 1 & 8 & | & 11 \\ 0 & 19 & | & 38 \end{bmatrix} \begin{array}{l} 2R1 + R2 \end{array} \Rightarrow$$

$$\begin{bmatrix} 1 & 8 & | & 11 \\ 0 & 1 & | & 2 \end{bmatrix} \tfrac{1}{19} R2 \Rightarrow$$

$$\begin{bmatrix} 1 & 0 & | & -5 \\ 0 & 1 & | & 2 \end{bmatrix} R1 - 8R2$$

Verify that the ordered pair $(-5, 2)$ satisfies both equations.
*Check:*

$$3x + 5y = -5 \; (1)$$
$$3(-5) + 5(2) = -5 \;\; ?$$
$$-15 + 10 = -5$$
$$-5 = -5 \;\; \text{True}$$

$$-2x + 3y = 16 \quad (2)$$
$$-2(-5) + 3(2) = 16 \;\; ?$$
$$10 + 6 = 16$$
$$16 = 16 \quad \text{True}$$

Solution set: $\{(-5, 2)\}$

**7.**
$$x + y + z = 1 \quad (1)$$
$$-x + y + z = 5 \quad (2)$$
$$y + 2z = 5 \quad (3)$$

Eliminate $x$ by adding equations (1) and (2) to get $2y + 2z = 6 \Rightarrow y + z = 3 \quad (4)$.

Now solve the system consisting of equations (3) and (4) by subtracting equation (4) from equation (3):

$$\begin{array}{r} y + 2z = 5 \quad (3) \\ y + z = 3 \quad (4) \\ \hline z = 2 \end{array}$$

Substitute the value for $z$ into equation (4) to solve for $y$: $y + 2 = 3 \Rightarrow y = 1$

Now, substitute the values for $y$ and $z$ into equation (1) to solve for $x$:
$$x + 1 + 2 = 1 \Rightarrow x = -2$$

Verify that the ordered triple $(-2, 1, 2)$ satisfies all three equations.

*Check*:

$$\begin{array}{l|l} x + y + z = 1 \;(1) & -x + y + z = 5 \quad (2) \\ -2 + 1 + 2 = 1 \;? & -(-2) + 1 + 2 = 5 \;? \\ \quad\quad 1 = 1 \;\text{True} & \quad\quad\quad 5 = 5 \;\text{True} \end{array}$$

$$y + 2z = 5 \quad (3)$$
$$1 + 2(2) = 5 \;?$$
$$\quad\quad 5 = 5 \;\text{True}$$

Solution set: $\{(-2, 1, 2)\}$

**9.**
$$7x + y - z = 4 \quad (1)$$
$$2x - 3y + z = 2 \quad (2)$$
$$-6x + 9y - 3z = -6 \quad (3)$$

$$D = \begin{vmatrix} 7 & 1 & -1 \\ 2 & -3 & 1 \\ -6 & 9 & -3 \end{vmatrix}$$

Adding 3 times row 2 to row 3, we have

$$D = \begin{vmatrix} 7 & 1 & -1 \\ 2 & -3 & 1 \\ 0 & 0 & 0 \end{vmatrix}$$

Since we have a row of zeros, $D = 0$ and we cannot use Cramer's rule.
Using the elimination method, we can add 3 times equation (2) to equation (3).

$$\begin{array}{r} 6x - 9y + 3z = 6 \quad (2) \\ -6x + 9y - 3z = -6 \quad (3) \\ \hline 0 = 0 \quad \text{True} \end{array}$$

This system will have infinitely many solutions.

Solve the system made up of equations (1) and (3) in terms of the arbitrary variable $y$. To eliminate $z$, multiply equation (1) by $-3$ and add the result to equation (3):

$$\begin{array}{r} -21x - 3y + 3z = -12 \\ -6x + 9y - 3z = -6 \\ \hline -27x + 6y = -18 \Rightarrow -27x = -6y - 18 \Rightarrow \end{array}$$
$$x = \frac{6y + 18}{27} = \frac{2y + 6}{9}$$

Now, express $z$ also in terms of $y$ by solving equation (1) for $y$ and substituting $\frac{2y+6}{9}$ for $x$ in the result.

$$7x + y - z = 4 \Rightarrow z = 7x + y - 4 \Rightarrow$$
$$z = 7\left(\frac{2y+6}{9}\right) + y - 4 = \frac{14y + 42}{9} + y - 4 = \frac{23y + 6}{9}$$

Solution set: $\left\{ \left( \frac{2y+6}{9}, y, \frac{23y+6}{9} \right) \right\}$

**11.** Let $x =$ the amount invested at 8%; $y =$ the amount invested at 11%; $z =$ the amount invested at 14%
The information in the problem gives the system:

$$\begin{cases} x + y + z = 5000 \\ z = x + y \Rightarrow \\ .08x + .11y + .14z = 595 \end{cases}$$

$$\begin{cases} x + y + z = 5000 \quad (1) \\ -x - y + z = 0 \quad (2) \\ .08x + .11y + .14z = 595 \quad (3) \end{cases}$$

This system has the augmented matrix

$$\begin{bmatrix} 1 & 1 & 1 & | & 5000 \\ -1 & -1 & 1 & | & 0 \\ .08 & .11 & .14 & | & 595 \end{bmatrix}$$

$$\begin{bmatrix} 1 & 1 & 1 & | & 5000 \\ 0 & 0 & 1 & | & 2500 \\ .08 & .11 & .14 & | & 595 \end{bmatrix} \frac{1}{2}(R1 + R2) \Rightarrow$$

$$\begin{bmatrix} 1 & 1 & 1 & | & 5000 \\ .08 & .11 & .14 & | & 595 \\ 0 & 0 & 1 & | & 2500 \end{bmatrix} R2 \leftrightarrow R3 \Rightarrow$$

$$\begin{bmatrix} 1 & 1 & 1 & | & 5000 \\ 8 & 11 & 14 & | & 59,500 \\ 0 & 0 & 1 & | & 2500 \end{bmatrix} 100R2 \Rightarrow$$

$$\begin{bmatrix} 1 & 1 & 1 & | & 5000 \\ 0 & 1 & 2 & | & 6500 \\ 0 & 0 & 1 & | & 2500 \end{bmatrix} \frac{1}{3}(-8R1 + R2) \Rightarrow$$

$$\begin{bmatrix} 1 & 1 & 1 & | & 5000 \\ 0 & 1 & 0 & | & 1500 \\ 0 & 0 & 1 & | & 2500 \end{bmatrix} R2 - 2R3 \Rightarrow$$

*(continued on next page)*

*(continued from page 281)*

$$\begin{bmatrix} 1 & 0 & 1 & | & 3500 \\ 0 & 1 & 0 & | & 1500 \\ 0 & 0 & 1 & | & 2500 \end{bmatrix} \begin{matrix} R1-R2 \\ \\ \\ \end{matrix} \Rightarrow$$

$$\begin{bmatrix} 1 & 0 & 0 & | & 1000 \\ 0 & 1 & 0 & | & 1500 \\ 0 & 0 & 1 & | & 2500 \end{bmatrix} \begin{matrix} R1-R3 \\ \\ \\ \end{matrix} \Rightarrow$$

$1000 was invested at 8%, $1500 was invested at 11%, and $2500 was invested at 14%.

**13.** $\begin{vmatrix} -1 & 2 & 4 \\ -3 & -2 & -3 \\ 2 & -1 & 5 \end{vmatrix}$

To expand by the first row, we will need to find $a_{11} \cdot M_{11} - a_{12} \cdot M_{12} + a_{13} \cdot M_{13}$.

$$M_{11} = \begin{vmatrix} -2 & -3 \\ -1 & 5 \end{vmatrix} = (-2)(5) - (-1)(-3) = -13$$

$$M_{12} = \begin{vmatrix} -3 & -3 \\ 2 & 5 \end{vmatrix} = (-3)(5) - (2)(-3) = -9$$

$$M_{13} = \begin{vmatrix} -3 & -2 \\ 2 & -1 \end{vmatrix} = (-3)(-1) - (2)(-2) = 7$$

$$a_{11} \cdot M_{11} - a_{12} \cdot M_{12} + a_{13} \cdot M_{13}$$
$$= (-1)(-13) - (2)(-9) + (4)(7) = 59$$

**15.** $\dfrac{2x^2 - 15x - 32}{(x-1)(x^2 + 6x + 8)} = \dfrac{2x^2 - 15x - 32}{(x-1)(x+2)(x+4)}$

$$= \dfrac{A}{x-1} + \dfrac{B}{x+2} + \dfrac{C}{x+4}$$

Multiply both sides by $(x-1)(x+2)(x+4)$:

$$2x^2 - 15x - 32$$
$$= A(x+2)(x+4) + B(x-1)(x+4)$$
$$+ C(x-1)(x+2)$$

Expanding and combining like terms on the right side of the equation, we have

$$2x^2 - 15x - 32$$
$$= A(x+2)(x+4) + B(x-1)(x+4)$$
$$+ C(x-1)(x+2)$$
$$= Ax^2 + 6Ax + 8A + Bx^2 + 3Bx - 4B$$
$$+ Cx^2 + Cx - 2C$$
$$= x^2(A + B + C) + x(6A + 3B + C)$$
$$+ (8A - 4B - 2C)$$

Equate the coefficients of like powers of $x$ on the two sides of the equation.

For the $x^2$-term, $2 = A + B + C$
For the $x$-term, $-15 = 6A + 3B + C$
For the constant term, $-32 = 8A - 4B - 2C$

Using the Gauss-Jordan method, we have

$$\begin{bmatrix} 1 & 1 & 1 & | & 2 \\ 6 & 3 & 1 & | & -15 \\ 8 & -4 & -2 & | & -32 \end{bmatrix} \Rightarrow$$

$$\begin{bmatrix} 1 & 1 & 1 & | & 2 \\ 0 & -3 & -5 & | & -27 \\ 0 & -12 & -10 & | & -48 \end{bmatrix} \begin{matrix} \\ -6R1+R2 \\ -8R1+R2 \end{matrix} \Rightarrow$$

$$\begin{bmatrix} 1 & 1 & 1 & | & 2 \\ 0 & -3 & -5 & | & -27 \\ 0 & 0 & 1 & | & 6 \end{bmatrix} \begin{matrix} \\ \\ \frac{1}{10}(-4R2+R3) \end{matrix} \Rightarrow$$

$$\begin{bmatrix} 1 & 1 & 1 & | & 2 \\ 0 & 1 & 0 & | & -1 \\ 0 & 0 & 1 & | & 6 \end{bmatrix} \begin{matrix} \\ -\frac{1}{3}(5R3+R2) \\ \end{matrix} \Rightarrow$$

$$\begin{bmatrix} 1 & 0 & 1 & | & 3 \\ 0 & 1 & 0 & | & -1 \\ 0 & 0 & 1 & | & 6 \end{bmatrix} \begin{matrix} R1-R2 \\ \\ \end{matrix} \Rightarrow$$

$$\begin{bmatrix} 1 & 0 & 0 & | & -3 \\ 0 & 1 & 0 & | & -1 \\ 0 & 0 & 1 & | & 6 \end{bmatrix} \begin{matrix} R1-R3 \\ \\ \end{matrix}$$

$A = -3$, $B = -1$, $C = 6$.
Thus

$$\dfrac{2x^2 - 15x - 32}{(x-1)(x^2 + 6x + 8)} = \dfrac{-3}{x-1} + \dfrac{-1}{x+2} + \dfrac{6}{x+4}$$

## Section 5.5: Nonlinear Systems of Equations

**1.** The system is $\begin{aligned} x^2 &= y - 1 \\ y &= 3x + 5 \end{aligned}$

The proposed solution set is $\{(-1, 2), (4, 17)\}$.

Check $(-1, 2)$.

| $x^2 = y - 1$ | $y = 3x + 5$ |
|---|---|
| $(-1)^2 = 2 - 1$ ? | $2 = 3(-1) + 5$ ? |
| $1 = 1$ | $2 = -3 + 5$ |
| | $2 = 2$ |

Check $(4, 17)$.

| $x^2 = y - 1$ | $y = 3x + 5$ |
|---|---|
| $4^2 = 17 - 1$ ? | $17 = 3(4) + 5$ ? |
| $16 = 16$ | $17 = 12 + 5$ |
| | $17 = 17$ |

Both solutions are valid.

**3.** The system is $\begin{aligned} x^2 + y^2 &= 5 \\ -3x + 4y &= 2 \end{aligned}$

The proposed solution set is
$\left\{ (-2,-1), \left(\frac{38}{25}, \frac{41}{25}\right) \right\}$.

Check $(-2,-1)$.

$$\begin{array}{ll} x^2 + y^2 = 5 & -3x + 4y = 2 \\ (-2)^2 + (-1)^2 = 5\ ? & -3(-2) + 4(-1) = 2\ ? \\ 4 + 1 = 5 & 6 + (-4) = 2 \\ 5 = 5 & 2 = 2 \end{array}$$

Check $\left(\frac{38}{25}, \frac{41}{25}\right)$.

$$\begin{array}{ll} x^2 + y^2 = 5 & -3x + 4y = 2 \\ \left(\frac{38}{25}\right)^2 + \left(\frac{41}{25}\right)^2 = 5\ ? & -3\left(\frac{38}{25}\right) + 4\left(\frac{41}{25}\right) = 2\ ? \\ \frac{1444}{625} + \frac{1681}{625} = 5 & -\frac{114}{25} + \frac{164}{25} = 2 \\ \frac{3125}{625} = 5 & \frac{50}{25} = 2 \\ 5 = 5 & 2 = 2 \end{array}$$

Both solutions are valid.

**5.** The system is $\begin{aligned} y &= \log x \\ x^2 - y^2 &= 4 \end{aligned}$

The proposed approximate solution set is
$\left\{ (2.0232821, .30605644) \right\}$.

To show this, we will recreate the solution with the graphing calculator. In order to get $x^2 - y^2 = 4$ to be displayed, solve for y and enter the relation as two functions.

$$x^2 - y^2 = 4 \Rightarrow y^2 = x^2 - 4 \Rightarrow y = \pm\sqrt{x^2 - 4}$$

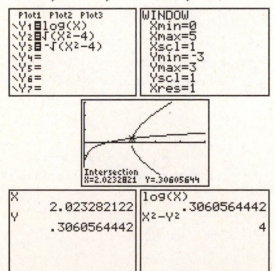

**7.** The system $\begin{aligned} x^2 - y &= 4 \\ x + y &= -2 \end{aligned}$ cannot have more than two solutions because a parabola and a line cannot intersect in more than two points.

In the solutions to Exercises 9–13, we will include both the algebraic solution and graphing calculator solution (like Example 1). For Exercises 15–41, we will include just the algebraic solution. Also, in the solutions to Exercises 9–41, we will omit the checking of solutions. Recall, though, when checking elements of the solution set, substitute the ordered pair(s) into both equations of the system.

**9.** $x^2 - y = 0$  (1)
   $x + y = 2$  (2)

Algebraic Solution:
Solving equation (2) for y, we have $y = 2 - x$.

Substitute this result into equation (1).
$$2 - x = x^2 \Rightarrow x^2 + x - 2 = 0$$
$$(x + 2)(x - 1) = 0 \Rightarrow x = -2 \text{ or } x = 1$$

If $x = -2$, then $y = 2 - (-2) = 4$. If $x = 1$, then $y = 2 - 1 = 1$.

Graphing Calculator Solution
$$x^2 - y = 0 \Rightarrow y = x^2 \text{ and } x + y = 2 \Rightarrow y = 2 - x$$

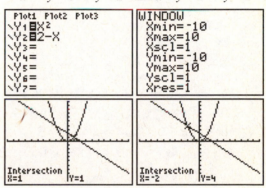

Solution set: $\left\{ (1,1), (-2,4) \right\}$

**11.** $y = x^2 - 2x + 1$  (1)
   $x - 3y = -1$     (2)

Algebraic Solution:
Solving equation (2) for y, we have $y = \frac{x+1}{3}$.

Substitute this result into equation (1).
$$\frac{x+1}{3} = x^2 - 2x + 1 \Rightarrow x + 1 = 3x^2 - 6x + 3 \Rightarrow$$
$$3x^2 - 7x + 2 = 0 \Rightarrow (3x - 1)(x - 2) = 0 \Rightarrow$$
$$x = \frac{1}{3} \text{ or } x = 2$$

If $x = 2$, then $y = \frac{2+1}{3} = 1$. If $x = \frac{1}{3}$, then
$$y = \frac{\frac{1}{3}+1}{3} = \frac{1+3}{9} = \frac{4}{9}.$$

(*continued on next page*)

*(continued from page 283)*

Graphing Calculator Solution

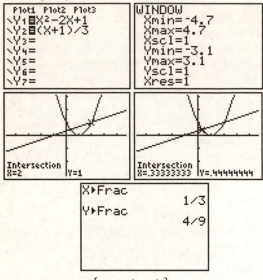

Solution set: $\left\{(2,1),\left(\frac{1}{3},\frac{4}{9}\right)\right\}$

**13.** $y = x^2 + 4x$ (1)

$2x - y = -8$ (2)

Algebraic Solution:

Solving equation (2) for *y*, we have $y = 2x + 8$. Substitute this result into equation

(1): $2x + 8 = x^2 + 4x \Rightarrow x^2 + 2x - 8 = 0 \Rightarrow$

$(x+4)(x-2) = 0 \Rightarrow x = -4$ or $x = 2$

If $x = -4$, then $y = 2(-4) + 8 = 0$. If $x = 2$, then

$y = 2(2) + 8 = 12$.

Graphing Calculator Solution

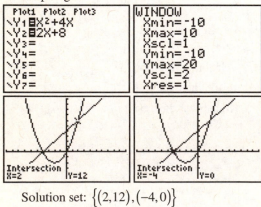

Solution set: $\{(2,12),(-4,0)\}$

**15.** $3x^2 + 2y^2 = 5$ (1)

$x - y = -2$ (2)

Solving equation (2) for *y*, we have $y = x + 2$.

Substitute this result into equation (1).

$$3x^2 + 2(x+2)^2 = 5$$
$$3x^2 + 2(x^2 + 4x + 4) = 5$$
$$3x^2 + 2x^2 + 8x + 8 = 5$$
$$5x^2 + 8x + 3 = 0$$
$$(5x+3)(x+1) = 0 \Rightarrow x = -\frac{3}{5} \text{ or } x = -1$$

If $x = -\frac{3}{5}$, then $y = -\frac{3}{5} + 2 = -\frac{3}{5} + \frac{10}{5} = \frac{7}{5}$. If

$x = -1$, then $y = -1 + 2 = 1$.

Solution set: $\left\{\left(-\frac{3}{5},\frac{7}{5}\right),(-1,1)\right\}$

**17.** $x^2 + y^2 = 8$ (1)

$x^2 - y^2 = 0$ (2)

Using the elimination method, we add equations (1) and (2).

$$\begin{array}{l} x^2 + y^2 = 8 \\ \underline{x^2 - y^2 = 0} \\ 2x^2 \quad\quad = 8 \Rightarrow x^2 = 4 \Rightarrow x = \pm 2 \end{array}$$

Find *y* by substituting back into equation (2).

If $x = 2$, then $2^2 - y^2 = 0 \Rightarrow 4 - y^2 = 0 \Rightarrow$

$y^2 = 4 \Rightarrow y = \pm 2$. If $x = -2$, then

$(-2)^2 - y^2 = 0 \Rightarrow 4 - y^2 = 0 \Rightarrow y^2 = 4 \Rightarrow$

$y = \pm 2$.

Solution set: $\{(2,2),(2,-2),(-2,2),(-2,-2)\}$

**19.** $5x^2 - y^2 = 0$ (1)

$3x^2 + 4y^2 = 0$ (2)

Using the elimination method, we multiply equation (1) by 4 and add to equation (2).

$$\begin{array}{l} 20x^2 - 4y^2 = 0 \\ \underline{3x^2 + 4y^2 = 0} \\ 23x^2 \quad\quad = 0 \Rightarrow x^2 = 0 \Rightarrow x = 0 \end{array}$$

Find *y* by substituting back into equation (1).

If $x = 0$, then

$5(0)^2 - y^2 = 0 \Rightarrow 0 - y^2 = 0 \Rightarrow y^2 = 0 \Rightarrow y = 0$.

Solution set: $\{(0,0)\}$

**21.** $3x^2 + y^2 = 3$ (1)
$4x^2 + 5y^2 = 26$ (2)
Using the elimination method, we multiply
equation (1) by $-5$ and add to equation (2).
$-15x^2 - 5y^2 = -15$
$\underline{4x^2 + 5y^2 = 26}$
$-11x^2 \quad = 11 \Rightarrow x^2 = -1 \Rightarrow x = \pm i$
Find $y$ by substituting back into equation (1).
If $x = i$, then $3(i)^2 + y^2 = 3 \Rightarrow$
$3(-1) + y^2 = 3 \Rightarrow -3 + y^2 = 3 \Rightarrow y^2 = 6 \Rightarrow$
$y = \pm\sqrt{6}$. If $x = -i$, then $3(-i)^2 + y^2 = 3 \Rightarrow$
$3(-1) + y^2 = 3 \Rightarrow -3 + y^2 = 3 \Rightarrow y^2 = 6 \Rightarrow$
$y = \pm\sqrt{6}$.
Solution set:
$$\left\{\left(i, \sqrt{6}\right), \left(-i, \sqrt{6}\right), \left(i, -\sqrt{6}\right), \left(-i, -\sqrt{6}\right)\right\}$$

**23.** $2x^2 + 3y^2 = 5$ (1)
$3x^2 - 4y^2 = -1$ (2)
Using the elimination method, we multiply
equation (1) by 4 and equation (2) by 3 and
then add the resulting equations.
$8x^2 + 12y^2 = 20$
$\underline{9x^2 - 12y^2 = -3}$
$17x^2 \quad = 17 \Rightarrow x^2 = 1 \Rightarrow x = \pm 1$
Find $y$ by substituting back into equation (1).
If $x = 1$, then $2(1)^2 + 3y^2 = 5 \Rightarrow$
$2 + 3y^2 = 5 \Rightarrow 3y^2 = 3 \Rightarrow y^2 = 1 \Rightarrow y = \pm 1$.
If $x = -1$, then $2(-1)^2 + 3y^2 = 5 \Rightarrow$
$2 + 3y^2 = 5 \Rightarrow 3y^2 = 3 \Rightarrow y^2 = 1 \Rightarrow y = \pm 1$.
Solution set: $\{(1, -1), (-1, 1), (1, 1), (-1, -1)\}$

**25.** $2x^2 + 2y^2 = 20$ (1)
$4x^2 + 4y^2 = 30$ (2)
Using the elimination method, we multiply
equation (1) by $-2$ and add to equation (2).
$-4x^2 - 4y^2 = -40$
$\underline{4x^2 + 4y^2 = 30}$
$0 = -10$
This is a false statement.
Solution set: $\varnothing$

**27.** $2x^2 - 3y^2 = 8$ (1)
$6x^2 + 5y^2 = 24$ (2)
Using the elimination method, we multiply
equation (1) by $-3$ and add to equation (2).
$-6x^2 + 9y^2 = -24$
$\underline{6x^2 + 5y^2 = 24}$
$14y^2 = 0 \Rightarrow y^2 = 0 \Rightarrow y = 0$
Find $x$ by substituting back into equation (2).
If $y = 0$, then $6x^2 + 5(0)^2 = 24 \Rightarrow$
$6x^2 + 0 = 24 \Rightarrow x^2 = 4 \Rightarrow x = \pm 2$.
Solution set: $\{(2, 0), (-2, 0)\}$

**29.** $xy = -15$ (1)
$4x + 3y = 3$ (2)
Solving equation (1) for $y$, we have $y = -\frac{15}{x}$.
Substitute this result into equation (2).
$4x + 3\left(-\frac{15}{x}\right) = 3 \Rightarrow 4x^2 - 45 = 3x \Rightarrow$
$4x^2 - 3x - 45 = 0 \Rightarrow (x + 3)(4x - 15) = 0 \Rightarrow$
$x = -3$ or $x = \frac{15}{4}$
If $x = -3$, then $y = -\frac{15}{-3} = 5$. If $x = \frac{15}{4}$, then
$y = -\frac{15}{\frac{15}{4}} = -4$.
Solution set: $\left\{(-3, 5), \left(\frac{15}{4}, -4\right)\right\}$

**31.** $2xy + 1 = 0$ (1)
$x + 16y = 2$ (2)
Solving equation (2) for $x$, we have
$x = 2 - 16y$.
Substitute this result into equation (1).
$2(-16y + 2)y + 1 = 0 \Rightarrow -32y^2 + 4y + 1 = 0 \Rightarrow$
$32y^2 - 4y - 1 = 0 \Rightarrow (8y + 1)(4y - 1) = 0 \Rightarrow$
$y = -\frac{1}{8}$ or $y = \frac{1}{4}$
If $y = -\frac{1}{8}$, then $x = 2 - 16\left(-\frac{1}{8}\right) = 2 + 2 = 4$. If
$y = \frac{1}{4}$, then $x = 2 - 16\left(\frac{1}{4}\right) = 2 - 4 = -2$.
Solution set: $\left\{\left(4, -\frac{1}{8}\right), \left(-2, \frac{1}{4}\right)\right\}$

**33.** $x^2 + 4y^2 = 25$ (1)

$xy = 6$ (2)

Solving equation (2) for $x$, we have $x = \frac{6}{y}$.

Substitute this result into equation (1).

$$\left(\tfrac{6}{y}\right)^2 + 4y^2 = 25 \Rightarrow \tfrac{36}{y^2} + 4y^2 = 25 \Rightarrow$$

$$36 + 4y^4 = 25y^2$$

$$4y^4 - 25y^2 + 36 = 0 \Rightarrow \left(y^2 - 4\right)\left(4y^2 - 9\right) = 0$$

$$y^2 = 4 \Rightarrow y = \pm 2 \quad \text{or} \quad y^2 = \tfrac{9}{4} \Rightarrow y = \pm \tfrac{3}{2}$$

If $y = 2$, then $x = \frac{6}{2} = 3$. If $y = -2$, then

$x = \frac{6}{-2} = -3$.

If $y = \frac{3}{2}$, then $x = \frac{6}{\frac{3}{2}} = 4$. If $y = -\frac{3}{2}$, then

$x = \frac{6}{-\frac{3}{2}} = -4$.

Solution set:

$$\left\{ (3,2), (-3,-2), \left(4, \tfrac{3}{2}\right), \left(-4, -\tfrac{3}{2}\right) \right\}$$

**35.** $x^2 - xy + y^2 = 5$ (1)

$2x^2 + xy - y^2 = 10$ (2)

Using the elimination method, we add
equations (1) and (2).

$$x^2 - xy + y^2 = 5$$
$$\underline{2x^2 + xy - y^2 = 10}$$
$$3x^2 \qquad = 15 \Rightarrow x^2 = 5 \Rightarrow x = \pm\sqrt{5}$$

Find $y$ by substituting back into equation (1).

If $x = \sqrt{5}$, then

$$\left(\sqrt{5}\right)^2 - \sqrt{5}\,y + y^2 = 5 \Rightarrow 5 - \sqrt{5}\,y + y^2 = 5 \Rightarrow$$

$$y^2 - \sqrt{5}\,y = 0 \Rightarrow y\left(y - \sqrt{5}\right) = 0$$

Thus, we have $y = 0$ or $y = \sqrt{5}$.

If $x = -\sqrt{5}$, then

$$\left(-\sqrt{5}\right)^2 - \left(-\sqrt{5}\right)y + y^2 = 5$$

$$5 + \sqrt{5}\,y + y^2 = 5 \Rightarrow y^2 + \sqrt{5}\,y = 0 \Rightarrow$$

$$y\left(y + \sqrt{5}\right) = 0 \Rightarrow y = 0 \text{ or } y = -\sqrt{5}$$

Thus, we have $y = 0$ or $y = -\sqrt{5}$.

Solution set:

$$\left\{ \left(\sqrt{5}, 0\right), \left(-\sqrt{5}, 0\right), \left(\sqrt{5}, \sqrt{5}\right), \left(-\sqrt{5}, -\sqrt{5}\right) \right\}$$

**37.** $x^2 + 2xy - y^2 = 14$ (1)

$x^2 - y^2 = -16$ (2)

This system can be solved using a combination
of the elimination and substitution methods.

Using the elimination method, multiply
equation (2) by $-1$ and add to equation (1).

$$x^2 + 2xy - y^2 = 14$$
$$\underline{-x^2 + \qquad y^2 = 16}$$
$$2xy \qquad = 30 \Rightarrow xy = 15\,(3)$$

Solve equation (3) for $y$.

$xy = 15 \Rightarrow y = \frac{15}{x}$ (4)

Find $x$ by substituting equation (4) into
equation (2).

$$x^2 - \left(\tfrac{15}{x}\right)^2 = -16 \Rightarrow x^2 - \tfrac{225}{x^2} = -16 \Rightarrow$$

$$x^4 - 225 = -16x^2 \Rightarrow$$

$$x^4 + 16x^2 - 225 = 0 \Rightarrow \left(x^2 - 9\right)\left(x^2 + 25\right) = 0$$

$x^2 = 9 \quad \text{or} \quad x^2 = -25$

$x = \pm 3 \qquad\quad x = \pm 5i$

If $x = 3$, then $y = \frac{15}{3} = 5$. If $x = -3$, then

$y = \frac{15}{-3} = -5$. If $x = 5i$, then $y = \frac{15}{5i} = \frac{3}{i} = -3i$.

If $x = -5i$, then $y = \frac{15}{-5i} = -\frac{3}{i} = 3i$.

Solution set:

$$\left\{ (3,5), (-3,-5), (5i,-3i), (-5i, 3i) \right\}$$

**39.** $x = |y|$ (1)

$x^2 + y^2 = 18$ (2)

If $x = |y|$, then $x^2 = y^2$ since $|y|^2 = y^2$.

Substitute $x^2 = y^2$ into equation (2).

$$y^2 + y^2 = 18 \Rightarrow 2y^2 = 18 \Rightarrow y^2 = 9 \Rightarrow y = \pm 3$$

If $y = 3$, then $x = |y| = |3| = 3$. If $y = -3$, then

$x = |y| = |-3| = 3$.

Solution set: $\left\{ (3,-3), (3,3) \right\}$

**41.** $2x^2 - y^2 = 4$ (1)

$|x| = |y|$ (2)

Since $|x|^2 = x^2$ and $|y|^2 = y^2$, we substitute

$y^2 = x^2$ into equation (1) and solve for $x$.

$$2x^2 - x^2 = 4 \Rightarrow x^2 = 4 \Rightarrow x = \pm 2$$

By substituting these values back into equation
(2) we have the following.

If $x = 2$, then $|2| = |y| \Rightarrow 2 = |y| \Rightarrow y = \pm 2$. If

$x = -2$, then $|-2| = |y| \Rightarrow 2 = |y| \Rightarrow y = \pm 2$.

Solution set: $\left\{ (2,2), (-2,-2), (2,-2), (-2,2) \right\}$

**43.** $y = \log(x+5)$

$y = x^2$

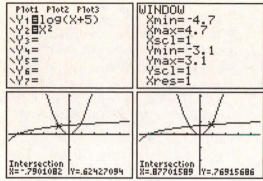

Solution set: $\{(-.79,.62),(.88,.77)\}$

**45.** $y = e^{x+1}$

$2x + y = 3 \Rightarrow y = 3 - 2x$

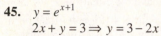

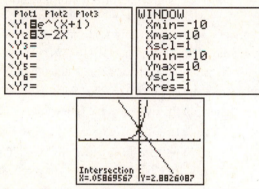

Solution set: $\{(.06, 2.88)\}$

**47.** Let $x$ and $y$ represent the numbers.
We obtain the following system.

$x + y = 17$ (1)

$xy = 42$ (2)

Solving equation (1) for $y$ we have $y = 17 - x$.
(3). Substituting this into equation (2) we have
$x(17 - x) = 42$. Solving this equation for $x$ we
have the following.

$$x(17 - x) = 42 \Rightarrow 17x - x^2 = 42 \Rightarrow$$

$$0 = x^2 - 17x + 42$$

$$(x - 3)(x - 14) = 0 \Rightarrow x = 3 \text{ or } x = 14$$

Using equation (3), if $x = 3$ then
$3 + y = 17 \Rightarrow y = 14$. If $x = 14$, then
$14 + y = 17 \Rightarrow y = 3$.

The two numbers are 14 and 3.

**49.** Let $x$ and $y$ represent the numbers.
We obtain the following system.

$x^2 + y^2 = 100$ (1)

$x^2 - y^2 = 28$ (2)

Adding equations (1) and (2) in order to
eliminate $y^2$, we have $2x^2 = 128$.
Solving this equation for $x$ we have

$$2x^2 = 128 \Rightarrow x^2 = 64 \Rightarrow x = \pm 8$$

Using equation (1), if $x = 8$ then

$$8^2 + y^2 = 100 \Rightarrow 64 + y^2 = 100 \Rightarrow$$

$$y^2 = 36 \Rightarrow y = \pm 6$$

If $x = -8$ then

$$(-8)^2 + y^2 = 100 \Rightarrow 64 + y^2 = 100$$

$$y^2 = 36 \Rightarrow y = \pm 6$$

The two numbers are 8 and 6, or 8 and −6, or
−8 and 6, or −8 and −6.

**51.** Let $x$ and $y$ represent the numbers.
We obtain the following system.

$\frac{x}{y} = \frac{9}{2}$   (1)

$xy = 162$ (2)

Solve equation (1) for $x$.

$$y\left(\frac{x}{y}\right) = y\left(\frac{9}{2}\right) \Rightarrow x = \frac{9}{2}y$$

Substitute $\frac{9}{2}y$ for $x$ in equation (2) and solve
for $y$.

$$\left(\frac{9}{2}y\right)y = 162 \Rightarrow \frac{9}{2}y^2 = 162 \Rightarrow$$

$$\frac{2}{9}\left(\frac{9}{2}y^2\right) = \frac{2}{9}(162) \Rightarrow y^2 = 36 \Rightarrow y = \pm 6$$

If $y = 6$, $x = \frac{9}{2}(6) = 27$. If $y = -6$,

$x = \frac{9}{2}(-6) = -27$.

The two numbers are either 27 and 6, or
−27 and −6.

**53.** Let $x$ = the length of the second side and
$y$ = the length of the third side.
We obtain the following system.

$x^2 + y^2 = 13^2 = 169$   (1)

$x = y + 7$        (2)

Substitute $y + 7$ for $x$ in equation (1) and
solve for $y$.

$$(y + 7)^2 + y^2 = 169$$

$$y^2 + 14y + 49 + y^2 = 169$$

$$2y^2 + 14y - 120 = 0$$

$$y^2 + 7y - 60 = 0$$

$$(y + 12)(y - 5) = 0 \Rightarrow y = -12 \text{ or } y = 5$$

We disregard the negative solution since a
length cannot be negative. If $y = 5$, then
$x = 5 + 7 = 12$.

The lengths of the two shorter sides are 5 m
and 12 m.

**55.** If the system formed by the following equations has a solution, the line and the circle intersect.

$3x - 2y = 9$  (1)

$x^2 + y^2 = 25$ (2)

Solving equation (1) for $x$ we have

$3x = 9 + 2y \Rightarrow x = \frac{9+2y}{3}$.  (3)

Substituting into equation (3) into equation (2)

we have $\left(\frac{9+2y}{3}\right)^2 + y^2 = 25$.

Solving this equation for $y$ we have the following.

$$\left(\frac{9+2y}{3}\right)^2 + y^2 = 25 \Rightarrow \frac{(9+2y)^2}{9} + y^2 = 25$$

$$(9+2y)^2 + 9y^2 = 225$$

$$81 + 36y + 4y^2 + 9y^2 = 225$$

$$13y^2 + 36y + 81 = 225$$

$$13y^2 + 36y - 144 = 0$$

Using the quadratic formula where $a = 13, b = 36$, and $c = -144$, we have the following.

$$y = \frac{-36 \pm \sqrt{(36)^2 - 4(13)(-144)}}{2(13)}$$

$$= \frac{-36 \pm \sqrt{1296 + 7488}}{26} = \frac{-36 \pm \sqrt{8784}}{26}$$

$$y = \frac{-36 - \sqrt{8784}}{26} \approx -4.989 \text{ and}$$

$$y = \frac{-36 + \sqrt{8784}}{26} \approx 2.220$$

If $y = \frac{-36 + \sqrt{8784}}{26}$, then

$$x = \frac{9 + 2\left(\frac{-36+\sqrt{8784}}{26}\right)}{3} = \frac{234 + 2\left(-36 + \sqrt{8784}\right)}{78}$$

$$= \frac{162 + 2\sqrt{8784}}{78} \approx 4.480$$

If $y = \frac{-36 - \sqrt{8784}}{26}$, then

$$x = \frac{9 + 2\left(\frac{-36-\sqrt{8784}}{26}\right)}{3} = \frac{234 + 2\left(-36 - \sqrt{8784}\right)}{78}$$

$$= \frac{162 - 2\sqrt{8784}}{78} \approx -.326$$

Thus, the circle and line do intersect, in fact twice, at approximately (4.48, 2.22) and (–.326, –4.99).

**57.** We must first find the solution to the following system.

$y = x^2$        (1)

$x^2 + y^2 = 90$ (2)

Substitute $y$ for $x^2$ in equation (2) and solve for $y$.

$$y + y^2 = 90 \Rightarrow y^2 + y - 90 = 0$$

$$(y+10)(y-9) = 0 \Rightarrow y = -10 \text{ or } y = 9$$

If $y = -10$, then $-10 = x^2 \Rightarrow x = \pm i\sqrt{10}$. Since we are seeking points of intersection, we reject these nonreal solutions. If $y = 9$, then $9 = x^2 \Rightarrow x = \pm 3$. The two points of intersection of the two graphs are (3, 9) and (–3, 9). The slope of the line through these two points is $m = \frac{9-9}{-3-3} = 0$, so this is the horizontal line with equation $y = 9$.

**59.** Let $x$ represent the length and width of the square base, and let $y$ represent the height. Using the formula for the volume of a box, $V = LWH$, we have the equation $x^2 y = 360$. (1). Since the height is 4 ft greater than both the length and width we have the equation $y = x + 4$. (2) Since the surface area consists of a square base and four rectangular sides we have the equation $x^2 + 4xy = 276$. (3) Substituting equation (2) into equation (3) we have $x^2 + 4x(x+4) = 276$. Solving this equation for $x$ we have the following.

$$x^2 + 4x(x+4) = 276 \Rightarrow x^2 + 4x^2 + 16x = 276$$

$$5x^2 + 16x = 276 \Rightarrow 5x^2 + 16x - 276 = 0$$

$$(5x+46)(x-6) = 0 \Rightarrow x = -\frac{46}{5} \text{ or } x = 6$$

We disregard the negative solution since a length (or width) cannot be negative. If $x = 6$, then $y = 6 + 4 = 10$. Thus, the length, which is equal to the width, is 6 ft and the height is 10 ft. Note: Since $6^2(10) = 360$, the conditions for the volume are satisfied.

**61.** The system is $\begin{array}{l} x^2 y = 75 \\ x^2 + 4xy = 85 \end{array}$

The first possible solution is $x = 5$ and $y = 3$.

| $x^2 y = 75$ | $x^2 + 4xy = 85$ |
|---|---|
| $5^2 \cdot 3 = 75$ ? | $5^2 + 4 \cdot 5 \cdot 3 = 85$ ? |
| $25 \cdot 3 = 75$ | $25 + 60 = 85$ |
| $75 = 75$ True | $85 = 85$ True |

Thus, the solution length = width = 5 in. and height = 3 in. is valid.

Ideally we should check the exact values in the system in order to determine if the approximate dimensions are valid. From Example 6 we have

$$x = \frac{-5 + \sqrt{5^2 - 4(1)(-60)}}{2(1)} = \frac{-5 + \sqrt{25 + 240}}{2} = \frac{-5 + \sqrt{265}}{2}.$$

The corresponding $y$-value would be

$$y = \frac{75}{x^2} = \frac{75}{\left(\frac{-5+\sqrt{265}}{2}\right)^2} = \frac{75}{\frac{\left(-5+\sqrt{265}\right)^2}{4}} = \frac{300}{\left(-5+\sqrt{265}\right)^2}.$$

The second possible solution is $x = \frac{-5+\sqrt{265}}{2}$

and $y = \frac{300}{\left(-5+\sqrt{265}\right)^2}$.

$$x^2 y = 75$$

$$\left(\frac{-5+\sqrt{265}}{2}\right)^2 \cdot \frac{300}{\left(-5+\sqrt{265}\right)^2} = 75 \ ?$$

$$\frac{\left(-5+\sqrt{265}\right)^2}{4} \cdot \frac{300}{\left(-5+\sqrt{265}\right)^2} = 75$$

$$75 = 75 \text{ True}$$

$$x^2 + 4xy = 85$$

$$\frac{\left(-5+\sqrt{265}\right)^2}{4} + \frac{600}{-5+\sqrt{265}} = 85 \ ?$$

$$\frac{\left(-5+\sqrt{265}\right)^3 + 2400}{4\left(-5+\sqrt{265}\right)} = 85$$

$$\frac{\left(-5+\sqrt{265}\right)^2 \left(-5+\sqrt{265}\right) + 2400}{4\left(-5+\sqrt{265}\right)} = 85$$

$$\frac{\left(25-10\sqrt{265}+265\right)\left(-5+\sqrt{265}\right)+2400}{4\left(-5+\sqrt{265}\right)} = 85$$

$$\frac{\left(290-10\sqrt{265}\right)\left(-5+\sqrt{265}\right)+2400}{4\left(-5+\sqrt{265}\right)} = 85$$

$$\frac{\left(-1450+340\sqrt{265}-2650\right)+2400}{4\left(-5+\sqrt{265}\right)} = 85$$

$$\frac{-4100+340\sqrt{265}+2400}{4\left(-5+\sqrt{265}\right)} = 85$$

$$\frac{-1700+340\sqrt{265}}{4\left(-5+\sqrt{265}\right)} = 85$$

$$\frac{340\left(-5+\sqrt{265}\right)}{4\left(-5+\sqrt{265}\right)} = 85$$

$$85 = 85 \text{ True}$$

**63.** supply: $p = \sqrt{.1q+9} - 2$
demand: $p = \sqrt{25-.1q}$

**(a)** Equilibrium occurs when supply equals demand, so solve the system formed by the supply and demand equations. This system can be solved by substitution. Substitute $\sqrt{.1q+9} - 2$ for $p$ in the demand equation and solve the resulting equation for $q$.

$$\sqrt{.1q+9} - 2 = \sqrt{25-.1q}$$

$$\left(\sqrt{.1q+9} - 2\right)^2 = \left(\sqrt{25-.1q}\right)^2$$

$$.1q + 9 - 4\sqrt{.1q+9} + 4 = 25 - .1q$$

$$.2q - 12 = 4\sqrt{.1q+9}$$

$$\left(.2q-12\right)^2 = \left(4\sqrt{.1q+9}\right)^2$$

$$.04q^2 - 4.8q + 144 = 16\left(.1q+9\right)$$

$$.04q^2 - 4.8q + 144 = 1.6q + 144$$

$$.04q^2 - 6.4q = 0$$

$$.04q\left(q-160\right) = 0 \Rightarrow$$

$$q = 0 \text{ or } q = 160$$

Disregard an equilibrium demand of 0. The equilibrium demand is 160 units.

**(b)** Substitute 160 for $q$ in either equation and solve for $p$.

$$p = \sqrt{.1(160)+9} - 2 = \sqrt{16+9} - 2$$

$$= \sqrt{25} - 2 = 5 - 2 = 3$$

The equilibrium price is \$3.

**65. (a)** The emission of carbon is increasing with time. The carbon emissions from the former USSR and Eastern Europe have surpassed the emissions of Western Europe.

**(b)** They were equal in 1963 when the levels were approximately 400 million metric tons.

**(c)**
$$W = E$$

$$375(1.008)^{(t-1950)} = 260(1.038)^{(t-1950)}$$

$$\ln\left[375(1.008)^{(t-1950)}\right]$$

$$= \ln\left[260(1.038)^{(t-1950)}\right]$$

$$\ln 375 + (t-1950)\ln 1.008$$

$$= \ln 260 + (t-1950)\ln 1.038$$

$$\ln 375 - \ln 260$$

$$= (t-1950)\ln 1.038 - (t-1950)\ln 1.008$$

$$\ln 375 - \ln 260$$

$$= (\ln 1.038 - \ln 1.008) \cdot (t-1950)$$

$$\frac{\ln 375 - \ln 260}{\ln 1.038 - \ln 1.008} = t - 1950$$

$$t = 1950 + \frac{\ln 375 - \ln 260}{\ln 1.038 - \ln 1.008} \approx 1962.49$$

If $t$ is approximately 1962.49,

$$W = 375(1.008)^{(1962.49-1950)} \approx 414.24.$$

In 1962, the emission levels were equal and were approximately 414 million metric tons.

**67.** Shift the graph of $y = x^2$ four units down to obtain the graph of $y = x^2 - 4$.

**69.** $x^2 - 4 = x - 1$ if $x \geq 1$ and $x^2 - 4 = 1 - x$ if $x < 1$

**71.** If $y = |x - 1|$ and $x = \frac{1+\sqrt{13}}{2}$, then

$$y = \left| \frac{1+\sqrt{13}}{2} - 1 \right| = \left| \frac{1+\sqrt{13}}{2} - \frac{2}{2} \right| = \left| \frac{-1+\sqrt{13}}{2} \right| = \frac{-1+\sqrt{13}}{2}.$$

If $y = 1 - x$ and $x = \frac{-1-\sqrt{21}}{2}$, then

$$y = 1 - \left( \frac{-1-\sqrt{21}}{2} \right) = \frac{2}{2} + \frac{1+\sqrt{21}}{2} = \frac{3+\sqrt{21}}{2}.$$

Solution set:

$$\left\{ \left( \frac{1+\sqrt{13}}{2}, \frac{-1+\sqrt{13}}{2} \right), \left( \frac{-1-\sqrt{21}}{2}, \frac{3+\sqrt{21}}{2} \right) \right\}$$

## Summary Exercises on Systems of Equations

As noted in the text, different methods of solving equations have been introduced. In the solutions to these exercises, we will present the solution using one method (or combination of methods). In general, the nonlinear systems cannot use the matrix, Gauss-Jordan, Cramer's rule methods. Only the methods of substitution or elimination should be considered for such systems.

**1.** $2x + 5y = 4$
$3x - 2y = -13$

$$D = \begin{vmatrix} 2 & 5 \\ 3 & -2 \end{vmatrix} = 2(-2) - 3(5) = -4 - 15 = -19,$$

$$D_x = \begin{vmatrix} 4 & 5 \\ -13 & -2 \end{vmatrix} = 4(-2) - (-13)(5)$$
$$= -8 - (-65) = 57$$

$$D_y = \begin{vmatrix} 2 & 4 \\ 3 & -13 \end{vmatrix} = 2(-13) - 3(4)$$
$$= -26 - 12 = -38 \Rightarrow$$

$$x = \frac{D_x}{D} = \frac{57}{-19} = -3 \text{ and } y = \frac{D_y}{D} = \frac{-38}{-19} = 2.$$

Solution set: $\{(-3, 2)\}$

**3.** $2x^2 + y^2 = 5$ (1)
$3x^2 + 2y^2 = 10$ (2)

Using the elimination method, we add $-2$ times equation (1) to equation (2).

$$\begin{array}{r} -4x^2 - 2y^2 = -10 \\ 3x^2 + 2y^2 = \phantom{-}10 \\ \hline -x^2 \phantom{+2y^2} = \phantom{-1}0 \Rightarrow x^2 = 0 \Rightarrow x = 0 \end{array}$$

If $x = 0$, then $2(0)^2 + y^2 = 5 \Rightarrow$

$$0 + y^2 = 5 \Rightarrow y^2 = 5 \Rightarrow y = \pm\sqrt{5}.$$

Solution set: $\left\{ \left(0, \sqrt{5}\right), \left(0, -\sqrt{5}\right) \right\}$

**5.** $6x - y = 5$ (1)
$xy = 4$ (2)

Solving equation (2) for $y$, we have $y = \frac{4}{x}$.

Substitute this result into equation (1).

$$6x - \frac{4}{x} = 5 \Rightarrow 6x^2 - 4 = 5x \Rightarrow$$

$$6x^2 - 5x - 4 = 0 \Rightarrow (2x+1)(3x-4) = 0 \Rightarrow \text{ If}$$
$$x = -\frac{1}{2} \text{ or } x = \frac{4}{3}$$

$x = \frac{4}{3}$, then $y = \frac{4}{\frac{4}{3}} = 3$. If $x = -\frac{1}{2}$, then

$$y = \frac{4}{-\frac{1}{2}} = -8.$$

Solution set: $\left\{ \left(\frac{4}{3}, 3\right), \left(-\frac{1}{2}, -8\right) \right\}$

**7.** $x + 2y + z = 5$ (1)
$y + 3z = 9$ (2)

This system has infinitely many solutions. We will express the solution set with $z$ as the arbitrary variable. Solving equation (2) for $y$, we have $y = 9 - 3z$. Substituting $y$ (in terms of $z$) into equation (1) and solving for $x$ we have the following.

$$x + 2(9 - 3z) + z = 5 \Rightarrow x + 18 - 6z + z = 5 \Rightarrow$$
$$x + 18 - 5z = 5 \Rightarrow x = -13 + 5z$$

Solution set: $\{(-13 + 5z, 9 - 3z, z)\}$

**9.** $3x + 6y - 9z = 1$
$2x + 4y - 6z = 1$
$3x + 4y + 5z = 0$

This system has the augmented matrix

$$\left[ \begin{array}{ccc|c} 3 & 6 & -9 & 1 \\ 2 & 4 & -6 & 1 \\ 3 & 4 & 5 & 0 \end{array} \right].$$

$$\left[ \begin{array}{ccc|c} 2 & 4 & -6 & 1 \\ 3 & 6 & -9 & 1 \\ 3 & 4 & 5 & 0 \end{array} \right] \begin{array}{l} \text{R1} \leftrightarrow \text{R2} \\ \\ \\ \end{array} \Rightarrow$$

$$\left[ \begin{array}{ccc|c} 1 & 2 & -3 & \frac{1}{2} \\ 3 & 6 & -9 & 1 \\ 3 & 4 & 5 & 0 \end{array} \right] \begin{array}{l} \frac{1}{2}\text{R1} \\ \\ \\ \end{array} \Rightarrow$$

$$\left[ \begin{array}{ccc|c} 1 & 2 & -3 & \frac{1}{2} \\ 0 & 0 & 0 & -\frac{1}{2} \\ 0 & -2 & 14 & -\frac{3}{2} \end{array} \right] \begin{array}{l} \\ -3\text{R1}+\text{R2} \\ -3\text{R1}+\text{R3} \end{array}$$

The second row of the augmented matrix corresponds to the statement $0 = -\frac{1}{2}$. This is a false statement. Thus the solution set is $\varnothing$.

**11.** $x^2 + y^2 = 4$ (1)

$y = x + 6$   (2)

Substitute equation (1) into equation (2).

$$x^2 + (x+6)^2 = 4$$

$$x^2 + x^2 + 12x + 36 = 4$$

$$2x^2 + 12x + 36 = 4$$

$$2x^2 + 12x + 32 = 0 \Rightarrow x^2 + 6x + 16 = 0$$

Using the quadratic formula where

$a = 1, b = 6$, and $c = 16$, we have the

following.

$$x = \frac{-6 \pm \sqrt{6^2 - 4(1)(16)}}{2(1)} = \frac{-6 \pm \sqrt{36 - 64}}{2}$$

$$= \frac{-6 \pm \sqrt{-28}}{2} = \frac{-6 \pm 2i\sqrt{7}}{2} = -3 \pm i\sqrt{7}$$

If $x = -3 + i\sqrt{7}$, then

$y = -3 + i\sqrt{7} + 6 = 3 + i\sqrt{7}$. If $x = -3 - i\sqrt{7}$,

then $y = -3 - i\sqrt{7} + 6 = 3 - i\sqrt{7}$.

Solution set:

$$\left\{ \left(-3 + i\sqrt{7}, 3 + i\sqrt{7}\right), \left(-3 - i\sqrt{7}, 3 - i\sqrt{7}\right) \right\}$$

**13.** $y + 1 = x^2 + 2x$  (1)

$y + 2x = 4$     (2)

Solving equation (2) for $y$, we have $y = 4 - 2x$.

Substitute this result into equation (1).

$$4 - 2x + 1 = x^2 + 2x \Rightarrow 5 - 2x = x^2 + 2x \Rightarrow$$

$$0 = x^2 + 4x - 5 \Rightarrow (x+5)(x-1) = 0 \Rightarrow$$

$$x = -5 \text{ or } x = 1$$

If $x = -5$, then $y = 4 - 2(-5) = 4 + 10 = 14$. If

$x = 1$, then $y = 4 - 2(1) = 4 - 2 = 2$.

Solution set: $\left\{ (1, 2), (-5, 14) \right\}$

**15.**  $2x + 3y + 4z = 3$

$-4x + 2y - 6z = 2$

$4x + 3z = 0$

This system has the augmented matrix

$$\begin{bmatrix} 2 & 3 & 4 & | & 3 \\ -4 & 2 & -6 & | & 2 \\ 4 & 0 & 3 & | & 0 \end{bmatrix}.$$

$$\begin{bmatrix} 1 & \frac{3}{2} & 2 & | & \frac{3}{2} \\ -4 & 2 & -6 & | & 2 \\ 4 & 0 & 3 & | & 0 \end{bmatrix} \begin{matrix} \frac{1}{2}R1 \\ \\ \end{matrix} \Rightarrow$$

$$\begin{bmatrix} 1 & \frac{3}{2} & 2 & | & \frac{3}{2} \\ 0 & 8 & 2 & | & 8 \\ 0 & -6 & -5 & | & -6 \end{bmatrix} \begin{matrix} \\ 4R1+R2 \Rightarrow \\ -4R1+R3 \end{matrix}$$

$$\begin{bmatrix} 1 & \frac{3}{2} & 2 & | & \frac{3}{2} \\ 0 & 1 & \frac{1}{4} & | & 1 \\ 0 & -6 & -5 & | & -6 \end{bmatrix} \begin{matrix} \\ \frac{1}{8}R2 \\ \end{matrix} \Rightarrow$$

$$\begin{bmatrix} 1 & 0 & \frac{13}{8} & | & 0 \\ 0 & 1 & \frac{1}{4} & | & 1 \\ 0 & 0 & -\frac{7}{2} & | & 0 \end{bmatrix} \begin{matrix} -\frac{3}{2}R2+R1 \\ \\ 6R2+R3 \end{matrix} \Rightarrow$$

$$\begin{bmatrix} 1 & 0 & \frac{13}{8} & | & 0 \\ 0 & 1 & \frac{1}{4} & | & 1 \\ 0 & 0 & 1 & | & 0 \end{bmatrix} \begin{matrix} \\ \\ -\frac{2}{7}R3 \end{matrix} \Rightarrow$$

$$\begin{bmatrix} 1 & 0 & 0 & | & 0 \\ 0 & 1 & 0 & | & 1 \\ 0 & 0 & 1 & | & 0 \end{bmatrix} \begin{matrix} -\frac{13}{8}R3+R1 \\ -\frac{1}{4}R3+R2 \\ \end{matrix}$$

Solution set: $\left\{ (0, 1, 0) \right\}$

**17.**  $-5x + 2y + z = 5$ (1)

$-3x - 2y - z = 3$ (2)

$-x + 6y = 1$ (3)

Add equations (1) and (2) in order to eliminate

$z$ to obtain $-8x = 8$ or $x = -1$.

Substitute $x = -1$ into equation (3) to solve

for $y$.

$$-(-1) + 6y = 1 \Rightarrow 1 + 6y = 1 \Rightarrow 6y = 0 \Rightarrow y = 0$$

Substitute $x = -1$ and $y = 0$ into equation (1)

to solve for $z$.

$$-5(-1) + 2(0) + z = 5 \Rightarrow 5 + 0 + z = 5 \Rightarrow$$

$$5 + z = 5 \Rightarrow z = 0$$

Solution set: $\left\{ (-1, 0, 0) \right\}$

**19.**  $2x^2 + y^2 = 9$   (1)

$3x - 2y = -6$ (2)

Solving equation (2) for $x$, we have

$3x = 2y - 6 \Rightarrow x = \frac{2y-6}{3}$.

Substitute this result into equation (1).

$$2\left(\frac{2y-6}{3}\right)^2 + y^2 = 9$$

$$2 \cdot \frac{(2y-6)^2}{9} + y^2 = 9$$

$$2 \cdot (2y-6)^2 + 9y^2 = 81$$

$$2\left(4y^2 - 24y + 36\right) + 9y^2 = 81$$

$$8y^2 - 48y + 72 + 9y^2 = 81$$

$$17y^2 - 48y + 72 = 81$$

$$17y^2 - 48y - 9 = 0$$

$$(y-3)(17y+3) = 0 \Rightarrow y - 3 \text{ or } y = -\frac{3}{17}$$

(*continued on next page*)

*(continued from page 291)*

If $y = 3$, then $x = \frac{2(3)-6}{3} = \frac{6-6}{3} = \frac{0}{3} = 0.$ If

$y = -\frac{3}{17}$, then $x = \frac{2\left(-\frac{3}{17}\right)-6}{3} = \frac{2(-3)-102}{51}$

$= \frac{-6-102}{51} = \frac{-108}{51} = -\frac{36}{17}.$

Solution set: $\left\{(0,3),\left(-\frac{36}{17},-\frac{3}{17}\right)\right\}$

**21.**
$$x + y - z = 0 \text{ (1)}$$
$$2y - z = 1 \text{ (2)}$$
$$2x + 3y - 4z = -4 \text{ (3)}$$

Add $-2$ times equation (1) to equation (3) in order to eliminate $x$.

$$\begin{aligned} -2x - 2y + 2z &= 0 \\ \underline{2x + 3y - 4z} &= \underline{-4} \\ y - 2z &= -4 \text{ (4)} \end{aligned}$$

Add $-2$ times equation (4) to equation (2) in order to eliminate $y$ and solve for $z$.

$$\begin{aligned} 2y - z &= 1 \\ \underline{-2y + 4z} &= \underline{8} \\ 3z &= 9 \Rightarrow z = 3 \end{aligned}$$

Substitute 3 for $z$ in equation (4) to obtain

$$y - 2(3) = -4 \Rightarrow y - 6 = -4 \Rightarrow y = 2.$$

Substitute 3 for $z$ and 2 for $y$ in equation (1) to obtain $x + 2 - 3 = 0 \Rightarrow x - 1 = 0 \Rightarrow x = 1.$

Solution set: $\{(1,2,3)\}$

**23.**
$$4x - z = -6 \text{ (1)}$$
$$\tfrac{3}{5}y + \tfrac{1}{2}z = 0 \text{ (2)}$$
$$\tfrac{1}{3}x + \tfrac{2}{3}z = -5 \text{ (3)}$$

Before applying one of the methods, it would be helpful to clear the fractions in equations (2) and (3) by multiplying equation (2) by 10 and equation (3) by 3 to obtain the following.

$$6y + 5z = 0 \text{ (4)}$$
$$x + 2z = -15 \text{ (5)}$$

The system formed by equations (1), (4), and (5) can be represented by the following augmented matrix.

$$\begin{bmatrix} 4 & 0 & -1 & -6 \\ 0 & 6 & 5 & 0 \\ 1 & 0 & 2 & -15 \end{bmatrix}$$

$$\begin{bmatrix} 1 & 0 & 2 & -15 \\ 0 & 6 & 5 & 0 \\ 4 & 0 & -1 & -6 \end{bmatrix} \begin{matrix} \text{R1} \leftrightarrow \text{R3} \\ \\ \end{matrix} \Rightarrow$$

$$\begin{bmatrix} 1 & 0 & 2 & -15 \\ 0 & 6 & 5 & 0 \\ 0 & 0 & -9 & 54 \end{bmatrix} \begin{matrix} \\ \\ -4\text{R1}+\text{R3} \end{matrix} \Rightarrow$$

$$\begin{bmatrix} 1 & 0 & 2 & -15 \\ 0 & 1 & \frac{5}{6} & 0 \\ 0 & 0 & -9 & 54 \end{bmatrix} \begin{matrix} \\ \frac{1}{6}\text{R2} \\ \\ \end{matrix} \Rightarrow$$

$$\begin{bmatrix} 1 & 0 & 2 & -15 \\ 0 & 1 & \frac{5}{6} & 0 \\ 0 & 0 & 1 & -6 \end{bmatrix} \begin{matrix} \\ \\ -\frac{1}{9}\text{R3} \end{matrix} \Rightarrow$$

$$\begin{bmatrix} 1 & 0 & 0 & -3 \\ 0 & 1 & 0 & 5 \\ 0 & 0 & 1 & -6 \end{bmatrix} \begin{matrix} -2\text{R3}+\text{R1} \\ -\frac{5}{6}\text{R3}+\text{R2} \\ \end{matrix}$$

Solution set: $\{(-3,5,-6)\}$

**25.**
$$x^2 + 3y^2 = 28 \text{ (1)}$$
$$y - x = -2 \text{ (2)}$$

Solving equation (2) for $y$, we have $y = x - 2$. Substitute this result into equation (1) and solve for $y$

$$x^2 + 3(x-2)^2 = 28$$
$$x^2 + 3(x^2 - 4x + 4) = 28$$
$$x^2 + 3x^2 - 12x + 12 = 28$$
$$4x^2 - 12x + 12 = 28$$
$$4x^2 - 12x - 16 = 0$$
$$x^2 - 3x - 4 = 0$$
$$(x+1)(x-4) = 0 \Rightarrow x = -1 \text{ or } x = 4$$

If $x = -1$, then $y = -1 - 2 = -3.$ If $x = 4$, then $y = 4 - 2 = 2.$

Solution set: $\{(4,2),(-1,-3)\}$

**27.**
$$2x^2 + 3y^2 = 20 \text{ (1)}$$
$$x^2 + 4y^2 = 5 \text{ (2)}$$

Using the elimination method, we add $-2$ times equation (2) to equation (1).

$$\begin{aligned} 2x^2 + 3y^2 &= 20 \\ \underline{-2x^2 - 8y^2} &= \underline{-10} \\ -5y^2 &= 10 \Rightarrow y^2 = -2 \Rightarrow y = \pm i\sqrt{2} \end{aligned}$$

Find $x$ by substituting back into equation (2).

If $y = i\sqrt{2}$, then $x^2 + 4\left(i\sqrt{2}\right)^2 = 5 \Rightarrow$

$x^2 + 4(-1)(2) = 5 \Rightarrow x^2 - 8 = 5 \Rightarrow x^2 = 13 \Rightarrow$

$x = \pm\sqrt{13}.$ If $y = -i\sqrt{2}$, then

$x^2 + 4\left(-i\sqrt{2}\right)^2 = 5 \Rightarrow x^2 + 4(-1)(2) = 5 \Rightarrow$

$x^2 - 8 = 5 \Rightarrow x^2 = 13 \Rightarrow x = \pm\sqrt{13}$

Solution set: $\left\{\left(\sqrt{13}, i\sqrt{2}\right), \left(-\sqrt{13}, i\sqrt{2}\right),\right.$

$\left.\left(\sqrt{13}, -i\sqrt{2}\right), \left(-\sqrt{13}, -i\sqrt{2}\right)\right\}$

**29.**
$$x + 2z = 9$$
$$y + z = 1$$
$$3x - 2y = 9$$

Adding $-3$ times row 1 to row 3, we have

$$D = \begin{vmatrix} 1 & 0 & 2 \\ 0 & 1 & 1 \\ 3 & -2 & 0 \end{vmatrix} = \begin{vmatrix} 1 & 0 & 2 \\ 0 & 1 & 1 \\ 0 & -2 & -6 \end{vmatrix}.$$

Expanding by column one, we have

$$D = 1 \cdot \begin{vmatrix} 1 & 1 \\ -2 & -6 \end{vmatrix} = -6 + 2 = -4.$$

Adding 2 times row 2 to row 3, we have

$$D_x = \begin{vmatrix} 9 & 0 & 2 \\ 1 & 1 & 1 \\ 9 & -2 & 0 \end{vmatrix} = \begin{vmatrix} 9 & 0 & 2 \\ 1 & 1 & 1 \\ 11 & 0 & 2 \end{vmatrix}.$$

Expanding by column two, we have

$$D_x = 1 \cdot \begin{vmatrix} 9 & 2 \\ 11 & 2 \end{vmatrix} = 18 - 22 = -4.$$

Adding $-3$ times row 1 to row 3 we have

$$D_y = \begin{vmatrix} 1 & 9 & 2 \\ 0 & 1 & 1 \\ 3 & 9 & 0 \end{vmatrix} = \begin{vmatrix} 1 & 9 & 2 \\ 0 & 1 & 1 \\ 0 & -18 & -6 \end{vmatrix}.$$

Expanding by column one, we have

$$D_y = 1 \cdot \begin{vmatrix} 1 & 1 \\ -18 & -6 \end{vmatrix} = -6 + 18 = 12.$$

Adding $-3$ times row 1 to row 3, we have

$$D_z = \begin{vmatrix} 1 & 0 & 9 \\ 0 & 1 & 1 \\ 3 & -2 & 9 \end{vmatrix} = \begin{vmatrix} 1 & 0 & 9 \\ 0 & 1 & 1 \\ 0 & -2 & -18 \end{vmatrix}.$$

Expanding by column one, we have

$$D_z = 1 \cdot \begin{vmatrix} 1 & 1 \\ -2 & -18 \end{vmatrix} = -18 + 2 = -16.$$

Thus, we have

$$x = \frac{D_x}{D} = \frac{-4}{-4} = 1, \quad y = \frac{D_y}{D} = \frac{12}{-4} = -3,$$

and $z = \frac{D_z}{D} = \frac{-16}{-4} = 4.$

Solution set: $\{(1, -3, 4)\}$

**31.**
$$-x + y \quad\quad = -1$$
$$x + \quad z = 4$$
$$6x - 3y + 2z = 10$$

This system has the augmented matrix

$$\begin{bmatrix} -1 & 1 & 0 & | & -1 \\ 1 & 0 & 1 & | & 4 \\ 6 & -3 & 2 & | & 10 \end{bmatrix}.$$

$$\begin{bmatrix} 1 & -1 & 0 & | & 1 \\ 1 & 0 & 1 & | & 4 \\ 6 & -3 & 2 & | & 10 \end{bmatrix} \begin{matrix} -1R1 \\ \\ \\ \end{matrix} \Rightarrow$$

$$\begin{bmatrix} 1 & -1 & 0 & | & 1 \\ 0 & 1 & 1 & | & 3 \\ 0 & 3 & 2 & | & 4 \end{bmatrix} \begin{matrix} \\ -1R1+R2 \\ -6R1+R3 \end{matrix} \Rightarrow$$

$$\begin{bmatrix} 1 & 0 & 1 & | & 4 \\ 0 & 1 & 1 & | & 3 \\ 0 & 0 & -1 & | & -5 \end{bmatrix} \begin{matrix} R2+R1 \\ \\ -3R2+R3 \end{matrix} \Rightarrow$$

$$\begin{bmatrix} 1 & 0 & 1 & | & 4 \\ 0 & 1 & 1 & | & 3 \\ 0 & 0 & 1 & | & 5 \end{bmatrix} \begin{matrix} \\ \\ -1R3 \end{matrix} \Rightarrow$$

$$\begin{bmatrix} 1 & 0 & 0 & | & -1 \\ 0 & 1 & 0 & | & -2 \\ 0 & 0 & 1 & | & 5 \end{bmatrix} \begin{matrix} -1R3+R1 \\ -1R3+R2 \\ \end{matrix}$$

Solution set: $\{(-1, -2, 5)\}$

**33.**
$$xy = -3 \quad (1)$$
$$x + y = -2 \quad (2)$$

Solving equation (1) for $y$, we have $y = -\frac{3}{x}$.

Substitute this result into equation (2).

$$x + \left(-\frac{3}{x}\right) = -2 \Rightarrow x^2 - 3 = -2x \Rightarrow$$
$$x^2 + 2x - 3 = 0 \Rightarrow (x + 3)(x - 1) = 0 \Rightarrow$$
$$x = -3 \text{ or } x = 1$$

If $x = -3$, then $y = -\frac{3}{-3} = 1$. If $x = 1$, then

$$y = -\frac{3}{1} = -3.$$

Solution set: $\{(1, -3), (-3, 1)\}$

**35.**
$$y = x^2 + 6x + 9 \, (1)$$
$$x + y = 3 \quad\quad (2)$$

Solve equation (2) for $y$ to obtain

$$x + y = 3 \Rightarrow y = 3 - x. \, (3)$$

Substitute equation (3) into equation (1) and solve for $x$.

$$3 - x = x^2 + 6x + 9$$
$$0 = x^2 + 7x + 6$$
$$(x + 6)(x + 1) = 0 \Rightarrow x = -6 \text{ or } x = -1$$

If $x = -6$, then $y = 3 - (-6) = 9$. If $x = -1$,

then $y = 3 - (-1) = 4.$

Solution set: $\{(-6, 9), (-1, 4)\}$

**37.**
$$2x - 3y \quad\quad = -2 \quad (1)$$
$$x + \; y - 4z = -16 \quad (2)$$
$$3x - 2y + \; z = \; 7 \quad (3)$$

Add 4 times equation (3) to equation (2) in order to eliminate $z$.
$$x + \; y - 4z = -16$$
$$\underline{12x - 8y + 4z = \; 28}$$
$$13x - 7y \quad\quad = \; 12 \quad (4)$$

Add 3 times equation (4) to $-7$ times equation (1) in order to eliminate $y$ and solve for $x$.
$$-14x + 21y = 14$$
$$\underline{39x - 21y = 36}$$
$$25x \quad\quad = 50 \Rightarrow x = 2$$

Substitute 2 for $x$ in equation (4) to obtain
$$13(2) - 7y = 12 \Rightarrow 26 - 7y = 12 \Rightarrow$$
$$-7y = -14 \Rightarrow y = 2$$

Substitute 2 for $x$ and 2 for $y$ in equation (2) to obtain the following.
$$2 + 2 - 4z = -16 \Rightarrow 4 - 4z = -16 \Rightarrow$$
$$-4z = -20 \Rightarrow z = 5$$
Solution set: $\{(2, 2, 5)\}$

**39.**
$$y = (x - 1)^2 + 2 \quad (1)$$
$$y = 2x - 1 \quad\quad (2)$$

Substitute $2x - 1$ for $y$ in equation (1) and solve for $x$:
$$2x - 1 = (x - 1)^2 + 2$$
$$2x - 1 = x^2 - 2x + 1 + 2$$
$$x^2 - 4x + 4 = 0 \Rightarrow (x - 2)^2 = 0 \Rightarrow x = 2$$

Substitute 2 for $x$ in equation (2) and solve for $y$: $y = 2(2) - 1 = 3$
Solution set: $\{(2, 3)\}$

## Section 5.6: Systems of Inequalities and Linear Programming

**1.** $x + 2y \le 6$

The boundary is the line $x + 2y = 6$, which can be graphed using the $x$-intercept 6 and $y$-intercept 3. The boundary is included in the graph, so draw a solid line. Solving for $y$, we have the following.
$$x + 2y \le 6 \Rightarrow 2y \le -x + 6 \Rightarrow y \le -\tfrac{1}{2}x + 3$$
Since $y$ is *less than* or equal to $-\tfrac{1}{2}x + 3$, the graph of the solution set is the line and the half-plane *below* the boundary.
We also can use $(0, 0)$ as a test point.

Since $0 + 2(0) \le 6 \Rightarrow 0 \le 6$ is a true statement, shade line and the side of the graph containing the test point $(0, 0)$ (the half-plane below the boundary).

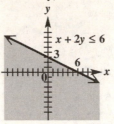

**3.** $2x + 3y \ge 4$

The boundary is the line $2x + 3y = 4$, which can be graphed using the $x$-intercept 2 and $y$-intercept $\tfrac{4}{3}$. The boundary is included in the graph, so draw a solid line. Solving for y, we have the following.
$$2x + 3y \ge 4 \Rightarrow 3y \ge -2x + 4 \Rightarrow y \ge -\tfrac{2}{3}x + \tfrac{4}{3}$$
Since $y$ is *greater than* or equal to $-\tfrac{2}{3}x + \tfrac{4}{3}$, the graph of the solution set is the line and the half-plane *above* the boundary. We also can use $(0, 0)$ as a test point. Since
$2(0) + 3(0) \ge 4 \Rightarrow 0 \ge 4$ is a false statement, shade the line and the side of the graph not containing the test point $(0,0)$ (the half-plane above the boundary).

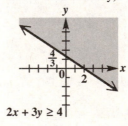

$2x + 3y \ge 4$

**5.** $3x - 5y > 6$

The boundary is the line $3x - 5y = 6$, which can be graphed using the $x$-intercept 2 and $y$-intercept $-\tfrac{6}{5}$. The boundary is not included in the graph, so draw a dashed line. Solving for $y$, we have the following  Since $y$ is *less than* $\tfrac{3}{5}x - \tfrac{6}{5}$, the graph of the solution set is the half-plane *below* the boundary. We also can use $(0, 0)$ as a test point. Since
$3(0) - 5(0) > 6 \Rightarrow 0 > 6$ is a false statement, shade the side of the graph not containing the test point $(0, 0)$ (the half-plane below the boundary).

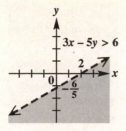

**7.** $5x \le 4y - 2$

The boundary is the line $5x = 4y - 2$, which can be graphed using the $x$-intercept $-\frac{2}{5}$ and $y$-intercept $\frac{1}{2}$. The boundary is included in the graph, so draw a solid line. Solving for $y$, we have the following.

$5x \le 4y - 2 \Rightarrow -4y \le -5x - 2 \Rightarrow y \ge \frac{5}{4}x + \frac{1}{2}$

Since $y$ is *greater than* or equal to $\frac{5}{4}x + \frac{1}{2}$, the graph of the solution set is the line and the half-plane *above* the boundary. We also can use (0, 0) as a test point. Since $5(0) \le 4(0) - 2 \Rightarrow 0 \le -2$ is a false statement, shade the line and the side of the graph not containing the test point (0, 0) (the half-plane above the boundary).

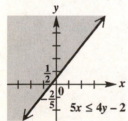

**9.** $x \le 3$

The boundary is the vertical line $x = 3$, which intersects the $x$-axis at 3. The boundary is included in the graph, so draw a solid line. Since $x \le 3$, it can easily be determined that we should shade to the left of the boundary. We also can use (0, 0) as a test point. Since $0 \le 3$ is a true statement, shade the line and the side of the graph containing the test point (0, 0) (the half-plane to the left of the boundary).

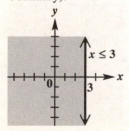

**11.** $y < 3x^2 + 2$

The boundary is the parabola $y = 3x^2 + 2$, which opens upwards. It has vertex $(0, 2)$, $y$-intercept 2, and no $x$-intercepts. Since the inequality symbol is <, draw a dashed curve. Since $y$ is *less than* $3x^2 + 2$, the graph of the solution set is the half-plane *below* the boundary. We also can use (0, 0) as a test point. Since $0 < 3(0)^2 + 2 \Rightarrow 0 < 2$ is a true statement, shade the region of the graph containing the test point (0, 0).

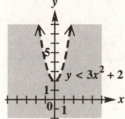

**13.** $y > (x-1)^2 + 2$

The boundary is the parabola $y = (x-1)^2 + 2$, which opens upwards. It has vertex $(1, 2)$, $y$-intercept 3, and no $x$-intercepts. Since the inequality symbol is >, draw a dashed curve. Since $y$ is *greater than* $(x-1)^2 + 2$, the graph of the solution set is the half-plane *above* the boundary. We also can use (0, 0) as a test point. Since $0 > (0-1)^2 + 2 \Rightarrow 0 > 3$ is a false statement, shade the region of the graph not containing the test point (0, 0).

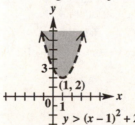

**15.** $x^2 + (y+3)^2 \le 16$

The boundary is a circle with center $(0, -3)$ and radius 4. Draw a solid circle to show that the boundary is included in the graph. Since $x^2 + (y+3)^2 \le 16$, it can easily be determined that points in the interior or on the boundary would satisfy this relation.

*(continued on next page)*

(*continued from page 295*)

We also can use (0, 0) as a test point. Since we have $0^2 + (0+3)^2 \le 16 \Rightarrow 9 \le 16$ is a true statement, shade the region of the graph containing the test point (0, 0).

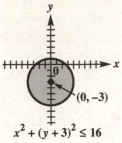

$$x^2 + (y+3)^2 \le 16$$

17. $y > 2^x + 1$

The boundary is an exponential function with *y*-intercept 2. The inequality symbol is >, wo draw a dashed curve. Since *y* is *greater than* $2^x + 1$, the graph of the solution set is the half-plane *above* the boundary. We also can use (0, 0) as a test point. Since $0 > 2^0 + 1 \Rightarrow 0 > 2$ is a false statement, shade the region of the graph not containing the test point (0, 0).

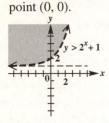

19. If an inequality is a "strict" inequality $(< \text{ or } >)$, then the boundary is not included and is represented by a dashed line. If an inequality is a "nonstrict" or "weak" inequality $(\le \text{ or } \ge)$, then the boundary is included and is represented by a solid line.

21. $Ax + By \ge C, B > 0$

Solving for *y* we have
$$Ax + By \ge C \Rightarrow By \ge -Ax + C \Rightarrow y \ge -\tfrac{A}{B}x + \tfrac{C}{B}$$
Since $B > 0$, the inequality symbol was not reversed when both sides are divided by *B*. Since $y \ge -\tfrac{A}{B}x + \tfrac{C}{B}$, you would shade above the line.

23. The graph of $(x-5)^2 + (y-2)^2 = 4$ is a circle with center (5, 2) and radius $r = \sqrt{4} = 2$. The graph of $(x-5)^2 + (y-2)^2 < 4$ is the region in the interior of this circle. The correct response is B.

25. The graph of $y \le 3x - 6$ is the region below the line with slope 3 and *y*-intercept – 6. This is graph C.

27. The graph of $y \le -3x - 6$ is the region below the line with slope –3 and *y*-intercept – 6. This is graph A.

29. $x + y \ge 0$
$2x - y \ge 3$

Graph $x + y = 0$ as a solid line through the origin with a slope of –1. Shade the region above this line. Graph $2x - y = 3$ as a solid line with *x*-intercept $\tfrac{3}{2}$ and *y*-intercept –3. Shade the region below this line. To find where the boundaries of the two lines intersect, solve the system $\begin{array}{l} x + y = 0 \ (1) \\ 2x - y = 3 \ (2) \end{array}$ by adding the two equations together to obtain $3x = 3 \Rightarrow x = 1$. Substituting this value into equation (1), we obtain $1 + y = 0 \Rightarrow y = -1$. Thus, the boundaries intersect at $(1, -1)$. The solution set is the common region, which is shaded in the final graph.

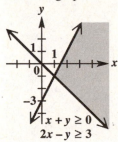

$$x + y \ge 0$$
$$2x - y \ge 3$$

For Exercises 31–61, we will omit finding the points in which the boundaries intersect.

31. $2x + y > 2$
$x - 3y < 6$

Graph $2x + y = 2$ as a dashed line with *y*-intercept 2 and *x*-intercept 1. Shade the region above this line. Graph $x - 3y = 6$ as a dashed line with *y*-intercept –2 and *x*-intercept 6. Shade the region above this line. The solution set is the common region, which is shaded in the final graph.

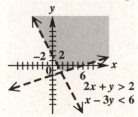

$$2x + y > 2$$
$$x - 3y < 6$$

**33.** $3x + 5y \leq 15$
$\quad\quad x - 3y \geq 9$

Graph $3x + 5y = 15$ as a solid line with $y$-intercept 3 and $x$-intercept 5. Shade the region below this line. Graph $x - 3y = 9$ as a solid line with $y$-intercept $-3$ and $x$-intercept 9. Shade the region below this line. The solution set is the common region, which is shaded in the final graph.

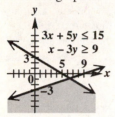

**35.** $4x - 3y \leq 12$
$\quad\quad y \leq x^2$

Graph $4x - 3y = 12$ as a solid line with $y$-intercept $-4$ and $x$-intercept 3. Shade the region above this line. Graph the solid parabola $y = x^2$. Shade the region outside of this parabola. The solution set is the intersection of these two regions, which is shaded in the final graph.

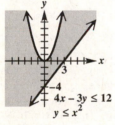

**37.** $x + 2y \leq 4$
$\quad\quad y \geq x^2 - 1$

Graph $x + 2y = 4$ as a solid line with $y$-intercept 2 and $x$-intercept 4. Shade the region below the line. Graph the solid parabola $y = x^2 - 1$. Shade the region inside of the parabola. The solution set is the common region, which is shaded in the final graph.

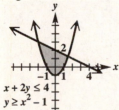

**39.** $y \leq (x + 2)^2$
$\quad\quad y \geq -2x^2$

Graph $y = (x + 2)^2$ as a solid parabola opening up with a vertex at $(-2, 0)$. Shade the region below the parabola. Graph $y = -2x^2$ as a solid parabola opening down with a vertex at the origin. Shade the region above the parabola. The solution set is the intersection of these two regions, which is shaded in the final graph.

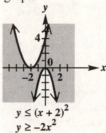

**41.** $x + y \leq 36$
$\quad\quad -4 \leq x \leq 4$

Graph $x + y = 36$ as a solid line with $y$-intercept 36 and $x$-intercept 36. Shade the region below this line. Graph the vertical lines $x = -4$ and $x = 4$ as solid lines. Shade the region between these lines. The solution set is the intersection of these two regions, which is shaded in the final graph.

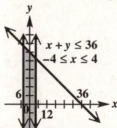

**43.** $y \geq x^2 + 4x + 4$
$\quad\quad y < -x^2$

Graph the solid parabola $y = x^2 + 4x + 4$ or $y = (x + 2)^2$, which has vertex $(-2, 0)$ and opens upward. Shade the region inside of this parabola. Graph the dashed parabola $y = -x^2$, which has vertex $(0, 0)$ and opens downward. Shade the region inside this parabola. These two regions have no points in common, so the system has no solution.

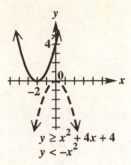

$y \geq x^2 + 4x + 4$
$y < -x^2$

**45.** $3x - 2y \geq 6$
$\quad x + y \leq -5$
$\qquad y \leq 4$

Graph $3x - 2y = 6$ as a solid line and shade the region below it. Graph $x + y = -5$ as a solid line and shade the region below it. Graph $y = 4$ as a solid horizontal line and shade the region below it. The solution set is the intersection of these three regions, which is shaded in the final graph.

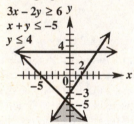

**47.** $-2 < x < 2$
$\quad y > 1$
$\quad x - y > 0$

Graph the vertical lines $x = -2$ and $x = 2$ as a dashed line. Shade the region between the two lines. Graph the horizontal line $y = 1$ as a dashed line. Shade the region above the line. Graph the line $x - y = 0$ as a dashed line through the origin with a slope of 1. Shade the region below this line. The solution set is the intersection of these three regions, which is shaded in the final graph.

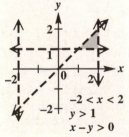

**49.** $x \leq 4$
$\quad x \geq 0$
$\quad y \geq 0$
$\quad x + 2y \geq 2$

Graph $x = 4$ as a solid vertical line. Shade the region to the left of this line. Graph $x = 0$ as a solid vertical line. (This is the $y$-axis.) Shade the region to the right of this line. Graph $y = 0$ as a solid horizontal line. (This is the $x$-axis.) Shade the region above the line. Graph $x + 2y = 2$ as a solid line with $x$-intercept 2 and $y$-intercept 1. Shade the region above the line. The solution set is the intersection of these four regions, which is shaded in the final graph.

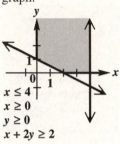

$x \leq 4$
$x \geq 0$
$y \geq 0$
$x + 2y \geq 2$

**51.** $2x + 3y \leq 12$
$\quad 2x + 3y > -6$
$\quad 3x + \ y < 4$
$\qquad x \geq 0$
$\qquad y \geq 0$

Graph $2x + 3y = 12$ as a solid line and shade the region below it. Graph $2x + 3y = 6$ as a dashed line and shade the region above it. Graph $3x + y = 4$ as a dashed line and shade the region below it. $x = 0$ is the $y$-axis. Shade the region to the right of it. $y = 0$ is the $x$-axis. Shade the region above it. The solution set is the intersection of these five regions, which is shaded in the final graph. The open circles at $(0, 4)$ and $\left(\frac{4}{3}, 0\right)$ indicate that those points are not included in the solution (due to the fact that the boundary line on which they lie, $3x + y = 4$, is not included).

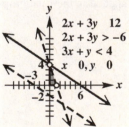

$2x + 3y \quad 12$
$2x + 3y > -6$
$3x + y < 4$
$x \quad 0, y \quad 0$

**53.** $y \leq \left(\frac{1}{2}\right)^x$

$y \geq 4$

Graph $y = \left(\frac{1}{2}\right)^x$ using a solid curve passing through the points $(-2, 4)$, $(-1, 2)$, $(0, 1)$, $\left(1, \frac{1}{2}\right)$, and $\left(2, \frac{1}{4}\right)$. Shade the region below this curve. Graph the solid horizontal line $y = 4$ and shade the region above it. The solution set consists of the intersection of these two regions, which is shaded in the final graph.

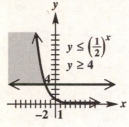

**55.** $y \leq \log x$

$y \geq |x - 2|$

Graph $y = \log x$ using a solid curve because $y \leq \log x$ is a nonstrict inequality. (Recall that "$\log x$" means $\log_{10} x$.) This graph contains the points $(.1, -1)$, $(1, 0)$, and $(10, 1)$. Use a calculator to approximate other points on the graph, such as $(2, .30)$ and $(4, .60)$. Because the symbol is $\leq$, shade the region *below* the curve. Now graph $y = |x - 2|$. Make this boundary solid because $y \geq |x - 2|$ is also a nonstrict inequality. This graph can be obtained by translating the graph of $y = |x|$ to the right 2 units. It contains points $(0, 2)$, $(2, 0)$, and $(4, 2)$. Because the symbol is $\geq$, shade the region *above* the absolute value graph. The solution set is the intersection of the two regions, which is shaded in the final graph.

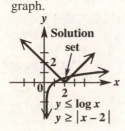

**57.** $y > x^3 + 1$

$y \geq -1$

Graph $y > x^3 + 1$ as a dashed curve which is the graph of $y = x^3$ translated up 1 unit. Shade the region above the curve. Graph $y = -1$ as a solid horizontal line through $(0, -1)$. Shade the area above the line. The solution set is the common region, which is shaded in the final graph.

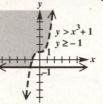

**59.** The upper line passes through $(0, 2)$ and $(4, 0)$, so $m = \frac{0 - 2}{4 - 0} = -\frac{1}{2}$, and the equation is

$y = -\frac{1}{2}x + 2 \Rightarrow x + 2y = 4$. The line is solid, so it is included in the inequality. The lower line passes through $(0, -3)$ and $(4, 0)$, so

$m = \frac{0 - (-3)}{4 - 0} = \frac{3}{4}$ and the equation is

$y = \frac{3}{4}x - 3 \Rightarrow 3x - 4y = 12$. The line is solid, so it is included in the inequality. The shaded region contains the point $(0, 0)$, so test $(0, 0)$ in each equation to determine the direction of the inequalities:

$0 + 2(0) = 0 < 4 \Rightarrow x + 2y \leq 4$,

$3(0) + 4(0) = 0 < 12 \Rightarrow 3x - 4y \leq 12$

The system is $\quad x + 2y \leq 4 \quad$.
$\qquad\qquad\quad 3x - 4y \leq 12$

**61.** The circle has center $(0, 0)$ and radius 4, so its equation is $x^2 + y^2 = 16$. The curve is solid, so it is included in the inequality. The horizontal line passes through $(0, 2)$, so its equation is $y = 2$. It is solid, so it is included in the inequality. The shaded region includes the point $(-1, 3)$, so test $(-1, 3)$ in each equation to determine the direction of the inequalities:

$(-1)^2 + 3^2 = 10 < 16 \Rightarrow x^2 + y^2 \leq 16$,

$3 > 2 \Rightarrow y > 2$. The system is $x^2 + y^2 \leq 16$.
$\qquad\qquad\qquad\qquad\qquad\qquad\qquad y \geq 2$

**63.** The graph is in the first quadrant, so $x > 0$ and $y > 0$. A circle with radius 2 centered at the origin has the equation $x^2 + y^2 = 4$. Since the region includes and is inside the circle, the inequality is $x^2 + y^2 \leq 4$. A line that passes through $(0, -1)$ and $(2, 2)$ has slope $m = \frac{2-(-1)}{2-0} = \frac{3}{2}$ and equation $y = \frac{3}{2}x - 1$. Since the region is above the line, but does not include the line, the inequality is $y > \frac{3}{2}x - 1$.

The system is
$$x > 0$$
$$y > 0$$
$$x^2 + y^2 \leq 4$$
$$y > \frac{3}{2}x - 1$$

**65.** $3x + 2y \geq 6$

Solving the inequality for $y$ we have
$$3x + 2y \geq 6 \Rightarrow 2y \geq -3x + 6 \Rightarrow y \geq -\frac{3}{2}x + 3.$$

Enter $Y_1 = (-3/2)x + 3$ and use a graphing calculator to shade the region above the line. Notice the icon to the left of $Y_1$.

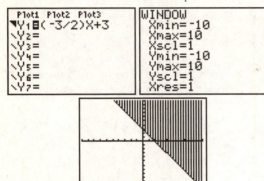

**67.** $x + y \geq 2$; $x + y \leq 6$

Solving each inequality for $y$, we have $y = -x + 2$ and $y = -x + 6$. Enter $Y_1 = -x + 2$ and $Y_2 = -x + 6$ and use the graphing calculator to shade the region above the graph of $y_1 = -x + 2$ and below the graph of $y_2 = -x + 6$.

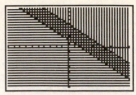

**69.**

| Point | Value of $3x + 5y$ | |
|-------|---------------------|---|
| $(1,1)$ | $3(1) + 5(1) = 8$ | ← Minimum |
| $(2,7)$ | $3(2) + 5(7) = 41$ | |
| $(5,10)$ | $3(5) + 5(10) = 65$ | ← Maximum |
| $(6,3)$ | $3(6) + 5(3) = 33$ | |

The maximum value is 65 at $(5,10)$. The minimum value is 8 at $(1, 1)$.

**71.**

| Point | Value of $3x + 5y$ | |
|-------|---------------------|---|
| $(1,0)$ | $3(1) + 5(0) = 3$ | ← Minimum |
| $(1,10)$ | $3(1) + 5(10) = 53$ | |
| $(7,9)$ | $3(7) + 5(9) = 66$ | ← Maximum |
| $(7,6)$ | $3(7) + 5(6) = 51$ | |

The maximum value is 66 at $(7,9)$. The minimum value is 3 at $(1,0)$.

**73.**

| Point | Value of $10y$ | |
|-------|----------------|---|
| $(1,0)$ | $10(0) = 0$ | ← Minimum |
| $(1,10)$ | $10(10) = 100$ | ← Maximum |
| $(7,9)$ | $10(9) = 90$ | |
| $(7,6)$ | $10(6) = 60$ | |

The maximum value is 100 at $(1,10)$. The minimum value is 0 at $(1,0)$.

**75.** Let $x$ = the number of Brand X pills; $y$ = the number of Brand Y pills.
The following table is helpful in organizing the information.

| | Number of Brand X pills ($x$) | Number of Brand Y pills ($y$) | Restrictions |
|---|---|---|---|
| Vitamin A | 3000 | 1000 | At least 6000 |
| Vitamin C | 45 | 50 | At least 195 |
| Vitamin D | 75 | 200 | At least 600 |

We have $3000x + 1000y \geq 6000$
$$45x + 50y \geq 195$$
$$75x + 200y \geq 600$$
$$x \geq 0, \ y \geq 0.$$

Graph $3000x + 1000y = 6000$ as a solid line with $x$-intercept 2 and $y$-intercept 6. Shade the region above the line. Graph $45x + 50y = 195$ as a solid line with $x$-intercept $4.\overline{3}$ and $y$-intercept 3.9. Shade the region above the line. Graph $75x + 200y = 600$ as a solid line with $x$-intercept 8 and $y$-intercept 3. Shade the region above the line. Graph $x = 0$ (the $y$-axis) as a solid line and shade the region to the right of it. Graph $y = 0$ (the $x$-axis) as a solid line and shade the region above it. The region of feasible solutions is the intersection of these five regions.

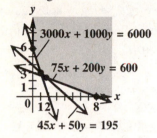

**77.** (a) Let $x$ = number of cartons of food; $y$ = number of cartons of clothing. The following table is helpful in organizing the information.

|  | Number of cartons of food ($x$) | Number of cartons of clothing ($y$) | Restrictions |
|---|---|---|---|
| Weight | 40 | 10 | Cannot Exceed 16,000 |
| Volume | 20 | 30 | No more than 18,000 |

We have $40x + 10y \leq 16,000$
$$20x + 30y \leq 18,000$$
$$x \geq 0, \ y \geq 0.$$

Maximize objective function, number of people $= 10x + 8y$.

Find the region of feasible solutions by graphing the system of inequalities that is made up of the constraints. To graph $40x + 10y \leq 16,000$, draw the line with $x$-intercept 400 and $y$-intercept 1600 as a solid line. Because the test point $(0,0)$ satisfies this inequality, shade the region *below* the line.

To graph $20x + 30y \leq 18,000$, draw the line with $x$-intercept 900 and $y$-intercept 600 as a solid line. Because the test point $(0,0)$ satisfies this inequality, shade the region *below* the line. The constraints $x \geq 0$ and $y \geq 0$ restrict the graph to the first quadrant. The graph of the feasible region is the intersection of the regions that are the graphs of the individual constraints.

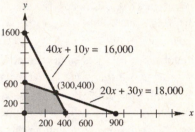

From the graph, observe that three are the vertices are $(0,0), (0,600),$ and $(400,0).$

The fourth vertex is the intersection point of the lines $40x + 10y = 16,000$ and $20x + 30y = 18,000.$ To find this point, solve the system $\begin{aligned} 40x + 10y &= 16,000 \\ 20x + 30y &= 18,000. \end{aligned}$

The first equation can be written as $4x + y = 1600 \Rightarrow y = 1600 - 4x.$

Substituting this equation into $20x + 30y = 18,000,$ we have

$$20x + 30(1600 - 4x) = 18,000$$
$$20x + 48,000 - 120x = 18,000$$
$$48,000 - 100x = 18,000$$
$$-100x = -30,000$$
$$x = 300$$

Substituting $x = 300$ into $y = 1600 - 4x,$ we have

$$y = 1600 - 4(300) = 1600 - 1200 = 400.$$

Thus, the fourth vertex is $(300, 400).$

Next, evaluate the objective function at each vertex.

| Point | Number of people $= 10x + 8y$ | |
|---|---|---|
| $(0,0)$ | $10(0) + 8(0) = 0$ | |
| $(0,600)$ | $10(0) + 8(600) = 4800$ | |
| $(300,400)$ | $10(300) + 8(400) = 6200$ | ← Maximum |
| $(400,0)$ | $10(400) + 8(0) = 4000$ | |

*(continued on next page)*

*(continued from page 301)*

The maximum value of $10x + 8y$ occurs at $(300, 400)$, so they should send 300 cartons of food and 400 cartons of clothes to maximize the number of people helped.

**(b)** The maximum number of people helped is 6200.

**79.** Let $x$ = number of cabinet A; $y$ = number of cabinet B. The cost constraint is $10x + 20y \leq 140$. The space constraint is $6x + 8y \leq 72$. Since the numbers of cabinets cannot be negative, we also have $x \geq 0$ and $y \geq 0$. We want to maximize the objective function, storage capacity $= 8x + 12y$. Find the region of feasible solutions by graphing the system of inequalities that is made up of the constraints. To graph $10x + 20y \leq 140$, draw the line with x-intercept 14 and y-intercept 7 as a solid line. Because the test point $(0, 0)$ satisfies this inequality, shade the region below the line. To graph $6x + 8y \leq 72$, draw the line with x-intercept 12 and y-intercept 9 as a solid line. Because the test point $(0, 0)$ satisfies this inequality, shade the region below the line. The constraints $x \geq 0$ and $y \geq 0$ restrict the graph to the first quadrant. The graph of the feasible region is the intersection of the regions that are the graphs of the individual constraints.

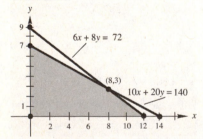

From the graph, observe that three vertices are $(0, 0), (0, 7),$ and $(12, 0)$. The fourth vertex is the intersection point of the lines $10x + 20y = 140$ and $6x + 8y = 72$. To find this point, solve the system $10x + 20y = 140$
$6x + 8y = 72$.

The first equation can be written as $x + 2y = 14 \Rightarrow x = 14 - 2y$. Substituting this equation into $6x + 8y = 72$, we have

$6(14 - 2y) + 8y = 72 \Rightarrow 84 - 12y + 8y = 72 \Rightarrow$
$84 - 4y = 72 \Rightarrow -4y = -12 \Rightarrow y = 3$

Substituting $y = 3$ into $x = 14 - 2y$, we have $x = 14 - 2(3) = 14 - 6 = 8$. Thus, the fourth vertex is $(8, 3)$. Next, evaluate the objective function at each vertex.

| Point | Storage Capacity $= 8x + 12y$ | |
|---|---|---|
| $(0, 0)$ | $8(0) + 12(0) = 0$ | |
| $(0, 7)$ | $8(0) + 12(7) = 84$ | |
| $(8, 3)$ | $8(8) + 12(3) = 100$ | ← Maximum |
| $(12, 0)$ | $8(12) + 12(0) = 96$ | |

The maximum value of $8x + 12y$ occurs at $(8, 3)$, so the office manager should buy 8 of cabinet A and 3 of cabinet B for a total storage capacity of 100 ft$^3$.

**81.** Let $x$ = number of servings of product A; $y$ = number of servings of product B. The Supplement I constraint is $3x + 2y \geq 15$. The Supplement II constraint is $2x + 4y \geq 15$. Since the numbers of servings cannot be negative, we also have $x \geq 0$ and $y \geq 0$. We want to minimize the objective function, cost $= .25x + .40y$. Find the region of feasible solutions by graphing the system of inequalities that is made up of the constraints. To graph $3x + 2y \geq 15$, draw the line with x-intercept 5 and y-intercept $\frac{15}{2} = 7\frac{1}{2}$ as a solid line. Because the test point $(0, 0)$ does not satisfy this inequality, shade the region *above* the line. To graph $2x + 4y \geq 15$, draw the line with x-intercept $\frac{15}{2} = 7\frac{1}{2}$ and y-intercept $\frac{15}{4} = 3\frac{3}{4}$ as a solid line. Because the test point $(0, 0)$ does not satisfy this inequality, shade the region *above* the line.

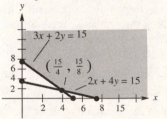

The constraints $x \geq 0$ and $y \geq 0$ restrict the graph to the first quadrant. The graph of the feasible region is the intersection of the regions that are the graphs of the individual constraints.

From the graph, observe that two vertices are $\left(0, \frac{15}{2}\right)$ and $\left(\frac{15}{2}, 0\right)$. The third vertex is the intersection point of the lines $3x + 2y = 15$ and $2x + 4y = 15$. To find this point, solve the system $3x + 2y = 15$
$\qquad\qquad 2x + 4y = 15$.

Multiply the first equation by $-2$ and add it to the second equation.

$$-6x - 4y = -30$$
$$\underline{\phantom{-6x} 2x + 4y = \phantom{-}15}$$
$$-4x \phantom{+4y} = -15 \Rightarrow x = \frac{15}{4}$$

Substituting this equation into $2x + 4y = 15$, we have $2\left(\frac{15}{4}\right) + 4y = 15 \Rightarrow \frac{15}{2} + 4y = 15 \Rightarrow$

$15 + 8y = 30 \Rightarrow 8y = 15 \Rightarrow y = \frac{15}{8}$. Thus, the

third vertex is $\left(\frac{15}{4}, \frac{15}{8}\right)$. Next, evaluate the objective function at each vertex.

| Point | Cost $= .25x + .40y$ |
|---|---|
| $\left(0, \frac{15}{2}\right)$ | $.25(0) + .40\left(\frac{15}{2}\right) = 3.00$ |
| $\left(\frac{15}{4}, \frac{15}{8}\right)$ | $.25\left(\frac{15}{4}\right) + .40\left(\frac{15}{8}\right) = 1.69$  ← Minimum |
| $\left(\frac{15}{2}, 0\right)$ | $.25\left(\frac{15}{2}\right) + .40(0) = 1.88$ |

The minimum cost is \$1.69 for $\frac{15}{4} = 3\frac{3}{4}$

servings of $A$ and $\frac{15}{8} = 1\frac{7}{8}$ servings of $B$.   .

## Section 5.7: Properties of Matrices

1. $\begin{bmatrix} -3 & a \\ b & 5 \end{bmatrix} = \begin{bmatrix} c & 0 \\ 4 & d \end{bmatrix}$

Since corresponding elements are equal, we have the following.
$a = 0, b = 4, c = -3, d = 5$

3. $\begin{bmatrix} x+2 & y-6 \\ z-3 & w+5 \end{bmatrix} = \begin{bmatrix} -2 & 8 \\ 0 & 3 \end{bmatrix}$

Since corresponding elements are equal, we have the following.
$x + 2 = -2 \Rightarrow x = -4$
$y - 6 = 8 \Rightarrow y = 14$
$z - 3 = 0 \Rightarrow z = 3$
$w + 5 = 3 \Rightarrow w = -2$

5. $\begin{bmatrix} 0 & 5 & x \\ -1 & 3 & y+2 \\ 4 & 1 & z \end{bmatrix} = \begin{bmatrix} 0 & w+3 & 6 \\ -1 & 3 & 0 \\ 4 & 1 & 8 \end{bmatrix}$

Since corresponding elements are equal, we have the following.
$x = 6$
$y + 2 = 0 \Rightarrow y = -2$
$z = 8$
$w + 3 = 5 \Rightarrow w = 2$

7. Since $\begin{bmatrix} x & y & z \end{bmatrix}$ is a $1 \times 3$ matrix and $\begin{bmatrix} 21 & 6 \end{bmatrix}$ is a $1 \times 2$ matrix, the statement cannot be true, hence we cannot find values of $x$, $y$, and $z$.

9. $\begin{bmatrix} -7+z & 4r & 8s \\ 6p & 2 & 5 \end{bmatrix} + \begin{bmatrix} -9 & 8r & 3 \\ 2 & 5 & 4 \end{bmatrix}$

$$= \begin{bmatrix} 2 & 36 & 27 \\ 20 & 7 & 12a \end{bmatrix} \Rightarrow$$

$\begin{bmatrix} -16+z & 12r & 8s+3 \\ 6p+2 & 7 & 9 \end{bmatrix} = \begin{bmatrix} 2 & 36 & 27 \\ 20 & 7 & 12a \end{bmatrix}$

Since corresponding elements are equal, we have
$-16 + z = 2 \Rightarrow z = 18$, $12r = 36 \Rightarrow r = 3$,
$8s + 3 = 27 \Rightarrow s = 3$, $6p + 2 = 20 \Rightarrow p = 3$,
$9 = 12a \Rightarrow a = \frac{3}{4}$

Thus, $z = 18$, $r = 3$, $s = 3$, $p = 3$, and $a = \frac{3}{4}$.

11. Two matrices are equal if they have the same <u>size</u> and if corresponding elements are <u>equal</u>.

13. $\begin{bmatrix} -4 & 8 \\ 2 & 3 \end{bmatrix}$

This matrix has 2 rows and 2 columns, so it is a $2 \times 2$ square matrix.

15. $\begin{bmatrix} -6 & 8 & 0 & 0 \\ 4 & 1 & 9 & 2 \\ 3 & -5 & 7 & 1 \end{bmatrix}$

This matrix has 3 rows and 4 columns, so it is a $3 \times 4$ matrix.

17. $\begin{bmatrix} 2 \\ 4 \end{bmatrix}$

This matrix has 2 rows and 1 column, so it is a $2 \times 1$ column matrix.

19. Answers will vary.

**21.** $\begin{bmatrix} -4 & 3 \\ 12 & -6 \end{bmatrix} + \begin{bmatrix} 2 & -8 \\ 5 & 10 \end{bmatrix} = \begin{bmatrix} -4+2 & 3+(-8) \\ 12+5 & -6+10 \end{bmatrix}$

$= \begin{bmatrix} -2 & -5 \\ 17 & 4 \end{bmatrix}$

**23.** $\begin{bmatrix} 6 & -9 & 2 \\ 4 & 1 & 3 \end{bmatrix} + \begin{bmatrix} -8 & 2 & 5 \\ 6 & -3 & 4 \end{bmatrix}$

$= \begin{bmatrix} 6+(-8) & -9+2 & 2+5 \\ 4+6 & 1+(-3) & 3+4 \end{bmatrix}$

$= \begin{bmatrix} -2 & -7 & 7 \\ 10 & -2 & 7 \end{bmatrix}$

**25.** Since $\begin{bmatrix} 2 & 4 & 6 \end{bmatrix}$ is a $1 \times 3$ matrix and $\begin{bmatrix} -2 \\ -4 \\ -6 \end{bmatrix}$ is

a $3 \times 1$ matrix, $\begin{bmatrix} 2 & 4 & 6 \end{bmatrix} + \begin{bmatrix} -2 \\ -4 \\ -6 \end{bmatrix}$ cannot be

added.

**27.** $\begin{bmatrix} -6 & 8 \\ 0 & 0 \end{bmatrix} - \begin{bmatrix} 0 & 0 \\ -4 & -2 \end{bmatrix} = \begin{bmatrix} -6-0 & 8-0 \\ 0-(-4) & 0-(-2) \end{bmatrix}$

$= \begin{bmatrix} -6 & 8 \\ 4 & 2 \end{bmatrix}$

**29.** $\begin{bmatrix} 12 \\ -1 \\ 3 \end{bmatrix} - \begin{bmatrix} 8 \\ 4 \\ -1 \end{bmatrix} = \begin{bmatrix} 12-8 \\ -1-4 \\ 3-(-1) \end{bmatrix} = \begin{bmatrix} 4 \\ -5 \\ 4 \end{bmatrix}$

**31.** Since $\begin{bmatrix} -4 & 3 \end{bmatrix}$ is a $1 \times 2$ matrix and

$\begin{bmatrix} 5 & 8 & 2 \end{bmatrix}$ is a $1 \times 3$ matrix,

$\begin{bmatrix} -4 & 3 \end{bmatrix} - \begin{bmatrix} 5 & 8 & 2 \end{bmatrix}$ cannot be subtracted.

**33.** $\begin{bmatrix} \sqrt{3} & -4 \\ 2 & -\sqrt{5} \\ -8 & \sqrt{8} \end{bmatrix} - \begin{bmatrix} 2\sqrt{3} & 9 \\ -2 & \sqrt{5} \\ -7 & 3\sqrt{2} \end{bmatrix}$

$= \begin{bmatrix} \sqrt{3}-2\sqrt{3} & -4-9 \\ 2-(-2) & -\sqrt{5}-\sqrt{5} \\ -8-(-7) & \sqrt{8}-3\sqrt{2} \end{bmatrix} = \begin{bmatrix} -\sqrt{3} & -13 \\ 4 & -2\sqrt{5} \\ -1 & -\sqrt{2} \end{bmatrix}$

**35.** $\begin{bmatrix} 3x+y & x-2y & 2x \\ 5x & 3y & x+y \end{bmatrix}$

$+ \begin{bmatrix} 2x & 3y & 5x+y \\ 3x+2y & x & 2x \end{bmatrix}$

$= \begin{bmatrix} (3x+y)+2x & (x-2y)+3y & 2x+(5x+y) \\ 5x+(3x+2y) & 3y+x & (x+y)+2x \end{bmatrix}$

$= \begin{bmatrix} 5x+y & x+y & 7x+y \\ 8x+2y & x+3y & 3x+y \end{bmatrix}$

In Exercises 37–39, $A = \begin{bmatrix} -2 & 4 \\ 0 & 3 \end{bmatrix}$ and $B = \begin{bmatrix} -6 & 2 \\ 4 & 0 \end{bmatrix}$.

**37.** $2A = 2 \begin{bmatrix} -2 & 4 \\ 0 & 3 \end{bmatrix} = \begin{bmatrix} 2(-2) & 2(4) \\ 2(0) & 2(3) \end{bmatrix} = \begin{bmatrix} -4 & 8 \\ 0 & 6 \end{bmatrix}$

**39.** $\frac{3}{2}B = \frac{3}{2} \begin{bmatrix} -6 & 2 \\ 4 & 0 \end{bmatrix} = \begin{bmatrix} \frac{3}{2}(-6) & \frac{3}{2}(2) \\ \frac{3}{2}(4) & \frac{3}{2}(0) \end{bmatrix} = \begin{bmatrix} -9 & 3 \\ 6 & 0 \end{bmatrix}$

**41.** $2A - B = 2 \begin{bmatrix} -2 & 4 \\ 0 & 3 \end{bmatrix} - \begin{bmatrix} -6 & 2 \\ 4 & 0 \end{bmatrix}$

$= \begin{bmatrix} -4 & 8 \\ 0 & 6 \end{bmatrix} - \begin{bmatrix} -6 & 2 \\ 4 & 0 \end{bmatrix}$

$= \begin{bmatrix} -4-(-6) & 8-2 \\ 0-4 & 6-0 \end{bmatrix} = \begin{bmatrix} 2 & 6 \\ -4 & 6 \end{bmatrix}$

**43.** $-A + \frac{1}{2}B = -\begin{bmatrix} -2 & 4 \\ 0 & 3 \end{bmatrix} + \frac{1}{2} \begin{bmatrix} -6 & 2 \\ 4 & 0 \end{bmatrix}$

$= \begin{bmatrix} 2 & -4 \\ 0 & -3 \end{bmatrix} + \begin{bmatrix} -3 & 1 \\ 2 & 0 \end{bmatrix}$

$= \begin{bmatrix} 2+(-3) & -4+1 \\ 0+2 & -3+0 \end{bmatrix} = \begin{bmatrix} -1 & -3 \\ 2 & -3 \end{bmatrix}$

**45.** $AB$ can be calculated and the result will be a $2 \times 5$ matrix.

Matrix $A$  Matrix $B$
$2 \times 3$      $3 \times 5$
matches
size of $AB$
$2 \times 5$

**47.** $BA$ cannot be calculated.

Matrix $B$  Matrix $A$
$3 \times 5$      $2 \times 3$
different

**49.** *BC* can be calculated and the result will be a $3 \times 2$ matrix.

Matrix *B*   Matrix *C*

$3 \times 5$ $\quad$ $5 \times 2$

matches

size of *CA*
$3 \times 2$

**51.** A $2 \times 2$ matrix multiplied by a $2 \times 1$ matrix results in a $2 \times 1$ matrix.

$$\begin{bmatrix} 1 & 2 \\ 3 & 4 \end{bmatrix}\begin{bmatrix} -1 \\ 7 \end{bmatrix} = \begin{bmatrix} 1(-1)+2(7) \\ 3(-1)+4(7) \end{bmatrix} = \begin{bmatrix} -1+14 \\ -3+28 \end{bmatrix} = \begin{bmatrix} 13 \\ 25 \end{bmatrix}$$

**53.** A $2 \times 3$ matrix multiplied by a $3 \times 1$ matrix results in a $2 \times 1$ matrix.

$$\begin{bmatrix} 3 & -4 & 1 \\ 5 & 0 & 2 \end{bmatrix}\begin{bmatrix} -1 \\ 4 \\ 2 \end{bmatrix} = \begin{bmatrix} 3(-1)+(-4)(4)+1(2) \\ 5(-1)+0(4)+2(2) \end{bmatrix} = \begin{bmatrix} -3+(-16)+2 \\ -5+0+4 \end{bmatrix} = \begin{bmatrix} -17 \\ -1 \end{bmatrix}$$

**55.** A $2 \times 3$ matrix multiplied by a $3 \times 2$ matrix results in a $2 \times 2$ matrix.

$$\begin{bmatrix} \sqrt{2} & \sqrt{2} & -\sqrt{18} \\ \sqrt{3} & \sqrt{27} & 0 \end{bmatrix}\begin{bmatrix} 8 & -10 \\ 9 & 12 \\ 0 & 2 \end{bmatrix} = \begin{bmatrix} \sqrt{2} & \sqrt{2} & -3\sqrt{2} \\ \sqrt{3} & 3\sqrt{3} & 0 \end{bmatrix}\begin{bmatrix} 8 & -10 \\ 9 & 12 \\ 0 & 2 \end{bmatrix}$$

$$= \begin{bmatrix} 8\sqrt{2}+9\sqrt{2} & -10\sqrt{2}+12\sqrt{2} \\ \quad -3\sqrt{2}(0) & \quad -3\sqrt{2}(2) \\ 8\sqrt{3}+3\sqrt{3}(9) & -10\sqrt{3}+3\sqrt{3}(12) \\ \quad +0(0) & \quad +0(2) \end{bmatrix} = \begin{bmatrix} 17\sqrt{2} & -4\sqrt{2} \\ 35\sqrt{3} & 26\sqrt{3} \end{bmatrix}$$

**57.** A $2 \times 2$ matrix multiplied by a $2 \times 2$ matrix results in a $2 \times 2$ matrix.

$$\begin{bmatrix} \sqrt{3} & 1 \\ 2\sqrt{5} & 3\sqrt{2} \end{bmatrix}\begin{bmatrix} \sqrt{3} & -\sqrt{6} \\ 4\sqrt{3} & 0 \end{bmatrix} = \begin{bmatrix} \sqrt{3}(\sqrt{3})+1(4\sqrt{3}) & \sqrt{3}(-\sqrt{6})+1(0) \\ 2\sqrt{5}(\sqrt{3})+3\sqrt{2}(4\sqrt{3}) & 2\sqrt{5}(-\sqrt{6})+3\sqrt{2}(0) \end{bmatrix}$$

$$= \begin{bmatrix} 3+4\sqrt{3} & -\sqrt{18} \\ 2\sqrt{15}+12\sqrt{6} & -2\sqrt{30} \end{bmatrix} = \begin{bmatrix} 3+4\sqrt{3} & -3\sqrt{2} \\ 2\sqrt{15}+12\sqrt{6} & -2\sqrt{30} \end{bmatrix}$$

**59.** $$\begin{bmatrix} -3 & 0 & 2 & 1 \\ 4 & 0 & 2 & 6 \end{bmatrix}\begin{bmatrix} -4 & 2 \\ 0 & 1 \end{bmatrix}$$

A $2 \times 4$ matrix cannot be multiplied by a $2 \times 2$ matrix because the number of columns of the first matrix (four) is not equal to the number of rows of the second matrix (two).

**61.** A $1 \times 3$ matrix multiplied by a $3 \times 3$ matrix results in a $1 \times 3$ matrix.

$$\begin{bmatrix} -2 & 4 & 1 \end{bmatrix}\begin{bmatrix} 3 & -2 & 4 \\ 2 & 1 & 0 \\ 0 & -1 & 4 \end{bmatrix} = \begin{bmatrix} -2(3)+4(2)+1(0) & -2(-2)+4(1)+1(-1) & -2(4)+4(0)+1(4) \end{bmatrix}$$

$$= \begin{bmatrix} -6+8+0 & 4+4+(-1) & -8+0+4 \end{bmatrix} = \begin{bmatrix} 2 & 7 & -4 \end{bmatrix}$$

**63.** A $3\times3$ matrix multiplied by a $3\times3$ matrix results in a $3\times3$ matrix.

$$\begin{bmatrix} -2 & -3 & -4 \\ 2 & -1 & 0 \\ 4 & -2 & 3 \end{bmatrix}\begin{bmatrix} 0 & 1 & 4 \\ 1 & 2 & -1 \\ 3 & 2 & -2 \end{bmatrix}$$

$$=\begin{bmatrix} -2(0)+(-3)(1)+(-4)(3) & -2(1)+(-3)(2)+(-4)(2) & -2(4)+(-3)(-1)+(-4)(-2) \\ 2(0)+(-1)(1)+0(3) & 2(1)+(-1)(2)+0(2) & 2(4)+(-1)(-1)+0(-2) \\ 4(0)+(-2)(1)+3(3) & 4(1)+(-2)(2)+3(2) & 4(4)+(-2)(-1)+3(-2) \end{bmatrix}$$

$$=\begin{bmatrix} 0+(-3)+(-12) & -2+(-6)+(-8) & -8+3+8 \\ 0+(-1)+0 & 2+(-2)+0 & 8+1+0 \\ 0+(-2)+9 & 4+(-4)+6 & 16+2+(-6) \end{bmatrix}=\begin{bmatrix} -15 & -16 & 3 \\ -1 & 0 & 9 \\ 7 & 6 & 12 \end{bmatrix}$$

In Exercises 65–71, $A=\begin{bmatrix} 4 & -2 \\ 3 & 1 \end{bmatrix}$, $B=\begin{bmatrix} 5 & 1 \\ 0 & -2 \\ 3 & 7 \end{bmatrix}$ and $C=\begin{bmatrix} -5 & 4 & 1 \\ 0 & 3 & 6 \end{bmatrix}$.

**65.** A $3\times2$ matrix multiplied by a $2\times2$ matrix results in a $3\times2$ matrix.

$$BA=\begin{bmatrix} 5 & 1 \\ 0 & -2 \\ 3 & 7 \end{bmatrix}\begin{bmatrix} 4 & -2 \\ 3 & 1 \end{bmatrix}=\begin{bmatrix} 5(4)+1(3) & 5(-2)+1(1) \\ 0(4)+(-2)(3) & 0(-2)+(-2)(1) \\ 3(4)+7(3) & 3(-2)+7(1) \end{bmatrix}=\begin{bmatrix} 20+3 & -10+1 \\ 0+(-6) & 0+(-2) \\ 12+21 & -6+7 \end{bmatrix}=\begin{bmatrix} 23 & -9 \\ -6 & -2 \\ 33 & 1 \end{bmatrix}$$

**67.** A $3\times2$ matrix multiplied by a $2\times3$ matrix results in a $3\times3$ matrix.

$$BC=\begin{bmatrix} 5 & 1 \\ 0 & -2 \\ 3 & 7 \end{bmatrix}\begin{bmatrix} -5 & 4 & 1 \\ 0 & 3 & 6 \end{bmatrix}=\begin{bmatrix} 5(-5)+1(0) & 5(4)+1(3) & 5(1)+1(6) \\ 0(-5)+(-2)(0) & 0(4)+(-2)(3) & 0(1)+(-2)(6) \\ 3(-5)+7(0) & 3(4)+7(3) & 3(1)+7(6) \end{bmatrix}$$

$$=\begin{bmatrix} -25+0 & 20+3 & 5+6 \\ 0+0 & 0+(-6) & 0+(-12) \\ -15+0 & 12+21 & 3+42 \end{bmatrix}=\begin{bmatrix} -25 & 23 & 11 \\ 0 & -6 & -12 \\ -15 & 33 & 45 \end{bmatrix}$$

**69.** $AB=\begin{bmatrix} 4 & -2 \\ 3 & 1 \end{bmatrix}\begin{bmatrix} 5 & 1 \\ 0 & -2 \\ 3 & 7 \end{bmatrix}$ is the product of a $2\times2$ matrix multiplied by a $3\times2$ matrix, which is not possible.

**71.** Since $A^2=AA$, we are finding the product of two $2\times2$ matrices, which results in a $2\times2$ matrix.

$$A^2=\begin{bmatrix} 4 & -2 \\ 3 & 1 \end{bmatrix}\begin{bmatrix} 4 & -2 \\ 3 & 1 \end{bmatrix}=\begin{bmatrix} 4(4)+(-2)(3) & 4(-2)+(-2)(1) \\ 3(4)+1(3) & 3(-2)+1(1) \end{bmatrix}=\begin{bmatrix} 16+(-6) & -8+(-2) \\ 12+3 & -6+1 \end{bmatrix}=\begin{bmatrix} 10 & -10 \\ 15 & -5 \end{bmatrix}$$

**73.** Since the answers to 65 and 69 are not equal, $BA\neq AB$.
  Since the answers to 67 and 68 are not equal, $BC\neq CB$.
  Since the answers to 66 and 70 are not equal, $AC\neq CA$.
  No, matrix multiplication is not commutative.

**75.** $A=\begin{bmatrix} 3 & 4 \\ -2 & 1 \end{bmatrix}$ and $B=\begin{bmatrix} 6 & 0 \\ 5 & -2 \end{bmatrix}$

  **(a)** $AB=\begin{bmatrix} 3 & 4 \\ -2 & 1 \end{bmatrix}\begin{bmatrix} 6 & 0 \\ 5 & -2 \end{bmatrix}=\begin{bmatrix} 3(6)+4(5) & 3(0)+4(-2) \\ -2(6)+1(5) & -2(0)+1(-2) \end{bmatrix}=\begin{bmatrix} 18+20 & 0+(-8) \\ -12+5 & 0+(-2) \end{bmatrix}=\begin{bmatrix} 38 & -8 \\ -7 & -2 \end{bmatrix}$

**(b)** $BA = \begin{bmatrix} 6 & 0 \\ 5 & -2 \end{bmatrix} \begin{bmatrix} 3 & 4 \\ -2 & 1 \end{bmatrix} = \begin{bmatrix} 6(3)+0(-2) & 6(4)+0(1) \\ 5(3)+(-2)(-2) & 5(4)+(-2)(1) \end{bmatrix} = \begin{bmatrix} 18+0 & 24+0 \\ 15+4 & 20+(-2) \end{bmatrix} = \begin{bmatrix} 18 & 24 \\ 19 & 18 \end{bmatrix}$

Note: $AB \neq BA$

**77.** $A = \begin{bmatrix} 0 & 1 & -1 \\ 0 & 1 & 0 \\ 0 & 0 & 1 \end{bmatrix}$ and $B = \begin{bmatrix} 1 & 0 & 0 \\ 0 & 1 & 0 \\ 0 & 0 & 1 \end{bmatrix}$

**(a)** $AB = \begin{bmatrix} 0 & 1 & -1 \\ 0 & 1 & 0 \\ 0 & 0 & 1 \end{bmatrix}\begin{bmatrix} 1 & 0 & 0 \\ 0 & 1 & 0 \\ 0 & 0 & 1 \end{bmatrix} = \begin{bmatrix} 0(1)+1(0)+(-1)(0) & 0(0)+1(1)+(-1)(0) & 0(0)+1(0)+(-1)(1) \\ 0(1)+1(0)+0(0) & 0(0)+1(1)+0(0) & 0(0)+1(0)+0(1) \\ 0(1)+0(0)+1(0) & 0(0)+0(1)+1(0) & 0(0)+0(0)+1(1) \end{bmatrix}$

$= \begin{bmatrix} 0+0+0 & 0+1+0 & 0+0+(-1) \\ 0+0+0 & 0+1+0 & 0+0+0 \\ 0+0+0 & 0+0+0 & 0+0+1 \end{bmatrix} = \begin{bmatrix} 0 & 1 & -1 \\ 0 & 1 & 0 \\ 0 & 0 & 1 \end{bmatrix}$

**(b)** $BA = \begin{bmatrix} 1 & 0 & 0 \\ 0 & 1 & 0 \\ 0 & 0 & 1 \end{bmatrix}\begin{bmatrix} 0 & 1 & -1 \\ 0 & 1 & 0 \\ 0 & 0 & 1 \end{bmatrix} = \begin{bmatrix} 1(0)+0(0)+0(0) & 1(1)+0(1)+0(0) & 1(-1)+0(0)+0(1) \\ 0(0)+1(0)+0(0) & 0(1)+1(1)+0(0) & 0(-1)+1(0)+0(1) \\ 0(0)+0(0)+1(0) & 0(1)+0(1)+1(0) & 0(-1)+0(0)+1(1) \end{bmatrix}$

$= \begin{bmatrix} 0+0+0 & 1+0+0 & -1+0+0 \\ 0+0+0 & 0+1+0 & 0+0+0 \\ 0+0+0 & 0+0+0 & 0+0+1 \end{bmatrix} = \begin{bmatrix} 0 & 1 & -1 \\ 0 & 1 & 0 \\ 0 & 0 & 1 \end{bmatrix}$

Note: $AB = BA = A$

**79.** In Exercise 77, $AB = A$ and $BA = A$. For this pair of matrices, B acts in the same way for multiplication as the number 1 acts for multiplication of real numbers.

**81. (a)** The sales figure information may be written as the $3 \times 3$ matrix where column 1 represents nonfat, column 2 represents regular, and column 3 represents supercreamy.

$\begin{matrix} \text{Location I} \\ \text{Location II} \\ \text{Location III} \end{matrix} \begin{bmatrix} 50 & 100 & 30 \\ 10 & 90 & 50 \\ 60 & 120 & 40 \end{bmatrix}.$

**(b)** The income per gallon information may be written as the $3 \times 1$ matrix $\begin{bmatrix} 12 \\ 10 \\ 15 \end{bmatrix}$.

Note: If the matrix in part (a) had been written with its rows and columns interchanged, then this income per gallon information would be written instead as a $1 \times 3$ matrix.

**(c)** $\begin{bmatrix} 50 & 100 & 30 \\ 10 & 90 & 50 \\ 60 & 120 & 40 \end{bmatrix}\begin{bmatrix} 12 \\ 10 \\ 15 \end{bmatrix} = \begin{bmatrix} 2050 \\ 1770 \\ 2520 \end{bmatrix}$

Note: This result may be written as a $1 \times 3$ matrix instead.

**(d)** $2050 + 1770 + 2520 = 6340$; The total daily income from the three locations is $6340.

**83.** Answers will vary depending on when numbers are rounded.

**(a)** $j_1 = 690, s_p = 210, a_1 = 2100$

1st year:

$j_2 = .33a_1 = .33(2100) = 693$
$s_2 = .18j_1 = .18(690) \approx 124.2 \approx 124$
$a_2 = .71s_1 + .94a_1$
$\quad = .71(210) + .94(2100)$
$\quad \approx 2123.1 \approx 2123$
$j_2 + s_2 + a_2 = 693 + 124 + 2123 = 2940$

2nd year:

$j_3 = .33a_2 = .33(2123) \approx 700.6$
$s_3 = .18j_2 = .18(693) \approx 124.7$
$a_3 = .71s_2 + .94a_2$
$\quad = .71(124) + .94(2123) \approx 2083.7$
$j_3 + s_3 + a_3 = 700.6 + 124.7 + 2083.7 = 2909$

*(continued on next page)*

(*continued from page 307*

3rd year:

$$j_4 = .33a_3 = .33(2084) \approx 687$$
$$s_4 = .18j_3 = .18(700) = 126$$
$$a_4 = .72s_3 + .94a_3$$
$$= .71(125) + .94(2084) = 2048$$
$$j_3 + s_3 + a_3 = 687 + 126 + 2048 = 2861$$

4th year:

$$j_5 = .33a_4 = .33(2048) \approx 676$$
$$s_5 = .18j_4 = .18(687) \approx 124$$
$$a_5 = .72s_4 + .94a_4$$
$$= .71(126) + .94(2048) = 2014$$
$$j_5 + s_5 + a_5 = 686 + 124 + 2048 = 2814$$

5th year:

$$j_6 = .33a_5 = .33(2014) \approx 664$$
$$s_6 = .18j_5 = .18(676) \approx 122$$
$$a_6 = .72s_5 + .94a_5$$
$$= .71(124) + .94(2014) \approx 1981$$
$$j_6 + s_6 + a_6 = 664 + 122 + 1981 = 2767$$

**(b)** The northern spotted owl will become extinct.

**(c)** $j_1 = 690,\ s_1 = 210,\ a_1 = 2100$

1st year:

$$j_2 = .33a_1 = .33(2100) = 693$$
$$s_2 = .3j_1 = .3(690) = 207$$
$$a_2 = .71s_1 + .94a_1$$
$$= .71(210) + .94(2100) \approx 2123$$
$$j_2 + s_2 + a_2 = 693 + 207 + 2123 = 3023$$

2nd year:

$$j_3 = .33a_2 = .33(2123) \approx 701$$
$$s_3 = .3j_2 = .3(693) \approx 208$$
$$a_3 = .71s_2 + .94a_2$$
$$= .71(207) = .94(2123) \approx 2143$$
$$j_3 + s_3 + a_3 = 701 + 208 + 2143 = 3052$$

3rd year:

$$j_4 = .33a_3 = .33(2143) \approx 707$$
$$s_4 = .3j_3 = .3(701) = 210$$
$$a_4 = .71s_3 + .94a_3$$
$$= .71(208) + .94(2143) \approx 2162$$
$$j_4 + s_4 + a_4 = 707 + 210 + 2162 = 3079$$

4th year:

$$j_5 = .33a_4 = .33(2162) \approx 714$$
$$s_5 = .3j_4 = .3(707) \approx 212$$
$$a_5 = .72s_4 + .94a_4$$
$$= .71(210) + .94(2162) \approx 2181$$
$$j_5 + s_5 + a_5 = 714 + 212 + 2181 = 3107$$

5th year:

$$j_6 = .71s_5 = .33a_5$$
$$= .33(2181) \approx 720$$
$$s_6 = .3j_5 = .3(714) \approx 214$$
$$a_6 = .71s_5 + .94a_5$$
$$= .71(212) + .94(2181) = 2201$$
$$j_6 + s_6 + a_6 = 720 + 214 + 2201 = 3135$$

**85.** Expanding $\begin{vmatrix} -x & 0 & .33 \\ .18 & -x & 0 \\ 0 & .71 & .94 - x \end{vmatrix}$ by row one, we have the following.

$$-x\begin{vmatrix} -x & 0 \\ .71 & .94 - x \end{vmatrix} - 0 + .33\begin{vmatrix} .18 & -x \\ 0 & .71 \end{vmatrix} = -x\left[-x(.94 - x) - .71(0)\right] + .33\left[(.18)(.71) - 0(-x)\right]$$

$$= -x\left(-.94x + x^2 - 0\right) + .33(.1278 - 0) = .94x^2 - x^3 + .042174$$

Thus, the polynomial is $-x^3 + .94x^2 + .042174$. Evaluating this polynomial with $x = .98359$, we have

$$-(.98359)^3 + .94(.98359)^2 + .042174 \approx .0000029.$$ Thus $.98359$ is an approximate zero.

In Exercises 87–93, $A = \begin{bmatrix} a_{11} & a_{12} \\ a_{21} & a_{22} \end{bmatrix}$, $B = \begin{bmatrix} b_{11} & b_{12} \\ b_{21} & b_{22} \end{bmatrix}$, and $C = \begin{bmatrix} c_{11} & c_{12} \\ c_{21} & c_{22} \end{bmatrix}$.

**87.** $A + B = B + A$

$$A + B = \begin{bmatrix} a_{11} & a_{12} \\ a_{21} & a_{22} \end{bmatrix} + \begin{bmatrix} b_{11} & b_{12} \\ b_{21} & b_{22} \end{bmatrix} = \begin{bmatrix} a_{11} + b_{11} & a_{12} + b_{12} \\ a_{21} + b_{21} & a_{22} + b_{22} \end{bmatrix} = \begin{bmatrix} b_{11} + a_{11} & b_{12} + a_{12} \\ b_{21} + a_{21} & b_{22} + a_{22} \end{bmatrix} = B + A$$

**89.** $(AB)C = \left(\begin{bmatrix} a_{11} & a_{12} \\ a_{21} & a_{22} \end{bmatrix} \begin{bmatrix} b_{11} & b_{12} \\ b_{21} & b_{22} \end{bmatrix}\right) C = \begin{bmatrix} a_{11}b_{11} + a_{12}b_{21} & a_{11}b_{12} + a_{12}b_{22} \\ a_{21}b_{11} + a_{22}b_{21} & a_{21}b_{12} + a_{22}b_{22} \end{bmatrix} \begin{bmatrix} c_{11} & c_{12} \\ c_{21} & c_{22} \end{bmatrix}$

$= \begin{bmatrix} (a_{11}b_{11} + a_{12}b_{21})c_{11} + (a_{11}b_{12} + a_{12}b_{22})c_{21} & (a_{11}b_{11} + a_{12}b_{21})c_{12} + (a_{11}b_{12} + a_{12}b_{22})c_{22} \\ (a_{21}b_{11} + a_{22}b_{21})c_{11} + (a_{21}b_{12} + a_{22}b_{22})c_{21} & (a_{21}b_{11} + a_{22}b_{21})c_{12} + (a_{21}b_{12} + a_{22}b_{22})c_{22} \end{bmatrix}$

$= \begin{bmatrix} a_{11}b_{11}c_{11} + a_{12}b_{21}c_{11} + a_{11}b_{12}c_{21} + a_{12}b_{22}c_{21} & a_{11}b_{11}c_{12} + a_{12}b_{21}c_{12} + a_{11}b_{12}c_{22} + a_{12}b_{22}c_{22} \\ a_{21}b_{11}c_{11} + a_{22}b_{21}c_{11} + a_{21}b_{12}c_{21} + a_{22}b_{22}c_{21} & a_{21}b_{11}c_{12} + a_{22}b_{21}c_{12} + a_{21}b_{12}c_{22} + a_{22}b_{22}c_{22} \end{bmatrix}$

$A(BC) = A\left(\begin{bmatrix} b_{11} & b_{12} \\ b_{21} & b_{22} \end{bmatrix} \begin{bmatrix} c_{11} & c_{12} \\ c_{21} & c_{22} \end{bmatrix}\right) = \begin{bmatrix} a_{11} & a_{12} \\ a_{21} & a_{22} \end{bmatrix} \begin{bmatrix} b_{11}c_{11} + b_{12}c_{21} & b_{11}c_{12} + b_{12}c_{22} \\ b_{21}c_{11} + b_{22}c_{21} & b_{21}c_{12} + b_{22}c_{22} \end{bmatrix}$

$= \begin{bmatrix} a_{11}(b_{11}c_{11} + b_{12}c_{21}) + a_{12}(b_{21}c_{11} + b_{22}c_{21}) & a_{11}(b_{11}c_{12} + b_{12}c_{22}) + a_{12}(b_{21}c_{12} + b_{22}c_{22}) \\ a_{21}(b_{11}c_{11} + b_{12}c_{21}) + a_{22}(b_{21}c_{11} + b_{22}c_{21}) & a_{21}(b_{11}c_{12} + b_{12}c_{22}) + a_{22}(b_{21}c_{12} + b_{22}c_{22}) \end{bmatrix}$

$= \begin{bmatrix} a_{11}b_{11}c_{11} + a_{11}b_{12}c_{21} + a_{12}b_{21}c_{11} + a_{12}b_{22}c_{21} & a_{11}b_{11}c_{12} + a_{11}b_{12}c_{22} + a_{12}b_{21}c_{12} + a_{12}b_{22}c_{22} \\ a_{21}b_{11}c_{11} + a_{21}b_{12}c_{21} + a_{22}b_{21}c_{11} + a_{22}b_{22}c_{21} & a_{21}b_{11}c_{12} + a_{21}b_{12}c_{22} + a_{22}b_{21}c_{12} + a_{22}b_{22}c_{22} \end{bmatrix}$

$= \begin{bmatrix} a_{11}b_{11}c_{11} + a_{12}b_{21}c_{11} + a_{11}b_{12}c_{21} + a_{12}b_{22}c_{21} & a_{11}b_{11}c_{12} + a_{12}b_{21}c_{12} + a_{11}b_{12}c_{22} + a_{12}b_{22}c_{22} \\ a_{21}b_{11}c_{11} + a_{22}b_{21}c_{11} + a_{21}b_{12}c_{21} + a_{22}b_{22}c_{21} & a_{21}b_{11}c_{12} + a_{22}b_{21}c_{12} + a_{21}b_{12}c_{22} + a_{22}b_{22}c_{22} \end{bmatrix}$

Since the final matrix for $(AB)C$ and $A(BC)$ are the same, we have obtained the desired results.

**91.** $c(A + B) = cA + cB,$ for any real number $c$.

$c(A + B) = c\left(\begin{bmatrix} a_{11} & a_{12} \\ a_{21} & a_{22} \end{bmatrix} + \begin{bmatrix} b_{11} & b_{12} \\ b_{21} & b_{22} \end{bmatrix}\right) = c\begin{bmatrix} a_{11} + b_{11} & a_{12} + b_{12} \\ a_{21} + b_{21} & a_{22} + b_{22} \end{bmatrix} = \begin{bmatrix} c(a_{11} + b_{11}) & c(a_{12} + b_{12}) \\ c(a_{21} + b_{21}) & c(a_{22} + b_{22}) \end{bmatrix}$

$= \begin{bmatrix} c \cdot a_{11} + c \cdot b_{11} & c \cdot a_{12} + c \cdot b_{12} \\ c \cdot a_{21} + c \cdot b_{21} & c \cdot a_{22} + c \cdot b_{22} \end{bmatrix} = \begin{bmatrix} c \cdot a_{11} & c \cdot a_{12} \\ c \cdot a_{21} & c \cdot a_{22} \end{bmatrix} + \begin{bmatrix} c \cdot b_{11} & c \cdot b_{12} \\ c \cdot b_{21} & c \cdot b_{22} \end{bmatrix}$

$= c\begin{bmatrix} a_{11} & a_{12} \\ a_{21} & a_{22} \end{bmatrix} + c\begin{bmatrix} b_{11} & b_{12} \\ b_{21} & b_{22} \end{bmatrix} = cA + cB$

**93.** $c(A)d = (cd)A,$ for any real numbers $c$ and $d$.

$c(A)d = \left(c\begin{bmatrix} a_{11} & a_{12} \\ a_{21} & a_{22} \end{bmatrix}\right)d = \begin{bmatrix} c \cdot a_{11} & c \cdot a_{12} \\ c \cdot a_{21} & c \cdot a_{22} \end{bmatrix}d = \begin{bmatrix} c \cdot a_{11} \cdot d & c \cdot a_{12} \cdot d \\ c \cdot a_{21} \cdot d & c \cdot a_{22} \cdot d \end{bmatrix} = \begin{bmatrix} c \cdot d \cdot a_{11} & c \cdot d \cdot a_{12} \\ c \cdot d \cdot a_{21} & c \cdot d \cdot a_{22} \end{bmatrix}$

$= \begin{bmatrix} (cd) \cdot a_{11} & (cd) \cdot a_{12} \\ (cd) \cdot a_{21} & (cd) \cdot a_{22} \end{bmatrix} = (cd)\begin{bmatrix} a_{11} & a_{12} \\ a_{21} & a_{22} \end{bmatrix} = (cd)A$

## Section 5.8: Matrix Inverses

**1.** $A = \begin{bmatrix} -2 & 4 & 0 \\ 3 & 5 & 9 \\ 0 & 8 & -6 \end{bmatrix}$ and $I_3 = \begin{bmatrix} 1 & 0 & 0 \\ 0 & 1 & 0 \\ 0 & 0 & 1 \end{bmatrix}$

$I_3 A = \begin{bmatrix} 1 & 0 & 0 \\ 0 & 1 & 0 \\ 0 & 0 & 1 \end{bmatrix}\begin{bmatrix} -2 & 4 & 0 \\ 3 & 5 & 9 \\ 0 & 8 & -6 \end{bmatrix} = \begin{bmatrix} 1(-2)+0(3)+0(0) & 1(4)+0(5)+0(8) & 1(0)+0(9)+0(-6) \\ 0(-2)+1(3)+0(0) & 0(4)+1(5)+0(8) & 0(0)+1(9)+0(-6) \\ 0(-2)+0(3)+1(0) & 0(4)+0(5)+1(8) & 0(0)+0(9)+1(-6) \end{bmatrix}$

$= \begin{bmatrix} -2+0+0 & 4+0+0 & 0+0+0 \\ 0+3+0 & 0+5+0 & 0+9+0 \\ 0+0+0 & 0+0+8 & 0+0+(-6) \end{bmatrix} = \begin{bmatrix} -2 & 4 & 0 \\ 3 & 5 & 9 \\ 0 & 8 & -6 \end{bmatrix}$

**3.** $\begin{bmatrix} 5 & 7 \\ 2 & 3 \end{bmatrix}\begin{bmatrix} 3 & -7 \\ -2 & 5 \end{bmatrix} = \begin{bmatrix} 5(3)+7(-2) & 5(-7)+7(5) \\ 2(3)+3(-2) & 2(-7)+3(5) \end{bmatrix} = \begin{bmatrix} 15+(-14) & -35+35 \\ 6+(-6) & (-14)+15 \end{bmatrix} = \begin{bmatrix} 1 & 0 \\ 0 & 1 \end{bmatrix}$

$\begin{bmatrix} 3 & -7 \\ -2 & 5 \end{bmatrix}\begin{bmatrix} 5 & 7 \\ 2 & 3 \end{bmatrix} = \begin{bmatrix} 3(5)+(-7)(2) & 3(7)+(-7)(3) \\ (-2)(5)+5(2) & (-2)(7)+5(3) \end{bmatrix} = \begin{bmatrix} 15+(-14) & 21+(-21) \\ (-10)+10 & (-14)+15 \end{bmatrix} = \begin{bmatrix} 1 & 0 \\ 0 & 1 \end{bmatrix}$

Since the products obtained by multiplying the matrices in either order are both the $2 \times 2$ identity matrix, the given matrices are inverses of each other.

**5.** $\begin{bmatrix} -1 & 2 \\ 3 & -5 \end{bmatrix}\begin{bmatrix} -5 & -2 \\ -3 & -1 \end{bmatrix} = \begin{bmatrix} -1(-5)+2(-3) & -1(-2)+2(-1) \\ 3(-5)+(-5)(-3) & 3(-2)+(-5)(-1) \end{bmatrix} = \begin{bmatrix} 5+(-6) & 2+(-2) \\ -15+15 & -6+5 \end{bmatrix} = \begin{bmatrix} -1 & 0 \\ 0 & -1 \end{bmatrix}$

Since this product is not the $2 \times 2$ identity matrix, the given matrices are not inverses of each other.

**7.** $\begin{bmatrix} 0 & 1 & 0 \\ 0 & 0 & -2 \\ 1 & -1 & 0 \end{bmatrix}\begin{bmatrix} 1 & 0 & 1 \\ 1 & 0 & 0 \\ 0 & -1 & 0 \end{bmatrix} = \begin{bmatrix} 0(1)+1(1)+0(0) & 0(0)+1(0)+0(-1) & 0(1)+1(0)+0(0) \\ 0(1)+0(1)+(-2)(0) & 0(0)+0(0)+(-2)(-1) & 0(1)+0(0)+(-2)(0) \\ 1(1)+(-1)(1)+0(0) & 1(0)+(-1)(0)+0(-1) & 1(1)+(-1)(0)+0(0) \end{bmatrix}$

$= \begin{bmatrix} 0+1+0 & 0+0+0 & 0+0+0 \\ 0+0+0 & 0+0+2 & 0+0+0 \\ 1+(-1)+0 & 0+0+0 & 1+0+0 \end{bmatrix} = \begin{bmatrix} 1 & 0 & 0 \\ 0 & 2 & 0 \\ 0 & 0 & 1 \end{bmatrix}$

Since this product is not the $3 \times 3$ identity matrix, the given matrices are not inverses of each other.

**9.** $\begin{bmatrix} -1 & -1 & -1 \\ 4 & 5 & 0 \\ 0 & 1 & -3 \end{bmatrix}\begin{bmatrix} 15 & 4 & -5 \\ -12 & -3 & 4 \\ -4 & -1 & 1 \end{bmatrix}$

$= \begin{bmatrix} -1(15)+(-1)(-12)+(-1)(-4) & -1(4)+(-1)(-3)+(-1)(-1) & -1(-5)+(-1)(4)+(-1)(1) \\ 4(15)+5(-12)+0(-4) & 4(4)+5(-3)+0(-1) & 4(-5)+5(4)+0(1) \\ 0(15)+1(-12)+(-3)(-4) & 0(4)+1(-3)+(-3)(-1) & 0(-5)+1(4)+(-3)(1) \end{bmatrix}$

$= \begin{bmatrix} -15+12+4 & -4+3+1 & 5+(-4)+(-1) \\ 60+(-60)+0 & 16+(-15)+0 & -20+20+0 \\ 0+(-12)+12 & 0+(-3)+3 & 0+4+(-3) \end{bmatrix} = \begin{bmatrix} 1 & 0 & 0 \\ 0 & 1 & 0 \\ 0 & 0 & 1 \end{bmatrix} = I_3$

$\begin{bmatrix} 15 & 4 & -5 \\ -12 & -3 & 4 \\ -4 & -1 & 1 \end{bmatrix}\begin{bmatrix} -1 & -1 & -1 \\ 4 & 5 & 0 \\ 0 & 1 & -3 \end{bmatrix}$

$= \begin{bmatrix} 15(-1)+4(4)+(-5)(0) & 15(-1)+4(5)+(-5)(1) & 15(-1)+4(0)+(-5)(-3) \\ -12(-1)+(-3)(4)+4(0) & -12(-1)+(-3)(5)+4(1) & -12(-1)+(-3)(0)+4(-3) \\ -4(-1)+(-1)(4)+1(0) & -4(-1)+(-1)(5)+1(1) & -4(-1)+(-1)(0)+1(-3) \end{bmatrix}$

$= \begin{bmatrix} -15+16+0 & -15+20+(-5) & -15+0+15 \\ 12+(-12)+0 & 12+(-15)+4 & 12+0+(-12) \\ 4+(-4)+0 & 4+(-5)+1 & 0+4+(-3) \end{bmatrix} = \begin{bmatrix} 1 & 0 & 0 \\ 0 & 1 & 0 \\ 0 & 0 & 1 \end{bmatrix} = I_3$

The given matrices are inverses of each other.

**11.** Find the inverse of $A = \begin{bmatrix} -1 & 2 \\ -2 & -1 \end{bmatrix}$, if it exists. Since $[A \mid I_2] = \begin{bmatrix} -1 & 2 & 1 & 0 \\ -2 & -1 & 0 & 1 \end{bmatrix}$, we have

$$\begin{bmatrix} 1 & -2 & -1 & 0 \\ -2 & -1 & 0 & 1 \end{bmatrix} \begin{matrix} -1R1 \\ \\ \end{matrix} \Rightarrow \begin{bmatrix} 1 & -2 & -1 & 0 \\ 0 & -5 & -2 & 1 \end{bmatrix} \begin{matrix} \\ 2R1+R2 \end{matrix} \Rightarrow \begin{bmatrix} 1 & -2 & -1 & 0 \\ 0 & 1 & \frac{2}{5} & -\frac{1}{5} \end{bmatrix} \begin{matrix} \\ -\frac{1}{5}R2 \end{matrix} \Rightarrow$$

$$\begin{bmatrix} 1 & 0 & -\frac{1}{5} & -\frac{2}{5} \\ 0 & 1 & \frac{2}{5} & -\frac{1}{5} \end{bmatrix} \begin{matrix} 2R2+R1 \\ \\ \end{matrix}. \text{ Thus, } A^{-1} = \begin{bmatrix} -\frac{1}{5} & -\frac{2}{5} \\ \frac{2}{5} & -\frac{1}{5} \end{bmatrix}.$$

**13.** Find the inverse of $A = \begin{bmatrix} -1 & -2 \\ 3 & 4 \end{bmatrix}$, if it exists. Since $\left[A \mid I_2\right] = \begin{bmatrix} -1 & -2 & 1 & 0 \\ 3 & 4 & 0 & 1 \end{bmatrix}$, we have

$$\begin{bmatrix} -1 & -2 & 1 & 0 \\ 0 & -2 & 3 & 1 \end{bmatrix} \begin{matrix} \\ 3R1+R2 \end{matrix} \Rightarrow \begin{bmatrix} 1 & 2 & -1 & 0 \\ 0 & -2 & 3 & 1 \end{bmatrix} \begin{matrix} -1R1 \\ \\ \end{matrix} \Rightarrow \begin{bmatrix} 1 & 0 & 2 & 1 \\ 0 & -2 & 3 & 1 \end{bmatrix} \begin{matrix} R2+R1 \\ \\ \end{matrix} \Rightarrow \begin{bmatrix} 1 & 0 & 2 & 1 \\ 0 & 1 & -\frac{3}{2} & -\frac{1}{2} \end{bmatrix} \begin{matrix} \\ -\frac{1}{2}R2 \end{matrix}$$

Thus, $A^{-1} = \begin{bmatrix} 2 & 1 \\ -\frac{3}{2} & -\frac{1}{2} \end{bmatrix}$.

**15.** Find the inverse of $\begin{bmatrix} 5 & 10 \\ -3 & -6 \end{bmatrix}$, if it exists. Since $\left[A \mid I_2\right] = \begin{bmatrix} 5 & 10 & 1 & 0 \\ -3 & -6 & 0 & 1 \end{bmatrix}$, we have

$$\begin{bmatrix} 1 & 2 & \frac{1}{5} & 0 \\ -3 & -6 & 0 & 1 \end{bmatrix} \begin{matrix} \frac{1}{5}R1 \\ \\ \end{matrix} \Rightarrow \begin{bmatrix} 1 & 2 & \frac{1}{5} & 0 \\ 0 & 0 & \frac{3}{5} & 1 \end{bmatrix} \begin{matrix} \\ 3R1+R2 \end{matrix}$$

At this point, the matrix should be changed so that the second-row, second-column element will be 1. Since that element is now 0, the desired transformation cannot be completed. Therefore, the inverse of the given matrix does not exist.

**17.** Find the inverse of $A = \begin{bmatrix} 1 & 0 & 1 \\ 0 & -1 & 0 \\ 2 & 1 & 1 \end{bmatrix}$, if it exists. Since $\left[A \mid I_3\right] = \begin{bmatrix} 1 & 0 & 1 & 1 & 0 & 0 \\ 0 & -1 & 0 & 0 & 1 & 0 \\ 2 & 1 & 1 & 0 & 0 & 1 \end{bmatrix}$ we have

$$\begin{bmatrix} 1 & 0 & 1 & 1 & 0 & 0 \\ 0 & -1 & 0 & 0 & 1 & 0 \\ 0 & 1 & -1 & -2 & 0 & 1 \end{bmatrix} \begin{matrix} \\ \\ -2R1+R3 \end{matrix} \Rightarrow \begin{bmatrix} 1 & 0 & 1 & 1 & 0 & 0 \\ 0 & 1 & 0 & 0 & -1 & 0 \\ 0 & 0 & -1 & -2 & 1 & 1 \end{bmatrix} \begin{matrix} \\ -1R2 \\ R2+R3 \end{matrix} \Rightarrow \begin{bmatrix} 1 & 0 & 1 & 1 & 0 & 0 \\ 0 & 1 & 0 & 0 & -1 & 0 \\ 0 & 0 & 1 & 2 & -1 & -1 \end{bmatrix} \begin{matrix} \\ \\ -1R3 \end{matrix} \Rightarrow$$

$$\begin{bmatrix} 1 & 0 & 0 & -1 & 1 & 1 \\ 0 & 1 & 0 & 0 & -1 & 0 \\ 0 & 0 & 1 & 2 & -1 & -1 \end{bmatrix} \begin{matrix} -1R3+R1 \\ \\ \end{matrix}. \text{ Thus, } A^{-1} = \begin{bmatrix} -1 & 1 & 1 \\ 0 & -1 & 0 \\ 2 & -1 & -1 \end{bmatrix}.$$

**19.** Find the inverse of $A = \begin{bmatrix} 1 & 3 & 3 \\ 1 & 4 & 3 \\ 1 & 3 & 4 \end{bmatrix}$, if it exists. Since $\left[A \mid I_3\right] = \begin{bmatrix} 1 & 3 & 3 & 1 & 0 & 0 \\ 1 & 4 & 3 & 0 & 1 & 0 \\ 1 & 3 & 4 & 0 & 0 & 1 \end{bmatrix}$, we have

$$\begin{bmatrix} 1 & 3 & 3 & 1 & 0 & 0 \\ 0 & 1 & 0 & -1 & 1 & 0 \\ 0 & 0 & 1 & -1 & 0 & 1 \end{bmatrix} \begin{matrix} \\ -1R1+R2 \\ -1R1+R3 \end{matrix} \Rightarrow \begin{bmatrix} 1 & 0 & 3 & 4 & -3 & 0 \\ 0 & 1 & 0 & -1 & 1 & 0 \\ 0 & 0 & 1 & -1 & 0 & 1 \end{bmatrix} \begin{matrix} -3R2+R1 \\ \\ \end{matrix} \Rightarrow \begin{bmatrix} 1 & 0 & 0 & 7 & -3 & -3 \\ 0 & 1 & 0 & -1 & 1 & 0 \\ 0 & 0 & 1 & -1 & 0 & 1 \end{bmatrix} \begin{matrix} -3R3+R1 \\ \\ \end{matrix}$$

Thus, $A^{-1} = \begin{bmatrix} 7 & -3 & -3 \\ -1 & 1 & 0 \\ -1 & 0 & 1 \end{bmatrix}$.

**21.** Find the inverse of $A = \begin{bmatrix} 2 & 2 & -4 \\ 2 & 6 & 0 \\ -3 & -3 & 5 \end{bmatrix}$, if it exists. Since $\begin{bmatrix} A \mid I_3 \end{bmatrix} = \begin{bmatrix} 2 & 2 & -4 & 1 & 0 & 0 \\ 2 & 6 & 0 & 0 & 1 & 0 \\ -3 & -3 & 5 & 0 & 0 & 1 \end{bmatrix}$,

$$\begin{bmatrix} 1 & 1 & -2 & \frac{1}{2} & 0 & 0 \\ 2 & 6 & 0 & 0 & 1 & 0 \\ -3 & -3 & 5 & 0 & 0 & 1 \end{bmatrix} \begin{matrix} \frac{1}{2}R1 \\ \\ \end{matrix} \Rightarrow \begin{bmatrix} 1 & 1 & -2 & \frac{1}{2} & 0 & 0 \\ 0 & 4 & 4 & -1 & 1 & 0 \\ 0 & 0 & -1 & \frac{3}{2} & 0 & 1 \end{bmatrix} \begin{matrix} \\ -2R1+R2 \\ 3R1+R3 \end{matrix}$$

$$\begin{bmatrix} 1 & 1 & -2 & \frac{1}{2} & 0 & 0 \\ 0 & 1 & 1 & -\frac{1}{4} & \frac{1}{4} & 0 \\ 0 & 0 & -1 & \frac{3}{2} & 0 & 1 \end{bmatrix} \begin{matrix} \\ \frac{1}{4}R2 \\ \end{matrix} \Rightarrow \begin{bmatrix} 1 & 0 & -3 & \frac{3}{4} & -\frac{1}{4} & 0 \\ 0 & 1 & 1 & -\frac{1}{4} & \frac{1}{4} & 0 \\ 0 & 0 & -1 & \frac{3}{2} & 0 & 1 \end{bmatrix} \begin{matrix} -1R2+R1 \\ \\ \end{matrix}$$

$$\begin{bmatrix} 1 & 0 & -3 & \frac{3}{4} & -\frac{1}{4} & 0 \\ 0 & 1 & 1 & -\frac{1}{4} & \frac{1}{4} & 0 \\ 0 & 0 & 1 & -\frac{3}{2} & 0 & -1 \end{bmatrix} \begin{matrix} \\ \\ -1R3 \end{matrix} \Rightarrow \begin{bmatrix} 1 & 0 & 0 & -\frac{15}{4} & -\frac{1}{4} & -3 \\ 0 & 1 & 0 & \frac{5}{4} & \frac{1}{4} & 1 \\ 0 & 0 & 1 & -\frac{3}{2} & 0 & -1 \end{bmatrix} \begin{matrix} 3R3+R1 \\ -1R3+R2. \\ \end{matrix}$$

Thus, $A^{-1} = \begin{bmatrix} -\frac{15}{4} & -\frac{1}{4} & -3 \\ \frac{5}{4} & \frac{1}{4} & 1 \\ -\frac{3}{2} & 0 & -1 \end{bmatrix}$.

**23.** Find the inverse of $A = \begin{bmatrix} 1 & 1 & 0 & 2 \\ 2 & -1 & 1 & -1 \\ 3 & 3 & 2 & -2 \\ 1 & 2 & 1 & 0 \end{bmatrix}$, if it exists.

$$\begin{bmatrix} A \mid I_4 \end{bmatrix} = \begin{bmatrix} 1 & 1 & 0 & 2 & 1 & 0 & 0 & 0 \\ 2 & -1 & 1 & -1 & 0 & 1 & 0 & 0 \\ 3 & 3 & 2 & -2 & 0 & 0 & 1 & 0 \\ 1 & 2 & 1 & 0 & 0 & 0 & 0 & 1 \end{bmatrix} \Rightarrow \begin{bmatrix} 1 & 1 & 0 & 2 & 1 & 0 & 0 & 0 \\ 0 & -3 & 1 & -5 & -2 & 1 & 0 & 0 \\ 0 & 0 & 2 & -8 & -3 & 0 & 1 & 0 \\ 0 & 1 & 1 & -2 & -1 & 0 & 0 & 1 \end{bmatrix} \begin{matrix} \\ -2R1+R2 \\ -3R1+R3 \\ -1R1+R4 \end{matrix}$$

$$\Rightarrow \begin{bmatrix} 1 & 1 & 0 & 2 & 1 & 0 & 0 & 0 \\ 0 & 1 & -\frac{1}{3} & \frac{5}{3} & \frac{2}{3} & -\frac{1}{3} & 0 & 0 \\ 0 & 0 & 2 & -8 & -3 & 0 & 1 & 0 \\ 0 & 1 & 1 & -2 & -1 & 0 & 0 & 1 \end{bmatrix} \begin{matrix} \\ -\frac{1}{3}R2 \\ \\ \end{matrix} \Rightarrow \begin{bmatrix} 1 & 0 & \frac{1}{3} & \frac{1}{3} & \frac{1}{3} & \frac{1}{3} & 0 & 0 \\ 0 & 1 & -\frac{1}{3} & \frac{5}{3} & \frac{2}{3} & -\frac{1}{3} & 0 & 0 \\ 0 & 0 & 2 & -8 & -3 & 0 & 1 & 0 \\ 0 & 0 & \frac{4}{3} & -\frac{11}{3} & -\frac{5}{3} & \frac{1}{3} & 0 & 1 \end{bmatrix} \begin{matrix} -1R2+R1 \\ \\ \\ -1R2+R4 \end{matrix}$$

$$\Rightarrow \begin{bmatrix} 1 & 0 & \frac{1}{3} & \frac{1}{3} & \frac{1}{3} & \frac{1}{3} & 0 & 0 \\ 0 & 1 & -\frac{1}{3} & \frac{5}{3} & \frac{2}{3} & -\frac{1}{3} & 0 & 0 \\ 0 & 0 & 1 & -4 & -\frac{3}{2} & 0 & \frac{1}{2} & 0 \\ 0 & 0 & \frac{4}{3} & -\frac{11}{3} & -\frac{5}{3} & \frac{1}{3} & 0 & 1 \end{bmatrix} \begin{matrix} \\ \\ \frac{1}{2}R3 \\ \end{matrix} \Rightarrow \begin{bmatrix} 1 & 0 & 0 & \frac{5}{3} & \frac{5}{6} & \frac{1}{3} & -\frac{1}{6} & 0 \\ 0 & 1 & 0 & \frac{1}{3} & \frac{1}{6} & -\frac{1}{3} & \frac{1}{6} & 0 \\ 0 & 0 & 1 & -4 & -\frac{3}{2} & 0 & \frac{1}{2} & 0 \\ 0 & 0 & 0 & \frac{5}{3} & \frac{1}{3} & \frac{1}{3} & -\frac{2}{3} & 1 \end{bmatrix} \begin{matrix} -\frac{1}{3}R3+R1 \\ \frac{1}{3}R3+R2 \\ \\ -\frac{4}{3}R3+R4 \end{matrix}$$

$$\Rightarrow \begin{bmatrix} 1 & 0 & 0 & \frac{5}{3} & \frac{5}{6} & \frac{1}{3} & -\frac{1}{6} & 0 \\ 0 & 1 & 0 & \frac{1}{3} & \frac{1}{6} & -\frac{1}{3} & \frac{1}{6} & 0 \\ 0 & 0 & 1 & -4 & -\frac{3}{2} & 0 & \frac{1}{2} & 0 \\ 0 & 0 & 0 & 1 & \frac{1}{5} & \frac{1}{5} & -\frac{2}{5} & \frac{3}{5} \end{bmatrix} \begin{matrix} \\ \\ \\ \frac{3}{5}R4 \end{matrix} \Rightarrow \begin{bmatrix} 1 & 0 & 0 & 0 & \frac{1}{2} & 0 & \frac{1}{2} & -1 \\ 0 & 1 & 0 & 0 & \frac{1}{10} & -\frac{2}{5} & \frac{3}{10} & -\frac{1}{5} \\ 0 & 0 & 1 & 0 & -\frac{7}{10} & \frac{4}{5} & -\frac{11}{10} & \frac{12}{5} \\ 0 & 0 & 0 & 1 & \frac{1}{5} & \frac{1}{5} & -\frac{2}{5} & \frac{3}{5} \end{bmatrix} \begin{matrix} -\frac{5}{3}R4+R1 \\ -\frac{1}{3}R4+R2 \\ 4R4+R3 \\ \end{matrix}$$

Thus, $A^{-1} = \begin{bmatrix} \frac{1}{2} & 0 & \frac{1}{2} & -1 \\ \frac{1}{10} & -\frac{2}{5} & \frac{3}{10} & -\frac{1}{5} \\ -\frac{7}{10} & \frac{4}{5} & -\frac{11}{10} & \frac{12}{5} \\ \frac{1}{5} & \frac{1}{5} & -\frac{2}{5} & \frac{3}{5} \end{bmatrix}$.

**25.** Since $A^{-1} = \begin{bmatrix} 5 & -9 \\ -1 & 2 \end{bmatrix}$, we need to find $A$, where $A = \left( A^{-1} \right)^{-1}$.

$$\left[ A^{-1} \mid I_2 \right] = \begin{bmatrix} 5 & -9 & 1 & 0 \\ -1 & 2 & 0 & 1 \end{bmatrix} \Rightarrow \begin{bmatrix} 1 & -\frac{9}{5} & \frac{1}{5} & 0 \\ -1 & 2 & 0 & 1 \end{bmatrix} \begin{smallmatrix} \frac{1}{5}R1 \end{smallmatrix} \Rightarrow \begin{bmatrix} 1 & -\frac{9}{5} & \frac{1}{5} & 0 \\ 0 & \frac{1}{5} & \frac{1}{5} & 1 \end{bmatrix} 1R1 + R2$$

$$\begin{bmatrix} 1 & -\frac{9}{5} & \frac{1}{5} & 0 \\ 0 & 1 & 1 & 5 \end{bmatrix} 5R2 \Rightarrow \begin{bmatrix} 1 & 0 & 2 & 9 \\ 0 & 1 & 1 & 5 \end{bmatrix} \begin{smallmatrix} \frac{9}{5}R2 + R1 \end{smallmatrix}$$

Thus, $A = \begin{bmatrix} 2 & 9 \\ 1 & 5 \end{bmatrix}$.

**27.** If $A = \begin{bmatrix} a & b \\ c & d \end{bmatrix}$, then $ad - bc = ad - cb$ is the

determinant of matrix $A$.

**29.** $A^{-1} = \begin{bmatrix} \frac{d}{|A|} & \frac{-b}{|A|} \\ \frac{-c}{|A|} & \frac{a}{|A|} \end{bmatrix} = \begin{bmatrix} \frac{1}{|A|}d & \frac{1}{|A|}(-b) \\ \frac{1}{|A|}(-c) & \frac{1}{|A|}a \end{bmatrix}$

$$= \frac{1}{|A|} \begin{bmatrix} d & -b \\ -c & a \end{bmatrix}$$

**31.** Given $A = \begin{bmatrix} 4 & 2 \\ 7 & 3 \end{bmatrix}$, $|A| = 12 - 14 = -2$. Thus

$$A^{-1} = \frac{1}{-2} \begin{bmatrix} 3 & -2 \\ -7 & 4 \end{bmatrix} = \begin{bmatrix} -\frac{3}{2} & 1 \\ \frac{7}{2} & -2 \end{bmatrix}.$$

**33.**  $-x + y = 1$
$\phantom{xxx}2x - y = 1$

Given $A = \begin{bmatrix} -1 & 1 \\ 2 & -1 \end{bmatrix}$, $X = \begin{bmatrix} x \\ y \end{bmatrix}$, and $B = \begin{bmatrix} 1 \\ 1 \end{bmatrix}$,

find $A^{-1}$. Since $\left[ A \mid I_2 \right] = \begin{bmatrix} -1 & 1 & 1 & 0 \\ 2 & -1 & 0 & 1 \end{bmatrix}$,

$$\begin{bmatrix} 1 & -1 & -1 & 0 \\ 2 & -1 & 0 & 1 \end{bmatrix} \begin{smallmatrix} -1R1 \end{smallmatrix} \Rightarrow$$

$$\begin{bmatrix} 1 & -1 & -1 & 0 \\ 0 & 1 & 2 & 1 \end{bmatrix} \begin{smallmatrix} -2R1 + R2 \end{smallmatrix} \Rightarrow$$

$$\begin{bmatrix} 1 & 0 & 1 & 1 \\ 0 & 1 & 2 & 1 \end{bmatrix} \begin{smallmatrix} R2 + R1 \end{smallmatrix}$$

Thus, $A^{-1} = \begin{bmatrix} 1 & 1 \\ 2 & 1 \end{bmatrix}$.

$$X = A^{-1}B = \begin{bmatrix} 1 & 1 \\ 2 & 1 \end{bmatrix} \begin{bmatrix} 1 \\ 1 \end{bmatrix} = \begin{bmatrix} 1(1) + 1(1) \\ 2(1) + 1(1) \end{bmatrix}$$

$$= \begin{bmatrix} 1+1 \\ 2+1 \end{bmatrix} = \begin{bmatrix} 2 \\ 3 \end{bmatrix}$$

Solution set: $\{(2,3)\}$

**35.**  $2x - y = -8$
$\phantom{xxx}3x + y = -2$

Given $A = \begin{bmatrix} 2 & -1 \\ 3 & 1 \end{bmatrix}$, $X = \begin{bmatrix} x \\ y \end{bmatrix}$, and $B = \begin{bmatrix} -8 \\ -2 \end{bmatrix}$,

find $A^{-1}$. We will use the relation given on page 588 to calculate $A^{-1}$ in the solution to this exercise.

If $A = \begin{bmatrix} a & b \\ c & d \end{bmatrix}$, then $A^{-1} = \begin{bmatrix} \frac{d}{ad-bc} & \frac{-b}{ad-bc} \\ \frac{-c}{ad-bc} & \frac{a}{ad-bc} \end{bmatrix}$.

Thus we have:

$$A^{-1} = \begin{bmatrix} \frac{1}{2(1)-(-1)(3)} & \frac{1}{2(1)-(-1)(3)} \\ \frac{-3}{2(1)-(-1)(3)} & \frac{2}{2(1)-(-1)(3)} \end{bmatrix}$$

$$= \begin{bmatrix} \frac{1}{2-(-3)} & \frac{1}{2-(-3)} \\ \frac{-3}{2-(-3)} & \frac{2}{2-(-3)} \end{bmatrix} = \begin{bmatrix} \frac{1}{5} & \frac{1}{5} \\ -\frac{3}{5} & \frac{2}{5} \end{bmatrix}$$

$$X = A^{-1}B = \begin{bmatrix} \frac{1}{5} & \frac{1}{5} \\ -\frac{3}{5} & \frac{2}{5} \end{bmatrix} \begin{bmatrix} -8 \\ -2 \end{bmatrix}$$

$$= \begin{bmatrix} \frac{1}{5}(-8) + \frac{1}{5}(-2) \\ -\frac{3}{5}(-8) + \frac{2}{5}(-2) \end{bmatrix} = \begin{bmatrix} -\frac{8}{5} + \left(-\frac{2}{5}\right) \\ \frac{24}{5} + \left(-\frac{4}{5}\right) \end{bmatrix}$$

$$= \begin{bmatrix} -\frac{10}{5} \\ \frac{20}{5} \end{bmatrix} = \begin{bmatrix} -2 \\ 4 \end{bmatrix}$$

Solution set: $\{(-2,4)\}$

**37.** $2x + 3y = -10$
$3x + 4y = -12$

Given $A = \begin{bmatrix} 2 & 3 \\ 3 & 4 \end{bmatrix}$, $X = \begin{bmatrix} x \\ y \end{bmatrix}$, and $B = \begin{bmatrix} -10 \\ -12 \end{bmatrix}$,

find $A^{-1}$. Since $\left[ A | I_2 \right] = \begin{bmatrix} 2 & 3 & | & 1 & 0 \\ 3 & 4 & | & 0 & 1 \end{bmatrix}$,

$\begin{bmatrix} 1 & \frac{3}{2} & | & \frac{1}{2} & 0 \\ 3 & 4 & | & 0 & 1 \end{bmatrix} \frac{1}{2}R1 \Rightarrow$

$\begin{bmatrix} 1 & \frac{3}{2} & | & \frac{1}{2} & 0 \\ 0 & -\frac{1}{2} & | & -\frac{3}{2} & 1 \end{bmatrix} -3R1 + R2 \Rightarrow$

$\begin{bmatrix} 1 & \frac{3}{2} & | & \frac{1}{2} & 0 \\ 0 & 1 & | & 3 & -2 \end{bmatrix} -2R2 \Rightarrow$

$\begin{bmatrix} 1 & 0 & | & -4 & 3 \\ 0 & 1 & | & 3 & -2 \end{bmatrix} -\frac{3}{2}R2 + R1$

Thus, $A^{-1} = \begin{bmatrix} -4 & 3 \\ 3 & -2 \end{bmatrix}$.

$X = A^{-1}B = \begin{bmatrix} -4 & 3 \\ 3 & -2 \end{bmatrix} \begin{bmatrix} -10 \\ -12 \end{bmatrix}$

$= \begin{bmatrix} -4(-10) + 3(-12) \\ 3(-10) + (-2)(-12) \end{bmatrix}$

$= \begin{bmatrix} 40 + (-36) \\ -30 + 24 \end{bmatrix} = \begin{bmatrix} 4 \\ -6 \end{bmatrix}$

Solution set: $\{(4, -6)\}$

**39.** $6x + 9y = 3$
$-8x + 3y = 6$

Given $A = \begin{bmatrix} 6 & 9 \\ -8 & 3 \end{bmatrix}$, $X = \begin{bmatrix} x \\ y \end{bmatrix}$, and $B = \begin{bmatrix} 3 \\ 6 \end{bmatrix}$,

find $A^{-1}$. We will use the results of the Relating Concepts exercises to calculate $A^{-1}$ in the solution to this exercise.

If $A = \begin{bmatrix} a & b \\ c & d \end{bmatrix}$, then $A^{-1} = \frac{1}{ad-bc}\begin{bmatrix} d & -b \\ -c & a \end{bmatrix}$.

Thus we have the following.

$A^{-1} = \frac{1}{6(3)-9(-8)}\begin{bmatrix} 3 & -9 \\ 8 & 6 \end{bmatrix} = \frac{1}{18-(-72)}\begin{bmatrix} 3 & -9 \\ 8 & 6 \end{bmatrix}$

$= \frac{1}{90}\begin{bmatrix} 3 & -9 \\ 8 & 6 \end{bmatrix} = \begin{bmatrix} \frac{3}{90} & \frac{-9}{90} \\ \frac{8}{90} & \frac{6}{90} \end{bmatrix} = \begin{bmatrix} \frac{1}{30} & -\frac{1}{10} \\ \frac{4}{45} & \frac{1}{15} \end{bmatrix}$

$X = A^{-1}B = \begin{bmatrix} \frac{1}{30} & -\frac{1}{10} \\ \frac{4}{45} & \frac{1}{15} \end{bmatrix}\begin{bmatrix} 3 \\ 6 \end{bmatrix}$

$= \begin{bmatrix} \frac{1}{30}(3) + \left(-\frac{1}{10}\right)(6) \\ \frac{4}{45}(3) + \frac{1}{15}(6) \end{bmatrix} = \begin{bmatrix} \frac{1}{10} + \left(-\frac{6}{10}\right) \\ \frac{4}{15} + \frac{2}{5} \end{bmatrix}$

$= \begin{bmatrix} -\frac{5}{10} \\ \frac{4}{15} + \frac{6}{15} \end{bmatrix} = \begin{bmatrix} -\frac{1}{2} \\ \frac{10}{15} \end{bmatrix} = \begin{bmatrix} -\frac{1}{2} \\ \frac{2}{3} \end{bmatrix}$

Solution set: $\left\{\left(-\frac{1}{2}, \frac{2}{3}\right)\right\}$

**41.** $.2x + .3y = -1.9$
$.7x - .2y = 4.6$

Given

$A = \begin{bmatrix} .2 & .3 \\ .7 & -.2 \end{bmatrix}$, $X = \begin{bmatrix} x \\ y \end{bmatrix}$, and $B = \begin{bmatrix} -1.9 \\ 4.6 \end{bmatrix}$,

find $A^{-1}$. Since $\left[ A | I_2 \right] = \begin{bmatrix} .2 & .3 & | & 1 & 0 \\ .7 & -.2 & | & 0 & 1 \end{bmatrix}$,

$\begin{bmatrix} 1 & 1.5 & | & 5 & 0 \\ .7 & -.2 & | & 0 & 1 \end{bmatrix} 5R1 \Rightarrow$

$\begin{bmatrix} 1 & 1.5 & | & 5 & 0 \\ 0 & -1.25 & | & -3.5 & 1 \end{bmatrix} -.7R1 + R2 \Rightarrow$

$\begin{bmatrix} 1 & 1.5 & | & 5 & 0 \\ 0 & 1 & | & 2.8 & -.8 \end{bmatrix} \frac{1}{-1.25}R2 \Rightarrow$

$\begin{bmatrix} 1 & 0 & | & .8 & 1.2 \\ 0 & 1 & | & 2.8 & -.8 \end{bmatrix} -1.5R2 + R1$

Thus, $A^{-1} = \begin{bmatrix} .8 & 1.2 \\ 2.8 & -.8 \end{bmatrix}$.

$X = A^{-1}B = \begin{bmatrix} .8 & 1.2 \\ 2.8 & -.8 \end{bmatrix}\begin{bmatrix} -1.9 \\ 4.6 \end{bmatrix}$

$= \begin{bmatrix} .8(-1.9) + 1.2(4.6) \\ 2.8(-1.9) + (-.8)(4.6) \end{bmatrix}$

$= \begin{bmatrix} -1.52 + 5.52 \\ -5.32 + (-3.68) \end{bmatrix} = \begin{bmatrix} 4 \\ -9 \end{bmatrix}$

Solution set: $\{(4, -9)\}$

**43.**  $x + y + z = 6$
$2x + 3y - z = 7$
$3x - y - z = 6$

Given $A = \begin{bmatrix} 1 & 1 & 1 \\ 2 & 3 & -1 \\ 3 & -1 & -1 \end{bmatrix}$, $X = \begin{bmatrix} x \\ y \\ z \end{bmatrix}$, and $B = \begin{bmatrix} 6 \\ 7 \\ 6 \end{bmatrix}$, find $A^{-1}$. Since $\begin{bmatrix} A | I_3 \end{bmatrix} = \begin{bmatrix} 1 & 1 & 1 & | & 1 & 0 & 0 \\ 2 & 3 & -1 & | & 0 & 1 & 0 \\ 3 & -1 & -1 & | & 0 & 0 & 1 \end{bmatrix}$,

$\begin{bmatrix} 1 & 1 & 1 & | & 1 & 0 & 0 \\ 0 & 1 & -3 & | & -2 & 1 & 0 \\ 0 & -4 & -4 & | & -3 & 0 & 1 \end{bmatrix} \begin{matrix} \\ -2R1+R2 \\ -3R1+R3 \end{matrix} \Rightarrow \begin{bmatrix} 1 & 0 & 4 & | & 3 & -1 & 0 \\ 0 & 1 & -3 & | & -2 & 1 & 0 \\ 0 & 0 & -16 & | & -11 & 4 & 1 \end{bmatrix} \begin{matrix} -R2+R1 \\ \\ 4R2+R3 \end{matrix}$

$\begin{bmatrix} 1 & 0 & 4 & | & 3 & -1 & 0 \\ 0 & 1 & -3 & | & -2 & 1 & 0 \\ 0 & 0 & 1 & | & \frac{11}{16} & -\frac{1}{4} & -\frac{1}{16} \end{bmatrix} \begin{matrix} \\ \\ -\frac{1}{16}R3 \end{matrix} \Rightarrow \begin{bmatrix} 1 & 0 & 0 & | & \frac{1}{4} & 0 & \frac{1}{4} \\ 0 & 1 & 0 & | & \frac{1}{16} & \frac{1}{4} & -\frac{3}{16} \\ 0 & 0 & 1 & | & \frac{11}{16} & -\frac{1}{4} & -\frac{1}{16} \end{bmatrix} \begin{matrix} -4R3+R1 \\ 3R3+R2 \end{matrix}$. Thus, $A^{-1} = \begin{bmatrix} \frac{1}{4} & 0 & \frac{1}{4} \\ \frac{1}{16} & \frac{1}{4} & -\frac{3}{16} \\ \frac{11}{16} & -\frac{1}{4} & -\frac{1}{16} \end{bmatrix}$.

$X = A^{-1}B = \begin{bmatrix} \frac{1}{4} & 0 & \frac{1}{4} \\ \frac{1}{16} & \frac{1}{4} & -\frac{3}{16} \\ \frac{11}{16} & -\frac{1}{4} & -\frac{1}{16} \end{bmatrix} \begin{bmatrix} 6 \\ 7 \\ 6 \end{bmatrix} = \begin{bmatrix} \frac{1}{4}(6) + 0(7) + \frac{1}{4}(6) \\ \frac{1}{16}(6) + \frac{1}{4}(7) + \left(-\frac{3}{16}\right)(6) \\ \frac{11}{16}(6) + \left(-\frac{1}{4}\right)(7) + \left(-\frac{1}{16}\right)(6) \end{bmatrix} = \begin{bmatrix} \frac{3}{2} + 0 + \frac{3}{2} \\ \frac{3}{8} + \frac{7}{4} + \left(-\frac{9}{8}\right) \\ \frac{33}{8} + \left(-\frac{7}{4}\right) + \left(-\frac{3}{8}\right) \end{bmatrix}$

$= \begin{bmatrix} \frac{6}{2} \\ \frac{7}{4} + \left(-\frac{6}{8}\right) \\ \frac{30}{8} + \left(-\frac{7}{4}\right) \end{bmatrix} = \begin{bmatrix} 3 \\ \frac{7}{4} + \left(-\frac{3}{4}\right) \\ \frac{15}{4} + \left(-\frac{7}{4}\right) \end{bmatrix} = \begin{bmatrix} 3 \\ \frac{4}{4} \\ \frac{8}{4} \end{bmatrix} = \begin{bmatrix} 3 \\ 1 \\ 2 \end{bmatrix}$

Solution set: $\{(3, 1, 2)\}$

**45.**  $x + 3y + 3z = 1$
$x + 4y + 3z = 0$
$x + 3y + 4z = -1$

We have $A = \begin{bmatrix} 1 & 3 & 3 \\ 1 & 4 & 3 \\ 1 & 3 & 4 \end{bmatrix}$, $X = \begin{bmatrix} x \\ y \\ z \end{bmatrix}$, $B = \begin{bmatrix} 1 \\ 0 \\ -1 \end{bmatrix}$, and $A^{-1} = \begin{bmatrix} 7 & -3 & -3 \\ -1 & 1 & 0 \\ -1 & 0 & 1 \end{bmatrix}$ (from Exercise 19).

$X = A^{-1}B = \begin{bmatrix} 7 & -3 & -3 \\ -1 & 1 & 0 \\ -1 & 0 & 1 \end{bmatrix} \begin{bmatrix} 1 \\ 0 \\ -1 \end{bmatrix} = \begin{bmatrix} 7(1) + (-3)(0) + (-3)(-1) \\ -1(1) + 1(0) + 0(-1) \\ -1(1) + 0(0) + 1(-1) \end{bmatrix} = \begin{bmatrix} 7 + 0 + 3 \\ -1 + 0 + 0 \\ -1 + 0 + (-1) \end{bmatrix} = \begin{bmatrix} 10 \\ -1 \\ -2 \end{bmatrix}$

Solution set: $\{(10, -1, -2)\}$

**47.** 
$$2x + 2y - 4z = 12$$
$$2x + 6y = 16$$
$$-3x - 3y + 5z = -20$$

We have $A = \begin{bmatrix} 2 & 2 & -4 \\ 2 & 6 & 0 \\ -3 & -3 & 5 \end{bmatrix}$, $X = \begin{bmatrix} x \\ y \\ z \end{bmatrix}$, $B = \begin{bmatrix} 12 \\ 16 \\ -20 \end{bmatrix}$, and $A^{-1} = \begin{bmatrix} -\frac{15}{4} & -\frac{1}{4} & -3 \\ \frac{5}{4} & \frac{1}{4} & 1 \\ -\frac{3}{2} & 0 & -1 \end{bmatrix}$ (from Exercise 21).

$$X = A^{-1}B = \begin{bmatrix} -\frac{15}{4} & -\frac{1}{4} & -3 \\ \frac{5}{4} & \frac{1}{4} & 1 \\ -\frac{3}{2} & 0 & -1 \end{bmatrix}\begin{bmatrix} 12 \\ 16 \\ -20 \end{bmatrix} = \begin{bmatrix} -\frac{15}{4}(12) + \left(-\frac{1}{4}\right)(16) + (-3)(-20) \\ \frac{5}{4}(12) + \frac{1}{4}(16) + 1(-20) \\ -\frac{3}{2}(12) + 0(16) + (-1)(-20) \end{bmatrix} = \begin{bmatrix} -45 + (-4) + 60 \\ 15 + 4 + (-20) \\ -18 + 0 + 20 \end{bmatrix} = \begin{bmatrix} 11 \\ -1 \\ 2 \end{bmatrix}$$

Solution set: $\{(11, -1, 2)\}$

**49.** 
$$x + y + 2w = 3$$
$$2x - y + z - w = 3$$
$$3x + 3y + 2z - 2w = 5$$
$$x + 2y + z = 3$$

We have $A = \begin{bmatrix} 1 & 1 & 0 & 2 \\ 2 & -1 & 1 & -1 \\ 3 & 3 & 2 & -2 \\ 1 & 2 & 1 & 0 \end{bmatrix}$, $X = \begin{bmatrix} x \\ y \\ z \\ w \end{bmatrix}$, $B = \begin{bmatrix} 3 \\ 3 \\ 5 \\ 3 \end{bmatrix}$, and $A^{-1} = \begin{bmatrix} \frac{1}{2} & 0 & \frac{1}{2} & -1 \\ \frac{1}{10} & -\frac{2}{5} & \frac{3}{10} & -\frac{1}{5} \\ -\frac{7}{10} & \frac{4}{5} & -\frac{11}{10} & \frac{12}{5} \\ \frac{1}{5} & \frac{1}{5} & -\frac{2}{5} & \frac{3}{5} \end{bmatrix}$ (from Exercise 23).

$$X = A^{-1}B = \begin{bmatrix} \frac{1}{2} & 0 & \frac{1}{2} & -1 \\ \frac{1}{10} & -\frac{2}{5} & \frac{3}{10} & -\frac{1}{5} \\ -\frac{7}{10} & \frac{4}{5} & -\frac{11}{10} & \frac{12}{5} \\ \frac{1}{5} & \frac{1}{5} & -\frac{2}{5} & \frac{3}{5} \end{bmatrix}\begin{bmatrix} 3 \\ 3 \\ 5 \\ 3 \end{bmatrix} = \begin{bmatrix} \frac{1}{2}(3) + 0(3) + \frac{1}{2}(5) + (-1)(3) \\ \frac{1}{10}(3) + \left(-\frac{2}{5}\right)(3) + \frac{3}{10}(5) + \left(-\frac{1}{5}\right)(3) \\ -\frac{7}{10}(3) + \frac{4}{5}(3) + \left(-\frac{11}{10}\right)(5) + \frac{12}{5}(3) \\ \frac{1}{5}(3) + \frac{1}{5}(3) + \left(-\frac{2}{5}\right)(5) + \frac{3}{5}(3) \end{bmatrix}$$

$$= \begin{bmatrix} \frac{3}{2} + 0 + \frac{5}{2} + (-3) \\ \frac{3}{10} + \left(-\frac{6}{5}\right) + \frac{3}{2} + \left(-\frac{3}{5}\right) \\ -\frac{21}{10} + \frac{12}{5} + \left(-\frac{11}{2}\right) + \frac{36}{5} \\ \frac{3}{5} + \frac{3}{5} + (-2) + \frac{9}{5} \end{bmatrix} = \begin{bmatrix} \frac{8}{2} + (-3) \\ \frac{3}{10} + \left(-\frac{9}{5}\right) + \frac{3}{2} \\ -\frac{21}{10} + \left(-\frac{11}{2}\right) + \frac{48}{5} \\ (-2) + \frac{15}{5} \end{bmatrix} = \begin{bmatrix} 4 + (-3) \\ \frac{3}{10} + \left(-\frac{18}{10}\right) + \frac{15}{10} \\ -\frac{21}{10} + \left(-\frac{55}{10}\right) + \frac{96}{10} \\ (-2) + 3 \end{bmatrix} = \begin{bmatrix} 1 \\ \frac{0}{10} \\ \frac{20}{10} \\ 1 \end{bmatrix} = \begin{bmatrix} 1 \\ 0 \\ 2 \\ 1 \end{bmatrix}$$

Solution set: $\{(1, 0, 2, 1)\}$

**51. (a)** 
$$602.7 = a + 5.543b + 37.14c$$
$$656.7 = a + 6.933b + 41.30c$$
$$778.5 = a + 7.638b + 45.62c$$

**(b)** $A = \begin{bmatrix} 1 & 5.543 & 37.14 \\ 1 & 6.933 & 41.30 \\ 1 & 7.638 & 45.62 \end{bmatrix}$, $X = \begin{bmatrix} a \\ b \\ c \end{bmatrix}$,

and $B = \begin{bmatrix} 602.7 \\ 656.7 \\ 778.5 \end{bmatrix}$

Using a graphing calculator with matrix capabilities, we obtain the following.

```
[A]-¹[B]
[[-490.547375]
 [-89         ]
 [42.71875    ]]
```

$X = A^{-1}B$ or $\begin{bmatrix} a \\ b \\ c \end{bmatrix} = \begin{bmatrix} -490.547375 \\ -89 \\ 42.71875 \end{bmatrix}$.

Thus, $a \approx -490.547$, $b = -89$, $c = 42.71875$.

**(c)** $S = -490.547 - 89A + 42.71875B$

**(d)** If $A = 7.752$ and $B = 47.38$, the predicted value of $S$ is given by the following.
$$S = -490.547 - 89(7.752)$$
$$+ 42.71875(47.38)$$
$$= 843.539375 \approx 843.5$$
The predicted value is approximately 843.5.

**(e)** If $A = 8.9$ and $B = 66.25$, the predicted value of $S$ is given by the following.
$$S = -490.547 - 89(8.9) + 42.71875(66.25)$$
$$= 1547.470188 \approx 1547.5$$
The predicted value is approximately 1547.5.
Using only three consecutive years to forecast six years into the future, it is probably not very accurate.

**53.** Answers will vary.

**55.** Given $A = \begin{bmatrix} \frac{2}{3} & .7 \\ 22 & \sqrt{3} \end{bmatrix}$, we obtain

$A^{-1} \approx \begin{bmatrix} -.1215875322 & .0491390161 \\ 1.544369078 & -.046799063 \end{bmatrix}$ using

a graphing calculator.

**57.** Given $A = \begin{bmatrix} \frac{1}{2} & \frac{1}{4} & \frac{1}{3} \\ 0 & \frac{1}{4} & \frac{1}{3} \\ \frac{1}{2} & \frac{1}{2} & \frac{1}{3} \end{bmatrix}$, we obtain

$A^{-1} = \begin{bmatrix} 2 & -2 & 0 \\ -4 & 0 & 4 \\ 3 & 3 & -3 \end{bmatrix}$

using a graphing calculator.

**59.** $2.1x + y = \sqrt{5}$
$\sqrt{2}x - 2y = 5$
Given
$A = \begin{bmatrix} 2.1 & 1 \\ \sqrt{2} & -2 \end{bmatrix}$, $X = \begin{bmatrix} x \\ y \end{bmatrix}$, and $B = \begin{bmatrix} \sqrt{5} \\ 5 \end{bmatrix}$,

using a graphing calculator, we have

$X = A^{-1}B = \begin{bmatrix} 1.68717058 \\ -1.306990242 \end{bmatrix}$. Thus, the

solution set is
$\{(1.68717058, -1.306990242)\}$.

**61.** $(\log 2)x + (\ln 3)y + (\ln 4)z = 1$
$(\ln 3)x + (\log 2)y + (\ln 8)z = 5$
$(\log 12)x + (\ln 4)y + (\ln 8)z = 9$

Given $A = \begin{bmatrix} \log 2 & \ln 3 & \ln 4 \\ \ln 3 & \log 2 & \ln 8 \\ \log 12 & \ln 4 & \ln 8 \end{bmatrix}$, $X = \begin{bmatrix} x \\ y \\ z \end{bmatrix}$,

and $B = \begin{bmatrix} 1 \\ 5 \\ 9 \end{bmatrix}$, using a graphing calculator, we

have $X = A^{-1}B = \begin{bmatrix} 13.58736702 \\ 3.929011993 \\ -5.342780076 \end{bmatrix}$.

Solution set:
$\{(13.58736702, 3.929011993, -5.342780076)\}$

**63.** Answers will vary.

**65.** Let $A = \begin{bmatrix} 1 & 0 \\ 1 & 1 \end{bmatrix}$ and $B = \begin{bmatrix} 1 & 1 \\ 0 & 1 \end{bmatrix}$. (Other

answers are possible.)
$AB = \begin{bmatrix} 1 & 0 \\ 1 & 1 \end{bmatrix}\begin{bmatrix} 1 & 1 \\ 0 & 1 \end{bmatrix}$
$= \begin{bmatrix} 1(1)+0(0) & 1(1)+0(1) \\ 1(1)+1(0) & 1(1)+1(1) \end{bmatrix}$
$= \begin{bmatrix} 1+0 & 1+0 \\ 1+0 & 1+1 \end{bmatrix} = \begin{bmatrix} 1 & 1 \\ 1 & 2 \end{bmatrix}$

Find $(AB)^{-1}$. Since $\begin{bmatrix} AB|I_2 \end{bmatrix} = \begin{bmatrix} 1 & 1 & 1 & 0 \\ 1 & 2 & 0 & 1 \end{bmatrix}$,

$\begin{bmatrix} 1 & 1 & 1 & 0 \\ 0 & 1 & -1 & 1 \end{bmatrix} \begin{matrix} \\ -R1+R2 \end{matrix} \Rightarrow$

$\begin{bmatrix} 1 & 0 & 2 & -1 \\ 0 & 1 & -1 & 1 \end{bmatrix} -R2+R1$

Thus, $(AB)^{-1} = \begin{bmatrix} 2 & -1 \\ -1 & 1 \end{bmatrix}$.

Now find $A^{-1}$. Since

$\begin{bmatrix} A|I_2 \end{bmatrix} = \begin{bmatrix} 1 & 0 & 1 & 0 \\ 1 & 1 & 0 & 1 \end{bmatrix} \Rightarrow$

$\begin{bmatrix} 1 & 0 & 1 & 0 \\ 0 & 1 & -1 & 1 \end{bmatrix} -R1+R2$

Thus, $A^{-1} = \begin{bmatrix} 1 & 0 \\ -1 & 1 \end{bmatrix}$.

Now find $B^{-1}$. Since

$$\left[B\,|\,I_2\right] = \begin{bmatrix} 1 & 1 & | & 1 & 0 \\ 0 & 1 & | & 0 & 1 \end{bmatrix} \Rightarrow$$

$$\begin{bmatrix} 1 & 0 & | & 1 & -1 \\ 0 & 1 & | & 0 & 1 \end{bmatrix} -R2 + R1$$

Thus, $B^{-1} = \begin{bmatrix} 1 & -1 \\ 0 & 1 \end{bmatrix}$.

Now find $A^{-1}B^{-1}$.

$$A^{-1}B^{-1} = \begin{bmatrix} 1 & 0 \\ -1 & 1 \end{bmatrix}\begin{bmatrix} 1 & -1 \\ 0 & 1 \end{bmatrix}$$

$$= \begin{bmatrix} 1(1)+0(0) & 1(-1)+0(1) \\ -1(1)+1(0) & -1(-1)+1(1) \end{bmatrix}$$

$$= \begin{bmatrix} 1+0 & -1+0 \\ -1+0 & 1+1 \end{bmatrix} = \begin{bmatrix} 1 & -1 \\ -1 & 2 \end{bmatrix}$$

Thus $(AB)^{-1} \neq A^{-1}B^{-1}$.

**67.** Given $A = \begin{bmatrix} a & 0 & 0 \\ 0 & b & 0 \\ 0 & 0 & c \end{bmatrix}$, and since $a$, $b$, and $c$

are all nonzero, $\frac{1}{a}, \frac{1}{b}$, and $\frac{1}{c}$ all exist. Thus,

$$\left[A\,|\,I\right] = \begin{bmatrix} a & 0 & 0 & | & 1 & 0 & 0 \\ 0 & b & 0 & | & 0 & 1 & 0 \\ 0 & 0 & c & | & 0 & 0 & 1 \end{bmatrix} \Rightarrow$$

$$\begin{bmatrix} 1 & 0 & 0 & | & \frac{1}{a} & 0 & 0 \\ 0 & 1 & 0 & | & 0 & \frac{1}{b} & 0 \\ 0 & 0 & 1 & | & 0 & 0 & \frac{1}{c} \end{bmatrix} \begin{matrix} \frac{1}{a}R1 \\ \frac{1}{b}R2 \\ \frac{1}{c}R3 \end{matrix}$$

and $A^{-1} = \begin{bmatrix} \frac{1}{a} & 0 & 0 \\ 0 & \frac{1}{b} & 0 \\ 0 & 0 & \frac{1}{c} \end{bmatrix}$.

**69.** Find the inverses of $I_n$, $-A$, and $kA$.

$I_n$ is its own inverse, since $I_n \cdot I_n = I_n$.

The inverse of $-A$ is $-\left(A^{-1}\right)$, since

$$(-A)\left[-\left(A^{-1}\right)\right] = (-1)(-1)\left(A \cdot A^{-1}\right) = I_n.$$

The inverse of $kA$ ($k$ a scalar) is $\frac{1}{k}A^{-1}$, since

$$(kA)\left(\tfrac{1}{k}A^{-1}\right) = \left(k \cdot \tfrac{1}{k}\right)\left(A \cdot A^{-1}\right) = I_n.$$

## Chapter 5: Review Exercises

**1.** $2x + 6y = 6$  (1)
$5x + 9y = 9$  (2)
Solving equation (1) for $x$, we have
$2x + 6y = 6 \Rightarrow 2x = 6 - 6y \Rightarrow x = 3 - 3y$  (3).
Substituting equation (3) into equation (2) and solving for $y$, we have the following.
$5(3 - 3y) + 9y = 9 \Rightarrow 15 - 15y + 9y = 9 \Rightarrow$
$\qquad 15 - 6y = 9 \Rightarrow -6y = -6 \Rightarrow y = 1$
Substituting 1 for $y$ in equation (3), we have
$x = 3 - 3(1) = 3 - 3 = 0$.
Solution set: $\{(0, 1)\}$

**3.** $x + 5y = 9$  (1)
$2x + 10y = 18$ (2)
Multiply equation (1) by $-2$ and add the resulting equation to equation (2).
$$\begin{array}{r} -2x - 10y = -18 \\ 2x + 10y = \phantom{-}18 \\ \hline 0 = 0 \end{array}$$
This is a true statement and implies that the system has infinitely many solutions. Solving equation (1) for $x$ we have $x = 9 - 5y$. Given $y$ as an arbitrary value, the solution set is
$\{(9 - 5y, y)\}$.

**5.** $y = -x + 3$  (1)
$2x + 2y = 1$ (2)
Substituting equation (1) into equation (2), we have $2x + 2(-x + 3) = 1$. Solving for $x$ we have $2x + 2(-x + 3) = 1 \Rightarrow$
$2x + (-2x) + 6 = 1 \Rightarrow 6 = 1$. This is a false statement and the solution is inconsistent. Thus, the solution set is $\varnothing$.

**7.** $3x - 2y = 0$ (1)
$9x + 8y = 7$ (2)
Multiply equation (1) by $-3$ and add the result to equation (2).
$$\begin{array}{r} -9x + 6y = 0 \\ 9x + 8y = 7 \\ \hline 14y = 7 \Rightarrow y = \tfrac{1}{2} \end{array}$$
Substituting $\frac{1}{2}$ for $y$ in equation (1) and solving for $x$, we have the following.
$3x - 2\left(\tfrac{1}{2}\right) = 0 \Rightarrow 3x - 1 = 0$
$\qquad 3x = 1 \Rightarrow x = \tfrac{1}{3}$
Solution set: $\left\{\left(\tfrac{1}{3}, \tfrac{1}{2}\right)\right\}$

**9.** $2x - 5y + 3z = -1$ (1)
$\phantom{2}x + 4y - 2z = 9$  (2)
$-x + 2y + 4z = 5$  (3)

First, we eliminate $x$. Multiply equation (2) by $-2$ and add the result to equation (1).

$\phantom{-}2x - 5y + 3z = -1$
$\underline{-2x - 8y + 4z = -18}$
$\phantom{-2x}-13y + 7z = -19$ (4)

Next, add equations (2) and (3) to obtain
$6y + 2z = 14$ (5). Now, we solve the system

$-13y + 7z = -19$ (4)
$\phantom{-1}6y + 2z = \phantom{-1}14$ (5)

Multiply equation (4) by 2, multiply equation (5) by $-7$, and add the resulting equations.

$-26y + 14z = -38$
$\underline{-42y - 14z = -98}$
$-68y \phantom{+14z}= -136 \Rightarrow y = 2$

Substituting this value into equation (4), we have $-13(2) + 7z = -19 \Rightarrow$

$-26 + 7z = -19 \Rightarrow 7z = 7 \Rightarrow z = 1$.

Substituting 2 for $y$ and 1 for $z$ in equation (2), we have $x + 4(2) - 2(1) = 9 \Rightarrow$

$x + 8 - 2 = 9 \Rightarrow x = 3$.

Solution set: $\{(3, 2, 1)\}$

**11.** $5x - y = 26$ (1)

$4y + 3z = -4$ (2)

$3x + 3z = 15$  (3)

Multiply equation (2) by $-1$ and add the result to equation (3) in order to eliminate $z$.

$\phantom{3x}- 4y - 3z = 4$
$\underline{3x \phantom{- 4y}+ 3z = 15}$
$3x - 4y \phantom{+ 3z}= 19$ (4)

Multiply equation (1) by $-4$ and add the result to equation (4).

$-20x + 4y = -104$
$\underline{\phantom{-20}3x - 4y = \phantom{-10}19}$
$-17x \phantom{+ 4y}= -85 \Rightarrow x = 5$

Substitute 5 for $x$ in equation (1) and solve for $y$ to obtain

$5(5) - y = 26 \Rightarrow 25 - y = 26 \Rightarrow y = -1$.

Substitute 5 for $x$ equation (3) and solve for $z$:

$3(5) + 3z = 15 \Rightarrow 15 + 3z = 15 \Rightarrow$

$3z = 0 \Rightarrow z = 0$

Solution set: $\{(5, -1, 0)\}$

**13.** One possible answer is $\begin{array}{l} x + y = 2 \\ x + y = 3 \end{array}$.

**15.** Let $x$ = amount of rice; $y$ = amount of soybeans.

We get the following system of equations. The first equation relates protein and the second relates calories.

$15x + 22.5y = 9.5$  (1)
$810x + 270y = 324$ (2)

Multiply equation (1) by $-12$ and add the result to equation (2).

$-180x - 270y = -114$
$\underline{\phantom{-1}810x + 270y = \phantom{-}324}$
$\phantom{-1}630x \phantom{+ 270y}= \phantom{-}210 \Rightarrow x = \frac{1}{3}$

Substitute $\frac{1}{3}$ for x in equation (1) and solve for $y$.

$15\left(\frac{1}{3}\right) + 22.5y = 9.5 \Rightarrow 5 + 22.5y = 9.5 \Rightarrow$

$22.5y = 4.5 \Rightarrow y = .20 = \frac{1}{5}$

$\frac{1}{3}$ cup of rice and $\frac{1}{5}$ cup of soybeans should be used.

**17.** Let $x$ = the number of blankets, $y$ = the number of rugs, $z$ = the number of skirts.

The following table is helpful in organizing the information.

|  | Number of blankets ($x$) | Number of rugs ($y$) | Number of skirts ($z$) | Available |
|---|---|---|---|---|
| Spinning yarn | 24 | 30 | 12 | 306 |
| Dying | 4 | 5 | 3 | 59 |
| Weaving | 15 | 18 | 9 | 201 |

We have the following system of equations.

$24x + 30y + 12z = 306 \Rightarrow 4x + 5y + 2z = 51$ (1)
$\phantom{24}4x + \phantom{3}5y + \phantom{1}3z = 59$  (2)
$15x + 18y + \phantom{1}9z = 201 \Rightarrow 5x + 6y + 3z = 67$ (3)

Multiply equation (1) by $-1$ and add the result to equation (2).

$-4x - 5y - 2z = -51$
$\underline{\phantom{-}4x + 5y + 3z = \phantom{-}59}$
$\phantom{-4x + 5y +}z = \phantom{-5}8$

Substitute 8 for z in equations (1) and (3) and simplify.

$4x + 5y + 2(8) = 51 \Rightarrow 4x + 5y + 16 = 51 \Rightarrow$
$\phantom{4x + 5y + 2(8)}4x + 5y = 35$  (4)
$5x + 6y + 3(8) = 67 \Rightarrow 5x + 6y + 24 = 67 \Rightarrow$
$\phantom{5x + 6y + 3(8)}5x + 6y = 43$  (5)

Multiply equation (4) by 5 and equation (5) by −4 and add the results.

$$20x + 25y = \phantom{-}175$$
$$\underline{-20x - 24y = -172}$$
$$\phantom{-20x - 2}y = \phantom{-17}3$$

Substituting 3 for y in equation (4) and solving for x, we obtain $4x + 5(3) = 35 \Rightarrow$

$$4x + 15 = 35 \Rightarrow 4x = 20 \Rightarrow x = 5$$

5 blankets, 3 rugs, and 8 skirts can be made.

**19.** $H = .491x + .468y + 11.2$
$H = -.981x + 1.872y + 26.4$

Substitute $H = 180$ and solve each equation for y so they can be entered into the graphing calculator.

$$180 = .491x + .468y + 11.2$$
$$168.8 - .491x = .468y \Rightarrow y = \tfrac{168.8 - .491x}{.468}$$
$$180 = -.981x + 1.872y + 26.4$$
$$153.6 + .981x = 1.872y \Rightarrow y = \tfrac{153.6 + .981x}{1.872}$$

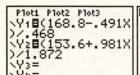

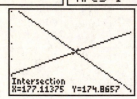

Thus, we have $x \approx 177.1$ and $y \approx 174.9$.

If an athlete's maximum heart rate is 180 beats per minute, then it will be about 177 beats per minute 5 seconds after stopping and 175 beats per second after stopping.

**21.** Since $y = ax^2 + bx + c$ and the points $(1, -2.3), (2, -1.3),$ and $(3, 4.5)$ are on the parabola, we have the following system of equations.

$$-2.3 = a(1)^2 + b(1) + c \Rightarrow a + b + c = -2.3 \ (1)$$
$$-1.3 = a(2)^2 + b(2) + c \Rightarrow 4a + 2b + c = -1.3 \ (2)$$
$$4.5 = a(3)^2 + b(3) + c \Rightarrow 9a + 3b + c = 4.5 \ (3)$$

First, eliminate c by multiplying equation (1) by −1 and adding the result to equation (2).

$$-a - \phantom{2}b - c = \phantom{-}2.3$$
$$\underline{4a + 2b + c = -1.3}$$
$$\phantom{-}3a + \phantom{2}b \phantom{+ c} = \phantom{-}1 \ (4)$$

Next, multiply equation (2) by −1 and add to equation (3).

$$-4a - 2b - c = 1.3$$
$$\underline{\phantom{-}9a + 3b + c = 4.5}$$
$$\phantom{-}5a + \phantom{2}b \phantom{+ c} = 5.8 \ (5)$$

Next, eliminate b by multiplying equation (4) by −1 and adding the result to equation (5).

$$-3a - b = -1$$
$$\underline{\phantom{-}5a + b = 5.8}$$
$$\phantom{-}2a \phantom{+ b} = 4.8 \Rightarrow a = 2.4$$

Substitute this value into equation (4) to obtain $3(2.4) + b = 1 \Rightarrow 7.2 + b = 1 \Rightarrow b = -6.2.$

Substitute 2.4 for a and −6.2 for b into equation (1) in order to solve for c.

$$(2.4) + (-6.2) + c = -2.3$$
$$-3.8 + c = -2.3 \Rightarrow c = 1.5$$

The equation of the parabola is
$$y = 2.4x^2 - 6.2x + 1.5 \text{ or}$$
$$Y_1 = 2.4X^2 - 6.2X + 1.5.$$

**23.** $3x - 4y + z = 2 \ (1)$
$\phantom{3x -} 2x + y = 1 \ (2)$

Solving equation (2) for y, we obtain $y = 1 - 2x$. Substitute $1 - 2x$ for y in equation (1) and solve for z.

$$3x - 4(1 - 2x) + z = 2 \Rightarrow 3x - 4 + 8x + z = 2 \Rightarrow$$
$$11x + z = 6 \Rightarrow z = 6 - 11x$$

Solution set: $\{(x, 1 - 2x, 6 - 11x)\}$

**25.** Writing $\begin{matrix} 5x + 2y = -10 \\ 3x - 5y = -6 \end{matrix}$ as an augmented

matrix we have $\left[\begin{array}{cc|c} 5 & 2 & -10 \\ 3 & -5 & -6 \end{array}\right].$

$$\left[\begin{array}{cc|c} 1 & \frac{2}{5} & -2 \\ 3 & -5 & -6 \end{array}\right] \frac{1}{5}R1 \Rightarrow$$

$$\left[\begin{array}{cc|c} 1 & \frac{2}{5} & -2 \\ 0 & -\frac{31}{5} & 0 \end{array}\right] -3R1 + R2 \Rightarrow$$

$$\left[\begin{array}{cc|c} 1 & \frac{2}{5} & -2 \\ 0 & 1 & 0 \end{array}\right] -\frac{5}{31}R2 \Rightarrow$$

$$\left[\begin{array}{cc|c} 1 & 0 & -2 \\ 0 & 1 & 0 \end{array}\right] -\frac{2}{5}R2 + R1$$

Solution set: $\{(-2, 0)\}$

**27.** Writing
$$2x - y + 4z = -1$$
$$-3x + 5y - z = 5$$
$$2x + 3y + 2z = 3$$
as an augmented

matrix we have $\begin{bmatrix} 2 & -1 & 4 & | & -1 \\ -3 & 5 & -1 & | & 5 \\ 2 & 3 & 2 & | & 3 \end{bmatrix}$.

$\begin{bmatrix} 1 & -\frac{1}{2} & 2 & | & -\frac{1}{2} \\ -3 & 5 & -1 & | & 5 \\ 2 & 3 & 2 & | & 3 \end{bmatrix} \frac{1}{2}R1 \quad \Rightarrow$

$\begin{bmatrix} 1 & -\frac{1}{2} & 2 & | & -\frac{1}{2} \\ 0 & \frac{7}{2} & 5 & | & \frac{7}{2} \\ 0 & 4 & -2 & | & 4 \end{bmatrix} \begin{matrix} \\ 3R1+R2 \\ -2R1+R3 \end{matrix} \Rightarrow$

$\begin{bmatrix} 1 & -\frac{1}{2} & 2 & | & -\frac{1}{2} \\ 0 & 1 & \frac{10}{7} & | & 1 \\ 0 & 4 & -2 & | & 4 \end{bmatrix} \frac{2}{7}R2 \quad \Rightarrow$

$\begin{bmatrix} 1 & 0 & \frac{19}{7} & | & 0 \\ 0 & 1 & \frac{10}{7} & | & 1 \\ 0 & 0 & -\frac{54}{7} & | & 0 \end{bmatrix} \begin{matrix} \frac{1}{2}R2+R1 \\ \\ -4R2+R3 \end{matrix} \Rightarrow$

$\begin{bmatrix} 1 & 0 & \frac{19}{7} & | & 0 \\ 0 & 1 & \frac{10}{7} & | & 1 \\ 0 & 0 & 1 & | & 0 \end{bmatrix} \begin{matrix} \\ \\ -\frac{7}{54}R2 \end{matrix} \Rightarrow$

$\begin{bmatrix} 1 & 0 & 0 & | & 0 \\ 0 & 1 & 0 & | & 1 \\ 0 & 0 & 1 & | & 0 \end{bmatrix} \begin{matrix} -\frac{19}{7}R3+R1 \\ -\frac{10}{7}R3+R2 \\ \end{matrix}$

Solution set: $\{(0,1,0)\}$

**29.** Let $x$ = number of pounds of \$4.60 tea;
$y$ = the number of pounds of \$5.75 tea;
$z$ = the number of pounds of \$6.50 tea.
From the information in the exercise, we obtain the system
$$x + y + z = 20$$
$$4.6x + 5.75y + 6.5z = 20(5.25).$$
$$x = y + z$$
Rewriting the system, we have
$$x + y + z = 20$$
$$4.6x + 5.75y + 6.5z = 105.$$
$$x - y - z = 0$$
With an augmented matrix of
$$\begin{bmatrix} 1 & 1 & 1 & | & 20 \\ 4.6 & 5.75 & 6.5 & | & 105 \\ 1 & -1 & -1 & | & 0 \end{bmatrix},$$
we solve by the Gauss-Jordan method.

$\begin{bmatrix} 1 & 1 & 1 & | & 20 \\ 0 & 1.15 & 1.9 & | & 13 \\ 0 & -2 & -2 & | & -20 \end{bmatrix} \begin{matrix} \\ -4.6R1+R2 \\ -1R1+R3 \end{matrix} \Rightarrow$

$\begin{bmatrix} 1 & 1 & 1 & | & 20 \\ 0 & 1 & \frac{38}{23} & | & \frac{260}{23} \\ 0 & -2 & -2 & | & -20 \end{bmatrix} \frac{20}{23}R2 \quad \Rightarrow$

$\begin{bmatrix} 1 & 0 & -\frac{15}{23} & | & \frac{200}{23} \\ 0 & 1 & \frac{38}{23} & | & \frac{260}{23} \\ 0 & 0 & \frac{30}{23} & | & \frac{60}{23} \end{bmatrix} \begin{matrix} -1R2+R1 \\ \\ 2R2+R3 \end{matrix} \Rightarrow$

$\begin{bmatrix} 1 & 0 & -\frac{15}{23} & | & \frac{200}{23} \\ 0 & 1 & \frac{38}{23} & | & \frac{260}{23} \\ 0 & 0 & 1 & | & 2 \end{bmatrix} \begin{matrix} \\ \\ \frac{23}{30}R3 \end{matrix} \Rightarrow$

$\begin{bmatrix} 1 & 0 & 0 & | & 10 \\ 0 & 1 & 0 & | & 8 \\ 0 & 0 & 1 & | & 2 \end{bmatrix} \begin{matrix} \frac{15}{23}R3+R1 \\ -\frac{38}{23}R3+R2 \\ \end{matrix}$

From the final matrix, $x = 10$, $y = 8$, and $z = 2$. Therefore, 10 lb of \$4.60 tea, 8 lb of \$5.75 tea, and 2 lb of \$6.50 tea should be used.

**31.** $y = .11x + 2.9$ (1)
$y = .19x + 1.4$ (2)
The given system can be solved by substitution. Substitute equation (1) into equation (2) to obtain $.11x + 2.9 = .19x + 1.4$. Solving this equation, we have
$1.5 = .08x \Rightarrow x = 18.75$.
Since $x = 0$ corresponds 1960, $x = 18.75$ implies during 1979 the male and female enrollments were the same.

The enrollment of males in this year is
$y = .11(18.75) + 2.9 = 4.9625$ million. The enrollment of females in this year (which is the same as the males) is
$y = .19(18.75) + 1.4 = 4.9625$ million. The total enrollment was therefore approximately 10 million.

**33.** $\begin{vmatrix} -1 & 8 \\ 2 & 9 \end{vmatrix} = -1(9) - 2(8) = -9 - 16 = -25$

**35.** Expanding $\begin{vmatrix} -1 & 2 & 3 \\ 4 & 0 & 3 \\ 5 & -1 & 2 \end{vmatrix}$ by the second column,

we have

$$-2\begin{vmatrix} 4 & 3 \\ 5 & 2 \end{vmatrix} + 0\begin{vmatrix} -1 & 3 \\ 5 & 2 \end{vmatrix} - (-1)\begin{vmatrix} -1 & 3 \\ 4 & 3 \end{vmatrix}$$
$$= -2\left[4(2) - 5(3)\right] + 0 + \left[(-1)(3) - (4)(3)\right]$$
$$= -2(8 - 15) + (-3 - 12) = -2(-7) + (-15)$$
$$= 14 - 15 = -1$$

**37.** $\begin{vmatrix} 3x & 7 \\ -x & 4 \end{vmatrix} = 8 \Rightarrow$

$$12x - (-7x) = 8 \Rightarrow 19x = 8 \Rightarrow x = \tfrac{8}{19}$$

Solution set: $\left\{\tfrac{8}{19}\right\}$

**39.** $3x + 7y = 2$
$5x - y = -22$

$$D = \begin{vmatrix} 3 & 7 \\ 5 & -1 \end{vmatrix} = 3(-1) - 5(7) = -3 - 35 = -38$$

$$D_x = \begin{vmatrix} 2 & 7 \\ -22 & -1 \end{vmatrix} = 2(-1) - (-22)(7)$$
$$= -2 - (-154) = 152$$

$$D_y = \begin{vmatrix} 3 & 2 \\ 5 & -22 \end{vmatrix} = 3(-22) - 5(2)$$
$$= -66 - 10 = -76$$

$$x = \frac{D_x}{D} = \frac{152}{-38} = -4 \text{ and } y = \frac{D_y}{D} = \frac{-76}{-38} = 2.$$

Solution set: $\left\{(-4, 2)\right\}$

**41.** Given $\begin{array}{l} 6x + y = -3 \quad (1) \\ 12x + 2y = 1 \quad (2) \end{array}$, we have

$$D = \begin{vmatrix} 6 & 1 \\ 12 & 2 \end{vmatrix} = 6(2) - 12(1) = 12 - 12 = 0.$$

Because $D = 0$, Cramer's rule does not apply. To determine whether the system is inconsistent or has infinitely many solutions, use the elimination method.

$\begin{array}{l} -12x - 2y = 6 \text{ Multiply equation (1) by } -2. \\ \underline{12x + 2y = 1} \\ \phantom{12x + 2y =} 0 = 7 \text{ False} \end{array}$

The system is inconsistent. Thus the solution set is $\varnothing$.

**43.** $x + y = -1$
$2y + z = 5$
$3x - 2z = -28$

Adding $-3$ times row 1 to row 3, we have

$$D = \begin{vmatrix} 1 & 1 & 0 \\ 0 & 2 & 1 \\ 3 & 0 & -2 \end{vmatrix} = \begin{vmatrix} 1 & 1 & 0 \\ 0 & 2 & 1 \\ 0 & -3 & -2 \end{vmatrix}$$

Expanding by column one, we have

$$D = \begin{vmatrix} 2 & 1 \\ -3 & -2 \end{vmatrix} = -4 - (-3) = -1$$

Adding $-2$ times row 1 to row 2, we have

$$D_x = \begin{vmatrix} -1 & 1 & 0 \\ 5 & 2 & 1 \\ -28 & 0 & -2 \end{vmatrix} = \begin{vmatrix} -1 & 1 & 0 \\ 7 & 0 & 1 \\ -28 & 0 & -2 \end{vmatrix}$$

Expanding about column two, we have

$$D_x = -(1)\begin{vmatrix} 7 & 1 \\ -28 & -2 \end{vmatrix} = -\left[-14 - (-28)\right] = -14$$

Adding 2 times row 2 to row 3, we have

$$D_y = \begin{vmatrix} 1 & -1 & 0 \\ 0 & 5 & 1 \\ 3 & -28 & -2 \end{vmatrix} = \begin{vmatrix} 1 & -1 & 0 \\ 0 & 5 & 1 \\ 3 & -18 & 0 \end{vmatrix}$$

Expanding by column three, we have

$$D_y = -(1)\begin{vmatrix} 1 & -1 \\ 3 & -18 \end{vmatrix}$$
$$= -1\left[-18 - (-3)\right] = -(-15) = 15$$

Adding $-3$ times row 1 to row 3, we have

$$D_z = \begin{vmatrix} 1 & 1 & -1 \\ 0 & 2 & 5 \\ 3 & 0 & -28 \end{vmatrix} = \begin{vmatrix} 1 & 1 & -1 \\ 0 & 2 & 5 \\ 0 & -3 & -25 \end{vmatrix}$$

Expanding by column one, we have

$$D_z = 1\begin{vmatrix} 2 & 5 \\ -3 & -25 \end{vmatrix} = -50 - (-15) = -35.$$

Thus, we have

$$x = \frac{D_x}{D} = \frac{-14}{-1} = 14, \quad y = \frac{D_y}{D} = \frac{15}{-1} = -15,$$
$$z = \frac{D_z}{D} = \frac{-35}{-1} = 35$$

Solution set: $\left\{(14, -15, 35)\right\}$

**45.** $\dfrac{2}{3x^2 - 5x + 2} = \dfrac{2}{(x - 1)(3x - 2)} = \dfrac{A}{x - 1} + \dfrac{B}{3x - 2}$

Multiply both sides by $(x - 1)(3x - 2)$ to get

$$2 = A(3x - 2) + B(x - 1). \; (1)$$

First substitute 1 for $x$ to get

$$2 = A\left[3(1) - 2\right] + B(1 - 1)$$
$$2 = A(1) + B(0) \Rightarrow A = 2$$

Replace $A$ with 2 in equation (1) and substitute $\frac{2}{3}$ for $x$ to get

$$2 = 2\left[3\left(\tfrac{2}{3}\right) - 2\right] + B\left(\tfrac{2}{3} - 1\right)$$

$$2 = 2(0) + B\left(-\tfrac{1}{3}\right) \Rightarrow 2 = -\tfrac{1}{3}B \Rightarrow B = -6$$

Thus, we have

$$\frac{2}{3x^2 - 5x + 2} = \frac{2}{x - 1} + \frac{-6}{3x - 2} \text{ or } \frac{2}{x - 1} - \frac{6}{3x - 2}$$

**47.** $\dfrac{5 - 2x}{\left(x^2 + 2\right)\left(x - 1\right)} = \dfrac{A}{x - 1} + \dfrac{Bx + C}{x^2 + 2}$

Multiply both sides by $\left(x^2 + 2\right)\left(x - 1\right)$ to get the following.

$$5 - 2x = A\left(x^2 + 2\right) + (Bx + C)(x - 1) \ (1)$$

First substitute 1 for $x$ to get the following.

$$5 - 2(1) = A\left(1^2 + 2\right) + \left[B(1) + C\right](1 - 1)$$

$$5 - 2 = A(1 + 2) + 0 \Rightarrow 3 = 3A \Rightarrow A = 1$$

Replace $A$ with 1 in equation (1) and substitute 0 for $x$ to get the following.

$$5 - 2(0) = \left(0^2 + 2\right) + \left[B(0) + C\right](0 - 1)$$

$$5 - 0 = (0 + 2) + C(-1) \Rightarrow 5 = 2 - C \Rightarrow$$

$$3 = -C \Rightarrow C = -3$$

$$5 - 2x = \left(x^2 + 2\right) + (Bx - 3)(x - 1) \ (2)$$

Substitute 2 (arbitrary value) in equation (2) for $x$ to get the following.

$$5 - 2(2) = \left(2^2 + 2\right) + \left[B(2) - 3\right](2 - 1)$$

$$5 - 4 = (4 + 2) + (2B - 3)(1)$$

$$1 = 6 + 2B - 3 \Rightarrow 1 = 3 + 2B$$

$$-2 = 2B \Rightarrow B = -1$$

Thus, we have

$$\frac{5 - 2x}{\left(x^2 + 2\right)(x - 1)} = \frac{1}{x - 1} + \frac{(-1)x + (-3)}{x^2 + 2} \text{ or }$$

$$\frac{1}{x - 1} - \frac{x + 3}{x^2 + 2}$$

**49.** $y = 2x + 10$ (1)

$x^2 + y = 13$ (2)

Substituting equation (1) into equation (2), we have $x^2 + (2x + 10) = 13$. Solving for $x$ we get the following.

$$x^2 + (2x + 10) = 13$$

$$x^2 + 2x - 3 = 0$$

$$(x + 3)(x - 1) = 0 \Rightarrow x = -3 \text{ or } x = 1$$

For each value of $x$, use equation (1) to find the corresponding value of $y$.

If $x = -3$, $y = 2(-3) + 10 = -6 + 10 = 4$.

If $x = 1$, $y = 2(1) + 10 = 2 + 10 = 12$.

Solution set: $\{(-3, 4), (1, 12)\}$

**51.** $x^2 + y^2 = 17$ (1)

$2x^2 - y^2 = 31$ (2)

Add equations (1) and (2) to obtain

$$3x^2 = 48 \Rightarrow x^2 = 16 \Rightarrow x = \pm 4.$$

For each value of $x$, use equation (1) to find the corresponding value of $y$.

If $x = -4$, $(-4)^2 + y^2 = 17 \Rightarrow 16 + y^2 = 17 \Rightarrow$

$y^2 = 1 \Rightarrow y = \pm 1$. If $x = 4$, $4^2 + y^2 = 17 \Rightarrow$

$16 + y^2 = 17 \Rightarrow y^2 = 1 \Rightarrow y = \pm 1.$

Solution set: $\{(-4, 1), (-4, -1), (4, -1), (4, 1)\}$

**53.** $xy = -10$ (1)

$x + 2y = 1$ (2)

Solving equation (2) for $x$ gives $x = 1 - 2y$ (3). Substituting this expression into equation (1), we obtain $(1 - 2y)y = -10$.

Solving for $y$ we have

$$(1 - 2y)y = -10 \Rightarrow y - 2y^2 = -10$$

$$0 = 2y^2 - y - 10$$

$$(y + 2)(2y - 5) = 0 \Rightarrow y = -2 \text{ or } y = \tfrac{5}{2}$$

For each value of $y$, use equation (3) to find the corresponding value of $x$.

If $y = -2$, $x = 1 - 2(-2) = 1 - (-4) = 5$.

If $y = \tfrac{5}{2}$, $x = 1 - 2\left(\tfrac{5}{2}\right) = 1 - 5 = -4$.

Solution set: $\left\{(5, -2), \left(-4, \tfrac{5}{2}\right)\right\}$

**55.** $x^2 + 2xy + y^2 = 4$ (1)

$x - 3y = -2$ (2)

Solve equation (2) for $x$: $x = 3y - 2$ (3).

Substituting this expression into equation (1), we obtain $(3y - 2)^2 + 2(3y - 2)y + y^2 = 4$.

Solving for $x$ we have the following.

$$(3y - 2)^2 + 2(3y - 2)y + y^2 = 4$$

$$9y^2 - 12y + 4 + 6y^2 - 4y + y^2 = 4$$

$$16y^2 - 16y + 4 = 4$$

$$16y^2 - 16y = 0$$

$$y^2 - y = 0$$

$$y(y - 1) = 0$$

$$y = 0 \text{ or } y = 1$$

(*continued on next page*)

(*continued from page 323*)

For each value of $y$, use equation (3) to find the corresponding value of $x$.

If $y = 0$, $x = 3(0) - 2 = 0 - 2 = -2$.

If $y = 1$, $x = 3(1) - 2 = 3 - 2 = 1$.

Solution set: $\{(-2, 0), (1, 1)\}$

**57.**    $3x - y = b$  (1)

$x^2 + y^2 = 25$ (2)

Solving equation (1) for $y$ we have $y = 3x - b$. Substitute this expression into equation (2).

$$x^2 + (3x - b)^2 = 25$$
$$x^2 + 9x^2 - 6bx + b^2 = 25$$
$$10x^2 - 6bx + (b^2 - 25) = 0$$

Recall that there is only one solution when the discriminant equals 0:

$$b^2 - 4ac = 0 \Rightarrow$$
$$(-6b)^2 - 4(10)(b^2 - 25) = 0$$
$$36b^2 - 40b^2 + 1000 = 0$$
$$-4b^2 = -1000$$
$$b^2 = 250 \Rightarrow b = \pm\sqrt{250}$$
$$b = \pm 5\sqrt{10}$$

**59.**    $x + y \le 6$

$2x - y \ge 3$

Graph the solid line $x + y = 6$, which has $x$-intercept 6 and $y$-intercept 6. Since $x + y \le 6 \Rightarrow y \le -x + 6$, shade the region below this line. Graph the solid line $2x - y = 3$, which has $x$-intercept $\frac{3}{2}$ and $y$-intercept –3. Since $2x - y \ge 3 \Rightarrow -y \ge -2x + 3 \Rightarrow y \le 2x - 3$, shade the region below this line.

The solution set is the intersection of these two regions, which is shaded in the final graph.

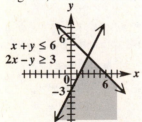

**61.** Find $x \ge 0$ and $y \ge 0$ such that $\begin{matrix} 3x + 2y \le 12 \\ 5x + y \ge 5 \end{matrix}$

and $2x + 4y$ is maximized. To graph $3x + 2y \le 12$, draw the line with $x$-intercept 4 and $y$-intercept 6 as a solid line. Because the test point $(0, 0)$ satisfies this inequality, shade the region *below* the line. To graph $5x + y \ge 5$, draw the line with $x$-intercept 1 and $y$-intercept 5 as a solid line. Because the test point $(0, 0)$ does not satisfy this inequality, shade the region *above* the line. The constraints $x \ge 0$ and $y \ge 0$ restrict the graph to the first quadrant. The graph of the feasible region is the intersection of the regions that are the graphs of the individual constraints. The four vertices are $(1, 0), (4, 0), (0, 5),$ and $(0, 6)$.

| Point | Value of $2x + 4y$ | |
|-------|-------------------|---|
| $(1, 0)$ | $2(1) + 4(0) = 2$ | |
| $(4, 0)$ | $2(4) + 4(0) = 8$ | |
| $(0, 5)$ | $2(0) + 4(5) = 20$ | |
| $(0, 6)$ | $2(0) + 4(6) = 24$ | ← Maximum |

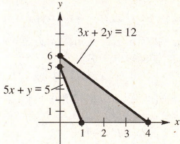

The maximum value is 24, which occurs at $(0, 6)$.

**63.** Let $x$ = number of servings of food A; $y$ = number of servings of food B. The protein constraint is $2x + 6y \ge 30$. The fat constraint is $4x + 2y \ge 20$. Also, $y \ge 2$ and since the numbers of servings cannot be negative, we also have $x \ge 0$. We want to minimize the objective function, cost $= .18x + .12y$. Find the region of feasible solutions by graphing the system of inequalities that is made up of the constraints. To graph $2x + 6y \ge 30$, draw the line with $x$-intercept 15 and $y$-intercept 5 as a solid line. Because the test point $(0, 0)$ does not satisfy this inequality, shade the region above the line.

To graph $4x + 2y \geq 20$, draw the line with x-intercept 5 and y-intercept 10 as a solid line. Because the test point $(0,0)$ does not satisfy this inequality, shade the region above the line. To graph $y \geq 2$, draw the horizontal line with y-intercept 2 as a solid line and shade the region above the line. The graph of the feasible region is the intersection of the regions that are the graphs of the individual constraints.

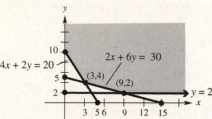

From the graph, observe that one vertex is $(0,10)$. The second occurs when y = 2 and

$$2x + 6y = 30 \Rightarrow 2x + 6(2) = 30 \Rightarrow 2x = 18 \Rightarrow x = 9.$$

Thus, $(9,2)$ is a vertex. The third vertex is the intersection point of the lines $2x + 6y = 30$ and $4x + 2y = 20$. To find this point, solve the system $2x + 6y = 30$
$\qquad\qquad\qquad\qquad\qquad 4x + 2y = 20.$

Multiply the first equation by –2 and add it to the second equation.

$$\begin{array}{r} -4x - 12y = -60 \\ 4x + 2y = 20 \\ \hline -10y = -40 \Rightarrow y = 4 \end{array}$$

Substituting 4 for y into $2x + 6y = 30$, we have the following.

$$2x + 6(4) = 30 \Rightarrow 2x + 24 = 30 \Rightarrow$$
$$2x = 6 \Rightarrow x = 3$$

Thus, the third vertex is $(3,4)$. Next, evaluate the objective function at each vertex.

| Point | Cost $= .18x + .12y$ | |
|-------|----------------------|---|
| $(0,10)$ | $.18(0) + .12(10) = 1.20$ | |
| $(3,4)$ | $.18(3) + .12(4) = 1.02$ | ← Minimum |
| $(9,2)$ | $.18(9) + .12(2) = 1.86$ | |

The minimum cost of $1.02 per serving will be produced by 3 units of food A and 4 units of food B.

**65.** $\begin{bmatrix} 5 & x+2 \\ -6y & z \end{bmatrix} = \begin{bmatrix} a & 3x-1 \\ 5y & 9 \end{bmatrix}$

$a = 5; \; x + 2 = 3x - 1 \Rightarrow 3 = 2x \Rightarrow x = \frac{3}{2};$
$-6y = 5y \Rightarrow 0 = 11y \Rightarrow y = 0; \; z = 9$

Thus, $a = 5,\; x = \frac{3}{2},\; y = 0,$ and $z = 9.$

**67.** $\begin{bmatrix} 3 \\ 2 \\ 5 \end{bmatrix} - \begin{bmatrix} 8 \\ -4 \\ 6 \end{bmatrix} + \begin{bmatrix} 1 \\ 0 \\ 2 \end{bmatrix} = \begin{bmatrix} 3-8 \\ 2-(-4) \\ 5-6 \end{bmatrix} + \begin{bmatrix} 1 \\ 0 \\ 2 \end{bmatrix}$

$\qquad = \begin{bmatrix} -5 \\ 6 \\ -1 \end{bmatrix} + \begin{bmatrix} 1 \\ 0 \\ 2 \end{bmatrix} = \begin{bmatrix} -5+1 \\ 6+0 \\ -1+2 \end{bmatrix}$

$\qquad = \begin{bmatrix} -4 \\ 6 \\ 1 \end{bmatrix}$

**69.** $\begin{bmatrix} 2 & 5 & 8 \\ 1 & 9 & 2 \end{bmatrix} - \begin{bmatrix} 3 & 4 \\ 7 & 1 \end{bmatrix}$

This operation is not possible because one matrix is $2 \times 3$ and the other $2 \times 2$. Matrices of different sizes cannot be added or subtracted.

**71.** $\begin{bmatrix} -1 & 0 \\ 2 & 5 \end{bmatrix} \begin{bmatrix} -3 & 4 \\ 2 & 8 \end{bmatrix}$

$\qquad = \begin{bmatrix} -1(-3)+0(2) & -1(4)+0(8) \\ 2(-3)+5(2) & 2(4)+5(8) \end{bmatrix}$

$\qquad = \begin{bmatrix} 3+0 & -4+0 \\ -6+10 & 8+40 \end{bmatrix} = \begin{bmatrix} 3 & -4 \\ 4 & 48 \end{bmatrix}$

**73.** The product of a $2 \times 3$ matrix and a $3 \times 2$ matrix is a $2 \times 2$ matrix.

$\begin{bmatrix} 3 & 2 & -1 \\ 4 & 0 & 6 \end{bmatrix} \begin{bmatrix} -2 & 0 \\ 0 & 2 \\ 3 & 1 \end{bmatrix} = \begin{bmatrix} 3(-2)+2(0)+(-1)(3) & 3(0)+2(2)+(-1)(1) \\ 4(-2)+0(0)+6(3) & 4(0)+0(2)+6(1) \end{bmatrix} = \begin{bmatrix} -6+0-3 & 0+4-1 \\ -8+0+18 & 0+0+6 \end{bmatrix} = \begin{bmatrix} -9 & 3 \\ 10 & 6 \end{bmatrix}$

**75.** The product of a $3 \times 3$ matrix and a $3 \times 3$ matrix is a $3 \times 3$ matrix.

$$\begin{bmatrix} -2 & 5 & 5 \\ 0 & 1 & 4 \\ 3 & -4 & -1 \end{bmatrix}\begin{bmatrix} 1 & 0 & -1 \\ -1 & 0 & 0 \\ 1 & 1 & -1 \end{bmatrix}$$

$$= \begin{bmatrix} -2(1)+5(-1)+5(1) & -2(0)+5(0)+5(1) & -2(-1)+5(0)+5(-1) \\ 0(1)+1(-1)+4(1) & 0(0)+1(0)+4(1) & 0(-1)+1(0)+4(-1) \\ 3(1)+(-4)(-1)+(-1)(1) & 3(0)+(-4)(0)+(-1)(1) & 3(-1)+(-4)(0)+(-1)(-1) \end{bmatrix}$$

$$= \begin{bmatrix} -2+(-5)+5 & 0+0+5 & 2+0+(-5) \\ 0+(-1)+4 & 0+0+4 & 0+0+(-4) \\ 3+4+(-1) & 0+0+(-1) & -3+0+1 \end{bmatrix} = \begin{bmatrix} -2 & 5 & -3 \\ 3 & 4 & -4 \\ 6 & -1 & -2 \end{bmatrix}$$

**77.** Find the inverse of $A = \begin{bmatrix} 2 & 1 \\ 5 & 3 \end{bmatrix}$, if it exists. We have $[A|I_2] = \begin{bmatrix} 2 & 1 & | & 1 & 0 \\ 5 & 3 & | & 0 & 1 \end{bmatrix}$, which yields

$$\begin{bmatrix} 1 & \frac{1}{2} & | & \frac{1}{2} & 0 \\ 5 & 3 & | & 0 & 1 \end{bmatrix}\begin{matrix} \frac{1}{2}R1 \\ \\ \end{matrix} \Rightarrow \begin{bmatrix} 1 & \frac{1}{2} & | & \frac{1}{2} & 0 \\ 0 & \frac{1}{2} & | & -\frac{5}{2} & 1 \end{bmatrix}\begin{matrix} \\ -5R1+R2 \end{matrix} \Rightarrow \begin{bmatrix} 1 & \frac{1}{2} & | & \frac{1}{2} & 0 \\ 0 & 1 & | & -5 & 2 \end{bmatrix}\begin{matrix} \\ 2R2 \end{matrix} \Rightarrow$$

$$\begin{bmatrix} 1 & 0 & | & 3 & -1 \\ 0 & 1 & | & -5 & 2 \end{bmatrix}\begin{matrix} -\frac{1}{2}R2+R1 \\ \end{matrix}$$

Thus, $A^{-1} = \begin{bmatrix} 3 & -1 \\ -5 & 2 \end{bmatrix}$.

**79.** Find the inverse of $A = \begin{bmatrix} 2 & -1 & 0 \\ 1 & 0 & 1 \\ 1 & -2 & 0 \end{bmatrix}$ if it exists. We have $\left[A \middle| I_3 \right] = \begin{bmatrix} 2 & -1 & 0 & | & 1 & 0 & 0 \\ 1 & 0 & 1 & | & 0 & 1 & 0 \\ 1 & -2 & 0 & | & 0 & 0 & 1 \end{bmatrix}$.

$$\begin{bmatrix} 1 & 0 & 1 & | & 0 & 1 & 0 \\ 2 & -1 & 0 & | & 1 & 0 & 0 \\ 1 & -2 & 0 & | & 0 & 0 & 1 \end{bmatrix}\begin{matrix} R1 \leftrightarrow R2 \\ \\ \end{matrix} \Rightarrow \begin{bmatrix} 1 & 0 & 1 & | & 0 & 1 & 0 \\ 0 & -1 & -2 & | & 1 & -2 & 0 \\ 0 & -2 & -1 & | & 0 & -1 & 1 \end{bmatrix}\begin{matrix} \\ -2R1+R2 \\ -1R1+R3 \end{matrix} \Rightarrow$$

$$\begin{bmatrix} 1 & 0 & 1 & | & 0 & 1 & 0 \\ 0 & 1 & 2 & | & -1 & 2 & 0 \\ 0 & -2 & -1 & | & 0 & -1 & 1 \end{bmatrix}\begin{matrix} \\ -1R2 \\ \end{matrix} \Rightarrow \begin{bmatrix} 1 & 0 & 1 & | & 0 & 1 & 0 \\ 0 & 1 & 2 & | & -1 & 2 & 0 \\ 0 & 0 & 3 & | & -2 & 3 & 1 \end{bmatrix}\begin{matrix} \\ \\ 2R2+R3 \end{matrix} \Rightarrow$$

$$\begin{bmatrix} 1 & 0 & 1 & | & 0 & 1 & 0 \\ 0 & 1 & 2 & | & -1 & 2 & 0 \\ 0 & 0 & 1 & | & -\frac{2}{3} & 1 & \frac{1}{3} \end{bmatrix}\begin{matrix} \\ \\ \frac{1}{3}R3 \end{matrix} \Rightarrow \begin{bmatrix} 1 & 0 & 0 & | & \frac{2}{3} & 0 & -\frac{1}{3} \\ 0 & 1 & 0 & | & \frac{1}{3} & 0 & -\frac{2}{3} \\ 0 & 0 & 1 & | & -\frac{2}{3} & 1 & \frac{1}{3} \end{bmatrix}\begin{matrix} -1R3+R1 \\ -2R3+R2 \\ \end{matrix}$$

Thus, $A^{-1} = \begin{bmatrix} \frac{2}{3} & 0 & -\frac{1}{3} \\ \frac{1}{3} & 0 & -\frac{2}{3} \\ -\frac{2}{3} & 1 & \frac{1}{3} \end{bmatrix}$.

$$3x+2y+\ z=-5$$

**81.** Given $\quad x-\ y+3z=-5\quad$ we have
$$2x+3y+\ z=0$$

$$A=\begin{bmatrix}3&2&1\\1&-1&3\\2&3&1\end{bmatrix},\ X=\begin{bmatrix}x\\y\\z\end{bmatrix},\ \text{and}\ B=\begin{bmatrix}-5\\-5\\0\end{bmatrix}.$$

Finding $A^{-1}$, we have

$$[A\,|\,I_3]=\begin{bmatrix}3&2&1&1&0&0\\1&-1&3&0&1&0\\2&3&1&0&0&1\end{bmatrix}\Rightarrow$$

$$\begin{bmatrix}1&-1&3&0&1&0\\3&2&1&1&0&0\\2&3&1&0&0&1\end{bmatrix}\begin{matrix}R1\leftrightarrow R2\\\\\end{matrix}\Rightarrow$$

$$\begin{bmatrix}1&-1&3&0&1&0\\0&5&-8&1&-3&0\\0&5&-5&0&-2&1\end{bmatrix}\begin{matrix}\\-3R1+R2\\-2R1+R2\end{matrix}\Rightarrow$$

$$\begin{bmatrix}1&-1&3&0&1&0\\0&1&-\frac85&\frac15&-\frac35&0\\0&5&-5&0&-2&1\end{bmatrix}\begin{matrix}\\\frac15R2\\\end{matrix}\Rightarrow$$

$$\begin{bmatrix}1&0&\frac75&\frac15&\frac25&0\\0&1&-\frac85&\frac15&-\frac35&0\\0&0&3&-1&1&1\end{bmatrix}\begin{matrix}R2+R1\\\\-5R2+R3\end{matrix}\Rightarrow$$

$$\begin{bmatrix}1&0&\frac75&\frac15&\frac25&0\\0&1&-\frac85&\frac15&-\frac35&0\\0&0&1&-\frac13&\frac13&\frac13\end{bmatrix}\begin{matrix}\\\\\frac13R3\end{matrix}\Rightarrow$$

$$\begin{bmatrix}1&0&0&\frac23&-\frac1{15}&-\frac7{15}\\0&1&0&-\frac13&-\frac1{15}&\frac8{15}\\0&0&1&-\frac13&\frac13&\frac13\end{bmatrix}\begin{matrix}-\frac75R3+R1\\\frac85R3+R2\\\end{matrix}$$

Thus, $A^{-1}=\begin{bmatrix}\frac23&-\frac1{15}&-\frac7{15}\\-\frac13&-\frac1{15}&\frac8{15}\\-\frac13&\frac13&\frac13\end{bmatrix}.$

Finally, we need to find $X=A^{-1}B$.

$$X=A^{-1}B=\begin{bmatrix}\frac23&-\frac1{15}&-\frac7{15}\\-\frac13&-\frac1{15}&\frac8{15}\\-\frac13&\frac13&\frac13\end{bmatrix}\begin{bmatrix}-5\\-5\\0\end{bmatrix}$$

$$=\begin{bmatrix}\frac23(-5)+\left(-\frac1{15}\right)(-5)+\left(-\frac7{15}\right)(0)\\-\frac13(-5)+\left(-\frac1{15}\right)(-5)+\frac8{15}(0)\\-\frac13(-5)+\frac13(-5)+\frac13(0)\end{bmatrix}$$

$$=\begin{bmatrix}-\frac{10}3+\frac13+0\\\frac53+\frac13+0\\\frac53+\left(-\frac53\right)+0\end{bmatrix}=\begin{bmatrix}-\frac93\\\frac63\\0\end{bmatrix}=\begin{bmatrix}-3\\2\\0\end{bmatrix}$$

Solution set: $\{(-3,2,0)\}$

## Chapter 5: Test

**1.** $3x-y=9\quad$ (1)

$x+2y=10\quad$ (2)

Solving equation (2) for $x$, we obtain $x=10-2y$ (3). Substituting this result into equation (1) and solving for $y$, we obtain
$$3(10-2y)-y=9\Rightarrow 30-6y-y=9\Rightarrow$$
$$-7y=-21\Rightarrow y=3$$

Substituting $y=3$ back into equation (3) in order to find $x$, we obtain
$$x=10-2(3)=10-6=4.$$

Solution set: $\{(4,3)\}$

**2.** $6x+9y=-21\quad$ (1)

$4x+6y=-14\quad$ (2)

Solving equation (1) for $x$, we obtain
$$6x=-9y-21\Rightarrow x=\frac{-9y-21}{6}\Rightarrow x=\frac{-3y-7}{2}\ (3).$$

Substituting this result into equation (2) and solving for $y$, we obtain the following.
$$4\left(\frac{-3y-7}{2}\right)+6y=-14$$
$$2(-3y-7)+6y=-14$$
$$-6y-14+6y=-14\Rightarrow -14=-14$$

This true statement implies there are infinitely many solutions. We can represent the solution set with $y$ as the arbitrary variable.

Solution set: $\left\{\left(\frac{-3y-7}{2},\ y\right)\right\}$

**3.** $\frac14x-\frac13y=-\frac{5}{12}\quad$ (1)

$\frac1{10}x+\frac15y=\frac12\quad$ (2)

To eliminate fractions, multiply equation (1) by 12 and equation (2) by 10.
$$3x-4y=-5\quad (3)$$
$$x+2y=5\quad (4)$$

Multiply equation (4) by 2 and add the result to equation (3).
$$3x-4y=-5$$
$$\underline{2x+4y=10}$$
$$5x\qquad =5\Rightarrow x=1$$

(*continued on next page*)

*(continued from page 327)*

Substituting $x = 1$ in equation (4) to find $y$, we obtain $1 + 2y = 5 \Rightarrow 2y = 4 \Rightarrow y = 2$.

Solution set: $\{(1, 2)\}$

**4.**
$$x - 2y = 4 \quad (1)$$
$$-2x + 4y = 6 \quad (2)$$

Multiply equation (1) by 2 and add the result to equation (2).
$$2x - 4y = 8$$
$$\underline{-2x + 4y = 6}$$
$$0 = 14$$

The system is inconsistent. The solution set is $\varnothing$.

**5.**
$$2x + y + z = 3 \quad (1)$$
$$x + 2y - z = 3 \quad (2)$$
$$3x - y + z = 5 \quad (3)$$

Eliminate $z$ first by adding equations (1) and (2) to obtain $3x + 3y = 6$ (4). Add equations (2) and (3) to eliminate $z$ and obtain $4x + y = 8$ (5). Multiply equation (5) by $-3$ and add the results to equation (4).
$$3x + 3y = 6$$
$$\underline{-12x - 3y = -24}$$
$$-9x \quad\quad = -18 \Rightarrow x = 2$$

Substituting $x = 2$ in equation (5) to find $y$, we have $4(2) + y = 8 \Rightarrow 8 + y = 8 \Rightarrow y = 0$.

Substituting $x = 2$ and $y = 0$ in equation (1) to find $z$, we have $2(2) + 0 + z = 3 \Rightarrow z = -1$.

Solution set: $\{(2, 0, -1)\}$

**6.** Writing $\begin{array}{l} 3x - 2y = 13 \\ 4x - y = 19 \end{array}$ as an augmented matrix,

we have $\left[\begin{array}{cc|c} 3 & -2 & 13 \\ 4 & -1 & 19 \end{array}\right].$

$$\left[\begin{array}{cc|c} 1 & -\frac{2}{3} & \frac{13}{3} \\ 4 & -1 & 19 \end{array}\right] \begin{array}{l} \frac{1}{3}R1 \\ \\ \end{array} \Rightarrow$$

$$\left[\begin{array}{cc|c} 1 & -\frac{2}{3} & \frac{13}{3} \\ 0 & \frac{5}{3} & \frac{5}{3} \end{array}\right] \begin{array}{l} \\ -4R1 + R2 \end{array} \Rightarrow$$

$$\left[\begin{array}{cc|c} 1 & -\frac{2}{3} & \frac{13}{3} \\ 0 & 1 & 1 \end{array}\right] \begin{array}{l} \\ \frac{3}{5}R2 \end{array} \Rightarrow \left[\begin{array}{cc|c} 1 & 0 & 5 \\ 0 & 1 & 1 \end{array}\right] \begin{array}{l} \frac{2}{3}R2 + R1 \\ \\ \end{array}$$

Solution set $\{(5, 1)\}$

**7.** Writing $\begin{array}{l} 3x - 4y + 2z = 15 \\ 2x - y + z = 13 \\ x + 2y - z = 5 \end{array}$ as an augmented

matrix, we have $\left[\begin{array}{ccc|c} 3 & -4 & 2 & 15 \\ 2 & -1 & 1 & 13 \\ 1 & 2 & -1 & 5 \end{array}\right].$

$$\left[\begin{array}{ccc|c} 1 & 2 & -1 & 5 \\ 2 & -1 & 1 & 13 \\ 3 & -4 & 2 & 15 \end{array}\right] \begin{array}{l} R1 \leftrightarrow R3 \\ \\ \\ \end{array} \Rightarrow$$

$$\left[\begin{array}{ccc|c} 1 & 2 & -1 & 5 \\ 0 & -5 & 3 & 3 \\ 0 & -10 & 5 & 0 \end{array}\right] \begin{array}{l} \\ -2R1 + R2 \\ -3R1 + R3 \end{array} \Rightarrow$$

$$\left[\begin{array}{ccc|c} 1 & 2 & -1 & 5 \\ 0 & 1 & -\frac{3}{5} & -\frac{3}{5} \\ 0 & -10 & 5 & 0 \end{array}\right] \begin{array}{l} \\ -\frac{1}{5}R2 \\ \\ \end{array} \Rightarrow$$

$$\left[\begin{array}{ccc|c} 1 & 0 & \frac{1}{5} & \frac{31}{5} \\ 0 & 1 & -\frac{3}{5} & -\frac{3}{5} \\ 0 & 0 & -1 & -6 \end{array}\right] \begin{array}{l} -2R2 + R1 \\ \\ 10R2 + R3 \end{array} \Rightarrow$$

$$\left[\begin{array}{ccc|c} 1 & 0 & \frac{1}{5} & \frac{31}{5} \\ 0 & 1 & -\frac{3}{5} & -\frac{3}{5} \\ 0 & 0 & 1 & 6 \end{array}\right] \begin{array}{l} \\ \\ -1R3 \end{array} \Rightarrow$$

$$\left[\begin{array}{ccc|c} 1 & 0 & 0 & 5 \\ 0 & 1 & 0 & 3 \\ 0 & 0 & 1 & 6 \end{array}\right] \begin{array}{l} -\frac{1}{5}R3 + R1 \\ \frac{3}{5}R3 + R2 \\ \\ \end{array}$$

Solution set: $\{(5, 3, 6)\}$

**8.** Since $y = ax^2 + bx + c$, and the points $(1, 5), (2, 3),$ and $(4, 11)$ are on the graph, we have the following equations which are then simplified.
$$5 = a(1)^2 + b(1) + c \Rightarrow a + b + c = 5 \quad (1)$$
$$3 = a(2)^2 + b(2) + c \Rightarrow 4a + 2b + c = 3 \quad (2)$$
$$11 = a(4)^2 + b(4) + c \Rightarrow 16a + 4b + c = 11 \quad (3)$$

First, eliminate $c$ by adding $-1$ times equation (1) to equation (2).
$$-a - b - c = -5$$
$$\underline{4a + 2b + c = 3}$$
$$3a + b \quad\quad = -2 \quad (4)$$

Eliminating $c$ again by adding $-1$ times equation (1) to equation (3), we have the following.
$$-a - b - c = -5$$
$$\underline{16a + 4b + c = 11}$$
$$15a + 3b \quad\quad = 6 \quad (5)$$

Next, eliminate $b$ by adding $-3$ times equation (4) to equation (5).

$$-9a - 3b = 6$$
$$\underline{15a + 3b = 6}$$
$$6a \qquad = 12 \Rightarrow a = 2$$

Substituting this value into equation (4), we obtain $3(2) + b = -2 \Rightarrow 6 + b = -2 \Rightarrow b = -8$.

Substituting 2 for $a$ and $-8$ for $b$ into equation (1), we have $2 + (-8) + c = 5 \Rightarrow$

$-6 + c = 5 \Rightarrow c = 11$.

The equation of the parabola is
$$y = 2x^2 - 8x + 11.$$

9. Let $x$ = number of units from Toronto;
$y$ = number of units from Montreal;
$z$ = number of units from Ottawa.
The information in the problem gives the system

$$x + y + z = 100$$
$$80x + 50y + 65z = 5990$$
$$x = z$$

Multiply the first equation by $-50$ and add to the second equation.

$$-50x - 50y - 50z = -5000$$
$$\underline{80x + 50y + 65z = \phantom{-}5990}$$
$$30x \phantom{+50y} + 15z = \phantom{-}990$$

Substitute $x$ for $z$ in this equation to obtain $30x + 15x = 990 \Rightarrow 45x = 990 \Rightarrow x = 22$.
If $x = 22$, then $z = 22$. Substitute 22 for $x$ and for $z$ in the first equation and solve for $y$:
$22 + y + 22 = 100 \Rightarrow y = 56$

22 units from Toronto, 56 units from Montreal, and 22 units from Ottawa were ordered.

10. $\begin{vmatrix} 6 & 8 \\ 2 & -7 \end{vmatrix} = 6(-7) - 2(8) = -58$

11. $\begin{vmatrix} 2 & 0 & 8 \\ -1 & 7 & 9 \\ 12 & 5 & -3 \end{vmatrix}$

This determinant may be evaluated by expanding about any row or any column. If we expand by the first row, we have the following.

$$\begin{vmatrix} 2 & 0 & 8 \\ -1 & 7 & 9 \\ 12 & 5 & -3 \end{vmatrix} = 2\begin{vmatrix} 7 & 9 \\ 5 & -3 \end{vmatrix} - 0\begin{vmatrix} -1 & 9 \\ 12 & -3 \end{vmatrix} + 8\begin{vmatrix} -1 & 7 \\ 12 & 5 \end{vmatrix}$$

$$= 2[7(-3) - 5(9)] - 0$$
$$\phantom{=} + 8[(-1)(5) - 12(7)]$$

$= 2(-21 - 45) + 8(-5 - 84)$
$= 2(-66) + 8(-89)$
$= -132 - 712 = -844$

12. $2x - 3y = -33$
$4x + 5y = 11$

$$D = \begin{vmatrix} 2 & -3 \\ 4 & 5 \end{vmatrix} = 2(5) - 4(-3) = 10 - (-12) = 22$$

$$D_x = \begin{vmatrix} -33 & -3 \\ 11 & 5 \end{vmatrix} = -33(5) - 11(-3)$$
$$= -165 - (-33) = -132$$

$$D_y = \begin{vmatrix} 2 & -33 \\ 4 & 11 \end{vmatrix} = 2(11) - 4(-33)$$
$$= 22 - (-132) = 154$$

$$x = \frac{D_x}{D} = \frac{-132}{22} = -6; \ y = \frac{D_y}{D} = \frac{154}{22} = 7$$

Solution set: $\{(-6, 7)\}$

13. $x + y - z = -4$
$2x - 3y - z = 5$
$x + 2y + 2z = 3$

Adding column 3 to columns 1 and 2, we have

$$D = \begin{vmatrix} 1 & 1 & -1 \\ 2 & -3 & -1 \\ 1 & 2 & 2 \end{vmatrix} = \begin{vmatrix} 0 & 0 & -1 \\ 1 & -4 & -1 \\ 3 & 4 & 2 \end{vmatrix}.$$

Expanding by row one, we have

$$D = -1\begin{vmatrix} 1 & -4 \\ 3 & 4 \end{vmatrix} = -[4 - (-12)] = -16.$$

Adding $-1$ times row 1 to row 2 and 2 times row 1 to row 3, we have the following.

$$D_x = \begin{vmatrix} -4 & 1 & -1 \\ 5 & -3 & -1 \\ 3 & 2 & 2 \end{vmatrix} = \begin{vmatrix} -4 & 1 & -1 \\ 9 & -4 & 0 \\ -5 & 4 & 0 \end{vmatrix}$$

Expanding by column three, we have

$$D_x = -1\begin{vmatrix} 9 & -4 \\ -5 & 4 \end{vmatrix} = -(36 - 20) = -16.$$

Adding $-1$ times row 1 to row 2 and 2 times row 1 to row 3, we have the following.

$$D_y = \begin{vmatrix} 1 & -4 & -1 \\ 2 & 5 & -1 \\ 1 & 3 & 2 \end{vmatrix} = \begin{vmatrix} 1 & -4 & -1 \\ 1 & 9 & 0 \\ 3 & -5 & 0 \end{vmatrix}.$$

Expanding by column three, we have

$$D_y = -1\begin{vmatrix} 1 & 9 \\ 3 & -5 \end{vmatrix} = -1(-5 - 27) = -(-32) = 32.$$

(*continued on next page*)

*(continued from page 329)*

Adding –2 times row 1 to row 2 and -1 times row 1 to row 3, we have

$$D_z = \begin{vmatrix} 1 & 1 & -4 \\ 2 & -3 & 5 \\ 1 & 2 & 3 \end{vmatrix} = \begin{vmatrix} 1 & 1 & -4 \\ 0 & -5 & 13 \\ 0 & 1 & 7 \end{vmatrix}.$$

Expanding by column one, we have

$$D_z = 1 \begin{vmatrix} -5 & 13 \\ 1 & 7 \end{vmatrix} = 1(-35 - 13) = -48.$$

Thus we have

$$x = \frac{D_x}{D} = \frac{-16}{-16} = 1, \ y = \frac{D_y}{D} = \frac{32}{-16} = -2,$$

and $z = \frac{D_z}{D} = \frac{-48}{-16} = 3.$

Solution set: $\{(1, -2, 3)\}$

**14.** $\dfrac{x+2}{x^3 + x^2 + x} = \dfrac{x+2}{x(x^2 + x + 1)} = \dfrac{x+2}{x(x+1)^2}$

$$= \frac{A}{x} + \frac{B}{x+1} + \frac{C}{(x+1)^2}$$

Multiply both sides by $x(x+1)^2$ to get

$x + 2 = A(x+1)^2 + Bx(x+1) + Cx$ (1).

Substituting 0 for $x$ in equation (1), we get

$0 + 2 = A(0+1)^2 + B(0)(0+1) + C(0) \Rightarrow A = 2$

Substituting 2 for $A$ and –1 for $x$ in equation (1), we get the following.

$$-1 + 2 = 2(-1+1)^2 + B(-1)(-1+1) + C(-1)$$
$$1 = -C \Rightarrow C = -1$$

We now have

$x + 2 = 2(x+1)^2 + Bx(x+1) + (-1)x$ (2).

Now substitute 1 for $x$ (arbitrary choice) in equation (2) to obtain the following.

$$1 + 2 = 2(1+1)^2 + B(1)(1+1) + (-1)(1)$$
$$3 = 2(4) + 2B - 1$$
$$3 = 8 + 2B - 1 \Rightarrow 3 = 7 + 2B$$
$$-4 = 2B \Rightarrow B = -2$$

Thus, $\dfrac{x+2}{x^3 + x^2 + x} = \dfrac{x+2}{x(x^2 + x + 1)} = \dfrac{x+2}{x(x+1)^2}$

$$= \frac{2}{x} + \frac{-2}{x+1} + \frac{-1}{(x+1)^2}.$$

**15.** The parabola passes through (0, 4), (1, 1), and (2, 0), so those points satisfy the equation $y = ax^2 + bx + c$. Substituting each ordered pair into the equation gives the system

$$4 = a(0)^2 + b(0) + c$$
$$1 = a(1)^2 + b(1) + c$$
$$0 = a(2)^2 + b(2) + c$$

which simplifies to

$$4 = \qquad c \quad (1)$$
$$1 = \ a + \ b + c \quad (2)$$
$$0 = 4a + 2b + c \quad (3)$$

Substitute $c = 4$ into equations (2) and (3), then solve the system consisting of those two equations for $a$ and $b$:

$$\begin{array}{l} 1 = \ a + \ b + 4 \\ 0 = 4a + 2b + 4 \end{array} \Rightarrow$$

$$-3 = a + b \qquad (4)$$
$$-4 = 4a + 2b \qquad (5)$$

From equation (4), we have $a = -b - 3$, so

$$-4 = 4(-b - 3) + 2b$$
$$-4 = -4b - 12 + 2b \Rightarrow 8 = -2b \Rightarrow -4 = b$$
$$a = -b - 3 \Rightarrow a = -(-4) - 3 \Rightarrow a = 1$$

Thus, the equation of the parabola is

$$y = x^2 - 4x + 4$$

The line passes through (–2, 0) and (1, 1), so $m = \frac{1-0}{1-(-2)} = \frac{1}{3}$, and the equation is

$$y - 0 = \tfrac{1}{3}(x+2) \Rightarrow 3y = x + 2 \Rightarrow x - 3y = -2.$$

The system is 
$$\begin{array}{l} y = x^2 - 4x + 4. \\ x - 3y = -2 \end{array}$$

**16.** $2x^2 + y^2 = 6 \qquad (1)$

$x^2 - 4y^2 = -15 \quad (2)$

Multiply equation (1) by 4 and add to equation (2), then solve for $x$.

$$\begin{array}{rl} 8x^2 + 4y^2 = & 24 \\ x^2 - 4y^2 = & -15 \\ \hline 9x^2 \qquad = & 9 \Rightarrow x^2 = 1 \Rightarrow x = \pm 1 \end{array}$$

Substitute these values into equation (1) and solve for $y$.

If $x = 1$, then $2(1)^2 + y^2 = 6 \Rightarrow 2 + y^2 = 6 \Rightarrow$ $y^2 = 4 \Rightarrow y = \pm 2$.

If $x = -1$, then $2(-1)^2 + y^2 = 6 \Rightarrow$ $2 + y^2 = 6 \Rightarrow y^2 = 4 \Rightarrow y = \pm 2$.

Solution set: $\{(1, 2), (-1, 2), (1, -2), (-1, -2)\}$

**17.** $x^2 + y^2 = 25$ (1)

$x + y = 7$ (2)

Solving equation (2) for $x$, we have

$x = 7 - y$ (3). Substituting this result into equation (1), we have

$$(7 - y)^2 + y^2 = 25$$
$$49 - 14y + y^2 + y^2 = 25$$
$$2y^2 - 14y + 24 = 0$$
$$y^2 - 7y + 12 = 0$$
$$(y - 3)(y - 4) = 0 \Rightarrow y = 3 \text{ or } y = 4$$

Substitute these values into equation (3) and solve for $x$. If $y = 3$, $x = 7 - 3 = 4$.

If $y = 4$, $x = 7 - 4 = 3$.

Solution set: $\{(3,4),(4,3)\}$

**18.** Let $x$ and $y$ be the numbers.

$x + y = -1$ (1)

$x^2 + y^2 = 61$ (2)

Solving equation (1) for $y$, we have

$y = -x - 1$ (3). Substituting this result into equation (2), we have

$$x^2 + (-x - 1)^2 = 61 \Rightarrow x^2 + x^2 + 2x + 1 = 61$$
$$2x^2 + 2x - 60 = 0 \Rightarrow x^2 + x - 30 = 0$$
$$(x + 6)(x - 5) = 0 \Rightarrow x = -6 \text{ or } x = 5$$

Substitute these values in equation (3) to find the corresponding values of $y$.

If $x = -6$, $y = -(-6) - 1 = 6 - 1 = 5$.

If $x = 5$, $y = -5 - 1 = -6$.

The same pair of numbers results from both cases. The numbers are 5 and –6.

**19.** $x - 3y \geq 6$

$y^2 \leq 16 - x^2$

Graph $x - 3y = 6$ as a solid line with $x$-intercept 6 and $y$-intercept of –2. Because the test point $(0,0)$ does not satisfies this inequality, shade the region *below* the line. Graph $y^2 = 16 - x^2$ or $x^2 + y^2 = 16$ as a solid circle with a center at the origin and radius 4. Shade the region, which is the interior of the circle. The solution set is the intersection of these two regions, which is the region shaded in the final graph.

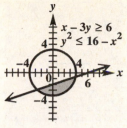

**20.** Find $x \geq 0$ and $y \geq 0$ such that

$x + 2y \leq 24$

$3x + 4y \leq 60$

and $2x + 3y$ is maximized.

Find the region of feasible solutions by graphing the system of inequalities that is made up of the constraints. To graph $x + 2y \leq 24$, draw the line with $x$-intercept 24 and $y$-intercept 12 as a solid line. Because the test point $(0,0)$ satisfies this inequality, shade the region *below* the line. To graph $3x + 4y \leq 60$, draw the line with $x$-intercept 20 and $y$-intercept 15 as a solid line. Because the test point $(0,0)$ satisfies this inequality, shade the region *below* the line. The constraints $x \geq 0$ and $y \geq 0$ restrict the graph to the first quadrant. The graph of the feasible region is the intersection of the regions that are the graphs of the individual constraints.

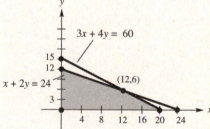

From the graph, observe that three vertices are $(0,0),(0,12)$, and $(20,0)$. A fourth vertex is the intersection point of the lines $x + y = 24$ and $3x + 4y = 60$. To find this point, solve the system $\begin{array}{l} x + 2y = 24 \quad (1) \\ 3x + 4y = 60 \quad (2) \end{array}$.

To eliminate $x$, multiply equation (1) by –3 and add the result to equation (2).

$$-3x - 6y = -72$$
$$\underline{3x + 4y = \phantom{0}60}$$
$$-2y = -12 \Rightarrow y = 6$$

Substituting this value into equation (1), we obtain $x + 2(6) = 24 \Rightarrow x + 12 = 24 \Rightarrow x = 12$.

The fourth vertex is $(12,6)$. Next, evaluate the objective function at each vertex.

*(continued on next page)*

(*continued from page 331*)

| Point | Profit $= 2x + 3y.$ | |
|-------|------------|---|
| $(0,0)$ | $2(0) + 3(0) = 0$ | |
| $(0,12)$ | $2(0) + 3(12) = 36$ | |
| $(20,0)$ | $2(20) + 3(0) = 40$ | |
| $(12,6)$ | $2(12) + 3(6) = 42$ | $\leftarrow$ Maximum |

The maximum value is 42 at $(12, 6)$.

**21.** Let $x =$ the number of VIP rings;
$y =$ the number of SST rings.
The constraints are $x + y \le 24$
$\qquad\qquad 3x + 2y \le 60$
$\qquad\qquad x \ge 0, \ y \ge 0.$

Maximize profit function, profit $= 30x + 40y.$

Find the region of feasible solutions by graphing the system of inequalities that is made up of the constraints. To graph $x + y \le 24,$ draw the line with $x$-intercept 24 and $y$-intercept 24 as a solid line. Because the test point $(0,0)$ satisfies this inequality, shade the region *below* the line. To graph $3x + 2y \le 60,$ draw the line with $x$-intercept 20 and $y$-intercept 30 as a solid line. Because the test point $(0,0)$ satisfies this inequality, shade the region *below* the line. The constraints $x \ge 0$ and $y \ge 0$ restrict the graph to the first quadrant. The graph of the feasible region is the intersection of the regions that are the graphs of the individual constraints.

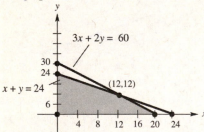

From the graph, observe that three vertices are $(0,0), (20,0),$ and $(0,24).$ A fourth vertex is the intersection point of the lines $x + y = 24$ and $3x + 2y = 60.$ To find this point, solve the system $\begin{array}{l} x + y = 24 \\ 3x + 2y = 60. \end{array}$

The first equation can be written as $x = 24 - y.$ Substituting this equation into $3x + 2y = 60,$ we have the following.

$3(24 - y) + y = 60 \Rightarrow 72 - 3y + 2y = 60 \Rightarrow$
$\qquad 72 - y = 60 \Rightarrow -y = -12 \Rightarrow y = 12$

Substituting $y = 12$ into $x = 24 - y,$ we have $x = 24 - 12 = 12.$ Thus, the four vertex is $(12,12).$ Next, evaluate the objective function at each vertex.

| Point | Profit $= 30x + 40y.$ | |
|-------|-------------|---|
| $(0,0)$ | $30(0) + 40(0) = 0$ | |
| $(20,0)$ | $30(20) + 40(0) = 600$ | |
| $(0,24)$ | $30(0) + 40(24) = 960$ | $\leftarrow$ Maximum |
| $(12,12)$ | $30(12) + 40(12) = 840$ | |

0 VIP rings and 24 SST rings should be made daily for a daily profit of $960.

**22.** $\begin{bmatrix} 5 & x+6 \\ 0 & 4 \end{bmatrix} = \begin{bmatrix} y-2 & 4-x \\ 0 & w+7 \end{bmatrix}$

All corresponding elements, position by position, of the two matrices must be equal.
$x + 6 = 4 - x \Rightarrow 2x = -2 \Rightarrow x = -1;$
$5 = y - 2 \Rightarrow y = 7; 4 = w + 7 \Rightarrow w = -3$
Thus, $x = -1, y = 7,$ and $w = -3.$

**23.** $3\begin{bmatrix} 2 & 3 \\ 1 & -4 \\ 5 & 9 \end{bmatrix} - \begin{bmatrix} -2 & 6 \\ 3 & -1 \\ 0 & 8 \end{bmatrix}$

$= \begin{bmatrix} 3(2) & 3(3) \\ 3(1) & 3(-4) \\ 3(5) & 3(9) \end{bmatrix} - \begin{bmatrix} -2 & 6 \\ 3 & -1 \\ 0 & 8 \end{bmatrix}$

$= \begin{bmatrix} 6 & 9 \\ 3 & -12 \\ 15 & 27 \end{bmatrix} + \begin{bmatrix} 2 & -6 \\ -3 & 1 \\ 0 & -8 \end{bmatrix}$

$= \begin{bmatrix} 6+2 & 9+(-6) \\ 3+(-3) & -12+1 \\ 15+0 & 27+(-8) \end{bmatrix} = \begin{bmatrix} 8 & 3 \\ 0 & -11 \\ 15 & 19 \end{bmatrix}$

**24.** $\begin{bmatrix} 1 \\ 2 \end{bmatrix} + \begin{bmatrix} 4 \\ -6 \end{bmatrix} + \begin{bmatrix} 2 & 8 \\ -7 & 5 \end{bmatrix}$

The first two matrices are $2 \times 1$ and the third is $2 \times 2.$ Only matrices of the same size can be added, so it is not possible to find this sum.

**25.** The product of a $2 \times 3$ matrix and a $3 \times 2$ matrix is a $2 \times 2$ matrix.

$$\begin{bmatrix} 2 & 1 & -3 \\ 4 & 0 & 5 \end{bmatrix} \begin{bmatrix} 1 & 3 \\ 2 & 4 \\ 3 & -2 \end{bmatrix} = \begin{bmatrix} 2(1)+1(2)+(-3)(3) & 2(3)+1(4)+(-3)(-2) \\ 4(1)+0(2)+5(3) & 4(3)+0(4)+5(-2) \end{bmatrix}$$

$$= \begin{bmatrix} 2+2+(-9) & 6+4+6 \\ 4+0+15 & 12+0+(-10) \end{bmatrix} = \begin{bmatrix} -5 & 16 \\ 19 & 2 \end{bmatrix}$$

**26.** $\begin{bmatrix} 2 & -4 \\ 3 & 5 \end{bmatrix} \begin{bmatrix} 4 \\ 2 \\ 7 \end{bmatrix}$

The first matrix is $2 \times 2$ and the second is $3 \times 1$. The first matrix has two columns and the second has three rows, so it is not possible to find this product.

**27.** There are associative, distributive, and identity properties that apply to multiplication of matrices, but matrix multiplication is not commutative. The correct choice is A.

**28.** Find the inverse of $A = \begin{bmatrix} -8 & 5 \\ 3 & -2 \end{bmatrix}$, if it exists.

The augmented matrix is

$$\left[ A \middle| I_2 \right] = \begin{bmatrix} -8 & 5 & 1 & 0 \\ 3 & -2 & 0 & 1 \end{bmatrix}.$$

$$\begin{bmatrix} 1 & -\frac{5}{8} & -\frac{1}{8} & 0 \\ 3 & -2 & 0 & 1 \end{bmatrix} \begin{matrix} -\frac{1}{8}R1 \\ \\ \end{matrix} \Rightarrow$$

$$\begin{bmatrix} 1 & -\frac{5}{8} & -\frac{1}{8} & 0 \\ 0 & -\frac{1}{8} & \frac{3}{8} & 1 \end{bmatrix} \begin{matrix} \\ -3R1+R2 \end{matrix} \Rightarrow$$

$$\begin{bmatrix} 1 & -\frac{5}{8} & -\frac{1}{8} & 0 \\ 0 & 1 & -3 & -8 \end{bmatrix} \begin{matrix} \\ -8R2 \end{matrix} \Rightarrow$$

$$\begin{bmatrix} 1 & 0 & -2 & -5 \\ 0 & 1 & -3 & -8 \end{bmatrix} \frac{5}{8}R2+R1$$

Thus, $A^{-1} = \begin{bmatrix} -2 & -5 \\ -3 & -8 \end{bmatrix}.$

**29.** Find the inverse of $A = \begin{bmatrix} 4 & 12 \\ 2 & 6 \end{bmatrix}$, if it exists.

The augmented matrix is

$$\left[ A \middle| I_2 \right] = \begin{bmatrix} 4 & 12 & 1 & 0 \\ 2 & 6 & 0 & 1 \end{bmatrix}.$$

$$\begin{bmatrix} 1 & 3 & \frac{1}{4} & 0 \\ 2 & 6 & 0 & 1 \end{bmatrix} \frac{1}{4}R1 \Rightarrow$$

$$\begin{bmatrix} 1 & 3 & \frac{1}{4} & 0 \\ 0 & 0 & -\frac{1}{2} & 1 \end{bmatrix} -2R1+R2$$

The second row, second column element is now 0, so the desired transformation cannot be completed. Therefore, the inverse of the given matrix does not exist.

**30.** Find the inverse of $A = \begin{bmatrix} 1 & 3 & 4 \\ 2 & 7 & 8 \\ -2 & -5 & -7 \end{bmatrix}$, if it exists. Performing row operations on the augmented matrix, we have the following.

$$\left[ A \middle| I_3 \right] = \begin{bmatrix} 1 & 3 & 4 & 1 & 0 & 0 \\ 2 & 7 & 8 & 0 & 1 & 0 \\ -2 & -5 & -7 & 0 & 0 & 1 \end{bmatrix} \Rightarrow$$

$$\begin{bmatrix} 1 & 3 & 4 & 1 & 0 & 0 \\ 0 & 1 & 0 & -2 & 1 & 0 \\ 0 & 1 & 1 & 2 & 0 & 1 \end{bmatrix} \begin{matrix} \\ -2R1+R2 \\ 2R1+R3 \end{matrix} \Rightarrow$$

$$\begin{bmatrix} 1 & 0 & 4 & 7 & -3 & 0 \\ 0 & 1 & 0 & -2 & 1 & 0 \\ 0 & 0 & 1 & 4 & -1 & 1 \end{bmatrix} \begin{matrix} -3R2+R1 \\ \\ -1R2+R3 \end{matrix} \Rightarrow$$

$$\begin{bmatrix} 1 & 0 & 0 & -9 & 1 & -4 \\ 0 & 1 & 0 & -2 & 1 & 0 \\ 0 & 0 & 1 & 4 & -1 & 1 \end{bmatrix} -4R3+R1$$

Thus, $A^{-1} = \begin{bmatrix} -9 & 1 & -4 \\ -2 & 1 & 0 \\ 4 & -1 & 1 \end{bmatrix}.$

**31.** The system $\begin{matrix} 2x+y=-6 \\ 3x-y=-29 \end{matrix}$ yields the matrix equation $AX = B$ where

$$A = \begin{bmatrix} 2 & 1 \\ 3 & -1 \end{bmatrix}, \ X = \begin{bmatrix} x \\ y \end{bmatrix}, \ \text{and} \ B = \begin{bmatrix} -6 \\ -29 \end{bmatrix}.$$

Find $A^{-1}$. The augmented matrix is

$$\left[ A \middle| I_2 \right] = \begin{bmatrix} 2 & 1 & 1 & 0 \\ 3 & -1 & 0 & 1 \end{bmatrix}.$$

*(continued on next page)*

(*continued from page 333*)

$$\begin{bmatrix} 1 & \frac{1}{2} & \Big| & \frac{1}{2} & 0 \\ 3 & -1 & \Big| & 0 & 1 \end{bmatrix} \frac{1}{2}R1 \Rightarrow$$

$$\begin{bmatrix} 1 & \frac{1}{2} & \Big| & \frac{1}{2} & 0 \\ 0 & -\frac{5}{2} & \Big| & -\frac{3}{2} & 1 \end{bmatrix} -3R1 + R2 \Rightarrow$$

$$\begin{bmatrix} 1 & 0 & \Big| & \frac{1}{5} & \frac{1}{5} \\ 0 & -\frac{5}{2} & \Big| & -\frac{3}{2} & 1 \end{bmatrix} \frac{1}{5}R2 + R1 \Rightarrow$$

$$\begin{bmatrix} 1 & 0 & \Big| & \frac{1}{5} & \frac{1}{5} \\ 0 & 1 & \Big| & \frac{3}{5} & -\frac{2}{5} \end{bmatrix} -\frac{2}{5}R2$$

Thus, $A^{-1} = \begin{bmatrix} \frac{1}{5} & \frac{1}{5} \\ \frac{3}{5} & -\frac{2}{5} \end{bmatrix}$. Since $X = A^{-1}B,$ we

have the following.

$$A^{-1}B = \begin{bmatrix} \frac{1}{5} & \frac{1}{5} \\ \frac{3}{5} & -\frac{2}{5} \end{bmatrix} \begin{bmatrix} -6 \\ -29 \end{bmatrix}$$

$$= \begin{bmatrix} \frac{1}{5}(-6) + \frac{1}{5}(-29) \\ \frac{3}{5}(-6) + \left(-\frac{2}{5}\right)(-29) \end{bmatrix}$$

$$= \begin{bmatrix} -\frac{6}{5} + \left(-\frac{29}{5}\right) \\ -\frac{18}{5} + \frac{58}{5} \end{bmatrix} = \begin{bmatrix} -\frac{35}{5} \\ \frac{40}{5} \end{bmatrix} = \begin{bmatrix} -7 \\ 8 \end{bmatrix}$$

Solution set: $\{(-7,8)\}$

**32.** The system $\begin{aligned} x + y &= 5 \\ y - 2z &= 23 \\ x + 3z &= -27 \end{aligned}$ yields the matrix

equation $AX = B$ where $A = \begin{bmatrix} 1 & 1 & 0 \\ 0 & 1 & -2 \\ 1 & 0 & 3 \end{bmatrix}$,

$X = \begin{bmatrix} x \\ y \\ z \end{bmatrix}$, and $B = \begin{bmatrix} 5 \\ 23 \\ -27 \end{bmatrix}$. Find $A^{-1}$. The

augmented matrix is

$$[A \mid I_3] = \begin{bmatrix} 1 & 1 & 0 & \Big| & 1 & 0 & 0 \\ 0 & 1 & -2 & \Big| & 0 & 1 & 0 \\ 1 & 0 & 3 & \Big| & 0 & 0 & 1 \end{bmatrix}.$$

$$\begin{bmatrix} 1 & 1 & 0 & \Big| & 1 & 0 & 0 \\ 0 & 1 & -2 & \Big| & 0 & 1 & 0 \\ 0 & -1 & 3 & \Big| & -1 & 0 & 1 \end{bmatrix} -R1 + R3 \Rightarrow$$

$$\begin{bmatrix} 1 & 0 & 2 & \Big| & 1 & -1 & 0 \\ 0 & 1 & -2 & \Big| & 0 & 1 & 0 \\ 0 & 0 & 1 & \Big| & -1 & 1 & 1 \end{bmatrix} \begin{matrix} -1R2 + R1 \\ \\ R2 + R3 \end{matrix} \Rightarrow$$

$$\begin{bmatrix} 1 & 0 & 0 & \Big| & 3 & -3 & -2 \\ 0 & 1 & 0 & \Big| & -2 & 3 & 2 \\ 0 & 0 & 1 & \Big| & -1 & 1 & 1 \end{bmatrix} \begin{matrix} -2R3 + R1 \\ 2R3 + R2 \\ \\ \end{matrix}$$

Thus, $A^{-1} = \begin{bmatrix} 3 & -3 & -2 \\ -2 & 3 & 2 \\ -1 & 1 & 1 \end{bmatrix}$. Since

$X = A^{-1}B,$ we have the following.

$$A^{-1}B = \begin{bmatrix} 3 & -3 & -2 \\ -2 & 3 & 2 \\ -1 & 1 & 1 \end{bmatrix} \begin{bmatrix} 5 \\ 23 \\ -27 \end{bmatrix}$$

$$= \begin{bmatrix} 3(5) + (-3)(23) + (-2)(-27) \\ -2(5) + 3(23) + 2(-27) \\ -1(5) + 1(23) + 1(-27) \end{bmatrix}$$

$$= \begin{bmatrix} 15 + (-69) + 54 \\ -10 + (69) + (-54) \\ -5 + 23 + (-27) \end{bmatrix} = \begin{bmatrix} 0 \\ 5 \\ -9 \end{bmatrix} = X$$

Solution set: $\{(0,5,-9)\}$